THIN FILM
PHENOMENA

THIN FILM PHENOMENA

KASTURI L. CHOPRA
Staff Scientist, Ledgemont Laboratory,
Kennecott Copper Corporation
Lexington, Massachusetts

Adjunct Professor of Mechanical Engineering
Northeastern University
Boston, Massachusetts

McGRAW-HILL BOOK COMPANY   New York   St. Louis
San Francisco   London   Sydney   Toronto   Mexico   Panama

THIN FILM PHENOMENA

10799

1234567890 MAMM 754321069

TO MY PARENTS

PREFACE

"It seems good for philosophers to move to fresh ways and systems; good for them to allow neither the voice of the detractor, nor the weight of ancient culture, nor the fullness of authority, to deter those who would declare their own views. In that way each age produces its own crop of new authors and new arts. . ." FERNEL, 16th Century

Because of their potential technical value and scientific curiosity in the properties of a two-dimensional solid, thin films have been extensively studied for over a century. However, until this decade sufficient technological progress has not been made to give reasonable scientific confidence to thin film research. The developments in this decade have made, directly or indirectly, significant contributions to many areas of basic and applied solid-state research.

Physical phenomena peculiar to thin films, and the basis for their study, are generally the consequence of their planar geometry, size, and unique structure. Epitaxial growth, the occurrence of metastable structures, size-limited electron and phonon transport processes in metals, insulators, and semiconductors, quantum-mechanical tunneling through normal and superconducting metal-insulator junctions, micro-magnetics, and plasma resonance absorption are some of the noteworthy contributions of thin film phenomena to solid-state physics. The technical interests which stimulated these studies have also been rewarded in the form of useful inventions such as a variety of active and passive microminiaturized components and devices, solar cells, radiation sources and detectors, magnetic memory devices, cryotrons, bolometers, interference filters, and reflection and antireflection coatings.

While progress in thin film research has been recorded, it is scattered in various scientific journals and thin film conference proceedings. Clearly, the presentation of a unified picture of the developments in a scientific field in the form of books is just as important as the scientific research itself. There is, at present, no such up-to-date book on the physics of thin films. The popular "know-how" book "Vacuum Deposition of Thin Films" by Holland, published in 1956, and the detailed textbook "Physik dünner Schichten" by Mayer, published in 1950 (vol. 1) and 1956 (vol. 2) are quite outdated and need major revisions to represent accurately the present status of our knowledge.

That a detailed and authoritative account of the enormous works on thin films in a single volume or by a single author is presently very difficult has led to the present-day trend of publishing edited books comprised of independent review articles on different subjects. The four volumes of "Physics of Thin Films" edited by Hass and Thun, and the NATO conference proceedings, "The Use of Thin Films in Physical Investigations," edited by Anderson, have many excellent articles. An unavoidable lack of coherence and completeness, however, is characteristic of such works.

The urgent need for a coherent, unified, and up-to-date account of thin film research inspired me to undertake the laborious task of writing this book. The book provides a comprehensive compilation of the known phenomena associated with the structural, mechanical, electrical, superconducting, magnetic, and optical behavior of films. Because of the necessary limitation on the size of a reasonable book, a detailed treatment and analysis of results is not possible, but sufficient description and literature references are provided to present a critical view of the salient features of the subjects. The experimental aspects of the basic phenomena are emphasized, although technical applications of these phenomena have also been mentioned. Wherever disorder exists in the literature, I have declared my own views and provided a reasonable synthesis.

It is natural that the contents of this book draw heavily from my own research interests and scientific contributions. This is reflected in the relatively more detailed treatment of the structure and growth, and transport properties of films. Since the preparation and characterization of films are vital parts of thin film research, Chaps. II and III are devoted to thin film technology. An extensive review of "Ferromagnetism in Films" (Chap. X) was contributed by Dr. Mitchell Cohen, to whom I am very grateful. In the chapters on Mechanical Effects and Optical Properties, respectively, considerable information is derived from the review articles by Hoffman, by Hass and coworkers, and by Heavens in the "Physics of Thin Films" series. My thanks are due to these authors and many others, and to several publishing companies for permission to use their published material freely. These sources are acknowledged in the text.

The order of presentation is primarily dictated by technical cohesiveness and is not chronological. Historical references are given where some confusion exists in literature. The enormous list of references (over 2,600), kept up to date until the middle of 1968, is only a fraction of the published literature but is adequately representative of the leading and informative articles in all areas discussed. The obviously doubtful and trivial literature has been eliminated. Since commonly used symbols have been employed for each subject, some overlap was unavoidable.

This work is intended for use as a reference and research book for graduate students, engineers, and research scientists. It is my earnest hope that it will succeed in inspiring new ideas and help coordinate work on thin films along more fruitful lines. If I have made some glaring errors or omissions, I hope the readers will bring them to my attention.

I am grateful to M. R. Randlett and S. K. Bahl for their untiring assistance in preparing and checking the entire manuscript at various stages, and to R. Johnson for editing the draft copy. Thanks are due to D. Barclay, M. R. Randlett, and R. L. Tennis for preparing most of the illustrations. I wish to thank my colleagues W. W. Harvey, I. L. Gelles, S. H. Gelles, R. H. Duff, J. Chen, J. Pack, and P. C. Clapp for reading parts of the manuscript. The burden of typing parts of the final manuscript was undertaken by Olga O'Brien, Joanne Johnson, Dorothy Smith, and Suzanne Archigian. It is a pleasure to thank E. W. Fletcher, the Director of Ledgemont Laboratory, for his encouragement. Finally, this book could not have been written without the constant help and understanding of my wife, Asha Chopra, who typed the first draft of the manuscript.

Lexington, Massachusetts   *K. L. Chopra*

CONTENTS

I INTRODUCTION

No, 'tis not so deep as a well, nor so wide as a
church-door, 'tis enough, 'twill serve.
                    SHAKESPEARE, *Romeo and Juliet*

While nonsolid films and the associated phenomenon of interference
colors have been studied for over three centuries, thin solid films were
probably first obtained by electrolysis in 1838. In the recorded
literature, however, Bunsen and Grove obtained metal films in 1852 by
means of a chemical reaction and by glow-discharge sputtering, respectively. Faraday obtained metal films in 1857 by the thermal evaporation
on explosion of a current-carrying metal wire. The usefulness of the
optical properties of metal films, and scientific curiosity about the
behavior of two-dimensional solids have been responsible for the
immense interest in the study of the science and technology of thin
films. The varied and irreproducible results often obtained on films led
most workers to conclude that vapor-deposited films represent a high
state of disorder and that no two films are alike. This chaotic state of the
knowledge has actually been a blessing in disguise since it sustained and
energized continued interest in thin film research.

1

The technology and understanding of films less than 1 micron thick have made tremendous advances in the last decade, primarily because of the industrial demand for reliable thin-film microelectronic devices to fulfill the urgent needs of the Sputnik era. This progress has brought maturity and much scientific confidence in the use of thin films for basic and applied research. In addition to major contributions to a variety of new and future scientifically based technologies, thin film studies have directly or indirectly advanced many new areas of research in solid-state physics and chemistry which are based on phenomena uniquely characteristic of the thickness, geometry, and structure of films. A discussion of these phenomena, of course, forms the central theme of this book.

This chapter will introduce the uninitiated reader to some of the outstanding scientific and technological achievements based on thin film research. The details and the appropriate references are provided in the main text.

Thin Film Technology (Chaps. II and III). The demand for clean ultrahigh-vacuum conditions during vacuum evaporation has been largely responsible for the growth of a highly specialized and scientifically based vacuum industry. In addition to the conventional oil pumps, a variety of high-speed, oil-free getter-ion, sublimation, and cryogenic pumps has emerged. Consequently, large deposition chambers maintained at $\sim 10^{-7}$ to 10^{-10} Torr are now commonplace in any laboratory.

A multitude of techniques have been developed to prepare poly-crystalline and nearly single-crystal films of all types of materials. Deposition rates may range from a fraction of an angstrom to thousands of angstroms per second. Among these deposition techniques, those of thermal evaporation by resistive and electron-bombardment heating, sputtering by means of glow discharge, rf and ion beams, and vapor deposition by a variety of chemical reactions have been perfected. With the introduction of electronic monitoring and control, thermal evaporation may be carried out at definite rates of evaporation from one or more sources, and definite rates of deposition onto one or more substrates, with a high (fraction of a percent) uniformity of film thickness over large surfaces (such as that of an artificial satellite).

Some of the laboratory deposition techniques have been carried over to the production stage in new industries. Notable examples of successful industrial processes are the use of thermal evaporation of Al for aluminization of foils, inert and reactive sputtering of Ta for thin-film microminiaturized resistors and capacitors with component densities exceeding $10^4/cm^2$, and chemical deposition of Nb_3Sn on foils for the winding of superconducting magnets.

Several microanalytical techniques have been improved and others invented to determine the composition and microstructure of thin films. Electron diffraction and electron microscopy, which owe their present high level of sophistication to the availability of thin films, are now responsible for much of the understanding of the structure of thin films. Among the various by-products in terms of new analytical techniques, that of the moiré fringes is significant since it is capable of revealing images of lattices and lattice defects down to a dimension of a few angstroms.

Structure and Growth (Chap. IV). The unique growth stages of a vapor-deposited film consist of a statistical process of nucleation of the vapor atoms, surface-diffusion-controlled growth of the three-dimensional nuclei, and the formation of a network structure and its subsequent filling. The most characteristic stage is that of the liquid-like coalescence of the nuclei to form the network structure. This stage plays a dominant role in the microstructure and epitaxial growth of films, and the introduction and annihilation of structural defects. Studies of these film-growth stages provide an insight into the basic crystal-growth processes.

The deposition parameters may be exploited to influence the kinetics of the film-growth stages and thereby obtain films with structures ranging from a completely disordered (amorphous) to a highly ordered (monocrystalline) form. Further, one can prepare films with an atomically smooth surface, or a rough film with an effective surface area hundreds of times larger than the geometrical area. The control of deposition conditions allows the preparation of stoichiometric films of multicomponent alloys and compounds. The phenomenon of epitaxy may be used to prepare suitably oriented single-crystal films of elements, alloys, and compounds on single-crystal native or foreign substrates.

The atomistic growth processes can be controlled to yield in many materials a multitude of metastabilized normal and abnormal structures which are impossible to obtain in the bulk form by any known physical method. Thus, for example, it is possible to prepare thick films of amorphous W and Zr, fcc Ta and Hf, cubic Co, amorphous Au-Co alloys, metastable binary alloys over a large solubility range, sphalerite CdS, wurtzite GaAs, cubic GeTe, superperiodic structures in certain binary alloys, and surface superstructures of a number of metals, insulators, and semiconductors. Thus, thin film techniques have opened a new and exotic dimension in materials research.

Mechanical Effects (Chap. V). Thin films are unusual specimens for the study of mechanical effects in materials in the presence of a high internal structural disorder. Vapor-deposited films are generally under enormous stresses ($\sim 10^9$ to 10^{10} dynes/cm^2) and further contain a high density of lattice defects which, even in the most favorable case of epitaxial films, amount to $\sim 10^{11}$ dislocations/cm^2. The level of the intrinstic stress is comparable with the yield strength of many bulk materials and should have a strong influence on the physical and mechanical stability and the properties of films. The high density of defects and the presence of free surfaces make it difficult to generate or move dislocations in a film. This condition results in the enhancement of the tensile strength of films up to 200 times the value in the corresponding bulk material, a value which is considerably larger than can be obtained by the severest cold-work treatment of the bulk material. Thus, thin films provide a medium for the study of the high yield strength and superplastic behavior of materials.

Transport Phenomena (Chaps. VI, VII, and VIII). Studies of the electronic properties of films have been largely stimulated by many attractive microelectronic-device applications. As a result of these studies, thin films of alloys, compounds, and mixtures used as resistors, and thin insulator films used in capacitors now find widespread applications in integrated circuits. The marked sensitivity of the surface conductivity of thin semiconducting films to a transverse electric field has been successfully exploited in the promising thin-film transistor (TFT). Single and multilayer thin piezoelectric films have been employed as hypersonic transducers in the gigacycle range.

Because of the surface scattering of carriers in films of thickness comparable with the mean free path (mfp), the electrical and thermal transport properties of metals and semiconductors are modified. This modification, called the "size effect," depends on the ratio of the film thickness to the mfp, which may be varied by changing the film thickness and temperature, and by applying a magnetic field. Such size-effect studies have provided tests for transport theories and have also offered convenient and, in some cases, direct methods of determining bulk-transport parameters such as the concentration and mobility of the carriers, the mfp's and their temperature and energy dependence for electron-phonon and electron-electron scattering, surface scattering coefficient, and Fermi surface topology. A variety of new phenomena of basic importance has emerged from these studies, e.g., the size-dependent specular scattering of electrons in metal films, thickness-dependent oscillatory variation of the transport parameters in semimetals due to the size quantization of the energy levels, and oscillatory variation of the transverse magnetoresistance of metal films.

Extensive theoretical and experimental investigations of conduction in thin insulating films have led to an understanding of the various transport mechanisms and also to the determination of the barrier parameters and barrier profile for an insulator and the insulator-electrode interfaces. By employing quantum-mechanical tunneling through thin insulators as a source of hot carriers, their injection, transfer, and collection properties have been studied. The possibility of using tunnel electrons to obtain a relatively temperature-insensitive and high-frequency-response tunnel (hot-electron) triode has been demonstrated. The characteristic dependence of the current through a tunnel junction on the energy-level diagram of the electrodes and, in the case of inelastic tunneling, on the excitable energy levels in the insulator forms the basis of a new, powerful technique of tunneling spectroscopy.

Superconductivity in Films (Chap. IX). Films of a thickness comparable with the penetration depth and coherence length of a superconductor are ideal specimens for studying the superconducting behavior (magnetization, critical current, and critical magnetic field) of type I and type II superconductors in the light of the various theories. Such studies have played a major role in the theoretical understanding of superconductivity and have made it possible to determine the superconductivity parameters of penetration depth, coherence length, and the Ginzburg-Landau coupling constant. The detection and measurement of quantized flux in a cylindrical film provided the most striking verification of the existence of electron or Cooper pairs in a superconductor.

The extension of the spectroscopic tool of quantum-mechanical tunneling through a thin insulator bounded by one or both superconducting electrodes provided the most direct measurement of the energy gap and the density-of-states function and thus an excellent verification of the Bardeen-Cooper-Schrieffer theory of superconductivity. The observation of the theoretically predicted potential-free, supercurrent, or Josephson tunneling of Cooper pairs through ultrathin insulators represents one of the most celebrated contributions of thin films to solid-state research. When a dc potential appears across a Josephson junction, microwave photons of definite frequency are emitted. Conversely, the absorption of such microwaves produces the supercurrent. A Josephson tunnel junction therefore acts as an emitter and detector of microwaves. The frequency of these microwaves has been used to determine the fundamental constant h/e with an unprecedented accuracy. The phase coherence obtained in a tunnel junction makes possible the study of electron quantum interference effects.

It is easy to suppress superconductivity, but impossible to enhance it appreciably. Observations of anomalously high enhancements of the transition temperature in highly disordered, amorphous-like structures of films of some materials offer much optimism for the discovery of new high-temperature superconductors.

The fact that the zero resistance state of a superconductor can be transformed reversibly to the finite resistance state, or switched to another part of the superconducting circuit by means of a suitable current or a magnetic field, forms the basic principle of a variety of promising superconductive devices. The high switching speed of the basic circuit, called a "cryotron," is achieved only by the use of thin films. Because of their simplicity of fabrication, small size, low power requirements (in microwatts), and lack of polarity dependence, with no need for connections, cryotrons are attractive switching, logic, and storage devices.

Ferromagnetism in Films (Chap. X). The attractive possibilities of utilizing the extremely fast magnetization-reversal processes in thin ferromagnetic films for computer circuitry have led to extensive research on the magnetic behavior of these films. These studies have contributed new magneto-optical and magnetoelectron-optical (Lorentz-microscopy) techniques, and various new phenomena which form a highly specialized field of micromagnetics.

The uniqueness of the thin film ferromagnetic phenomena is due to the geometry and microstructure of films. Owing to the pinning of an increasing number of spins at the surface of a ferromagnetic film, the saturation magnetization decreases rapidly below about 30 Å thickness. The surface pinning makes it possible to excite spin waves across the film thickness, an interesting phenomenon which yields a direct measurement of the ferromagnetic exchange constant.

The planar geometry of a ferromagnetic film produces a very high demagnetizing field perpendicular to the plane as compared with that in the plane of the film. This anisotropy becomes uniaxial when the film is deposited in the presence of a magnetic field so that the magnetization is aligned with the field direction. This direction of magnetization can be reversed to the energetically equivalent direction in a very short time by means of a small magnetic field. This switching process is the attractive feature for computer applications.

The static and dynamic behavior of the magnetization of a ferromagnetic film is dominated by the various magnetic anisotropies and ripple or dispersion in the uniaxial anisotropy in the form of quasiperiodic local angular deviations of a few degrees from the average direction. The magnetization reversal involves the rotation of domains and some motion of the domain boundaries or walls. In contrast to bulk ferromagnets, thin ferromagnetic films have domains extending throughout the film thickness. The domain walls are of Néel or cross-tie type instead of the Bloch type formed in bulk materials. This difference again arises from the high magnetostatic energy normal to the film plane so that the magnetization rotates in the film plane. Studies of these various features have contributed knowledge to thin film computer technology, basic ferromagnetism, and the structure of films.

Optical Properties (Chap. XI). Since the early use of metal films in mirrors and interferometers, interest in the applications of thin metal and dielectric films in optical devices has dominated research in this field. The ingenious exploitation of the interference phenomenon in thin films has led to the development of sophisticated multilayer systems with nearly ideal reflection, antireflection, polarization, narrow- and wideband reflection and transmission filtering, and absorptance and emittance properties. Such optical coatings now find routine and indispensable industrial applications in optical instruments.

The ideally suited, clean, smooth surfaces of vapor-deposited films have been utilized to determine the optical constants of numerous materials. Thin film absorption in the infrared and ultraviolet and photoemission studies have yielded valuable information on the basic electronic parameters of materials. The film geometry makes it possible to excite surface modes in an ionic lattice and affect plasma-radiation emission and absorption in metals. Studies of these absorption phenomena have thrown light on the lattice dynamics of the materials.

Future Developments. With rapid technological advances in the preparation of films with controlled, reproducible, and well-defined structures, thin films are expected to play an increasingly important role in the studies of a variety of solid-state phenomena of basic and practical interest. The properties of a large variety of new and exotic materials obtained by thin film techniques will undoubtedly draw considerable attention in the future.

A multitude of thin-film optical, magnetic, electronic, and super-conductive devices have been successfully operated. With increasing flexibility and diversity of the application-oriented industry, a natural course of growth and selection of new and promising devices based on complete and careful utilization of the science and technology of thin films will dominate future developments. This remark is best illustrated by the development by Weimer et al. [1] of a television camera with a completely integrated self-scanned solid-state image sensor, employing more than 10^5 thin-film components obtained by simple vacuum deposition techniques in a single pump-down cycle. Figure 1 shows the

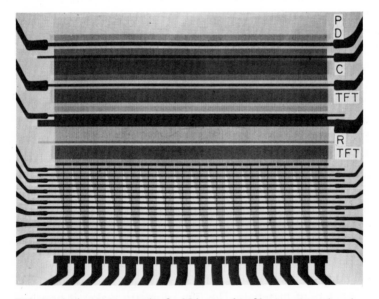

FIG. 1. A photomicrograph of a 256-stage thin-film transistor decoder connected to a 256-element line-scan sensor. The notations stand for 256 elements each of photoconductors (P), diodes (D), capacitors (C), CdSe thin-film transistors (TFT), and resistors (R). The center-to-center spacing between elements is about 52 μ. (*Courtesy of P. K. Weimer et al.* [1])

photomicrograph of a 256-stage thin-film transistor decoder for television scanning, driven by two 16-stage shift registers, and the associated photoconductors, diodes, capacitors, and resistors. This example represents perhaps the best exploitation of both the technology and science of thin films for an integrated microelectronic circuit of the future.

REFERENCE

1. P. K. Weimer, G. Sadasiv, J. E. Meyer, L. Meray-Hovrath, and W. S. Pike, WESCON 67, 13/3 August, 1967; *Proc. IEEE,* **55**:1591 (1967).

II THIN FILM DEPOSITION TECHNOLOGY

1. INTRODUCTION

This chapter describes various methods and the related technology for preparing thin solid films for research, development, and production purposes. Although this subject cannot be dealt with in depth in this book, the summary provided here should help the reader to evaluate various techniques critically. A general description with references to important reviews and representative publications is presented.

The deposition techniques for thin films may be broadly classified under three headings: thermal evaporation (physical vapor deposition), cathodic sputtering, and chemical deposition. A comparative summary of the various methods within these categories is presented in Table V at the end of Sec. 4.

2. THERMAL EVAPORATION

2.1 General Considerations

Solid materials vaporize when heated to sufficiently high temperatures. The condensation of the vapor onto a cooler substrate yields thin solid films. Thin films of carbon deposited by evaporation inside an electric bulb were probably first observed by Edison. Historically [1], however, the deposition of a metal film from a wire exploded by a high current density is attributed to Faraday, and by vacuum evaporation, to Nahrwold, and Pohl and Pringsheim. The deposition by the thermal-evaporation method is simple and very convenient and is at present the most widely used. Excellent and detailed reviews of the "know-how" of the subject are given by Holland [2, 3].

Because of collisions with ambient gas atoms, a fraction of the vapor atoms proportional to exp $(-d/l)$ will be scattered and hence randomized in direction in a distance d during their transfer through the gas. Here l is the mean free path (mfp) of gas atoms which for air molecules at $25°C$ and pressures of 10^{-4} and 10^{-6} Torr (1 Torr = 1 mm Hg), respectively, is about 45 and 4,500 cm. Thus, pressures lower than 10^{-5} Torr are necessary to ensure a straight-line path for most of the emitted vapor atoms, for substrate-to-source distances of $\sim$10 to 50 cm in a vacuum evaporator.

The rate of free evaporation of vapor atoms from a clean surface of unit area in vacuum is given by the Langmuir expression

$$m_e = 5.83 \times 10^{-2} p_e \sqrt{\frac{M}{T}} \quad \text{g/(cm}^2)(\text{sec}) \tag{1}$$

where p_e ($<10^{-2}$) is the equilibrium vapor pressure (in Torr) of the evaporant under saturated-vapor conditions at a temperature T, and M is the molecular weight of the vapor species. Alternatively, we may write the evaporation rate as

$$N_e = 3.513 \times 10^{22} p_e \sqrt{\frac{1}{MT}} \quad \text{molecules/(cm}^2)(\text{sec}) \tag{2}$$

Note, however, that the rate of deposition of the vapor on a substrate depends on the source geometry, its position relative to the substrate, and the condensation coefficient.

Holland [2] has discussed thoroughly the theoretical distribution of vapor from a point, a wire, a small surface, an extended strip, and from cylindrical and ring types of sources. For the ideal case of deposition from a clean, uniformly emitting point source onto a plane receiver, the rate of deposition varies as $\cos\theta/r^2$ (Knudsen cosine law), where r is the radial distance of the receiver from the source and θ is the angle between the radial vector and the normal to the receiver direction. If t_0 and t are the thicknesses of deposits at the receiver vertically below the source at a distance h, and at a horizontal distance x from the vertical line, respectively, then the deposit distribution (assuming the same condensation coefficient) is given by

$$\frac{t}{t_0} = \frac{1}{\left[1 + (x/h)^2\right]^{3/2}} \tag{3}$$

For evaporation from a small area onto a parallel plane receiver, the deposition rate is proportional to $\cos^2\theta/r^2$, and the thickness distribution is given by

$$\frac{t}{t_0} = \frac{1}{\left[1 + (x/h)^2\right]^{2}} \tag{4}$$

In both cases, the thickness decreases by about 10 percent for $x = h/4$. More complicated expressions for the distribution of the deposit result for other types of sources.

A parameter of interest in understanding the influence of ambient gases on the properties of films is the impingement rate of gas atoms or molecules. This is given by the kinetic theory of gases under equilibrium conditions as

$$N_g = 3.513 \times 10^{22} \frac{p_g}{\sqrt{M_g T_g}} \quad \text{molecules/(cm}^2\text{)(sec)}$$

which is the same expression as Eq. (2), with subscript g for gas. Table I lists the mfp and impingement rate of air molecules at different pressures. Note that at constant gas and evaporation temperatures, the

Table I. Some Facts about the Residual Air at 25°C in a Typical Vacuum Used for Film Deposition

Pressure,* Torr	Mean free path, cm (between collisions)	Collisions/sec (between molecules)	Molecules/(cm^2)(sec) (striking surface)	Monolayers/sec†
10^{-2}	0.5	9×10^4	3.8×10^{18}	4,400
10^{-4}	51	900	3.8×10^{16}	44
10^{-5}	510	90	3.8×10^{15}	4.4
10^{-7}	5.1×10^4	0.9	3.8×10^{13}	4.4×10^{-2}
10^{-9}	5.1×10^6	9×10^{-3}	3.8×10^{11}	4.4×10^{-4}

*1 Torr = 1 mm Hg.
†Assuming the condensation coefficient is unity.

ratio N_g/N_e is proportional to p_g/p_e. Values of N_g given in Table 1 show that under the commonly employed experimental conditions of vacuum ($\sim 10^{-5}$ Torr) and deposition rates (~ 1 Å/sec), the impingement rate of gas atoms is relatively quite large, so that if the sticking coefficient of gas atoms is not negligibly small, a considerable amount of gas sorption could occur.

2.2 Evaporation Methods

Thermal evaporation may be achieved directly or indirectly (via a support) by a variety of physical methods. Several variants are described below.

(1) Resistive Heating. This method consists of heating the material with a resistively heated filament or boat, generally made of refractory metals such as W, Mo, Ta, and Nb, with or without ceramic coatings. Crucibles of quartz, graphite, alumina, beryllia, and zirconia are used with indirect heating. The choice of the support material is primarily determined by the evaporation temperature and resistance to alloying and/or chemical reaction with the evaporant. With the exception of highly reactive materials such as Si, Al, Co, Fe, and Ni, most materials present no problem with evaporation from suitable supports. New materials are constantly being developed to overcome difficulties with reactive evaporants. A comprehensive list of materials, their physical properties of interest for thermal deposition, possible support materials, and applicable deposition techniques other than by evaporation is given in the Appendix (Sec. 7).

Vapor sources of various types, geometries, and sizes can be easily constructed or obtained commercially. Some forms of these sources are illustrated in Fig. 1. The baffled sources are indispensable for materials which spatter. Among other interesting spatter-free sources are those of Vergara et al. [4] utilizing evaporation and subsequent reevaporation from another hot surface, and of Gretz [5] for controlled magnetic feeding of degassed material particles into the evaporation furnace inside an ultrahigh-vacuum ($\sim 10^{-10}$ Torr) ion-microscope system. A satisfactory support material [6] for evaporation of Al is an intermetallic composite ceramic (50 percent mixture of boron nitride and titanium diboride, available from Union Carbide). Some sublimation sources have been described by Card and Galen [7]. The deposition of radioactive materials is described by Parker and Grunditz [8].

It should be noted that a point vapor source (Knudsen source) rarely yields the ideal cosine distribution. Most Knudsen sources are highly directional. A free molecular flow with cosine distribution is expected when vapor emission occurs through an aperture of dimensions ~5 mm for 10^{-2} Torr vapor pressure. A multiple Knudsen source [Fig. 1(c)] used by the author gives a uniform deposit due to the superimposition of different beams. Preuss and Alt-Anthony [9] established marked preferential evaporation (beaming effect) from the open ends of both the basket and the cylindrical filaments. For a more opened mouth, 45° conical spiral, an approximate cosine distribution is realized.

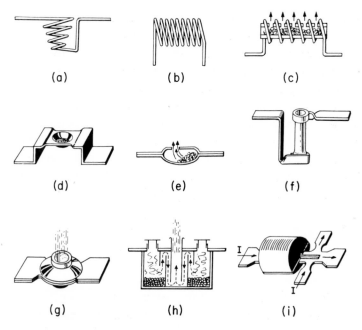

(a)   (b)   (c)

(d)   (e)   (f)

(g)   (h)   (i)

FIG. 1. Some thermal-evaporation sources: (*a*) basket, (*b*) spiral, (*c*) multiple-vapor-beam source consisting of a quartz tube with several holes ~ 1 mm diameter and heated by a tungsten spiral (*Chopra, unpublished*), (*d*) dimpled boat, (*e*) asymmetric oven-type point source, (*f*) howitzer, (*g*) crucible heater, (*h*) baffled chimney (also called Drumheller source), (*i*) dual boat, flash-evaporation source. Here the liquid mixture of components is kept in the covered section of the boat and a controlled amount is allowed to pass through a narrow aperture into the open boat for flash evaporation at a much higher temperature maintained by a separate, high-resistance electrical path (*after Dale, Res. Develop., May, 1967, p. 60*).

Sublimation. If a material has a sufficiently high vapor pressure before melting occurs, it will sublime, and the condensed vapors form a film. Since the rates of sublimation for most materials are small, this method does not find widespread applications. However, a useful application has been obtained in sublimation of resistively heated Nichrome wires. Huijer et al. [10] showed that as a result of diffusion of Cr at elevated temperatures, an equilibrium state is reached under which the Nichrome components sublimate at relative rates equal to their relative concentration in the alloy, thereby making it possible to obtain films of the same composition as the evaporant. Calculations and experiments show that equilibrium is reached when Nichrome wire is held at 1330°C for 3 hr.

Multiple-component Evaporation. When multicomponent alloys or compounds are thermally evaporated, the components may evaporate at different rates because of their different vapor pressures, their different tendencies to react with the support material, and possible thermal decomposition of the parent material. These factors will obviously produce nonstoichiometric films. Due to vapor-pressure difference alone, the ratio of the evaporation rates of components A and B of an alloy may be obtained from Langmuir's expression [Eq. (1)] by assuming that the vapor pressure of each component in the alloy is depressed compared with that in the pure state by an amount proportional to the relative concentrations (called Raoult's law). It is given by

$$\frac{N_A}{N_B} = \frac{C_A p_A}{C_B p_B} \sqrt{\frac{M_B}{M_A}} \tag{5}$$

where the C's are the atomic fractions of the components. Generally, this relation is not obeyed because of strong interaction between the components of the alloy, and one must introduce an activity coefficient [11] as a function of C to correct for the departure.

A point of technical interest [12] is that the higher the evaporation temperature of an alloy of widely different vapor-pressure components is, the closer is the composition of the film to the bulk alloy. By using inert support materials and a suitable evaporation temperature, a large number of alloys and compounds can actually be deposited with only slight deviations from the original composition. Large deviations occur, however, for compounds which dissociate readily on heating.

A satisfactory method of preparing alloys and compounds with precisely controlled compositions is to evaporate each component from a separate source with reaction and homogeneity brought about at the substrate at an elevated temperature. Such a method has been used by many workers (13-16) and is referred to as the "two-source" or "three-temperature" or "Günther" [17] (in the case of semiconducting compounds) technique.

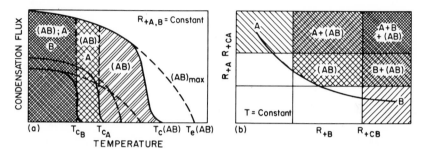

FIG. 2. (*a*) Schematic representation of the variation of the condensation flux as a function of the substrate temperature for a constant flux of incident vapor of type *A, B,* and *AB* molecules. The subscript with plus sign indicates the incident flux; the subscripts *c* and *e* refer to the critical (for condensation) and the equilibrium values, respectively. The dotted curves are the equilibrium curves which take into account the adatom densities of the other components. (*b*) Condensation diagrams for *A* and *B* incident vapor at a constant substrate temperature. Regions of different compositions obtained are shown. The region where only pure *AB* compound is formed is obtained by the straight lines representing critical flux conditions and the solid curve given by $R_{+A} \times R_{+B}$ = constant. (*Günther* [17])

Günther [17] analyzed the condensation of multicomponent vapors on the basis of the nucleation theory (Chap. IV), according to which a critical flux exists for every substrate temperature (and vice versa) at which spontaneous nucleation occurs. After nucleation, the condensation flux quickly approaches a maximum value. The schematic representation of the dependence of the condensation flux on the incident fluxes and the substrate temperature is shown in Fig. 2(*a*) and (*b*), respectively. The vapor density is generally low enough to neglect collisions between particles of the *A* and *B* components in the vapor phase. However, attractive interaction takes place between these particles in the adsorbed stage on the substrate surface, which may lead to the formation of a general molecule $A + B \rightarrow A_n B_m$ of a stable compound.

As seen in Fig. 2, a broad stoichiometric interval exists for both temperature and incident fluxes of A and B vapors within which an exactly stoichiometric compound is condensed. Thus, in principle, the three-temperature method allows deposition of stoichiometric films of any material when one component has a significantly higher vapor pressure than the other. The applications of this technique to II-VI, IV-VI, III-V, and other compounds are reviewed by Günther.

(2)   Flash Evaporation. A rapid evaporation of a multicomponent alloy or compound, which tends to distill fractionally, may be obtained [18,19] by continuously dropping fine particles of the material onto a hot surface so that numerous discrete evaporations occur. Alternatively, a mixture of the components in powder form may be fed into the evaporator. The arrangement used is shown in Fig. 3(a). Fine powder ($\sim$100 to 200 mesh) is fed into a heated Ta or Ir boat by mechanically or ultrasonically agitating the feed chute. This system has been used for preparing films of cermets [20], Bi_2Te_3 [21], III-V compounds [22], etc. The Ta boat is held at 1300 to 1400°C for III-V compounds, and the substrate is heated to $\sim$200 to 550°C for obtaining single-phase compounds.

There is evidence [23] that flash-evaporated films in some cases can show considerable deviation from the original composition. Further, since the large amount of rapidly released gas produces spattering of particles from the evaporant, this method is not easily controllable.

(3)   Arc Evaporation. By striking an arc between two electrodes of a conducting material, sufficiently high temperatures can be generated to evaporate refractory materials such as Nb and Ta. This method, widely used for evaporation of carbon for electron-microscope specimens, employs a standard dc arc-welding generator connected to the electrodes with a capacitor across the electrodes [see Fig. 3(b)]. The arc, initiated by bringing the electrodes together, may last $\sim$1/10 sec. Deposition rates $\sim$50 Å/sec or higher can be obtained [24, 25] for refractory metals, but the process is not easily reproducible.

(4)   Exploding-wire Technique. This technique [26,27] consists of exploding a wire by a sudden resistive heating of the wire with a transient high current density approaching 10^6 A/cm^2. This is achieved by discharging a bank of condensers ($\sim$10 to 100 μF), charged to a voltage $\sim$1 to 10 kV, through a metallic wire [Fig. 3(c)]. Thus, a catastrophic destruction and vaporization of the wire at some region takes place.

Kul'gavchuk and Novoskol'tseva [28] studied the kinetics of the explosion stages by x-ray techniques. But the evaporation kinetics remain obscure. The equivalent source temperature as deduced from the measured velocity of the vapor atoms is estimated to be as high as $10^{6\circ}$K. Mattox et al. [27] obtained equivalent film-deposition rates $\sim 10^6$ Å/sec by exploding 10- to 20-mil-diameter Cu and Au wires in less than 100 μsec, using 1,000 J of energy. Thin films so formed showed regions of defects due to the condensation of microparticles spattered during the explosion.

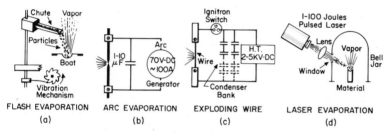

FIG. 3. Diagrams of some special, high-thermal-evaporation-rate sources. (*a*) Flash evaporation, (*b*) arc evaporation, (*c*) exploding-wire (with high current density), and (*d*) laser evaporation. The techniques are described in the text.

(5) Laser Evaporation. The enormous intensity of a laser may be used to heat and vaporize materials by keeping the laser source outside the vacuum system and focusing the beam onto the surface of the material to be evaporated [Fig. 3(*d*)]. This promising method [29,30] has not yet been fully exploited. Since the laser penetration depth is small ($\sim$100 Å), evaporation takes place at the surface only. Degassing and explosion of the specimen can be minimized by using material in a fine powder form ($\sim$10 μ).

Schwarz and Tourtellotte [30] used a non-Q-spoiled glass neodymium laser to deliver 80 to 150 J of energy per burst with a duration of 2 to 4 msec to evaporate $BaTiO_3$, $SrTiO_3$, ZnS, and Sb_2S_3. Deposition of several thousand angstroms per burst was obtained corresponding to a rate $\sim 10^6$ Å/sec. The actual rate may be higher, since most of the evaporation probably takes place in the initial 10 to 100 μsec. The temperature of the vapor source is estimated to be $\sim$20,000°K (corresponding to an energy of 1 to 1.5 eV). The emitted vapors were observed to be positively charged carrying a current up to 1 mA.

Note that methods (2) to (4) (Fig. 3) have among them the common feature of high evaporation rate due to a high source temperature. This means that the vapors have a high kinetic energy and are expected to carry a fair amount of electrostatic charges due to thermal ionization. Because the influence of these features on the structure and growth of films is of fundamental interest (Chap. IV, Sec. 2.3), these methods have special attraction.

(6) RF Heating. The rf or induction heating [31-33] may be supplied to the evaporant directly or indirectly from the crucible material. By suitable arrangement of the rf coils, the induction-heated material can be levitated, thereby eliminating the possibility of contamination of the film by the support material. Ames et al. [6] made good use of rf heating to prevent migration of Al out of the crucible by evaporating it from the surface of a (BN + TiB$_2$) crucible. They used a crucible thin enough near the surface to offer no shielding of the rf field and thus allow heating of the Al melt by the 200 kc/sec field. However, the crucible was thick enough elsewhere to screen the melt from excessive coupling with the rf field, thus minimizing turbulence.

(7) Electron-bombardment Heating. The simple resistive heating of an evaporation source suffers from the disadvantages of possible contamination from the support material and the limitations of the input power, which make it difficult to evaporate high-melting-point materials. These drawbacks may be overcome by an efficient source of heating by electron bombardment of the material. In principle, this type of source is capable of evaporating any material at rates ranging from fractions of an angstrom to microns per second. Thermal decomposition and structural changes of some chemical compounds may occur because of the intense heat and/or energetic electron bombardment. These effects are frequently exploited to form films of the decomposed products and are discussed in Sec. 4.

The simplest electron-bombardment arrangement consists of a heated W filament to supply electrons which are accelerated by applying a positive potential to the material for evaporation. The electrons lose their energy in the material very rapidly, their range being determined by their energy and the atomic number of the material. Thus the surface of the material becomes a molten drop and evaporates. In contrast to this simple work-accelerated electron-beam arrangement [34-38], one may use electron optics [39,40] to focus the beam and direct it onto the material for evaporation. A multitude of electron-beam source designs

(Fig. 4) are possible and have been employed by experimenters and commercial producers. The simplest sources are shown in Fig. 4(a) and (b); the latter, designed by Chopra and Randlett [37], is a demountable system. The components are mounted on dovetailed H_2-fired lava blocks and can be moved relative to each other to allow adjustments for focusing and beam current. The filament is shielded from the vapor by the grounded Ta shield, which also acts to focus the electron beam electrostatically to about 3 mm diameter. The vapor traverses the electron beam and is partially charged. This aspect is of interest in understanding the role of charges on the structure and growth of films. The material to be evaporated is placed in a water-cooled Cu pedestal or can be fed into the electron beam by an automatic mechanical or electrical device. Since the melting and evaporation of the material are confined to the surface, the support pedestal presents no contamination problems.

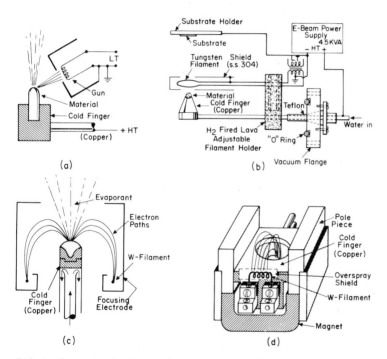

FIG. 4. Some simple electron-beam sources for thermal evaporation. (a) and (b) are work-accelerated guns designed by Robin-Kandaré et al. [36], and Chopra and Randlett [37], respectively; (c) uses transverse electrostatic focusing (*Unvala and Booker* [40]), and (d) magnetic focusing (*Varian Associates*).

The arrangement [40] shown in Fig. 4(c) allows a transverse focusing by the electrostatic shield surrounding the filament and the evaporant, and has been used to deposit films of Si at a rate $\sim3\mu$/min. Magnetic bending and focusing of the beam are readily achieved by suitably designed permanent (e.g., ceramic magnets), or wirewound magnets yielding a field ~100 to 300 G. A commercial design by Varian Associates is shown in Fig. 4(d). The ease with which an electron beam can be deflected by a magnetic and/or electrostatic field may be exploited to scan or to program the same beam to perform sequential depositions from different source materials. Such systems are in common use at present.

The electron-source filament is generally heated by a low-tension (~6 to 12 V, 100 A) transformer with a center tap that is grounded or, if floating, it should be capable of withstanding high-voltage isolation. The high tension supplies 0 to 20 kV and currents of up to 500 mA. Generally, 2- to 4-kW power supplies suffice for most laboratory projects. The maximum current density is limited by the space charge and is given by

$$I = 2.3 \times 10^{-6} \frac{V^{3/2}}{d^2} \quad \text{amp/cm}^2 \tag{6}$$

so that a higher accelerating voltage or a shorter cathode-anode distance d is required to obtain a high current density. If the vacuum is poor either to start with or as a result of thermal desorption of gases from the evaporant, positive ions will be formed by the ionization of the gas (thermally produced positive ions of the evaporant may form a negligible part) and will neutralize the space charge. Precautions are necessary to prevent sudden current transients due to discharges in a poor vacuum. Note that the energetic electrons lose most of the energy at the end of their range in the material. Thus, most of the heating takes place at a depth (determined by the accelerating voltage) below the surface. If excessive gases are desorbed internally, the molten material will be splashed. It is therefore essential to degas the material by slow heating for some time.

3. CATHODIC SPUTTERING

3.1 Sputtering Process

The ejection of atoms from the surface of a material (the target) by bombardment with energetic particles is called "sputtering." If ejection is due to positive-ion bombardment, it is referred to as "cathodic sputtering." The ejected or sputtered atoms can be condensed on a substrate to form a thin film. The sputtering phenomenon has been known since 1852 [41] and exploited for deposition of films. Because of the high pressure of gas used and high sensitivity to contamination in commonly used glow-discharge sputtering, the technique has generally been termed "dirty." But improved technology and new variants of sputtering arrangements have now revived low- and high-pressure sputtering as a versatile and powerful deposition technique for both research and production purposes.

The sputtering process has been subjected to numerous experimental and theoretical investigations, and excellent reviews of the subject have been published in the literature [42-47]. The practical application of the sputtering process to thin film deposition and the properties of sputtered films have also been reviewed by several workers [48-50]; particular mention should be made of the comprehensive reports by Holland [2] and Maissel [50]. We shall present here a summary of the observed salient features of the sputtering phenomenon primarily as an aid to understanding sputtering as a deposition technique.

1. The sputtering yield [51,52], defined as the average number of atoms ejected from the target per incident ion, increases with the increasing energy of ions and their mass (relative to the atomic weight of the target). Typical yield variations with the bombardment energy for Cu, Ni, and Mo [53] are shown in Fig. 5. A sputtering threshold exists between 5 and 25 eV for most metals, the lower limit being set by the sublimation energy. The questions of the existence and the nature of the threshold are not satisfactorily settled at present. The yield increases very rapidly for energies beyond threshold, shows a small region of linear proportionality, and then approaches saturation. Moreover, at very high energies, the yield decreases because of the increasing penetration depth and hence increasing energy losses below the surface, with the consequence that not all ejected atoms are able to reach the surface to escape. The saturation region occurs at higher energies for heavier

bombarding particles. Thus, Xe^+ bombardment of Au shows saturation above 100 keV, whereas the Ar^+ bombardment curve is saturated at less than 20 keV.

The yield depends on the angle of incidence (see the data for Cu and Mo in Fig. 5) and increases approximately as $(\cos\theta)^{-1}$, where θ is the angle between the normal to the target surface and the beam direction [47].

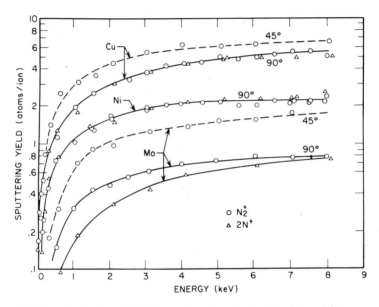

FIG. 5. Sputtering yield (atoms ejected per ion) of Cu, Ni, and Mo as a function of the bombarding energy of the N^+ and $2N^+$ ions. The variation of the yield for a 45° incidence of N_2^+ ions is shown (dotted curve) for Cu and Mo. (*Bader et al.* [53])

2. The yield as a function of the atomic number of the target displays an undulatory behavior, and the periodicity of the undulations corresponds to the groups of elements of the periodic chart of the elements. The yield increases strongly with decreasing heat of vaporization. Yields for Ar^+ bombardment of several materials are listed in Table II, and those for a Cu target bombarded with different types of ions are compiled in Table III.

3. The yield of a single-crystal target increases with decreasing transparency of the crystal in the direction of the ion beam. Atoms sputtered from such targets tend to be ejected predominantly along

preferred directions of close packing, which have the least transparency [54]. Ejection patterns are observed at all temperatures only if the bombarding energy is below a critical value. For higher energies, patterns appear only above a certain target temperature.

4. The yield is rather insensitive [55] to the target temperature except at very high temperatures (above ~600°C for Cu and Au) when it shows an apparent rapid increase due to accompanying thermal evaporation.

Table II. (a) Sputtering Yields for Various Materials Bombarded by Ar+ (Compiled by Maissel [50])

Target	Bombarding energy, kV					
	0.2	0.6	1	2	5	10
Ag	1.6	3.4				8.8
Al	0.35	1.2				
Au	1.1	2.8				
Co	0.6	1.4				
	0.7	1.3				
Cu	1.1	2.3	3.2	4.3	5.5	6.6
Fe	0.5	1.3	1.4	2.0*	2.5*	
Ge	0.5	1.2	1.5	2.0	3.0	
Mo	0.4	0.9	1.1			2.2
Nb	0.25	0.65				
Ni	0.7	1.5	2.1			
Os	0.4	0.95				
Pd	1.0	2.4				
Pt	0.6	1.6				
Re	0.4	0.9				
Rh	0.55	1.5				
Si	0.2	0.5	0.6	0.9	1.4	
Ta	0.3	0.6				
Th	0.3	0.7				
Ti	0.2	0.6				
U	0.35	1.0				
W	0.3	0.6				
Zr	0.3	0.75				
GaSb (111)	0.4	0.9	1.2			
SiC		1.8				

*Type 304 stainless steel.

(b) Sputtering Yields[†] in Molecules per Ion for Various Insulators Bombarded by Argon Using RF (Davidse and Maissel [96])

Target	Mean ion energy, kV		
	1.1	2.0	2.9
SiO_2	0.16	0.39	0.50
Pyrex 7740	0.15	0.33	0.43
Al_2O_3	0.05	0.12	0.17

[†]The values were obtained by comparison with Ge and Si, whose dc yields in this energy range are known.

Table III. Sputtering Yield (Normal Incidence and Normal Ejection) for Polycrystalline Copper Target Bombarded by 1-kV Ions of Different Masses (Collated from Data Compiled in Ref. [50])

Positive ion	Approximate mass	Yield
He	4	0.08 (at 600 V)
N	14	1.5
Ne	20.2	1.8
N_2	28	2.0
Ar	39.9	3.2
Kr	83.8	3.4
Xe	131.3	3.6
Hg	200.6	1.3 (at 400 V)

5. The ejected atoms have considerable energies [56,57]. A typical spectrum is shown in Fig. 6(a). The distribution is Maxwellian below the peak. With increasing bombarding energy, the peak shifts only slightly toward higher energies, because of the counteracting effect of increased penetration of more energetic ions. The long tail after the peak continues to increase in length and area with increasing energy. The velocities

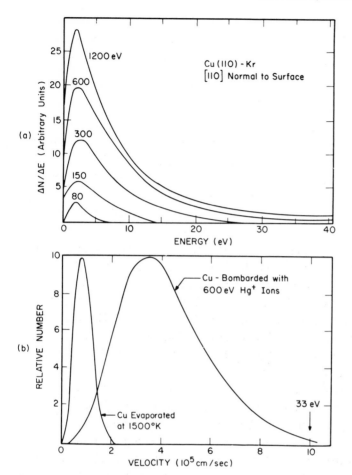

FIG. 6. (*a*) Energy distributions of Cu atoms ejected in the [110] direction from a (110) Cu target for several energies of the bombarding Kr ions. (*b*) Velocity distribution of sputtered and evaporated Cu atoms. Cu bombarded with 600-eV Hg⁺, and Cu evaporated at 1500°K. (*Stuart and Wehner* [56])

corresponding to the energies of sputtered atoms are considerably higher than those of the evaporated atoms, as seen in Fig. 6(*b*). The sputtering rate of 10^{18} atoms/(cm²)(sec) from a Cu target by Hg⁺ ions corresponds to evaporation at 1500°K. The Maxwellian distribution of thermally evaporated atoms has a peak at 0.77×10^5 cm/sec (0.2 eV) and a cutoff at $\sim 2 \times 10^5$ cm/sec (1.3 eV), whereas corresponding values for sputtered atoms are 3.5×10^5 cm/sec (4.0 eV) and 12×10^5 cm/sec (48 eV), respectively.

The peak energy of the ejected atoms increases [55] with the angle of ejection with respect to the target-surface normal. However, a decrease occurs for very large angles ($>60°$), probably because the increased depth of the escape factor is dominant over the enhancement factor resulting from the smaller energy required for oblique ejection.

6.   The ejected particles are considered to be primarily atoms or clusters of atoms. There is evidence that in sputtering of Cu and Ag at high energies, more than 10 percent of the ejected particles are diatomic molecules [58]. Molecular ejection is also indicated [48] for a GaSb target. Comas and Cooper [59] carried out mass analysis of species from (110), (111), and ($\bar{1}\bar{1}\bar{1}$) GaAs crystals ejected by 0- to 140-eV Ar ions. For each plane, approximately 99.4 percent of the collected ions were neutral Ga and As atomic species; the balance were neutral GaAs molecules. No neutral Ga_2, As_2, or $(GaAs)_2$ molecules, or negative $\bar{Ga}$, $\bar{As}$, or $(Ga\bar{As})$ ions were detected. These examples show that the type of the ejected species depends on the target material.

7.   The ejected atoms leave the surface in an excited or ionized state and show characteristic recombination light emission. The low-energy particles are rapidly deexcited by resonance and Auger processes near the surface. About 1 percent of the ejected particles from metals, Ge, and Si are charged; both species, positive and negative, are emitted [44]. This fraction increases rapidly with increasing energy of H_2^+, D_2^+, and He_4^+. A part of the ions are reflected projectiles and the rest are ejected atoms. A considerable number of secondary electrons are generated because of their kinetic ejection [60] via interaction of the energetic projectile with the valence electrons of the target, potential ejection [61] via neutralization of the ion at the surface, and Auger (excited) emission. The fraction of secondary electrons per Ar^+ and He^+ ion bombardment of a W surface is ~0.1 and 0.24, respectively, for energies ~200 to 1,000 eV.

Let us now say a few words about the theories of sputtering. The hot-spot evaporation theory, formulated by von Hippel [62] and investigated by Townes [63], has yielded to the overwhelming experimental evidence, due largely to the work of Wehner [51,54], that sputtering is not an energy but rather a momentum–transfer process, as originally suggested by Stark [64]. The early momentum-transfer or collision theory of Kingdon and Langmuir [65] has seen numerous developments. Henschke's [66] billiard-ball model, valid at low energies, considers sputtering as the result of double or triple collisions of the ion with the lattice atoms followed by its back reflection by the lattice.

More sophisticated theories, such as those of Keywell [67] and others [68,69], consider sputtering as essentially a radiation-damage phenomenon. Accordingly, the incident ion displaces a number of atoms (referred to as "knock-ons") during its passage through the material and thus loses its energy, or "cools." Some fraction of the knocked-on atoms will diffuse to the surface and emerge as sputtered atoms. The knocked-on atoms may also have sufficient energy to produce additional displaced atoms which can contribute to the total sputtering yield. The exact details of the interaction of an ion with the target atom, the momentum transfer, and the collision mean free path clearly depend on its energy. With increasing energy (E), the interaction changes from billiard-ball to Coulombic and then to a nucleonic model. Pease's [69] theory gives the following expression for the sputtering yield:

$$Y = An^{2/3} \frac{E}{4E_d} \left\{ 1 + \left[\frac{\ln (\bar{E}/E_s)}{\ln 2} \right]^{1/2} \right\} \qquad (7)$$

where A is the cross section for imparting an energy greater than E_d, the energy required to displace an atom from its lattice site, $\bar{E}$ is the mean energy of the struck atom, E_s is the sublimation energy, and n is the number of atoms per unit volume. Note that in this theory, a threshold is set by the displacement energy E_d.

The ejection of atoms along close-packed directions [54] can be understood on the basis of focused collision sequence (focusons) along close-packed directions as postulated by Silsbee [70]. The applicability of a focuson process for low-energy sputtering has, however, been questioned by Harrison et al. [71], and Lehmann and Sigmund [72], who have shown that ejection patterns can result simply from the regular crystal structure at the surface of a solid.

3.2 Glow-discharge Sputtering

A cheap and simple source of ions for sputtering is provided by the well-known phenomenon of glow discharge due to an applied electric field between two electrodes in a gas at low pressures. A typical dc current-voltage characteristic of such a diode structure and its visual appearance are shown in Fig. 7(a) and (b), respectively. The gas breaks down to conduct electricity when a certain minimum voltage is reached.

The attendant glow discharge maintains itself at a constant voltage and is referred to as "normal glow." The region where both voltage and current increase together is called the "abnormal glow." A luminous layer which covers the cathode partially in the normal glow and entirely in the abnormal glow is known as the "cathode glow." A fairly well defined region of relatively low luminosity known as "Crookes" or "cathode" dark space is adjacent to it. This is followed by a bright "negative glow" region, after which ill-defined regions of the Faraday dark space and the "positive column" can be seen.

The cathode dark space is the most important region. Most of the applied voltage is dropped (called "cathode fall") across it. Ions and electrons created at the breakdown are accelerated across this region. The energetic electrons produce more ions by collisions with the gas atoms in the negative glow, and the energetic ions strike the cathode to produce sputtering and emit secondary electrons which are essential for

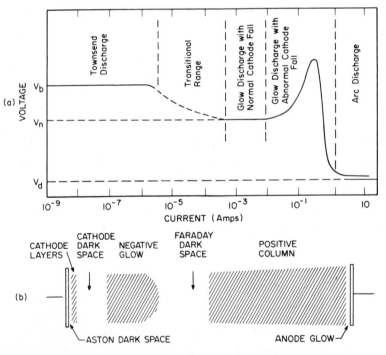

FIG. 7. (*a*) Typical current-voltage characteristics of an electrical discharge through gas at low pressures. (*b*) A visual representation of the principal regions in a low-pressure dc glow discharge.

sustaining the glow. The thickness d of the cathode dark space is inversely proportional to the pressure p of the gas (Paschen's law) such that for Ar gas the product $pd = 0.3$ Torr-cm. The number of collisions of the electrons traversing this region is the same at all pressures and is about 8 for Ar gas.

Effective sputtering is possible only when both the number of ions and their energy are large and controllable. This is conveniently effected in the abnormal-glow discharge region. The ion energy is less than or equal to the cathode fall, depending on whether or not it collides with gas atoms during its transit.

Several factors which influence the operation of glow discharge as a technique for sputter deposition of films will now be considered.

(1) Pressure. As the gas pressure is increased, the discharge current increases, the voltage falls, and the cathode dark space decreases. The number of ions increases (approximately proportional to p^2), but their energies decrease. Since the yield increases proportionally with the number of ions, but decreases with decreasing ion energy linearly or less than linearly in the practical range of a few kilovolts, a net increase in the total number of atoms ejected results. There is, however, an upper limit to it, since with increasing pressure, ejected atoms suffer more collisions and are thus prevented from reaching the anode. For example [62], at 0.1 Torr, less than 10 percent of the ejected atoms may reach the anode. The sputtering yield of Ni[73], bombarded by 150-eV Ar ions, is fairly constant (~0.47) for pressures down to 20 mTorr and thereafter shows an apparent drop. Thus, the optimum pressure range for glow-discharge sputtering is between 25 and 75 mTorr (or microns).

(2) Deposit Distribution. Because of collisions with the ambient gas atoms at high pressures, the sputtered atoms are diffusely scattered during transit and therefore reach the anode with randomized directions and energies. As a result of the diffuse nature of material transport, the atoms deposit at places not necessarily in the line of sight of the cathode. Also note that, because of collisions, the energetic ions hit the cathode at high oblique angles, which actually is helpful in increasing the yield.

At constant pressure and constant applied voltage, the deposition rate is low at large distances from the cathode and shows a decided maximum at the center. As this distance is decreased, a more uniform deposit first results which then becomes annular in nature with a maximum thickness on a circle slightly smaller than the target. The optimum conditions of deposition with uniformity of deposit extending to about half the area of the target are obtained when the cathode-anode distance is about twice the length of the cathode dark space.

As the anode (or any other physical obstruction) approaches the cathode dark space, the glow discharge extinguishes and no sputtering occurs. This fact is advantageously utilized in preventing glow discharge between the back side of the cathode or the high-tension lead wires and the neighboring support materials. It is achieved by connecting the support materials and any auxiliary electrode such as Al foil wrapped around the high-tension lead wires to anode potential, and by keeping the distance between the cathode and the nearest anode less than the cathode dark space.

Ions and electrons in a plasma do not recombine efficiently because the difference in their masses makes it difficult to conserve momentum. Thus, low-temperature plasmas may diffuse at substantial distances from the cathode. Recombination, however, takes place much more easily at walls and pointed and contaminated regions of the glow-discharge geometry, which act as sinks for a diffusing plasma. The recombination, resulting in a neutral molecule, releases considerable energy as heat. The plasma-sink regions also distort the uniformity of the glow discharge and hence the sputtering rate.

(3) Current and Voltage Dependence. The sputtering rate is proportional to the current for a constant voltage which is thus a very convenient control parameter. The voltage dependence is nonlinear (see Fig. 5), but for a certain range of applied voltages, depending on the gas and the target material, the glow-discharge sputtering may be operated in a linear range so that the sputtering rate is proportional to the product of current and voltage. Typical conditions employed for plane cathode sputtering are 1 to 5 kV potential with a current density 1 to 10 mA/cm^2 obtained from a rectified power supply capable of supplying up to 500 mA. A high-wattage current-limiting series resistor is essential to prevent arcing.

(4) Cathode. A plane cathode of area about twice that required for a uniform deposit is used. The cathode material may be a plate, foil, or electroplated deposit onto a suitable (normally low-yield materials such as stainless steel) support target material. Because of the bombardment of ions, the cathode gets hot. The temperature increases rapidly to approach an equilibrium value. Both the rate of temperature rise and the maximum temperature attained depend on the power dissipated at the cathode, the thermal characteristics (such as conductivity and emissivity) of the cathode, the gas pressure, etc. Typically, at 1 kV and 1 mA/cm^2 the temperature of an Au target rises to 200 to 300°C in about 1 min operation. These temperatures do not significantly alter the sputtering yield, but they have other undesirable effects such as that of heating the substrate, or heating the gas resulting in changes in density and discharge conditions. Although a heavy, high-thermal-conductivity cathode may have only tolerable temperature rise, it is generally desirable to cool the cathode with running water or some other cooling fluid. Several cooling systems are described in the literature [74].

In addition to plane cathodes, wire [75], cylindrical [76], and concave [2] cathodes have also been studied. The wire geometry which is useful for deposition inside a cylindrical substrate enhances ion bombardment and thus gives deposition rates that are considerably higher than those obtained by plane cathodes. The concave cathode concentrates bombardment between the boundary and center of the cathode, thus producing a deposit roughly annular in shape with little deposit in the middle. Mention should also be made of the multiple-cathode designs [74,77-79] which allow simultaneous or sequential sputtering from various cathodes.

(5) Contamination Problem. Even if a leakproof sputtering system is initially pumped down to a high vacuum ($\sim 10^{-8}$ Torr) and then sputtering gas of high purity (the commercially available high-purity Ar generally requires passing through a cold trap to remove oil and water vapors) is admitted, contaminants may still appear from (1) the outgassing as a result of plasma-discharge heating of the walls and contaminated components of the sputtering chamber which are not adequately grounded or shielded, and (2) the decomposition of oil vapors as a result of back streaming from the diffusion pump operated at high pressures. Since only fluid pumps (mechanical and/or diffusion) can be used for pumping a high-pressure sputtering chamber, the use of an optically dense baffle, preferably cooled, and some provision for throttling of the diffusion pump are highly advisable.

Mass-spectrometer analysis [80,81] of the composition of the background gases before, during, and after sputtering shows an immediate decrease in the concentration of reactive gases such as O_2, N_2, and water vapors, but a sharp increase of H_2 occurs during sputtering. This is possibly due to cracking of higher-mass hydrocarbons in the glow discharge. The presence of H_2 influences the sputtering yield considerably [81]. Analysis by flash photolysis (Chap. III, Sec. 2.1) of the gas content of both glow-discharge and ion-beam low-pressure sputtered films of Mo prepared in the author's laboratory confirms the presence of large amounts (>3 percent) of H_2 in both high- and low-pressure ($\sim 10^{-5}$ Torr) sputtered films.

In addition to chemical sorption of the ambient gases by the film, Winters and Kay [230] have shown that an impact-activated sorption of the sputtering gas occurs which increases rapidly in the case of Ni films sputtered with Ar ions of energy greater than 100 eV. Depending on the deposition conditions, a concentration of 10^{-1} to 10^{-4} Ar atoms per Ni atom was found.

(6) Deposition Control. One of the chief advantages of the sputtering technique is that the rate of deposition remains constant with time, provided the current density and voltage do not vary, a condition that is easily attained by using an automatic pressure controller and a regulated power supply. One can also control the rate by controlling the discharge-current density electronically. A quartz-crystal oscillator (discussed later) may be used for monitoring and controlling rate by means of a feedback mechanism to control the discharge current. It is important, however, to position the monitor so that it does not disturb the plasma and is well shielded. A small magnet may be used to deflect the ions.

3.3 Sputtering Variants

Several systems have been employed for deposition of films by sputtering. These sputtering variants are shown schematically in Fig. 8. The simplest and most widely used one (thoroughly discussed in Ref. 2, among others) utilizes the glow discharge between two electrodes and is commonly referred to as a diode arrangement (*a*). The substrate in such a diode system is normally placed on the anode and kept at anode potential. It may be left floating, in which case it will acquire a negative potential of several volts relative to the anode and will thus attract impurities and gas ions, leading to contamination of the film. On the other hand, if the substrate is held at a large negative potential, the film will be subjected to steady ion bombardment throughout its growth, a process which effectively "cleans" the film of adsorbed gases otherwise trapped in it as impurities. This technique (*b*), employed by Maissel and Schaible [82], is called "bias sputtering." The idea of sputter deposition with simultaneous sputter cleaning by bombardment was originally conceived by Frerichs [83], and his arrangement is shown in (*c*). Here,

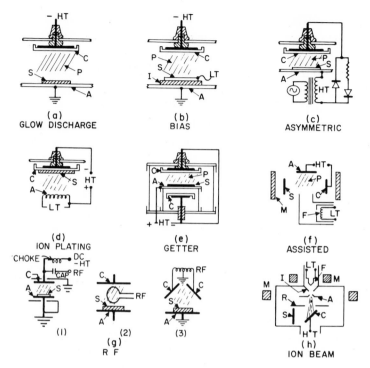

FIG. 8. Schematic arrangements for several sputtering variants; (*a*) dc glow discharge (diode sputtering); (*b*) dc bias; (*c*) ac asymmetric (or bias); (*d*) ion plating (deposition by evaporation with partial dc sputtering); (*e*) getter sputtering; (*f*) thermionically and/or magnetically assisted glow discharge; (*g*) rf sputtering with (1) capacitor coupling for sputtering of metals and the inductor coupling for isolating dc from rf power supply, (2) indirect or induced rf coupling, and (3) push-pull double cathode; (*h*) duoplasmatron ion-beam source. The notations are P, plasma or discharge; C, cathode; A, anode; S, substrate; LT, low tension; HT, high tension; F, filament (electron source); M, magnet; V, vapor; R, reference; I, intermediate electrode, insulating in (*b*).

an asymmetric alternating rather than direct current is applied between cathode and substrate so that more material is deposited on one half-cycle than is removed by reverse sputtering in the other half-cycle. Bombardment removes not only the adsorbed gases (yielding a purer film) but also the initial oxide layers which are responsible for good bonding of the film. Good bonding may be retained by precoating with a positive bias.

The ac sputtering method is, however, more complicated and less efficient than bias sputtering. The purer quality of the films in both cases has been established by measuring the electrical resistance and super-conducting properties of the films. Increasing the bias voltage rapidly reduces the resistivity of sputtered Ta films from several hundred at zero bias to less than 40 $\mu\Omega$-cm in films $\sim$1,000 Å thick for a bias of about -200 V. The high resistivity of the unbiased film has been attributed to $\sim$20 percent impurity concentration, which seems anomalously large. The reduction of impurities may not be the only resistivity-reducing factor in bias sputtering. The bombardment of the film with charged ions during growth is expected (Chap. IV, Sec. 2.3) to have a considerable influence on film structure and will increase the crystallite size, thus reducing the grain-boundary scattering contribution to resistivity. Structural studies of bias-sputtered films are therefore of interest.

The effect of sputter cleaning of an evaporated film is utilized in a system (d) called "ion plating" [84] (a misnomer). The films are obtained by evaporation from a filament and the deposit is simultaneously bombarded with accelerated gas ions. The electrons bombard the filament and partially sustain evaporation.

The efficient gettering action of films for reactive gases during deposition was put to advantage by Theuerer and Hauser [85] in technique (e), called "getter sputtering." Here, two cathodes of the material to be sputtered are symmetrically located with respect to a Ni anode can. The lower cathode is used to maximize the gettering action of active gases where they enter near the bottom of the can at $\sim$100 mTorr pressure. After sputtering for a few minutes, the sputtered material from the second cathode is allowed to deposit on the substrate. The effectiveness of this method in reducing the reactive-gas content as judged by the closeness of the superconducting transition temperature of some transition-metal films to the bulk value has been established.

3.4 Low-pressure Sputtering

The (a) to (e) arrangements essentially yield useful sputtering rates only in the 20- to 100-mTorr Ar pressure range since the density of ions required for sputtering falls rapidly with decreasing pressure. The decreasing influence of gas atoms, the lower concentration of trapped gas atoms, and the controlled direction and higher mean energy of the ejected atoms striking the substrate owing to smaller collision losses are some of the desirable features of low-pressure sputtering. Reasonable sputtering rates at low pressures may be obtained by increasing the ionization of the sputtering gas by (1) increased ionizing efficiency of the available electrons, (2) increased supply of ionizing electrons, and (3) an ion-beam source.

(1) Magnetic Field. The ionizing efficiency may be increased very conveniently by increasing the path length of the ionizing electrons, for example, by a transverse magnetic field normal to the electric field. Such a field will, however, concentrate discharge at one side of a planar cathode and thus will reduce the uniformity of the deposit. This disadvantage may be overcome by using [86] a cylindrical cathode and a magnetic field parallel to its axis. Much more uniform deposits, particularly on three-dimensional substrates, may be obtained by deposition on a substrate placed inside a hollow cathode. The "hollow-cathode effect" allows the anode to be placed outside without interfering with the discharge. A hollow-cathode discharge supported by an external magnetic field with an internal coaxial anode (inverted magnetron) was used by Gill and Kay [87] to sputter at $\sim 10^{-5}$ Torr.

A longitudinal magnetic field (50 to 100 G) is generally convenient to use for a diode geometry. Such a field has no effect on electrons moving parallel to the field, but it helps to concentrate the diffuse plasma by preventing lateral motion and also increases the path length of randomly moving electrons. Consequently, the current density is considerably increased and reasonable sputtering rates may be obtained at pressures down to a few millitorr. The benefits of both longitudinal and transverse field were realized by Kay [88] by using two solenoids in opposition, which give components of both field directions. An increase in the sputtering rate by a factor of 30 was obtained.

(2) Assisted (Triode) Sputtering. Auxiliary electrons may be supplied thermionically from a filament. Both the total ionization and the ionization efficiency are increased by accelerating the electrons by means of a third electrode and injecting them into the plasma. This assisted sputtering (f) process, also called "triode sputtering," is further enhanced by the presence of a magnetic field inclined to the lines of force between the cathode and the anode. This system, first employed by Ivanov et al. [89], has become, in a variety of forms and electrode configurations [90], a standard sputtering unit for most laboratories. Such a system has been used by the author for sputtering noble metals at ~ 20 Å/min at about 1 mTorr. A significant advantage of the system is the fine control of current density and hence the sputtering rate by the easily varied magnetic field.

(3) RF Sputtering. Ionization of the gas can be produced by suitable electromagnetic radiations such as rf, uv, x-rays, and γ-rays. High ionization yield may be obtained conveniently by the use of rf of several megacycles. The rf may be applied directly to the anode through a capacitor [g(1), for metal sputtering], or via a high-frequency coil inside [g(2)] or outside the discharge vessel. Gawehn [91] used the electrodeless technique with rf of 1.8 to 12 Mc/sec to sputter at low pressures. Note that rf may be used additionally in any one of the (a) to (f) sputtering arrangements to obtain increased sputtering rates at lower pressures.

A significant development employing rf-excited discharge to sputter insulators was announced by Anderson et al. [92]. Direct-current sputtering of insulators is not possible because of the buildup of positive surface charges which would repel the energetic ions. Methods to neutralize this surface charge by injecting electrons [93] from a gun, or by placing a metal screen [94] over the cathode surface, thus producing a conductive sputtered metal film, have been devised. These techniques obviously are not "clean." A high-frequency alternating potential may be used to neutralize the insulator surface periodically with plasma electrons, which must be switched in a period that is short compared with the time positive ions require to travel from the edge of the ion sheath to the surface of the insulator. The fast electrons can still respond to the alternating field. The surface potential changes at a rate $dV/dt = I_i/C$, where I_i is the ion current and C the capacitance of the insulator. For $I_i = 20$ mA/cm^2, and $C = 20$ $\mu\mu$F/cm^2, $dV/dt = 10^9$ V/sec. At 10 Mc/sec, the potential changes by 100 V/cycle. Thus, both the amplitude and frequency of rf should be high to yield high rates.

Davidse and Maissel [95] described the application of the rf technique to sputtering of insulator films. A transmitter or a standard rf heating power supply (1 to 20 Mc/sec, 2 kW at 2 kV) may be used to couple directly to the insulator target mounted on a metal plate. The 13.56 Mc/sec frequency is allowed by the U.S. Federal Communications Commission for commercial uses. Two cathodes in a push-pull arrangement may be used [g (3)]. The direct coupling of rf is somewhat sensitive to the design of the feedthroughs for the insulator cathode and to the rf power input. The reflective losses from the power supply into the sputtering apparatus necessitate water cooling of the leads.

An applied magnetic field assists in supporting and stabilizing the high-frequency discharge and makes operation possible at low pressures of a few millitorr. Quartz and various glasses may be readily deposited [95,96] to form films at rates of up to 30 Å/sec under typical operating conditions of a 13.56 Mc/sec, rf power of 800 W at 3,000 V peak to peak and an axial magnetic field of 110 G.

Radio-frequency sputtering is a versatile technique and has several other useful applications. (1) If the rf power supply is coupled capacitively to a metal electrode [see Fig. *g* (1)], metals can also be rf-sputtered. The capacitor prevents a dc current from flowing in the circuit and thus allows the buildup of the negative bias on the metal electrode that is necessary if sputtering is to occur. The self-biasing effect is caused by the difference in electron and ion mobilities. With this method, metal films of many materials can be deposited by sputtering at rates ~1,000 Å/min for a rf power of ~1 kW at pressures ~5μ. (2) The high rf sputtering rates of metals and insulators at relatively low pressures can be further enhanced by employing simultaneous dc (1 to 5 kV) and rf (~1 kW) sputtering [97]. The superimposed rf and dc method further allows a flexibility of operation with conducting and insulating materials over a large range of the inert-gas pressure. The superimposition of rf and dc power supplies is made possible by using a filtering network of an inductor and capacitor [see Fig. *g*(1)] between the power supplies and the cathode. The capacitor protects the rf power supply from dc, while the inductor protects the dc power supply from the rf source. (3) We have already mentioned the advantage of bias sputtering. If rf potential is applied to the back of an insulating substrate, a dc bias will develop on the front of the substrate. This is a convenient technique for obtaining bias without a conducting electrode, and it may also be used for sputter-cleaning the substrate prior to deposition. (4) Reactive sputtering in a diode arrangement is limited to making films at low rates. This is limited by the fact that a small amount of the reactive gas must be used so that the reaction at the cathode does not decrease sputtering rate because of the formation of an insulating layer or poisoning of the cathode. Since insulators can be rf-sputtered, rf sputtering is ideally suited for high-rate reactive sputtering. (5) If the filmed substrates are placed on the cathode, rf sputtering may be used to etch metals, semiconductors, and insulators. If certain patterns are required, they may be generated by photoresist materials (e.g., KTFR, Kodak Thin Film Resist). Thereafter, the bare film material and the photoresist material on the pattern can be etched away simultaneously by rf sputtering. The universality and the absence of undercutting (in contrast to chemical etching) are the chief advantages of rf sputtering. The universality characteristic is, however, a disadvantage since it does not allow a selective or restricted etching of materials. This technique has been successfully used for etching of resistor patterns in cermet films with a definition significantly better than that obtained by conventional chemical-etching techniques.

(4) Ion-beam Sputtering. The systems described above are ineffective below 10^{-3} Torr because of the scarcity of ions. By producing ions in a high-pressure chamber and then extracting them into a differentially pumped vacuum chamber through suitable apertures with the help of suitable electron and ion optics, a beam of ions may be obtained for sputtering in vacuum. Such an arrangement is called a "duoplasmatron," and the principle of extraction is attributed to von Ardenne [98]. Although the von Ardenne type of ion source [99-104] is theoretically very efficient, in practice many complicated ion-optical problems limit the total beam current density. Densities $\sim$50 mA/cm^2 for H$_2$ ions at potentials exceeding 10 kV have been obtained for beam sizes of a fraction of a millimeter. Such sources have found a wide variety of applications but are not well suited for deposition of films by sputtering which would require a high-current beam spread over a large area. Some sources have, however, been used in measuring sputtering yield of materials.

Chopra and Randlett [105] designed a duoplasmatron ion-beam source capable of yielding currents of $\sim$500 mA for a beam size of $\sim$1 cm^2. This source, shown schematically in Fig. 8(*h*), produces Ar ions in a small chamber at $\sim$50 mTorr by means of an arc discharge. The ions are extracted into a vacuum chamber through apertures by confining the plasma in the narrow aperture by means of suitable magnetic and electric fields. By placing the anode aperture in the low-pressure side of the system, the electrons are forced to move through the apertures, resulting in a high ionization efficiency of the gas. A well-defined and collimated ion beam extracted into the separately pumped vacuum chamber maintained at $\sim$10^{-5} Torr impinges upon the water-cooled cathode (target) at 45°. Metal, semiconductor, or insulator targets may be sputtered. The rate of sputtering is approximately proportional to the ion current for a constant accelerating voltage and therefore may be conveniently controlled. Typical deposition rates onto a substrate 8 cm away from the target are 40 Å/min for quartz and 500 Å/min for Ag, using 50 mA/cm^2 current density and an accelerating potential of 5 kV.

The observed sputtering of insulators at reasonable rates is the result of the leakage of the positive charge to the backing electrode directly or via neutralization by valence electrons extracted from the surface of the insulator by the ions. Sputtering yields for both metals and insulators are consistent with the data in the literature (Tables II and III). The simple Chopra-Randlett ion source may be used for both sputter deposition and sputter etching.

By placing the anode at a substantial distance from an electron-beam source, accelerated electrons may be used very effectively for ionization of the ambient gas. Such a "tetrode" structure was used by Gaydou [106]. Although this arrangement resembles a duoplasmatron, it produces a broad ion-glow discharge along the path of the electron beam rather than an ion beam. The technique is, however, capable of supplying large ion currents (~ 1 A) so that reasonable sputtering rates at pressures of $\sim 10^{-4}$ Torr can be obtained.

3.5 Reactive Sputtering

The high chemical reactivity of the atomic form of stable diatomic species and their ions which are formed in a glow discharge can be used advantageously in producing thin film carbides, nitrides [80], oxides [107-114], hydrides [115], sulfides [116], etc. This method is conveniently realized by introducing a mixture (preferably premixed) of the inert sputtering gas and a small quantity of the reactant in gas form in the conventional glow discharge or rf sputtering arrangement.

This subject has been reviewed by Holland [2] and Schwartz [80]. Of particular interest are the studies on the formation of carbide and nitride of Ta (reviewed in Ref. 80), oxides of various metals (see, for instance, Smith and Ayling [107]), epitaxial growth of oxides of Al and Ta [111] (by both reactive evaporation and sputtering), mixed oxides by sequential sputtering or co-sputtering of Pb and Si [112], Al-Si alloy [113], and ferrites [114].

The chemical reaction for reactive sputtering may take place at the cathode, during transport of ejected atoms, at the film surface, or a combination of these possibilities. Clearly, the role of each process depends much on the pressure and the chemical activity of the reacting species. At low partial pressures of the reactant, a reaction at the film surface during growth is expected, as supported by the studies of Krikorian and Sneed [111]. They found the growth rates of the oxide film to increase rapidly with substrate temperature. The comparison with reactivity evaporated films showed rates several times higher for reactively sputtered films. Kinetics of oxidation and stoichiometry of Cu films were studied by Perny et al. [110]. Details of the mechanism of reactive sputtering are by no means clearly established because of insufficient basic studies carried out in this area. The major problem of this technique is the difficulty of controlling the stoichiometry of the products. Nevertheless, the method is very promising in producing acceptable dielectric films.

If the metal film is biased positively with respect to the anode, it may be "anodized" with bombarding oxygen ions in a glow discharge. This process, called "gaseous anodization," was used by Miles and Smith [117] to oxidize Ta, Cr, Be, Ge, and Si. Pinhole-free Al oxide films grown at a rate of ~22 Å/V for positive potential up to 60 V were obtained by Tibol and Hull [118] in a low-pressure plasma of oxygen ions. This rate is about 50 percent higher than that obtained in a typical nonsolvent electrolyte.

Radio-frequency plasma was used by Sterling and Swann [119] to promote a reaction between SiH_4 and H_2O vapors to deposit Si, and between SiH_4 and NH_3 to form durable and insulating Si_3N_4 films. The glow-discharge breakdown of tetraethylorthosilicate vapors was used by Alt et al. [120] to deposit silicon oxide films. Hydrocarbon molecules can be decomposed in the presence of a glow discharge, and the radicals thus formed may then combine to form larger polymer molecules and condense as polymer films [121-124] with different carbon-hydrogen compositions depending on the growth conditions. This subject is discussed later.

3.6 Sputtering of Multicomponent Materials

An important advantage of the sputtering process is that the composition of a sputtered film is the same as that of the cathode provided that (1) the cathode temperature is not too high for the rapid compensation of the higher-sputtering-yield material at the surface by volume diffusion from the bulk; (2) the cathode does not decompose; (3) the surface of the cathode does not alter chemically, say by oxidation; and (4) the sticking coefficients for the components on the substrate are the same. If condition (1) is satisfied, the cathode is sputtered away layer by layer. However, the cathode surface may be appreciably altered by penetration of the energetic ions. For example, Gillam [125] found that 400-eV ions of He, Ar, and Xe sputtered Cu selectively from the surface of $AuCu_3$ alloy, leaving a 40-Å-thick surface layer of a new and uniform composition. With subsequent sputtering, the altered layer was maintained and the deposited film had the same composition as that of the altered layer as checked by electron diffraction. The depth of this altered layer should clearly depend on the various sputtering parameters.

Evidence for the correspondence between sputtered film and the cathode compositions has been reported for various materials, such as type 347 stainless steel and brass [126], a number of Al alloys [127], and group III-V intermetallic films [128]. In the last mentioned case, Moulton found no evidence of elemental materials and suggested that compounds sputtered either molecularly or as polyatomic aggregates. Electron-diffraction studies [129] of sputtered Bi_2Te_3, PbTe, and InSb films also indicate no compositional change. It must be emphasized that, although changes in composition on sputtering appear to be small, precise determination of the film composition by sensitive analytical techniques has not yet been reported in the literature.

4. CHEMICAL METHODS

4.1 Introduction

Chemical-vapor-transport methods for deposition of thin and thick coatings have been developed from the early observations of Bunsen [130] on chemical reduction of $FeCl_3$ by steam to produce Fe_2O_3. Chemical methods may be broadly classified in two categories: (1) electroplating and (2) chemical vapor deposition (CVD). Both have widespread applications of major technical and commercial value to an industrialized nation. These techniques enable coating thicknesses to be varied from angstroms to fractions of a millimeter in a well-controlled fashion. Although the impurities and their effects vary with the material to be deposited, most refractory metals and several nonmetals are obtained chemically in a purer form than by conventional metallurgical practices.

The vast subject of chemical deposition is treated in various texts and reference books. The comprehensive works of Milazzo [131], Lowenheim, [132] and Potter [133] on electroplating, of Young [134] and Potter [133] on anodic oxidation, and of Schäfer [135] and Powell et al. [136] on chemical vapor deposition are recommended as references. The review articles on gas-phase deposition of organic and inorganic insulating films by Gregor [137], on the chemical-vapor deposition of single-crystal semiconductor films by Joyce [138], and on Si heteroepitaxy by CVD by LaChapelle et al. [139] are of particular interest.

A brief description of the various chemical methods is given in the following sections.

4.2 Electrodeposition

(1) Electrolytic Deposition. The first application of the principles of electrolysis to the deposition of metal films is the subject of controversy [140] but probably took place in 1838. According to the laws of electrolysis, the weight of the material deposited is proportional to the amount of electricity passed, 1 gram equivalent of the material being deposited by 96,490 C. The metallic ions in the electrolyte migrate toward the cathode under the influence of the applied electric field, which can be very high ($\sim 10^7$ V/cm) between the cathode surface and the ions in the double layer. The process by which a metal ion is incorporated into the cathode lattice is a subject of some discussion. According to the extensive studies of Bockris and coworkers [141], the cation crosses the double layer, loses part of its water of hydration, and then becomes adsorbed on the metal surface as an adion. The adion (corresponding to adatom in vapor deposition) diffuses over the surface, loses the rest of its water of hydration, and is finally incorporated as an ion into the metal lattice.

The experimental arrangement requires a suitable electrolyte through which current is passed between two electrodes. The deposition rate is proportional to the electrolysis time and the current density, and may be varied over a wide range from about an angstrom to several microns per second. The rates can be influenced markedly by the geometry of the cathode, the temperature, and the agitation of the electrolyte. The structure [142] of the deposit depends sensitively on the rate of deposition and the characteristics (such as pH value) of the electrolyte, and may vary from a highly disordered to a well-oriented single crystal. The occurrence of epitaxy in electrodeposits at room temperature is of considerable interest, though not understood at present. It is probably the result of considerable mobility of the adions due to the high field present at the surface.

(2) Electroless Deposition. Electrolytic action may be achieved without an external potential source by a chemical-reduction process such as that used in the well-known technique of silvering glass dewars. This technique, called "electroless deposition" [132], has been employed to deposit Ni, Co, and Pb films by reduction of their chlorides by sodium hypophosphite. For nonmetallic surfaces, it may be necessary to use a sensitizer such as 0.1 percent stannous chloride. The rate of film growth in this method depends greatly on the reaction temperature and is generally difficult to control.

(3) Anodic Oxidation. A large number of metals (called "valve" metals) tend to form a protective oxide film of limited thickness when exposed to oxygen. By anodic polarization of these metals in a suitable aqueous solution (which does not dissolve the oxide), a protective high-resistance film can be grown. The anodization process involves the migration of ions of oxygen, metal, or both, depending on the material, through the existing oxide film. The details of the ion-transport process, however, are still a subject of some debate [134].

Growth rate of an anodic film depends on the current density and the temperature of the electrolyte. If the current is held constant, the growth increases linearly with time. For a constant voltage, the growth increases rapidly in several minutes to approach an asymptotic value. For current densities of $\sim$2 mA/cm^2, the rates are similar to those found with electroplating (e.g., 10 Å/sec for Ta [143]). The maximum (actually, asymptotic) value of thickness developed, however, depends on the applied anodization voltage, the electrolyte, and its temperature. Electrolytes and the practical values of the oxide thickness per volt obtained for some metals are listed in Table IV. The upper limit of the film thickness is determined by the breakdown potential, which occurs at about 1,100 V for Al (giving 15-μ-thick film) and 700 V for Ta (giving 11-μ-thick film). In some cases [134], the limit is imposed by a high-field recrystallization of the oxide, leading to its breakdown.

A simple experimental arrangement consists of an electrolytic cell with a cathode of the same material as the anode, or of platinum. Anodization is carried out at constant current, constant voltage, or by a sequence of constant current and constant voltage. The former is generally preferred by scientific workers. Simple electrical circuits may be designed to perform these functions (see Ref. 134 for more details). The quality of the oxide films and their adhesion to the parent metal depend much on the surface of the metal. A smooth, electrolytically polished surface is essential for obtaining adherent, pinhole-free films.

The anodic films are generally amorphous [144], although crystalline [145,146] films (e.g., TiO_2 on Ti) may be obtained under certain anodization conditions. A suitable heat treatment of the amorphous anodic and thermal oxide films can yield single-crystal films [147], which is not necessarily the result of epitaxial growth.

Table IV.   Thickness of Some Anodic Oxide Films for 1 V of Anodization at Room Temperature

Material	Anodization solution	Oxide thickness, Å/V
Al	Ammonium phosphate, tartrate, or citrate	13-14
Nb	Ammonium tartrate	22
Ta	Ammonium borate	16
Ti	Ammonium tartrate	22
Si	0.04 mole KNO_3 in N-methylacetamide	4.6

Anodic oxidation has been extensively utilized in producing ultrathin and thick oxide films for tunnel devices, capacitors, protective layers, etc. The anodization process is applicable only to certain metals. Wood et al. [148] have shown, however, that it can be extended to other metals such as Mo and V, by alloying them with relatively low proportions of valve metals.

A serious disadvantage of anodization in aqueous solutions is that it results, in some cases, in the incorporation of water and OH ions into the films, producing deleterious effects on their dielectric behavior [146]. Gaseous anodization (already discussed) overcomes this problem.

4.3 Chemical Vapor Deposition (CVD)

When a volatile compound of the substance to be deposited is vaporized, and the vapor is thermally decomposed or reacted with other gases, vapors, or liquids at the substrate to yield nonvolatile reaction products which deposit atomistically (atom by atom) on the substrate, the process is called chemical vapor deposition [135,136]. Since a large variety of chemical reactions are available, CVD is a versatile and flexible technique in producing deposits of pure metals, semiconductors, and insulators. A very significant application of the CVD processes is the preparation of single-crystal metal oxides, notably the ferrites, garnets, sapphire, MgO, etc. (see Chap. IV, Table VIII).

Most CVD processes operate in the range of a few Torr to above atmospheric pressure of the reactants. Although no electric currents or fields are required in a CVD process, the application of a transverse electric field to substrate during deposition is known [149] to increase the growth rate of Si, Ge, and GaAs. The same effect is observed in the growth of thermal oxides [150]. Thus, an applied field may prove to be an important deposition-control parameter in CVD techniques.

Several types of reactions are classified under the CVD heading. We shall present representative examples to illustrate each type. Some of the commonly used experimental systems for these reactions are shown schematically in Fig. 9.

(1) Pyrolysis (Thermal Decomposition).   The thermal decomposition of a compound to yield a deposit of the stable residue is called pyrolysis. Organometallic compounds decompose at low temperatures ($<600°C$), whereas metal halides, particularly the iodides, decompose above about $600°C$ to yield metallic deposits. Electron-beam-bombardment heating [151] is best suited for pyrolysis of metal compounds.

Pyrolysis of silane (SiH_4) and germane (GeH_4) is employed [152-156] to produce epitaxial layers of Si and Ge. The silane process, depicted in Fig. 9(*a*), produces silane by the reaction of an alkali-metal hydride in an organic solvent. One of the reactions is

$$SiCl_4 (sol) + LiAlH_4 (sol) = SiH_4 (g) + LiCl (s) + AlCl_3 (s)$$

with tetraethylene glycol dimethyl ether (TGDE) as the solvent. Silane at a pressure of less than 8 Torr decomposes completely above $777°C$. The Si deposit can be grown epitaxially on Si. Using ultrahigh-vacuum techniques, Joyce and Bradley [156] obtained epitaxial deposits over the range 800 to $1350°C$. The reaction is immeasurably slow below $800°C$.

Many of the organic silicates can be thermally decomposed to give substantially pure SiO_2 films of dielectric properties close to those of silica. This deposition is usually accomplished by passing an inert carrier gas carrying vapor from the silicates over a heated substrate. The vapor feed is picked up by bubbling the carrier gas through several centimeters of fluid in a gas wash bottle. The tetraethoxysilane (TEOS) [157] vapors decompose in the reactor in the 600 to 900°C range and deposit a film of SiO_2 on the substrate. The equipment involved is shown in Fig. 9(*b*). The unreacted material carried with the silane is removed by cold traps and the absorption column is used to purify silane.

Deposition of Al_2O_3 films has been obtained by pyrolysis of aluminum triethoxide [158] in vacuo at 550°C, and also of aluminum triiso-propoxide [159] [$Al\ (OC_3H_7)_3$]. The latter compound, studied by Aboaf [159], yielded deposition rates of ~100 Å/min at 420°C with 10 liters/min gas flow.

Yoshioka and Takayanagi [160] described the silane-hydrazine process involving pyrolysis of $SiH_4:N_2H_4$ of mole ratio of 0.1 at temperatures between 550 and 1150°C to deposit films of Si_3N_4. Growth rates of 10^2 to 10^3 Å/min were obtained at temperatures above 700°C.

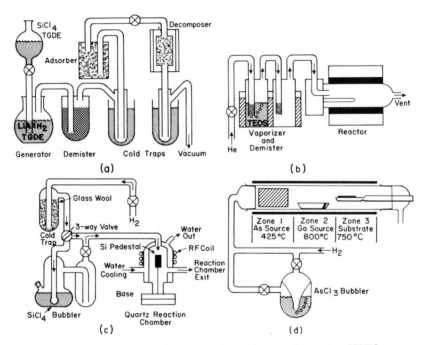

FIG. 9. Typical arrangements for some chemical-vapor deposition (CVD) processes: (*a*) silane (or germane) pyrolysis process for epitaxial Si (or Ge); (*b*) pyrolysis of Si $(OC_2H_5)_4$ (tetraethylorthosilicate or TEOS) to produce SiO_2 (*Smits* [157]); (*c*) hydrogen reduction of chlorides of Si and Ge to produce epitaxial Si and Ge (*after Ref. 136, p. 614*); (*d*) transfer reaction of As with Ga to produce epitaxial GaAs (*Knight et al.* [171]).

(2) **Hydrogen Reduction.** Hydrogen reduction may be thought of as a pyrolysis reaction which is facilitated by the removal of one or more of the gaseous products of decomposition. The reaction temperature is lowered by several hundred degrees below that needed in the absence of hydrogen. Hydrogen reduction of metal halides is frequently used for depositing metal films. Hydrogen may also be used in other types of deposition reactions such as pyrolysis of organometallic compounds where reduction is not the primary objective but improves the deposit characteristics.

The reduction of $SiCl_4$ by CO_2 in the presence of H_2 is a widely used method to obtain SiO_2 at deposition rates of ~0.5 μ/min at 1200°C. This method is called the "CO_2" or the "Phillips" process [161].

The reduction method has found widespread use in preparing device-quality epitaxial Si (and Ge) films. A simple experimental arrangement is shown in Fig. 9(c). Both $SiCl_4$ [162,163] and $SiHCl_3$ [164] have been studied. The appropriate chloride is maintained at a controlled temperature and is picked up by a stream of dry H_2 or He to give a 2 to 30 Torr partial pressure of chloride in the reactor, the whole system being at atmospheric pressure. Substrate heating is carried out by rf induction in the substrate (high-purity Si) or its pedestal. Deposition is carried out between 1100 and 1300°C with growth rates of ~200 Å/sec. This method is particularly suited [165] for the introduction of dopants (e.g., PH_3, B_2H_6) by feeding them into the reactor from a separate source.

(3) **Halide Disproportionation.** The basis of this process is the control of the equilibrium that exists between Si (or Ge), the tetraiodide, and the diiodide. At 1100°C, SiI_4 attacks the source Si to put SiI_2 into the vapor phase. At 900°C, the reverse reaction takes place, resulting in the deposition of the epitaxial Si on the substrate by disproportionation of the SiI_2 vapors. Both open-tube [166] (dynamic) and closed-tube [167] (static) systems have been used. The former is not suited for a controlled equilibrium process, and the latter makes it difficult to clean the substrate.

If multilayered structures with different dopants are required, separate sources of Si must be individually introduced or removed from the etching phase of the reaction. Wajda and Glang [168] have discussed practical solutions to this problem.

(4) **Transfer Reactions.** Deposition can be obtained in a heterogeneous chemical-reaction system at equilibrium by setting up a

temperature differential in the system to disturb the equilibrium. Among the more useful applications of this method are the epitaxial growth of III-V compounds. The technique may be illustrated by reference to the case of GaAs.

Two different types of processes exist. In the first, GaAs itself is used as a source material for the epitaxial layer, and it may be transported by a number of reagents. For example, the normal iodide disproportionation type of reaction may be used [169], with source and seed maintained at about 750 and 700°C, respectively. Hydrogen halides (e.g., HCl) may also be used [170] as transport agents. Both these systems have been employed using closed-tube (static-equilibrium) and open-tube (dynamic-equilibrium) methods.

The second approach, which permits a much greater control over growth parameters and impurity levels, uses the elements and their halides as source materials for transport reaction. The system used by Knight et al. [171] for this purpose [Fig. 9(d)] uses conveniently available high-purity $AsCl_3$ and Ga as reagents. Hydrogen is bubbled through the $AsCl_3$, maintained at 20°C, and is then passed through the furnace. The initial reaction in zone 1 reduces $AsCl_3$ to As, which is completely absorbed by the Ga source in zone 2 at 800°C until saturation occurs at 2.25 atomic percent arsenic. During this period, no free As condenses beyond the furnace, but Ga is transported as a lower chloride. Gallium arsenide substrates are vapor-etched if they are maintained at 900°C, but epitaxial growth occurs when the temperature is lowered to 750°C.

(5) Polymerization. The deposition of insulating polymer films by the decomposition of some hydrocarbon molecules in a glow discharge and rf discharge has already been mentioned. The polymerization process probably results in loss of hydrogen or dissociation by breaking the carbon chain in a hydrocarbon. The process can also be accomplished by a high-temperature electron bombardment, or exposure to ionizing radiations, e.g., ultraviolet light (in which case it is called a "photolytic process"). Several workers have used electron-beam bombardment to produce and study properties of polymer films of DC 704 pump fluid, siloxanes [172-174], styrene [175] and related monomers, and butadiene [176]. White [177] studied polymer films of vapors of butadiene, styrene, acrylonitrile, and methyl isopropenyl ketone by electron bombardment as well as by photolytic process and found them to be essentially similar for both preparative techniques. Solid poly-*p*-xylene films have been formed [178] by pyrolysis of *p*-xylene at 600°C, resulting in monomers or dimers of *p*-xylene.

Although details of the polymerization process are not well established, the usefulness of the polymer films as coherent, pinhole-free, and stable insulators that are strongly adherent to substrates is widely recognized. The preparation and properties of polymer films are extensively reviewed elsewhere [3,137]. It should be noted that, by replacing some or all of the carbon atoms in organic polymers by Si, P, N, B, As, etc., inorganic polymers may be produced which are expected to be stable at relatively higher temperatures. These materials have not yet received much attention.

Attention should also be drawn to a simple and powerful technique [179] for obtaining epitaxial films of homopolymers (*a*) by direct isothermal immersion of a single-crystal substrate into a suitable solution of the polymer at a temperature of ~100 to 200°C, and (*b*) by allowing the evaporation of the solvent from the solute placed on the substrate.

4.4 Miscellaneous Methods

1. A brief mention should be made of the well-known Langmuir-Blodgett technique for depositing monolayer films of fatty acids. If, for example, a glass plate is raised through barium stearate spread on water, a well-defined 48-Å-thick monolayer with the hydrocarbon surface oriented outward is formed. One may then dip the plate again into the film-covered surface and obtain a second monolayer (same thickness) "back to back" or with like orientation, depending on the direction of dipping of the plate. Such monolayers can be built up by as many as one hundred layers. The films are pinhole-free and good insulators, but are fragile.

2. Photoconducting and electroluminescent films of II-VI compounds can be prepared by a chemical-spray method. Reagents such as thiourea, selenourea, and thioacetamide interact with salt solutions of heavy metals and form precipitates of II-VI compounds when heated at a suitable pH value. Films up to 20 μ thickness can be prepared but contain trapped nonmetallic or organic enclosures, which makes a subsequent forming process necessary.

A comparative summary of various deposition techniques is presented in Table V.

Table V. Some Characteristics of the Various Thin-film-deposition Techniques

Technique	Materials	Deposition rate*	Rate control	Vacuum, μ	Epitaxy	Adhesion	Remarks
Evaporation:							
Resistive heating	Most	L, M	Yes	<0.1	Yes	Fair	Simple, support alloying and decomposition in certain cases
Multisource	Alloys, compounds	L, M	Yes	<0.1	Yes	Fair	Stoichiometry control possible
RF heating	Conducting	L, M	Poor	<0.1	Yes	Fair	Levitation eliminates support contamination
Electron beam	Most	L–H	Poor	<0.1	Yes	Good	Versatile, gas bursts in certain cases
Flash	Alloys, compounds	M, H	No	<0.1	Yes	Good	Nearly stoichiometric films, gas bursts, spattering
Arc	Conducting	H	No	<1	Possible	Fair	Crude, spattering, very high rate
Laser	Most	H	No	<0.1	Yes	Fair	Very high rate, spattering, promising method
Exploding wire	Conducting wires	H	No	<1	Possible	Fair	High rate, crude, spattering
Sputtering:							
Glow discharge (diode)	Conducting	L, M	Yes	20–60	Yes	Good	Simple, widely used

Method	Materials					Remarks
Ion plating	Conducting	L, M	Yes	20-60	Yes	Good	Evaporation and sputtering in poor vacuum
Asymmetric ac and dc bias	Conducting	L, M	Yes	20-60	Yes	Good	Contamination reduced
Getter	Getter metals	L, M	Yes	20-60	Yes	Good	Reactive gases removed
Triode (assisted ionization)	Conducting	L, M	Yes	~1-60	Yes	Good	Versatile, low pressure, confined plasma
RF	Most	L-H	Poor	~1	Possible	Good	Useful, plasma effects complicated
Ion beam	Most	L-H	Poor	<0.1	Yes	Good	Sophisticated, well-defined conditions, low pressure, substrate out of plasma
Reactive	Oxides, hydrides, carbides, nitrides, sulfides	L, M	No	1-60	Yes	Excellent	Powerful, many variables control difficult
Chemical: Electrolysis	Metals	M, H	Yes		Yes	Fair	Cheap, production method
Anodization	Oxides of valve metals	L, M	Yes		Yes	Excellent	Controlled thickness of thin oxide films obtained, solution contaminants
Pyrolysis	Most	H	No	~1,000	Yes	Fair	Production method, high-temperature system, contamination problems

Table V. Some Characteristics of the Various Thin-film-deposition Techniques *(Continued)*

Technique	Materials	Deposition rate*	Rate control	Vacuum, μ	Epitaxy	Adhesion	Remarks
Hydrogen reduction	Most	H	No	~1,000	Yes	Fair	Production method for metals, high-temperature system, contamination problems
Disproportiona-tion	Products of some halide compounds	H	No	~1,000	Yes	Fair	High-temperature contamination problem
Transfer reaction	Depends on reaction	H	No	~1,000	Yes	Fair	High-temperature system, contamination
Polymerization	Hydrocarbons	H	No		Yes	Fair	Cheap, good films, but thermally unstable
Electroless solution	Polymers	H	No		Yes	Good	Cheap, good films

*L, M, H stand for low (<0.1 Å/sec), medium (0.1 to 20 Å/sec), and high (>20 Å/sec) rates of deposition, respectively. The values are chosen arbitrarily.

5. VACUUM-DEPOSITION APPARATUS

The preparation of films by the deposition techniques described in the preceding section requires a variety of ancillary techniques, equipment, and jigs. This section outlines a brief "know-how" on this important aspect of thin film technology. Deposition-rate control and thickness-monitor arrangements are described in the next chapter.

5.1 Vacuum Systems

The performance of a vacuum system with a chamber or bell jar ("evaporator") is the single most important consideration for vacuum-deposition techniques. This consideration arises from the fact that the structure and properties of a film, depending on the material, may be influenced profoundly by the ultimate vacuum and residual gases and their partial pressures. As we have already seen in Table I, the residual air at 10^{-5} Torr forms about 4.4 monolayers ($\sim$10 Å) per second if the gas atoms have a sticking coefficient of unity. Although the sticking coefficient for the first monolayer is very small, considerable sorption of the gas can occur and be detrimental to the properties of films. It is therefore necessary to employ as good a vacuum condition as possible. In practice, however, several factors, e.g., cost and pumping time, limit the generally employed vacuum in the range 10^{-5} to 10^{-8} Torr. The ultrahigh-vacuum (uhv) range below 10^{-8} Torr is easily accessible with recent advancements in vacuum technology and is being adopted increasingly.

Vacuum technology has undergone major developments primarily as a result of the demand of thin film technology for better vacuum. The subject is covered extensively in several recent books [180,181]. Excellent and detailed reviews on evaporators are given by Holland [2,3] and Caswell [182]. The various types of pumps, their ultimate pressure, and the various types of gauges employed to measure pressure are summarized in Fig. 10. The following practical points should be of interest to a thin-film technologist.

1. Basically, a diffusion-pump (mercury or oil, although high-speed oil pumps are most commonly used) bell-jar system, and a getter-ion-pump bell-jar system are the two types used as evaporators. Typical representatives of these, used in the author's laboratory, are shown schematically in Figs. 11 and 12 with the various components and accessories essential for a reasonable system. The list is by no means

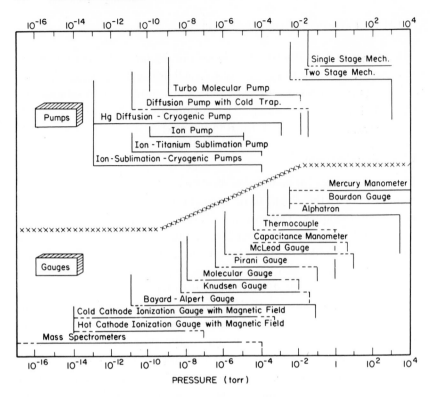

FIG. 10. Chart showing approximate ranges (possible extensions shown by the dotted lines) of pressures for different types of pumps and gauges. (*Partly taken from Dushman* [180])

comprehensive but should give the reader an idea of the requirements of an evaporator. A high-speed oil-diffusion pump system ($\sim$1,000 liters/sec after baffling in a 19-in. system) is relatively simple and cheap. It allows rapid 1-hr cycling from atmosphere to $\sim$$10^{-6}$ to 10^{-7} Torr for a bell jar of reasonable size and is therefore indispensable for routine applications. By careful baking and degassing of the evaporator components, pressures of $\sim$$10^{-8}$ Torr are obtained in several hours. On the other hand, an ion pump backed by sorption pumps readily yields $\sim$$10^{-8}$ Torr but is more expensive and has a relatively slower speed and longer cycle time. It is, however, free from oil, which can back-stream in a diffusion-pump system. With recent improvements and innovations such as Orbion and electrostatic getter-ion pumps, ion pumps with typical speeds of $\sim$400 liters/sec are finding routine applications despite their high cost. With the addition of titanium sublimation and/or a cryogenic pump (employing

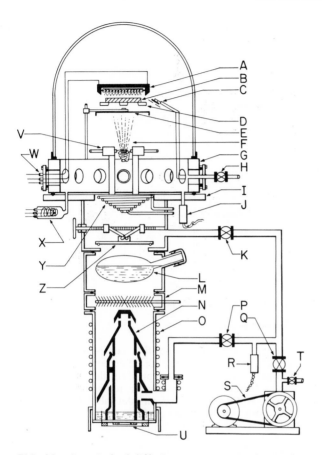

FIG. 11. A typical oil-diffusion-pump evaporation station used in the author's laboratory. The notations stand for A, quartz iodine lamp heater; B, substrate; C, quartz-crystal rate controller and deposition monitor; D, substrate mask; E, shutter (mechanical or electromagnetic); F, vapors from evaporation source; G, adapter collar between the bell jar and the pump baseplate flange; H, air-inlet valve; I, baseplate flange; J, Pirani or thermocouple gauge; K, roughing valve; L, liquid air trap; M, cooled chevron baffles; N, diffusion pump; O, cooling coils; P and Q, backing valves; R, Pirani gauge; S, forepump with air-inlet valve T; U, diffusion-pump heater; V, filament holders; W, multiple feedthrough; X, ionization gauge; Y, Meissner trap; Z, baffle valve.

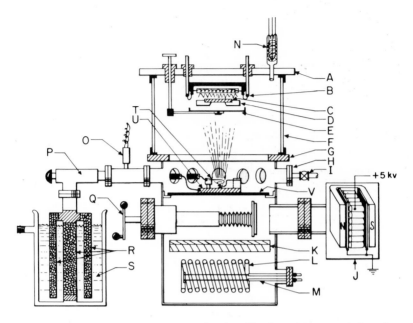

FIG. 12. A typical ion-pump system for thin-film deposition, consisting of
sputter-ion pump (J), titanium sublimation pump (M) with a cold envelope
(L), cryogenic pump (K), and sorption pump (R). A, top plate for the open
bell jar; B, C, D, and E are heater, substrate, mask, and shutter, respectively;
F and G are bell jar and lower plate; H, Cu-seal flange; I, air-inlet valve; N
and O are gauges; P and Q are valves; R, Zeolite holder cooled with liquid
nitrogen (S); T and U are filament holders; V, pump shield.

liquid N_2 or He), pressures in the range of $\sim10^{-12}$ Torr may be easily
obtained. The getter-ion pumps are, however, selective in pumping gases;
the pumping speeds are very high for reactive gases, but rather low for
argon. This feature, together with their requirement for a low starting
pressure ($\sim10^{-4}$ Torr), makes getter-ion pumps impractical for dynamic
sputtering systems.

2. The ultimate pressure in an evaporator depends not only on the
pumps employed but also on the rate of desorption of the evaporator
hardware and degassing of the evaporant. These rates are sufficiently
high for a commonly employed system to warrant the use of pumps with
speeds no less than 200 liters/sec. High-vapor-pressure materials such as
brass should not be used. Polished stainless steel (type 304) and
aluminum are some of the materials with lowest desorption rates. Since
baking or predegassing is an essential practice for obtaining uhv, the

materials used in a system should be carefully selected to perform well at an optimum baking temperature of ~400°C.

At elevated temperatures, a considerable amount of water vapor diffuses from the interior of Pyrex glass to its surface. At temperatures above 350°C, glass decomposes in vacuo, liberating water vapors and alkali ions (e.g., Na^+, K^+, among others), which are a serious source of contamination for electronic films. Moreover, considerable permeation of Ne, H_2, and He through glass occurs to limit the upper limit of the vacuum attained. Thus, a glass bell jar is not desirable for uhv and bakeable systems. Note that the permeation of gases through metals is negligible, with the important exception of hydrogen, which is of serious concern for most systems below 10^{-10} Torr.

Among the various elastomers used for gaskets and O rings, the fluoroelastomer called Viton A (Du Pont trademark) has the lowest permeability for the principal atmospheric gases and low (but still 10^3 times more than that for stainless steel) desorption rate after degassing at 150°C. It decomposes at temperatures above 150°C, however, liberating mainly HF. Thus, a bakeable system should have O-ring seals made of metals (e.g., Cu, Al, Au).

3. Oil vapors reach a vacuum system from the rotary pump (the vapor pressure of oils used is $\sim 10^{-3}$ Torr at 40°C) during the roughing cycle, and from the diffusion pump by back streaming, and also as a result of the high vapor pressure of light hydrocarbons even at low temperatures. Optically dense, cryogenic chevron baffles are absolutely essential even though they do not completely eliminate the oil vapors. Traps with a molecular sieve (e.g., Zeolite) preferably cooled by liquid nitrogen are also essential between the rotary pump and the vacuum system. The vapor pressure of diffusion-pump fluids (DC 704 or 705) is $\sim 10^{-8}$ to 10^{-10} Torr at 15°C, so that a monolayer of oil will be adsorbed under equilibrium conditions in 10^2 to 10^4 sec if every molecule sticks on first impact. Oil vapors or hydrocarbons present in an ion-pump system will decompose in the glow discharge to deposit a polymer film.

The residual gases in a kinetic system usually consist of H_2O, CO_2, N_2, H_2, and organic vapors of different molecular weights. During deposition, considerable change in the partial pressures of residual gases may occur because of desorption from the evaporant, interaction of the gases with the hot filament and the depositing film, chemical reactivity of the glow-discharge process, etc. Water vapor may react with a growing metal film to form its oxide and liberate H_2. The reaction with a heated filament produces H_2, CO, CO_2, CH_4, and other hydrocarbon products.

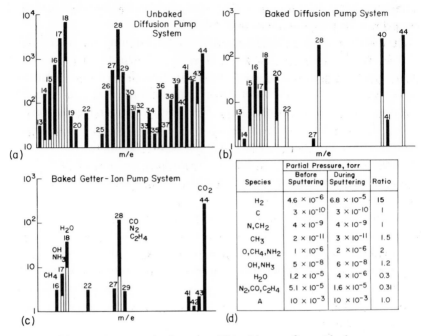

The table in part (d):

Species	Partial Pressure, torr		Ratio
	Before Sputtering	During Sputtering	
H_2	4.6×10^{-6}	6.8×10^{-5}	15
C	3×10^{-10}	3×10^{-10}	1
N, CH_2	4×10^{-9}	4×10^{-9}	1
CH_3	2×10^{-11}	3×10^{-11}	1.5
O, CH_4, NH_2	1×10^{-6}	2×10^{-6}	2
OH, NH_3	5×10^{-8}	6×10^{-8}	1.2
H_2O	1.2×10^{-5}	4×10^{-6}	0.3
N_2, CO, C_2H_4	5.1×10^{-5}	1.6×10^{-5}	0.31
A	10×10^{-3}	10×10^{-3}	1.0

FIG. 13. Mass spectrograms in the m/e = 12 to 44 range for residual gases evolved in the systems with a substrate heater in operation (white areas), and for gases evolved (dark areas) during evaporation from a tungsten helix in the following cases: (*a*) Sn in an unbaked pump system, (*b*) Sn in a baked (at 350°C) diffusion-pump system, (*c*) 83:17 Ni-Fe in a baked getter-ion-pump system. The level 10 represents a partial pressure of 2×10^{-6} Torr (*Nicholls* [183]). For comparison, residual gas composition before and during sputtering of Ta in Ar-N_2 dynamic flow system at a total pressure of 10×10^{-3} Torr of Ar, 5 kV and 0.15 mA/cm² is listed in (*d*) (*Schwartz* [80]).

Mass spectrograms [183] of residual gases of 12 to 44 m/e values observed before and during evaporation runs in a typical baked and unbaked diffusion-pump system and a getter-ion-pump system of commonly used designs are displayed in Fig. 13. Also shown in the table of Fig. 13(*d*) is a relative change in the partial pressure of gases during reactive sputtering. It is clear that some residual gases and other decomposition products such as OH and CO are enhanced. As compared with a baked system, an unbaked diffusion-pump system exhibits a considerably higher variety of constituents with a tenfold increase in the level of some components. The evolution of large quantities of H_2 in sputtering, and to a lesser extent in evaporation, is noteworthy.

It is clear that the amount of residual gases in an average evaporator is large enough to be of concern if good-quality films are required. The use of a Meissner trap at 77°K can play a paramount role in reducing the influence of water vapors. A liquid-helium cryogenic pump or a cold finger in a vacuum system is, of course, the solution for drastically reducing the residual pressure of all constituents.

4. The pressure in the evaporator is measured by using one or more of the gauges listed in Fig. 10. A Pirani, thermocouple, or capacitance manometer gauge is commonly used for the high-pressure range (down to 1 μ) and an ionization gauge for lower pressures. The former types are indispensable for sputtering. Furthermore, the electrical output of such gauges (which have a fast response and expanded scale) can be employed to control the gas pressure during sputtering in conjunction with a servosystem.

Measured pressures in high-vacuum systems depend on the location of the gauge, their pumping action, and the relative location of different gauges (electrostatic-charge interference effects). Extreme care is needed in measuring and interpreting pressures below 10^{-8} Torr. It is generally not recognized that unshielded or poorly shielded gauges and ion pumps contribute a large amount of stray electrostatic charge that strongly influences the growth of films (Chap. IV, Sec. 2.3).

5.2 Substrate-deposition Technology

This section deals with various problems and techniques related to the process of obtaining suitable thin films on a substrate. The discussion is again of a practical "know-how" nature. Only general references will be given.

1. **Substrate Materials.** The nature and surface finish of the substrate are extremely important because they greatly influence the properties of films deposited onto them. Glass, quartz, and ceramic substrates are commonly used for polycrystalline films. Single-crystal substrates of alkali halides, mica, MgO, Si, Ge, etc. are used for epitaxial growth. The most commonly used glass slides as substrates are of Pyrex, which, like other glasses and ceramics, is noncrystalline and has a composition of 80.5 percent SiO_2, 12.9 percent B_2O_3, 3.8 percent Na_2O, 2.2 percent Al_2O_3, and 0.4 percent K_2O. Fused quartz and Vycor glasses have higher silicate content (~96 percent) and are stable at temperatures of ~500°C. The alkali content of different glasses varies and is important

from the point of view of failures of devices caused by the movement of alkali ions in high electric fields. An exhaustive review of the properties of glass surfaces is given by Holland [184].

Soft glass, Pyrex, and quartz surfaces [185,186] exhibit large (several microns) smooth areas with occasional bowl-like depressions [Fig. 14(a)]. The figure also shows macroscopic-size ($\sim 1 \mu$) scratches attributable to rolling and polishing of poor-quality glass slides. These imperfections may be reduced by optical polishing and fire polishing to less than 50 Å.

If suitable materials are used, the cleaved surface of single-crystal substrates can be obtained with smoothness on an atomic scale. This is particularly true of mica. The surface of a freshly cleaved alkali-halide crystal varies considerably and shows a large number of cleavage steps of atomic heights and various types of other surface defects. The deposition of ultrathin metal films decorates [187-189] these defects and makes them visible by electron microscopy, as shown in Fig. 14(b). Surface defects may be reduced by polishing [190] of NaCl on a moist cloth followed by quick rinsing in water and methanol.

Mica, alkali halides, and MgO [191] crystals can be easily cleaved. By producing a crevice with a sharp knife, mica sheets can be pulled apart in air or vacuum. Alkali halides can be split apart by a stroke of a sharp knife actuated by a hammer, pendulum, or spring action. Mechanical arragements for cleaving crystals in vacuum and subsequent deposition of the film are described in the literature [189,192]. Cleaving of mica and NaCl leaves very fine dust on the surface, which presents a serious drawback for vacuum-cleaved surfaces since the dust particles cannot be easily removed and may act as nucleating centers.

2. Substrate Cleaning. Glass substrates must be thoroughly cleaned before deposition, and a variety of procedures exist for this purpose. References to substrate materials, their surfaces, and the cleaning processes are given by Gaffee [193]. The following procedure used in the author's laboratory is found adequate. The gross contaminants are first removed by a lukewarm, ultrasonically agitated, ionic detergent. Note that a hot detergent solution may produce [194] nonuniform etching of soda-lime glasses. The glass is then rinsed thoroughly several times in deionized water and later subjected to a vapor degreaser using pure alcohol. The cleaned glass may be stored immersed in pure alcohol and occasionally agitated ultrasonically. Before use, the glass is dried by blowing with dry nitrogen.

Additional methods for cleaning glass and cleaved surfaces utilize heat and electron and ion bombardment. Fire polishing or even heating to near the softening point produces a clean glass surface. Thermal treatment yields etched, rough surfaces [Fig. 14(c)] on alkali-halide crystals above about 400°C. Atomically smooth surfaces are obtained for metals by lengthy annealing close to their melting point [195].

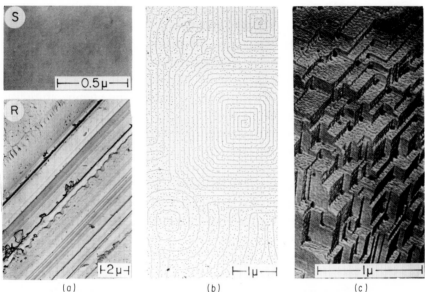

(a) (b) (c)

FIG. 14. Micrographs of the surface structures of (a) glass slide (*Chopra, unpublished*), (b) freshly cleaved rock salt after decoration with Au deposit (*courtesy of H. Bethge and W. Keller*), (c) thermally etched rock salt after heating at 400°C (*Sella and Trillat* [189]).

An efficient method to remove contaminants and oxide layers from a substrate is to sputter the surface by ionic bombardment, as described in detail by Holland [2]. An extensive investigation of the cleanliness and surface smoothness obtained by sputtering was made by Farnsworth et al. [196] and other investigators [197-199] using the low-energy electron diffraction (LEED) technique. The general conclusions are that an atomically smooth surface can be realized by sputtering if it is preceded by degassing and extended annealing. It must be noted that low bombardment energies (~0.5 kV) and low current densities (~100 $\mu A/cm^2$) should be used to prevent surface damage. Furthermore, glow discharge should be confined to the surface for cleaning and extraneous sputtering should be eliminated.

The use of hot substrates is required for many thin film depositions. A quartz iodine radiation lamp (an extended source), a Nichrome wire heater sandwiched between mica sheets, and W or Ta strip heaters are convenient heat sources employed in the author's laboratory. An electron-bombardment heater for obtaining temperatures up to 1200°C in uhv ($\sim 10^{-10}$ Torr) has been described by Naber [200]. Ideally, a uniform temperature is ensured only if the substrate is confined to a blackbody chamber. By cementing the substrate with graphite or silver paint to a recessed platform in a heated (or cooled) high-purity Cu block, equilibrium temperatures very close to that of the block are easily obtained. The heating and controlling of the substrate temperature have been discussed by several authors [201]. Conventional temperature controllers using thin-film thermocouples (Chap. VI, Sec. 6.2) deposited directly on the substrate are recommended for an accurate control of the surface temperature.

3. Uniform and Nonuniform Deposits. It is necessary to interpose a mechanically or magnetically controlled shutter between the evaporation source and the substrate for the obvious reason of controlling deposition. A rotating shutter of suitable shape can be utilized to intercept the vapor beam nonuniformly to produce wedge-shaped films [202] and spherical surfaces [203,204]. With increasing size of the substrate, deposition from a small vapor source will yield an increasingly nonuniform deposit. This may be corrected by a rotating shutter to intercept vapors at different places for different periods of time.

Uniform coating of large surfaces is generally done by using a large number of ring sources (equivalent to a rotating small source) or by rotation of the substrate. For example, Strong [205] satisfactorily coated a 200-in.-diameter telescope mirror with Al from 175 coils located on five concentric circles with radii of 20, 40, 60, 70, and 100 in. The problem of uniformity of deposits for extended sources has been discussed by Holland [2] and Behrndt [206,207]. A rotating substrate is expected to be equivalent to a ring source with respect to the thickness distribution and has been utilized by several investigators [206-213] to coat large surfaces such as that of a satellite for outer space. A conventional arrangement is to use a planetary motion in which substrates arranged on a circle concentric to the source rotate around the vapor source as well as their axes. Under optimum positioning,

4-in.-diameter substrates have been coated with a thickness variation of ±0.25 percent [206]. Note that a rotating substrate also reduces structural effects that may arise because of the large angle of incidence of vapors (Chap. IV, Sec. 2.3).

4. Masks and Connections. By laying a suitable mask over the substrate during deposition, films in some pattern or shape can be obtained. A moving mask may be used to grow films laterally. Resolution of the pattern is determined by the definition of the mask and its proximity to the substrate. Preparation of masks from thin (~100 μ) metal foils by selective etching or photoetching techniques is described in great detail by Gaffee [193]. Resolution of lines of about 10 μ can be obtained. A wire-grill mask [214] placed very close to the substrate yields a resolution of ~1 μ. Instead of using masks during deposition, the required patterns may be photoetched or cut with a microelectron beam. This is, of course, an important area of technology for microelectronics and is extensively dealt with elsewhere [193]. Lessor [215] has discussed the design of suitable crossovers when superimposed films are deposited.

Electrical connections to thin films for measurements, or interconnections within thin-film circuits can be obtained in a variety of ways which are thoroughly discussed by several authors [216,217]. Indium, In alloys (with Sn, Pb, Ag, etc.) of different melting points, Sn-Ag alloys, and 60:40 Sn-Pb solders can be used for soldering wires onto films using a low-wattage (~10 to 20W) fine-tip soldering gun. Microwelding techniques, such as thermocompression bonding, welding in a stream of hot H_2, or ultrasonic welding, which are widely used in commercial production of semiconductor circuitry, are readily applicable to thin films.

5. Multiple-film Deposition. Deposition of films of the same material on a large number of substrates (batch process), and a large number of different types of films on a variety of substrates during the same pump-down cycle (sequential process) is commonly required for research and development applications. A variety of mechanical arrangements (jigs) to perform these processes in a sequential or a batch manner have been described in the literature [218-226]. The designs have largely been dictated by the specific needs of the investigators.

Batch coating is simple and can be performed by mounting substrates on a carrousel holder which may be continuously rotated or moved step by step. An arrangement for sequential coating is more versatile and more useful. A typical design [221-225] consists of a carrousel substrate holder, a vapor-source turret, and a carrousel mask holder which allows masks to be raised and brought in contact with the substrate with a register accuracy of about 20 μ. The system should be made of stainless steel and use unlubricated (MoS_2 powder impregnated with DC 705 silicon oil may be used) stainless-steel ball bearings. Among the other accessories needed are jigs for loading the substrates into the substrate holder, heaters to heat the substrates, and suitably placed deposition-rate monitors.

A simple and unique system (Fig. 15) designed in the author's laboratory makes it possible to prepare films sequentially from any one of five vapor sources, onto any one of five substrates maintained at any temperature from 23 to 450°C, using any one of the five masks. Helical gears are used to effect translatory motion of the substrate and mask carriages via sprocket and chain linkages, and the rotary motion of the

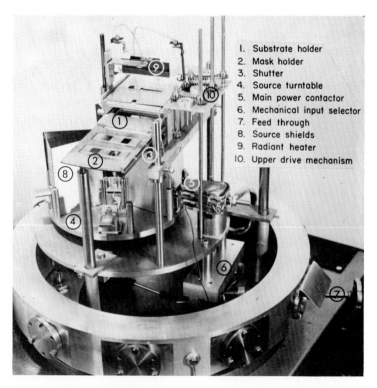

1. Substrate holder
2. Mask holder
3. Shutter
4. Source turntable
5. Main power contactor
6. Mechanical input selector
7. Feed through
8. Source shields
9. Radiant heater
10. Upper drive mechanism

FIG. 15. Photograph of a mechanical system for multiple-film deposition. (*Randlett, Stroberg, and Chopra* [226])

turntable carrying individually shielded vapor sources. Their motion is obtained by mating with a helical gear mounted at the end of a shaft of a feedthrough capable of push-pull, rotation, and 60° conical movements. The feedthrough consists of a Viton, O-ring-sealed, spherical ball with a shaft through the ball sealed to it by two O rings. The carriages have indented holder positions for precision registry. The various parts can easily be removed for cleaning purposes. The substrates are heated at the place of deposition by means of a quartz-iodine lamp. Deposition is monitored by a fixed quartz-crystal oscillator. A unique copper roller with silver contacts capable of handling up to 300 A is used for making electrical contacts for the vapor-source heaters. This system has been operated routinely at 10^{-6} to 10^{-7} Torr in a standard 19-in. diffusion-pump system.

Continuous-coating machines with moving-belt-type-substrate holders and differentially pumped chambers or loading-lock arrangements have been designed and operated [227,228] successfully for deposition with both evaporation and sputtering techniques. Such machines are, of course, only of commercial interest. The reader may appreciate that many of the laboratory deposition techniques can be transformed to pilot-plant production with suitable engineering modifications. For example, a continuous deposition system using several ion-beam sputtering modules has been proposed by Chopra and Randlett [229]. This machine is capable of sequential deposition of any combination of metal and nonmetal films.

6. CONCLUSIONS

A large variety of methods for depositing thin solid films at rates ranging from a fraction of an angstrom to 10^6 Å/sec have been successfully utilized. Several of these methods, notably thermal evaporation and sputtering, permit automatic monitoring and controlling of the rate of deposition and film thickness. The fact that the properties of films depend on various deposition parameters must be considered in the choice of a particular deposition technique. These parameters include the rate of deposition, temperature or kinetic energy of the vapor atoms, angle of incidence of vapor, electrostatic charges carried by vapors, the substrate and vapor-source materials, substrate (deposition) temperature, composition of the ambient atmosphere, and thermal accommodation and condensation coefficients of the vapor atoms and the ambient gas atoms. These parameters may vary from technique to technique. Their influence on the film properties is not yet well established in all cases. Consequently, a clear-cut comparison of films obtained by different techniques is not feasible at present. Much work is therefore needed in the future to characterize the important deposition techniques. This knowledge is essential to answer significant questions on the variations of the structure, growth, epitaxial conditions, and properties of films prepared by different techniques.

7. APPENDIX

Miscellaneous Technical Information for Preparation of Films of Various Materials

Material	ρ, g/cc	mp, °C	Temp, °C for vp = 10^{-4} Torr	Evaporation sources		Other techniques*	Remarks
				Filament	Boat/Crucible		
Ag	10.5	961	1105	Ta, W, Mo, Nb, Fe, Ni, Cr	Mo, Ta	ED,SP	Does not wet W
Al	2.7	660	1280	W, Ta	Graphite, Al_2O_3 coated boats, (BN + 50% TiB_2) boat	SP	Alloys with refractory metals; reacts with carbon and oxide crucibles
Al_2O_3	3.6	2020	1600	W, Ta	W, Ta	EB,SP(rf), AN	Yields some Al and AlO
As	5.7	Subl.	480	Graphite	Al_2O_3,C, BeO	EB,SP	Sublimes, toxic
Au	19.3	1063	1405	W, Mo	Al_2O_3,C,W,Mo	EB,SP	Wets but reacts with Ta, possibly alloy formation. Partially wets W, Mo
B	2.3	2550	2400		C	EB	Material explodes on rapid cooling
BN	2.3	Subl. ~3000				EB	Thermal shock requires slow heat-up
B_4C	2.5	2450	2650	No	No	EB,SP	Thermal shock requires slow heat-up

69

7. APPENDIX (Continued)

Material	ρ, g/cc	mp,°C	Temp,°C for $vp = 10^{-4}$ Torr	Evaporation sources		Other techniques*	Remarks
				Filament	Boat/Crucible		
Ba	3.5	725	735	W,Ta,Mo,Nb, Ni,Fe,Cr	Ta,Mo,W	EB,SP	Violently reacts with ceramics; wets sources
BaTiO$_3$	Tetr. 6.0, hex. 5.8		Decomposes	Flash evaporation (Ir boat); two-source		SP(rf)	Yields some free Ba
Be	1.9	1284	1270	Ta,W,Mo	C,BeO,Ta,W,Mo	EB	Wets sources. Vapors, powders, and oxides toxic
Bi	9.8	271	790	Chromel, W, Ta,Mo,Nb	Al$_2$O$_3$,C,Fe,Mo, Ta,W		Wets Chromel. Vapors toxic
Bi$_2$O$_3$	8.9	820	1840 at 10μ	Pt		SP (react)	Readily sputters in O$_2$-Ar mixture
Bi$_2$Te$_3$	7.7	573		W,Ta,Mo		SP	
C	1-2	3700	2410	Arc evaporation	Graphite	EB,SP	Solid "sparks" from point source. Density variable
Ca	1.5	842	732	W	Al$_2$O$_3$		Deposits corrode in air
CaF$_2$	3.2	1360		Ta,W	Ta,W		Rate control important. Sublimes
Cd	8.6	321	450	Chromel,Nb, Ta,Mo,W,Ni, Fe	Mo,Ta,fused quartz		Wets Nb and Chromel
CdS	4.8	1750	920	W	Graphite, quartz, C	EB, SP(rf)	EV film contains free Cd; sticking coeff. strongly affected by substrate temp
				Two-source			

Material	Density	T1	T2	Source A	Source B	Technique	Remarks
CdTe	6.2	1041		W	Al₂O₃,C	EB,SP	
Ce	6.9	804	1650	W	W		
CeF₃	6.16	1460					
CeO₂	6.9	1600	2310		W	EB	Attacks boat. Rate control important. Does not decompose on evaporation from W
Co	8.9	1495	1530	Nb,W	Al₂O₃,BeO	EB,SP	Forms low-melting eutectic with refractory metals
Cr	6.9	1890	1430	W	Al₂O₃,BeO	EB,SP	Can be sublimed
Cr/Ni	8.2	1350	1490	W,Ta	Ta,W,Mo, Al₂O₃	EB,SP	Smooth deposit from resistance sources
Cu	8.9	1083	1032	Pt or Pt/Ir coil, Nb,Mo, Ta,W	Mo,Ta,W, Al₂O₃	EB,SP,CVD	Alloys with Ni, Fe, Chromel. Does not wet Mo, W, Ta
Fe	7.9	1535	1207	W	Al₂O₃,W,BeO	EB,SP	Alloys with W if Fe exceeds 1/3 total weight of filament
Ga	5.9	30	1180	Fused quartz	Al₂O₃,BeO quartz		Alloys with metals
GaAs		1238		Two-source, flash evaporation (Ta boat)		SP,CVD	Compound decomposes
GaP		1350		Ta,W, Flash evaporation (Ta boat)		EB,CVD	Does not decompose. Rate control important
GaSb				Flash evaporation (Ta boat)		EB,CVD	
Ge	5.3	959		Ta,Mo,W	Mo,Ta,W, Al₂O₃,C	EB,SP,CVD	Wets Ta, Mo, and W

7. APPENDIX (Continued)

Material	ρ, g/cc	mp, °C	Temp, °C for vp = 10^{-4} Torr	Evaporation sources		Other techniques*	Remarks
				Filament	Boat/Crucible		
GeTe	6-17	Subl.		W,Mo		SP	Wets W
In	7.4	157	1015	W,Fe	W,Mo		
InAs		940		Two-source, flash evaporation (Ta boat)		SP,CVD	Compound decomposes
InP			1070	Ta,W, flash evaporation		SP,CVD	Deposits P rich
InSb		535		Two-source, flash evaporation		SP,CVD	Compound can be reconstituted by deposition on substrate at 900°C
Ir	22.4	2454	2380		Graphite	EB	
Li	0.5	186	680		Fused quartz, steel, Fe	EB	Oxide destroys fused quartz
LiF	2.6	870	1180		Mo	EB	Rate control important for optical films
Mg	1.7	651	600	W,Ni,Fe,Ta, Mo,Nb, Chromel	Mo,Ta,C		Sublimes at high rate
MgF$_2$	2.9-3.2	1266	1540	W,Ta,Mo	C	EB	Rate control important for optical films
Mn	7.2	1260	1020	W,Ta,Mo,Nb	Al$_2$O$_3$	EB	Wets resistance sources
Mo	10.2	2622	2390	No	No	EB,SP	Very volatile if oxidized to form MoO$_3$

		1000	1000 (10μ)	Mo			
Na₃AlF₆ (cryolite)							Hygroscopic
NaCl	2.2	801	1014		Ta,W,C	Furnace	Hygroscopic film
NaF	2.8	980	1200		Ta,W	Furnace	Soluble film
Nb	8.5	2500	2550	W	W	EB,SP	Used as source material. Reacts with W, reducing source life
Nd₂O₃	7.2	~1900		W (thick)	W	SP(rf)	
Ni	8.9	1455	1535		Al₂O₃,BeO	EB,SP	Forms low-melt eutectic with filaments
NiFe	8.7	1395	1580		Al₂O₃	EB,SP	Ni content low in film. Use 84% Ni source
Pb	11.3	328	870	Fe,Ni,W, Chromel	Al₂O₃,W,Mo, Ta,Fe	EB	Does not wet W, Ta, Mo, Nb; toxic
PbS	7.5	1114		W,Mo			
PbSe	8.1	1065		W			
PbSnO	8.1	1115	905	Pt,Ir	Pt,Ir	SP,CVD	Disproportionates
PbTe	8.1	917	1050	W	Al₂O₃	EB,SP	
Pd	12.0	1550	1465	W	Al₂O₃,BeO	EB,SP	
Pt	21.5	1774	2020	W		EB,SP,ED	Alloys with metals
Rh	12.1	2149	1980	W		EB,SP,ED	Very low pressure for W source
Sb	6.7	630	1480	Chromel, Ta,Mo	Ta,C,Al₂O₃, Mo		Wets Chromel

7. APPENDIX (Continued)

Material	ρ, g/cc	mp, °C	Temp, °C for vp = 10^{-4} Torr	Evaporation sources		Other techniques*	Remarks
				Filament	Boat/Crucible		
Sb_2S_3	4.12	550	550 at 10μ	Mo			
Se	4.3	234	437	Chromel,Fe, Mo,Nb, 304 stainless steel	To,Ta,Al_2O_3		Very volatile, may contaminate vacuum system. Wets filament metals quoted
Si	2.4	1420	1610	SiC	BeO,C	EB,SP,CVD	Difficult to prepare Si films free from SiO
SiC	3.217	2700				EB,reaction with methane	
SiO	2.1	Softens, no sharp mp	1250	W,Ta,Mo	W,Ta	EB,furnace	Pinholes reduced by low rate
SiO_2	2.1	Softens	Influenced by composition	Decomposes		EB,SP(rf), CVD	Yields SiO at poor pressure
Sn	5.7	232	1270	Chromel,Ta, Mo	Mo,Ta,Al_2O_3,C		Wets Chromel, Mo, and Ta
SnO_2	6.4	Decomposes			W	SP	Sublimes
Ta	16.6	2996	2860			EB,SP	Getters oxygen
Ta_2O_5	8.7	1470	1920	Ta,W	Ta,W	EB,SP(rf)	Forms smooth spectral film, high dielectric constant

Material	Density	Melting/Boiling Temp		Source Materials		Technique	Remarks
Te	6.2	452	550	W,Ta,Mo,Nb, Ni,Fe	Al_2O_3,fused quartz		Poisons system; toxic vapors. Wets without alloying all sources
Th	11.7	1827	1831	W	W	EB,SP	Getters gas, oxide film on melt
Ti	4.5	1800	1715	W,Ta	W,graphite	EB,SP(rf)	Decomposes into TiO, Ti; must be reoxidized
TiO_2	4.3	~1850		Ta,W	Ta,W		Decomposes into TiO, Ti, TiO_2
Ti_2O_3	4.6	2130 (decomp.)		Ta,Mo			
Tl	11.9	302	740	Ni,Fe,Nb,Ta	Quartz,Al_2O_3,Ta		Decomposes into TiO, Ti, TiO_2
U	18.7	1132	1855	W	W	EB	Wets sources without alloying
V	5.9	1710	1820	W,Mo	W,Mo	EB	Films oxidize
W	19.3	3370	3030	No	No	EB,SP	Wets Mo. Reacts with Ta, W
W_2C	15.7	2860		No	Graphite	EB,SP	Forms volatile oxides. Films oxidize
WO_3	12.1	1473	1460	W	Ta,W	EB	Smooth, conductive, hard film
Zn	7.1	419	520	W,Ta,Mo,Nb	Al_2O_3,Fe,C, porcelain		Yields some $(WO_3)_x$, $x = 3, 4, 5$
ZnS	3.9	1900	1265	Ta,Mo	Ta,Mo,C	Furnace	Wets refractory metals; no reaction
Zr	6.4	1857	2260	W	W	EB,SP	Partially decomposes. Sticking coeff. varies greatly with substrate temp
ZrO_2	5.9	~2700		W	W	EB,SP(rf)	Requires low pressure $<10^{-6}$ Torr to prevent film oxidation
							Yields some ZrO

REFERENCES

1. M. Faraday, *Phil. Trans. Roy. Soc. London,* **147**:145 (1857); R. Nahrwold, *Wied. Ann.,* **31**:467 (1887); R. Pohl and P. Pringsheim, *Verhandl. Deut. Physik. Ges.,* **14**:546 (1912).
2. L. Holland, "Vacuum Deposition of Thin Films," John Wiley & Sons, Inc., New York, 1956.
3. L. Holland, in "Thin Film Microelectronics" (L. Holland, ed.),chap. IV, John Wiley & Sons, Inc., New York, 1956.
4. W. C. Vergara, H. M. Greenhouse, and N. C. Nicholas, *Rev. Sci. Instr.,* **34**:520 (1963).
5. R. D. Gretz, *Rev. Sci. Instr.,* **38**:112 (1967).
6. I. Ames, L. H. Kaplan, and P. A. Roland, *Rev. Sci. Instr.,* **37**:1737 (1966).
7. F. E. Card and J. J. Galen, *Rev. Sci. Instr.,* **32**:858 (1961).
8. W. Parker and Y. Grunditz, *Nucl. Instr. Methods,* **22**:73 (1963).
9. L. E. Preuss and C. Alt-Anthony, *Trans. 6th Natl. Vacuum Symp.,* p. 228, Pergamon Press, New York, 1959.
10. P. Huijer, W. T. Langendam, and J. A. Lely, *Philips Tech. Rev.,* **34**:144 (1963); also, see T. K. Lakshmanan, *Trans. 8th Natl. Vacuum Symp.,* p. 868, Pergamon Press, New York, 1961.
11. G. Zinsmeister, *Vakuum-tech.,* **8**:223 (1964).
12. L. S. Palatnik, G. V. Fedorov, and P. N. Bogatov, *Phys. Metals Metallog.,* **21**:89 (1966).
13. For example, R. F. Bis, A. S. Rodolakis, and J. N. Zemel, *Rev. Sci. Instr.,* **36**:1626 (1965); A. E. Feuersanger, A. K. Hagenlocher, and A. L. Solomon, *J. Electrochem. Soc.,* **111**:1387 (1964).
14. For example, J. deKlerk and E. F. Kelly, *Rev. Sci. Instr.,* **36**:506 (1965); also, F. A. Pizzarello, *J. Appl. Phys.,* **35**:2730 (1964).
15. For example, K. H. Behrndt and R. W. Love, *Vacuum,* **12**:1 (1962).
16. For example, J. E. Davey and T. Pankey, *J. Appl. Phys.,* **35**:2203 (1964).
17. K. G. Günther, in "The Use of Thin Films in Physical Investigations" (J. C. Anderson, ed.), p. 213, Academic Press Inc., New York, 1966; also, *Z. Naturforsch.,* **13a**:1081 (1958); in "Compound Semiconductors" (Willardson and Goering, eds.), vol. 1, Reinhold Publishing Corporation, New York, 1962.
18. L. Harris and B. M. Siegel, *J. Appl. Phys.,* **19**:739 (1948).
19. W. R. Beam and T. Takahashi, *Rev. Sci. Instr.,* **35**:1623 (1964).
20. W. Himes, B. F. Stout, and R. E. Thun, *Trans. 9th Natl. Vacuum Symp.,* p. 144, The Macmillan Company, New York, 1962.
21. E. K. Müller, B. J. Nicholson, and G. L'E. Turner, *J. Electrochem. Soc.,* **110**:969 (1963).
22. J. L. Richards, in "The Use of Thin Films for Physical Investigations" (J. C. Anderson, ed.), p. 71, Academic Press Inc., New York, 1966.
23. S. G. Ellis, *J. Appl. Phys.,* **38**:2906 (1967).
24. M. S. P. Lucas, H. A. Owen, Jr., W. C. Stewart, and C. R. Vail, *Rev. Sci. Instr.,* **32**:203 (1961).
25. M. Kikuchi, S. Nagakura, H. Ohmura, and S. Oketani, *Japan. J. Appl. Phys.,* **36**:3686 (1965).
26. W. M. Conn, *Phys. Rev.,* **79**:213 (1950).
27. D. M. Mattox, A. W. Mullendore, and F. N. Rebarchik, *J. Vacuum Sci. Tech.,* **4**:123 (1967).

28. V. M. Kul'gavchuk and G. A. Novoskol'tseva, *Soviet Phys. Tech. Phys.*, **11**:406 (1966).
29. H. M. Smith and A. F. Turner, *Appl. Opt.*, **4**:147 (1965).
30. H. Schwarz and H. A. Tourtellotte, *13th Natl. Vacuum Symp.*, p. 87, The Macmillan Company, New York, 1966; also, P. D. Zavitsanos and W. E. Sauer, *J. Electrochem. Soc.*, **115**:109 (1968).
31. J. A. Turner, J. K. Birtwistle, and G. R. Hoffman, *J. Sci. Instr.*, **40**:557 (1963).
32. E. A. Roth, E. A. Margerum, and J. A. Amick, *Rev. Sci. Instr.*, **33**:686 (1962).
33. J. Van Audenhove, *Rev. Sci. Instr.*, **36**:383 (1965).
34. G. Siddall and B. A. Probyn, *Trans. 8th Natl. Vacuum Symp.*, p. 1017, Pergamon Press, New York, 1961.
35. O. S. Heavens, *J. Sci. Instr.*, **36**:95 (1959).
36. S. Robin-Kandaré, M. H. Damany, and L. Tertian, *J. Phys. Radium,* **20**:504 (1959).
37. K. L. Chopra and M. R. Randlett, *Rev. Sci. Instr.*, **37**:1421 (1966).
38. D. H. Blackburn and W. Haller, *Rev. Sci. Instr.*, **36**:901 (1965).
39. R. E. Thun and J. B. Ramsey, *Trans. 6th Natl. Vacuum Symp.*, p. 192, Pergamon Press, New York, 1959.
40. B. A. Unvala and G. R. Booker, *Phil. Mag.,* **9**:691 (1964).
41. W. R. Grove, *Phil. Trans. Roy. Soc. London*, **142**:87 (1852).
42. G. K. Wehner, *Advan. Electron. Electron Phys.*, **7**:239 (1955).
43. W. J. Moore, *Am. Scientist,* **48**:109 (1960); J. S. Colligon, *Vacuum,* **11**:272 (1961).
44. E. Kay, *Advan. Electron. Electron Phys.*, **17**:245 (1962).
45. M. Karminsky, "Atomic and Ionic Impact Phenomena on Metal Surfaces," Academic Press Inc., New York, 1965.
46. H. S. W. Massey and E. H. S. Burhop, "Electrons and Ion Impact Phenomena," Oxford University Press, Fairlawn, N.J., 1956.
47. P. K. Rol, D. Onderdelinden, and J. Kistemaker, *Proc. 3d Intern. Congr. Vacuum Tech.,* Stuttgart, vol. 1, p. 75, Pergamon Press, New York, 1965.
48. S. P. Wolsky, *Trans. 10th Natl. Vacuum Symp.*, p. 309, The Macmillan Company, New York, 1963.
49. M. H. Francombe, *Trans. 10th Natl. Vacuum Symp.*, p. 316, The Macmillan Company, New York, 1963.
50. L. I. Maissel, in "Physics of Thin Films" (G. Hass and R. E. Thun, eds.), vol. 3, p. 61, Academic Press Inc., New York, 1966.
51. G. K. Wehner, *Phys. Rev.,* **108**:35 (1957); **112**:1120 (1958); also, Ref. 73.
52. O. Almén and G. Bruce, *Nucl. Instr. Methods,* **2**:257 (1961).
53. M. Bader, F. Witteborn, and T. W. Snouse, *NASA Tech. Rept.* TR-R105, 1961.
54. G. K. Wehner, *J. Appl. Phys.,* **26**: 1056 (1955); G. S. Anderson, G. K. Wehner, and H. J. Olin, *J. Appl. Phys.,* **34**:3492 (1963); M. Koedam, *Physica,* **25**:742 (1959); M. Koedam and A. Hoogendoorn, *Physica,* **26**:351 (1960).
55. R. S. Nelson and B. J. Sheldon, *Harwell Unclassified Rept.* AERE R-4694.
56. G. K. Wehner, *Phys. Rev.,* **114**:1270 (1960); N. Laegreid and G. K. Wehner, *J. Appl. Phys.,* **32**:365 (1961); R. V. Stuart and G. K. Wehner, *Trans. 9th Natl. Vacuum Symp.,* 1962, p. 160.
57. V. I. Veksler, *Soviet Phys. JETP,* **11**:235 (1960).
58. M. Kaminsky, *Phys. Rev.,* **126**:1267 (1962); J. R. Woodyard and C. B. Cooper, *J. Appl. Phys.,* **35**:1107 (1964).
59. J. Comas and C. B. Cooper, *J. Appl. Phys.,* **38**:2956 (1967).

60. N. N. Petrov, *Soviet Phys. Solid State*, **2**:857, 1182 (1960); O. Roos, *Z. Physik*, **147**:210 (1957).
61. H. D. Hagstrum, *Phys. Rev.*, **96**:325, 336 (1954); **104**:1516 (1956); **122**:83 (1961); *J. Phys. Chem. Solids*, **14**:33 (1960); *J. Appl. Phys.*, **32**:1015, 1020 (1961).
62. A. von Hippel, *Ann. Physik*, **81**:1043 (1926).
63. C. H. Townes, *Phys. Rev.*, **65**:319 (1944).
64. F. Stark, *Z. Elektrochem.*, **15**:509 (1909).
65. K. H. Kingdon and I. Langmuir, *Phys. Rev.*, **22**:148 (1923).
66. E. B. Henschke, *J. Appl. Phys.*, **33**:1773 (1962); E. Landberg, *Phys. Rev.*, **111**:91 (1958).
67. F. Keywell, *Phys. Rev.*, **97**:1611 (1955).
68. D. E. Harrison, Jr., *Phys. Rev.*, **102**:1473 (1956); *J. Chem. Phys.*, **32**:1336 (1960).
69. R. S. Pease, *Nuovo Cimento*, Suppl. 13, 1960.
70. R. H. Silsbee, *J. Appl. Phys.*, **28**:1246 (1957).
71. D. E. Harrison, J. P. Johnson, and N. S. Levy, *Appl. Phys. Letters*, **8**:33 (1966).
72. C. Lehmann and P. Sigmund, *Phys. Stat. Solidi*, **16**:507 (1966).
73. N. Laegreid and G. K. Wehner, *J. Appl. Phys.*, **32**:365 (1961).
74. For example, L. I. Maissel and J. H. Vaughn, *Vacuum*, **13**:421 (1963).
75. H. Fetz, *Z. Physik*, **119**:590 (1942).
76. W. R. Sinclair and F. G. Peters, *J. Vacuum Sci. Tech.*, **2**:178 (1965); E. Kay, U.S. Patent 3,354,074.
77. J. G. Simmons and L. I. Maissel, *Rev. Sci. Instr.*, **32**:642 (1961).
78. W. R. Sinclair and F. G. Peters, *Rev. Sci. Instr.*, **33**:744 (1962).
79. R. S. Humphries, *Rev. Sci. Instr.*, **37**:1734 (1966).
80. N. Schwartz, *Trans. 10th Natl. Vacuum Symp.*, p. 325, The Macmillan Company, New York, 1963.
81. J. Sosniak, *J. Vacuum Sci. Tech.*, **4**:87 (1967); E. A. Stern and H. L. Caswell, *J. Vacuum Sci. Tech.*, **4**:128 (1967).
82. L. I. Maissel and P. M. Schaible, *J. Appl. Phys.*, **36**:237 (1965).
83. R. Frerichs, *J. Appl. Phys.*, **33**:1898 (1962).
84. G. M. Mattox, *Electrochem. Tech.*, **2**:295 (1964).
85. H. C. Theuerer and J. J. Hauser, *J. Appl. Phys.*, **35**:554 (1964).
86. F. M. Penning and J. H. A. Moubis, *Proc. Konikl. Ned. Akad. Wetenschap.*, **43**:41 (1940).
87. W. D. Gill and E. Kay, *Rev. Sci. Instr.*, **36**:277 (1965).
88. E. Kay, *J. Appl. Phys.*, **34**:760 (1963).
89. R. D. Ivanov, G. V. Spivak, and G. K. Kislova, *Izv. Akad. Nauk SSSR Ser. Fiz.*, **25**:1524 (1961).
90. For example, J. Edgecumbe, L. G. Rosner, and D. E. Anderson, *J. Appl. Phys.*, **35**:2198 (1964).
91. H. Gawehn, *Z. Angew. Phys.*, **14**:126 (1962).
92. G. S. Anderson, W. N. Mayer, and G. K. Wehner, *J. Appl. Phys.*, **33**:2991 (1962).
93. R. L. Hines and R. Wallor, *J. Appl. Phys.*, **32**:202 (1961).
94. G. V. Spivak, A. I. Krokhina, T. V. Yavorskaya, and Y. A. Durasova, *Dokl. Akad. Nauk SSSR*, **114**:1001 (1957).
95. P. D. Davidse and L. I. Maissel, *J. Appl. Phys.*, **37**:574 (1966); P. D. Davidse, *Vacuum*, **17**:139 (1967).

96. P. D. Davidse and L. I. Maissel, *J. Vacuum Sci. Tech.*, **4**:33 (1967).
97. F. Vratny, *J. Electrochem. Soc.*, **114**:505 (1967).
98. M. von Ardenne, "Tabellen der Elektronenphysik, Ionenphysik und Über-mikroskopie," vol. 1, p. 653, VEB Deutscher Verlag der Wissenschaften, Berlin, 1956.
99. J. Kistemaker and H. L. Duowes-Dekker, *Physica*, **16**:195 (1950).
100. A. T. Finkelstein, *Rev. Sci. Instr.*, **11**:94 (1940).
101. C. D. Moak, H. E. Banta, J. N. Thurston, J. W. Johnson, and R. F. King, *Rev. Sci. Instr.*, **30**:694 (1959).
102. C. E. Carlston and F. D. Magnuson, *Rev. Sci. Instr.*, **33**:905 (1962).
103. H. J. Liebl and R. F. K. Herzog, *J. Appl. Phys.*, **34**:2893 (1963).
104. H. Tawara, *Japan. J. Appl. Phys.*, **3**:342 (1964); L. J. Christensen and E. J. Zaharis, *Rev. Sci. Instr.*, **37**:1571 (1966).
105. K. L. Chopra and M. R. Randlett, *Rev. Sci. Instr.*, **38**:1147 (1967); U.S. Patent 3,408,283.
106. F. Gaydou, *Vacuum*, **17**:325 (1967).
107. E. E. Smith and S. G. Ayling, *Proc. Electron. Components Conf.*, 1962, p. 82.
108. C. S. Murphy, *Electron. Reliability Micromin.*, **2**:235 (1963).
109. P. Lloyd, *Solid-State Electron.*, **3**:74 (1961).
110. G. Perny and B. Laville-Saint-Martin, *J. Phys. Radium*, **25**:993 (1964); G. Perny et al., *J. Phys. Radium*, **25**:5 (1964).
111. E. Krikorian and R. J. Sneed, *J. Appl. Phys.*, **37**:3674 (1966); E. Krikorian, *12th Natl. Vacuum Symp. Abstracts*, 1966, p. 175.
112. W. R. Sinclair and F. G. Peters, *J. Am. Ceram. Soc.*, **46**:20 (1963).
113. J. C. Williams, W. R. Sinclair, and S. E. Koonce, *J. Am. Ceram. Soc.*, **46**:161 (1963).
114. A. W. Simpson, unpublished (Plassey Co. Internal Report, 1962).
115. A. Güntherschulze, *Z. Physik*, **36**:563 (1926).
116. T. K. Lakshmanan and J. M. Mitchell, *Trans. 10th Natl. Vacuum Symp.*, p. 335, The Macmillan Company, New York, 1963.
117. J. L. Miles and P. H. Smith, *J. Electrochem. Soc.*, **110**:1240 (1963).
118. G. J. Tibol and R. W. Hull, *J. Electrochem. Soc.*, **111**:1368 (1964).
119. H. F. Sterling and R. C. G. Swann, *Solid-State Electron.*, **8**:653 (1965).
120. L. L. Alt, S. W. Ing., Jr., and K. W. Laendle, *J. Electrochem. Soc.*, **110**:465 (1963).
121. H. Pagnia, *Phys. Stat. Solidi*, **1**:90 (1961).
122. E. M. DaSilva and R. E. Miller, *Electrochem. Tech.*, **2**:147 (1964).
123. R. A. Connell and L. V. Gregor, *J. Electrochem. Soc.*, **112**:1198 (1965).
124. J. Goodman, *J. Polymer Sci.*, **44**:551 (1960).
125. E. Gillam, *Phys. Chem. Solids*, **11**:55 (1959).
126. T. F. Fisher and C. E. Weber, *J. Appl. Phys.*, **23**:181 (1952).
127. R. Hanau, *Phys. Rev.*, **76**:153 (1949).
128. C. Moulton, *Nature*, **195**:793 (1962).
129. M. H. Francombe, J. J. Flood, and G. L'E. Turner, *Proc. 5th Intern. Conf. Electron Microscopy*, paper DD-8, Academic Press Inc., New York, 1962.
130. R. Bunsen, *J. Prakt. Chem.*, **56**:53 (1952).
131. G. Milazzo, "Electrochemistry," Elsevier Publishing Company, Amsterdam, 1963.
132. F. A. Lowenheim (ed.), "Modern Electroplating," John Wiley & Sons, Inc., New York, 1963.
133. E. C. Potter, "Electrochemistry," Cleaver-Hume, London, 1956.

134. L. Young, "Anodic Oxide Films," Academic Press Inc., New York, 1961.
135. H. Schäfer, "Chemical Transport Reactions," Academic Press Inc., New York, 1964.
136. C. F. Powell, J. H. Oxley, and J. M. Blocher, Jr. (eds.), "Vapor Deposition," John Wiley & Sons, Inc., New York, 1966.
137. L. V. Gregor, in "Physics of Thin Films" (G. Hass and R. E. Thun, eds.), vol. 3, p. 130, Academic Press Inc., New York, 1966.
138. B. A. Joyce, in "The Use of Thin Films for Physical Investigations" (J. C. Anderson, ed.), p. 87, Academic Press Inc., New York, 1966.
139. T. J. LaChapelle, A. Miller, and F. L. Morritz, in "Progress in Solid State Chemistry" (H. Reiss, ed.), vol. 3, p. 1, Pergamon Press, New York, 1967.
140. G. Gore, "The Art of Electro-metallurgy," 3d ed., Longmans, Green & Co., Inc., New York, 1877.
141. See the review by J. O'M. Bockris and A. Damjanovic, in "Modern Aspects of Electrochemistry" (J. O'M. Bockris and B. E. Conway, eds.), vol. 3, p. 224, Butterworths, Washington, D.C., 1964.
142. K. R. Lawless, *J. Vacuum Sci. Tech.*, 2:24 (1965). A detailed review in "Physics of Thin Films" (G. Hass and R. E. Thun, eds.), vol. 4, p. 191, Academic Press Inc., New York, 1967.
143. D. S. Campbell, in "The Use of Thin Films for Physical Investigations" (J. C. Anderson, ed.), p. 11, Academic Press Inc., New York, 1966.
144. D. J. Stirland and R. W. Bicknell, *J. Electrochem. Soc.*, 106:481 (1959).
145. M. L. Levin, *Trans. Faraday Soc.*, 54:935 (1958).
146. D. A. Vermilyea, in "Non-crystalline Solids" (V. D. Frechette, ed.), p. 328, John Wiley & Sons, Inc., New York, 1958.
147. R. L. Schwoebel, *J. Appl. Phys.*, 34:2784 (1963); P. E. Doherty and R. S. Davis, *J. Appl. Phys.*, 34:619 (1963).
148. G. C. Wood, C. Pearson, A. J. Brock, and S. W. Khoo, *J. Electrochem. Soc.*, 114:145 (1967).
149. Y. Tarui, H. Teshima, K. Okura, and A. Minamiya, *J. Electrochem. Soc.*, 110:1167 (1963).
150. P. J. Jorgenson, *J. Chem. Phys.*, 37:874 (1962).
151. For example, A. G. Baker and W. C. Morris, *Rev. Sci. Instr.*, 32:458 (1961); R. W. Christy, *J. Appl. Phys.*, 33:1884 (1962); O. G. Fritz, Jr., *J. Appl. Phys.*, 35:2272 (1964).
152. S. R. Bhola and A. Mayer, *RCA Rev.*, 24:511 (1963).
153. W. A. Kagdis, *J. Electrochem. Soc.*, 109:71C (1962).
154. C. Lewis, H. C. Kelly, M. B. Giusto, and S. Johnson, *J. Electrochem. Soc.*, 108:1114 (1961).
155. S. E. Mayer and D. E. Shea, *J. Electrochem. Soc.*, 111:550 (1964).
156. B. A. Joyce and R. R. Bradley, *J. Electrochem. Soc.*, 110:1235 (1963).
157. F. M. Smits, *Proc. IRE*, 46:1049 (1958); J. Klerer, *J. Electrochem. Soc.*, 108:1070 (1961).
158. N. N. Tvorogov, *Zh. Fiz. Khim.*, 34:2203 (1961).
159. J. A. Aboaf, *J. Electrochem. Soc.*, 114:948 (1967).
160. S. Yoshioka and S. Takayanagi, *J. Electrochem. Soc.*, 114:962 (1967).
161. W. Steinmaier and J. Bloem, *J. Electrochem. Soc.*, 111:206 (1964).
162. M. J. Rand, *J. Electrochem. Soc.*, 110:184C (1963); K. J. Miller and M. J. Grieco, *J. Electrochem. Soc.*, 110:1252 (1963).

163. H. C. Theuerer, *J. Electrochem. Soc.*, **108**:649 (1961); *J. Electrochem. Soc.*, **109**:742 (1962).
164. J. M. Charig and B. A. Joyce, *J. Electrochem. Soc.*, **109**:957 (1962).
165. S. K. Tung and R. A. Porter, *J. Electrochem. Soc.*, **110**:217C (1963).
166. J. J. Oberly and A. Adams, *J. Electrochem. Soc.*, **109**:210C (1962).
167. E. S. Wajda, B. W. Kippenhan, and W. H. White, *IBM J. Res. Develop.*, **4**:288 (1960).
168. E. S. Wajda and R. Glang, in "Metallurgy of Elemental and Compound Semiconductors" (R. O. Grubel, ed.), p. 229, Interscience Publishers, Inc., New York, 1961.
169. F. A. Pizzarello, *J. Electrochem. Soc.*, **110**:1059 (1963).
170. R. R. Moest and B. R. Shupp, *J. Electrochem. Soc.*, **109**:1061 (1962).
171. J. R. Knight, D. Effer, and P. R. Evans, *Solid-State Electron.*, **8**:178 (1965).
172. A. E. Ennos, *Brit. J. Appl. Phys.*, **5**:27 (1954).
173. R. W. Christy, *J. Appl. Phys.*, **31**:1680 (1960); *J. Appl. Phys.*, **35**:2179 (1964).
174. L. Holland and L. Laurenson, *Vacuum,* **14**:325 (1964).
175. M. Stuart, *Nature,* **199**:59 (1963).
176. I. Haller and P. White, *J. Phys. Chem.*, **67**:1784 (1963).
177. P. White, *J. Phys. Chem.*, **67**:2493 (1963); *Insulation,* September, 1963; May, 1967.
178. D. J. Valley and J. S. Wagener, *IEEE Trans. Component Pts.*, **CP-11**:205 (1964).
179. See Ref. 267, chap. IV.
180. S. Dushman, "Scientific Foundations of Vacuum Technique" (J. M. Lafferty, ed.), 2d ed., John Wiley & Sons, Inc., New York, 1962.
181. C. M. van Atta, "Vacuum Science and Engineering," McGraw-Hill Book Company, New York, 1965.
182. H. L. Caswell, in "Physics of Thin Films" (G. Hass, ed.), vol. 1, Academic Press Inc., New York, 1963.
183. For example, A. A. Nicholls, *Proc. 3d Intern. Congr. Vacuum Tech.*, *Stuttgart,* vol. 2, p. 21, Pergamon Press, New York, 1965.
184. L. Holland, "The Properties of Glass Surface," John Wiley & Sons, Inc., New York, 1964.
185. C. J. Calbick, *Trans. 8th Natl. Vacuum Symp.*, vol. 2, p. 1013, Pergamon Press, New York, 1961.
186. W. C. Levengood and T. S. Vong, *J. Appl. Phys.*, **31**:1416 (1960).
187. G. A. Bassett, J. W. Menter, and D. W. Pashley, in "Structure and Properties of Thin Films," p. 11, John Wiley & Sons, Inc., New York, 1959.
188. H. Bethge, *Phys. Stat. Solidi,* **2**:3,775 (1962).
189. C. Sella and J. J. Trillat, in "Single-crystal Films" (M. H. Francombe and H. Sato, eds.), p. 201, Pergamon Press, New York, 1964.
190. L. O. Brockway and R. B. Marcus, *J. Appl. Phys.*, **34**:921 (1963).
191. G. W. Gobeli and F. G. Allen, *J. Phys. Chem. Solids,* **14**:23 (1960); R. B. Marcus, *J. Appl. Phys.*, **37**:3121 (1966).
192. A. Catlin and R. A. Draughn, *Rev. Sci. Instr.*, **35**:1609 (1964).
193. D. J. Gaffee, in "Thin Film Microelectronics" (L. Holland, ed.), p. 252, John Wiley & Sons, Inc., New York, 1966.
194. G. R. Stilwell and D. B. Dove, *J. Appl. Phys.*, **34**:1941 (1963).

195. See reviews by R. W. Roberts, *Brit. J. Appl. Phys.*, **14**:537 (1963); J. Moll, *Vide*, **18**:248 (1963).
196. H. E. Farnsworth, R. E. Schlier, T. H. George, and R. M. Burger, *J. Appl. Phys.*, **29**:1150 (1958); H. E. Farnsworth, in "The Surface Chemistry of Metals and Semiconductors" (H. Gatos, ed.), p. 21, John Wiley & Sons, Inc., New York, 1959.
197. D. Haneman, *Phys. Rev.*, **119**:563 (1960).
198. S. Nielson, *Brit. J. Appl. Phys.*, **12**:603 (1961).
199. H. D. Hagstrum and C. D'Amico, *J. Appl. Phys.*, **31**:715 (1960).
200. C. T. Naber, *Rev. Sci. Instr.*, **38**:1161 (1967).
201. For example, M. M. Hanson, P. E. Oberg, and C. H. Tolman, *J. Vacuum Sci. Tech.*, **3**:277 (1966).
202. B. O'Brian and T. A. Russel, *J. Opt. Soc. Am.*, **24**:54 (1934).
203. J. Strong, "Procedures in Experimental Physics," Prentice-Hall, Inc., Englewood Cliffs, N.J., 1938.
204. P. Giacomo, B. Roizen-Dossier, and S. Roizen, *J. Physique*, **25**:285 (1964).
205. J. Strong, *J. Phys. Radium*, **11**:441 (1950).
206. K. H. Behrndt and D. W. Doughty, *J. Vacuum Sci. Tech.*, **4**:199 (1967).
207. K. H. Behrndt, in "Physics of Thin Films" (G. Hass and R. E. Thun, eds.), vol. 3, p. 1, Academic Press Inc., New York, 1966.
208. K. Hefft, R. Kern, G. Nöldeke, and A. Steudel, *Z. Physik*, **175**:391 (1963).
209. W. Steckelmacher, J. M. Parisot, L. Holland, and T. Putner, *Vacuum*, **9**:171 (1959).
210. P. Giacomo, *Rev. Opt.*, **35**:317, 442 (1956).
211. P. H. Lisberger and J. Ring, *Opt. Acta*, **2**:42 (1955).
212. C. Dufour, *Vide*, **5**:837 (1950).
213. P. Prugne and P. Leger, *J. Phys. Radium*, **13**:129A (1952).
214. P. K. Weimer, *Proc. IRE*, **50**:1462 (1962).
215. A. E. Lessor, *IEEE Trans. Component Pts.*, **CP-11 (2)**:48 (1964).
216. E. Keonjian, "Microelectronics: Theory, Design, and Fabrication," McGraw-Hill Book Company, New York, 1963.
217. F. Z. Keister, R. D. Engquist, and J. M. Holley, *IEEE Trans. Component Pts.*, **CP-11 (1)**:33 (1964).
218. J. P. Hoekstra and P. White, *Rev. Sci. Instr.*, **32**:362 (1961).
219. K. Taylor, *Trans. 8th Natl. Vacuum Symp.*, p. 981, Pergamon Press, New York, 1961.
220. H. L. Caswell and J. R. Priest, *Trans. 9th Natl. Vacuum Symp.*, p. 138, The Macmillan Company, New York, 1962.
221. A. Duthie, S. Humphery, and B. A. Probyn, *Electron. Eng.*, **35**:430 (1963).
222. F. A. Nunn and D. S. Campbell, *J. Sci. Instr.*, **40**:337 (1963).
223. H. Zerbst, *Vakuum-tech.*, **12**:173 (1963).
224. D. A. Tandeski, M. M. Hanson, and P. E. Oberg, *Vacuum*, **14**:3 (1964).
225. P. M. Chirlian, V. A. Marsocci, H. W. Phair, and W. V. Kraszewski, *Rev. Sci. Instr.*, **35**:1718 (1964).
226. M. Randlett, E. Stroberg, and K. L. Chopra, *Rev. Sci. Instr.*, **37**:1378 (1966).
227. W. Shockley and E. Geissinger, *IRE Trans. Component Pts.*, **11**:34 (1964).
228. S. S. Charschan and H. Westgaard, *Electrochem. Tech.*, **2**:1 (1964).
229. K. L. Chopra and M. R. Randlett, Ledgemont Laboratory, Kennecott Copper Corporation, U.S. Patent 3,409,529.
230. H. F. Winters and E. Kay, *J. Appl. Phys.*, **38**:3928 (1967).

III THICKNESS MEASUREMENT AND ANALYTICAL TECHNIQUES

As a continuation of our description of thin film technology in the previous chapter, this chapter is devoted to methods for monitoring and measuring thickness, structure, and composition of films. Specific methods and techniques for measuring physical quantities related to mechanical, electrical, magnetic, and optical properties are dealt with elsewhere in this book.

1. THICKNESS MEASUREMENT

Thickness is the single most significant film parameter. It may be measured either by in-situ monitoring of the rate of deposition, or after the film is taken out of the deposition chamber. Techniques of the first type, often referred to as "monitor" methods, generally allow both monitoring and controlling of the deposition rate and film thickness. Some of these methods are shown schematically in Fig. 1 and are described in the following section. Instead of describing the methods of

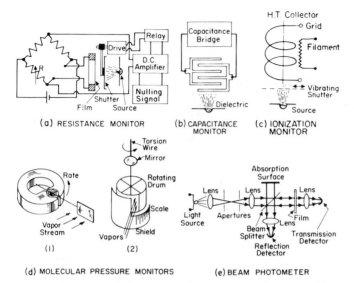

FIG. 1. Diagrams of some in-situ deposition-rate monitors: (*a*) Wheatstone-bridge electrical-resistance monitor and controller for metal films; (*b*) capacitance monitor for insulator films; (*c*) ionization monitor and rate controller; (*d*) microbalance (1) and impingement momentum (2) rate monitors [the microbalance (1) can also be used for weight measurements] ; (*e*) beam photometer to monitor the optical density of films of all types of materials.

the two types separately, we shall group them together but classify them according to the underlying physical principles.

Several reviews [1-5] of the subject have appeared in the literature. Following the preceding chapter, we shall confine ourselves to a practical description rather than a detailed analysis of the techniques. It may be noted that any physical quantity related to film thickness can in principle be used to measure film thickness. Such methods are, of course, very useful when the physical quantity concerned needs also to be known. The methods chosen here are commonly employed because of their convenience, simplicity, and reliability.

1.1   Electrical Methods

(1)   Film Resistance. This simple method, applicable to metallic and low-resistivity semiconductor films, rests on the fact that the resistance is related to the film thickness and the mean free path (mfp) of charge carriers (see Chaps. VI and VII). Whatever the theoretically expected relation may be, an empirical thickness-resistivity relation describing polycrystalline films (with zero surface scattering coefficient) may be found experimentally and used to deduce the unknown film thickness. The situation is complex in ultrathin, structurally discontinuous films (<100 Å), so that no reliance can be put on the resistance method in this thickness region. The resistivity of semiconductor films is very sensitive to deposition conditions, and as a consequence, the resistance method is applicable only for comparison of film thicknesses rather than for absolute measurements.

Resistance of metal films can be measured very easily by making the film one arm of a dc or ac Wheatstone bridge [Fig. 1(a)]. For a given value of the ratio arms of the bridge, the film resistance is proportional to the variable resistance, which can be measured by automatically recording the potential across it. Several variations [6-8] of the bridge circuit can be employed for recording the resistance changes from several hundred megohms to a fraction of an ohm by automatically changing the value of the ratio arms. The variable resistance may be set to a value corresponding to a certain film thickness. When the bridge is balanced, a relay may be energized to actuate a magnetically controlled shutter to interrupt the vapor beam and then shut off power to the vapor source.

By using suitable components in the bridge circuit, an accuracy in the resistance of ~1 percent is readily obtained. Using a sensitive dc amplifier, the author has measured resistance changes of ~0.01 percent. The accuracy of the film thickness is, however, limited by the reproducibility of the resistivity-thickness relation used and may not be better than 5 percent.

Simultaneous measurement of the resistivity ρ and Hall coefficient R_H in thin metal films can be utilized [9] to determine film thickness from the graphic solution of their size-effect equations (Chap. VI).

Electrical resistivity of metal and semiconductor films may be measured by other techniques. An excellent method, which can be used for in-situ measurement of both ρ and R_H of metal and semiconductor films, is due to van der Pauw [10]. Using the reciprocity theorem, he showed that if A, B, C, and D are any four sufficiently small, ohmic contacts arranged successively on the circumference of a film of an arbitrary shape, then

$$\rho = 4.53t \frac{R_{AB,CD} + R_{BC,DA}}{2} f\left(\frac{R_{AB,CD}}{R_{BC,AD}}\right) \quad \text{ohm-cm} \tag{1}$$

and

$$R_H = 10^8 \frac{t}{H} \Delta R_{BD,AC} = 10^8 \frac{t}{H} \Delta R_{AC,BD} \equiv \mu_H \rho \quad \text{cm}^3/\text{C} \tag{2}$$

where t is film thickness in centimeters and μ_H is the Hall mobility in units of $\text{cm}^2/\text{V-sec}$. The notation $R_{AB,CD}$ denotes the resistance between A and B when current is passed between C and D, and ΔR's are the resistance changes due to an applied magnetic field H in gauss. The correction factor f has been calculated by van der Pauw and is nearly unity when $R_{AB,CD} \approx R_{BC,AD}$.

Lange [11] has analyzed an infinite-sheet method and a circular-sheet method which do not require contacts to the edge of the sample. By placing the four contacts near the center of the specimen, R_H is again given by Eq. (2), provided the interspacing between points is small compared with the size of the specimen.

The most widely used method for measuring resistivity in the $\sim 10^{-3}$ to 10^{+3} ohm-cm range is due to Valdes [12] and employs a collinear four-point probe with equal separation between the points. If I is the current passing between the outer points and V is the potential drop between the inner points with a point spacing s, then the resistivity for a specimen of infinite volume is given by

$$\rho = 2\pi s \frac{V}{I} \tag{3}$$

If the specimen dimensions are comparable to the point spacing, then the resistivity for various geometries and different types of boundaries is given by

$$\rho = 2\pi s \frac{V}{I} f\left(\frac{L}{s}\right) \tag{4}$$

where L is the distance between the nearest probe and the limiting boundary. Valdes [12] and Uhlir [13] have calculated the function f for several possible cases. Note that the four points in the probe need not be either equidistant or collinear; in this case, however, the resistivity expressions are more complicated.

A useful simplification of the collinear four-probe relation [Eq. (3)] obtained for a thin specimen ($t \ll s/2$) is

$$\rho = 4.53 \frac{V}{I} t \tag{5}$$

which is clearly a convenient relation to use. Furthermore, $\rho/t = 4.53 \ V/I$ gives the sheet resistivity of the film. The probe can be made from thin (~0.05-cm-diameter) tungsten wires which are sharpened electrolytically and fixed in a retractable plexiglass header by a suitable cement. A commonly used point spacing is $s = 0.159$ cm, which makes the $2\pi s$ factor unity in Eq. (4) for a bulk specimen.

The four-probe method measures the average resistivity of the film on a substrate provided the film is either isolated from the substrate or its resistivity is much lower than that of the substrate. A three-probe called the "spreading-resistance technique" [14,15] is of particular interest since it can be used for films of the same (no isolation junction) or opposite type of conductivity as the substrate, as well as for measuring the resistivity profiles of multilayered semiconductor structures. The technique is based on the fact that for a flat circular voltage probe making ohmic contact to a semi-infinite conducting material, practically all the potential drop occurs within a distance of a few probe radii of the probe center. If I is the current through contacts of radius s, the resistivity is related to the voltage drop V by

$$\rho = 4 \frac{sV}{I} \tag{6}$$

If the medium is not semi-infinite, the voltage drop also depends on the thickness of the film, or films in a multilayer structure. This dependence forms the basis of the method for the determination of the resistivity profile.

The experimental arrangement for the spreading-resistance technique uses three collinear probes, two of which are connected to a current source, and the potential difference is measured between one of these and the third probe. It is important to maintain constant pressure on the one probe common to the current source and the voltmeter. A spring-loaded tungsten-carbide probe is commonly used.

Electrical resistivity of metal films can also be measured without making contact with the film by measuring the decay of eddy currents (Chap. VI, Sec. 4.7) induced in the film [16]. One may measure the effective ac impedance of the film and therefrom deduce the resistivity. Johnson and Johnson [17] have described a method to determine the effective impedance as a function of the distance beneath the surface of a specimen. This is achieved by varying the current frequency and hence the skin depth. The frequency to be used depends on the resistivity and thickness of the specimen. For metal films less than 1 μ thick at room temperature, a frequency of $\sim 10^2$ Gc would be required to yield a skin depth smaller than or comparable with the film thickness.

(2) Capacitance Monitors. The thickness of dielectric films may be determined by directly monitoring the electrical capacitance of a capacitor configuration of the type shown in Fig. 1(*b*). Keister and Scapple [18] constructed a capacitance monitor plate by depositing Al films and then photoetching the comb pattern. Using a 1- by 3-in. silica monitor plate with 56 lines, each of width $\sim$ 0.0075 in., a capacitance of 65 pF was obtained. A 10-μ-thick deposit of SiO increased the capacitance by about 13 percent.

Riddle [19] measured the rate of evaporation by measuring changes in the capacitance of a parallel-plate condenser due to changes in the dielectric constant resulting from the presence of the vapor of the evaporant. The method is not very sensitive and requires careful measurements. It may be subject to spurious effects due to stray electrical charges in the vapor and in the vacuum chamber.

(3) Ionization Monitors. By ionizing the vapor from the evaporant and measuring the resultant ion current, the evaporation rate can be monitored and controlled via a feedback servomechanism. This method, used for detection of molecular beams as early as 1928 [20], was first adapted to thin films by Haase [21] and Metzger [22], and has since been utilized by a number of other workers [23-26].

The ion current is proportional to the total number of vapor atoms and their ionization probability. The linear relation between the ion current and deposition rate is an important advantage of the technique. An empirical observation by Engel and Steenbeck [27] suggested that the ion current was proportional to the number of electrons per atom or molecule of the vapor beam. This relation, also verified by Schwarz [24] for Al, Ni, and Cr, allows a convenient calibration for other materials by extrapolation.

Basically, an ionization monitor is a nude, triode-type, ionization gauge used for measuring pressures. It is necessary, however, to heat the anode and the collector to prevent deposition of the vapors onto them, or at least to allow their periodic cleaning. As shown schematically in Fig. 1(c), it consists of a thermal source (filament) for electrons which are accelerated by the spiral wire grid (anode) at a positive potential ($\sim$250 V) and injected to ionize the vapor in the region between the anode and collector wire. The collector at a negative potential collects the ions. The measurement of small currents (fraction of a microampere for a typical operation) can be made by a variety of instruments, but the interpretation of the results presents several problems. First, the vapor may carry negative charges (particularly serious in electron-beam evaporation). These charges may be suppressed in the case of filament evaporation by applying a positive dc voltage to the evaporating source. Second, a considerable number of residual gas atoms present at pressures of $\sim$$10^{-6}$ Torr give an appreciable background ion current. This may be minimized by modulating the vapor beam by using a vibrating shutter (say at 20 to 30 cycles) and monitoring the modulated component of the ion current. Another method to overcome this problem, employed by Dufour and Zega [25], used two anodes and two collectors symmetrically disposed on either side of the common cathode (called "double-ionization gauge system"). One of them was exposed to the modulated vapor beam. The difference signal derived from the ion currents of the two gauges provides an accurate measure of the vapor ion current.

Simple and demountable ionization monitors can be operated in ultrahigh-vacuum conditions and at high temperatures. Brownell et al. [26] have given a detailed description and electrical circuitry for a small, automatic, high-stability ionization monitor for measuring and controlling the rate of deposition. It allows films of thicknesses from 50 to

1,500 Å to be deposited automatically by electron bombardment at selected rates between 0.1 and 10 Å/sec with a reproducibility of ± 10 Å.

The vapor ions produced in the ionization-rate monitor could be used [25,28] as a source in a mass spectrometer, which would provide selective response to particular vapor species from mixed materials. Crawford [28] suggested the possibility of absolute monitoring of the vapor beam using the Paul-type quadrupole mass filter. Another variant of interest was investigated by Brooks and Herschbach [29], who used the ionized vapor to neutralize the space charge partly and thus modify the space-charge-limited current flow in a diode. This method can yield high sensitivity for electronic detection and may find an application in thin film monitoring.

1.2 Microbalance Monitors

(1) Microbalances. These monitors are termed the "gravimetric" or "momentum" type depending on whether they measure the weight or the momentum of the impinging vapor, respectively. Gravimetric methods are among the earliest and most convenient to use for determining film thickness.

Various types of balances, such as pivotal, torsion fiber, quartz or tungsten helical spiral, and magnetic suspension, (all of which employ null-balance principles using mechanical, optical, electromagnetic, or electrostatic detection methods), have found applications as monitors. References to the major developments in vacuum microbalance techniques since 1960 have been published in the proceedings of the annual symposia on the subject. A review of the techniques used before 1956 is given by Behrndt [30]. The detection sensitivities of various balances range from 1 to 10^{-2} μg. A sensitivity of 0.1 μg, which is conveniently obtained in a commercial pivotal balance, corresponds to about a 1-Å-thick Ag film of 1 cm^2 area. The electrical signal from the microbalance detection system may be used to monitor and control the deposition rate.

Microbalance monitors using the movements of moving-coil current meters (1 to 20 μA) have been described by Campbell and Blackburn [31], Hayes and Roberts [32], and Houde [33]. One arrangement, shown schematically in Fig. 1(d.1), uses a lightweight large-area vane with a counterweight on the other side of the meter movement for obtaining a high sensitivity. Depending on the relative orientation of the vane and the vapor stream, the arrangement may be used to measure total weight (average thickness), the force due to the vapor momentum transfer (rate of deposition), or both. For example, if the vane is vertical and the vapor stream impinges horizontally, only rate can be measured. By impinging vapor onto the vane inclined to the horizontal and vertical positions, the components of forces at right angles can be measured so that both rate and the total weight can be determined.

Neugebauer [34] used a simple torsion pendulum [Fig. 1(d.2)] to measure the momentum transfer rate. A light Al-foil cylinder 5 cm in diameter and 5 cm long, suspended by a 0.0025-cm-diameter W torsion wire 25 cm long, intercepted a vapor beam on one side. The resulting momentum exchange caused a rotation about the axis of the cylinder. A sensitivity of ~1.6 Å/min per degree rotation for Sn vapor was obtained. The rotation clearly depends on the number, mass, and velocity of the impinging vapors.

(2) Quartz-crystal Monitor. A sensitive and rugged microbalance is based on measuring changes in the resonant frequency of a quartz-crystal oscillator with mass loading when operated in a particular mode of vibration. A quartz-crystal monitor for monitoring and controlling the rates of both the deposition and the evaporation of metals, nonmetals, and multicomponent films has become universally accepted and is, at present, the single most important monitor for thin film technology.

The use of a quartz crystal as a thin-film monitor was first proposed by Sauerbrey [35,36], who made an extensive investigation of the various parameters of the monitor. It has been employed by numerous workers, and reference is made here to a few of the published reports [37-42].

The monitor utilizes the thickness shear mode of a piezoelectric quartz crystal. Here, the major crystal surfaces are antinodal, and mass added on either one or both sides shifts the resonance frequency irrespective of the thickness, density, elastic constants, or stiffness of the added material. A $35°\ 20'$ quartz-crystal cut, called the AT cut, is generally used for the monitor because of its low temperature coefficient ($\pm\ 5\ \times\ 10^{-6}$ between -20 and $+60°C$) for the resonant frequency. The frequency of the fundamental resonance of a thickness mode for an AT cut crystal is given by [43]

$$f = \frac{1}{2d}\left(\frac{C}{\tilde{\rho}_q}\right)^{1/2} = \frac{N}{d} = \frac{1,670}{d} \quad \text{mm kc/sec} \tag{7}$$

where d is the crystal thickness, $\tilde{\rho}_q$ its density, C its shear elastic constant, and $N \equiv (C/4\tilde{\rho}_q)^{1/2} = 1,670$ mm kc/sec.

Warner and Stockbridge [41] showed that a change in frequency Δf, due to a deposit of mass m, added to the area A of the antinodal surface of a mechanical resonator, is given by

$$\Delta f = -f\frac{Km}{\tilde{\rho}_q Ad} \tag{8}$$

where the constant $K \approx 1$ and the negative sign implies a decrease in the frequency. Combining Eqs. (7) and (8), we obtain

$$\Delta f = -\frac{f^2 K}{N\tilde{\rho}_q}\frac{m}{A} = -C_f \frac{m}{A} = -C_f t\tilde{\rho}_{\text{film}} \tag{9}$$

where $C_f \equiv f^2 K/N\tilde{\rho}_q$ is a constant of the crystal, and $m = At\tilde{\rho}_{\text{film}}$ assuming a uniform film of thickness t and a constant density $\tilde{\rho}_{\text{film}}$. Thus $t = \Delta f/C_f\tilde{\rho}_{\text{film}}$ yields average film thickness. Table I lists values of C_f along with other parameters of quartz crystals of different frequencies.

Note that Δf is proportional to f^2, and so higher-frequency crystals yield higher sensitivity. On the other hand, since the linear relation between Δf and m [Eq. (8)] is valid only if t is very small compared with d and Δf is less than $0.01 f$, the nonlinear region is reached for smaller thicknesses with increasing frequency. Thus, one must compromise between the sensitivity and the maximum film thickness for which a crystal may be used in the linear range. Frequencies of 5 or 6 Mc/sec are therefore commonly used. At 5 Mc/sec, a 1-cps change corresponds to a weight increase of about 1.8×10^{-8} g/cm^2, which for a Fe film corresponds to 0.12 monolayers. In this case, a deposit of 25,000 monolayers can be obtained for a 1 percent change in f. When the deposit has exceeded the linear range, a new crystal should be used. If

Table I. Constants and Sensitivities of AT-cut Crystals of a Range of Frequencies for Use as Thickness Monitors

Resonance frequency, Mc/sec	Crystal thickness mm	Mass change per unit area for $f = 1$ cps, g/cm^2	Frequency change per unit density of 1-Å-thick film, cps	Max. thickness of a unit density for $\Delta f/f = 1\%$, μ
1	1.67	4.42×10^{-7}	0.022	44.2
3	0.56	4.91×10^{-8}	0.203	14.73
6	0.28	1.23×10^{-8}	0.815	7.38
10	0.167	4.42×10^{-9}	2.26	4.42
15	0.11	1.96×10^{-9}	5.1	2.94

the electrodes are precoated with Al, the deposit can easily be removed by dissolving the underlayer in NaOH solution, and the same crystal may be used again.

The sensitivity of the crystal does not increase appreciably by depositing over an area larger than the electrode, except when the conductive coating also increases the active area of the electrode. For a deposit covering areas smaller than the electrode area, the mass sensitivity decreases slightly [36].

The standard method of operation of an AT-cut crystal is by perpendicular excitation using metal electrodes on the central area of each face. The contacts to the electrode may be secured by spring clips [see Fig. 2(a)]. The electrode resistance should be low since the crystal presents a low resistance at series resonance. Contacts may be improved by bonding using a soft solder or metal paste. The crystal is operated at series resonance where parallel capacitance has a small effect. However, the capacitance should be kept small to maintain high-frequency stability and a high Q value. The use of an oscillator circuit with neither electrode

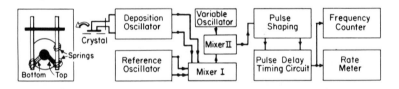

(a) FILM THICKNESS MONITOR

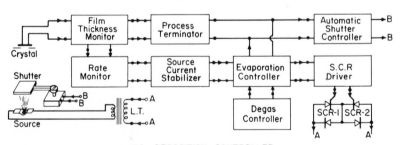

(b) DEPOSITION CONTROLLER

FIG. 2. (*a*) Block diagram of a quartz-crystal deposition-rate and film-thickness monitor. A spring-loaded crystal holder for parallel excitation is shown. (*b*) Block diagram of an evaporation controller using the output of a quartz-crystal monitor to control the power input to the thermal-vapor source via the silicon-controlled rectifier driver.

connected to ground is advantageous from this point of view. However, precautions must then be taken to prevent any effects due to collection of electrostatic charges. At antiresonance, the capacitance across the crystal has a profound effect on the frequency. This mode is therefore not used for monitoring.

The maximum sensitivity of a quartz-crystal monitor is limited by variations in the crystal frequency due to the temperature, oscillator drive level, and changes in the oscillator circuit. Although sensitivities of $\sim 10^{-12}$ g/cm^2 have been attained [41] under extremely careful conditions, a sensitivity $\sim 10^{-8}$ g/cm^2 is a more practical figure. Since the crystal must see the vapor source, its temperature will rise because of the source radiation. The temperature will also rise because of the heat of condensation liberated during deposition of the vapor. It is necessary to use suitable radiation shields so that only the active area is exposed to the source. Water cooling of the crystal holder can easily be incorporated and may be necessary for excessive heating due to lengthy deposition times from extended sources.

The frequency and frequency changes of a quartz-crystal oscillator circuit can be measured by using a suitable pulse-analog or pulse-digital counter with a moderate accuracy of $\pm$ 1 cps for a 6 Mc/sec crystal. The rate of change can be measured by feeding the counter output to the pulse-differentiating unit. Frequency changes of 1 part in 10^{10} can be measured under ideal conditions.

A cheaper method to measure the frequency change by a meter readout is to drive one intermediate frequency by beating the frequency of the monitor crystal with that of another quartz crystal of a slightly different frequency, and then obtaining another intermediate frequency by beating the first intermediate frequency with a variable-frequency oscillator. The output is amplified, rectified, and read on a meter. The rate of frequency change can be obtained by using an RC differentiating circuit. Steckelmacher et al. [5,42] have described in detail one type of such a circuit. A block diagram of the electrical circuit is shown in Fig. 2(a). The components are self-explanatory.

The crystal monitor may be used not only to monitor the deposition rate but also to control conveniently the evaporation rate from a vapor source. The block diagram of an electric circuit used for this purpose is shown in Fig. 2(b). The crystal-monitor output is used to drive two back-to-back silicon-controlled rectifiers (SCR) which are capable of controlling large power for heating the evaporation filament. The details of a circuit are given by Steckelmacher [5].

Two crystal oscillators may be used together to obtain controlled rate of deposition as well as rate of evaporation from a source. The latter is important for studies in which the effect of the vapor temperature is being investigated. In the arrangement used by Chopra and Randlett [44] for this purpose [Fig. 3(a)], one crystal oscillator controls the evaporation rate. By moving the substrate and the second crystal together, relative to the vapor source, the required rate of deposition can be obtained.

An interesting application [45] of a two-crystal-oscillator arrangement [Fig. 3(b)] is to control the relative evaporation rate of two different sources to obtain alloys or compounds of definite compositions. The output of the two monitors is mixed and after discrimination is

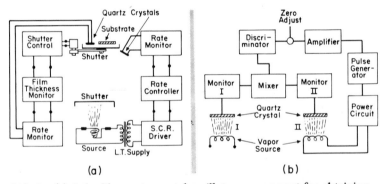

(a)   (b)

FIG. 3. (a) A double quartz-crystal oscillator arrangement for obtaining films at a controlled rate of deposition as well as controlled rate of source evaporation (*Chopra and Randlett* [44]). (b) A dual monitor arrangement to control the relative composition of a deposit from two vapor sources (*Gerber* [45]).

compared with a preset signal. The difference or an error signal from this preset value is then used to control the evaporation rate of one of the vapor sources by a servomechanism, thereby allowing any desired composition of the two materials to be attained.

1.3 Mechanical Method (Stylus)

Variations in the movements of a mechanical stylus can be amplified electronically so that step heights and surface irregularities of ~10 Å can be measured. This system, called "Talysurf" (manufactured by the Rank-Taylor-Hobson division of the Rank Organization, England), has become quite popular for measuring film thicknesses and surface irregularities. The stylus in this case consists of a diamond with a rounded ($\sim$0.7- to 1.3-μ-diameter) or four-sided pyramid tip fastened to a lever arm. The arm is delicately balanced so that the load on the stylus is very small. The vertical movement of the stylus is detected with a transducer, amplified 10^5 to 10^6 times, and then fed into a recorder. Under the pressure of the stylus, the films of soft materials deform considerably, but a suitable choice of the tip and counterbalance arrangement can largely surmount this difficulty. The results obtained by this technique have been compared with other methods by various authors [46], and satisfactory agreement has been reported.

1.4 Radiation-absorption and Radiation-emission Methods

The thickness dependence of the absorption of light, x-rays, α-rays, and β-rays may be utilized to determine the thickness and the thickness distribution of a film.

Several workers [47-51] have used the optical-absorption (also called optical-density) method to determine thickness and have verified the exponential dependence of absorption on film thickness for continuous films. The absorption in discontinuous films is strongly influenced by the granular nature of films, but it may still be used for relative measurement of the average film thickness. The measurement techniques for optical absorption are fairly standard and straightforward. A typical beam photometer for in situ reflection and transmission studies is shown schematically in Fig. 1(e). The optical absorption can also be conveniently used for monitoring during deposition by glow-discharge sputtering [51].

Absorption or scattering of x-rays [52,53], α-rays [54,55], and β-rays [56-59] emitted from radioactive isotopes have been employed for thickness measurement. Thicknesses from less than 100 Å to about 1 mm may be measured using α- and β-ray absorption. Electron-gun sources can also be used to provide fast electrons for absorption studies provided Bragg reflections are absent (that is, material should not have preferred orientations). Such a measurement can be conveniently performed inside an electron microscope [60,61]. The maximum thickness is limited to several thousand angstroms for 100-keV electrons. It is a particularly useful technique for thin insulator films [60].

When radiations are incident on a surface, some scattering takes place. The back scattering of β-rays depends on the atomic number, density, and thickness of the scattering material, and its measurement therefore allows [56,58,59] a determination of the film density and thickness. This simple technique employs a β-source such as C^{14} (0.16 MeV) or Pm^{147} (0.22 MeV) of strength ~50 μCi and conventional Geiger-tube detectors. A system for measuring the scattered electrons (and hence film thickness) in the Debye-Scherrer cone has been described by Behrens [62].

If the evaporant is radioactive, the activity of the deposited film will be proportional to its thickness. Antal and Weber [63] used this technique for Bi^{210} films. Preuss and coworkers [64-66] used Au^{198} and Cr^{51} evaporants to radiograph the vapor source and measure the deposit distribution.

X-ray emission, also known as x-ray fluorescence, is an important and most frequently used nondestructive method for determining the mass of the material components of the film, and hence its thickness and the chemical composition Secondary x-rays may be excited in several ways and measured with conventional x-ray spectrometers. One technique [67-70] is to excite them in the substrate using white x-ray radiation, or an energetic electron beam and measure their attenuation in the film material. For thinner films, the excited characteristic x-rays of the film material itself may be measured [70-80]. Such measurements can conveniently be made with a standard electron-microprobe analyzer [76-80]. The observed intensity of radiations emitted from the surface of the film normalized to that obtained from its bulk is proportional to thickness for small thicknesses. Sweeney et al. [76] and Chopra et al. [77] studied several metal films and found a linear relation up to about 2,500 and 5,000 Å for Ag and Al films, using 29-kV electrons. Chopra et al. also verified the validity of this method for superimposed films of Ag and Al, and thus established the usefulness of this technique for multicomponent films.

The measurable lower limit of thickness is determined by the detection sensitivity and is about 10 Å. The thickness can be determined to within ± 10 Å. With increasing film thickness, the x-ray yield approaches that obtained from the surface of a bulk material. The limit of the thickness dependence is determined by the electron-range value in the particular material. Castaing [79] and Katz and Penfold [80] have given suitable relations for the electron range in a material. Cockett and Davis [78] showed that film excitation is a more accurate method for thicknesses up to about a quarter of the electron range while the substrate-excitation method is better for thicker films.

Characteristic x-rays of a material may also be generated by proton bombardment and used to measure thin film density. Khan et al. [81] and Christensen et al. [82] used a 100-kV proton beam to excite x-rays in Al, Cu, and Yb films. The x-ray intensity increases linearly for small thicknesses and approaches saturation for thick films. The maximum useful thickness for density measurement was found to be ~4,000 Å for Al and 1,000 Å for Yb.

1.5 Optical-interference Methods

Several methods for determining the optical constants of films described in Chap. XI involve thickness as a parameter. If optical constants are known, the thickness can be calculated. Among these methods are the photometric and spectrophotometer techniques, which are based on the optical-interference phenomenon and find widespread applications for measurements and control of multilayer dielectric films and semiconductor films.

(1) Photometric Method. The photometric method for measuring the optical density of a film has already been mentioned. If a transparent or slightly absorbing film is deposited on a transparent substrate of a different refractive index, the optical reflectance and transmittance behavior of the film-substrate combination shows an oscillatory behavior with increasing film thickness because of interference effects (Chap. XI, Sec. 5). Reflectance is reduced or enhanced depending on the relative values of the indices of the film and substrate material. Film thickness is determined from the maxima and minima of the reflectance which occur at intervals given by

$$2m \frac{\lambda}{4} = n_f t \tag{10}$$

where n_f is the film refractive index, t the film thickness, λ the wavelength of light, and m the order of the maximum or minimum. For example, if $\lambda = 1$ μ, then 1 μ thickness of an SiO film ($n_f = 2.0$) deposited on glass ($n_g = 1.5$) substrate is obtained after four maxima or minima are traversed.

The maximum and minimum reflectance in the above-mentioned example are given by Eqs. (14) and (15) of Chap. XI and are 20.7 and 4 percent respectively. Thus, relatively sharp reflectance changes occur and form the basis of this method for monitoring and controlling the deposition of multilayer dielectric films for thin-film optical devices (Chap. XI). A number of workers [83-92] have described beam photometers for this purpose. High-intensity laser sources may be used [87] as a source of monochromatic light.

Figure 4 illustrates the observed variation of the reflectance and transmittance of alternating ZnS and MgF$_2$ $\lambda/4$ films. We may note that with an increasing number of layers the amplitude of the oscillatory variation decreases, thereby reducing the sensitivity of the method.

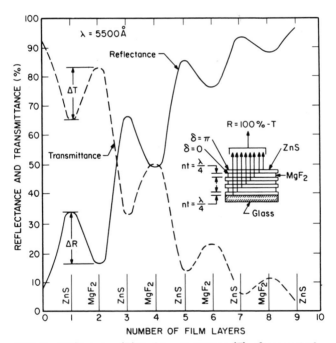

FIG. 4. Reflectance (R) and transmittance (T) of successively increasing number of multilayers consisting of alternating dielectric films of MgF$_2$ and ZnS (*Steckelmacher et al.* [86]).

Several approaches to overcome this difficulty have been described in the literature [89-92]. Lindberg and Irland [89] used two photocells in series opposition. One cell monitored the reflected intensity from the sample and the other was illuminated from a variable-intensity light source to provide a bucking or nulling signal so that only the oscillatory part is obtained and amplified. This procedure allowed monitoring of up to 15 layers.

Giacomo and Jacquinot [90] employed an arrangement in which the wavelength of the light from the dispersion system of a monochromator is modulated by reflecting it from a vibrating ($\sim$100 cps) concave mirror. The intensity T of the light transmitted through the film thus has a component of the same frequency as the vibrating mirror and an amplitude proportional to $dT/d\lambda$, the vanishing of which determines accurately the position of the maxima or minima of transmission. Traub [91] used a rotating disk with two filters to obtain two wavelengths which are symmetric to the wavelength at which $\lambda/4$ thickness is desired. Thus, the sum of the two light signals transmitted through the film passes through zero when the $\lambda/4$ optical thickness is obtained. Sawaki et al. [92] used an arrangement to split the light beam transmitted through the film into two components which were measured by two photocells after each component was passed through a filter of a wavelength symmetrically distributed with respect to the desired value. By changing the filters, this method allows monitoring over a wide wavelength range. A reproducibility of $\pm\lambda/65$ for the passband was found by these authors.

(2) Spectrophotometric Method. If light is incident at an angle θ from a medium of index n_0 onto a film of index n_1 and thickness t, deposited onto a substrate of index n_2, with n_1 lying between n_0 and n_2, the reflected light will show an interference maximum for a wavelength λ when the path difference $2n_1 t \cos\theta$ between the successive beams reflected at each surface is equal to $m\lambda$, where m is an integer. If n_1 is greater than n_0 and n_2, the reflected intensity will show a minimum (dark band results) when $2n_1 t \cos\theta = m\lambda$ and a maximum if $2n_1 t \cos\theta = (m - 1/2)\lambda$.

An interference maximum will produce a characteristic hue of the film. When white light is used, the reflected light will show maxima for various wavelengths for which the interference condition is satisfied. This is the basis of the visual method of monitoring film thickness and was realized by Newton as early as 1675. The color-judgment method is, of course, subjective, and fails for thick films when many maxima in the reflected spectrum make the light appear white. A spectrophotometer [93] may be employed to measure the transmitted or reflected intensity as a function of the wavelength and thus record positions of maxima and minima. If mth-order maximum occurs at λ_1, and the $(m + 1)$th order at λ_2, we have for normal incidence (assuming the same index for λ_1 and λ_2)

$$2n_1 t = m\lambda_1 = (m + 1)\lambda_2 \tag{11a}$$

Therefore,

$$2n_1 t = \frac{\lambda_1\lambda_2}{\lambda_1 - \lambda_2} \tag{11b}$$

Thus, the method determines t if n_1 is known, or vice versa, provided the index does not vary rapidly with the wavelength. If the substrate is absorbing, the phase-change relation becomes quite complicated [2].

The spectrophotometric method employing a double-beam recording spectrophotometer has been used [94-97] extensively for thickness measurement of epitaxial semiconductor films deposited on a substrate of different index (or the same material with a different carrier concentration). A precision of better than 1 percent for SiO_2 films as thin as 400 Å has been obtained [96] with this method.

(3) Interference Fringes. When two reflecting surfaces are brought into close proximity, interference fringes are produced, the measurement of which makes possible a direct determination of the film thickness and surface topography with high accuracy. Wiener [98] was the first to use interference fringes to measure film thickness. The interference-fringe methods have been developed to a remarkable degree by Tolansky [99,100] and are now accepted as the absolute-standard methods.

Two types of fringes are utilized for thickness measurements. The Fizeau fringes of equal thickness are obtained in an optical apparatus of the type shown in Fig. 5(a.1). The interferometer consists of two slightly inclined optical flats, one of them supporting the film, which forms a step on the substrate. When the second optical flat is brought in contact

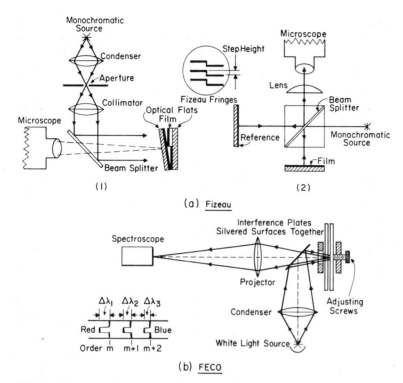

(a) _Fizeau_

(b) _FECO_

FIG. 5. Diagrams of some optical systems of interferometers for measurement of film thickness. (*a*) Fizeau method of fringes of constant thickness using a multiple-beam (1) and a two-beam (2) arrangement. (*b*) Fringes of equal chromatic order method (FECO) (*after Scott et al.* [102]).

with the film surface, and the interferometer is illuminated with a parallel monochromatic beam at normal incidence and viewed with a low-power microscope, dark fringes can be observed which trace out the points of equal air-gap thickness. The two adjacent fringes are separated by $\lambda/2$. If the surfaces of the optical flats are highly reflecting (the upper flat must, however, possess an observable transmission) and very close to each other, the reflected fringe system consists of very fine dark lines against a white background with a fringe width which can be made as small as $\lambda/100$. By adjusting the relative positions of the flats to form a wedge-shaped air gap, the fringes can be made to run in straight lines perpendicular to the steps on the opaque film. The fringes show a displacement as they pass over the film step edge. This displacement expressed as a fraction of the $\lambda/2$ fringe spacing gives the film thickness, and it can be measured to about a tenth of a fringe. Note that it is necessary to coat the film as well as the exposed glass surface with the same reflecting layer in order that phase changes on reflection from the two sides of the step will be the same.

Instead of using multiple-beam interference as in the preceding case, one may use an interference-objective [101] arrangement [Fig. 5(a.2)] with a beam splitter to produce two-beam Fizeau interference fringes. While this arrangement yields relatively broader fringes (and hence less accuracy), the optical flat does not physically touch the film so that it does not get scratched. Furthermore, a reflecting overlayer or a step is not required, which makes this technique very useful and convenient to use. Thicknesses as low as 100 Å with an accuracy of $\sim$ 20 Å are routinely measured with this technique in the author's laboratory.

If, instead of the air wedge, two parallel plates illuminated with white light are used, fringes will occur at wavelengths for which t/λ is constant so that the resulting spectrum is banded. This is called the method of "fringes of equal chromatic order (FECO)." Heavily silvered plates are again necessary to obtain fine fringes. In reflection, fringes appear dark on bright background. In transmission, the complementary pattern is seen. If one of the plates is covered by a film, a displacement $\Delta\lambda$ is seen in the mth-order fringe from which film thickness can be calculated as

$$t = \frac{m}{2} \Delta\lambda \qquad (12)$$

Scott et al. [102] used a 32-mm microscope objective to produce an enlarged (3×) image of the interference pattern at the slit of a constant-deviation-type spectroscope [Fig. 5(b)]. Thicknesses down to 15 Å were determined with an accuracy of ± 5 Å. The observed fringes were about one-quarter the width of those obtained with optical flats as a result of the silvered surfaces being parallel so that no fringe broadening from the wedge effect occurs.

Lockyer [103] proposed a simple modification of the multiple-beam FECO interferometer which effectively increases the wavelength used, and thus makes the method applicable to thicker films. The method utilizes the "vernier" principle through a reinforced fringe pattern which is obtained with the overlap of two fringe patterns, when two monochromatic optical sources have a wavelength ratio close to a simple rational fraction. Thus, if a mercury green line ($\lambda = 5{,}461$ Å) and a blue line ($\lambda = 4{,}358$ Å) are used together, the fringes will coincide at every fourth green fringe and every fifth blue fringe. The reinforced fringes form their distinct pattern of effective $\lambda/2 = 10{,}908$ Å. Similarly, the use of 5,876- and 5,016-Å lines from a helium-discharge source would give an effective $\lambda/2 = 17{,}592$ Å, so that relatively thicker films could be accurately measured.

Multiple-beam interferometers are not convenient for in situ studies. Greenland and Billington [104,105] described an arrangement to monitor the growth of dielectric films. Interference is established by employing a beam incident at an angle at which total reflection occurs at the interface between dielectric and vacuum. The dark fringes were observed spectroscopically. A precision of 5 to 10 Å may be obtained by using a total reflection prism of the type described by de Jong [106].

(4) X-ray Interference Fringes. When a thin sheet of monochromatic x-rays is incident on a sufficiently thin, smooth, and flat film, so that the glancing angle is in the region of the critical angle for total reflection, x-ray interference can be observed [107] between those fractions of the incident beam reflected at the air-film interface and the film-substrate interface. The separation of the interference fringes can be used [108,109] to determine thickness of films as thin as 300 Å. Sauro et al. [110] observed a secondary interference pattern which also results from interference between the components of scattered radiation originating at the two surfaces of the film. Because of the wider fringe spacing available, this secondary interference structure allows a more accurate determination of film thickness. These authors used the technique to measure thickness of Cu films ranging from 250 to 1,000 Å.

1.6 Summary of Methods

We have described a number of techniques for the measurement of film thickness. Some of these techniques can be used for in-situ monitoring and controlling of the deposition of films, and others are well suited for scanning the film and thus determining the thickness distribution and surface roughness. The methods to be used in a particular case will be dictated by the type of deposit, deposition technique, and particular use of the film. A critical summary of the characteristics of the commonly used methods is presented in Table II.

Table II. Some Common Methods for Deposition-rate, Film-thickness, and Thickness-profile Measurements

Method	Materials	Maximum sensitivity	Maximum thickness	Control automation	Remarks
Resistance	Metals	$\sim 1\%$	$\sim 1\ \mu$	Yes	Convenient, empirical thickness-resistance relation required
Ionization gauge	All	< 1 Å/sec	None	Yes	Good rate monitor, compensation for residual gas pressure required
Microbalance (gravimetric)	All	~ 1 Å/cm^2	None	Possible	Simple, rate and thickness monitor
Momentum meter	All	~ 1 Å/min per degree rotation	None	Possible	Simple, relative-rate monitor
Quartz-crystal oscillator	All	$\ll 1$ Å/cm^2	$\sim 1\ \mu$ (depends on density)	Yes	Most useful, simple, rate, thickness, and vapor distribution monitor
Stylus	All	~ 20 Å	None	No	Rapid, absolute, thickness and thickness profile
X-ray emission	All (elements)	~ 100 Å	Depends on excitation agency	No	Simple, relative, thickness and thickness profile. Electron-probe microanalyzer well suited

Method					Remarks
Absorption of x-, α-, and β-rays	All	Depends on absorption coefficient	$\sim 100\,\mu$	No	Simple, rapid scan, thickness profile
Optical density	All	1%	Transparent films	Possible	Relative, rapid, continuous scan, thickness profile
Photometric	Dielectric	$\lambda/300$	Many μ	Yes	Indispensable for multilayer $\lambda/4$ dielectric films
Color of films	Dielectric	$\sim 100\,\text{Å}$	$\sim 1\,\mu$	No	Subjective method
Spectro-photometric	Dielectric	$\lambda/100$	Many μ	No	Useful for thick semiconductor and dielectric films
Interfer-ometric	All	$\sim 2\,\text{Å}$	Several μ	No	Most accurate, absolute, highly reflecting surface or overcoat required in a two-beam method where sensitivity is limited to $\sim 50\,\text{Å}$
Polarimetric	Dielectric (primarily)	$< 1\,\text{Å}$		No	Tedious, extensive calculations

2. ANALYTICAL TECHNIQUES

In order to understand fully the behavior of thin solid films, it is essential to characterize them physically, chemically, and structurally. The various physical properties and their measurements are discussed in the corresponding chapters. This section is concerned with the analytical techniques for chemical characterization of the composition and its gradient, impurities, absorbed gases, and structural analysis encompassing crystal structure, texture, surface structure, and surface and volume structural defects.

A large number of analytical techniques have been developed or modified for thin-film specimens which typically consist of 1 to 10 cm^2 surface area and 10 to 1,000 μg of the material. A detailed discussion of these techniques is beyond the scope of this book, but a brief description shall be presented.

2.1 Chemical Analysis

The methods used in general chemical analysis may be listed in the following groups:

1. Conventional analytical techniques such as gravimetric, volumetric, polarography, radioactivation, and chromatography
2. Optical spectrography (arc, spark, and flame)
3. Calorimetry, and absorption and emission spectrophotometry
4. X-ray fluorescence and electron-probe x-ray microanalyzer
5. X-ray and electron diffraction
6. Mass spectrometry

Most of these techniques are dealt with in depth in standard reference books [111-116]. The annual reviews of *Analytical Chemistry* are the most comprehensive sources of the latest information on, and applications of, these techniques. One or more of these techniques are routinely available in an average thin film laboratory. The gravimetric and volumetric methods are sensitive down to a fraction of a percent of the concentration of the element under investigation. Spectrographic and spectrophotometric methods are capable of detecting down to 10^{-4} weight percent of the trace element. The x-ray microprobe analyzer [115-117] using a 2-μ-diameter beam of 1- to 10-Å wavelength electrons for generation of x-rays has been claimed (for the Hitachi instrument) to be able to detect materials as small as 10^{-13} g, and to measure concentrations of $\sim 10^{-4}$ percent. X-ray and electron-diffraction techniques are of a more qualitative nature and are capable of detecting trace elements with a concentration of a fraction of a percent. Perhaps the most sensitive technique is mass spectrometry [118,119], with which metals can be measured down to a concentration of 10^{-5} to 10^{-7} weight percent and nonmetals down to 10^{-3} to 10^{-5} percent. Thus, we have a variety of techniques spanning a large range of sensitivities.

Some special techniques deserve a brief mention. For example, Guldner [120] devised a method for quantitative analysis of the gas content of metal films. The film is vaporized in ultrahigh vacuum by means of a xenon-flash discharge lamp which provides an estimated temperature of 5000 to 8000°C. This temperature is adequate to dissociate all stable compounds of most metals. The gases evolved in dissociation are collected in the vacuum chamber and analyzed by a cryogenic chromatograph. Guldner's results on sputtered Ta films obtained by this technique compare well with those for concentration of O_2 analyzed by vacuum fusion, and of N_2 by Kjeldahl and other methods. The method is capable of detecting all gases and has been used to measure the concentration of N_2 in sputtered Ta films as low as 0.03 atomic percent. Winters and Kay [121] used a laser to induce thermal desorption of gases from films and then analyzed the gases with a mass spectrometer.

A systematic utilization of the spectrophotometric methods can yield much information on the composition and structure of films, particularly dielectric films. An excellent example is provided by the comprehensive infrared transmission and reflection studies of the absorption spectrum of various glass films performed by Pliskin and coworkers (see the review by Pliskin et al. [122]).

Double-focusing mass spectrometers using a combination of electrostatic and magnetic fields to obtain both velocity and direction focusing are commonly employed to obtain trace analysis of microgram samples of thin films at a fraction of a part per million level. The specimen is converted into ions for mass analysis by techniques such as a spark source [123,124], laser evaporation [125,126], and ion bombardment [127] (sputtering). A laser source of suitable power is a very efficient method of producing ions of several microamperes from metal surfaces.

Theoretical aspects and experimental investigations of laser vaporization of Au and Al films were reported by Board and Townsend [126]. They used a He-Ne pulsed laser which produced a 10-μ-diameter area of interaction with a penetration depth of ~100 to 1,000 Å. A review of the use of mass spectrometry for thin film analysis is given by Willardson [128].

The sputter-ion source is of special interest to thin films. Here, an ion beam generated in a duoplasmatron-ion source is accelerated to about 10 keV and focused onto the sample surfaces. The surface atoms are removed by sputtering, and the ionized sputtered species are mass analyzed. This technique allows a continuous examination of the specimen composition as a function of the depth. Due to a large variation in the sputtering yield of various elements, the sensitivity of detection also varies markedly. The technique has been employed by Herzog [129] for analysis of tantalum oxide and nitride films.

2.2 Structural Analysis

The techniques employed for structural analysis of thin films may be classified under two groups, one dealing primarily with the "surface" structure and the other with "volume and surface" structure. This arbitrary classification is obviously meaningless for ultrathin films. The two groups consist of the following techniques.

1. Optical interference, light-figure reflectograms, low-energy electron diffraction (LEED), field-emission and field-ion microscopy, sputter-ion microscopy, electron-reflection diffraction, and electron microscopy

2. X-ray microscopy or topography, x-ray and transmission electron diffraction, and transmission electron microscopy

We shall now present a brief account of each technique, with special emphasis on electron diffraction and electron microscopy, which are the most important and indispensable tools for thin film investigations.

2.3 Surface Structure

(1) Optical Methods. We have already mentioned that multiple-beam interferometry developed to a high level of sophistication by Tolansky [99,100] is useful in the study of surface smoothness. Careful use of the method allows determination of surface roughness on an atomic scale [130].

A technique to determine the crystallographic features and orientation of the surface of an epitaxial or partially oriented semiconducting film is provided by the corresponding symmetry of the observed reflected (in the backward direction) light pattern, also called "light-figure reflectogram" [131]. For a mirrorlike surface, a light preferential etch would suffice to produce etch figures that would yield reflection patterns. The method has been successfully used for epitaxial Ge [132] and CdTe [133] films. A significant advantage of the technique lies in the possibility [134] of identifying the surface polarity of the films of compound materials because the shape of the etch pattern depends on the polarity. The method is simple and inexpensive and may be used in situ, particularly for epitaxial growth by the CVD techniques (Chap. II).

(2) Low-energy Electron Diffraction (LEED). Low-energy (5 to 500 eV) electron diffraction is extremely sensitive to surface atom arrangements [135]. Owing to this high sensitivity, clean surfaces are essential for reliable LEED studies. A primary need for such studies is an ultrahigh vacuum in the 10^{-9} to 10^{-10} Torr. range so that contamination of the specimen surface by gas adsorption takes a sufficiently long time ($\sim$15 to 150 min) to enable a "clean" experiment to be performed. Readily available ultrahigh-vacuum equipment and the technical advances made in very efficient detection of the low-energy electrons directly on a fluorescent screen have brought about the renaissance of this powerful technique. Consequently, LEED is becoming a very popular technique and is finding increasing applications in surface studies. The reader is referred to the detailed reviews by Germer [136], Lander [137], MacRae [138], and Schlier and Farnsworth [139].

The diffraction equipment consists of an electron source, which is usually a low-work-function thermionic emitter, a means of collimating and then focusing the low-energy electrons onto a crystal, and some means of detecting the diffracted electrons. The electrons are focused on the crystal with electrostatic lenses and are diffracted in the backward direction. The beam size at the crystal is usually of the order of 0.5 mm in diameter, but beam sizes smaller than this can be achieved under special conditions. Note that this beam size is considerably larger than that in high-energy-electron sources and results from the space-charge limitations for workable beam currents of the order of 1 μA at these low energies.

The low-energy diffracted electrons, after passing through three semicircular grids having high transparency of about 85 percent, are accelerated to a fluorescent screen to produce a visual pattern that can be observed or photographed through the front window. One function of the grids is to shield the incident and diffracted beams from the 4 kV on the fluorescent screen. The two grids closest to the screen suppress or filter out electrons that have lost energy after interaction with the target crystal. By biasing these grids slightly negative with respect to the cathode, only the elastically scattered electrons, which produce the diffraction patterns, can reach the fluorescent screen. The first grid can be kept at the same potential as the crystal to provide a field-free region in the vicinity of the crystal, but it is often biased positive with respect to the crystal to improve beam focus at energies less than 50 eV as well as to deflect diffracted electrons not subtended by the grid system onto the fluorescent screen. The entire diffraction pattern is obtained on the screen in this way. The resultant distortion of the pattern does not complicate interpretation.

Diffraction patterns may be obtained from single-crystal surfaces of metals, semiconductors, and insulators. A gradual accumulation of charge on the surface of an insulator changes both the energy of the incident beam and the focusing conditions. This problem may be minimized [138] by alternately (e.g., at 60 cps) changing the energy of the incident beam from the desired low energy to a higher energy. At the high energy the secondary electron-emission ratio for most insulators is greater than unity and the charge accumulated on the crystal during the operating cycle can be dissipated. The time constant of the phosphor being greater than 1/60 sec, the LEED pattern is displayed continuously.

Because of the strong interaction of the low-energy electrons with atoms, the diffraction process is characteristic of the first three or so atomic layers at the surface. The positions of the diffraction spots are directly related to the two-dimensional symmetry and dimensions of the structural unit cell. In some respects, the diffraction pattern resembles that obtained from crossed optical-diffraction gratings. The plane grating formula $n\lambda = d(\sin\phi_r - \sin\phi_i)$, where λ is the wavelength, d the spacing between lines, and ϕ_r and ϕ_i are the angles between the surface normal and the reflected and incident beams, respectively, holds for both cases. The rows of atoms on the surface are the analogs of the lines in a grating.

We may also describe diffraction in terms of the reciprocal lattice, which for a two-dimensional lattice consists of rods having a spacing of $1/d$. Diffraction spots arise where the Ewald sphere of radius $1/\lambda$ ($\lambda = \sqrt{150/V}$ Å) intersects the reciprocal lattice. The Ewald sphere will always intersect at least one rod in the present case. With increasing electron energy, the Ewald sphere increases its radius, intersecting the rods at different positions, and the resultant diffraction beams move toward the center of the pattern.

It is clear from the reciprocal-lattice description of the diffraction process that a polycrystalline surface consisting of a continuous array of reciprocal rods would yield diffraction beams for all angles so that no discernible pattern is produced. Thus, the technique is restricted to single crystals or crystals with a preferred orientation. This property of the technique forms a useful tool for detecting surface contamination on an atomic scale. The single-crystal nature of the surface and the crystal structure of the surface may be determined from the diffraction patterns and their symmetry. The method is therefore valuable in studying the initial stages of epitaxial growth of a deposit. However, as expected, little information may be obtained about extended structural defects. If facets or inclined planes are present in the deposit surface, extra diffraction beams are produced which can be easily separated from those produced by the original surface by increasing the energy of incident electrons. This process allows a study of the size of the crystallites and their orientation.

The intensities of diffraction patterns may in principle be utilized to determine the actual position of atoms by the well-known Fourier-synthesis techniques applied in the case of x-ray diffraction. This is, however, not yet feasible for LEED because of insufficient knowledge of the effective scattering factors, which are complicated here by multiple scattering, and the differences in the real and imaginary parts of the atomic scattering factors between the different atoms.

The LEED technique has been employed to study surfaces of mica, MgO, alkali halides, Ge, Si, III-V compound semiconductors, NiO, Au, Ag, etc. Early studies indicated that the arrangement of surface atoms at the surface of a clean noncovalent material was the same as in bulk. On the other hand, covalent crystal surfaces showed an atomic arrangement that often bears little resemblance to their bulk structure. The situation is, however, changing rapidly at present, and there appears to be conclusive evidence for the existence of superstructures on many materials. Since this subject is reviewed in Chap. IV (Sec. 4.4), no literature references or further discussion is given here. It may be noted, however, that the surface superstructures may play a significant role in epitaxial growth, and from that point of view, the contribution of this technique to our knowledge may be invaluable.

(3) Auger-electron Spectroscopy. When a material is bombarded by low-energy electrons, Auger (inelastically scattered) electrons are emitted [137] with energies which are characteristic of the surface atoms of the material. The energy analysis of Auger electrons thus presents itself as a powerful spectroscopic technique for chemical analysis of solid surfaces, particularly those of light elements which are expected to have a high Auger-electron yield.

Harris [204] used electronic differentiation of the energy-distribution function and demonstrated the high sensitivity of this technique; Weber and Peria [205] subsequently showed that a standard three-grid LEED system may be effectively employed for measurement of the derivative of the energy-distribution function. Thus, both the chemical analysis of a film or surface and its crystal structure (with LEED, using the elastically scattered electrons) may be monitored with the same arrangement. The study of Weber and Peria showed that the Auger electron could be detected with a sensitivity of less than one-tenth of a monolayer of Cs and Si. Palmberg and Rhodin [206] monitored the energy spectra and the intensities of the peaks of metal surfaces during deposition of a second metal. The results indicated that the surface region which contributed to the Auger spectrum was only a few monolayers thick. The mean escape depth of Auger electrons in Ag was estimated to be 4 and 8 Å for energies of 72 and 362 eV, respectively.

(4) Field-emission, Field-ion, and Sputter-ion Microscopy. In the field-emission microscope, first devised by Müller [140,141], the electron emission of the surface under high electric field is measured. The sample in the form of a fine hemispherical point ($\sim$1,000 Å radius) is made the cathode, and a potential of a few thousand volts is applied to an anode to produce a field $\sim10^7$ V/cm at the surface of the point. The field- or tunnel-emitted electron current from the surface is projected onto a fluorescent screen where it portrays a magnified image of the point. The magnification, given by the ratio of the sample-screen separation to the radius of curvature of the point, can be made as high as

10^6, corresponding to a resolution along the surface of 20 Å. The emission current from any small area of the point is an exponential function of both the field intensity and the potential barrier for tunneling. These factors make emission very sensitive to steps, gross structural features, and adsorption of foreign atoms at the surface.

Field-emission patterns display some symmetry related to that of the crystal structure [142]. Adsorbed gases obliterate the patterns, and it is possible to detect a few percent of a monolayer coverage (e.g., on Si [143]). The technique is well suited and has been utilized by several workers [144-147] to study the nucleation stage of films. Note that by using an ac rather than a dc field, the cathode may be cleaned by field desorption of contaminant gases and also studied simultaneously [143].

The magnified image of the surface can also be produced by using the field-desorbed ions themselves [148]. The ion current may be enhanced by introducing an inert gas into the system. As a result of the high field in the vicinity of the point, the gas molecules are ionized just outside the surface. They are then drawn to the fluorescent screen where they reproduce on an enlarged scale the field distribution at the surface of the point. The resolution of the technique is improved by the reduced thermal velocity of the ions at low temperatures. With this technique of field-ion microscopy it is possible to resolve monatomic steps and even single atoms. It has been applied successfully to study surfaces, particularly of metals, and is undoubtedly an important tool for studying adsorption and initial nucleation stages of epitaxial growth. More extensive use of the technique is, however, still to be made.

An interesting, low-resolution type of ion microscope was described by Castaing and Slodzian [149,150]. The specimen surface is sputtered with a primary beam of rare-gas ions of energy ~10 kV. The ions of the sputtered species are then mass-analyzed by a double-focusing mass spectrometer and detected by accelerating them onto a fluorescent screen. The spatial distribution of a given mass component at the same surface is thus imaged on the screen by scanning. Although this ion microprobe has a low resolution of a fraction of a micron, it is expected to be a useful technique for analyzing the surface composition of films.

(5) Reflection Electron Diffraction. The high-energy electron-diffraction (HEED) method uses fast electrons (30 to 100 kV) for diffraction by reflection or transmission [151-156]. The penetration depth of fast electrons in a material is appreciable. However, the actual penetration depth perpendicular to the specimen surface for a reflection-diffraction condition, for which a grazing angle of incidence must be used owing to the small Bragg angle ($\sim 1°$, since λ is very small, e.g., 0.037 Å for 100-kV electrons), is much less. Consequently, reflection diffraction has a sensitivity for detecting surface films which closely approaches that of the LEED technique. For example, reflection-diffraction patterns were observed by Newman and Pashley [157] for an average thickness of 0.5 Å of Cu films (actually the film consisted of nuclei of height ~ 12 Å and diameter ~ 50 Å). Further, Sewell and Cohen [158] have demonstrated that the results obtained by an ultrahigh-vacuum HEED on oxidation of a Ni surface are quite similar to those obtained by the LEED technique.

The diffraction patterns yield information on the texture and crystallinity of the surface. If no atomic ordering exists on the surface position at which the electron beam impinges, no reinforcement of diffracted electrons takes place and a continuous radial falloff of the intensity from the primary beam is obtained. With increasing atomic order, broad halos and then sharp rings appear, the latter corresponding to a large number of randomly oriented crystallites. If the surface has one crystallographic orientation (or the crystallite size is as large as the electron-beam size, $\sim 10\ \mu$), then the diffracted beams consist of a series of well-defined rays and produce a pattern of spots in the diffraction pattern. If crystallites tend to have a particular zone axis, oriented in a common direction called the "fiber axis" (the orientation is called "fibrous"), the diffraction pattern consists of arced rings of a definite symmetry. The extent of the arcs will depend on the spread of the fibrous orientation and the crystallite size.

If the surface is atomically smooth, the reflection-diffraction spots become streaked or elongated in a direction at right angles to the shadow edge because of the very limited penetration of the electrons beneath the surface. The elongation is approximately equal to the inverse of the penetration depth. On the basis of the reciprocal-lattice concept, a thin crystal has a reciprocal lattice consisting of spikes perpendicular to the planes containing the thickness. Owing to the very small value of the wavelength of the electrons, the radius of the Ewald sphere is quite large so that it intersects long sections of the spikes to yield elongated spots.

With decreasing film thickness and increasing surface smoothness, the spikes merge to form lines [Fig. 6(a)].

If the surface is smooth, the electron beam is appreciably refracted because of the inner potential of the specimen [151,153], resulting in a deflection by a discrete amount for each spot. The diffracted beams are deflected directly toward the shadow edge, or at an angle to this edge if there are crystallographic facets on the crystal surface. Undulations in the surface cause asymmetrical streaks in the pattern. The amount of the refractive shift decreases as the distance between the diffracted electron beam and the shadow edge increases. Thus, the semicircular arcs of a diffraction ring from a very smooth polycrystalline surface become slightly U-shaped.

When the electron beam penetrates through surface asperities, transmission diffraction patterns are observed. Thus, with increasing surface roughness, the streaks change to a mixture of streaks and spots [Fig. 6(b)] and then to spots [Fig. 6(c)]. The transmission pattern consists of spots for a single-crystal surface and rings for a polycrystalline

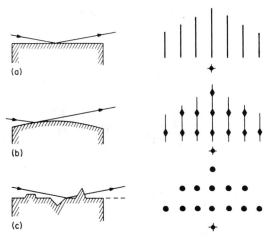

FIG. 6. The influence of surface smoothness on the formation of reflection-diffraction patterns from a single-crystal surface. The pattern changes from streaks for an atomically smooth surface (*a*) to a mixture of streaks and spots with increasing roughness (*b*). When surface asperities are large spot patterns characteristic of transmission diffraction are observed (*c*).

surface. Typical reflection and transmission diffraction patterns of films of some materials with different types of microstructure are shown in Fig. 7.

Both reflection and transmission diffraction patterns are utilized to determine crystal structure following the standard method of x-ray diffraction analysis. Interplanar spacings (d values) are obtained from the relation $d = \lambda L/r$, where r is the radius of the diffraction ring or the

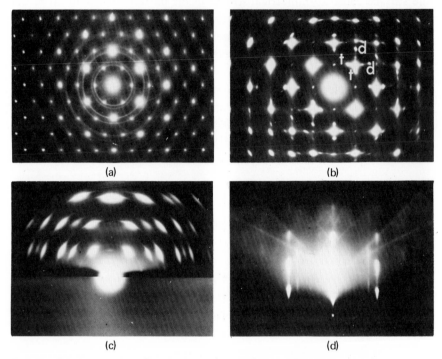

(a) (b)

(c) (d)

FIG. 7. Some examples of transmission (a and b) and reflection (c and d) electron-diffraction patterns of thin films. (a) Mixed polycrystalline and predominantly (111) oriented monocrystalline Au film (~ 500 Å thick) evaporated onto (100) Ag at $\sim 100°$C. The epitaxial Ag film was obtained by evaporation onto cleaved NaCl at 200°C. Note the occurrence of the forbidden reflections by the interference of multiply-diffracted beams. (b) Epitaxial (100) Ag film (~ 500 Å thick) sputtered onto cleaved NaCl at room temperature. The sharp extra reflections in the vicinity of t and d markers are due to twins and double diffraction, respectively. (c) A fibrous structure ~ 1-μ-thick Mo film, ion-beam sputtered onto glass at room temperature (*Chopra, unpublished*). (d) Epitaxial β-SiC film ($\sim 1,000$ Å thick) prepared by chemical reaction of (110) Si with unsaturated hydrocarbons at $\sim 950°$C. The strong streaks and the Kikuchi lines are due to β-SiC, and the inner weak streaks are due to Si (*courtesy of I. H. Khan*).

distance of the diffraction spot from the center of the pattern, and L is the effective specimen-to-plate distance, called the camera constant. These numbers are then compared with the tabulated lists [159] of d values for known structures until an identification is made. With an internal calibration (to determine L) and for crystallites larger than $\sim$100 Å, lattice constant may be determined with an accuracy of $\sim$1 percent.

The reflection-diffraction technique is much less accurate than the transmission technique, owing to such factors as elongation of spots resulting from refraction and limited penetration, the uncertainty of the camera constant due to the specimen size in the direction of the beam, and the effects due to the charge-up of insulating substrates. The charge-up is a serious factor in producing considerable deflection of the pattern and can be minimized by spraying the insulator surface with low-energy ($\sim$200 eV) electrons during examination. This spray of electrons neutralizes the positive charge on the insulating substrate produced by the secondary emission of electrons. The reflection-diffraction technique, however, offers the advantage of the possibility of determining the orientation of the surface relative to the electron beam by observing the symmetry of the patterns on rotation of the specimen surface in the plane of the beam.

A convenient means of varying the local contrasts in the diffraction patterns, thus allowing the position of very weak beams from ultrathin deposits to be recorded, is provided by a "scanning-electron diffraction" system developed by Grigson and coworkers [160-163]. This technique, while applicable for both transmission and reflection diffraction, is particularly advantageous for plotting two-dimensional spot patterns obtained by reflection [163]. The only shortcoming of this technique is that it takes too long to perform a dynamic in situ experiment.

Electron microscopy based on reflection diffraction has found little application in the study of thin films, primarily because the resolution of the technique (about 300 to 400 Å at best) is poorer than that of transmission microscopy. Halliday and Newman [164] have refined the technique by using a diffracted beam for imaging rather than diffusely scattered electrons. The resolution is thereby improved to 80 Å. The technique may therefore be useful for the study of surfaces of thick films.

(6) Replica Electron Microscopy. An easy technique for obtaining information about surface texture is the electron-microscopic examination of shadowed replicas of the surface. A review of the various replication techniques is given by Bradley [165]. The basic principle of the technique involves creating an impression of a surface with a thin film and making this impression visible in the electron microscope by providing diffraction contrast differences with the help of a suitable shadowing material. The impression may be obtained by coating the surface under study with a dilute solution of a plastic such as collodion in amyl acetate or Formvar dissolved in ethylene dichloride. When dry, the plastic film can be peeled from the surface. A common method is to deposit obliquely on the surface a material which scatters electrons heavily. This material is evaporated from a point source at a large angle to the average surface normal. Surface regions of varying gradations receive different thicknesses of the deposit, and asperities produce shadows. A supporting film is deposited over the "shadowed" surface, and the shadowed replica is removed for examination. It is clear that, given the shadowing angle and the length of the shadow, the height of the asperity can be easily determined trigonometrically.

The resolution of the replica is limited by the granularity of the shadowing material and is nominally $\gtrsim 20$ Å. A large number of materials such as high-atomic-number and high-melting-point metals (e.g., Pt and Pd, which yield films of very small grain size) and various oxides are suitable. The support film needed for the physical stability of the replica must be a poor electron scatterer; carbon is extensively used. A convenient procedure is to deposit both films simultaneously by using a mixed C-Pt source. When replicas are not easily removed by dissolution from the replicated surface, a surface coating of a water-soluble material (such as Victawet) can be used prior to replication. Victawet has a low melting point so that it may also be mixed with the C-Pt source since its evaporation would occur first.

If the surface to be examined has structural defects such as cleavage steps and dislocations, preferential nucleation of the condensed vapors of a high-mobility deposit occurs at the defects. Thus, an ultrathin deposit, equivalent to a few angstroms of Au, Ag, etc., would decorate these defects and make them visible in the electron micrograph (see Fig. 14, Chap. II). Several workers [166-169] have observed this effect on single-crystal alkali halides, MgO, and mica surfaces. The extensive work of Bethge and other workers in this field is reviewed elsewhere [168].

The decoration technique is obviously limited to the evaluation of the defect structure of the substrate surface only.

2.4 Volume Structure

(1) X-ray Diffraction. X-ray diffraction methods have not been applied to the study of thin films as extensively as electron-diffraction methods. This stems from the fact that the diffracted intensities are considerably smaller ($\sim 10^3$ times) for x-rays than for electrons. X-ray diffraction methods have, however, advantages of relative ease and convenience, large diffraction angles making accurate measurements of spot patterns possible, the diffraction patterns representing the average throughout the film due to increased penetration, and simultaneous display of diffraction patterns from the film and the substrate.

Improved detection methods for x-rays, the availability of commercial monochromators, and intense microfocus x-ray sources have made x-ray diffraction methods applicable to films as thin as 100 Å [170]. The methods are generally applied to films thicker than several hundred angstroms. Several workers [170-174] have described x-ray-diffraction arrangements suited to the study of thin films. The electron microprobe (x-ray microanalyzer), for example, is ideally suited for x-ray analysis of films.

Borie [170] studied the profiles of x-ray-diffraction maxima and deduced information on strains in thin copper oxide films. A comprehensive work on the position and profiles of diffraction maxima by in situ measurements on Cu and Au films deposited at low temperatures and subsequently annealed is due to Vook and Witt [175]. These authors used the profile data to calculate strains and crystallite sizes. The lattice parameters perpendicular to the film surface, and intrinsic stacking faults and twins were derived from the analysis of the peak positions of the various (hkl) reflections. Strains of a fraction of a percent and lattice constants to an accuracy of ~ 300 ppm were measured. Similar studies have also been made by other workers [176-179]. The reader is referred to the original papers for further details of the analysis of the x-ray diffraction data.

The profile of the diffraction maxima for epitaxial films is a qualitative measure of the perfection and the alignment of the crystallites [180] and its measurement presents itself as a convenient and

rapid evaluation method. The method may be made very sensitive to misorientation by the use of a double-crystal diffractometer to obtain the profile of a particular diffraction spot. This technique employs a monochromator to provide a diffracted beam which is further diffracted from the film surface oscillating about the mean diffraction position. The diffraction curves so obtained are also called "rocking curves." If the monochromator is a single crystal of the same material, a parallel arrangement of the specimen and the monochromator eliminates dispersion due to the finite x-ray line width. A discussion of the results of rocking curves from epitaxial GaAs and Ge is given by Holloway [181].

It should be noted that the diffraction curves broaden as the crystallite size decreases. The x-ray line broadening is commonly used to determine the crystallite size d, which is given by $d = \lambda/(D \cos \theta)$, where λ is the wavelength, θ is the Bragg angle, and D is the angular width at half the maximum intensity. Note that, because of the much smaller value of λ for fast electrons, the line broadening of electron-diffraction patterns is considerably reduced and is measurable only for crystallites much smaller than 100 Å.

(2) X-ray Microscopy (Topographic Methods). X-ray microscopy or topographic methods in which images of imperfections are formed by diffraction contrasts have been widely used for bulk single crystals. The methods are also well suited for thick epitaxial films and have been exploited by several workers [181-185].

The topographic methods generally employ two types of arrangements, one using Laue (transmission) diffraction devised by Lang [186] and the other employing Bragg (reflection) diffraction [187]. In both cases, the collimated x-ray beam is incident on the specimen, and a particular diffracted beam is allowed to reach the photographic plate. Both the specimen and the photographic plate, which are parallel to each other, are given a parallel scanning motion. In a modified arrangement for the Laue case, Schwuttke [185] used a scanning-oscillator technique in which, in addition to scanning, the crystal and the film also oscillate simultaneously around the normal to the plane containing incident and diffracted beams. The angle of oscillation covers the whole reflecting range of the crystal. This modification increases the intensity considerably, allowing rapid nondestructive testing of specimens.

The topographs reveal imperfections throughout the crystals and over large surface areas, and stereo pairs may be made for three-dimensional analysis. The method is superior to any other x-ray technique for obtaining sharp images of dislocations. The resolution is in excess of 100 Å. The topographs represent diffraction contrasts from several microns of the depth of an epitaxial layer. For thin layers, the defects of the substrate will also be seen. Clearly, the Laue method is applicable only if the substrate and the film are very thin, and thus the Bragg case is more useful for epitaxial studies.

Holloway and coworkers [182,183] used the reflection technique for examining defects in epitaxial GaAs films grown on GaAs and Ge [see Fig. 27(c), Chap. IV]. Schwuttke [185] employed the Laue case to study the surface of Si wafers, the homoepitaxial growth of Si films, the interfacial dislocations, and elastic strains present in these films.

(3) Transmission Electron-diffraction and Electron-microscope Methods. The texture, crystal structure, orientation of crystallites, and structural defects associated with a thin-film specimen can be determined by using the electron microscope both as an instrument for imaging a thin material at a high magnification and as a means for obtaining its diffraction patterns. The combination of the two approaches with several variants is widely used. The diffraction theory of contrast associated with various types of structural defects is now sufficiently well advanced to allow quantitative information to be obtained from the electron micrographs. Experimental and theoretical aspects of electron-diffraction methods are treated extensively in various books [151-156], and their application to epitaxial films is reviewed in excellent articles by Pashley [188,189] and Stowell [190]. We shall present here an elementary account of the type of information one may get from these techniques. For theoretical analysis or interpretation of the electron micrographs, the reader is referred to the general references cited above.

(a) Electron Diffraction. We have already described the reflection-electron-diffraction technique. Transmission electron diffraction is clearly restricted to specimens of thickness smaller than the penetration depth of the fast electrons employed. The thickness limit is generally of the order of 1,000 to 2,000 Å for 100-keV electrons. As in the reflection case, the symmetry of diffraction spots and spatial distribution of diffraction spots or rings may be used to determine the orientation and

crystal structure of the specimen. The diffraction pattern formed on the fluorescent screen of a microscope is actually an enlarged projection of the reciprocal-lattice "points" intersected by the Ewald sphere. Thus, if the reciprocal lattice is known, the complete diffraction pattern can be envisaged. The expected diffraction patterns for some directions of different crystal structures are shown in Fig. 8.

The accuracy for lattice-constant determination is generally ~1 percent. The standard technique does not allow interplanar spacings larger than ~20 Å to be resolved. In special low-angle diffraction techniques [191-192] used to resolve spacings of ~4,000 Å, one or both of the condenser lenses are overfocused and the intermediate lens is used to magnify the diffraction patterns from its object plane to that of the

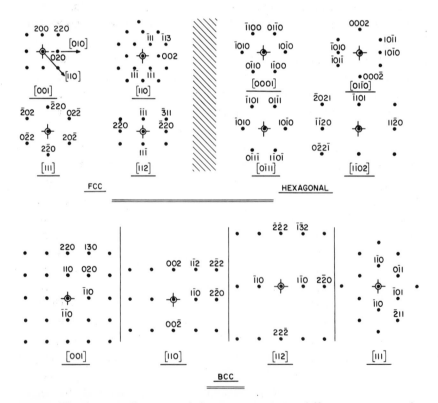

FIG. 8. The theoretically expected electron-transmission diffraction patterns of crystals of different structures with the direct electron beam along certain [hkl] reciprocal lattice directions.

projector lens. The projector lens produces the final magnification, and the objective lens is kept off.

If sharp Kikuchi lines are present, their center of symmetry allows a very accurate determination of the lattice constants. The orientation is also revealed since, unlike spots, Kikuchi lines depend solely on the orientation of the specimen and are independent of the direction of the incident electrons. Kikuchi lines appear when rather stringent diffraction conditions are satisfied, and their appearance may be taken to imply perfectness of crystallites and the parallelness of the crystallite surfaces [151] (for transmission).

A particular advantage in transmission diffraction may be realized by obtaining diffraction (as well as a micrograph) from a small and selected area of the specimen, which may be a single crystallite. This is achieved by inserting a physical aperture in the first image plane (which utilizes the magnification of the objective lens for viewing the aperture). The minimum size of the selected area is generally limited by the beam size (a few microns). Cockayne et al. [193] described a modification with which diffraction patterns from crystallites as small as 100 Å may be obtained. This technique is obviously of considerable interest.

Another sensitive and important variant of transmission-electron-diffraction arrangement for in situ studies is provided by the scanning electron-diffraction camera devised by Grigson and coworkers [160-163].

Electron-diffraction patterns are generally complicated by the presence of "extra" reflections and sharp "satellite" spots located symmetrically around some primary spots. Since a thin crystal has a reciprocal lattice of long spikes perpendicular to the plane of the crystal, the Ewald sphere may intersect not only the first but also the second layer of reciprocal-lattice spikes. The latter will give rise to an outer zone of extra spots, displaced relative to the inner zone of normal spots. If the crystal is extremely thin, the reciprocal-lattice spikes may become continuous from layer to layer, and extra spots will arise over the whole of the observed pattern. Diffuse streaking patterns may arise from thermal effects. This subject has been discussed by Honjo [194].

Any diffracted beam can act as the primary beam for further diffraction (dynamical effects), as seen in pattern (a) of Fig. 7. The high intensities of the electron-diffraction beams make such rediffraction effects quite significant and commonplace in electron-diffraction patterns. These dynamical effects modify the spot intensities and also give rise to reflections forbidden by the structure factors. From simple geometric considerations of reflection, it is clear that if reflections $(h_1 k_1 l_1)$ and $(h_2 k_2 l_2)$ occur, then reflections $(h_1 \pm h_2, k_1 \pm k_2, l_1 \pm l_2)$ can also occur. Dynamical-refraction effects may also give rise to forbidden or satellite reflections which make definite angles with the reflection radius.

Twins present in a crystal give extra diffraction spots. When a twinned crystal is in an orientation where both matrix and twin are diffracting strongly, it is common for double diffraction to occur across the twin boundary [190]. An example is shown in pattern (b) of Fig. 7. The double-diffraction spots usually cannot be given rational indices. More complicated double-diffraction spots may appear when overlapping zones are involved in diffraction conditions.

(b) Electron Microscopy. Transmission-electron-microscopy technique yields images at very high magnifications with a resolution less than 10 Å, thus allowing a detailed study of the localized regions of a specimen. It is the only technique with which images of lattice defects with a resolution below 10 Å may be easily formed.

Electron microscopy may be applied to thin film problems in several modes. Shadowed replicas (described already) may be used to study the surface structure. The film may be deposited on a thin electron-transparent substrate such as cleaved mica, MoS_2, MgO, and Bi_2Te_3 so that both the substrate and film can be examined together. This is a particularly useful technique for deposition inside the electron microscope and studying the initial processes of a film. A number of workers [195-201] have used this technique but generally without controlled deposition parameters and ultrahigh-vacuum conditions so as to yield reliable quantitative results. Poppa [201] used ultrahigh vacuum ($\sim 10^{-9}$ Torr) for the specimen chamber and a quartz crystal to monitor the deposition rate. He also demonstrated the use of glow discharge inside the electron microscope to clean the substrate and also to sputter-etch the film for studying the different stages of formation of a film.

The more commonly used method for electron microscopy of films is to remove the film from the substrate by dissolving away the substrate in a suitable solution (e.g., NaCl in water, mica in dilute HF, and MgO in hot HF) and then examine the film inside the microscope. For an ultrathin film, a supporting carbon film is deposited before removal. It is interesting to point out [202] that most clean metal films (prepared under the best possible conditions) as thin as 20 Å average thickness deposited on rock salt float off in water in large self-supporting pieces which are quite suitable for examination.

A powerful mode of operation of the electron microscope is that of forming dark-field images. A diffracted beam is used to form the image rather than the undiffracted or undeviated beam (bright-field image). The contrast in a bright-field image arises from the depletion of electrons from the undeviated beam in areas which are strongly diffracting (the image is therefore negative). In a dark-field image, those areas of the specimen which diffract electrons into the beam used for image formation will appear bright. This is achieved by inserting an aperture to allow a diffracted beam corresponding to a particular diffraction spot to pass through. The chosen diffraction spot may be moved to the center of the electron-optical-imaging system to reduce spherical aberration of the image by electromagnetically deflecting the central beam. This technique obviously allows the distribution of regions of a particular orientation to be determined in great detail. The region in a diffraction condition will appear bright on a dark background. In this way, twins and multiply-positioned nuclei or crystallites (which arise when the rotational symmetry about the normals to substrate and deposit are different) can be distinguished. The dark-field mode is also a valuable method for studying the distribution of variously oriented nuclei that may occur during the initial stages of epitaxial growth.

The single most important application of transmission electron microscopy is to form images of lattice defects such as stacking faults, microtwins, dislocations, vacancy clusters, and grain and domain boundaries. Point defects such as single vacancies or interstitials have not been observed in films. The resolution is thought to be inadequate, and also the contrast is probably insufficient to lead to the observable effect. However, aggregates of point defects, or features (e.g., dislocation loops, tetrahedra) resulting from the aggregation, have been observed.

The observation of lattice defects and other crystallographic features is made possible by the process of diffraction contrast. Basically, small variations in crystal-structure orientation or composition cause appreciable variations in the intensity of the locally diffracted beams. This results in local variations in the intensity of the transmitted undeviated beam, which is used to form the normal, bright-field image. The number of observable defects is determined by the diffraction contrast and the magnification of the image. At 20,000 magnification, for example, a surface density of defects of ~10^7/cm^2 must be present for one defect to be observable in the microscope image-viewing-screen area of ~35 cm^2. Dislocations with considerably lower densities may be observed if the dislocations lie parallel to the film surface and extend for long distances.

A detailed interpretation of the electron-diffraction contrasts is possible by the application of a dynamical theory which takes into account interference effects between all diffraction beams within the specimen. The subject is reviewed in standard treatises [155,156]. It should be emphasized that, although most defects can be easily observed and probably recognized, the interpretation, in some cases, can be made only by specialists in the field.

The Bragg angles for electron diffraction are very small, and distortion of a crystal by bending just a fraction of a degree is therefore ample to modify diffraction conditions. A bent crystal may be considered as a wedge-shaped crystal which should yield an interference-fringe pattern similar to that observed in a multiple-beam optical interferometer. Each fringe represents the locus of points which reflect strongly into a particular diffracted beam, i.e., the locus of points for which the reflecting planes are equally inclined to the incident electron beam. The fringes are known as extinction contours due to bending, and are commonly observed in thin-film micrographs.

A stacking fault lying in a plane inclined to that of the film gives rise to a fringe pattern in the electron image. By comparing the bright- and dark-field images of the stacking fault and from the diffraction pattern, it is possible to determine the sense of inclination and the nature of the fault [190]. A wide twin gives rise to an image bounded by fringes corresponding to the wedge-shaped interfaces. A narrow twin, which is essentially equivalent to overlapping stacking faults, produces fringes similar to those of stacking faults. A distinction between a stacking fault and a microtwin on the basis of the fringe pattern alone is therefore extremely difficult. The microtwins may be identified by their character-istic extra diffraction spots and also by the dark-field-image technique.

A very important technique for observing and identifying lattice defects, originally used by Hashimoto and Uyeda [203] and later extensively and successfully exploited by Bassett and Pashley and coworkers [188-190] (among others), is that of the moiré patterns. Moiré patterns or fringes are produced by the interference of waves diffracted by the periodic matching or mismatching of the overlapping lattices of two crystals of the same or different spacing. When the two crystals have parallel diffracting planes with spacings d_1 and d_2 a "parallel" moiré pattern of spacing $D = |d_1 d_2/(d_1-d_2)|$ is formed. If the two planes have the same spacing d but are rotated by a small angle θ relative to each other, the resulting "rotation" moiré pattern has a spacing $D = d/\theta$ The net effect is to produce a new lattice of highly magnified lattice spacing, thus making it quite easy to observe defects.

The moiré-pattern technique is valuable for the study of defects and, in particular, the kinetics of their incorporation and annihilation in epitaxial films deposited on suitable substrates. Pashley and coworkers have made use of the cleaved MoS_2 substrates for such studies with in situ epitaxial growth of fcc metals. Some examples of the moiré patterns of various structural defects in films are shown in Fig. 26 of Chap. IV.

3. CONCLUSIONS

It is satisfying to conclude that a large variety of techniques for monitoring thickness, controlling deposition, and analysis of thin films of all materials have been devised and successfully applied. Technologically speaking, it is therefore possible to deposit a film of x angstroms under x different but defined conditions. The structure of such a film, although not yet completely controllable (as we shall see in the following chapter), can nevertheless be defined to a large extent. Despite all these advances, however, it is very disconcerting to remark that thin film workers have so far hardly utilized even the minimum required deposition controls and analytical techniques for obtaining a well-characterized film. Consequently, it is not uncommon to find conflicting results and widely divergent views on the structure and properties of films. These remarks will unfortunately echo time and again in the rest of the book.

It must also be emphasized that any further breakthroughs in the understanding of some of the basic phenomena such as nucleation, epitaxy, and abnormal structures will be possible only by the employment of controlled deposition techniques and the simultaneous use of in-situ LEED, HEED, and electron-microscopy techniques.

REFERENCES

1. H. Mayer, "Physik dünner Schichten," vol. 1, Wissenschaftliche Verlag, Stuttgart, 1950.
2. O. S. Heavens, "Optical Properties of Thin Solid Films," Dover Publications, Inc., New York, 1965.
3. L. Holland, "Vacuum Deposition of Thin Films," John Wiley & Sons, Inc., New York, 1956.
4. K. H. Behrndt, in "Physics of Thin Films" (G. Hass and R. E. Thun, eds.), vol. 3, p. 1, Academic Press Inc., New York, 1966.
5. W. Steckelmacher, in "Thin Film Microelectronics" (L. Holland, ed.), p. 193, John Wiley & Sons, Inc., New York, 1966.
6. J. A. Bennett and T. P. Flanagan, *J. Sci. Instr.,* **37**:143 (1960).
7. W. Steckelmacher, J. English, H. H. A. Bath, D. Haynes, and J. T. Holden, *Trans. 10th Natl. Vacuum Symp.,* p. 515, The Macmillan Company, New York, 1963.
8. J. A. Turner, J. K. Birtwistle, and G. R. Hoffman, *J. Sci. Instr.,* **40**:557 (1963).
9. W. F. Leonard and R. L. Ramey, *J. Appl. Phys.,* **35**:2963 (1964).
10. L. J. van der Pauw, *Philips Res. Rept.,* **13**:1 (1958).
11. J. Lange, *J. Appl. Phys.,* **35**:2659 (1964).
12. L. B. Valdes, *Proc. IRE,* **42**:420 (1954).
13. A. Uhlir, Jr., *Bell System Tech. J.,* **34**:105 (1955).
14. P. A. Schumann, Jr., and J. F. Hallenbach, Jr., *J. Electrochem. Soc.,* **110**:538 (1963); R. G. Mazur and D. H. Dickey, *J. Electrochem. Soc.,* **113**:255 (1966).
15. E. E. Gardner, in "Measurement Techniques for Thin Films" (N. Schwartz and B. Schwartz, eds.), p. 240, Electrochemical Society, 1967.
16. For example, P. Cotti, *Phys. Letters,* **4**:114 (1963).
17. E. W. Johnson and H. H. Johnson, *Rev. Sci. Instr.,* **35**:1510 (1964).
18. F. Z. Keister and R. Y. Scapple, *Trans. 9th Natl. Vacuum Symp.,* p. 116, The Macmillan Company, New York, 1962.
19. G. C. Riddle, *Proc. 4th Symp. Electron Beam Technol.,* p. 340, Alloyd Electronics Corp., Cambridge, Mass., 1962.
20. See the review article by J. G. King and R. Zacharias, in "Advances in Electronics and Electron Physics" (L. Marton, ed.), vol. 8, p. 1, Academic Press Inc., New York, 1956.
21. O. Haase, *Z. Naturforsch.,* **12a**:941 (1957).
22. F. Metzger, *Helv. Phys. Acta,* **16**:323 (1943).
23. G. R. Giedd and M. H. Perkins, *Rev. Sci. Instr.,* **31**:773 (1960).
24. H. Schwarz, *Trans. 7th Natl. Vacuum Symp.,* p. 326, Pergamon Press, New York, 1960; *Rev. Sci. Instr.,* **32**:194 (1961).

25. C. Dufour and B. Zega, *Vide,* **18**:180 (1963).
26. R. B. Brownell, W. D. McLennan, R. L. Ramey, and E. J. White, *Rev. Sci. Instr.,* **35**:1147 (1964).
27. A. V. Engel and M. Steenbeck, "Electrical Gas Discharges," vol. 1, p. 33, Springer-Verlag OHG, Berlin, 1932; also, J. L. Cobine, "Gaseous Conductors," Dover Publications, Inc., New York, 1958.
28. C. K. Crawford, *Proc. 4th Symp. Electron Beam Technol.,* p. 59, Alloyd Electronics Corp., Cambridge, Mass., 1962.
29. P. R. Brooks and D. R. Herschbach, *Rev. Sci. Instr.,* **35**:1528 (1964).
30. K. H. Behrndt, *Z. Angew. Phys.,* **8**:453 (1956).
31. D. S. Campbell and H. Blackburn, *Trans. 7th Natl. Vacuum Symp.,* p. 313, Pergamon Press, New York, 1960. Also, A. R. Beavitt, *J. Sci. Instr.,* **43**:182 (1965).
32. R. E. Hayes and A. R. V. Roberts, *J. Sci. Instr.,* **39**:428 (1962).
33. A. L. Houde, in "Vacuum Microbalance Techniques," vol. 3, p. 109, Plenum Press, 1963.
34. C. A. Neugebauer, *J. Appl. Phys.,* **35**:3599 (1964).
35. G. Sauerbrey, *Z. Physik,* **155**:206 (1959).
36. G. Sauerbrey, *Arch. Elektron. Übertr.,* **18**:617 (1964).
37. P. Lostis, *Rev. Opt.,* **38**:1 (1959).
38. K. H. Behrndt and R. W. Love, *Vacuum,* **12**:1 (1962).
39. S. J. Lins and H. S. Kukuk, *Trans. 7th Natl. Vacuum Symp.,* p. 333, Pergamon Press, New York, 1960.
40. D. McKeown, *Rev. Sci. Instr.,* **32**:133 (1961).
41. A. W. Warner and C. D. Stockbridge, *J. Appl. Phys.,* **34**:437 (1963).
42. W. Steckelmacher et al., *Trans. 10th Natl. Vacuum Symp.,* p. 415, The Macmillan Company, New York, 1963.
43. W. P. Mason, "Piezoelectric Crystals and Their Application to Ultrasonics," D. Van Nostrand Company, Inc., Princeton, N.J., 1956.
44. K. L. Chopra and M. R. Randlett, *J. Appl. Phys.,* **39**:1874 (1968).
45. R. Gerber, *Rev. Sci. Instr.,* **38**:77 (1967).
46. N. Schwartz and R. Brown, *Trans. 9th Natl. Vacuum Symp.,* p. 836, The Macmillan Company, New York, 1962; M. D. Silver and E. T. K. Chow, *J. Vacuum Sci. Tech.,* **2**:203 (1965).
47. T. A. Anastasio, *Rev. Sci. Instr.,* **34**:740 (1963).
48. A. W. Agar, *Brit. J. Appl. Phys.,* **8**:35 (1957).
49. K. H. Behrndt, *Trans. 9th Natl. Vacuum Symp.,* p. 111, The Macmillan Company, New York, 1962; K. H. Behrndt and D. W. Doughty, *J. Vacuum Sci. Tech.,* **3**:264 (1966).
50. P. F. Varadi and J. R. Suffredini, *Rev. Sci. Instr.,* **36**:1331 (1965).
51. H. Mase, S. Nakaya, and Y. Hatta, *J. Appl. Phys.,* **38**:2960 (1967).
52. H. Friedman and L. S. Birks, *Rev. Sci. Instr.,* **17**:99 (1946).
53. A. Eisenstein, *J. Appl. Phys.,* **17**:874 (1946).
54. W. H. T. Davison, *J. Sci. Instr.,* **34**:418 (1957).
55. M. de Croes, W. Parker, and K. Sevier, *Nucl. Instr. Methods,* **7**:10 (1960).
56. J. F. Cameron, in "Progress in Non-destructive Testing" (E. G. Stanford and J. H. Fearons, eds.), vol. 2, p. 91, Heywood, London, 1960.
57. W. Dietzsch, *Isotopen Tech.,* **1**:66, 98 (1961).

58. Yu. S. Zavalvskii, G. I. Shor, A. D. Stukin, and E. D. Stukin, *Instr. Exptl. Tech. USSR English Transl.*, **1**:141 (1963).
59. D. Kantelhardt and O. Schott, *Z. Angew. Phys.*, **15**:307 (1963).
60. S. Ohh and K. G. Carroll, *J. Appl. Phys.*, **30**:1620 (1959).
61. L. Reimer, *Z. Angew. Physik*, **9**:34 (1957).
62. G. Behrens, *Z. Physik*, **162**:180 (1961).
63. J. J. Antal and A. H. Weber, *Rev. Sci. Instr.*, **23**:424 (1952).
64. L. E. Preuss, *J. Appl. Phys.*, **24**:1401 (1953).
65. C. Bugenis and L. E. Preuss, *Trans. 10th Natl. Vacuum Symp.*, p. 374, The Macmillan Company, New York, 1963.
66. L. E. Preuss and C. E. Alt, *Vakuum-Tech.*, **9**:99, 121, 155 (1960).
67. H. F. Beeghly, *J. Electrochem. Soc.*, **97**:152 (1950).
68. P. D. Zemany and H. A. Liebhafsky, *J. Electrochem. Soc.*, **103**:157 (1956).
69. H. A. Liebhafsky and P. D. Zemany, *Ann. Chem.*, **28**:455 (1956).
70. B. W. Schumacher and S. S. Mitra, *Electron. Reliability Micromin.*, **1**:321 (1962).
71. P. K. Koh and E. Caugherty, *J. Appl. Phys.*, **23**:427 (1952).
72. T. N. Rhodin, *Ann. Chem.*, **27**:1857 (1955).
73. P. J. Brown, *J. Sci. Instr.*, **37**:394 (1960).
74. R. Weyl, *Z. Angew. Physik*, **13**:283 (1961).
75. J. F. Finnegan and P. R. Gould, *Trans. 9th Natl. Vacuum Symp.*, p. 129, The Macmillan Company, New York, 1962.
76. W. E. Sweeney, R. E. Seebold, and L. S. Birks, *J. Appl. Phys.*, **31**:1061 (1960).
77. K. L. Chopra, M. R. Randlett, and S. L. Bender, *Rev. Sci. Instr.*, **39**:1755 (1968).
78. G. H. Cockett and C. D. Davis, *Brit. J. Appl. Phys.*, **14**:813 (1963).
79. R. Castaing, *Advan. Electron. Electron Phys.*, **13**:317 (1960).
80. L. Katz and A. S. Penfold, *Rev. Mod. Phys.*, **24**:28 (1952).
81. J. M. Khan, D. L. Potter, and R. D. Worley, *J. Appl. Phys.*, **37**:564 (1966).
82. L. J. Christensen, J. M. Khan, and W. F. Brunner, *Rev. Sci. Instr.*, **38**:20 (1967).
83. C. Dufour, *Vide*, **3**:480 (1948); *J. Phys. Radium*, **11**:353 (1950).
84. C. Fremont, *Rev. Sci. Instr.*, **20**:620 (1949).
85. A. F. Bogenschütz, F. Bergmann, and J. Jentzsch, *Z. Angew. Phys.*, **14**:469 (1962).
86. W. Steckelmacher, J. M. Parisot, L. Holland, and T. Putner, *Vacuum*, **9**:171 (1959).
87. D. L. Perry, *Appl. Opt.*, **4**:987 (1965).
88. A. Andant, *J. Phys. Radium*, **11**:351 (1950).
89. V. L. Lindberg and M. J. Irland, *J. Opt. Soc. Am.*, **45**:328 (1955).
90. P. Giacomo and P. Jacquinot, *J. Phys. Radium*, **13**:59A (1952); P. Giacomo, *Rev. Opt.*, **35**:317, 442 (1956).
91. A. C. Traub, *J. Opt. Soc. Am.*, **46**:999 (1956).
92. T. Sawaki, M. Iwata, S. Katsube, and K. Hara, *J. Physique*, **25**:258 (1964).
93. P. Jacquinot, *Rev. Opt.*, **21**:15 (1942).
94. W. G. Spitzer and M. Tanenbaum, *J. Appl. Phys.*, **32**:744 (1961).
95. M. P. Albert and J. F. Combs, *J. Electrochem. Soc.*, **109**:709 (1962).

96. N. Goldsmith and L. A. Murray, *Solid-State Electron.*, **9**:331 (1966).
97. D. J. Dumin, *Rev. Sci. Instr.*, **38**:1107 (1967).
98. O. Wiener, *Wied. Ann.*, **31**:629 (1887).
99. S. Tolansky, "Multiple Beam Interferometry of Surfaces and Films," Oxford University Press, Fair Lawn, N.J., 1948.
100. S. Tolansky, "Surface Microtopography," John Wiley & Sons, Inc., New York, 1960.
101. A. C. Terrell, *The Microscope and Crystal Front,* vol. 14, July-August, 1964.
102. G. D. Scott, T. A. McLauchlan, and R. S. Sennett, *J. Appl. Phys.*, **21**:843 (1950).
103. C. Lockyer, *J. Sci. Instr.*, **44**:393 (1967).
104. K. M. Greenland, *Vacuum,* **2**:216 (1952).
105. K. M. Greenland and C. Billington, *Proc. Phys. Soc. London,* **B63**:359 (1950).
106. A. N. de Jong, *Opt. Acta,* **10**:115 (1963).
107. H. Kiessig, *Naturwiss.,* **18**:847 (1930).
108. W. Hink and W. Petzold, *Z. Angew. Phys.,* **10**:135 (1958).
109. Y. Fujiki, *J. Phys. Soc. Japan,* **14**:1308 (1959).
110. J. Sauro, I. Fankuchen, and N. Wainfan, *Phys. Rev.,* **132**:1544 (1963).
111. E. B. Sandell, "Colorimetric Determination of Traces of Metals," Interscience Publishers, Inc., New York, 1959.
112. G. E. F. Lundell, H. A. Bright, and J. I. Hoffman, "Applied Inorganic Analysis," John Wiley & Sons, Inc., New York, 1962.
113. I. M. Kolthoff and P. J. Elving (eds.), "Treatise on Analytical Chemistry," Interscience Publishers, Inc., New York, 1964.
114. *Analytical Chemistry,* Annual Reviews, American Chemical Society.
115. H. A. Liebhafsky et al., "X-ray Absorption and Emission in Analytical Chemistry," John Wiley & Sons, Inc., New York, 1960.
116. H. H. Paltee, V. E. Cosslett, and A. Engstrom (eds.), "X-ray Optics and X-ray Microanalysis," Academic Press Inc., New York, 1963.
117. For other references, see E. Fuchs, *Rev. Sci. Instr.,* **37**:623 (1966).
118. R. W. Kiser, "An Introduction to Mass Spectrometry and Its Applications," Prentice-Hall, Inc., Englewood Cliffs, N.J., 1965.
119. H. H. Willard, L. L. Merritt, Jr., and J. A. Dean, "Instrumental Methods of Analysis," D. Van Nostrand Company, Inc., Princeton, N.J., 1965.
120. W. G. Guldner, *Anal. Chem.,* **35**:1744 (1963).
121. H. F. Winters and E. Kay, *J. Appl. Phys.,* **38**:3928 (1967).
122. W. A. Pliskin, in "Measurement Techniques for Thin Films" (B. Schwartz and N. Schwartz, eds.), p. 280, Electrochemical Society, 1967; W. A. Pliskin, D. R. Kerr, and J. A. Perri, in "Physics of Thin Films" (G. Hass and R. E. Thun, eds.), vol. 4, p. 257, Academic Press Inc., New York, 1967.
123. G. G. Sweeney, W. M. Hickam, and L. B. Crider, *Proc. 14th Ann. Conf. Mass Spectrometry Allied Topics,* 1966, p. 165.
124. D. L. Malm, *Proc. 14th Ann. Conf. Mass Spectrometry Allied Topics,* 1966, p. 138.
125. R. E. Honig, *Proc. 12th Ann. Conf. Mass Spectrometry Allied Topics,* 1964, p. 233.
126. K. Board and W. G. Townsend, *Microelec. Reliability,* **5**:251 (1966).

127. H. J. Liebl and R. F. K. Herzog, *J. Appl. Phys.*, **34**:2893 (1963).
128. R. K. Willardson, in "Measurement Techniques for Thin Films" (B. Schwartz and N. Schwartz, eds.), p. 58, Electrochemical Society, 1967.
129. R. F. K. Herzog (unpublished).
130. W. F. Koehler and W. C. White, *J. Opt. Soc. Am.*, **45**:1011 (1955); D. R. Herriott, *J. Opt. Soc. Am.*, **51**:1142 (1961).
131. G. A. Wolff and J. D. Broder, *Acta Cryst.*, **12**:313 (1959).
132. J. G. Gualtieri and A. J. Kerecman, *Rev. Sci. Instr.*, **34**:108 (1963).
133. M. Weinstein, G. A. Wolff, and B. N. Das, *Appl. Phys. Letters*, **6**:73 (1965).
134. For example, A. N. Mariano and G. A. Wolff, *Z. Krist.*, **126**:244 (1968).
135. C. J. Davisson and L. H. Germer, *Phys. Rev.*, **30**:705 (1927).
136. L. H. Germer, *Sci. Am.*, **212**:32 (1965).
137. J. J. Lander, in "Progress in Solid State Chemistry," vol. 2, p. 26, Pergamon Press, New York, 1965.
138. A. U. MacRae, *Science*, **139**:379 (1963); also, in "The Use of Thin Films for Physical Investigations" (J. C. Anderson, ed.), p. 149, Academic Press Inc., New York, 1966.
139. R. E. Schlier and H. E. Farnsworth, *Advan. Catalysis*, **9**:434 (1957).
140. E. W. Müller, *Z. Physik*, **106**:541 (1937); *Phys. Rev.*, **102**:618 (1956).
141. See the review by R. H. Good, Jr., and E. W. Müller, in "Handbuch der Physik," vol. 21, p. 176, Springer-Verlag OHG, Berlin, 1956.
142. F. G. Allen, *J. Phys. Chem. Solids*, **19**:87 (1961).
143. F. G. Allen, J. Eisinger, H. D. Hagstrum, and J. T. Law, *J. Appl. Phys.*, **30**:1563 (1959).
144. R. D. Gretz and G. M. Pound, in "Condensation and Evaporation of Solids" (E. Rutner, P. Goldfinger, and J. P. Hirth, eds.), p. 575, Gordon and Breach, Science Publishers, Inc., New York, 1964.
145. R. D. Gretz, *Surface Sci.*, **5**:255 (1966).
146. A. J. Melmed, *J. Appl. Phys.*, **37**:275 (1966).
147. A. J. Melmed, *J. Chem. Phys.*, **43**:3057 (1965).
148. E. W. Müller, *Z. Physik*, **126**:642 (1949); *Science*, **149**:591 (1965).
149. R. Castaing and G. Slodzian, *J. Microscopie*, **1**:395 (1962).
150. R. Castaing, "Advances in Mass Spectrometry," vol. 3, p. 91, The Institute of Petroleum, London, 1964.
151. Z. G. Pinsker, "Electron Diffraction," Butterworth & Co. (Publishers) Ltd., London, 1953.
152. B. K. Vainstein, "Structure Analysis by Electron Diffraction," Pergamon Press, New York, 1964.
153. R. D. Heidenreich, "Fundamentals of Transmission Electron Microscopy," Interscience Publishers, Inc., New York, 1964.
154. D. H. Kay, "Techniques for Electron Microscopy," F. A. Davis Company, Philadelphia, 1965.
155. P. B. Hirsch, A. Howie, R. B. Nicholson, D. W. Pashley, and M. J. Whelan, "Electron Microscopy of Thin Crystals," Butterworths, Washington, D.C., 1965.
156. S. Amelinckx, "The Direct Observation of Dislocations," Solid State Physics Supplement 6, Academic Press Inc., New York, 1964.
157. R. C. Newman and D. W. Pashley, *Phil. Mag.*, **46**:927 (1955).

158. P. B. Sewell and M. Cohen, *Appl. Phys. Letters,* **7**:32 (1965).
159. ASTM Index X-ray Powder Data File (revised annually).
160. C. W. B. Grigson, *Nature,* **192**:647 (1961).
161. C. W. B. Grigson, *Rev. Sci. Instr.,* **36**:1587 (1965).
162. P. N. Denbigh and C. W. B. Grigson, *J. Sci. Instr.,* **42**:305 (1965); P. N. Denbigh and D. B. Dove, *J. Appl. Phys.,* **38**:99 (1967).
163. M. F. Tompsett and C. W. B. Grigson, *Nature,* **206**:923 (1965).
164. J. S. Halliday and R. C. Newman, *Brit. J. Appl. Phys.,* **11**:158 (1960).
165. D. E. Bradley, in "Techniques for Electron Microscopy" (D. H. Kay, ed.), chap. 5, F. A. Davis Company, Philadelphia, 1965.
166. G. A. Bassett, *Phil. Mag.,* **3**:1042 (1958).
167. C. Sella and J. J. Trillat, in "Single Crystal Films" (M. H. Francombe and H. Sato, eds.), p. 201, Pergamon Press, New York, 1964.
168. H. Bethge, *Phys. Stat. Solidi,* **2**:3, 775 (1962).
169. J. J. Gilman, *J. Appl. Phys.,* **30**:1584 (1959), and the earlier references cited.
170. B. Borie, *Acta Cryst.,* **13**:542 (1960); B. Borie, C. J. Sparks, and J. V. Cathcart, *Acta Met.,* **10**:691 (1962).
171. D. A. Brine and R. A. Young, *Trans. 7th Natl. Vacuum Symp.,* p. 250, Pergamon Press, New York, 1960.
172. M. J. Hall and M. W. Thompson, *Brit. J. Appl. Phys.,* **12**:495 (1961).
173. P. E. Lighty, D. Shanefield, S. Weissman, and A. Shrier, *J. Appl. Phys.,* **34**:2233 (1963).
174. R. W. Vook and F. R. L. Schoening, *Rev. Sci. Instr.,* **34**:792 (1963).
175. R. W. Vook and F. Witt, *J. Vacuum Sci. Tech.,* **2**:49 (1965); **2**:243 (1965).
176. T. B. Light and C. N. J. Wagner, *J. Vacuum Sci. Tech.,* **3**:1 (1966).
177. C. N. J. Wagner, *Acta. Met.,* **5**:427 (1957); J. B. Cohen and C. N. J. Wagner, *J. Appl. Phys.,* **33**:2073 (1962).
178. R. E. Smallman and K. H. Westmacott, *Phil. Mag.,* **2**:669 (1957).
179. D. E. Mikkola and J. B. Cohen, *J. Appl. Phys.,* **33**:892 (1962).
180. For example, W. H. Zachariasen, "Theory of X-ray Diffraction in Crystals," John Wiley & Sons, Inc., New York, 1945.
181. H. Holloway, in "The Use of Thin Films for Physical Investigations" (J. C. Anderson, ed.), p. 111, Academic Press Inc., New York, 1966.
182. H. Holloway, K. Wollmann, and A. S. Joseph, *Phil. Mag.,* **11**:263 (1965).
183. H. Holloway and L. C. Bobb, *J. Appl. Phys.,* **38**:2893 (1967).
184. G. H. Schwuttke and V. Sils, *J. Appl. Phys.,* **34**:3127 (1963).
185. G. H. Schwuttke, *Proc. 3d Intern. Congr. Vacuum Tech. Stuttgart,* vol. 2, p. 301, Pergamon Press, New York, 1965.
186. A. R. Lang, *J. Appl. Phys.,* **29**:597 (1958); **30**:1748 (1959).
187. C. S. Barrett and T. B. Massalski, "Structure of Metals," 3d ed., McGraw-Hill Book Company, New York, 1966.
188. D. W. Pashley, in "Modern Developments in Electron Microscopy" (B. M. Siegel, ed.), p. 149, Academic Press Inc., New York, 1964.
189. D. W. Pashley, *Advan. Phys.,* **14**:327 (1965).
190. M. J. Stowell, in "The Use of Thin Films for Physical Investigations" (J. C. Anderson, ed.), p. 131, Academic Press Inc., New York, 1966.
191. H. Mahl and W. Weitsch, *Z. Naturforsch.,* **15a**:1051 (1960).

192. R. H. Wade and J. Silcox, *Appl. Phys. Letters,* 8:7 (1966); *Phys. Stat.Solidi,* 19:57, 63 (1967).
193. D. J. H. Cockayne, P. Goodman, J. C. Mills, and A. F. Moodie, *Rev. Sci. Instr.,* 38:1097 (1967).
194. G. Honjo, in "Single Crystal Films" (M. H. Francombe and H. Sato, eds.), p. 189, Pergamon Press, New York, 1964.
195. T. A. McLauchlan, R. S. Sennett, and G. D. Scott, *Can. J. Res.,* A28:530 (1950).
196. G. A. Bassett, *Phil. Mag.,* 3:1042 (1958).
197. D. W. Pashley, M. J. Stowell, M. H. Jacobs, and T. J. Law, *Phil. Mag.,* 10:127 (1964).
198. I. M. Watt, *Proc. European Regional Conf. Electron Microscopy, Delft,* 1960, p. 341.
199. A. E. Curzon and K. Kimoto, *J. Sci. Instr.,* 40:601 (1963).
200. K. J. Hanszen, *Z. Naturforsch.,* 19a:820 (1964).
201. H. Poppa, *Phil. Mag.,* 7:1013 (1962); *Z. Naturforsch.,* 19a:835 (1964); *J. Vacuum Sci. Tech.,* 2:42 (1965).
202. G. G. Sumner, *Surface Sci.,* 4:313 (1966).
203. H. Hashimoto and R. Uyeda, *Acta Cryst.,* 10:143 (1957).
204. L. A. Harris, *J. Appl. Phys.,* 39:1419 (1968).
205. R. E. Weber and W. T. Peria, *J. Appl. Phys.,* 38:4355 (1967).
206. P. W. Palmberg and T. N. Rhodin, *J. Appl. Phys.,* 39:2425 (1968).

IV NUCLEATION, GROWTH, AND STRUCTURE OF FILMS

Thin films are most commonly prepared by the condensation of atoms from the vapor phase of a material. At the earliest stage of observation, atomistic condensation takes place in the form of three-dimensional nuclei which then grow to form a continuous film by diffusion-controlled processes. The manner in which a thin film comes into being is, indeed, unique. The novel structural behavior and properties of films can largely be ascribed to this growth process, which is therefore of basic importance to the science and technology of thin films. This chapter reviews the significant theoretical and experimental results related to the understanding of the various stages of growth and the film structure.

1. NUCLEATION

This, the birth stage of a film, is essentially a problem of vapor-solid-phase transformation. Several theoretical approaches to this problem are reviewed in the literature [1-6]. The reviews by Hirth and Pound [2] and Sigsbee and Pound [4] are recommended for reference.

In the following sections, we shall first present some classical concepts of the condensation process followed by a brief and elementary report on heterogeneous nucleation and the experimental results on the condensation process.

1.1 Condensation Process

The condensation of a vapor atom is determined by its interaction with the impinged surface in the following manner. The impinging atom is attracted to the surface by the instantaneous dipole and quadrupole moments of the surface atoms. As a result, the atom loses its velocity component normal to the surface in a short time, provided the incident kinetic energy is not too high. The vapor atom is then physically adsorbed (called "adatom"), but it may or may not be completely thermally equilibrated. It may move over the surface by jumping from one potential well to the other because of thermal activation from the surface and/or its own kinetic energy parallel to the surface. The adatom has a finite stay or residence time on the surface during which it may interact with other adatoms to form a stable cluster and be chemically adsorbed (incorporated into the surface) with the release of the heat of condensation. If not adsorbed, the adatom reevaporates or desorbs into the vapor phase. Therefore, condensation is the net result of an equilibrium between the adsorption and desorption processes.

The probability that an impinging atom will be incorporated into the substrate (surface) is called the "condensation" or "sticking" coefficient. It is measured by the ratio of the amount of material condensed on a surface to the total amount impinged. The degree of thermal equilibration is described by the accommodation coefficient α_T, defined by

$$\alpha_T = \frac{T_I - T_R}{T_I - T_S} = \frac{E_I - E_R}{E_I - E_S} \tag{1}$$

Here T's and E's correspond, respectively, to the equivalent rms temperatures and equivalent kinetic energies of the incident (I), reflected or reevaporated (R) atoms, and the substrate (S).

The problem of capturing an incident atom and its energy exchange through the van der Waals forces has been treated by several workers (see for example, the review by Massey and Burhop [7]). Several theoretical investigations [8-11] considered the problem of a head-on collision of an atom with a one-dimensional lattice of spring-connected masses, and showed that, for nearly equal masses of the impinging atom and the substrate lattice atom, a unity condensation coefficient should be obtained for kinetic energies up to 25 times the desorption energy Q_{des}. Since Q_{des} for metal vapor incident on metal substrates is of the order of 1 to 4 eV, vapor atoms of equivalent beam temperatures of the order of a million degrees should readily be physically adsorbed. The capture of impinging atoms is less complete for a three-dimensional lattice [12], principally because of its greater stiffness. If the impinging atom is considerably lighter than the substrate atom, or if it has a high kinetic energy, the sticking coefficient can be appreciably less than unity.

The mean relaxation time τ_e required for an adatom to equilibrate thermally with the substrate is estimated [13] to be less than $2/\nu$, where ν is the adatom surface-vibrational frequency. According to McCarrol and Ehrlich [14], the captured atom will lose all but a few percent of E_I within three lattice oscillations for comparable masses of the impinging and the impinged atoms. The adatom moves over the substrate surface and will have a mean stay time, before being desorbed, given by

$$\tau_s = \frac{1}{\nu} \exp\left(\frac{Q_{des}}{kT}\right) \tag{2}$$

Thus,

$$\tau_e \approx 2\tau_s \exp\left(-\frac{Q_{des}}{kT}\right) \tag{3}$$

At high binding energies ($Q_{des} \gg kT$), τ_s is very large and τ_e is small. That is, thermal equilibrium occurs rapidly. The adatom can then be considered as localized and will diffuse by discrete jumps. On the other hand, if $Q_{des} \sim kT$, the adatoms do not equilibrate rapidly and thus remain "hot," generally resulting in a less than unity condensation coefficient. The mobile adatoms in this case can be appropriately regarded as forming a two-dimensional gas, so that the kinetic theory of gases may be applied to determine their movements. It must be emphasized that, for mathematical convenience, nucleation theories assume thermodynamical equilibrium for the adsorbate, and neglect the role of the kinetic energy in the condensation process.

During its stay time, an equilibrated adatom diffuses over the substrate surface to a diffusion distance $\bar{X}$ given by the Einstein relation for the Brownian movement as

$$\bar{X} = (2D_s \tau_s)^{1/2} = (2\nu\tau_s)^{1/2} a \exp\left(-\frac{Q_d}{2kT}\right) \tag{4}$$

$$= 2^{1/2} a \exp\left(\frac{Q_{des} - Q_d}{2kT}\right) \tag{4a}$$

where a is the jump distance between the adsorption sites on the surface, Q_d is the activation energy for a surface-diffusion jump, and the surface-diffusion coefficient $D_s = a^2 \nu \exp(-Q_d/kT)$.

It is clear that Q_{des} and Q_d play an important role in the condensation process. Their magnitudes are therefore of interest and are listed for some systems in Table I. The values of these energies depend very sensitively on the conditions of the surface, as illustrated by the values for Cd on clean and contaminated Ag. Although a precise relation between Q_d and Q_{des} is not known, it is commonly observed that $Q_d \sim 1/4 \, Q_{des}$.

1.2 Langmuir-Frenkel Theory of Condensation

Langmuir [8] and Frenkel [15] formulated a condensation model* in which the adsorbed atoms move over the surface during their lifetimes to

---

*This model considers vapor → solid transformation. At high deposition temperatures, a vapor → liquid (amorphous) → solid condensation mode may occur (discussed later). Semenoff [16] suggested that heterogeneous nucleation always proceeds by formation of an amorphous film followed by nucleation of crystallites within the amorphous film.

Table I. Experimental Values for the Binding Energy Q_{des} and the Activation Energy for Surface Diffusion Q_d for Some Typical Systems

Condensate	Substrate	Q_{des}, eV	Q_d, eV	Ref.
Ag	NaCl		0.2	27
Ag	NaCl		0.15 (evaporation) 0.10 (sputtering)	52
Al	NaCl	0.6		6
	Mica	0.9		
Ba	W	3.8	0.65	11
Cd	Ag (fresh film)	1.6		6
	Ag, glass	0.24		
Cu	Glass	0.14		6
Cs	W	2.8	0.61	11
Hg	Ag	0.11		6
Pt	NaCl		0.18	110
W	W	3.8	0.65	11

form pairs which, in turn, act as condensation centers for other atoms. If one assumes a steady state between the relative rates of impingement and desorption, atom pairs are formed on a surface at a temperature T, if a critical density R_c of the beam given by

$$R_c = \frac{\nu}{4A} \exp\left(-\frac{\mu}{kT}\right) \tag{5}$$

is obtained. Here, A is the cross section for capture of an atom, and μ is the sum of the adsorption energy of a single atom to the surface and the dissociation energy of a pair of atoms. Although this concept of a critical beam density is in accord with the experimental observations of Cockcroft [17] (condensation of Cd on Cu surface) and others [18], its relationship with the beam and substrate temperatures is by no means as simple as expressed by Eq. (5), since a nucleation barrier exists for condensation to occur and it depends sensitively on the temperature, chemical nature, structure, and cleanliness of the surface. The value of R_c drops markedly immediately after the initial nucleation has occurred.

The Langmuir-Frenkel theory was extended by Zinsmeister [19] by considering equilibrium between the growth and decay of adatom clusters. The qualitative conclusions of the theory are consistent with general experience, as expected. The occurrence of decay processes, however, needs experimental verification.

1.3 Theories of Nucleation

We shall now examine the conditions for growth of a cluster and the rate of formation of such critical clusters as a function of the deposition parameters and practical consequences thereof.

(1) **Capillarity Theory.** A homogeneous-nucleation theory which takes into account the total free energy of formation of a cluster of adatoms was postulated by Volmer and Weber [20], and Becker and Döring [21] (see the review by Holomon and Turnbull [22]). It was later extended to heterogeneous nucleation by Volmer [20] and to the particular shapes of clusters in a thin film case by Pound et al. [23,2]. In this theory, clusters (also called embryos, or subcritical nuclei) are formed by collisions of adatoms on the substrate surface, and in the vapor phase if supersaturation is sufficiently high. They develop initially with an increase in free energy (Fig. 1) until a critical size is reached,

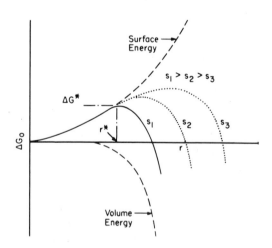

FIG. 1. Total free energy (surface plus volume free energies) of cluster formation vs. size (r) for different supersaturations (s).

above which growth continues with a decrease in free energy. If one assumes that bulk thermodynamical quantities can be ascribed to the properties of clusters, the Gibbs' free energy of formation of a spherical cluster of radius r, given by the sum of the surface energy (to create a surface) and the volume energy of condensation, is

$$\Delta G_0 = 4\pi r^2 \sigma_{cv} + \frac{4}{3}\pi r^3 \Delta G_v \tag{6}$$

Here, σ_{cv} is the condensate-vapor interfacial free energy and $\Delta G_v = (-kT/V)\ln(p/p_e)$ is the Gibbs' free-energy difference per unit volume of the phase of molecular volume V condensed from the supersaturated vapor of pressure p to the equilibrium pressure p_e ($p/p_e \equiv S$ is the supersaturation). A plot of ΔG_0 vs. r (Fig. 1) shows a maximum at

$$r^* = -\frac{2\sigma_{cv}}{\Delta G_v} = \frac{2\sigma_{cv} V}{kT \ln(p/p_e)} \tag{7}$$

This is the radius of the critical nucleus. Clusters smaller than r^* are unstable, while larger aggregates grow to become stable deposits.

If the critical nucleus is cap-shaped, its contact angle θ in a condensate(c)-vapor(v)-substrate(s) system is determined by Young's equation for the minimization of the surface free energies

$$\sigma_{cv} \cos\theta = \sigma_{sv} - \sigma_{sc} \tag{8}$$

By rewriting Gibbs' free-energy term ΔG_v in terms of the various surface energies, one obtains

$$\Delta G_0 = \frac{1}{3}\pi r^3 \Delta G_v (2 - 3\cos\theta + \cos^3\theta) + 2(1 - \cos\theta)\pi r^2 \sigma_{cv}$$
$$+ \pi r^2 \sin\theta (\sigma_{sc} - \sigma_{sv}) \tag{9}$$

The value of r^* is still given by Eq. (7) for any value of $\theta > 0$. The critical value of ΔG_0 is now given by

$$\Delta G^* = \frac{16}{3} \pi \frac{\sigma_{cv}^{\;3}}{\Delta G_v^{\;2}} \phi(\theta)$$ (10)

where $\phi(\theta) = \frac{1}{4}(2 - 3\cos\theta + \cos^3\theta)$.

Note from Fig. 1 that with increasing supersaturation (e.g., with higher-melting-point materials and lower substrate temperatures) r^* decreases; that is, a larger number of smaller-sized stable nuclei are formed. This variation is, however, insignificant since, on the basis of the present theory, the calculated value of r^* for most cases is of atomic dimensions. For example, using bulk values of σ_{cv} and deposition rates of 1 Å/sec at 300°K, r^* for Ag is calculated to be 2.2 Å.

When $\theta = 0$ (complete wetting which holds if $\sigma_{sv} \geq \sigma_{sc} + \sigma_{cv}$), $\Delta G^* = 0$, so that there is no activation barrier to the nucleation of the condensed phase. For $\theta = 180$ (no wetting), $\phi(\theta) = 1$, and the foreign interface is inactive in the nucleation process. *Note that the value of r^* is not affected by the values of θ.* On the other hand, ΔG^* depends on θ. Chakraverty and Pound [24] showed that for $\theta < 45°$, and $50° < \theta < 105°$, clusters will form, with a smaller ΔG^*, at steps rather than on the flat surfaces. Thus, the concentration of critical nuclei at steps would be increased relative to that on flat surfaces, yielding the well-known surface-decoration effect (see Fig. 14b, Chap. II). Electrostatic charges present on the vapor atoms or adsorption sites may also lower ΔG^*, thus facilitating the condensation process. Impurities may facilitate or retard condensation depending on whether ΔG^* is lowered or raised.

The nucleation rate I is proportional to the product of the concentration, $N^* = N_0 \exp(-\Delta G^*/kT)$, of the critical nuclei, and the rate Γ at which molecules join the critical nuclei by a diffusion process.[†] Here, N_0 is the density of adsorption sites. Thus,

$$I = Z(2\pi r^* \sin\theta)\Gamma N^*$$ (11)

where $2\pi r^* \sin\theta$ is the periphery of the critical nucleus and Z is the Zeldovich correction factor to take into account the departure from

---

†The effective catchment area ($\sim\pi\bar{X}^2$) for a diffusional process is much larger than the physical area for direct impingement.

equilibrium due to nucleation and the fact that some of the nuclei would decay. This factor is about 10^{-2} for both cap-shaped and disklike nuclei.

Now, the diffusion process yields

$$\Gamma \approx N_1 a_0 \nu' \exp\left(-\frac{Q_d}{kT}\right) \qquad (12)$$

where a_0 is the separation between adsorption sites, ν' is a frequency of the order 10^{12} per sec, Q_d is the activation energy for surface diffusion, and N_1 is the density of adatoms defined by $N_1 = R\tau_s$, so that

$$N_1 \approx \frac{R}{\nu} \exp\left(\frac{Q_{des}}{kT}\right) \qquad (13)$$

Here, R is the rate of incidence of single atoms from the vapor (called "impingement flux") and Q_{des} is the energy of desorption of single atoms from the substrate. If we assume $\nu \sim \nu'$, we obtain the following expression for I by using Eqs. (11), (12), and (13),

$$I = \frac{4\pi\sigma_{cv}}{\Delta G_v} \sin\theta R a_0 N_0 \exp\left(\frac{Q_{des} - Q_d - \Delta G^*}{kT}\right) \qquad (14)$$

(2) Statistical or Atomistic Theory. The size of the critical nucleus for metals, as suggested by the capillarity theory, is typically of atomic dimensions involving a few atoms. Consequently, the applicability of the capillarity model and the use of bulk thermodynamical quantities for such small clusters are questionable. These difficulties may be overcome by writing the partition functions and potential energies for the reacting species and products [25]. An approximate analysis which considered the energies and bonds of nucleation clusters treated as macromolecules was given by Walton and Rhodin [5,26,27]. Accordingly, at low substrate temperatures or very high supersaturations, the critical nucleus may be a single atom which will form a pair with another atom by random occurrence to become a stable cluster and grow spontaneously.

The stability of a pair is derived from an assumed one bond per atom. At higher substrate temperatures, a pair of atoms may no longer be a stable cluster. The next smallest stable cluster is one that has a minimum of two bonds per atom, which may be achieved by putting atoms at the corners of a triangle configuration. A four-atom configuration with two bonds per atom would be a square.

The nucleation rate in this theory is again proportional to $N^*\Gamma$. If one assumes that the vibrational partition functions are unity, and takes E_n as the energy required to dissociate the n-adatoms cluster into n single atoms adsorbed on the surface, the general expression for the nucleation rate of critical nuclei with n^* atoms is

$$I = Ra_0{}^2 N_0 \left(\frac{R}{\nu N_0}\right)^{n^*} \exp\left[\frac{(n^* + 1)Q_{des} - Q_d + E_{n^*}}{kT}\right] \tag{15}$$

The terms used here have already been defined. Note that, with increasing size (i.e., n^*) of the critical nucleus, the nucleation rate and temperature dependence increase as $R^{n^* + 1}$.

A stability transition from an n^* to an $n^* + 1$ atom cluster with increasing substrate temperature may be found by equating the two corresponding expressions in Eq. (15). For example, a transition from a 2- to a 3-atom cluster occurs at

$$T = -\frac{Q_{des} + \frac{1}{2}E_3}{k \ln (R/\nu N_0)} \tag{16}$$

The significance of this transition temperature for epitaxial growth is discussed later.

In view of the similarity of the basic principles, it is not surprising to find a similarity between Eqs. (14) and (15) of the capillarity and the atomistic models, respectively. The difference lies in the use of a continuously varying surface energy and hence changing size of the cluster in the former as compared with discontinuous changes in the binding energy of the adatoms with changing size of the cluster in the latter case. The discontinuities of the atomistic model are of course more realistic for a small cluster. By applying macroscopic data to the atomistic model, Lewis [28,29] made a detailed theoretical comparison of the results of nucleation process derived from the two theories. He showed that, because of the difference in the value of supersaturation for small nuclei, the capillarity model predicts a relatively larger size for the critical nucleus and a lower nucleation rate. The fact that the idealized shapes of the capillarity model give higher cluster energies, smaller critical nuclei, and a higher nucleation rate compensates for this difference. In general, however, the two models exhibit a wide agreement with each other.

(3) Miscellaneous Models

1. It is possible to use the powerful method of Monte Carlo calculations to analyze the condensation process via nucleation involving relatively few atoms. By ascribing simple sets of behavior (movement) rules to atoms, the initial results [30] of such calculations show qualitatively the expected agglomeration or clustering of adatoms. Further refinements of this method will no doubt receive more attention in the future.

2. The more complex problem of the nucleation of a binary condensed phase is of much technical interest since this is directly related to the deposition of alloy and compound films. Reiss [31] analyzed this problem theoretically to a limited extent. The details of his results fall beyond our scope. However, it is noteworthy that Günther [32] made use of the simple critical condensation concept to analyze qualitatively the formation of different phases from codeposited vapor atoms (Sec. 2.2, Chap. II).

(4) Further Deductions of the Nucleation Theories

1. Because of the exponential dependence on ΔG^* in the capillarity theory and on E_{n*} in the atomistic theory, the nucleation rate is very sensitive to changes in supersaturation. In fact, the rate changes nearly discontinuously from a negligible to a very high value over a narrow range of values above a certain "critical" supersaturation. Further, since the effective supersaturation is a rapidly varying function of the substrate temperature, small changes in the latter cause several orders of magnitude variation in the critical supersaturation.

2. The nucleation rate in the capillarity theory is proportional to the impingement flux, and to its square (for the smallest cluster) in the atomistic theory.

3. The nucleation rates given in the two theories are for the case of incoherent nucleation. If coherent nucleation occurs (as is the case in pseudomorphs; see Sec. 4.1), the surface energy and hence the shape of the cluster in the capillarity model will be determined by the nucleating crystal faces.

4. Both theories predict a similar type of temperature dependence [Eqs. (14) and (15)]. By measuring critical nucleation conditions and the nucleation rate as a function of temperature, an Arrhenius type of plot should yield a straight line (or lines). By assuming a suitable value of θ, and the validity of the thermodynamical relation in Eq. (10), the value of $(Q_{des} - Q_d)$ may be obtained. This parameter may also be obtained from the atomistic theory provided some values of n^* and E_{n^*} are assumed. Thus, both theories require the assumption of empirical parameters to obtain values of $(Q_{des} - Q_d)$ which are reasonable but not unique.

5. The nucleation rates given by the capillarity and atomistic theories are steady-state rates, valid *only* until the density of nuclei reaches a maximum and the average distance between nuclei corresponds to the mean diffusion distance. Thereafter, the nuclei grow by capturing adatoms by diffusion. This depletion process does not allow further nucleation to occur. The saturation density of nuclei is *independent* of the impingement rate provided (*a*) the impinging atoms are equilibrated instantaneously, (*b*) the effect of the momentum of vapor atoms is not significant, and (*c*) the impingement rate is smaller than the adatom diffusion rate so that equilibrium conditions exist. Under these conditions, the saturation density decreases with substrate temperature† as

$$N_s = N_0 \exp\left(-\frac{Q_{des} - Q_d}{kT}\right) \tag{17}$$

6. The steady state is attained after an "induction" period which is the sum of the times necessary for adatoms to attain the population in kinetic equilibrium with the vapor and that for forming the equilibrium

---

†Implies effective temperature, which includes effect of factors such as heat of condensation and the kinetic energy of the incident vapor atoms.

population of different-sized embryos. After this period, the nucleation density increases rapidly to approach saturation at a rate which depends on the supersaturation.

1.4 Experimental Results

The experiments on condensation phenomena are generally concerned with the study of the onset of condensation, the sticking coefficient, and the nucleation rate as related to impingement rate and temperature of the substrate.

(1) Sticking Coefficient. The onset of condensation has been variously determined: visually, with electron microscopy and field-ion microscopy, or mass spectrometrically. A very rapid, spontaneous type of condensation is generally found to occur when the "critical" supersaturation condition is fulfilled. The sticking coefficient is expected to decrease with increasing substrate temperature and decreasing binding energy of the adsorbate to the substrate. It should increase with increasing deposit and approach unity for the case of self-deposition. These conclusions are in general qualitative agreement with the various observations reviewed by Wexler [6], and Hirth and Pound [2]. Some measured values of the sticking coefficient for several deposits at different temperatures and film thicknesses are listed in Table II to illustrate these conclusions.

Extensive measurements of the sticking coefficient of Sb, Cd, and Au deposits on Cu, Al, mica, and glass substrates have been carried out by Devienne [33]. A number of workers [33-38,333] have studied Au and Ag deposits on a variety of substrates. The sticking coefficient of Au on Ag, for example, is found to be considerably less than unity (as low as 0.5) in the initial growth stages and approaches unity at film thicknesses of ~250 Å. The varied thickness dependence of the sticking coefficient in this case is generally attributed to the presence of surface contamination.

Careful, ultrahigh-vacuum, in situ experiments of Henning [333] established that the sticking coefficient of Au is unity on air-contaminated NaCl but is considerably lower (as low as 0.3 at 300°C) for clean NaCl substrates. The coefficient was found to increase with film thickness (approaching unity at ~500 Å) as well as with increasing thermal etching of the clean NaCl substrate in high vacuum. According to Henning, the observed thickness and temperature dependence of the

Table II. Illustration of the Effects of Deposit Thickness (for Cd), Surface Temperature (for Au), and Lattice Mismatch (for Ag) on the Condensation Coefficient of Some Vapor Atoms
Data on Cd and Au deposits taken from Devienne [33] and for Ag from Yang et al. [35]

Condensate	Surface	Surface temp. °C	Deposit, Å	Condensation coefficient
Cd	Cu	25	0.8	0.037
			4.9	0.26
			6	0.24
			42.4	0.60
Au	Glass,Cu,Al	25	Detectable	0.90-0.99
	Cu	350		0.84
	Glass	360		0.50
	Al	320		0.72
	Al	345		0.37
Ag	Ag (0)*	20	Detectable	1.0
	Au (0.18)*			0.99
	Pt (3.96)*			0.86
	Ni (13.7)*			0.64
	Glass			0.31

*Lattice mismatch, percent, with respect to Ag.

sticking coefficient may be understood on the basis of the island structure (geometrical) of the deposit. The larger the number of islands and hence the smaller the interisland distance for surface diffusion, the higher is the probability of capture of the adatoms. In order to explain the role of surface contamination, such a model should take into account both the geometrical distribution of the islands and the adatom mobility.

The thickness dependence of the coefficient varies considerably for different substrate-vapor combinations and the condition of the substrate surface. Yang et al. [35] found the coefficient for Ag deposited on various metal substrates to decrease with increasing lattice mismatch (see the data in Table II). Whether this effect is due to interfacial strain, change of contact angle, or a systematic change of impurity adsorption is not clear.

Gas contamination of the surface affects condensation considerably. Hudson and Sandejas [39] observed that the sticking coefficient, which is nearly unity for Cd on an atomically clean W substrate at 10^{-10} Torr, decreases appreciably when the W surface is exposed to 10^{-5} Torr pressure, and the critical supersaturation increases drastically by many orders of magnitude. It is well known that nucleation is enhanced by the presence of surface defects and metallic impurities. Further, enhanced nucleation has been observed for Sn and In [40,41] on glass exposed to a partial pressure of O_2, for Cu [42] on W exposed to partial pressures of $\sim 10^{-8}$ Torr of O_2 and N_2, and for Au [43] on air-cleaved rather than vacuum-cleaved NaCl. Melmed [42] established from the field-emission microscope studies of Cu that enhanced nucleation was accompanied by increased Q_d and hence Q_{des}. Gretz [44] observed a significantly higher critical supersaturation for a clean rather than contaminated surface. It is thus quite clear that no general statement about the role of contaminants can be made since it depends entirely on whether the binding energy is increased or decreased.

Systems which exhibit sticking coefficients less than unity suggest thermal nonequilibrium or nonaccommodation. The unusually high (many powers of 10) supersaturation required for condensation in these cases led Sears and Cahn [45] to suggest that adatoms remain "hot" during adsorption so that their effective temperature, rather than the lower equilibrium substrate temperature, should be used to calculate the effective supersaturation. This hypothesis has been questioned [2,4] but without due justification. The results of Hudson [46] on Cd and Zn support thermal nonaccommodation. On the other hand, those of Hruska and Pound [47], and Shade [48], who varied the temperature of the Cd vapor beam and found it to have negligible effect on the critical beam flux for deposition, suggest perfect thermal accommodation.

The "hot-adatom" concept needs careful consideration. It is, of course, not compatible with the thermodynamical equilibrium assumed by the nucleation theories. Refined and clear-cut experiments must be performed to determine the validity of this concept.

(2) Observations on Nucleation. Since the predicted size of the critical nucleus (for condensation) is of atomic dimensions, its direct observation by any presently known technique is not possible. The smallest nucleus of an Au deposit observed [49] by electron microscopy is $\sim$5 Å diameter containing about 20 atoms. Clearly, this is already a postnucleation or nucleation-growth stage, and probably all other known electron- and ion-microscopy studies of nucleation also represent this stage of growth. We shall refer to such nuclei as "islands."

The occurrence of three-dimensional islands in ultrathin films of metals, insulators, and semiconductors prepared by evaporation [50,51] and metal films prepared by sputtering [52] and electrolysis [53] has been well established by electron microscopy. Similar results are obtained by simultaneous deposition and examination of metal films inside the electron microscope under both poor-vacuum [50,54-58] ($\sim$10^{-5} Torr) and ultrahigh-vacuum [58] ($\sim$10^{-8} Torr) conditions. The occurrence of three-dimensional islands is the most direct proof of the existence of a nucleation barrier for condensation as well as that of the growth of islands primarily by surface-diffusion mechanisms rather than by direct capture of the vapor atoms.

Although nucleation and subsequent growth have been extensively studied only in films of a few metals, the process is qualitatively quite general for any method of atomistic deposition of a material onto any substrate. The quantitative features are, however, expected to vary markedly with different materials. Among the other representative examples where three-dimensional nucleation has been established and studied are the epitaxial growth of evaporated and electroplated Au [59,60] on Ag films, Sn [61] on carbon films, AgBr [62] on Ag films, various tarnishing layers [63], oxidation of Cu [64], Ni [65], and Ta [66], epitaxial II-VI [67] compound films, and auto expitaxial films of Si by pyrolysis [68] as well as]hysical-evaporation [69] techniques.

The occurrence of three-dimensional nucleation in most deposit-substrate systems is an established experimental fact. A two-dimensional or monolayer type of coverage *may* occur in any of the following cases: (1) Assuming the validity of the capillarity model, if $\sigma_{sv} \geq \sigma_{cv} + \sigma_{sc}$ [that is, $\theta = 0$ from Eq. (8)], then the nucleation barrier is absent and wetting is possible. As seen later, θ is an empirical parameter of finite value, and it may bear no relation to the bulk values of the contact angle. Consequently, it is not possible to predict the possibility of a monolayer coverage from the known bulk surface-energy values. An excellent example of this case is provided by the surprising observation [68,69] of the nucleation type of growth of Si on Si substrate which is a continuation of the same material. One may be tempted to explain this observation by proposing nucleation at traces of contaminants. This is, however, rejected by the observed epitaxial growth of the deposit. It must then be concluded that θ may not be zero even for the case of self-deposition. The condensation process thus appears to be statistical in nature and occurs via a nucleation barrier irrespective of the substrate. (2) If the nucleation barrier is small and the adatom surface mobility is high, large (laterally) platelet-like islands will be formed. These islands will give the appearance of "extended" monolayer coverage. Such a deposit will give rise to the streaking of the reflection electron-diffraction patterns as seen in the observations of Pb [70] films on (111) Ag, and of self-deposited alkali halides [71]. (3) If nucleation occurs at or very near (negligible surface diffusion) the position of impingement of the vapor atom, an "effective" monolayer-by-monolayer deposition takes place. This condition is satisfied when the number of nucleation centers is $\sim 10^{15}/cm^2$, which may result from very high supersaturation (low substrate temperature, high-melting-point vapor source) and/or suitable choice of a substrate (e.g., Bi_2O_3 for Au films).

Some interesting observations on the nucleation stage may now be noted. (1) The saturation density of uniformly distributed islands for a number of metals of comparable melting point deposited on ionic, amorphous, and single-crystal insulating substrates at room temperatures is $\sim 10^{10}$ to $10^{12}/cm^2$. That is, the islands are approximately 100 to 1,000 Å apart, and vapor atoms condensing anywhere in between therefore move to join the nearest island. The islands appear to be uniformly distributed on a defect-free surface. The size distribution [57] of the islands exhibits a strongly peaked, bell-shaped curve. In some cases [49,52] of low supersaturation, the island density is initially low and increases very rapidly with further deposition to approach a saturation value. (2) After reaching saturation, the island density decreases by mutual coalescence with increasing deposition rate and substrate temperature, and with subsequent film growth. The rate of decrease is more rapid with sputtered than with evaporated films, and with increasing smoothness of the substrate and its temperature [52]. (3) A time lag or an "induction" period (see, for example, Poppa [58]) between the time of deposition and the observation of condensation occurs under low-supersaturation conditions. It is due partly to the time taken for nuclei to reach the minimum observable size and partly to the reasons given already. (4) The nucleation-growth stage is affected [72-75] if the substrate is irradiated with energetic electrons, since electrostatic charges should facilitate condensation by lowering the nucleation barrier. Further, Chopra [72,73] showed that the presence of positive or negative electrostatic charges on the islands may increase their surface area and enhance their surface diffusion (discussed later). (5) Because of a lower nucleation barrier and an increased binding energy, preferential and enhanced (by a factor of up to 100) nucleation occurs at surface defects, ledges, and monoatomic steps, thereby resulting in their decoration. This decoration effect offers a powerful technique for studying surface imperfections [43,76,77] and for determining the polarity of a compound surface. (6) Condensation appears to occur at random sites. A further discussion of the nature of condensation centers [78-81] is given later.

The critical dependence of nucleation on supersaturation observed by early workers was clearly demonstrated by Yang et al. [35], who studied the nucleation of a Na atomic beam on polycrystalline Ag, Pt, Cu, Ni, and CsCl over a range of beam fluxes and substrate temperatures. The size of the critical nucleus deduced from the data was of atomic dimensions. The calculated thermodynamical parameters indicated poor agreement with the bulk values. Field-emission microscopy (resolution of 20 Å or more) was used by Moazed and Pound [82] and Gretz and Pound [83] to study the temperature dependence of critical nucleation of various metals condensed on W tips at $\sim 10^{-10}$ Torr.

The results of the various nucleation studies reported in the literature as deduced on the basis of the capillarity theory are summarized by Hirth and Moazed [334] and are listed in Table III. It is clear from the data that (1) the critical size of the nucleus is of atomic dimensions, and (2) the value of θ required for obtaining reasonable values of thermodynamical parameters bears no relation to the bulk contact angle for the system.

Walton et al. [27], using the electron microscope, measured both the incidence rate and temperature dependence of the nucleation rate of Ag deposited on in situ cleaved (100) NaCl in ultrahigh vacuum. The results (Fig. 2) show a change in the slope of the straight line at $\sim 250°C$ which the authors interpreted as representing a transition from the surface

Table III. Summary of Nucleation Data
Compiled from the literature by Hirth and Moazed [334]

Condensate	Substrate	T, °K	σ, ergs/cm^2	$(\Delta Q_{des} - \Delta Q_d)$ kcal/mole	θ, deg	$-\Delta G^*$, cal/cc	r^*, Å
Ag	W	300	1,100		~ 84	2,900	2
Ag	W	300	1,100	~34.5	78	3,170	1.5
Ag	W	75	1,100	~34.5	78	3,300	1.5
Zn	W	373	750	~23	44	685	5
Zn	W	75	750	~23	44	730	5
Cd	W	75	600	~23	20	340	9
Cd	W	300	600	~23	20	350	9
Ni	W	75	1,900	~46	120	8,300	1.1
Ni	W	300	1,900	~46	120	8,390	1.1
Au	W	300	1,200	~46	~180	4,160	1.4
Cd	NaCl	167	600		~135	1,208	2.2
Cd	LiF	167	600		~135	1,215	2.2
Na	CsCl	300	310	2.3	~125	500	3
Cd	Cu	300	600		~141	410	8
H$_2$O	Hg	300	73	6.8	~ 15	12	220
Zn	Glass	300	750	3.1	~ 50	1,090	3.3
Zn	Glass	685	750	3.1	~ 50	15	240
Zn	Mica	300	750			1,100	3.3
Zn	LiF	300	750			1,120	3.3
Ag	NaCl	500	1,200	6.6	110	2,800	2
Mg	Glass	500				3,000	12
Sb	Mg$_2$Sb$_2$	600				55	51
Mg$_3$Sb$_2$	Glass	770				500	5

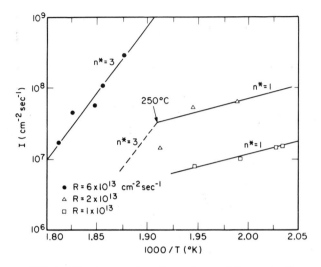

FIG. 2. The temperature dependence of the nucleation rate of Ag on cleaved NaCl for different impingement rates. (*Walton et al.* [27])

diffusion of clusters of one atom to those with three atoms. The slopes yield Q_{des} = 0.4 eV, Q_d = 0.2 eV, and ΔE_3 = 2.1 eV for critical nucleation. The occurrence of this temperature-dependent transition has been cited in support of the statistical or atomistic model of Walton and Rhodin. This is by no means an unambiguous explanation since Q_d or Q_{des} may not be constant but may rather increase [51,58] with temperature for similar systems, and this dependence can offer an alternative explanation.

Poppa [58] made careful in situ studies of the nucleation density of Bi and Ag on amorphous carbon films using a high-vacuum (5×10^{-8} Torr) electron microscope with heating stage and an in situ deposition-rate controller. Poppa found that both the capillarity and atomistic theories can be used to explain the qualitative features of his results. The quantitative results revealed the inadequacies of both theories. For example, the atomistic theory yielded too high a value of a (the jump distance) for Bi, while the nucleation rate at low temperatures did not fit the capillarity theory. The two theories yielded desorption energies that differ by a factor of 1.7. The data on the effect of the impingement flux on the saturation nucleation density were too limited to allow reasonable conclusions. Poppa's experiments represent the best effort made so far to verify the nucleation theories. Further refinements in the experiments including a vapor source of controlled temperature would help to lend more confidence to the nucleation data.

Despite some optimistic reviews in the literature [3,4,334], we must conclude that, because of both the experimental difficulties and the uncertainty of the validity of the assumed parameters in the nucleation theories, an unambiguous, comparative, quantitative test of the theories has not yet been made. The present nucleation theories are by no means complete or perfect since they are unable to explain a variety of nucleation phenomena, including the phenomenon of epitaxy (Sec. 5).

The basic concepts developed to understand the nucleation of a deposit have found useful applications in thin film technology. Some of these applications are: (1) The high mobility of adatoms on the substrate surface is the basis of the three-source method (Chap. II) for deposition of compound films of controlled stoichiometry. (2) By changing the substrate temperature, the condensation mechanism may be changed to produce amorphous and metastable structures of some materials (Sec. 4.2). (3) Preferential or enhanced condensation of vapor atoms is possible by prenucleating the substrate with a suitable material to increase the binding energy of the particular adsorbate. (4) The prenucleation technique may also be used for producing ultrathin continuous films if the adsorbate and the prenucleation material are chosen so as to wet each other completely. (5) The decoration effect may be exploited as a sensitive technique for identifying surface ledges, monoatomic steps, and polarity of the compound surface (in conjunction with etching).

(3) Condensation Centers. According to Cabrera [78], nucleation could occur preferentially on the emergent points of dislocations. Some evidence of this was obtained by Oudar and Bénard [63] in the formation of Cu_2S layers on polygonization boundaries in Cu, and by Phillips [62] on the formation of Ag_2S on stacking faults and partial dislocations in Ag films.

On the other hand, it is well established that the nuclei are randomly and isotropically distributed and their number is considerably smaller than the available atomic adsorption sites but considerably larger than the possible density of line defects on the surface of a typical single-crystal substrate used for epitaxial growth. These facts suggest

[52,79] that the nucleation process is homogeneous and that it occurs as a result of statistical fluctuations in the supersaturation. Stirland's [80] observation that nucleation centers of Au deposited on two matched surfaces of cleaved rock salt bear no correlation to each other further supports the statistical nature of the condensation process.

If defects are indeed the nucleation sites, then the observed nucleation density can be explained only by assuming nucleation to occur at point defects since only such defects have a density of $\sim 10^{10}$ to $10^{12}/\text{cm}^2$ just below the surface. The observed exponential decrease of the nucleation density with the inverse of temperature, however, runs contrary to the temperature dependence of the density of point defects. Further, it is not clear how the density of point defects could be approximately the same for a variety of substrates of different perfections.

Some unusual observations on the nature of condensation centers have been made by Distler et al. [81,350]. They have reported that the decoration of the cleavage surface of NaCl and its orienting influence can also occur through an intermediate amorphous layer (e.g., carbon less than 200 Å thick, Parlodion plastic from 100 Å to several microns thick). Similar long-range activity has been observed for cleaved mica and for single crystals of silicon and quartz. The orienting influence through the intermediate layers has been studied for PbS on mica and for polyethylene, Au, and CdS on NaCl substrates.

Distler et al. have proposed that these observations are due to the long-range effect of some active centers. Since decoration has been found by these authors to persist to much longer distances than the orienting influence, two types of active centers have been proposed, one for nucleation and the other for epitaxy. The great range of the long-range effects indicates that these cannot be associated with point defects. These centers may be the result of electrostatic, dipole, or van der Waals interaction of an aggregate of defects. The electric-field-induced changes in the growth of films discovered by Chopra [72] (discussed later) have been cited to suggest that these active centers must be charged so as to interact with the field.

The aforementioned observations are in sharp disagreement with the commonly known fact (from both high-energy and low-energy electron diffraction) that epitaxial growth on a surface is completely destroyed by the presence of even a monolayer of a polycrystalline or an amorphous deposit. It is also difficult to see how charge centers can act "at a distance" through a thick carbon film which is a reasonably good conductor at ~200 Å thickness. Further, the density of the charge centers must vary considerably from one single-crystal surface to another, but, by contrast, the observed density of nucleation centers does not vary much.

In view of the need to confirm the controversial observation of Distler et al., a careful and systematic study of epitaxial growth through intermediate amorphous layers was undertaken by Chopra.* He studied the epitaxial growth of Au, Ag, PbS, and SnS on rock salt and mica substrates held at temperatures ~100 to 300°C. The epitaxial deposits were obtained by thermal evaporation at rates ~1 to 10 Å/sec. Parts of the substrates were coated with amorphous deposits of carbon, quartz, SiO, and $BaTiO_3$ of thickness ranging from a few to 200 angstroms. While epitaxial growth was observed on the virgin parts of the single crystal substrates in all cases, *only* polycrystalline deposits were obtained on the amorphous (intermediate) regions. When a high density of steps and ledges was present on cleaved rock salt surfaces, oriented nucleation near these defects was observed, presumably due to the presence of virgin regions shadowed by these defects. The observations of Chopra strongly support the generally accepted hypothesis that epitaxial growth is an interfacial phenomenon and that the orienting influence of the substrate is governed by and confined to its surface atoms only.

(4) Condensate Temperature. Another serious question was raised by Semenoff [16], and Palatnik and coworkers (see reviews by Palatnik [84,85]) concerning the validity of the direct vapor-to-solid transformation assumed for the condensation mechanism in the nucleation theories. Based on the general electron-microscope observations that the islands in metal films of Bi, Sn, Pb, Ag, Au, Cu, Al, etc., deposited on amorphous substrates at a temperature $T_S > 2/3\ T_m$ (T_m = the bulk melting point of the vapor source) are spherical, Palatnik et al. postulated that film condensation occurs via vapor→liquid→solid transformation. For $T_S < 2/3\ T_m$, the transformation is directly vapor to solid. In some cases (such as Sb), however, the mechanism of condensation reverts to vapor→liquid→amorphous solid for $T_S < 1/3\ T_m$. A support for the

*Chopra [354].

vapor→liquid→solid condensation mechanism is provided by the observed frozen-in high-temperature metastable phases of some solids in their condensates, as expected from Ostwald's rule (discussed later). This point is further elaborated by Behrndt [86].

The significance of the $2/3\,T_m$ and $1/3\,T_m$ values was discussed by Komnik [84]. The $1/3\,T_m$ limit is essentially that below which the adatoms do not have sufficient mobility to produce ordered structure. It therefore corresponds to an annealing temperature, and the condensation mechanism may properly be denoted vapor→disordered solid. The $2/3\,T_m$ is, however, a significant parameter since it is nearly the melting point of ultrathin films of various metals. This follows from the fact that the melting point T_r of a spherical crystallite (or an island) of radius r is depressed because of the increase of vapor pressure inside the curved surface. It is given by the Thomson-Frenkel relation [87]

$$T_r = T_m \exp\left(-\frac{2\sigma V}{Lr}\right) \tag{18}$$

where σ is the solid-liquid interfacial energy, L is the latent heat of fusion of the solid, and V is the molecular volume of the solid. Equation (18) yields for small depression $\Delta T = T_m - T_r$,

$$\Delta T \approx \frac{2\sigma}{r}\frac{T_m}{L}V \tag{18a}$$

Note that Eq. (18a) can also be derived easily by equating the total thermodynamical potential of the particle to zero at T_r. That is,

$$(\Delta U - T_r \Delta S)\,\Delta V - \sigma \Delta A = 0 \tag{18b}$$

where ΔU and ΔS are the changes in the internal energy and entropy (per unit volume) during melting, and ΔV and ΔA are the changes in the volume and surface area of one of the phases during melting. Since for a bulk metal $\Delta S = \Delta U/T_m = L/T_m$, Eq. (18b) yields Eq. (18a).

Komnik pointed out that if we consider the smallest value of r as that given by the radius of the critical nucleus [Eq. (7)], then by substitution we obtain

$$\frac{T_m}{T_r} = 1 + \frac{\sigma}{\sigma_{cv}} \frac{kT_m}{L} \ln \frac{p}{p_e} \tag{18c}$$

Komnik noted that, for a large number of metals with close-packed structures, the value of kT_m/L is nearly the same and is approximately unity. Further, for such cases, the value of $\sigma/\sigma_{cv} \sim 0.1$. Thus, in order to explain the observed [88,89] experimental results that $T_m/T_r \sim 1.5$ for many metals in the earliest observable stage of growth, the required supersaturation p/p_e for nucleation to occur must be $\sim 10^2$. This is a reasonable theoretical value, but it is much smaller than the observed values of critical supersaturation.

The predictions of Eq. (18) have not been tested for crystallites of definite sizes and shapes. However, the depression of the melting point of Bi, Pb, Sn, and Ag films has been measured [88,89] as a function of increasing average film thickness (and hence increasing grain size). The results are in reasonable accord with Eq. (18).

The thickness dependence of the melting point of Ag and Cu films observed by Gladkich et al. [89] is shown in Fig. 3 and is compared with Eq. (18). The finite melting point at negligible thicknesses obtained by extrapolation must be ascribed to the finite island size of the thinnest observable film.

On chemical adsorption, the liberated heat of condensation (~ 1 eV) may raise the temperature of small nuclei considerably. This rise obviously depends on the various heat-loss factors, e.g., thermal conductivity and emissivity. Using bulk parameters for heat loss* (an

*A recent theoretical paper of interest in this connection is due to Harrington [*J. Appl. Phys.,* **39**:3699 (1968)]. Using kinetic theory, he showed that the thermal conductivity decreases near the surface of a metal and becomes zero at the surface if there is no heat flow across the surface. When the metal surface is exposed to blackbody radiation, the absorption of radiation occurs within a depth equal to the mean free path of electrons in the metal and results in an exponential temperature distribution. The effective thermal conductivity in the surface region may be reduced by a factor $\sim 10^5$ to 10^6 due to the small value of the loss of energy of electron per elastic collision.

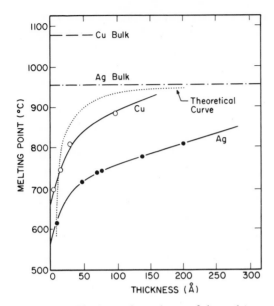

FIG. 3. Thickness dependence of the melting point of Cu and Ag films deposited on a carbon film substrate at 200°C inside an electron-diffraction camera at 10^{-9} Torr. The dotted curve is calculated from Eq. (18) for Ag, using $\sigma_{Ag} = 126$ ergs/cm^2. (*Gladkich et al.* [89])

unrealistic assumption for ultrathin films), Gafner [90] showed that the temperature rises to a high value in the form of a transient ($\sim10^{-9}$ sec) and decays rapidly to a small steady-state value. Yoda [91] measured the spectral temperature of evaporated Ag films by measuring the surface radiation. He found a temperature rise in ultrathin (~100 Å) films of up to 400°C for a deposition rate of ~3 Å/sec. With increasing film thickness, the temperature rise decreased.

The temperature rise during condensation may be high enough to melt ultrathin deposits of some metals, and thus condensation would occur via vapor→liquid→solid transformation. The observed [92] increasing sphericity of the islands of Au and Ag films on thin mica and NaCl with increasing condensation rates is a consequence of the increasing deposit temperature.

2. GROWTH PROCESSES

2.1 General Description

The preceding section discussed nucleation and the formation of three-dimensional islands (agglomeration). This section deals with the subsequent growth and intergrowth of the islands (agglomerated structure) to form a continuous film and the influence of various deposition parameters and other physical agencies on this process.

The growth sequence of a film described below was originally deduced by Andrade [93] in 1938 from the observed optical-transmission behavior of Ag films. This deduction is in remarkable agreement with the electron-microscope observations first made by Uyeda [94] and later in detail by Levinstein [95] and others [96,97,49-58]. Electron-microscope studies may be performed in situ, as first used by Bassett [50]. This is most convenient since rapid growth changes can be followed only by photographing electron micrographs on the fluorescent screen of the microscope from outside the vacuum using a movie camera. Moving pictures with frame speeds of up to 32 frames/sec, and sequential photographs of the growth process have been obtained by several workers. Poor vacuum inside the microscope, uncertain effects due to contamination by decomposed hydrocarbons and the electron beam, and difficulty of controlling deposition parameters hamper this technique. The best arrangement to date is that of Poppa [57,58], who employed ultrahigh vacuum (5×10^{-8} Torr) for the specimen chamber and a rate monitor for film deposition. Fortunately, however, the growth sequence observed in an in situ experiment is not significantly different from that obtained from micrographs of different growth stages of films prepared in a separate system and detached from the substrate for electron microscopy. This second technique has therefore found widespread use since it is possible not only to prepare films under controlled conditions but also to employ other physical measurements such as electrical resistivity to supplement the electron-microscope data (see Chopra and coworkers [98,99]). A summary of the results of in situ studies of Bassett [50] and Pashley and coworkers [54,100] and a comprehensive review of the whole subject is given by Pashley [100].

The characteristic sequential growth stages are: (1) Randomly distributed, three-dimensional nuclei are first formed and rapidly approach a saturation density with a small amount of deposit. These nuclei then grow to form observable islands whose shapes are determined by interfacial energies and deposition conditions. The growth is diffusion-controlled; that is, adatoms and subcritical clusters diffuse over the substrate surface and are captured by the stable islands. (2) As islands increase their size by further deposition and come closer to each other, the larger ones appear to grow by coalescence of the smaller ones. Island density decreases monotonically at a rate determined by the deposition conditions. This stage, arbitrarily denoted here as coalescence I, involves considerable mass transfer by diffusion between the islands. Although not completely unambiguously verified, physical movement of the small islands (which may not be observable by electron microscope)

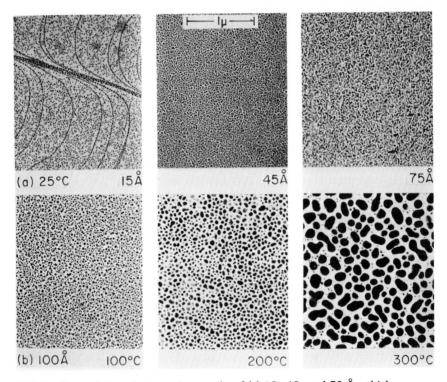

FIG. 4. Transmission electron micrographs of (*a*) 15-, 45-, and 75-Å – thick argon-sputtered Au films deposited on NaCl at 25°C at a rate of ~1 Å/sec. (*b*) 100-Å-thick Au films on NaCl at 100, 200, and 300°C evaporated at a rate of ~1 Å/sec. (*Chopra, unpublished*)

over the surface at elevated temperatures is very likely. The disappearance of small islands is generally quite rapid (in less than a fraction of a second). If a 10-Å-radius island ($\sim 10^3$ atoms) coalesces into a larger island in 0.1 sec with a contact area of $\sim 10^{-14}$ cm^2, the mass transport takes place at $\sim 10^{18}$ atoms/(cm^2)(sec). (3) When the island distribution reaches a "critical" state, a rapid large-scale coalescence of the islands results in a connected network structure, and the islands are flattened to increase surface coverage. This process, coalescence II, is very rapid initially but slows down considerably on formation of the network. The network contains a large number of empty channels, which in some cases of epitaxial growth are holes of crystallographic profiles. Occasionally, they are long, uniformly wide, homogeneously distributed, and possess approximately matching curvatures over small areas (see Fig. 4). (4) The final stage of growth is a slow process of filling the empty channels which requires a considerable amount of deposit. Wherever large surface areas are vacated by coalescence to form a composite structure, secondary nucleation occurs. These nuclei generally grow and coalesce very slowly with further deposition. This effect is particularly marked when the secondary nuclei are completely surrounded by the deposited material.

Electron micrographs of sputtered Au films of three different thicknesses deposited on NaCl at 25°C, and of evaporated 100-Å-thick Au films deposited at three temperatures are shown in Fig. 4(a) and (b), respectively, to illustrate the growth sequence.

It must be emphasized that the above sequence is qualitatively common to other types of vapor-deposited films prepared by a variety of techniques. The kinetics of each stage may, however, vary markedly, depending on the deposition parameters and the deposit-substrate combination. These differences may be described by such qualitative terms as agglomeration and mobility. The increasing tendency to form larger islands (and hence smaller in density) is termed "increasing agglomeration" and is the result of higher surface "mobility" of adatoms and subcritical and critical clusters. The term "mobility" cannot, unfortunately, be defined quantitatively since it is influenced by a large number of physical parameters. A reasonable measure of the mobility is provided by the rate of change of the island density or the interisland spacing with respect to a physical variable such as film thickness or substrate temperature. Thus, higher mobility (and hence higher agglomeration) is obtained for films of low-melting-point materials, on smooth and inert substrates, increasing substrate temperatures, increasing rate and kinetic energy of the deposit vapor, etc. The high surface-mobility deposits of Au and Ag, for example, condensed on Pyrex glass slides at 25°C become electrically conducting at an average thickness (given by the weight per cm^2) of ~50 to 60 Å. On the other hand, films of W, Ta, Ge, Si, various metal oxides, etc., deposited on several substrates at 25°C exhibit agglomeration on a very fine scale and reach continuity at an average thickness of several angstroms. The mobility in these cases is significantly enhanced at elevated temperatures of the substrate, with a corresponding effect on the growth sequence.

2.2 Liquid-like Coalescence

(1) Experimental Observations. The growth sequence described above proceeds quite rapidly for high-adatom-mobility cases and involves significant changes in the shape of the islands and the network as a result of considerable mass transfer during coalescence stages I and II. An in situ continuous experiment gives the general impression of a "liquid-like" behavior, and the growth process is therefore termed "liquid-like coalescence." The islands are certainly not liquid since they yield normal single-crystal or crystalline diffraction patterns at all stages of the growth, indicating that the coalescence phenomena are of solid-state origin.

The coalescence phenomena have a profound effect on the structure and properties of the resultant film since recrystallization, grain growth, orientation changes, incorporation and removal of defects, etc., occur as a consequence of coalescence. The dynamics of the coalescence phenomena are not completely understood at present. Pashley et al. [54,100] discussed coalescence in terms of "sintering" of two particles in contact with each other. Before we discuss their model, some other controversial observations on the coalescence II stage should be noted. (1) Whatever the coalescence mechanism may be, its general qualitative features are not influenced drastically by poor vacuum and substrate contamination. (2) Since growth sequences are essentially similar for both in situ and detached films, coalescence is not caused by the electron beam, as suggested by Dove [101]. The coalescence may, however, be considerably modified by the presence of electrostatic charges and applied electric field (discussed later). (3) Once initiated, the coalescence II stage requires only a small amount of additional material to complete the network stage. The rapidity of the coalescence of discrete islands is illustrated in Fig. 5 by the successive micrographs of Ag films, each taken with the addition of about 5 Å (an average amount) of the deposit at the critical stage, which was predetermined from the electrical-resistance curves. Examples of the electrical-resistance curves for Ag films displayed

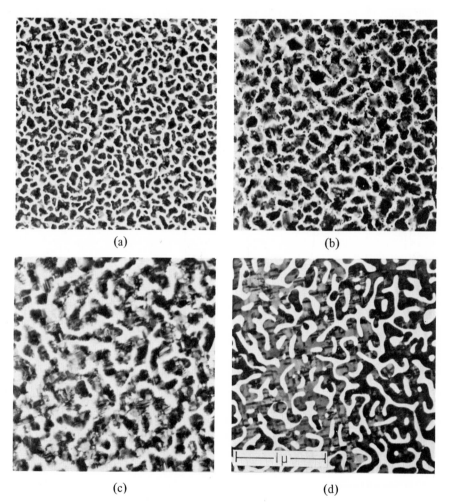

(a)   (b)

(c)   (d)

FIG. 5.   A sequence (*a*) to (*d*) of electron micrographs of Ag films of 100, 105, 110, and 115 Å average thickness, respectively, evaporated onto NaCl at 100°C, demonstrating the rapidity of the coalescence process leading to a network structure. (*Chopra, unpublished*)

in Fig. 6(a) show the sharpness of the coalescence transition from a discontinuous to an electrically continuous structure. At 100°C, the resistance falls by several orders of magnitude for a nominal addition of ~10 Å of the deposit beyond the critical thickness. Both the width of the transition and the critical thickness t_c (Fig. 6) increase with substrate temperature. Note that, once initiated, the electrical-coalescence curve continues to fall [the dotted curve in Fig. 6(a)] at a much lower rate after the vapor supply is removed. This process may last as long as several seconds for noble-metal films at both room temperature and 77°K. (4) The formation of pronounced bridges between discrete islands preceding coalescence was reported by Adamsky and LeBlanc [49], Poppa [57], and Chopra [52]. Pashley et al. [54], however, failed to detect these bridges and suggested that if contamination (e.g., adsorbed hydrocarbon layers) formed a skin over the islands, condensation on this skin may form bridges on the deposit material. This explanation is questionable

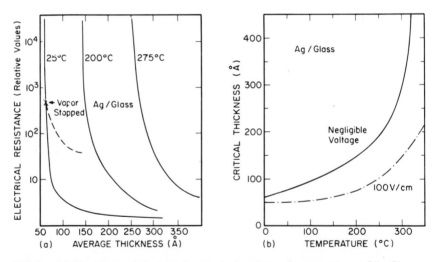

FIG. 6. (a) Electrical-resistance behavior during the coalescence stage of Ag films evaporated at a rate of 1 to 5 Å/sec on glass substrates held at temperatures of 25, 200, and 275°C. The continuation of the fall of resistance (continuation of the coalescence process) with time on intercepting the vapor source is indicated by the dotted part of the 25°C curve. (b) The temperature dependence of the critical thickness (at which electrical continuity is attained) for evaporated Ag films on glass. The dotted curve is for films prepared with a field of ~100 V/cm applied laterally during film deposition. (*Taken partly from Refs. 72 and 99*)

since bridge formation has been observed in at least as clean a system as that used by Pashley et al., and further, the bridge formation is found [52] to be more pronounced in sputtered Ag films on mica at elevated temperature ($\sim$350°C) where contamination should be reduced. The bridge formation is probably dependent on the nature of the film-substrate interface and other factors such as electrostatic charges present on the islands. It should be pointed out that bridges or necks form quite extensively across the channels in a network film, as observed by Bassett [50] and confirmed by Pashley et al.

Bassett [50] observed small translatory movements of Ag islands on graphite as well as amorphous carbon. Pashley et al. [54] reported no evidence of such motion for Ag and Au islands on MoS_2 substrate. However, both Bassett and Pashley et al. observed that islands rotate slightly (angles of up to 10° are possible) during growth, particularly during coalescence when two slightly misaligned nuclei rotate into perfect alignment. Pashley et al. assert that islands are not mobile and that all apparent movement is due to the liquid-like coalescence. Although it is reasonable to assume that large stable islands do not move except as a result of spreading and electron-beam irradiation, [72,73,102] some movements of small, nearly critical size islands are essential to explain the observed rapid decrease in the density of discrete islands in the early growth stages and the rapid interisland coalescence at the later stage. This may be a result of factors such as momentum transfer on impingement by the incident energetic vapor, and attractive electrostatic forces between islands possessing charges. The experimental observations that coalescence can be induced and accelerated by a laterally applied electric field [72] during deposition, by irradiation of the substrate by electrostatic charges [73-75], and by ultrasonic agitation [103] of the substrate suggest the occurrence of physical movements of small islands.

(2) Coalescence Model. The various coalescence effects have been interpreted by Pashley et al. [54] in terms of the surface mobility of the deposit atoms over the surfaces of the deposit islands. Coalescence is believed to occur *only* when islands touch each other physically, the mechanism being analogous to that of sintering [104,105] of two spheres. When two spheres of radii r touch each other at a point, the curvature of radius R of the neck so formed at the point of contact produces a driving force $2\sigma/R$ (σ is the surface energy) to transfer the material from the spheres to fill up the neck. The transport of material may take place by volume or surface diffusion. The growth of the radius $x (< 0.3r)$ of the neck in time t is given by the relation

$$\frac{x^n}{r^m} = A(T) t \qquad (19)$$

where $n = 5$ and $m = 2$ for volume diffusion, and $n = 7$ and $m = 3$ for surface diffusion (the choice of the values of n and m for the two cases is *not* unambiguous [105]), T is the temperature, and $A(T)$ is a function which contains transport parameters. For $T = 400°K$, and assuming the bulk value of $A(T)$ for Au spheres, the calculated time required to obtain a neck radius $x = 0.1r$ by surface diffusion is ~10^{-7} and 10^{-3} sec for $r = 100$ and $1,000$ Å respectively. Similarly, for volume diffusion, the respective times are 2×10^{-3} and 2.0 sec. These authors conclude that, since experimentally observed times for 1,000-Å islands are about a fraction of a second, surface diffusion is the important transfer mechanism.

The general behavior of necks and channels has been explained by Pashley et al. on the basis of mass transport resulting from the tendency to minimize the surface energy. The rapid formation of necks and then a slow growth after reaching a critical size follows from Eq. (19) since the growth rate dx/dt is proportional to $1/t^{0.85}$ for surface diffusion. A neck once formed continues to grow up to several seconds even if the vapor supply is stopped (by means of a mechanical shutter) because of the effort to minimize curvature. Resumption of vapor deposition causes the necks to grow just as though there had been no interruption of the vapor supply. This observation was explained by assuming that the mobile vapor atoms migrate to deposit preferentially in the regions of high negative curvature, such as any necks. Thus, the authors suggest that the formation and initial growth of a neck occur predominantly by transport of previously deposited material, but the latter stages of growth occur predominantly as a result of the preferential deposition of freshly arrived material at positions of high curvature. The same general arguments are applied to the case of channel filling except that the channel-filling process should not slow down as rapidly as for the neck formation since the radius of curvature (and hence the driving force) does not change very much as the channel bridge widens.

The surface-energy and diffusion-controlled mass-transport mechanisms undoubtedly play important roles during the liquid-like coalescence of islands in physical contact with each other. However, some other driving as well as restraining forces, such as may arise from the presence of electrostatic charges on the islands, can affect the coalescence stage preceding the sintering process. This is further supported by the fact that the sintering mechanism is not able to explain the following observations: (1) the observed liquid-like coalescence of Ag [106,269] on NaCl, mica, and MgO, of Au [108] on NaCl, and alkali metals on metal substrates [109] at temperatures (~77°K) low enough for thermal diffusion to be negligible; (2) the slower rate of filling of channels at higher substrate temperatures; (3) the widely varying stabilities of similarly irregular shaped necks and channels [see Fig. 5(d)] with very high curvatures at some points; (4) a large range of time constants for filling of the visually similar types of necks and channels; (5) the existence [61] of strikingly uniformly wide channels [see Fig. 26(c)] and thus correlation of the curvatures of neighboring channels; (6) the high stability of the secondary nuclei against coalescence; and (7) the continuation of the coalescence process for several seconds at 77°K and further the fact that the coalescence rate can be accelerated by an applied electric field.

2.3 Influence of Deposition Parameters

(1) General Aspects. The influence of the deposition parameters on film growth may be understood in terms of their effects on the sticking coefficient, the nucleation density, and the surface mobility of adatoms. Agglomeration of a film increases with increasing surface mobility and decreasing nucleation density. An increasing agglomeration implies that the film will reach continuity at a higher average thickness and will have a larger grain size and smaller number of frozen-in structural defects.

The initial saturation nucleation density under thermodynamical-equilibrium conditions is determined by the substrate-vapor system and is not expected to depend on the deposition rate. This is, however, not valid when equilibrium conditions are not obtained as in the case of a very fast deposition rate (arrival rate higher than the diffusion rate), the presence of electrostatic charges on the vapor atom and/or on the surface, structural defects on the surface, the penetration of the energetic vapor atoms into the surface, and the resulting creation of surface defects, etc. All these factors bring about an increase in the initial nucleation density and hence lesser subsequent agglomeration. Adsorbed impurities also influence the nucleation density. For instance, by using prenucleation centers, such as Bi_2O_3 on glass, the nucleation density of Au films is considerably increased and thus these films attain electrical continuity at a thickness ~ 20 Å as compared with ~ 60 Å on clean glass. This method is therefore useful in obtaining ultrathin, continuous film.

An important parameter which determines agglomeration and film growth is the surface migration or mobility of an adatom. If the mobility is random in direction, the adatoms and the embryos will execute random walk on the surface until they are either reevaporated or chemisorbed on the surface. Under equilibrium conditions, discrete islands will be formed which conform to a distribution such that the average interisland distance corresponds to the mean diffusion distance of the random-walk process. The mobility increases with decreasing activation energy of surface diffusion, and with increasing effective temperature or kinetic energy of the adatom during migration, and also with the substrate temperature and surface smoothness. The effect of the kinetic energy of an adatom on film growth is generally neglected in theoretical treatments. Chopra and Randlett [103] considered the role of the kinetic energy as a diffusion mechanism. Assuming a lateral velocity of $\sim 10^5$ cm/sec (say for Au vapor atom at $1,000°$K) and no frictional losses for movement, an adatom will impart on impingement a velocity of ~ 1 cm/sec to a 100-Å embryo containing 10^5 atoms. The embryo can move a considerable distance with the imparted momentum until impinged upon by another vapor atom or stopped by another island. Because of the random impingement of vapor atoms, the momentum-transfer-induced diffusion is random and cannot be distinguished from that due to thermal activation.

The preceding description of the mobility process suggests higher agglomeration in a postnucleation growth stage from (1) higher deposition temperature; (2) higher kinetic energy of the vapor atoms, which for thermal evaporation implies higher evaporation rate; and (3) an increasing angle of incidence of the vapor. These conclusions assume a constant condensation coefficient and an atomically smooth substrate.

The degree of agglomeration is easily observed by electron micros-copy. The average critical thickness t_c at which electrical continuity of the film occurs rapidly is also a good measure of agglomeration. Furthermore, the slope of the resistance curve of the coalescence II stage determines the rate of coalescence. The few reported experimental results present no consistent picture of the dependence of the growth stage on deposition parameters, primarily because of the lack of controlled conditions. However, the detailed electrical and electron-microscope investigations of the growth of Au and Ag films by Chopra and Randlett [99], using quartz-crystal oscillators to control and monitor both the deposition and the source-evaporation rates, are in agreement with the preceding conclusions. A further discussion of the results is given below.

With increasing surface smoothness of an inert substrate, the t_c increases, as illustrated by the following approximate values for Au films deposited at 25°C:

Substrate t_c, Å	Bi_2O_3 25	Glass 40	Mica 50	Rock salt 60

The island density N or the interisland spacing d_I (radius of the capture zone) has been found by various workers [27,52,58,110] to follow an $\exp(-1/T)$ dependence. A plot of $\log d_I$ vs. $1/T$ for evaporated and sputtered Ag/mica films is shown in Fig. 7. The slopes of the straight lines so obtained yield Q_d, if we assume that Eq. (4) is valid for the island-growth process and that τ_s is constant. The large increase of Q_d (from 0.15 to 0.9 eV for evaporated films) beyond ~520°K is not clearly understood, but it may correspond to the activation energy for surface diffusion of larger clusters of adatoms rather than individual adatoms. This is reasonable since the minimum size of the stable clusters is expected to increase with temperature. Poppa [58] also observed a temperature-dependent slope of the $\log N$ vs. $1/T$ curves for Ag films and ascribed it to changes in τ_s. We may note that it is not possible in these experiments to distinguish between the contributions from changes in Q_d and τ_s.

The critical thickness t_c increases [111,112,52,99] with increasing substrate temperature [Fig. 6(b)] and follows an $e^{-Q/kT}$ dependence.

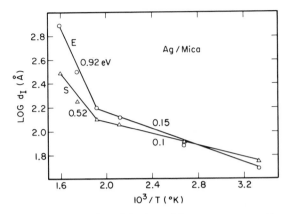

FIG. 7. A plot of the log of the average separation distance between the islands of 100-Å-thick argon-sputtered (S) and evaporated (E) Ag/mica films vs. the reciprocal of the substrate temperature. Rate of deposition ~1 Å/sec. (*Chopra* [52])

The activation energy Q of growth by coalescence was found by Chopra and Randlett for Ag on glass to be 0.26 and 0.85 eV for temperatures below and above 450°K, respectively.

Activation energy for surface migration of adatoms provided by the substrate temperature may also be obtained from forced vibrations of the lattice. In order to effect the diffusion of adatoms during their short stay time, we need to employ a hypersonic source of vibrations. But, the channel filling during coalescence II stage is sufficiently slow (for noble metals) to be influenced by lattice vibrations in the kilo- to megacycle frequency range. This effect was observed by Chopra and Randlett [103] in Ag and Au films deposited on NaCl coupled to a quartz crystal oscillating at 6 Mc/sec. Comparison of electron micrographs of films deposited on thin slices of NaCl with and without ultrasonic agitation established that the substrate agitation results in enhancement of agglomeration and increased preferential crystallographic orientation of the islands. In addition to substituting for the substrate temperature, ultrasonic agitation may allow influencing of the coalescence process along definite crystallographic directions.

Sennett and Scott [96] observed increasing agglomeration with decreasing deposition rate of Ag films on Formvar. Campbell et al. [113] observed a higher concentration of the islands in LiF and Au films deposited on carbon with increasing deposition rate. Neither of these studies attempted to keep the evaporation rate (vapor-source temperature) constant. Chopra and Randlett [99] employed arrangements to control both the rate of source evaporation and the rate of film deposition and observed the agglomeration (and, therefore, the critical thickness) in Au and Ag films deposited onto glass and NaCl to increase with the increasing deposition rate. At very high rates ($>$ 20 Å/sec), the trend reversed, which may be ascribed to the inequilibrium conditions resulting from the fact that the adatoms form stable clusters by collisions before they are able to diffuse to equilibrium positions. A reversal in the agglomeration behavior can also occur at very low deposition rates in poor vacuum conditions because of the dominant influence of the residual gas contamination effect.

(2) Kinetic-energy Effect. By using a mechanical velocity selector, Levinstein [95] studied the deposition of Au and Ag vapor atoms of well-defined kinetic energy. Although he observed no effect on the structure and growth of films, the result is questionable since the deposition rate was not held constant. The effect of higher kinetic energy in increasing agglomeration and inducing oriented growth at relatively lower temperatures is established by various studies of Chopra et al. [52,99,106]. The average temperatures for epitaxial growth of sputtered Ag [52,106], Au [29,114], and reactively sputtered oxides of Al and Ta [115] are known to be considerably lower than those of films of the same materials formed by evaporation. That the observed lowering of the epitaxial temperature is due completely to the vapor-atom kinetic energy being equivalent to a higher substrate temperature is not clear since a similar effect may also result from the desorption and cleaning of the surface by the energetic atoms. The situation is further complicated by the influence of charged particles present in the sputtered species. The improvement of epitaxial growth of low-pressure ($\sim$1 μ) sputtered Ag film at $\sim$77°K with an increasing number of the average-kinetic-energy vapor atoms suggests the direct influence of the kinetic energy.

The more rapid rate of decrease of the number of islands by coalescence with deposit thickness (Fig. 8) for sputtered films as compared with evaporated films is a convincing case for the role of the kinetic energy in enhancing agglomeration. The higher agglomeration for sputtered films is also verified by electron micrographs. As the deposit thickness increases, the density of sputtered islands approaches a constant value, and thereafter, the islands flatten to obtain earlier continuity of a sputtered film as compared with an evaporated film. The flattening of larger islands may be due to the electrostatic charges (discussed later), which are significantly more abundant in sputtered than in evaporated films. If sputtered atoms create surface defects on impingement, thereby increasing nucleation density, the above discussion does not hold since agglomeration will then be considerably reduced.

The evaporation rate of a thermal source increases very rapidly with a small change in the source temperature, thereby allowing only a limited range of variation of the kinetic energy of vapor atoms. The kinetic-energy effect is, however, observable by electron microscopy, as verified for Ag and Au films by Chopra and Randlett [99].

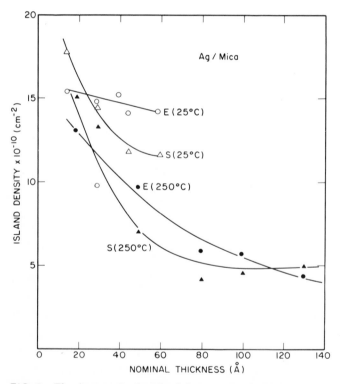

FIG. 8. The decrease in the island density with further deposition of evaporated (E) and sputtered (S) Ag films on mica at 25 and 250°C. (*Chopra* [52])

Further evidence for the kinetic-energy effect is clearly provided by the observed [116-119] penetration and formation of compounds and alloys at the film-substrate interface by the energetic sputtered and evaporated atoms at temperatures (as low as 20°K) low enough for thermal diffusion to be negligible. The observed [120] decrease in the grain size of Au films with increasing pressure of the inert gas (Ar) present during evaporation is also indicative of the role of the kinetic energy since increasing numbers of collisions with the gas atoms decrease the energy and assist rapid thermal equilibration of the vapor atoms with the substrate to produce a fine-grained structure.

(3) Oblique Deposition. Oblique impingement, that is, vapor deposition at a nonnormal angle of incidence, should increase the velocity component of adatoms migrating on the surface and thus yield high mobility effects. Electron-microscope studies of the growth of evaporated Au and Ag films by Chopra and Randlett showed increasing

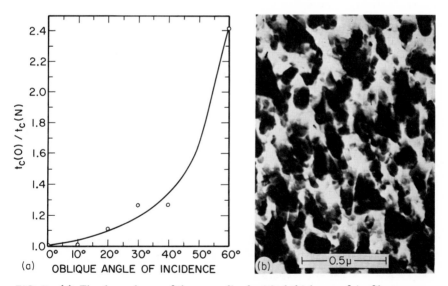

(a) OBLIQUE ANGLE OF INCIDENCE

(b) ⊢——0.5μ——⊣

FIG. 9. (*a*) The dependence of the normalized critical thickness of Ag films, evaporated on glass at 25°C and at a rate of ~1 Å/sec, on the angle of incidence of the impinging vapor. O stands for oblique and N for normal (to the substrate) incidence. (*b*) Electron-transmission micrograph showing anisotropic, columnar growth of a 1,000-Å-thick Al film deposited on NaCl at 25°C and at an angle of 80° away from the normal to the substrate. (*Chopra and Randlett* [99])

agglomeration with the angle of incidence. The early growth and distribution of islands was found to be isotropic in the film plane for angles of incidence of up to 80°. With increasing size of the islands, the self-shadowing becomes pronounced, and columnar growth (elongated growth normal to the substrate) in the direction of the incident vapor occurs. Consequently, the critical thickness for electrical continuity of the films increases [Fig. 9(a)] very rapidly at higher angles of incidence. The surface area of such films increases [121] drastically beyond a certain angle of incidence and is consistent with the picture of the columnar growth. Further, the observed anisotropy in mechanical stress [122] and magnetic properties [123], and poor reflectance and enhanced optical absorption [124,125] of obliquely deposited films also suggest anisotropic growth.

It is clear that the anisotropic, columnar growth is controlled by the vapor supply (and hence direction) and the adatom surface mobility. The higher the initial mobility, the less pronounced is the anisotropic growth. Gold and silver films deposited at elevated temperatures therefore show little anisotropic growth even at large angles of incidence, while the low-mobility Al adatoms condensed on glass and rock salt at room temperature show marked columnar growth [99,124,126] [Fig. 9(b)].

Kooy and Nieuwenhuizen [126] developed a replica technique for electron microscopy of the cross section of thick ($\sim 1 \mu$) Al and Si films deposited on glass. The micrographs showed the expected columnar growth on oblique incidence and further revealed columnar grain growth in Al films. The grain size increased with thickness, with values up to 10^2 times larger than for normal incidence. The elimination of a large number of grain boundaries to produce a single, large grain may be the result of enhanced agglomeration at oblique incidence (see also Sec. 4.5).

Wade and Silcox [127] studied small-angle diffuse electron scattering from evaporated Pd, Ni, and Permalloy films. The interpretation of these results suggested that films deposited at normal incidence at slow rates of ~ 1 Å/sec consist of separated parallel rods with their long axis perpendicular to the film. At faster rates, ~ 6 Å/sec, a more continuous film was obtained, presumably because of increased mobility. With oblique incidence, the long axis of the rods was inclined to the normal of the substrate toward the direction of the beam, but not necessarily coincident with it.

(4) Electrostatic Effects. We have already discussed the kinetics of the coalescence II stage. Chopra [128,72] discovered that a dc electric field of ~100 V/cm applied laterally during the deposition of high-adatom-mobility metal films induces coalescence II at a lower average film thickness than is normally observed. The transmission electron micrographs and diffraction patterns of Ag films deposited on NaCl with and without the laterally applied field are shown in Fig. 10. The applied field appears to flatten the discrete islands, increase their surface area, and force them to coalesce. This results in a well-oriented crystallographic network with parallel orientation to the NaCl surface, which is in contrast to partial orientation obtained under the deposition conditions employed for the films without the field. The applied field reduces the critical thickness of the film, the reduction being very marked at higher deposition temperatures, as shown by the dotted curve of Fig. 6.

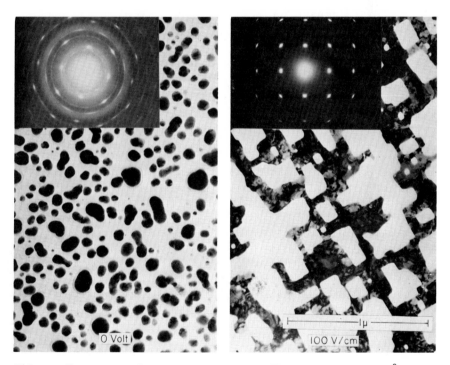

FIG. 10. Transmission electron micrographs and diffraction patterns of 50-Å Ag films on NaCl at 200°C with 100 V/cm and without lateral field applied during deposition. (*Chopra* [128])

The increasing field enlarges the surface area of the islands up to a limit beyond which degradation of the film results because of electrical arcing. The field effect is primarily electrostatic (negligible current) in nature since the field is effective only just prior to coalescence (discontinuous film) and it may be removed after coalescence sets in. The observed effect cannot be due to Joule heating, since this would lead to agglomeration and hence increased discontinuity. Similar results may be obtained by an ac or a dc field with or without direct electrode connection to the film. The growth behavior changes considerably by Joule heating if excessive current is allowed to pass through the film during its network stage.

The electrical resistance of the continuous film prepared with applied field is, in general, lower than that of the film without the field. As the field is increased, the resistivity decreases to approach a constant value close to the bulk value. The decrease is more marked at higher substrate temperatures. The reduction in the resistivity is considered to be due to the decrease of frozen-in structural defects in the induced coalescence case. This conclusion is best supported by the observed increase in the resistivity ratio $(\rho_{293}/\rho_{4.2°K})$ of epitaxial Ag films prepared with applied field. Resistivity-ratio values as high as 1,200 have been obtained for ~10-μ-thick epitaxial Ag films, as compared with ~130 for the high-purity starting bulk material.

Other workers [124,125,352] verified the field effect. Electric fields as small as ~10 V/cm were found to affect the coalescence stage of In films [129]. An interesting consequence of the applied field is that significant changes in the structure and growth behavior of oriented films can be obtained [67,130-134] (Sec. 5.1). Note that if the islands in a discontinuous film were electrically charged to varying degrees by irradiation or by contact potential at the heterogeneous substrate interface, strong, random electric fields between islands will exist which should influence the growth of films in a similar way.

Chopra explained the field effect on the basis of the interaction of the field with the electrically charged islands, the charges being derived from the ionized-vapor material and/or the potential at the substrate interfaces. The presence of a charge q on a spherical particle of radius r increases the free energy, which now becomes the sum of surface energy $(4\pi r^2 \sigma$; σ is surface tension) and the electrostatic energy $(q^2/8\pi\epsilon_0 r$, where

ϵ_0 is the permittivity constant). The increase in the total energy would be accommodated by an increase in the surface area; i.e., the sphere would change to an oblate spheroid, the exact shape being determined by the balance of various free energies. On further increase, the particle may break up. This sequence can be easily demonstrated by charging a drop of mercury. For $r = 100$ Å, $q = 10^3$ unit charges, the electrostatic contribution is nearly the same as the surface-tension energy, resulting in considerable flattening of islands and enhanced surface diffusion. If q is proportional to the number of atoms in the islands, flattening will become more marked with larger islands.

Additionally, we must consider the electrostatic forces between two islands of radii r_1 and r_2 carrying charges q_1 and q_2, and separated by a distance d. Taking image forces into consideration, the net force is given by

$$F = \frac{q_1 q_2}{4\pi\epsilon_0 d^2} - \frac{q_1^2 r_2 d}{4\pi\epsilon_0 \left(r_1^2 - d^2\right)^2} - \frac{q_2^2 r_1^2 d}{4\pi\epsilon_0 \left(r_2^2 - d^2\right)^2} + \cdots \tag{20}$$

Thus, if either charge is much larger than the other, or the radii are nearly equal to the separation distance, the attractive image forces dominate irrespective of the sign of charges [101]. If $r_1 = 100$ Å, $q_1 = 100$ unit charges, $r_2 = 20$ Å, $q_2 = $ a unit charge, and $d = 200$ Å, the net attractive force is $\approx 10^{-4}$ dynes and the shear stress on the small particle is $\approx 10^{10}$ dynes/cm^2 = 10^4 atm, which is large enough to move the small particle to coalesce with the larger one. It must be noted that, for equally charged particles, the repulsive forces are also of the same magnitude, which would prevent coalescence.

If no electric field is applied, the islands have an equilibrium distribution. This equilibrium is disturbed by the redistribution or transfer of charges due to an applied field. The resultant gradient of charges may initiate coalescence between two islands which would then set up a continuing inequilibrium situation so that a rapid coalescence retarded by repulsive coulomb forces would occur throughout the film.

A similar effect should occur by spatially nonuniform irradiation of the islands with charged particles. This was observed by Chopra with epitaxial Pb films by using the electron beam of the electron microscope for both imaging and irradiation. In order to observe the slow effects of charging without seriously affecting the imaging performance of the beam, the detached films were either slightly oxidized or floated off with traces of rock salt. Lead was chosen because of its high thermal mobility at relatively low temperatures. On irradiation, effects of large-scale movement of small islands (<100 Å) over distances up to 1,000 Å, coalescence as well as break-up of islands, flattening of islands, and continuous adjustments of the crystallographic profile of islands were observed and recorded on movie film. Although the observed process appears liquid-like, the film certainly maintains its single-crystal structure as observed by the diffraction patterns. Further, the dark-field electron microscopy established the crystallographic continuity on coalescence of two islands. Although elevated film temperature amplifies the movements observed, the effect cannot be attributed to the melting of the film. A deliberate melting of the film produces spherical islands and all movement ceases. The electron-beam-induced movements of islands have also been observed by Pócza [102].

Electron irradiation of the substrate before or during deposition produces coalescence of islands at an earlier stage and affects the further growth of films [74-76]. Since condensation is facilitated by the presence of charges (either sign) on the substrate, an increase in the saturation nucleation density is expected. Whether earlier coalescence is the result of the increased nucleation density or a combination of nucleation and electrostatic effects described above is not established at present because of the limited available data.

The author† has found that the uv illumination of the substrate during deposition of metal films influences the growth stages. Knight and Jha [335] studied the illumination effect in detail and established that an effective higher-substrate temperature-like growth occurs when the wavelength of the incident light corresponds to the absorption band of the metal. The effect has been attributed to the interaction between the dipoles created by absorption.

3. SOME ASPECTS OF THE PHYSICAL STRUCTURE OF FILMS

(1) Crystallite Size. The crystallite size of a vapor-deposited polycrystalline film increases with the increasing surface mobility of adatoms

† Unpublished.

and clusters during deposition. Therefore, the crystallite size is expected to increase with increasing substrate (deposition) temperature, rate of deposition, velocity of the vapor atoms parallel to the surface, inertness and smoothness of the surface, etc. The variations of the crystallite size with some of these parameters are shown schematically in Fig. 11. It must be recognized that (1) the effective substrate temperature for deposition is a relative value with respect to the melting point of the vapor source; (2) contamination and residual gases can influence the crystallite size considerably; (3) at very high deposition, the adatoms interact strongly with each other to become chemisorbed with little surface migration, thereby resulting in a fine-grained deposit; and (4) the higher the surface mobility, the more marked is the film-thickness dependence of the crystallite size.

Annealing at temperatures higher than the deposition temperature increases the crystallite size, but the growth effect is significantly different from that obtained by using the same temperature during

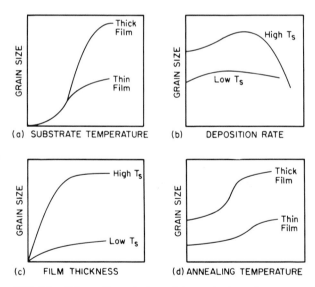

FIG. 11. Schematic variation of the grain size of pure metal films as a function of (a) substrate temperature during deposition; (b) rate of deposition; (c) film thickness; and (d) annealing temperature subsequent to deposition.

deposition [135]. The difference arises from the high-activation-energy process of thermal diffusion of surface atoms of a condensate in the former case as compared with the orderly process of condensation of mobile adatoms in the latter case.

Although no systematic experimental studies of the crystallite size of films have been published in literature, a qualitative behavior in general agreement with the above discussion is observed. This is exemplified by the data of Fleet [136] on the film-thickness and substrate-temperature dependence of the crystallite size of Ni films. The crystallite sizes of pure films of low-melting-point materials at room temperature are ~100 Å or more, while those of high-melting-point materials are considerably less, and those of several oxides, and covalent bond materials such as Ge and Si, are sufficiently small (≤ 10 Å) to characterize these films as amorphous if recrystallization takes place spontaneously via nucleation.

The crystallite size of films decreases with decreasing deposition temperatures. It is noteworthy, however, that most metal films deposited at helium temperatures under clean conditions remain crystalline with crystallite size of ~20 to 40 Å. Palatnik et al. [137,138], using microbeam electron-diffraction techniques, analyzed Al films of 60 to 200 Å thickness deposited on cool substrates. They found that "block crystallites" extended throughout the thickness of the film. They also established that most blocks have a large orientation difference with their neighbors.

The anomalously large (~5,000 Å as compared with 50 Å for normal incidence) crystallite size on the surface of the obliquely deposited Al films is an interesting observation [126] illustrating the high mobility of the adatoms. Since the grain boundaries appear to be aligned along the vapor-incidence direction, grain growth by a recrystallization process is unlikely. An extensive grain-boundary movement is also not expected. The growth behavior apparently results from a preferential condensation and high mobility of the vapor atoms.

Epitaxially grown films are sometimes referred to as "single-crystal" films. This is not a correct description in the strict sense since such films generally consist of a number of large crystallites of dimensions (~10^{-4} cm) determined by the system and the deposition parameters. The crystallites, however, have nearly the same crystallographic orientation, so that the film consists of blocks of crystallites with very low angle grain boundaries.

Structural changes in films on annealing have been investigated by several workers using electron-microscopy [139-144] and x-ray-diffraction [145,146] techniques. Generally, the crystallite size increases as a result of a sintering process of the crystallites in contact. The driving force is provided by the tendency to minimize the total surface energy. If the film is thin, the process of sintering produces discrete islands, and thereafter no change in the crystallite size is observed, except, of course, a decrease in the size by evaporation if the temperature is sufficiently high.

Chopra [144] studied the annealing behavior of Ag and W films on carbon and thin mica (with a carbon coating on the deposit to prevent electrostatic-charge effects) substrates. The crystallite size of W films on mica increased rapidly by an order of magnitude at temperatures above ~700°C. Silver films showed grain growth above ~100°C. Keith [146] observed a change of crystallite size from 40 to 400 Å for Cu films deposited at 90°K and warmed to room temperature. In the presence of oxygen, however, the size increased to only 60 Å. As the temperature at which rapid grain growth takes place varies with the nature and smoothness of the substrate, provided the film is deposited at temperatures at which some adatom mobility exists, the annealing temperature of such films is not simply related to the Debye temperature of the vapor source material. If a film is deposited at sufficiently low temperatures at which the adatom mobility is negligible, subsequent annealing yields a monotonic grain growth.

(2) Surface Roughness. The condition for a minimum surface energy for a film may be met by an ideally flat surface. Deposited films, however, show a considerable departure from the ideally flat surface. The surface smoothness of a film is determined by the statistical process of nucleation and growth and the adatom surface mobility. If a film is deposited at low temperatures so that the atomic mobility is small, the surface roughness or average deviation Δt (and hence the surface area of a random deposit) from the average film thickness t is statistical in nature and is given by the Poisson distribution as

$$\Delta t = \sqrt{t} \tag{21}$$

As the surface mobility of adatoms increases (e.g., at higher substrate temperatures), the condensation can occur preferentially at the surface concavities and thus tend to smooth the surface. If, on the other hand, the condensate has a tendency to grow preferentially along certain crystal faces, because of either the large anisotropy of surface energy or the presence of faceted roughness on the substrate, the roughness of the film surface will be enhanced. A further enhancement would occur for oblique deposition owing to the shadowing of the neighboring area by the growing crystal faces. It should be pointed out that oblique deposition causes roughness *even* in the absence of facets in the deposit, but it is reduced considerably at elevated temperatures.

Under suitable deposition conditions, atomically smooth surfaces can be obtained for epitaxial films of various metals and for films with fine-grained and amorphous structures. The presence of Kikuchi lines [98,147,148] (see Fig. 7(d), Chap. III), and strong refraction effects in the reflection diffraction patterns, and the interference fringe pattern accompanying x-ray diffraction lines [149] are some of the criteria that may be used to establish the atomic smoothness of a surface. Figure 12 shows the surface micrographs of: an atomically smooth sputtered Au film; a fast-evaporated ($>10^3$ Å/min) matte Ag film; a faceted, sputtered Bi_2Te_3 film; and an epitaxial polyoxymethylene film with rodlike crystallite growth.

A quantitative measure of roughness, the roughness factor, is the ratio of the real to the geometric area. The real surface area of a thin film is generally measured by the well-known BET (Braunner-Emmet-Teller) method in which the volume (which is proportional to the surface area) of a suitable gas adsorbed on the film is measured.   Several investigations show this factor to approach unity for epitaxial film and films deposited at elevated temperatures.   Values as high as 100 are obtained for films condensed at low temperatures or at oblique incidence.

The extensive studies of Beeck et al. [150,151] on Pd, Pt, Co, and Ni films, of Porter and Tompkins [152] on Fe films, and of Brennan et al. [153] on Ni, Fe, Rh, W, Ta, and Mo films, show the general behavior that unannealed or slightly annealed films display a surface area increasing linearly with film thickness. The roughness increases markedly when films are deposited in the presence of ambient gases, presumably as a result of the rapid thermalization of the vapor atom leading to the formation of stable, individual clusters in the medium, and on the substrate without significant surface mobility. O'Connor and Uhlig

FIG. 12. Electron micrographs of a variety of surfaces of different films: (*a*) atomically smooth 1,000-Å sputtered Au film on NaCl; (*b*) a rough Ag film evaporated on mica at 300°C, roughness caused by a deposition rate exceeding 3,000 Å/min (*Chopra* [92]); (*c*) a faceted growth with platelet crystals in a 3-μ-thick Bi_2Te_3 film sputtered on glass at 400°C (*courtesy of M. H. Francombe* [315]); and (*d*) a rodlike crystallite growth of polyoxymethylene film on (100) KCl ([110] direction shown by an arrow) from nitrobenzene solution held at 120°C (*courtesy of Koutsky et al.* [267]).

[154] found that the roughness of Fe films decreased with increasing temperature of the evaporation source.

The increasing roughness with decreasing substrate temperature for 300- to 850-Å-thick (average) evaporated Cu films studied by Allen et al.

[155] is illustrated in Fig. 13. The linear increase of surface area with the mass (or thickness) of the film clearly suggests a porous film structure with its relatively large internal surface area accessible to the adsorbing gas even in the lowest layers of the film. Allen et al. [155] proposed an explanation of the large internal surface area in terms of a structural model according to which the film consists of rod-shaped elements of approximately equal length growing upward in certain crystallographic directions. These rods are separated from each other by a distance sufficient for the diffusion of adatoms. On annealing, the crystallites grow bigger because of surface migration of the atoms, thus decreasing the surface area.

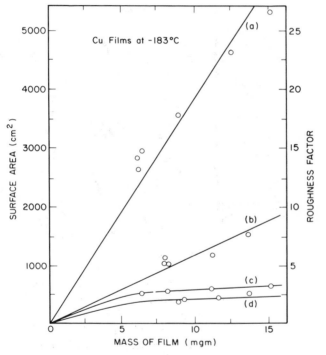

FIG. 13. Surface area and roughness factor vs. mass thickness of evaporated Cu films. (*a*) Deposited and kept at −183°C during measurement; (*b*) deposited at −183°C, annealed for 2 hr at 18°C, and cooled again to −183°C for measurement; (*c*) deposited at 18°C, then cooled to −183°C for measurement; (*d*) deposited at 100°C, then cooled to −183°C for measurement. (*Allen et al.* [155])

Swaine and Plumb [121] studied the effect of oblique incidence of vapor on the surface area of films. They observed catastrophic increase in surface area at large angles of incidence. This result is due to the film porosity which, as discussed in Sec. 2.3.3, arises from the shadowing effect.

(3) Density of Thin Films. Density is an important parameter of physical structure. It must be known for the determination of film thickness by gravimetric and volume emission and absorption methods. Density is generally determined by measuring the mass per unit area and the interferometric film thickness. If a film is discontinuous or porous because of the deposition conditions employed, the film density is expected to be lower than the bulk value.

Several workers have measured the densities of films. A general behavior, namely, a decrease in density with decreasing film thickness, has been observed. Discrepancies exist, however, as to the thickness at which density approaches the bulk value, but these may be ascribed primarily to differences in the deposition conditions and measurement techniques employed by different observers. For example, Blois and Rieser [156] found the densities of Cu and Ag films to approach bulk values at thicknesses of 3,800 and 1,200 Å, respectively. On the other hand, Wainfan et al. [157] obtained bulk density for Cu films at thicknesses exceeding 300 Å.

Using the sensitive in situ quartz-crystal microbalance, the measurements by Wolter [158] on Al, Au, Ag, Cu, and Cr films, by Edgecumbe [159] on Au, Al, KCl, and CsI films, and by Hartman [160] on Al, Au, and Ag films showed reasonably consistent thickness dependence of the density. In particular, Hartman obtained bulk values for Au and Ag films at thicknesses down to 200 Å. His data on Al films (Fig. 14) show constant density at thicknesses above 525 Å. The observed density of 2.58 is, however, lower than the bulk value. The result has been tentatively ascribed to the presence of a surface oxide layer of a lower density, so that the effective density of the film is lowered.

4. CRYSTALLOGRAPHIC STRUCTURE OF FILMS

Generally speaking, thin films of most materials assume the same crystal structure as the bulk material. However, the structural order, e.g., size and orientation of the crystallites, departs considerably from that of

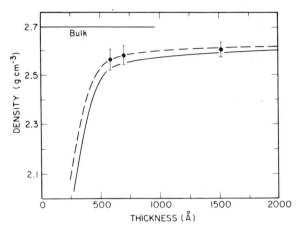

FIG. 14. Thickness dependence of the density of Al films evaporated on glass at 25°C. Data points with error flags refer to interferometric measurements on films deposited on the quartz crystal (the thickness monitor). The dotted curve is corrected for the oxide layer thickness. (*Hartman* [160])

the bulk. Depending on the adatom mobility during deposition, the structural order of atomistically deposited films can be varied from a highly disordered (amorphous-like) to a well-ordered (epitaxial growth on a single-crystal substrate) state. In contrast to these "normal" films, those of a large number of materials prepared under suitable conditions exhibit structures markedly different from those of the bulk. These "abnormal" structures assume various forms, such as amorphous, superstructures, and metastable, unstable, and stable polymorphs. These structures, strictly thin-film phenomena, provide new materials for solid-state research, and their growth may shed new light on the mechanism of epitaxial growth, which has defied any reasonable explanation so far.

This section deals with various observations of abnormal and normal structures. The order of presentation is designed to emphasize the abnormal features of the initial growth of a film. The extensive reviews of the normal, ordered structures given by Mayer [1], Pashley [161,162,100], and Palatnik and Papirov [85] should be consulted for details. Some brief and general reviews of the structure and growth of films are also given by other workers [163-165].

Before the various structures of films are discussed, a basically important question about the possible effect of the film thickness or crystallite size and the substrate structure on the lattice parameter of a film needs to be examined.

4.1 Lattice Constant of Thin Films

The lattice constant of a thick, normal film is generally found to be the same as that of the bulk material. Departures from the bulk value have, however, been found in ultrathin deposits of some materials, and they are discussed below in relation to the influence of the film or crystal size and the nature of the substrate.

(1) Size Effect. Since the atoms on the surface of a crystal in equilibrium have a different environment, one expects the atomic arrangement to be different from that in the bulk. If we apply the surface-energy concept to a spherical crystallite of diameter D and surface energy σ, the crystallite has an internal pressure equal to $4\sigma/D$. This should lead to a change δa in the lattice spacing a, given by

$$\frac{\delta a}{a} = -\frac{4}{3}\frac{\sigma}{ED} \tag{22}$$

where E is the bulk modulus of the material. Thus, depending on the sign of σ, an increase or decrease in the lattice constant is expected.

A decrease in the lattice constants of evaporated alkali-halide films with decreasing crystallite size was observed by Finch and Fordham [166], and Boswell [167]. The latter author observed a $1/D$ dependence for LiF films, with 1 percent increase in the lattice constant for crystallites ranging from 30 to 100 Å in size. The results with other alkali-halide films were essentially the same. On the other hand, Halliday et al. [168] observed an increase (negative σ) in the lattice constant of LiF films. The negative value of σ is consistent with the theoretical estimate [169] of surface energy. Rymer [170] defended the work of Halliday et al. but criticized Boswell's calculations on the ground that the electron-diffraction theory was not applicable to crystals of dimensions less than 50 Å.

The lattice constant of small crystallites of vacuum-prepared MgO was found by Cimino et al. [171] to increase, showing a $1/D$ dependence.

The opposite variation observed for MgO prepared in air was attributed to the interfacial strains from an adsorbed hydroxide layer.

Careful and precise x-ray-diffraction measurements of the lattice constant of 100 to 1,000-Å-thick, partially oriented, Sn films deposited on glass were made by Vook et al. [149]. Film thickness was determined from the x-ray interference-fringe pattern. The lattice constant was found to increase with decreasing film thickness (Fig. 15) and showed a $1/\sqrt{t}$ dependence (t = film thickness). The crystallite size perpendicular to the film surface was found to be the same as the film thickness, while that parallel to the surface was ~700 Å for films of thickness greater than 100 Å. Thus, the observed variation may be described as a thickness effect. A similar increase (~0.1 percent) in the lattice parameter of $\{111\}$ planes parallel to the surface of annealed Au films (100 to 850 Å thick) evaporated onto glass was observed by Vook and Otooni [349].

An unambiguous verification of the existence of a thickness dependence of the lattice constant of a material can be made only for a strain-free film obtained by monolayer-by-monolayer deposition rather than by a three-dimensional nucleation process. Furthermore, it is

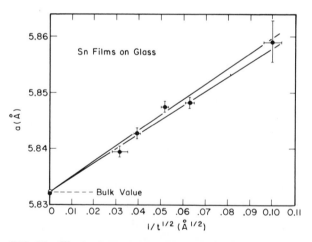

FIG. 15. The lattice parameter a perpendicular to the surface of a ⟨100⟩ textured Sn film deposited on glass at a rate of 30 Å/sec vs. $1/\sqrt{t}$ (t = film thickness). Lattice constant was determined from the (200), (400), and (600) x-ray reflections. (*Courtesy of R. W. Vook* [149])

necessary to employ special techniques such as that of moire fringes (Chap. III, Sec. 2.4) to measure the lattice constant accurately. A thickness-dependent increase in the strain and hence the lattice constant of discrete Pt nuclei electrodeposited onto Au was established by Jesser et al. [336]. The strain was determined from the spacing of the moiré fringes. The calculated lattice constant was found to be inversely proportional to the radius of the nuclei, and it increased by about 1 percent as the radius decreased from ~180 to ~40 Å. The results were found to be in agreement with the van der Merwe theory of elastic strain in epitaxial overgrowths (Chap. V, Sec. 2.3).

(2) **Surface Pseudomorphism.** If an ideal monolayer-by-monolayer deposition is obtained, it is safe to assume that the structure of the first one or two monolayers may be influenced by the substrate lattice. The *constraining* of the deposit lattice to match the substrate lattice is called "surface pseudomorphism." The constrained, new lattice may be called a "pseudomorph." It must be noted that, in many cases, the substrate lattice may *sustain* superstructures, and also *induce* the epitaxial growth of normal and abnormal structures of the deposit (these cases are discussed later). These are, however, *not* examples of pseudomorphism.

Finch and Quarrell [172] introduced the term "basal plane pseudo-morphism" to explain the observed abnormal lattice constants of sputtered Mg, Al, and Zn films, and ZnO film formed on Zn surface. Several other cases of pseudomorphism were subsequently reported in literature. For example, Cochrane [173] observed the transition of electrolytically deposited Ni and Co on (110) Cu single-crystal substrates from pseudomorphic (copper-like) to the normal structures at thicknesses above 100 Å. Most of such reports received either no further confirmation or were contradicted by other careful experiments such as those on ZnO [174] films, and electrolytic deposits of Ni and Co [175]. The analysis of the diffraction patterns of pseudomorphic Al deposited on Pt obtained by Finch and Quarrell was shown by Shishakov [176] and Pashley [161] to be erroneous.

Evidence that the first few monolayers of Cu on W are pseudomorphic was deduced by Jones [177] from his observation that the field-emission patterns of the deposit under ultrahigh-vacuum conditions at 78°K were similar to those of the substrate. This result is in complete disagreement with the LEED observations of epitaxial growth in ultrahigh vacuum of

Cu on Ti [178] and W [179] and of Ni on Cu [180], which show only the normal structures for the initial deposit. Taylor [179] observed partial alloying of Cu and W at room temperature, and this may explain Jones' observations. Normal structures for the epitaxial growth of Cu [181], Ag [182], and Pb [183] on W tips have also been observed by field-emission microscopy. In some cases [182, 184] (e.g., Ag on W), however, there appears to be evidence that the first few adlayers are not crystalline. Such results need further confirmation.

A critical review of the early experimental results by Pashley [161] and Palatnik and Papirov [85] concluded that the evidence presented in literature did not support the concept of pseudomorphism. One may, on the other hand, criticize these investigations on the ground that the necessary accuracy for measurement of the lattice constant was not available nor were the systems employed suitable. According to the theory of Frank and van der Merwe [289] (later extended by van der Merwe [290]), the deposit lattice may be strained to fit the substrate lattice only over narrow ranges of difference in lattice parameter and deposit thickness. If the misfit (defined as the percentage difference of the lattice parameters of the deposit and the substrate with respect to that of the deposit) is very small (~0.2 percent) and the interfacial bonding is strong (i.e., the shear modulus of the interface is comparable with those of deposit and substrate), then it is predicted that the deposit will strain to fit the substrate exactly until the thickness of the deposit exceeds several hundred angstroms. If the misfit is 4 percent, and the interfacial bonding is strong, then the deposit is expected to be pseudomorphic until a thickness of several angstroms is reached. When the deposit thickness exceeds these critical values, some misfit will be accommodated by misfit dislocations [291]. If the misfit is greater than ~12 percent, then much of the misfit will be accommodated by misfit dislocations for all deposit thicknesses.

For a film consisting of hemispherical islands on an infinitely thick substrate, Cabrera and others [337] have shown that, if there is to be a coherent, pseudomorphic interface between the island and the substrate, the island radius should be less than a critical value R_c that is given approximately by

$$R_c = \frac{-3b\left(1 + G_0 a_s^2/4G_s a_0^2\right) \ln\left[(2\beta - 2\beta^2)/e\right]}{2(1 + G_0/G_s)f} \tag{23}$$

where $\beta = 2\pi G_0 f/(1 - \nu)(1 + G_0/G_s)G_i$; $f = (a_s - a_0)/a_0$; a_s and a_0 are the lattice constants of substrate and overgrowth; G_s, G_0, and G_i are the shear moduli of substrate, overgrowth, and interface between overgrowth and substrate; and b is the Burgers vector of the misfit dislocations. The Poisson ratios of overgrowth, substrate, and interface are assumed equal and are represented by ν. In the derivation of this expression for R_c it was assumed that the misfit dislocations were in edge orientation and that their Burgers vectors lay in the interface. This assumption was made because these dislocations are best able to accommodate the misfit between one lattice and another.

Let us consider the example [341] of the Fe-Cu interface. Using $G_i = 7.8 \times 10^{11}$ dynes/cm^2, $a_0 = 3.571$ Å, $a_s = 3.615$ Å, $f = 0.012$, $b = a_s/\sqrt{2}$, $\nu = 1/3$, $G_0 = 1.2 \times 10^{12}$ dynes/cm^2, $G_s = 4.6 \times 10^{11}$ dynes/cm^2, we obtain $R_c = 870$ Å. Similarly, the values of R_c for Au/Ag [338] and Pt/Au [339] are 460 and 5 Å, respectively. The choice of suitable systems for the observation of pseudomorphism is clearly indicated.

Jesser and Matthews and others have carried out a series of electron-diffraction, electron-microscopy, and moiré-fringe experiments on Au/Ag [338], Pt/Au [339], Cr/Ni [340], Fe/Cu [341], Co/Cu [342], and Ni/Cu [343] systems. The deposits in these systems grow approximately as monolayers and are therefore well suited for such studies. These authors established that in all these cases: (1) the deposits were pseudomorphic; (2) pseudomorphism continues up to a critical average thickness of the deposit or the radius of the islands (the latter generally being much higher) in reasonable agreement with Eq. (23); (3) with increasing deposit thickness as well as misfit, misfit dislocations were generated; and (4) pseudomorphism was not due to any inter-diffusion or interface alloying since, for example, in case of the Au/Ag system the alloying (and also the elevated temperature) will increase the lattice misfit. Thus, these studies have provided excellent verification of both the concept of pseudomorphism and van der Merwe's theory of accommodation of the misfit at the interface between two different lattices.

4.2 Disordered and Amorphous Structures

A number of workers [185] have observed broad, halo-type electron-diffraction patterns of the first few monolayers (mass equivalent) of the deposits of several metals and have interpreted them as due to the amorphous structure of the deposits. These results have generally been attributed to the contamination effects, but there is now mounting evidence that this may not be true for all cases, for example, the observations of amorphous structure by Kehoe [186] in Sn/Cu, and in Sn/(111) Ag by Newmann (referred to by Pashley [161]) using an ultrahigh-vacuum electron-diffraction camera, by Grigson and Dove [187] in Al and Fe films (~10 Å thick) on carbon in a scanning electron-diffraction camera, and by Melmed [182] and Gretz [184] in Ag/W in an ultrahigh-vacuum field-emission microscope deserve careful considerations.

Since structural order in a film is determined largely by the surface mobility of the adatoms, a highly disordered amorphous-like structure would be produced if mobility is negligible so that the atoms condense at or near the point of impingement. This condition is obtained by vapor deposition at room temperature of various materials such as *pure* elements of C, S, Ge, Si, Se, and Te, and many compounds of Se and Te, and high-melting-point nonmetallic compounds (e.g., oxides, titanates, niobates, stannates). Amorphous structures (exhibiting rapid recrystallization by nucleation, at a well-defined temperature, accompanied by the release of heat of transformation ~1 eV) are obtained by the following mobility-reducing processes.

(1) Impurity Stabilization. If vapor atoms are deposited in the presence of "mobility-inhibiting" gaseous impurities, fine-grained and, in some cases, amorphous structures are obtained. A partial pressure ($\sim 10^{-4}$ to 10^{-5} Torr) of O_2 decreases [40,41] the mobility of oxidizable vapor material. The clusters formed during the condensation process are probably covered with an oxide layer which prevents further coalescence and hence grain growth. Abeles et al. [188] used this technique to produce amorphous-like thin deposits of Al, Pb, Ga, In, and Sn which exhibit enhanced superconductivity phenomena (Chap. IX, Sec. 3.3). The amorphous structures observed by Chopra et al. [189,190] in argon ion-beam sputtered films of W, Mo, Ta, Hf, Zr, and Re may be stabilized by traces (~1 atomic percent) of N_2. Amorphous Nb films obtained by Hutchinson [191] were attributed to stabilization by oxygen.

(2) Vapor Quenching (VQ). Atomic mobility is reduced by condensing vapor atoms onto a substrate cold enough to reduce thermal diffusion considerably and also to absorb heat of condensation without an appreciable rise in temperature. This process, "vapor quenching," resembles "splat cooling" [192] (and its modification "plasma-jet spraying" [193]) in which small liquid droplets are quenched rapidly at rates of $\sim 10^6$ degrees per second to produce amorphous and metastable structures. This technique was used by Hilsch and Buckel and coworkers [194-197], and others (see the reviews by Hilsch [194] and Buckel [195] to obtain high-temperature superconducting, amorphous Bi, Ga, As, [196] Sb, and Be [198] films by depositing at liquid-helium temperatures. Most nonmetallic compounds, such as sulfides and halides, yield amorphous deposits when deposited at temperatures below 77°K.

It may be noted that, while elements with covalent bonds yield amorphous deposits, pure-metal deposits remain crystalline by vapor quenching at helium temperatures [197]. The crystallite size of metal deposit is, however, decreased considerably (< 50 Å) but is still large enough to yield crystalline diffraction patterns. The small crystallite size implies a highly disordered structure, a conclusion which is supported by the high residual resistivity of VQ metal films. The difference in the behavior of covalent-bond elements and metals may be understood [195] on the basis that, because of the small (four) coordination number of the former, relatively large displacements of the randomly condensed atoms are necessary to form the crystalline structure. On the other hand, metals crystallize in close-packed structures which are not too different from the randomly packed condensed atoms. Moreover, because of smaller desorption and adsorption energies of metal condensates as compared with covalent-bond condensates, the metal deposits have a considerably higher atomic mobility even at low temperatures, so that significant recrystallization takes place. Thus, because of the lack of mobility, a highly disordered, amorphous-like structure is frozen into covalent-bond systems.

(3) Codeposit Quenching. A codeposition of two incompatible systems onto a cold substrate produces the most effective mobility-inhibiting process, which is very useful for producing amorphous structures at low temperatures below 77°K. A mixture of a metal and nonmetal (e.g., Ag + 16 percent SiO [199], Cu + Bi [200]) or two incompatible metals with high concentration [(Sn + 10 percent Cu) [196,201]] may be used. Chopra [202] showed that amorphous structures (e.g., Sn + 40 percent Cu, Ag + Cu) may also be obtained by co-sputtered deposition onto glass at $77°$ K. Deposition by sputtering provides a convenient method of obtaining controlled composition of the constituents.

Detailed studies by Mader and coworkers [203-205] on various two-metal systems, in particular Cu-Ag and Co-Au, established that amorphous structures are obtained in these cases when (1) an equilibrium diagram shows limited terminal solubilities of the two components; (2) the composition range corresponds to the center portions of the miscibility gap, e.g., the Cu-Ag system in the range from Cu + 35 percent Ag to Cu + 65 percent Ag, and for the whole range in the Cu-Mg system; (3) a size difference greater than 10 percent exists between the incompatible component atoms which do not form solid solutions; and (4) the substrate surface is amorphous or atomically rough to reduce mobility.

The composition range of the amorphous structure increases with increasing size factor. A transformation to the crystalline state occurs in many systems at temperatures in the 0.30 to 0.35 T_m range, where T_m is the average melting point of the two components. This temperature range bears no simple relation to the Debye temperature of the materials. Thus, amorphous structures stable at or above room temperature may be formed by a suitable choice of materials.

We have used the term "amorphous" to denote noncrystalline structures. Amorphous (no translational periodicity over several inter-atomic spacings) and fine-grained (<20 Å) structures cannot be distinguished from each other on the basis of electron-diffraction patterns since the patterns in both cases consist of diffuse halos. Richter and Steeb [206], and Fujime [207] attempted to deduce radial distribution curves from the intensities of halo patterns. Fujime's extensive studies on vapor-quenched Bi and Ga indicated that the structure is amorphous. In view of the large inaccuracies in electron-diffraction intensity measurements, these results are not unambiguous. The deduced coordination number is closer to the solid phase than to the liquid phase in these cases. The amorphous nature of the structure is strongly suggested by the observed rapid recrystallization at definite temperatures in cases of VQ films of Bi, Ga, and Sb, and binary metal systems. A fine-grained structure is expected to yield a continuous recrystallization. During the amorphous-to-crystalline transformation, heat is produced [208,209] which is nearly half the heat of fusion for the bulk. This observation suggests that these disordered phases are amorphous. On the other hand, the radial distribution analysis of Cu-Ag VQ films [351] indicates that these films are fine-grained rather than amorphous.

A list of various thin-film materials which exhibit "amorphous" structures is compiled in Table IV.

4.3 Abnormal Metastable Crystalline Structures

The formation of thin-film metastable structures in polycrystalline and, in some cases, epitaxial forms is made possible for several materials by means of the following growth processes uniquely characteristic of a film.

(1) Amorphous-Crystalline Transformation. Buckel [195] noted that the metastable amorphous state in VQ films of Bi, Ga, and Sb is a consequence of the Ostwald rule [210] according to which a system undergoing reaction proceeds from a less stable state through a series of increasingly more stable intermediate states to reach the final equilibrium state. If we assume [16] that the liquid (or liquid-like) phase is formed as the first step during condensation and is stabilized by quenching, then the Ostwald rule suggests that on annealing we should observe various metastable high-temperature phases of the material.

Although condensation via liquid phase may be in question under certain deposition conditions, observation of many cases of metastable structures provides remarkable examples of the Ostwald rule. For example, Ga deposited at $2°K$ is amorphous; it transforms to a γ phase at $15°K$ and the normal structure at $70°K$. The irreversible transformations seen by electron-diffraction patterns are also verified by the electrical-resistivity changes, particularly the occurrence of superconductivity in different phases at different temperatures. In fact, the superconducting energy-gap measurements of impurity-stabilized amorphous Ga (deposited at room temperature) suggest [211] that three metastable modifications, e.g., an unidentified high-temperature phase, and γ and β phases coexist in these films. Careful electron-diffraction studies are obviously necessary to corroborate their existence. Similarly, annealing of VQ films of Bi [196] at $15°K$ yields hitherto unidentified, relatively denser crystalline phase.

Let us emphasize that, depending on the system, the transformations via metastable structures may be very rapid and thus difficult to detect with electron diffraction. Careful studies of changes in the electrical

200

Table IV. Abnormal Structures of Some Thin-film Materials

Single-crystal substrates are required for obtaining the abnormal structures of semiconducting compounds and halides of Cs and Tl, while both amorphous and single-crystal substrates may be used for the rest. The data on temperature and thickness limits are approximate. Ev, Sp, El. Dep., and Sol stand for evaporation, sputtering, electrodeposition, and from solution, respectively

Deposit	Normal structure	Abnormal structure	Deposition technique	Temp. range, °C	Thickness limit, Å	Ref.
Ge, Si	cub. diam.	Amorphous	Ev, Sp	<150	No	General observations
As, B, Bi, C, Ga, Se, Te		Amorphous	Ev	≤RT	No	General observations
Many semiconducting compounds, halides, codeposited mixtures of incompatible systems		Amorphous	Ev	~−180	No	General observations
Fe, Al		Amorphous	Ev	RT	<100	187
W, Mo, Ta, Nb, V	bcc	Amorphous	Sp	~100	No	189, 191, 220
Hf, Zr, Re	hcp	Amorphous	Sp	~100	No	189
W, Mo, Ta, Nb, V	bcc	fcc	Sp	~150-400	~1 μ	189
Mo, Ta	bcc	fcc	Ev	~150-400	~100	220, 191
Hf, Zr, Re	hcp	fcc	Sp		~100	189
Ta	bcc	β (tetr.)	Sp		No	221

Ni	fcc	hcp	Ev	RT	<50	215
			El. Dep.	RT	1,000	343
Co	hcp	fcc	Ev	RT	No	213, 214
			El. Dep.	RT	~1 μ	344
Cu	fcc	hcp	Ev	RT		226
Au, Ag	fcc	hcp	El. Dep.	RT		224, 225
Cs and Tl halides	CsCl	NaCl	Ev, Sol	RT		229
CdS, CdSe	Wurtzite	Sphal.	Ev	200	<1 μ	67, 232-237
ZnTe, ZnSe, ZnS, CdTe, HgSe, HgTe	Sphal.	Wurtzite	Ev	>200	<1 μ	67, 232-237
SnS, SnSe	Orthorhomb.	NaCl	Ev	>200	~1,000	240
HgS	Cinnabar	NaCl	Ev	>200	~1,000	134
GeTe	Rhomb.	NaCl	Ev	>200	<1 μ	134
Bi_2Te_3	Rhomb.	β phase	Ev	280-420		314
GaAs, GaP	Sphal.	Wurtzite	Ev	>450	No	238, 239
InSb	Sphal.	Wurtzite	Ev			317
Pseudomorphs:						
Cr/Ni	bcc	fcc	Ev	RT-350	10 (70 Å nuclei)	340
Fe/Cu	fcc	bcc	Ev	RT	20 (250 Å nuclei)	341
Co/Cu	hcp	fcc	Ev	RT	20 (<375 Å nuclei)	342
			El. Dep.		~1 μ	344
Ni/Co	fcc	hcp	El. Dep.	RT	1,000	343

resistivity on annealing are better suited to establish the occurrence of metastable structures in such systems.

Among other examples of the Ostwald rule are the metastable growth of white Sn [212] and fcc Co [213-215,342,344] phases in films of the respective materials deposited at or near room temperature. Both structures are known to be the high-temperature (>24°C for Sn and >420°C for Co) bulk phases. With increasing deposition temperature, mixed fcc and hcp phases of Co are observed.

Shiojiri [216] studied crystallization of amorphous TiO_2 films prepared by vacuum evaporation onto rock salt at room temperature. Titanium dioxide exists in three polymorphs: tetragonal rutile, anatase, and unstable orthorhombic brookite. This author observed that a 50-keV electron beam of 100 mA/cm^2 set off rapid localized crystallization of the film, yielding polycrystalline rutile coexisting with large crystallites of the anatase phase. Chopra and Bahl [217] observed that electron irradiation (inside the electron microscope) of an amorphous GeTe film obained by vapor deposition at room temperature also initiates rapid localized crystallization. The crystalline regions (Fig. 16) are either spherulites or dendrites which are created by a random choice and coexist with the amorphous regions. The dendrites show predominantly single-crystal growth of an abnormal NaCl structure, while other regions show predominantly the normal (rhombohedral) structure, as verified by the diffraction patterns and the dark-field electron micrographs of the dendritic region. The crystallization appears to be triggered by the high electric field set up by the charge gradients on the insulating amorphous film. Once triggered, the heat released on recrystallization accelerates the transformation process. Just how the oriented structure is obtained with the help of the electron beam is not clear. Note that recrystallization can also be obtained by means of an elevated annealing temperature (~150°C for GeTe films), leading to uniform grain growth over the whole film and a predominantly polycrystalline normal structure. These examples point out the complexity of the amorphous-to-crystalline transformations, and also the potential use of the electron beam for recrystallization and metastabilization of abnormal and polymorphic structures.

(2) Codeposit Quenching of Metastable Alloys. Since the vapor phase has no miscibility restrictions, this process of fabricating a solid by

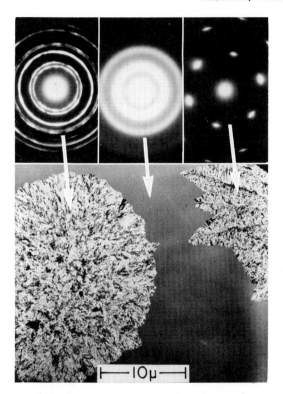

FIG. 16. Electron micrograph of an electron-beam-irradiated 500-Å-thick GeTe film evaporated on NaCl at 25°C, showing the coexistence of amorphous, spherulitic, and dendritic regions. The selected area diffraction patterns show that spherulites are rhombohedral and the dendrites are NaCl structure. (*Chopra and Bahl* [217])

quenching atom by atom may be utilized to produce random and homogeneous solid solutions of systems which have essentially no solubility in the solid or even liquid state. If the phase diagram shows complete solubility, the proper crystalline solid solution with disordered structure is obtained. If the terminal solubility is limited, there is a range of compositions beyond the equilibrium solubility limit in which films with structures of the terminal solid solutions are formed. This possibility of extension of the range was utilized by Kneller [218] to produce metastable solid solutions of Co-Cu and Fe-Cu at and above room temperature. The extent of this range of metastable crystalline solubility depends on the ratio of the atomic radii of the component. The range is very large ($\sim$50 percent) for size ratios smaller than 1.1. This leads to complete crystalline solid miscibility when the components have the same crystal structure. The composition range decreases for larger size ratios, approaching 10 percent for a ratio of 1.2.

The most interesting composition range is that in the insolubility range for size ratios exceeding 1.1. As described earlier, these cases yield amorphous structures which, on annealing to a temperature $T_1 \sim 0.30$ to $0.35\ T_m$, transform rapidly and irreversibly to a single-phase metastable alloy (of fcc structure in all the cases studied by Mader [203]). This annealing temperature depends on the substrate, deposition temperature, the size ratio, the position of the composition in the miscibility gap, etc. Higher stability of the amorphous phase and higher T_1 are obtained for amorphous substrates (lower atomic mobility), lower deposition temperature, larger size ratio, and by sputtering rather than evaporation. For example, deposition of a Cu-Ag film on a cleaved NaCl substrate at 78°K yields mixed amorphous and metastable fcc, but deposition on an Ag substrate bypasses the amorphous state. Deposition of Cu + 50 percent Ag film at room temperature on cleaved MoS_2 yields the metastable fcc directly, which grows epitaxially with a perfect parallel orientation. Note that the deposition temperature required to yield the same microstructure is generally considerably lower than the annealing temperature subsequent to deposition. The annealing temperature for extraordinary systems of Co-Ag and Au-Mg, which are completely insoluble in each other even in the liquid state, is much lower ($\sim$0.12 T_m) relative to the partially miscible systems because of a higher system entropy.

On further annealing, to a temperature of $\sim$0.5 T_m, the metastable alloy decomposes to form a two-phase system. The various structural changes of a Co + 30 percent Au film are shown in Fig. 17. The corresponding resistivity changes are shown in Fig. 17 of Chap. VI. The annealing temperature for this second stage depends on various parameters discussed above for the annealing temperature for the first stage. If the system forms intermetallic compounds (e.g., $Cu_2 Mg$ and $Cu Mg_2$ for Cu + Mg) in the composition range used, the amorphous phase anneals directly to the compound structures.

The kinetics of the two crystallization stages have been studied in detail by Mader et al. [204] on evaporated films and by Chopra [202] on sputtered films. A fair understanding can be obtained by simulating the annealing process in two dimensions by vibrating a tray of hard spheres [219]. The stability of the amorphous phase is due to the result of different size spheres locking each other in place. The first annealing stage is an activated diffusional process while the second stage resembles a spinodal decomposition of a concentrated solid solution into a two-phase structure. The first stage obviously involves smaller movements of the atoms such that the system strives to set up a short-range order to form a metastable alloy phase. With increased activation energy available at higher temperatures, long-range atomic movements lead to decomposition. This transformation process is another example of the Ostwald rule.

It is clear from the above discussion that, by choosing suitable components which do not dissolve appreciably in each other under equilibrium conditions, metastable alloys with new and unique combinations

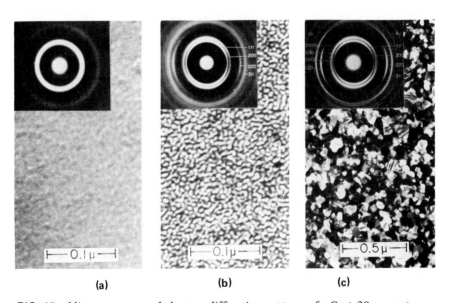

(a)   (b)   (c)

FIG. 17. Microstructure and electron-diffraction patterns of a Co + 30 percent Au film deposited at 80°K. (*a*) Film warmed up to room temperature, amorphous. (*b*) Film warmed up to 470°K, single-phase crystalline. (*c*) Film heated to 650°K, two-phase equilibrium state. (*Courtesy of S. Mader* [205])

of physical properties may be obtained. The structural behavior of a few binary-alloy systems is summarized in Table V.

(3) Deposition-parameter-controlled and Nucleated Metastable Structures. Various cases of metastable structures have been found which depend on deposition parameters such as film thickness, rate of deposition, deposition technique, nature of substrate, substrate temperature, a lateral dc field applied during deposition, and plasma conditions. The mechanisms of stabilization in any of these cases are not well understood and present a challenging basic problem.

The existence of abnormal structures, β Be (a high-temperature phase), hcp Ni, fcc V, and complex cubic Cr in evaporated ultrathin (<20 Å) films *only*, was reported by Bublik and Pines [215]. They attributed the stabilization of the abnormal close-packed structures to the high-surface-energy contribution [Eq. (6)] by the small crystallites to the total free energy of an ultrathin film. Thickness-limited (~100 Å), abnormal fcc structures were observed in evaporated Ta and Mo films by Denbigh and Marcus [220]. But the crystallite size in epitaxial, fcc, Ta films was large enough (~100 to 200 Å) to make any size-effect hypothesis questionable.

Sputtered Ta films show [221] abnormal β Ta (tetragonal) phase under certain plasma-density conditions [222] during sputtering. Chopra et al. [189,190] observed stable, abnormal fcc structures in argon ion-beam-sputtered thick films of W, Mo, and Ta (normally bcc) and Hf, Zr, and Re (normally hcp). The fcc structure persists in Mo films for thicknesses up to $2\,\mu$, beyond which a mixture with the normal structure occurs. These phases are obtained on amorphous and single-crystal substrates at elevated temperatures (~200 to $400°$C, depending on the metal). Epitaxial growth of the fcc phases is obtained on NaCl in most cases at temperatures of $\sim400°$C. If deposited at room temperature, the films of these metals yield amorphous phases which on annealing inside the electron microscope show an easily observable transformation to the normal structure via the abnormal fcc structure, as illustrated in Fig. 18.

Chopra et al. undertook various experiments to show that the fcc structures were not impurity compounds such as oxides, carbides, or nitrides, which have similar structures. The stabilization process in clean sputtering conditions, although not clear, is apparently intimately connected with the characteristics of the sputtering process, e.g., the

Table V. Approximate Composition Ranges and First and Second Crystallization Temperatures (T_1 and T_2, Respectively) for Single-phase Metastable and Amorphous Alloys of Some Systems Codeposited at 80°K (After Mader [203]) T_m refers to the average of the melting points of the components

System A-B	Atom-size ratio r_B/r_A	Metastable (fcc) single-phase rel. composition, atomic %	Amorphous rel. composition, atomic %	T_1, °K	T_1/T_m	T_2/T_m
Co-Cu	1.02	100	No			0.45
Co-Au	1.11	<25 (both)	>25 (both)	~430		0.36
Cu-Au	1.12	100 (disordered)	No			
Cu-Ag	1.13	<37 (both)	>37 (both)	~370	0.28	0.38
Co-Ag	1.15	No	100	~180	0.12	0.24
Cu-Sn	1.18		>30	~150	0.1	
Cu-Mg	1.25	<18 Cu (fcc) <10 Mg (hcp)	>18 Cu >10 Mg	~400	0.35	
Fe-Au*	1.11		20-60 Au	~200	0.13	
Gd-Au*	1.13		50-80 Gd	~480	0.3	

*S. Mader, unpublished.

207

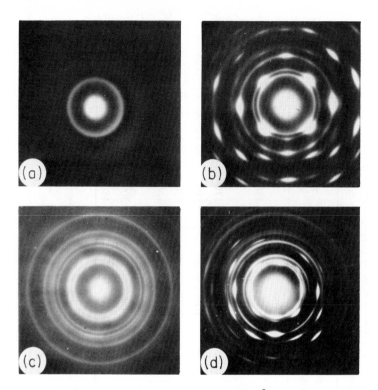

FIG. 18. Electron-diffraction patterns of ~500-Å-thick Zr
films ion-beam-sputtered onto NaCl at (*a*) 23°C, amorphous;
(*b*) 250°C, fcc; (*c*) 450°C, hcp; (*d*) fcc annealed to 675°C
in vacuum, fcc + hcp Zr. (*Chopra et al.* [190])

high kinetic energy of atoms, ions, and atomic forms of gas impurities.
By varying the plasma density, either the normal or the abnormal
structure can be produced, an observation also made by Cook [222].
The fcc phase in Nb films observed by Hutchinson and Olsen [223] was
attributed to stabilization by oxygen. If the fcc structures of the various
transition metals are stabilized by impurities, then it is clear that the
presence of an impurity atom must be sufficiently effective to build a
structure with a large number of surrounding vacant octahedral
positions. Just how it occurs is an open question.

Abnormal hcp phase has been reported for evaporated Ag [224] and
Au [224,225] films, epitaxial electrodeposit of Cu [226] on β-brass

substrate, and oriented electrodeposit of Ni [227] on Cu and Co substrates. It is likely that the hcp phase is stabilized by impurities. In some cases, according to Pashley [100], the interpretation of the diffraction pattern may be questionable. For example, Pashley et al. [228] showed that the extra rings attributed by Davey and Deiter [225] to an hcp phase of Au can be explained by double diffraction at twin boundaries.

The stabilizing influence of the substrate is suggested by the observed [229] abnormal NaCl structure of evaporated films of halides of Cs and Tl (normal CsCl structure) deposited on cleaved surfaces of certain alkali halides. When deposited on an amorphous substrate at room temperature, CsI, TlCl, TlBr, and Tl I showed normal structure, but CsCl showed a mixed normal and abnormal one [230]. On annealing of the amorphous halide films deposited at $-180°C$ on an amorphous substrate, Blackman and Khan [231] observed a mixture of the normal and abnormal phases for all cases except for CsCl, which had a completely abnormal structure.

Polycrystalline films of CdS, CdTe, CdSe, ZnSe, ZnS, ZnTe, GaP, and GaAs generally grow with mixed wurtzite and sphalerite polymorphs of these materials. Either of the two polymorphs may, however, be obtained by means of epitaxial growth on NaCl, mica, etc., by choosing a suitable deposition temperature [67,232-236]. The wurtzite phase appears at relatively lower substrate temperatures. The excess of the metal atoms increases the stability of the hexagonal phase in CdS [67] and CdSe [236] films. Chopra and Khan [67] reported that the higher-temperature sphalerite phase can be stabilized and grown epitaxially at a temperature at which normally the wurtzite phase grows by applying a lateral electric field of ~300 V/cm during deposition of the film on NaCl substrate. The stabilization process appears to be related to the enhanced atomic mobility through an interaction of the electric field with the electrostatic charges present on the islands in the initial-growth stages of the film (Sec. 2.3). Weinstein et al. [237] obtained many microns thick epitaxial films of metastable hexagonal CdTe and sphalerite CdS phases by chemical-vapor deposition onto CdS substrates at temperatures of $\sim500°C$.

Abnormal hcp phases were observed by Davey and Pankey in thick ($\sim6,000$ Å) epitaxial films of GaAs [238] on (111) planes of Ge and

GaAs and of GaP [239] on Ge. The normal zinc blende structure is obtained for epitaxial GaAs at ~400°C, while the hcp phase with (10$\bar{1}$0) plane parallel to the substrate surface is obtained between 450 and 500°C. A diagram of the observed electron-diffraction patterns and their identification is shown in Fig. 19.

A number of materials (e.g., SnS, SnSe, PbSnS, HgS, HgSe, and GeTe) have slightly distorted NaCl structures. When grown epitaxially (on NaCl, mica, or other single-crystal substrates at temperatures of ~200°C), an "abnormal" NaCl structure with parallel orientation is obtained [240] in all cases. The perfection of the epitaxy may depend on the substrate. For example, the rhombohedral GeTe grows epitaxially more favorably on

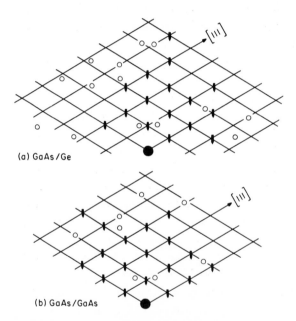

(a) GaAs/Ge

(b) GaAs/GaAs

FIG. 19. Schematic of reflection electron-diffraction patterns ([001] azimuth) of two-source evaporated, ~6,000-Å-thick GaAs films deposited at ~460°C and a rate of ~16 to 50 Å/min. (a) GaAs on (111) Ge, (b) GaAs on Ga face of GaAs. The line ($\dagger$) spots are identified as of an hcp modification, and the weak extra spots (o) of the cubic zinc blende phase. (*Courtesy of J. Davey and T. Pankey* [238])

mica than on rock salt. Mixed structures, distorted and undistorted, are obtained by deposition onto heated glass substrates.

Mariano and Chopra [240] pointed out that since the bulk, NaCl structures of PbS, PbSe, and SnTe transform under hydrostatic pressures to exactly the same orthorhombic phase (distorted NaCl) as that of SnS, SnSe, and PbSnS at atmospheric pressure, a small lattice dilation of the latter group may be responsible for stabilizing the NaCl structures.

Single-crystal fcc or hcp films of Co can be grown [343] epitaxially up to about 1 μ thickness onto single-crystal Cu substrates by electrodeposition. The phase formed during deposition has been shown to be dependent upon the bath composition, the current density, and in particular, the pH of the plating-bath solution. Similarly, hcp modification of Ni can be grown [344] epitaxially to thicknesses of ~1,000 Å by electrodeposition in suitable baths onto Cu {110} surfaces and epitaxial hcp deposits of Co {100}.

Goddard and Wright [343] established that, if electrodeposition of Co is carried out in a magnetic field, there is a critical value of pH and correspondingly a critical film thickness above which the metastable fcc phase transforms to the hcp phase. The critical transition thickness depends on the particular crystallographic orientation of the single-crystal substrate and the strength of the applied field. The critical thickness decreases rapidly with increasing field and is about several hundred angstroms at a field of ~17 kOe. Cisman et al. [345] observed that the applied magnetic field enhanced the hcp structure of electrodeposited Ni-Fe films. Above about 15 Oe, however, the cubic rather than hexagonal structure was favored.

(4) Pseudomorphs and Superstructures. We have already mentioned that a substrate can help stabilize abnormal structures of the deposit. In some cases, the substrate lattice can *constrain* the lattice of the deposit to produce a pseudomorph (discussed in detail in Sec. 4.1.2). Further, a number of substrates are known to *sustain* superstructures on the surface. This subject is discussed later (Sec. 4.4).

(5) Conclusions. We have compiled in Table IV a list of common materials which exhibit new and abnormal structures in thin film form. Although the number of examples is limited, some general conclusions are possible. For example: (1) most materials can be obtained in amorphous structures; (2) the distorted NaCl structures can be transformed to undistorted NaCl structures; (3) the wurtzite compounds can be prepared in sphalerite structures and vice versa; (4) the bcc and hcp structures have a tendency to assume fcc phases.

We have described various physical conditions which are necessary to stabilize the new structures. These conditions vary from material to material and only in some cases may be common to a group of similar materials. A summary of these conditions and typical examples of materials (given in the parentheses) to which they are applicable are given in the following.

1. Film thickness: (pseudomorphic Cr/Ni, Fe/Cu, Co/Cu)
2. Substrate: (Pb and Sn chalcogenides, Cs and Tl halides, super-structures of Ge, Si, Au, MgO, Al_2O_3)
3. Deposition temperature: (Co, W, Mo, Hf, Zr, some II-VI and IV-VI compounds)
4. Electrolyte concentration and pH: (electrolytic deposits of Co, Ni, Ni-Fe)
5. Sputtering current density: (β Ta, fcc Mo and W)
6. Impurities: (Nb, W, Ta)
7. Fine grain size: (Bi, Ga, Sb, Be, incompatible alloys and compounds such as Co-Au)
8. Electron irradiation: (GeTe, TiO_2)
9. Electric field: (CdS, SnS, SnSe)
10. Magnetic field: (electrolytic deposits of Co, Ni)
11. Film stress: (Ni)
12. Nonstoichiometry: (some II-VI, IV-VI, and III-V compounds)

Some other characteristics of the metastable structures may now be mentioned. Once formed, abnormal, metastable thin-film structures are generally stable with aging. Detachment of the film from the substrate does not affect the stability of the metastable structure. We term them metastable since these can be transformed to the normal structures by means of one of the physical agencies such as temperature, electric field, electron irradiation, or magnetic field. The critical thickness up to which a certain metastable structure may grow depends on the system and the deposition condition, and may range from several angstroms to several microns. Under favorable conditions, films several microns (or more) thick are easily obtainable. The metastable structures are not limited to any particular grain size; both large ($\sim 1~\mu$) and small (~ 100 Å) grain sizes have been observed.

(6) Metastabilization Mechanisms. The preceding summary makes it quite clear that the stabilization processes are complex and varied. Since no detailed quantitative studies have been reported in the literature on any one system, only a phenomenological understanding is possible. Ostwald's rule offers the best phenomenological explanation of the occurrence of metastable structures. According to this rule, these structures are also the high-temperature and/or high-pressure polymorphs of the materials.

In order to provide a basic understanding of the metastabilization processes, we must follow an atomistic treatment of the relative stability of the different structures of a material. The solid modification of a substance for which the free energy is a relative minimum under the prevailing conditions is metastable, but not necessarily stable. Basically, the problem consists of determining the Gibbs free energy and its variation with the stabilizing factors for the different modifications. This problem is exceedingly difficult since the energy differences between different modifications of a material are, in many cases, a very small fraction of the cohesive energy—of a magnitude at best comparable with the accuracy of the theory of the cohesive energy itself. This difficulty is further augmented by the fact that metastable structures in films appear to be nucleated; that is, these phases seem to exist in the smallest ($\sim$20 Å) observable nuclei of the deposit. The applicability of the thermodynamical concepts to such small nuclei is at best doubtful.

Although inherent difficulties exist in a rigorous theoretical approach, we can apply qualitative considerations to the total energy of the small nuclei and show that the metastabilizing factors *can* provide the energy difference between the normal and the abnormal phases. As mentioned above, this energy difference is very small. For example, the energy difference between the face-centered cubic and hexagonal phases of cobalt is less than 10^9 ergs/g-atom. The total energy of an island of the material deposited on a substrate may be considered as being the sum of interatomic forces (lattice energy), surface energy, thermal energy, strain energy, surface potential (due to electrostatic charges on the deposit, and/or ionized substrate), interfacial potential energy, magnetization energy (for magnetic materials under a magnetic field), etc.

In most thin film cases, the strain energy, and the magnetization energy (as also the relatively smaller contributions due to magneto-strictive strain energy, demagnetization energy, magnetocrystalline anisotropy energy) are not large enough to have significant effect on the crystal structure of the film. On the other hand, small lattice changes due to stresses in a film may make the normal structure unstable. This is probably the case for those materials in which the slightly distorted NaCl structure reverts to an undistorted one.

The existence of a well-defined temperature of transformation from metastable to normal structure in a number of cases suggests that the thermal energy plays an important role. Both the surface-energy and surface-potential terms depend inversely on the size of the island and therefore become increasingly more significant for smaller islands. In electrolysis, changes in the pH and concentration of the solution have marked effect on the surface potential and the interfacial potential energy which would explain the sensitive dependence of the stability of the new structures on these parameters. Thus, various new energy contributions *can* be large enough to change the equilibrium and stability of the normal structure in a small island.

4.4   Two-dimensional Superstructures

It is commonly assumed that a thermodynamically stable crystal surface (or a film) exhibits the structure of the bulk solids. This cannot be justified in the presence of a sharp surface discontinuity. We must expect ordered surface-atom displacements relative to the bulk lattice both laterally and vertically to minimize the surface energy. The surface atoms must rearrange to form two-dimensional (2D) superstructures related in some way to the bulk structure, but ready to respond to changing external conditions by adopting various ordered structures. If superstructures exist, then the order-disorder transitions must be common in thin films so that different superstructures can occur on the same surface. This is also suggested [241,242] by consideration of the extension of the order-disorder transition phenomenon in bulk materials.

The recent revival of low-energy electron-diffraction (LEED) studies has verified the existence of superstructures in films and surfaces of several metals, insulators, and semiconductors and has opened a new field of 2D crystallography of the unexpectedly large and varied class of superstructures. Present LEED studies primarily yield the crystallographic symmetry of the surface lattice. A structural analysis must await better understanding of the theory of low-energy electron diffraction.

Nevertheless, the widespread presence of superstructures must have significant consequences in relation to nucleation and growth, and the physical properties of the condensates. The limited evidence available lends support to this contention, although much remains to be explored.

A list of some of the superstructures is presented in Table VI. An excellent comprehensive review of the subject is given by Lander [243].

Table VI.   Superstructures of Some Bulk and Thin-film Surfaces

Deposit	Substrate	Superstructure	Existence Temp, °C	Ref.
Si	{111} Si	(7 x 7)	>600	243, 251
		(7 x 7) + (5 x 5)	400-600	252
Si	{100} Si	(2 x 2)	>350	252
Si	{311} Si	(3 x 1)	600	316
Si	(111) Ge	(7 x 7)	600	316
Al	(111) Si	($\sqrt{3}$ x $\sqrt{3}$) (and others)		247
In	(111) Si	(7 x 7) (and others)		243
	(100) Ge	(4 x 4)		243
	(111) Ge	(8 x 8)	>300	243
Au	(100) MgO	(5 x 1)	200	256
	(100) Au	(5 x 1)	200-400	254
		(6 x 6)	350-550	254
		Ring pattern	>750	254
Ag	(100) MgO	Normal		256
	(100) Ag	(2 x 2)	600-750	254
		Ring pattern	>750	254
	(100) Pt	(2 x 1)	300-500	254
		(5 x 5)	350-500	254
		Ring pattern	>600	254
	(001) α Al_2O_3	(6 x 6)	1000	249

(See also Chap. III for more references on LEED.) A few examples of interest and the important conclusions deduced therefrom will now be discussed. Most studies have been done on the surfaces of cleaved single crystals. Under suitable clean conditions, similar results have been obtained with evaporated films of a number of materials. It is essential for observation of the LEED pattern that the substrate be single-crystal or highly oriented.

The superstructures are considered to be generated by the repetition in two dimensions parallel to the surface of a 2D mesh (corresponding to the 3D unit-cell term) which is highly anisotropic in the perpendicular direction. Consequently, 3D centers of inversion are not possible. The superstructure is supported by the substrate structure and is therefore referred to its 3D unit cell. The superstructures are designated: Si (111)–7 (or 7×7); or Ni (110)–2×1–O; or Si (111)–$\sqrt{3}$–Al. The notation implies: Substrate surface–Lengths of the two principal vectors of the unit mesh of the superstructure relative to the unit cell of the substrate lattice–Absorbate, if any. Thus, $\sqrt{3}$ indicates $\sqrt{3}$ times larger lengths of the unit cell obtained by a rotation of $30°$ between the two basic vectors.

Various studies of Ni and Ni + O systems made by Farnsworth and coworkers [244], and others (reviewed by MacRae [245]) established that no superstructures exist for Ni (111), (110), or (100) surfaces. The atoms on the (111) surface appear to be displaced outward by 5 percent [245]. In the presence of oxygen, numerous superstructures are readily obtained which must involve surface migration of Ni atoms over hundreds of angstroms at room temperature. For example, with increasing coverage of oxygen, (2×1), (5×2), (3×1), and (5×1) structures are obtained on Ni (110). Similarly, (2×2) and $(\sqrt{2} \times \sqrt{2})$ are formed on Ni (100), and (2×2) and $(\sqrt{3} \times \sqrt{3})$ on Ni (111). Some of these structures show order-disorder transitions at definite temperatures. When nearly one monolayer coverage of NiO is obtained, the 2D structures transform to nucleate three-dimensional NiO crystallites of normal structure with exposed (100) surfaces in all cases. On the (100) and (111) surfaces, the oxide crystallites have matching planes at the interfaces.

Superstructures developed during oxidation of several other metals, Cu [244], Pt, and Rh [246], have been found. Insulator surfaces, such as MgO [247,248], α Al$_2$O$_3$ [249], mica, and alkali halides [248], also exhibit superstructures. For example, a (6×6) structure on α Al$_2$O$_3$ appears on annealing above 1,000°C and reverts to the normal structure on subsequent annealing at 700 to 900°C.

Extensive studies have been made on cleaved Ge and Si surfaces and vacuum-evaporated Si films. The works of Farnsworth and coworkers [244,250], extended considerably by Lander and coworkers [243], established that the cleaved surface of Si (111) exhibits a superstructure with twofold rotational symmetry (on a substrate of threefold symmetry). Upon heat treatment in vacuum, the superstructure is quickly and irreversibly converted to Si (111)–7 at ~700°C [to Ge (111)–8 at ~300°C for Ge]. Annealing the 7×7 structure often produced a (5×5) structure for silicon. The Si (100) exhibits (2×2) structure on annealing, with occasional fourth-order disorder. Jona [251] studied the annealing behavior of an amorphous Si film deposited on Si {111} at room temperature [Fig. 20(A)]. Recrystallization occurs above 150°C, producing normal-lattice LEED spots along with superlattice 1/7- and 6/7-order spots. At about 350°C, pure normal spots were observed, and at higher temperatures, first diffuse half-order spots appeared which progressively resolved themselves into forming the (7×7) structure. Similarly, the deposit on Si {100} also showed the transitional diffusion half-order spots which on further annealing become sharp to yield (2×2) structure. The sequence of events is formally similar to some order-disorder transitions in bulk alloys.

Thomas and Francombe [252] studied the epitaxial growth of Si films several microns thick on (111) and (100) surfaces of Si at different temperatures of deposition. They observed the (7×7) structure on Si (111) between 600 and 1,000°C, a mixed (5×5) and (7×7) below 600°C [Fig. 20(B)], and an amorphous structure below 350°C. On a (100) surface, epitaxy occurred above 300°C with a (2×2) structure at higher temperatures, and a new, more ordered form of the (2×2) structure between 650 and 700°C.

The above description suggests that (7×7) and (2×2) are the stable superstructures of clean, annealed (111) and (100) surfaces, respectively, of silicon. The (5×5), (4×4), and other structures observed by some workers may be metastable with their stability depending on the particular system used by the investigator. That the temperature range of

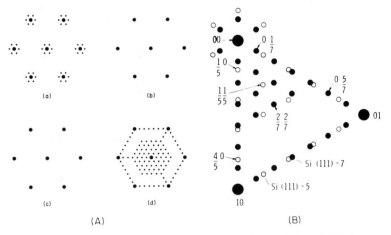

FIG. 20. (*A*) Schematic LEED patterns observed during annealing of an amorphous film of Si on {111} Si. Large circles represent the normal, small circles the superlattice spots. (*a*) Intermediate phase (150 to 350°C), (*b*) normal Si phase above 350°C, (*c*) transition stage with antiphase domains, (*d*) final 7×7 structure. (*Courtesy of F. Jona* [251]). (*B*) Schematic of a sextent of the LEED pattern of an epitaxial Si film evaporated on Si {111} at 550°C, showing mixed (5×5) and (7×7) structures. (*Thomas and Francombe* [252])

stability of different superstructures varies in different reports suggests that the deposition systems are not the same in all cases. Thomas and Francombe [252] observed a correlation between the development of stacking faults in epitaxial (111) Si films deposited below 600°C with the occurrence of the mixed (5×5) and (7×7) structures. They attributed the formation of stacking faults to the formation of the (5×5) domains. Since stacking faults in epitaxial Si films are known [100] to arise from the presence of surface impurities or defects, the (5×5) structure is probably stabilized by surface impurities and imperfections.

Results of investigations of changes in superstructure of the principal surfaces of Si and Ge due to chemical reaction with a variety of elements undertaken at Bell Telephone Laboratories have been summarized by Lander [243]. In many cases, a large number of superstructures are obtained. For example, with Cs and halogens deposited on Si (111), translational periodicity factors such as $\sqrt{3}$, 2, $2\sqrt{3}$, 4, 5, 6, 7, 8, 12, 18, and 24 were observed. Similarly, at least 5 phases were observed with Al, and 8 with In deposits.

The existence of ordered and disordered surface structures with well-defined temperature ranges of stability have been established for Pt [253], Ag [254,255], Au, and Pd [254] single-crystal surfaces. Palmberg and Rhodin [256] confirmed the existence of (1×5) structure on evaporated (100) epitaxial Au films, but found no superstructure for (100) Ag epitaxial films. It was suggested that the Au (1×5) structure may be consistent with the presence of a thin layer of an hcp phase. The results of different authors lead to the conclusion that some of the reproducible superstructures on metal surfaces are indeed characteristic of a clean surface rather than being impurity-stabilized. This stems from the fact that the presence of alloys and impurities produces varied and irreproducible (from specimen to specimen) structures.

The presence of superstructures on clean surfaces of a variety of materials is thus well established. Some of the remarkable properties of the superstructures are: (1) They exist in a variety of ordered and disordered forms within definite temperature ranges and are stable, reproducible, and reversible in many cases. (2) Higher-order (i.e., long-range, less-dense) structures are obtained with increasing temperatures and require enormous movements of surface atoms to form perfect domains of the superstructure thousands of angstroms on edge. (3) Ordering of structures appears to be nucleated at some favorable sites, spreads by cooperative interactions, and is very sensitive to ambient conditions. (4) When thickness exceeds a monolayer or two, the superstructures exhibit instability. (5) There is a correspondence (registry) between the superstructure lattice and the substrate lattice. The "registry" may be obtained with translational and/or rotational movements which must clearly be performed only in the directions parallel to the surface. There are some uncertain cases, however, where registry in one of the directions is absent. (6) Rotational symmetry is maintained in superstructures even when much of the translational symmetry matching is discarded. The occurrence of Si (111)−7−In, for example, with a poor matching of symmetry, *strongly rules out the necessity of matching systems for epitaxial growth.* (7) Examples of the case where the unit mesh is smaller than the unit cell of the substrate are not known.

It is evident from the registry condition that superstructures must be related to the periodic arrangement of bonding forces between surface atoms and substrate atoms. *The type or origin of bonding is apparently not very significant.* What causes superstructures to nucleate and possess a certain symmetry is far from clear. A vacancy model which considers domains of surface vacancies has been suggested. The model, although plausible in the light of adsorption data, still remains hypothetical.

4.5 Fiber Texture (Oriented Overgrowth)

It is well known that crystallites in vapor and electrolytically deposited films tend to assume a preferred orientation, called "fiber texture." It may be a single or double fiber texture, depending on whether the crystallites have one or two degrees of preferred orientations. A fiber texture may develop in a film at various stages, e.g., nucleation, coalescence, growth, epitaxial growth, oblique growth, annealing. It is accordingly called nucleation, nucleation-growth, growth, epitaxial, oblique, and annealing texture, respectively. Most studies on nucleation texture have been concerned with the epitaxial growth where the substrate has a dominating influence on the orientation (discussed in the following section). The fiber-texture data on films deposited on amorphous and inert substrates generally reported in the literature refer primarily to the initial-growth stage.

Bauer [257] and Palatnik and Papirov [85] have compiled an extensive list of the fiber textures of various evaporated and sputtered films and have reviewed their interpretation critically. Fiber texture in electrodeposits is reviewed in several books [1,85,258], and it is apparent from these references that a wide variety of textures is possible for any one system, depending on the deposition parameters employed. Consequently, disagreements between different observers are not uncommon. One cannot generalize the growth behavior of fiber textures for different materials, although regularities for some materials of one class of crystal structure can be noted as follows.

Evaporated films of fcc metals on amorphous substrates exhibit an initial-growth $\langle 111 \rangle$ texture which changes to $\langle 110 \rangle$ in thicker films. The $\langle 110 \rangle$ texture is also observed for films deposited in the presence of inert gases. An interesting exception is Al films which show $\langle 100 \rangle$ texture. The $\langle 111 \rangle$ and $\langle 100 \rangle$ orientations of Ge (diamond structure) resemble those of the fcc metals. Films with NaCl (e.g., various halides) as well as tetragonal structures exhibit the $\langle 100 \rangle$ textures. The low-melting-point hcp metals and wurtzite-structure compounds show pronounced [0001] orientation starting from the nucleation stage. Let us emphasize that the textures in all cases are modified strongly by the deposition parameters which influence the atomic mobility, presence of electrostatic charges and applied electric field [67,131,132], presence and adsorption of ambient gases, nature and smoothness of substrate surface, and angle of incidence of vapors.

The situation in electrodeposits is much more complex. Both the degree of orientation and the nature of the texture can be altered drastically by plating conditions. It is noteworthy that, when free to assume its own texture, the fiber axis stands parallel to the direction of current flow. The variety of textures obtained in electrodeposits is illustrated by the example [258] of Cu, which shows [110], [100], [111], [110] + [100], and [111] + [100] fiber textures.

The fiber axis of films deposited at normal incidence is very close to the substrate normal; small deviations on both sides of the normal are observed in some cases. With oblique deposition, the fiber axis tilts toward the vapor direction, though not necessarily coincident with it. This angle-of-incidence effect is generally absent in the nucleation stage, and for materials or conditions yielding high surface mobility of adatoms, with the notable exception of low-melting-point hcp metals. It is more pronounced with low-mobility deposits such as Al (in the presence of oxygen) and alkali halides. The hcp systems with [0001] fiber orientation display [257,259] a monotonically increasing tilt of the fiber axis with increasing angle of incidence, as shown in Fig. 21 for Mg and Zn films. This behavior has been exploited in producing CdS-transducer films with different orientations and hence excitation modes (Chap. VIII, Sec. 3).

Phenomenological models to understand the various textures proposed by a number of workers are reviewed by Bauer [257]. No single model has withstood a critical test, primarily because of the neglect of the simultaneous role of the many deposition variables. In one of the more basic and successful approaches, due to Bauer, the textures are determined by the crystal habits, the strongest texture tendency being for the crystals with the simplest habits. The crystal habits are simple for NaCl and hcp structures, but complicated for other structures, which explains the diversity of their textures. According to the Gibbs-Wulff theorem, a crystal in equilibrium with its vapor is bounded by a small

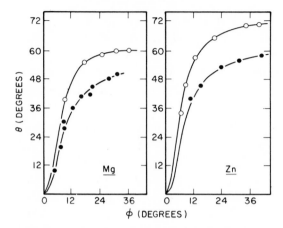

FIG. 21. Dependence of the angle (θ) deviation of the hexagonal axis of the texture in Mg and Zn films on angle (ϕ) of incidence of vapor atoms with respect to the substrate normal. Films were deposited at room temperature ($\bullet$), and at $-70°$C ($\circ$). (*Mel'nikov et al.* [259])

number of equilibrium planes. If h_i is the central distance of an equilibrium plane and σ_i is the specific free energy of this plane, then

$$\frac{\sigma_i}{h_i} = \text{constant} \tag{24}$$

Clearly, if σ_i is altered by impurities, nonstoichiometry, adsorption, electrostatic charges, etc., the crystal habits will be changed.

Several empirical rules have been put forward to explain texture formation. According to the principle of geometric selection [260], in the growth of randomly oriented nuclei, only those grains will survive in which the direction of maximum rate of growth approximately coincides with the normal to the crystallization front. This rule, stated variously by Finch et al., Evan and Wilman, and Dixit, suggests that, if the surface mobility of adatoms during deposition is high, the densest-packed planes will lie parallel to the substrate surface. On the other hand, if the mobility is low, the atoms will condense with a dense packing but with random orientation. On annealing, the most probable $\langle 111 \rangle$ texture should first appear in this case. There exists, therefore, a correlation between mobility and texture tendency which explains Levinstein's [95] crude classification of materials in terms of their melting points, since mobility generally increases (but not always) with decreasing melting point.

The oblique texture may originate from surface roughess or crystal habits. The preferential growth of certain planes once initiated is further enhanced by the effects of geometrical shadowing, angular dependence of the sticking coefficient, and the selective supply of the obliquely incident vapor atoms. The increased tendency to exhibit fiber texture in obliquely deposited films on a smooth surface may arise from the higher mobility of the obliquely incident vapor atoms and hence enhanced crystallization [126].

In conclusion, we should note that the existence of a fiber texture must have strong influence on the properties of films, as illustrated by the many examples of anisotropic behavior of the properties mentioned in this text. There is, however, a notable lack of systematic work in this area.

4.6 Alloy Superlattices

The ease with which epitaxial films of alloys and compounds of varying compositions can be prepared has been advantageously exploited in studying the phenomenon of long-period superlattices in binary alloys of noble metals. Electron-microscope studies of these superstructures provide indirect information about the Fermi surface of alloys. The review article by Sato [261] on the subject should be consulted for the original references.

At the equiatomic composition Cu Au, the structure of this alloy at high temperatures is "disordered," while at low temperatures, it forms a superlattice called "Cu Au I." Between these phases, the long-period superlattice Cu Au II exists. The Cu Au I is a regular tetragonal-symmetry superlattice in an fcc lattice. The unit cell of Cu Au II consists of Cu Au I cells arranged perpendicular to the c axis such that after every five cells the atom layer alternates and becomes superperiodic. Thus the unit cell divides into two antiphase domains, the size of the domain being an intrinsic and structure-insensitive property.

Long-period superlattices have been observed in epitaxial films of alloys such as Cu-Pt, Cu-Pd, Au-Zn, Au-Cd, and Ag-Mg. As the composition of Cu in Cu Au alloy is increased, the cell period increases to 10 for 31 atomic percent. When Cu_3Au composition is obtained, the long-period structure vanishes. However, the superstructure can be restored by adding multivalent impurities like Al. The superperiod decreases with increasing electron/atom ratio of the system.

Based on their extensive studies of the effect of various impurities, Sato and coworkers have concluded that a superperiod is created in an ordered alloy to lower the electron energy. This process is opposed by the antiphase boundary energy resulting from the antiphase domains so formed. The superperiodic structure appears when the decrease in the electron energy is greater than the antiphase boundary energy. The various observations on superlattices are in accord with this theory.

5. EPITAXIAL-GROWTH PHENOMENON

Royer [262] introduced the term epitaxy ("arrangement on") to denote the phenomenon of the oriented growth of one substance on the crystal surface of a foreign substance. The first observed examples of such growth occurred naturally on minerals. Frankenheim [263] was apparently the first to successfully obtain epitaxy of alkali-halide deposits from a solution onto a cleaved mica surface. The growth of alkali halides upon each other was studied by Barker [264] and was repeated and extended by Royer [262]. Since then, epitaxial growth of films of metals, insulators, and semiconductors has been the subject of extensive investigations. The oriented growth of a material over itself is now termed "auto-epitaxy," and over another material "hetero-epitaxy" or simply epitaxy. It is clear that hetero-epitaxy changes over to auto-epitaxy when films are thick enough for the initial substrate to have no orienting influence. Another type of epitaxial growth of a solid film over a liquid surface of another material is called "rheotaxy," which does not concern us here, but it is of interest to remark that oriented overgrowth can be obtained on liquid surfaces. For example [265], (111) planes of Au and Si grow parallel to the surface of liquid mercury and molten glass, respectively.

The experimental and theoretical aspects of thin film epitaxy have been thoroughly reviewed by Pashley [100,161] and Palatnik and

Papirov [85]. These authors have also compiled comprehensive lists of the various cases of epitaxial growth and the lattice mismatch of the deposit-substrate systems.

The following section describes the role of various parameters which affect the oriented growth of films deposited on single-crystal substrates. The emphasis is on determining the general trends, if they exist within a broad frame of reference, and comparing them with the theories of epitaxy. The experimental details and interpretation of the epitaxial relationships cannot be presented in a limited space. The reader should consult the original references. Most studies of epitaxial phenomena have been carried out on films of alkali halides on themselves, on mica and metal substrates, and of metal films on themselves, on alkali halides, and on mica substrates. Our conclusions are primarily based on the information provided by electron-diffraction studies of a deposit-substrate epitaxial system in the nucleation-growth stage of a thin deposit.

5.1 Influence of Substrate and Deposition Conditions

(1) Substrate. The single-crystal substrate has a dominant influence on the oriented growth of the deposit. Epitaxy can occur between two substances of completely different crystal structures and of different types of chemical bonds. For example, epitaxial growth is observed for materials of various structures on cubic alkali halides as well as monoclinic mica, of Au [266] on various faces of Ge, and of a variety of long-chain polymers [267] on alkali halides, etc. The resulting orientation of the deposit, however, depends on the orientation and the crystal structure of the substrate. Thus fcc metals grow with parallel (both planes and directions coincident) orientation on (100), (110), and (111) surfaces of NaCl, but with (111) plane parallel to a (001) mica cleavage plane forming a hexagonal arrangement of atoms and ions.

Some symmetry exists, though not always obvious, between the contacting planes of the two materials. The symmetry relation is generally complicated and may be produced by generalized translational and/or rotational movements of the two lattices. The preservation of rotational symmetry rather than purely translational symmetry is observed more often. As a result of a particular symmetry relation, the deposit plane may have two or more equivalent orientations, so that the deposit will consist of a mixture of distinct orientations. Thus, double and quadruple positioning is obtained for the fcc deposits on mica and NaCl, respectively.

A small lattice misfit (defined as a percentage difference between the lattice constants of the deposit and the substrate planes in contact) is *neither a necessary nor a sufficient condition* for the occurrence of epitaxy. The maximum allowed misfit of 15 percent for epitaxy deduced by Royer [262] from his limited observations of alkali-halide deposits on mica (called the Royer rule of epitaxy) was shown by Schulz [71] not to be obeyed by many systems. The following extreme examples illustrate the point:

System	CsI/LiF	RbBr/Ag	Sn/Ag	Cu/NaCl	Ni/NaCl	Ag/PbS
Misfit, %	+ 90	+ 105 and + 18	+ 101	− 36	− 38	− 31

Since both positive and negative, large and small, and anisotropic misfits are tolerated, it seems meaningless to quote misfits to justify observations of good or poor epitaxial growth. Unfortunately, this continues to be a common practice in literature. But we must emphasize that this does not mean that misfit is not an important parameter. On the contrary, this is expected to be the key to the understanding of epitaxy, but the *relative positioning* of the atoms rather than the *best geometrical fit* of the two lattices is the significant variable.

Freshly deposited, cleaved (in air or vacuum), mechanically polished and chemically etched surfaces are some of the types of surfaces commonly used successfully for epitaxial-growth studies. Faceted and rough surfaces would obviously yield multiple orientations. Atomically smooth surfaces enhance adatom mobility, resulting in increased perfection of epitaxy. The observed nearly perfect orientation of the preferentially condensed nuclei at the cleavage steps suggests that the orienting influence of the substrate is not altered by such a defect. In fact, the dark-field micrographs taken by the author show that these nuclei are oriented even in the normally unfavorable deposition conditions. Color centers on NaCl produced by x-ray irradiation enhance the epitaxial influence of the substrate [353].

Although suitable atomic bonding of the contacting planes must occur for epitaxy, a clear dependence on the nature and strength of bonding still needs to be established. Qualitatively, in this regard, the tendency and perfection of epitaxy and the substrate temperature required for epitaxy indicate some systematic variations which are not understood at present. For instance, as compared with Au, the epitaxy of Ag on NaCl and mica occurs at relatively lower temperatures and is generally more perfect. Both Au and Ag show a higher tendency for orientation on NaCl than on mica.

(2) Substrate Temperature. Brück [268] examined the effect of substrate temperature on the epitaxy of fcc metals on NaCl and concluded the existence of a critical temperature, subsequently called the "epitaxial temperature," above which epitaxy is perfect and below which it is imperfect. Although an epitaxial temperature does exist for a system, it is not just a characteristic of the deposit-substrate system but also strongly depends on the deposition conditions. Consequently, workers using different conditions have reported widely varying epitaxial temperatures for the same system. For example, the epitaxy of Au/MgO [269], Ag/NaCl [106] and Cu/W [107], Sn/Cu [270] is normally obtained above room temperature but, under certain conditons, may be obtained at temperatures as low as $77°K$.

An increasing substrate temperature may improve epitaxy by (1) aiding the desorption of adsorbed surface contaminants; (2) lowering supersaturation, thus allowing the dilute gas of adatoms sufficient time to reach the equilibrium positions; (3) providing activation energy for adatoms to occupy the positions of potential minima; (4) enhancing recrystallization due to the coalescence of islands by increasing surface and volume diffusion; and (5) assisting a possible ionization of surface atoms. Contributions (3) and (4) may also be obtained by ultrasonic agitation of the substrate during deposition, as demonstrated by Chopra and Randlett [103] for Au and Ag films on rock salt.

The factors mentioned above are, in turn, dependent on other deposition conditions, making it difficult to define a precise value of the epitaxial temperature. However, despite these various influences, comparable values of the epitaxial temperature for the same system under similar conditions have been reported by different observers. As a result of the confidence in the reliability of an epitaxial temperature peculiar to a system, it is possible to conclude that the epitaxial temperature T_e of a deposit shows definite substrate dependence for substrates of similar structures, as exemplified by the following data for (100) epitaxial growth of Ag [295]:

System T_e, °C	Ag/LiF 340	Ag/NaCl 150	Ag/KCl 130	Ag/KI 80

(3) Deposition Rate. Studies of several workers [271-274], particularly Sloope and Tiller [271, 272] on evaporated Ge films on CaF_2, NaCl, NaF, and MgO, and Ag on cleaved NaCl, established that, for a given deposition rate R, a minimum epitaxial temperature T_e exists related to R by $R \leq Ae^{-Q_d/kT_e}$. Here, A is a constant and Q_d is the energy of surface diffusion. This relation follows from elementary considerations that an adatom should have sufficient time to jump to an equilibrium position of an ordered state by surface diffusion before it interacts with another adatom. For a given system with fixed atomic mobility, this is satisfied if the deposition rate of a monolayer should be less than the jump frequency of adatoms. A similar relation also follows from the Walton-Rhodin theory [Eq. (16)] of nucleation, where epitaxial temperature is the temperature at which a transition occurs from clusters with a single bond to those with a double bond.

Figure 22 shows the deposition-rate dependence of the temperature required for polycrystalline-to-monocrystalline and for amorphous-to-crystalline transformations, in films of Ge evaporated onto CaF_2. The above mentioned exponential relation is obeyed, which yields $Q_d \sim 1.4$ eV for evaporation at 10^{-7} to 10^{-9} Torr. At higher pressures ($\sim10^{-5}$ Torr), the epitaxial temperature increases and Q_d decreases to 1.15 eV. In contrast to these values, Krikorian and Sneed [273] obtained $Q_d = 0.4$ and 1 eV for evaporated and sputtered Ge films (on CaF_2), respectively.

The amorphous-to-crystalline transformation (Fig. 22) also obeys an exponential relation (justified by Sloope and Tiller again on the basis of a diffusion-controlled process), yielding an activation energy of 1.5 eV. In sharp contrast, however, this transformation is found by other workers [275] *not* to depend on the deposition rate. The results of Krikorian and Sneed indicate a sharp deposition-rate dependence only below 250 Å/min and a well-defined "triple point" between the amorphous, polycrystalline, and monocrystalline phases. This rather

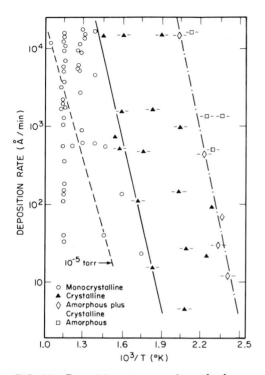

FIG. 22. Deposition rate vs. reciprocal substrate temperature showing the transitions from the amorphous to crystalline to monocrystalline structures for a Ge film evaporated onto polished (111) CaF_2 substrates at 10^{-7} to 10^{-9} Torr. The dotted curve is for films prepared at 10^{-5} Torr. (*Sloope and Tiller* [271])

unusual and unexpected observation of a triple point is not consistent with the existing data in the literature. The consistency of Sloope and Tiller's data with the triple point as suggested by Behrndt [86] is based on an unjustified extrapolation of their data. Further, Behrndt's comparison between this triple point and that for a vapor-liquid-solid transformation is physically unrealistic. A solid-film system exhibiting such a triple point cannot be in thermodynamical equilibrium. Instead of a sharp triple point, it is generally agreed among most workers that amorphous, polycrystalline, and single-crystal phases coexist in a broad temperature range [275].

(4) Contamination. A monolayer of an amorphous deposit (e.g., carbon) on a single-crystal substrate is known to hinder epitaxial growth of a deposit. Distler's [350] observations that the orienting influence of the substrate can be transmitted across thick amorphous plastic and carbon deposits through electrostatic effects is questionable and has not been verified [354] (see Sec. 1.4.3).

If the substrate is exposed to air or a poor vacuum ($> 10^{-7}$ Torr), some adsorption of the ambient gases must take place. The LEED observations [243] show that some gases (e.g., O_2 and N_2) adsorb epitaxially, generally with a superstructure. We have already mentioned that, depending on its influence on the desorption energy, adsorption of a gas may decrease or increase the mobility of adatoms and hence affect the epitaxial temperature. The lowering of the epitaxial temperature of several metals by depositing on vacuum-cleaved ($\sim 10^{-4}$ to 10^{-5} Torr) rather than air-cleaved rock salt was first observed by Ino et al. [276] and then verified by others [43, 277]. The effect is marked, as indicated by the following data [43] on epitaxial temperatures $T_e(A)$ and $T_e(V)$ for parallel orientation of several metals on air-cleaved and vacuum-cleaved (100) NaCl, respectively:

System	Au	Ag	Al	Ni	Cr	Fe	Cu	Ge
$T_e(A)$, °C	400	150	440	370	500	500	300	500
$T_e(V)$, °C	200	RT	300	200	300-350	300-350	100	350-400

When, on the other hand, rock salt was cleaved in ultrahigh vacuum (uhv) ($\sim 10^{-9}$ Torr) before deposition, Ino et al. [276] observed that the parallel orientation for Cu, Au, and Ag deteriorated, but no such deterioration occurred for Al and Ni. Further, these authors observed that, instead of the parallel (100), the (111) orientation was prominent for Cu, Au, and Ag deposited on NaCl cleaved in good or poor vacuum. Matthews and Grünbaum [277], on the other hand, observed parallel orientation for initial layers of Au on uhv cleaved NaCl at 360°C. With increasing thickness, eight (111) orientations appeared, and above a thickness of several hundred angstroms, only (111) orientations could be observed. On air-exposed NaCl substrate, the (111) orientations appeared initially but failed to develop sufficiently so that the parallel orientation predominated with further deposition. These experiments suggest clearly the need for suitable contaminants for epitaxial growth of some materials on NaCl substrates.

Harsdorff and coworkers [278, 279] made a series of detailed studies of the effect of gas contaminants on the perfection of epitaxy. They found the perfection of epitaxy to be a periodic function of temperature, which they interpreted as a stage-by-stage thermal desorption of adsorbed gases, in particular, water vapor. Bauer et al. [280] questioned this explanation since the experimental results do not suggest a thermal-desorption process in the temperature range used for epitaxy. Exposure of the NaCl substrate to water vapor was found by Adam and Harsdorff [279], among others [277], to be beneficial for epitaxy, most probably as a consequence of changes [76] in the surface of rock salt due to recrystallization.

(5) Film Thickness. We have already referred to changes in the orientation of Au films with increasing thickness. According to Matthews [281], the orientation changes occur by surface diffusion and migration [282] of the grain boundaries during coalescence of the two differently oriented nuclei. In this process, the more abundant and/or larger-sized nuclei dominate the overall orientation. For example, for both vacuum-cleaved and air-contaminated NaCl, the (100) nuclei predominate initially. However, the (111) nuclei, which appear with further growth, grow more rapidly in the former case while the (100) grow more rapidly in the latter case. Thus, (111) and (100) orientations prevail, respectively, with increasing thickness. In either case, a small nucleus may rotate on coalescence and become a twin of the large one, thereby explaining the occurrence of a large number of twins in fcc metal films.

On the basis of coalescence-induced recrystallization, Matthews [283] demonstrated a technique for growing epitaxial films of Au and Fe by first evaporating a thin layer (~20 Å) at a very high rate (> 1,000 Å/min), followed by slow evaporation. It was argued that rapid deposition yields a large number of initial (100) nuclei, resulting in an earlier coalescence and hence preservation of the (100) growth. That the saturation density of the initial nuclei is increased is questionable [99, 284] and further, the reported observation could easily be due to higher substrate temperature attained during rapid evaporation.

Systematic thickness effects on orientation have not been studied. Deposition conditions vary with increasing deposition, a fact that may be used to explain the observations of both improvement and deterioration of epitaxy with increasing thickness. Twinning occurs on all four (111) planes with increasing thickness of most of the fcc metals. Atomically smooth surfaces obtained in high-mobility cubic metal films are maintained with further deposition. Marked faceting, resulting in a rough surface, develops for materials with highly anisotropic surface energies.

(6) **Electrostatic Effects.** We have already indicated that an applied electric field can enhance coalescence, improve epitaxial growth, and stabilize crystallographic structures other than the normal. A field applied normal to the substrate is also known [285] to enhance the growth rate of epitaxial Ge, Si, and GaAs films obtained by chemical-vapor-deposition technique. Chopra's [72] explanation of the field effect suggested that effects similar to that of the applied field can also be obtained by electrostatically charging either the substrate and/or the deposit during vapor deposition. These predictions have been verified by the observed earlier coalescence and improved epitaxial growth in Au films [286], and orientation changes in evaporated NaCl [131, 287], and sputtered Ni films [132].

(7) **Deposition Methods.** Given suitable deposition conditions and a single-crystal substrate, epitaxial growth of elements, compounds, and alloys should be feasible by any deposition method. This contention is amply supported by a variety of examples listed in Tables VII, VIII, and IX. Each deposition technique, however, has special characteristics which should be reflected in the nucleation, growth, and epitaxy of films. For example, the observed lowering of the epitaxial temperature for several materials by sputter deposition is most likely due to the high kinetic energy of the vapor atoms and/or electrostatic charges on the deposit and the substrate. This effect may be the result of both the high adatom mobility and substrate cleaning by desorption. The relative contributions of the two factors are, however, not established. The kinetic energy and the electrostatic charges are also known to influence the growth of evaporated films. Here, the electrostatic charges present in the vapor atoms depend on the ionizable impurities in the source, and the vaporization temperature.

The epitaxial growth of electrodeposits of metals at room temperature is most probably due to the higher kinetic energy and the surface mobility of the adsorbed metal ions. A brief description of the

Table VII. Some Examples of Epitaxial Growth of Metal Films Deposited by Different Techniques on Various Air-exposed Substrates

The listed minimum epitaxial temperatures refer to one particular work; the values may vary widely with the deposition conditions employed. Where no references are given, the information is either well known, or is taken from the author's work. Ev, evaporation; Sp, sputtering; El. Dep., electrodeposition.

Deposit	Substrate	Orientation dep ∥ sub	Deposition method	Epitaxial temp, °C	Reference
Au	NaCl (100)	(111) ∥ (100) / [$1\bar{1}0$] ∥ [011]	Ev, Sp	150	
	NaCl	P-P*	Ev	400	295
			Sp	RT	114
	LiF (100)	P-P*	Ev	440	295
	KCl (100)			380	
	KI (100)			184	
	CaF$_2$ (111)			420	
	MgO (100)			400	
	Ge (100), (110)			50	266
	Mica	(111) ∥ (001) / [$1\bar{1}0$] ∥ [100]	Ev	400	
			Sp	150	
	MoS$_2$		Ev	180	
	MgO		Ev	−190	269
Ag	NaCl (100)	P-P*	Ev	200	105
			Sp	−180	
Al	NaCl (100)	(111) ∥ (100) / [$1\bar{1}0$] ∥ [011]	Ev	100	
		P-P*	Ev	300	

Table VII. Some Examples of Epitaxial Growth of Metal Films Deposited by Different Techniques on Various Air-exposed Substrates *(Continued)*

Deposit	Substrate	Orientation dep ∥ sub	Deposition method	Epitaxial temp, °C	Reference
Pd	NaCl (100)	P-P*	Ev	250	
Fe	NaCl (100)	P-45°† P-P*	Ev	400	
	Mica	$(111) \parallel (001)$ $[1\bar{1}0] \parallel [100]$	Ev	400	
Cr, Mo, Nb, Ta	MgO	P-P*	Ev, Sp	500	
Zn, Cd	NaCl, mica	Depends on deposition conditions	Ev	100	
Ni ⎫ Cu ⎬ Au ⎪ Ag, Pd ⎭	Ag (100) film	P-P*	Ev	100 0 −100 −196	318
Fe	Ag (100)	P-P*	El. Dep.	RT	85 (page 111)
Fe ⎫ Cr ⎬ Fe-Ni ⎪ Fe-Co ⎭	Cu (100)	$\{(110) \parallel (100)$ $[1\bar{1}1] \parallel [011]$	El. Dep.	RT	85 (page 111)

*P-P means parallel planes with parallel directions.

†P-45° means parallel planes with corresponding directions inclined at 45° to each other.

234

Table VIII. Epitaxial Growth of Films of Some Insulating Materials

Parallel orientation means parallel to both surface and direction. Ev, CVD, Sol, and Sp stand for evaporation, chemical–vapor deposition, from the solution, and sputtering, respectively

Deposit	Substrate	Orientation dep ‖ sub	Deposition method	Substrate temp. °C	Ref.
Thin (<100 Å) oxides	Ni, Cu, Fe, Zn, Cd	Generally, parallel	Heated in air	Variable	161
MgO	Mg	(111) [1$\bar{1}$0] ‖ (0001) [10$\bar{1}$0]	Heated in air	400-500	319
Al_2O_3	Al	Parallel	Heated in air	>500	320
MgO	MgO	Parallel	CVD	950-1100	321
Al_2O_3	Sapphire	Parallel (hex. planes)	CVD	1550-1800	322
Al_2O_3, Ta_2O_5	Sapphire	Parallel	React. Sp.	~500	115
FeO, NiO, CoO	MgO	Parallel	CVD	600-700	323
$X\,Fe_2O_4$ ferrites (X = Ni, Co, Fe, or Mn)	MgO, Pt, Al_2O_3, $MgAl_2O_4$	Parallel	CVD	~700 800-1050	324
$Y_3Fe_5O_{12}$, $Gd_3Fe_5O_{12}$	$Y_3Al_5O_{12}$	Parallel	CVD	~1200	325
$BaTiO_3$, $SrTiO_3$ $CaTiO_3$, $NaNbO_3$ $BaSnO_3$, $BaCeO_3$	LiF, NaF, Au (100)	Parallel	Flash Ev	500-700	326
NaCl, KCl, NaBr	Ag (001) film	(111) [1$\bar{1}$0] ‖ (001) [100] (001) [100] ‖ (001) [110]	Ev Sol	RT	71
NaCl, KCl, NaBr	Ag (111) film	(111) [1$\bar{1}$0] ‖ (111) [2$\bar{1}\bar{1}$]	Ev and Sol	RT	71
LiF	Ag (111) film	Parallel			71
Alkali halides (except Cs and Tl)	Mica	(111) [1$\bar{1}$0] ‖ (001) [100] [1$\bar{1}$0]	Ev	RT	71
Homopolymers	Alkali halides (001)	Chain axis ‖ (001) ⟨110⟩	Sol	~100	267

235

Table IX. Some Examples of Epitaxial Growth of Semiconductor Films
The orientation of the deposit is parallel to that of the substrate in most
cases. The minimum epitaxial temperatures are given, primarily to indicate
approximate values

Deposit	Substrate	Deposition method	Epitaxial temp, °C	Reference
Ge	Ge (100), (111)	Sp	225	273
		Ev	300	
	CaF$_2$ (111) Polished	Sp, Ev	225	273
		Ev	600-700	271
	Ge	CVD	600	285
	Sapphire (0001)	CVD	750-900	327
Si	Si (100)	Ev	RT	251
		Ev	380	328
	Si (111)	Ev	450	328
		Ev	550	252
	Si	CVD	1100	69
	Sapphire (0001)	CVD	1200	329
GaAs	Ge (100), (111)	Fl. Ev	475	330
	Ge, GaAs	CVD	700-850	307
GaP	Ge	Fl. Ev	540	330
	GaAs (111)	CVD		331
GaSb $\rbrace$ InP $\rbrace$ InAs $\rbrace$ InSb $\rbrace$	Ge	Fl. Ev	500 300 500 300	330
CdS	NaCl, mica	Ev	25-300	
SiC	Si (100), (111) (110)	CVD	800-1000	148
Bi$_2$Te$_3$	NaCl (100), (111)	Sp	200	315
Pb (or Sn) S (Se, Te)	Mica (111)	Ev	200	
	NaCl (100), (111)	Ev	200	

ion-migration process is given in Sec. 4.2, Chap. II. In the case of anodic oxidation, the mobilities must be low because of the strong affinity of oxygen with the metal ions, and consequently amorphous or fine-grained structure is obtained in the oxide films. A subsequent annealing of these films can produce oriented growth [320].

Nonstoichiometry, common to thermal and chemical-vapor deposition methods, is known to affect (generally assist) the epitaxial growth of semiconducting compounds. The high temperatures and contaminating ambient generally prevalent in a chemical-deposition method strongly influence the structure and growth of epitaxial films. The epitaxial temperatures are generally appreciably higher than those obtained by physical evaporation of the same deposit, but the chemical techniques conveniently allow epitaxial deposition of most materials with controlled stoichiometry. Excellent examples are provided by the epitaxy of various complicated metal oxides (see Table VIII).

One may use a chemical reaction to obtain epitaxial growth of compounds and alloys. For example, an Ag film may be grown epitaxially and then a Se or Te film may be evaporated onto the heated Ag film. The chemical reaction between the two films results in good epitaxial growth of a compound film. Furthermore, the composition of the compound can be controlled conveniently by electronic monitoring of the deposits. This simple technique can be extended to other systems.

A controlled, lateral epitaxial growth from a large, single crystallite without the subsequent influence of the substrate may be obtained [346-348] by lateral deposition through a moving mask. First, a thin film is deposited on a suitable substrate through a fixed mask which has a narrow tapered position with an acute angle. Then, further deposition is carried out through a movable (at speeds of $\sim 100\ \mu$/min) slotted mask placed between the fixed mask and the evaporation source. Clark and Alibozek [347] sputtered Si onto Si through a moving-mask arrangement and obtained epitaxial growth at a temperature of $\sim 200°$C, which is significantly lower than is normally required. Braunstein et al. [348] combined the moving-mask method with the VLS technique, using Au-Si eutectic liquid alloy film as a predeposit, to obtain lateral ($\sim 100\ \mu$) oriented growth of Si films on amorphous quartz at $\sim 850°$C.

(8) Summary. The varied dependence of the epitaxial growth of films on several parameters points out the difficulty of comparing results from different observers. Neveretheless, it is evident that the necessary requirements for obtaining epitaxial growth by various deposition techniques are a single-crystal substrate and a high atomic-mobility supplied by the substrate temperature or some other physical agency.

The list of epitaxially grown thin-film materials is lengthy and increasing rapidly. Tables VII, VIII, and IX are a brief compilation of some interesting examples of metals, insulators, and semiconductors (see Chap VII, Sec. 3.3 for more references on semiconductor films), respectively, to illustrate the feasibility of epitaxial growth of a wide variety of materials deposited on various substrates by the major deposition techniques. The lattice misfits (see Refs. 85 and 161) are omitted since they convey no significant meaning as such. Furthermore, representative values of the epitaxial temperatures are given only as a guide to the beginner in the field. But it must be recognized that these values are peculiar to the particular surface and the deposition conditions used by the investigator. The listed orientations are the predominant ones at the epitaxial temperature. Many systems exhibit more than one orientation in different temperature ranges and under different deposition conditions.

5.2 Theories of Epitaxy

We can conclude from the preceding section that any theory of epitaxy must explain the following experimental facts: (1) A particular value of lattice misfit or the best-fit geometrical matching of lattice is neither a necessary nor a sufficient condition. (2) Pseudomorphism or constraining of the lattice of the deposit by the substrate, if present, is not required for obtaining epitaxial growth. (3) Epitaxy appears at the earliest stages of film growth by three-dimensional nucleation, although orientation changes with subsequent growth may also result by reorientation or recrystallization on coalescence of differently oriented nuclei. (4) For a given material and suitable deposition conditions such as a low rate and high deposition temperature, epitaxy, or at least partial epitaxy, is generally possible on single-crystal substrates of a variety of crystal structures and different types of atomic bonds. The orientation, degree of perfection, and the deposition conditions required for epitaxy do, however, vary with different substrate materials for the same deposit and vice versa.

(1) Royer Hypothesis. Several geometrical and phenomenological theories of epitaxy have been proposed. Some may be rejected outright, while others are too limited to explain the above observations. Three rules for epitaxy proposed by Royer [262], and now clearly only of historical interest, are: (1) Lattice planes exist with elementary or multiple networks which are identical in form and are of nearly the same dimension in the two lattices. (2) Where ionic crystals are involved, the ions of the substrate should take up positions which corresponding ions of the substrate of the same polarity would have occupied had the substrate continued to grow. (3) The substrate and overgrowth crystals should have the same type of bonding.

Menzer [288] proposed that an oriented layer corresponding to a good fit occurs during the initial stages of growth, and subsequent growth gives rise to different orientations. Accordingly, the orientation of initial deposits of Ag on NaCl should be (221). But this orientation is not observed, so that the hypothesis is not supported.

(2) van der Merwe Theory. This theory, initiated by Frank and van der Merwe [289] and investigated in detail by the latter author [290], assumes that the initial growth stage of an oriented deposit is the formation of a monolayer of regular atomic order with a spacing determined by the energy of the substrate-deposit interface. The lowest-energy state of the system is shown to be the one with a thin film constrained to fit the substrate. This occurs when the lattice misfit is less than a critical value, about 4 percent in an average case. For small misfits of several percent, the difference in the spacings of the film and the substrate can be reduced to zero with a homogeneous lateral strain. Above a misfit of about 12 percent in an average case, a cross grid of misfit dislocations will occur at the interface even in a monolayer deposit, provided that a suitable activation energy for their generation is available. Thus, for a low misfit, an initially pseudomorphic film will grow which, beyond a critical thickness, will undergo a transition process to yield an oriented and strain-free deposit of normal structure.

The occurrence of pseudomorphism (Sec. 4.1) and misfit dislocations [291] in a number of systems has been verified recently. Nevertheless, two-dimensional deposition and pseudomorphism are of little significance for most epitaxial systems for which the absolute value of misfit is not the important parameter for epitaxy. Furthermore, epitaxy apparently takes place even in the smallest observable nuclei (~10 to 20 Å across), the interface of which is too small for misfit dislocations to form. This is, therefore clearly not a theory of epitaxy but rather a theory of accommodation of misfit between two lattices.

(3) Brück-Engel Theory.   The epitaxial growth of metals on alkali halides was explained by Brück [268] by means of a rule according to which the sum of the distances between the atoms (ions) of a metal and halogen ions must be minimal. This rule is illustrated in Fig. 23, where the possible positions (according to this rule) of the metal atoms relative to the net of Cl ions on the (100) face of NaCl are compared with the observed, and the geometrically lowest misfit positions. It is clear that on (100) NaCl, parallel orientations suggested by the rule are more favorable positions. For an orientation of the type (111) metal ∥ (100) NaCl,

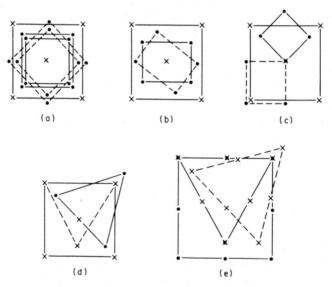

FIG. 23. Relative geometrical arrangement of metal atoms (solid circles) and Cl ions (crosses) in various metal–NaCl (100) epitaxial systems based on Brück's rule. (*a*) (100) Ag, (*b*) (110) Fe, (*c*) (100) Fe, (*d*) (111) Au and Al, and (*e*) (111) oriented NaCl on (100) Ag film. The solid-line con- figurations are actually observed, and the dotted ones correspond to the best geometrical fit of lattices. (*After Palatnik and Papirov* [85])

Brück's rule suggests two nearly equivalent positions, which may explain the occurrence of both orientations.

It may be noted from Fig. 23(d) and (e) that the orientation of (111) Ag on (100) NaCl is not the same as that of (111) NaCl on (100) Ag. Dankov [292] explained this difference on the basis of the requirement of electrostatic-charge neutrality conditions which are obviously different for metal atoms of the same size depositing on a network of Na and Cl ions in one case, and for Na and Cl ions of different size and opposite charges trying to find an equilibrium position on the metal lattice.

A criticism [293] of Brück's rule that it should apply only to a unit cell rather than the whole nucleus is not justified since epitaxy is certainly an atomistic rather than a geometric process. Engel [294] interpreted the Brück rule physically to mean that the minimum sum of the distances between the positive ions of the deposit and the negative ions of the substrate is the direct result of the coulombic forces. Accordingly, the dominant orientation is the one for which the net attractive coulombic forces are maximum. Sample calculations showed that the parallel orientation was energetically favorable for an fcc metal on an alkali-halide surface.

Engel assumed that the ionization energy of a metal atom adsorbed on the surface of ionic crystals may be considerably reduced so that ionization takes place at relatively low temperatures. Epitaxial growth then occurs at a temperature at or above which the electrostatic interaction with the ionized metal surface is sufficiently strong to mobilize atoms to take up equilibrium positions for an oriented structure. This definition of epitaxial temperature is necessarily qualitative in view of the uncertainty of the surface ionization energy. But it is interesting to note that, consistent with this model, the epitaxial temperatures of metals on alkali halides increase with the increasing bulk ionization potential, e.g., in the series Ag, Cu, Al, and Au.

Göttsche [295] developed concepts of epitaxial growth based on the polarization of metallic atoms on the surfaces of ionic salts, combined with electrostatic interactions between the various charged atoms at the interface. He suggested that the orientation of the nuclei may be strongly influenced by the closest-packing direction of ions of like signs lying on the substrate surfaces, since the distance between ions is minimum along such directions.

The dependence of epitaxial temperature on the nature of the substrate may be understood in terms of the differences in the polarizability of the atoms. Polarizability increases in the sequence Li, Na, and K for the metallic ions, and in the sequence F, Cl, Br, and I for the halogen ions. An increasing polarization is believed to increase attractive forces between the metallic deposit and the alkali-halide surface, thereby lowering the epitaxial temperature. Thus, the epitaxial temperature on LiF should be higher than on NaCl, which is in agreement with the observations.

An empirical criterion for the determination of the epitaxial correlation of metals and salts has been pointed out by Papirov and Palatnik [296], among others. Accordingly, orientation on salts is such that the two directions with a maximum close packing of particles of like sign in the salt lattice are parallel to the two directions with close-packed atoms in the metal lattice. The correlation between the directions of closest packing of atoms in the two epitaxial systems is in accord with many examples of epitaxy. This result may be interpreted as indicating the importance of the electrostatic interactions in determining epitaxy. Thus, although the original Brück-Engel hypothesis is qualitative and limited in scope, the underlying concept may be useful for a detailed theory of epitaxy. In this connection, the theory of interfacial adhesion based on electrostatic interactions between lattices (Chap. V, Sec. 4.2) is of considerable interest.

For the sake of completeness, let us mention that the possible influence of the van der Waals forces in orientation effects has been considered by Dixit [297] and by Dankov [292]. These theories are, however, rather limited in concepts and are not pursued here further.

(4) Nucleation Theories. The smallest observable nuclei appear to be oriented. Whether this is due to the growth of the initially oriented *critical* nuclei, or the growth of a dominant orientation resulting from the coalescence of the variously oriented critical nuclei cannot be answered at present. It is, however, very likely that epitaxy is nucleated. To explain the phenomenon of epitaxy, any nucleation theory must predict the occurrence, or at least the predominance, of critical nuclei of a particular orientation in the earliest stages of growth. Unfortunately, the existing nucleation theories cannot predict but can only qualitatively suggest the conditions which may lead to the realization of this prediction.

According to the capillarity theory of nucleation, epitaxy will occur for that orientation which gives lower interfacial free energies, and hence a lower free energy of formation for the critical nucleus, and a much higher nucleation rate than any other orientation. At sufficiently high supersaturations (high deposition rate and/or low substrate temperature) many orientation relationships may have appreciable nucleation rates, thus destroying the epitaxy.

In the Walton-Rhodin atomistic model, orientation is believed to result from the arrangement of atoms in the smallest cluster (critical nucleus plus one atom). As supersaturation decreases from a very high value, the size of the critical nucleus increases discretely from one atom to two atoms and so on. The smallest cluster leading to orientation is the one with three adsorbed atoms. A minimum number of bonds is achieved for a three-atom cluster forming a (111) orientation for an fcc metal. The next size of cluster of four atoms would form a (100) orientation, and of five atoms a (110) orientation. Since the concentration of a cluster decreases rapidly with increasing size, and further, since the smallest-sized clusters correspond to the more close-packed orientation, critical nuclei which lead to the close-packed orientations occur in greater numbers and are favored, and for this reason these orientations prevail in the deposit. With increasing substrate temperature (decreasing supersaturation), a transition from (111) to (100) orientation is expected on the basis of this theory. Only the epitaxy of Au and Al films shows a transition in agreement with this prediction. Clearly, the theory does not favor epitaxial growth of one orientation only at high temperatures, which is in contradiction to the general observations.

Both nucleation theories explain the role of the substrate lattice in determining epitaxial orientation by assuming that the substrate energetically favors the adsorption of one geometrical arrangement of atoms in the cluster or nuclei over another. The lack of reliable data on the parameters of interfacial energies and the cluster binding energy for an atomic size cluster makes it impossible to justify this assumption. Further, this assumption is unfortunate since it is indeed the central fact which needs to be explained by any formal theory of epitaxy. The failure of the nucleation theories of epitaxy is dramatized by the fact that neither can explain even the simplest examples of epitaxy, e.g., that of the fcc metals films with parallel azimuthal orientation on rock salt, and (111) orientation on mica.

(5) Summary. Although a qualitative understanding of the occurrence of oriented growth can be obtained from empirical rules and nucleation concepts, no existing theory satisfactorily explains and predicts the epitaxial relation between a condensate and the substrate. A theory of epitaxy required to explain the preferential adsorption of certain orientations must take into account interactions between the substrate and condensate atoms. The Brück-Engel theory suggests one reasonable approach in this direction.

6. STRUCTURAL DEFECTS IN THIN FILMS

Electron-microscopic examination of thin ($<1,000$ Å), single-crystal epitaxial films of fcc metals has revealed a variety of structural defects, such as dislocation lines, stacking faults, microtwins and twin boundaries, multiple positioning boundaries, grain boundaries, and minor defects arising from aggregation of point defects (e.g., dislocations loops, stacking-fault tetrahedra, and small dotlike defects). Some of these defects can be seen in the typical transmission electron micrograph of an epitaxial Ag/NaCl film shown in Fig. 24.

Little work has been done on structural imperfections in polycrystalline films. Grain boundaries are expected to form a major contribution to imperfections in this case. The evidence of structural defects reported in the literature has been primarily obtained for evaporated fcc metal films up to 1,000 Å thick grown epitaxially on NaCl, mica, and MoS_2 substrates. Among the various studies [51, 161, 282, 298-302], those of Bassett, Pashley, and coworkers [161, 282], who exploited the moiré-pattern technique, are the most systematic. Significant contributions to the observations and interpretation of defects in metal films have been made by Matthews [299] and Jacobs et al. [282]. This section contains a brief summary of the observations and interpretation of structural defects based on the reviews by Pashley [161] and Stowell [298]. The reader should consult these reviews for details of the results and the pertinent references. Only leading references are given here.

Moiré-pattern studies indicate that no lattice defects are present in the small discrete nuclei of the epitaxial deposits of the fcc metals but that they are incorporated into the film during and after coalescence of the

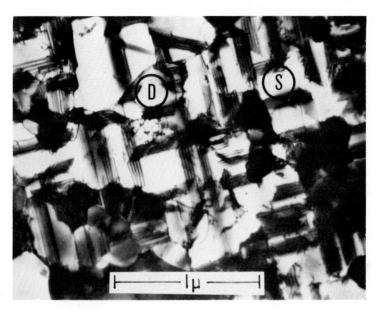

FIG. 24. A typical transmission electron micrograph of ~1,000-Å-thick epitaxial Ag film evaporated onto cleaved NaCl at 200°C, showing stacking fault (S) fringe arrays, dislocations (D), and other defects. (*Chopra, unpublished*)

nuclei with each other. Dislocations are the most frequently encountered lattice defect in such films. They consist of lines passing from one side of the film to the other by the shortest distance, and their density is about 10^{10} to 10^{11} lines/cm^2. Dislocations with appreciable lengths running nearly parallel to the plane of the film are very rare in thin films (<500 Å), but there may be a tendency for them to form in thicker films.

Dislocations can be formed by the following mechanisms:

1. Since nuclei form at random points on the substrate surface, misorientations and/or misfit displacements of the lattices of nuclei or islands exist. On coalescence, dislocations are incorporated at the boundary of the two islands. If the islands are small, they rotate and translate to accommodate small misfits. Consequently, dislocations are permanently incorporated only when the islands become too large to move to minimize the misfit. Thus, the dislocation density (Fig. 25) rises rapidly when the final-stage coalescence of large islands takes place to

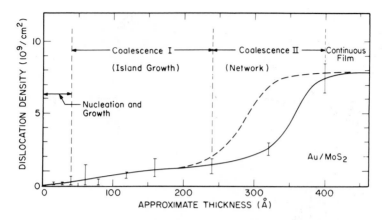

FIG. 25. The density of dislocations (per unit area of the substrate) in Au films evaporated onto cleaved MoS_2 at 300°C, at various stages of growth. The dotted curve includes an estimate of the incipient dislocations present in the holes and channels. (*Jacobs et al.* [282])

make the film continuous. The formation of a dislocation on coalescence of islands is illustrated in Fig. 26(*a*) by the moiré-pattern technique.

2. If coalescence of more than two islands takes place simultaneously so as to minimize the elastic strain at each interface, the islands may join to leave a hole between them. Stresses in the film may generate incipient dislocations in the hole. It is a geometrical necessity that real dislocations are formed as these holes fill in, since the incipient dislocations are effectively dislocations with a hollow core. The incipient-dislocation contribution is significant, as shown by the data in Fig. 25.

Dislocation densities in thicker films have not been studied. It is clear, however, that if the misfit is the major cause of dislocations, the density of dislocations should decrease when self-deposition or auto-epitaxy occurs. This is supported by the observation that dislocations can be largely absent in carefully prepared deposits of silicon on silicon.

3. An interface misfit-dislocation network may be set up to relieve strain at the film-substrate interface. Evidence of this mechanism has been obtained in various systems, and an extensive review of this subject is given by Matthews [291] (see Sec. 4.1.2 for further discussion).

4. When islands containing stacking faults bounded by the surfaces of the island coalesce, partial dislocations must now bound these faults in the continuous film [299].

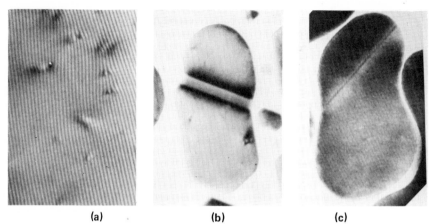

FIG. 26.   Moiré images of lattice defects formed during coalescence of islands. (*a*) Dislocations in (111) Au film on MoS$_2$ ($\times$ 550,000) (*courtesy of T. J. Law*); (*b*) wide twins in a 200-Å-thick (111) Ag film on MoS$_2$ ($\times$ 255,000); (*c*) stacking faults in a 200-Å-thick (111) Ag film on MoS$_2$ ($\times$ 590,000) (*courtesy of M. J. Stowell* [298]). The magnifications given are the values of original photographs. The moiré-fringe width in each case is 17 Å.

5. Dislocations (and other defects) may be formed because of point-defect aggregation during growth provided that the vacancies are relatively immobile. Deposition at very high rates and sufficiently low temperatures might cause many vacancies to become trapped, and these could aggregate slowly.

6. The extension of the substrate imperfections into the film is a possible mechanism [300]. It cannot be the major mechanism in most systems, however, because the dislocation density in the deposit is usually several orders of magnitude greater than in the substrate surface ($\sim 10^5$/cm^2 for NaCl). Moreover, the nucleation process on the substrate surface is known to be random; that is, it largely ignores the presence of surface dislocations. The only condition under which the extension of substrate dislocations can take place is if the nuclei are constrained (pseudomorphism) by the substrate. It should also be pointed out that surface impurities can have a significant effect in introducing dislocations and other defects in the deposit. This cannot, however, be an important mechanism since dislocations are formed in the deposit even under apparently clean surface conditions.

We have mentioned above that vacancies may become trapped during rapid quenching of the supersaturated vapor. Single vacancies have not yet been observed with the electron microscope, possibly because of the lack of sufficient diffraction contrast and/or resolution. However, minor defects such as dislocation loops, stacking-fault tetrahedra, and small triangular defects which can occur because of vacancy collapse have been observed. Phillips [300] measured about 10^{14} loops/cm^3, of 100- to 300-Å size, in a 650-Å-thick epitaxial Ag film on rock salt. If each loop is indeed a collapsed disk of vacancies, one has to postulate a vacancy concentration of at least 1.5×10^{-5} in the film before collapse takes place. This value suggests that vacancies are frozen in by an effectively high-temperature quenching of the vapor atoms. This is by no means an unreasonably high value for a thin film condensation process. Much higher ($\sim$1 percent) values are suggested by the annealing kinetics of the electrical resistivity of metal films (Chap. VI, Sec. 3.7). The observed dotlike features may be either unresolved dislocation loops, vacancy aggregates, aggregates of impurity atoms trapped in the film during preparation, or decomposed hydrocarbon contamination of the film during electron bombardment inside the electron microscope.

The electron micrographs (of epitaxial films of fcc metals) show a high density of planar defects on the {111} planes. Many of these defects are joined together in characteristic arrays as seen in Fig. 24. The observed fringe patterns may be due to both stacking faults and microtwins which can be distinguished by electron-diffraction patterns and dark-field images (Chap. III, Sec. 2.4). It is easily shown that both stacking faults and twins exist in these films. Simple and complex (extended to two or more planes) stacking faults as well as simple and multiple [301] twins are commonly found in these films.

Twins and stacking faults occurring on the inclined {111} planes of fcc metal films are formed during coalescence of islands and occur at the junction plane between the islands as shown in Figs. 26(b) and (c). Complex stacking faults are formed when more than two islands coalesce. Stacking faults are more stable than dislocations in the small islands since a fault which completely traverses an island introduces no elastic strain. Jacobs et al. [282] found that, as further coalescence occurs, the faults are forced either to terminate or to extend appreciably so that all the faults are eventually eliminated by the time a continuous film is produced. Thus the fault density passes through a maximum value of about $2 \times 10^{10}/cm^2$ when the area of coverage of the substrate is in the region of 50 percent. Note that faults may also be formed because of the dissociation of the whole dislocations into two partials. Fewer twins than stacking faults are formed, and it is difficult to distinguish the very fine twins from stacking faults. All the twins are also eliminated as the film becomes continuous.

Brockway et al. [302] studied the microstructure of Cu films grown epitaxially on NaCl and KCl and observed a high ($\sim 10^9/cm^2$) density of dislocations and some narrow stacking faults. On annealing at $\sim 630°C$ in H_2, the dislocation density decreased to $10^7/cm^2$ and stacking faults of ~ 2 μ width with faulted area of $\sim 5 \times 10^4$ cm^2/cc were observed. By contrast, similar annealing of films detached from the substrate introduced no faults but increased the dislocation density a hundredfold. The authors concluded that faults are formed under compressive stresses during cooling of the film on the substrate. The formation of complex faults due to stressing of films was first proposed by Phillips [300].

An effectively incoherent twin boundary imperfection is provided by the multiple-positioning boundary, which is a common type of imperfection in epitaxial films because of the occurrence of equivalent deposit orientations in many systems. The most studied example of this imperfection is the double-positioning boundary produced by the two equivalent (111) orientations of a triangular crystallite of an fcc metal on a substrate of hexagonal symmetry (e.g., Au on mica). Double-positioning boundaries are inevitably produced as islands of two equivalent orientations grow together. If liquid-like coalescence of the two double-positioned islands takes place, the boundary migrates, probably under the driving force of the boundary energy, and moves out of the compound island, leaving it in one position. The smaller island generally takes on the orientation of the larger one. This recrystallization effect may be understood in terms of the thermally activated diffusion of point defects at the boundary.

Dislocations, stacking faults, and twins are also observed in epitaxial semiconductor films. The only systematic studies [303-306] of imperfections in semiconducting films have been carried out on thick ($\sim$ a few microns) auto-epitaxial deposits of silicon. A large number of dislocations are generally present in hetero-epitaxial deposits. A well-defined triangular defect [Fig. 27(a)] occurring in Si films deposited on (111) Si surfaces by various techniques such as evaporation, sputtering, and chemical disproportionation consists of three inclined stacking faults forming a triangular-based tetrahedron with its apices coinciding with the substrate-deposit interface. Similar types of faults [e.g., a square-based tetrahedron on (100) surfaces] are formed on other Si surfaces. The large stacking-fault triangles in Fig. 27(a) are at tetrahedra which nucleated during the initial growth of the film. The smaller ones nucleated during the second deposition carried out subsequent to cooling of the substrate and then reheating it. In this case, more tetrahedra nucleated, probably because of the effect of contamination during cooling. As the film thickness increases, most overlapping tetrahedra

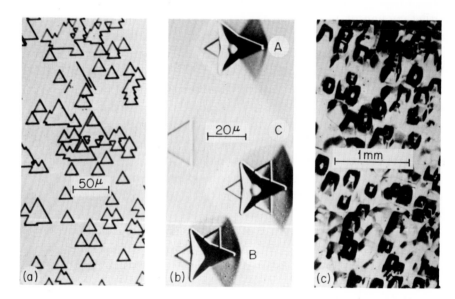

FIG. 27. (a) Optical micrographs of etched ~10-μ-thick Si films prepared by chemical-vapor deposition on (111) Si surfaces showing stacking faults; (b) tripyramid type of defect accompanied by stacking-fault tetrahedra in three stages of growth in the sequence A to C form a hexagram (*courtesy of S. Mendelson* [303]); (c) an x-ray reflection topograph of hillock-like features due to twins on ~1-μ-thick epitaxial GaAs film formed on {100} GaAs by chemical-vapor deposition at ~1,000°C (*courtesy of H. Holloway* [307]).

blend into large polyhedra; the opposite type may retain a stacking fault between them.

The twinned regions consist of simple twin lamella on inclined {111} planes, with thickness ranging from a very few atom-plane spacings to large fractions of a micron. A more complex arrangement is the tripyramid defect in which three oblique pyramids grow together by deposition within the crater as shown in the sequence of Fig. 27(*b*). The tripyramids shown here are accompanied by stacking-fault tetrahedra at various stages of growth to form hexagrams. Investigations of these defects led Chu and Gavaler [305] to suggest that the tripyramids consist of an arrangement of doubly twinned material. Booker [306] showed that each segment of the tripyramid is doubly twinned with respect to the substrate orientation. The central region is similar to the double-positioning structure in fcc metal films. The coalescence of this region with neighboring nuclei seems to produce the tripyramid.

Many studies conclude that the origin of the tripyramid defects is associated with small traces of contamination on the substrate surface. Booker and Unvala [69] studied the growth stages of evaporated Si films and found a very high fault density in the regions where the first islands form and a relatively low density between these islands. This was explained by the fact that mobile Si atoms act as getters and carry the contamination from the surface to the stable nuclei.

Detailed mechanisms and crystallographic models for the production of the stacking faults in Si have been put forward by many workers and are reviewed by Mendelson [303]. It is generally agreed that the physical displacement of the deposit atoms by impurity atoms, and the prevalent twinning of the deposit layers are chiefly responsible for the observed stacking-fault configuration in epitaxial Si films.

Epitaxial films of III-V semiconductors show a variety of surface features. Holloway and coworkers [307] studied x-ray topographs of GaAs film surfaces grown epitaxially on GaAs (both polar faces) and Ge substrates by chemical-vapor deposition. The {100} surfaces of both homo- and hetero-epitaxial GaAs films revealed a variety of hillock-like features [Fig. 27(c)] which were ascribed to the formation of twins at or near the epitaxial interface. During growth of the film, the twins can become buried, leaving a matrix-oriented surface. The occurrence of the twin features correlated with the presence of an unidentified contaminant. An important observation was made that neither the conditions for epitaxial growth nor the structural perfection of the deposit depended in any way on the polarity of the GaAs surface, provided vigorous precautions were taken to exclude oxygen and water vapor from the system.

Sloope and Tiller [308] studied the microstructure of epitaxial Ge films evaporated onto (111) CaF_2 at $\sim600°C$. Electron micrographs showed increasing density and complexity of defects with increasing deposition rate and decreasing substrate temperature. The density of defects consisting of dislocations, stacking faults, and twins was $\sim10^9$ to $10^{11}/cm^2$, which remained unchanged for thicknesses up to 4,000 Å. The density of defects decreased by a factor of up to 10^3 over the 0.5- to 17-μ film-thickness range.

Complex stacking-fault defects obtained in Si films have also been found in epitaxial films of Ge [309, 310] and other semiconductors such as GaAs [311] and CdTe [312]. The microstructure of Ge and Si films has been reviewed by Newman [313].

It is clear from the preceding discussion that the mechanisms responsible for introducing lattice defects in epitaxial films are directly related to the dynamics of coalescence of the three-dimensional islands and crystallites in the films. The coalescence stage in vapor deposition is critical and holds the key to understanding and controlling structural defects. An early and enhanced coalescence of islands brought about, for example, by an applied electric field or electron-beam irradiation during film deposition, or by annealing is expected to reduce the density of structural defects. The experimental results are in general accord with this conclusion.

7. CONCLUDING REMARKS

One should not expect an exact quantitative description of the wide variety of growth effects in thin films in view of the variable but significant influence of a multitude of parameters involved. Nevertheless, some understanding of these effects is clearly necessary to exploit thin film specimens effectively for physical investigations and device applications. Unfortunately, many thin film structural phenomena remain obscure, primarily because of the lack of properly organized and well-defined experiments which may exploit one or more of the analytical techniques to monitor the deposition process. Some of the problems which need attention are: more details of the condensation process (e.g., energy exchange between the vapor atom and the substrate, the high mobility of adatoms on some surfaces even at sufficiently low temperatures, the nature and density of the condensation centers, the dynamics of the coalescence processes), the high chemical activity and long-range rearrangements of atoms on some clean surfaces, the origin of nucleation fiber textures in films deposited on amorphous substrates, the phenomenon of epitaxy, the processes of metastabilization of new and abnormal crystal structures in films, and the incorporation and annihilation of lattice defects in thick films. These studies must recognize that the clean surfaces of a substrate can be more reactive and differently ordered than the bulk materials.

It is ironic that the phenomenon of epitaxy, which, because of its basic and practical importance, has been so thoroughly investigated experimentally, still remains a mystery. It is clear that both the statistical nucleation concepts and microscopic interactions between the atoms of the substrate and the deposit must be taken into account by any successful theory of epitaxy.

The atomistic growth process of a vapor-deposited film allows the preparation of an infinite variety of multicomponent alloys and compounds with normal and abnormal structures which are often impossible to obtain in the bulk form. In many cases, these materials can be grown epitaxially. Thus, a promising and exciting new dimension has been added to materials technology.

Solid-state diffusion is a subject of much basic and practical interest. Thin superimposed films are ideally suited for such studies. Almost all the literature in this area has been primarily concerned with the identification of various alloy phases formed by interdiffusion. The subject is thoroughly reviewed by Weaver and Hill [332]. Clearly, it should be possible to obtain a significant amount of quantitative information on diffusion by using superimposed films [355].

REFERENCES

1. H. Mayer, "Physik dünner Schichten," vol. 2, Wissenschaftliche Verlagsgesellschaft, Stuttgart, 1955.
2. J. P. Hirth and G. M. Pound, "Condensation and Evaporation–Nucleation and Growth Kinetics," The Macmillan Company, New York, 1963.
3. G. M. Pound and J. P. Hirth, in "Condensation and Evaporation of Solids" (E. Rutner, P. Goldfinger, and J. P. Hirth, eds.), p. 475, Gordon and Breach, Science Publishers, Inc., New York, 1964.
4. R. A. Sigsbee and G. M. Pound, *Advan. Colloid Interface Sci.,* 1:335 (1967).
5. T. N. Rhodin and D. Walton, similar reviews in "Single Crystal Films," p. 31, Pergamon Press, New York, 1964; "Metal Surfaces," p. 259, American Society for Metals, Metals Park, 1963; "The Use of Thin Films in Physical Investigations" (J. C. Anderson, ed.), p. 187, Academic Press Inc., New York, 1966.
6. S. Wexler, *Rev. Mod. Phys.,* 30:402 (1958).
7. H. S. W. Massey and E. H. S. Burhop, "Electronic and Ionic Impact Phenomena," chap. 9, Oxford University Press, London, 1952.
8. I. Langmuir, *Phys. Rev.,* 8:149 (1916).
9. N. Cabrera, *Discussions Faraday Soc.,* 28:16 (1959).
10. R. W. Zwanzig, *J. Chem. Phys.,* 32:1173 (1960).
11. G. Ehrlich, in "Structure and Properties of Thin Films" (C. A. Neugebauer, J. B. Newkirk, and D. A. Vermilyea, eds.), p. 423, John Wiley & Sons, Inc., New York, 1959.
12. F. O. Goodman, *Phys. Chem. Solids,* 23:1269, 1491 (1962); 24:1451 (1963).
13. J. E. Lennard-Jones, *Proc. Roy. Soc. London,* A163:127 (1937).
14. B. McCarrol and G. Ehrlich, *J. Chem. Phys.,* 38:523 (1963); in "Condensation and Evaporation of Solids" (Rutner et al., eds.), p. 521, Gordon and Breach, Science Publishers, Inc., New York, 1964.
15. J. Frenkel, *Z. Physik,* 26:117 (1924).
16. N. Semenoff, *Z. Phys. Chem.,* B7:741 (1930).
17. J. D. Cockcroft, *Proc. Roy. Soc. London,* A119:293 (1928).
18. For example, see T. N. Rhodin, *Discussions Faraday Soc.,* 5:215 (1949); P. Benjamin and C. Weaver, *Proc. Roy. Soc. London,* A252:418 (1959).
19. G. Zinsmeister, in "Basic Problems in Thin Film Physics" (R. Niedermayer and H. Mayer, eds.), p. 33, Vandenhoeck and Ruprecht, Göttingen, 1966.
20. M. Volmer and A. Weber, *Z. Phys. Chem.,* 119:277 (1925); M. Volmer, "Kinetik der Phasenbildung," Theodor Steinkopf Verlagsbuchhandlung, Dresden-Leipzig, 1939.
21. R. Becker and W. Döring, *Ann. Physik,* 24:719 (1935).
22. See the review by J. H. Holomon and D. Turnbull, *Prog. Metal Phys.,* 4:333 (1953).
23. G. M. Pound, M. T. Simnad, and L. Yang, *J. Chem. Phys.,* 22:1215 (1954).
24. B. K. Chakraverty and G. M. Pound, *Acta Met.,* 12:851 (1964).
25. J. P. Hirth, *Acta Met.,* 7:755 (1959).
26. D. Walton, *J. Chem. Phys.,* 37:2182 (1962).
27. D. Walton, T. N. Rhodin, and R. Rollins, *J. Chem. Phys.,* 38:2695 (1963).

28. B. Lewis, *Thin Solid Films,* 1:85 (1967).
29. B. Lewis and D. S. Campbell, *J. Vacuum Sci. Tech.,* 4:209 (1967).
30. A. J. W. Moore, *J.Australian Inst. Metals,* 11:220 (1967); a similar approach is treated by R. Kikuchi, *Hughes Res. Lab. Rept.* 233, 1964.
31. H. Reiss, *J. Chem. Phys.,* 18:840 (1950).
32. K. G. Günther, *Z. Naturforsch.,* 13a:1081 (1959); K. G. Günther and H. Freller, *Z. Naturforsch.,* 16a:279 (1961).
33. F. M. Devienne, *Compt. Rend.,* 231:740 (1950); 234:30 (1952); *J. Phys. Radium,* 13:53 (1952); 14:257 (1953); *Vacuum,* 3:392 (1953); *Compt. Rend.,* 238:2397 (1954).
34. H. Frauenfelder, *Helv. Phys. Acta,* 23:347 (1950).
35. L. Yang, C. E. Birchenall, G. M. Pound, and M. J. Simnad, *Acta Met.,* 2:462 (1954); L. Yang, M. J. Simnad, and G. M. Pound, *Acta Met.,* 2:470 (1954).
36. S. Aisenberg, *Trans. 10th Natl. Vacuum Symp.,* p. 457, The Macmillan Company, New York, 1963.
37. R. L. Schwoebel, *Surface Sci.,* 2:356 (1964).
38. L. Bachman and J. J. Shin, *J. Appl. Phys.,* 37:242 (1966).
39. J. B. Hudson and J. S. Sandejas, *J. Vacuum Sci. Tech.,* 4:230 (1967); also, *Abstracts, 14th Natl. Vacuum Symp.,* 1967, p. 41.
40. H. L. Caswell, *J. Appl. Phys.,* 32:105, 2641 (1961).
41. J. B. Preece, H. Wilman, and C. T. H. Stoddart, *Phil. Mag.,* 16:447 (1967).
42. A. J. Melmed, *J. Appl. Phys.,* 37:242 (1966).
43. C. Sella and J. J. Trillat, in "Single Crystal Films" (M. H. Francombe and H. Sato, eds.), p. 201, Pergamon Press, New York, 1964.
44. R. D. Gretz, Ph.D. Thesis, Carnegie Institute of Technology, 1963.
45. G. W. Sears and J. W. Cahn, *J. Chem. Phys.,* 33:494 (1960).
46. J. B. Hudson, *J. Chem. Phys.,* 36:887 (1962); in "Condensation and Evaporation of Solids" (Rutner et al., eds.), p. 639, Gordon and Breach, Science Publishers, Inc., New York, 1964.
47. S. J. Hruska and G. M. Pound, *Trans. AIME,* 230:1406 (1964).
48. R. W. Shade, *J. Chem. Phys.,* 40:915 (1964).
49. R. F. Adamsky and R. E. LeBlanc, *J. Vacuum Sci. Tech.,* 2:79 (1965).
50. G. A. Bassett, *Phil. Mag.,* 3:1042 (1958); in "Condensation and Evaporation of Solids" (Rutner et al., eds.), p. 599, Gordon and Breach, Science Publishers, Inc., New York, 1964.
51. G. A. Bassett, J. W. Menter, and D. W. Pashley, *Proc. Roy. Soc. London,* A246:345 (1959); in "Structure and Properties of Thin Films" (Neugebauer et al., eds.), p. 11, John Wiley & Sons, Inc., New York, 1959.
52. K. L. Chopra, *J. Appl. Phys.,* 37:3405 (1966).
53. K. R. Lawless, *J. Vacuum Sci. Tech.,* 2:24 (1956); in "Physics of Thin Films" (G. Hass and R. E. Thun, eds.), vol. 4, p. 191, Academic Press Inc., New York, 1967.
54. D. W. Pashley, M. J. Stowell, M. H. Jacobs, and T. J. Law, *Phil. Mag.,* 10:127 (1964).
55. A. E. Curzon and K. Kimoto, *J. Sci. Instr.,* 40:601 (1963).
56. K-J. Hanszen, *Z. Naturforsch.,* 19a:820 (1964).

57. H. Poppa, *Z. Naturforsch.*, **19a**:835 (1964); *J. Vacuum Sci. Tech.*, **2**:42 (1965).
58. H. Poppa, *J. Appl. Phys.*, **38**:3883 (1967).
59. E. W. Dickson, M. H. Jacobs, and D. W. Pashley, *Phil. Mag.*, **11**:575 (1965).
60. J. G. Allpress and J. V. Sanders, *Phil. Mag.*, **9**:645 (1964).
61. J. van der Waterbeemd, *Philips Res. Rept.*, **21**:27 (1966).
62. V. A. Phillips, *Phil. Mag.*, **5**:571 (1960); *J. Appl. Phys.*, **33**:712 (1962); D. W. Pashley, *Mem. Sci. Rev. Met.*, **62**:93 (1965).
63. J. Bénard, *Acta Met.*, **8**:272 (1960); *Met. Rev.*, **9**:473 (1964); J. Oudar and J. Bénard, *Acta Met.*, **7**:295 (1959).
64. K. R. Lawless and D. F. Mitchell, *Mem. Sci. Rev. Met.*, **62**:27, 39 (1965).
65. A. U. MacRae, *Appl. Phys. Letters*, **2**:88 (1963).
66. R. E. Pawel, J. V. Cathcart, and J. J. Campbell, *Acta Met.*, **10**:149 (1962).
67. K. L. Chopra and I. H. Khan, *Surface Sci.*, **6**:33 (1967).
68. J. M. Charig and B. A. Joyce, *J. Electrochem. Soc.*, **109**:957 (1962).
69. G. R. Booker and B. A. Unvala, *Phil. Mag.*, **11**:11 (1965).
70. R. C. Newman, *Phil. Mag.*, **2**:750 (1957); also, E. Grünbaum, *Proc. Phys. Soc. London*, **72**:459 (1958).
71. L. G. Schulz, *Acta Cryst.*, **4**:483 (1951), 487 (1951); **5**:264 (1952).
72. K. L. Chopra, *J. Appl. Phys.*, **37**:2249 (1966).
73. K. L. Chopra, in "Basic Problems in Thin Film Physics" (R. Niedermayer and H. Mayer, eds.), p. 89, Vandenhoeck and Ruprecht, Göttingen, 1966.
74. D. J. Stirland, *Appl. Phys. Letters*, **8**:326 (1966).
75. R. M. Hill, *Nature*, **210**:512 (1966).
76. Reviews by H. Bethge, *Phys. Stat. Solidi*, **2**:3,775 (1962), also, H. Bethge, *Proc. Intern. Conf. Crystal Growth, Boston*, 1966, p. 623; K. W. Keller, *Proc. Intern. Conf. Crystal Growth, Boston*, 1966, p. 629.
77. A review by D. J. Stirland in "The Use of Thin Films in Physical Investigations" (J. C. Anderson, ed.), p. 163, Academic Press Inc., New York, 1966.
78. N. Cabrera, *J. Chim. Phys.*, **53**:675 (1956).
79. For example, J. W. Faust, *Acta Met.*, **11**:1077 (1963); F. Grønlund, *J. Chim. Phys.*, **53**:660 (1956).
80. D. J. Stirland, *Phil. Mag.*, **13**:1181 (1966); D. J. Stirland and D. S. Campbell, *J. Vacuum Sci. Tech.*, **3**:258 (1966).
81. G. I. Distler, in *7th Intern. Crystallography Congress and Symposium on Crystal Growth, Moscow*, 1966, p. A288; *Rost Kristallov*, **8**:108 (1966).
82. K. L. Moazed and G. M. Pound, *Trans. AIME*, **230**:234 (1964).
83. R. D. Gretz and G. M. Pound, in "Evaporation and Condensation of Solids" (Rutner et al., eds.), p. 575, Gordon and Breach, Science Publishers, Inc., 1964.
84. L. S. Palatnik and Yu. F. Komnik, *Soviet Phys. "Doklady" English Transl.*, **5**:1072 (1960); L. S. Palatnik, G. V. Fedorov, and P. N. Bogatov, *Soviet Phys.-Solid State English Transl.*, **7**:2141 (1966); Yu. F. Komnik, *Soviet Phys.-Solid State English Transl.*, **6**:2309 (1965); a review by L. S. Palatnik in "Basic Problems in Thin Film Physics" (R. Niedermayer and H. Mayer, eds.), p. 92, Vandenhoeck and Ruprecht, Göttingen, 1966.

85. L. S. Palatnik and I. I. Papirov, "Oriented Crystallization" (in Russian), Moscow, 1964.
86. K. H. Behrndt, *J. Appl. Phys.*, 37:3841 (1966).
87. W. Thomson, *Phil. Mag.*, 42:448 (1871); J. Frenkel, "Kinetic Theory of Liquids," Dover Publications, Inc., New York, 1955.
88. M. Takagi, *J. Phys. Soc. Japan*, 9:359 (1954); M. Blackman and A. E. Curzon, in "Structure and Properties of Thin Films" (Neugebauer et al., eds.), p. 217, John Wiley & Sons, Inc., New York, 1959; L. S. Palatnik and B. T. Boiko, *Phys. Metals Metallog.*, 11:119 (1961).
89. N. T. Gladkich, R. Niedermayer, and K. Spiegel, *Phys. Stat. Solidi*, 15:181 (1966).
90. G. Gafner, *Phil. Mag.*, 5:1041 (1960).
91. E. Yoda, *Proc. 6th Intern. Congr. Electron Microscopy, Kyoto*, 1966, p. 517; also M. V. Belous and C. M. Wayman, *J. Appl. Phys.*, 38:5119 (1967).
92. K. L. Chopra, *Bull. Am. Phys. Soc.*, 11(ser. 11):75 (1966).
93. E. N. Andrade, *Trans. Faraday Soc.*, 31:1137 (1935).
94. R. Uyeda, *Proc. Phys. Math. Soc. Japan*, 24:809 (1942).
95. H. Levinstein, *J. Appl. Phys.*, 20:306 (1949).
96. R. S. Sennett and G. D. Scott, *J. Opt. Soc. Am.*, 40:203 (1950).
97. S. Ogawa, D. Watanabe, and F. E. Fujita, *J. Phys. Soc. Japan*, 10:429 (1955).
98. K. L. Chopra and L. C. Bobb, in "Single Crystal Films" (Francombe and Sato, eds.), p. 371, Pergamon Press, New York, 1964.
99. K. L. Chopra and M. R. Randlett, *J. Appl. Phys.*, 39:1874 (1968).
100. D. W. Pashley (review), *Advan. Phys.*, 14:361 (1965).
101. D. B. Dove, *J. Appl. Phys.*, 35:2785 (1964).
102. J. Pócza (unpublished).
103. K. L. Chopra and M. R. Randlett, *Appl. Phys. Letters*, 11:202 (1967).
104. G. C. Kuczynski, *Trans. AIME*, 185:169 (1949); W. D. Kingery and M. Berg, *J. Appl. Phys.*, 26:1205 (1955).
105. F. A. Nichols, *J. Appl. Phys.*, 37:2805 (1966).
106. K. L. Chopra and M. R. Randlett, *Appl. Phys. Letters*, 10:241 (1966).
107. A. J. Melmed, *J. Appl. Phys.*, 36:3585 (1965).
108. E. F. Wassermann and R. L. Hines, *J. Appl. Phys.*, 38:196 (1967).
109. R. Nossek, *Z. Physik*, 142:321 (1955).
110. G. G. Sumner, *Phil. Mag.*, 12:767 (1965).
111. E. T. S. Appleyard and J. R. Bristow, *Proc. Roy. Soc. London*, A172:530 (1939).
112. A. J. Learn and R. S. Spriggs, *J. Appl. Phys.*, 34:3012 (1963).
113. D. S. Campbell, D. J. Stirland, and H. Blackburn, *Phil. Mag.*, 7:1099 (1962).
114. D. S. Campbell and D. J. Stirland, *Phil. Mag.*, 9:703 (1964).
115. E. Krikorian, *12th Natl. Vacuum Symp. Abstracts*, 1966, p. 175; E. Krikorian and R. J. Sneed, *J. Appl. Phys.*, 37:3674 (1966).
116. R. C. Newman and D. W. Pashley, *Phil. Mag.*, 46:917 (1955).
117. P. Hilsch and R. Hilsch, *Naturwiss.*, 48:549 (1961).
118. S. Shirai, Y. Fukuda, and M. Nomura, *J. Phys. Soc. Japan*, 16:1989 (1961).
119. E. Klokholm and C. Chiou, *Acta Met.*, 14:565 (1966).
120. R. Aziz and G. D. Scott, *Can. J. Phys.*, 34:73 (1956).

121. J. W. Swaine and R. C. Plumb, *J. Appl. Phys.*, **33**:2378 (1962).
122. J. D. Finegan and R. W. Hoffman, *Trans. 8th Natl. Vacuum Symp.*, p. 935, Pergamon Press, New York, 1961.
123. D. O. Smith, M. S. Cohen, and G. P. Weiss, *J. Appl. Phys.*, **31**:1755 (1960).
124. H. Koenig and G. Helwig, *Optik,* **6**:111 (1950).
125. L. Holland, *J. Opt. Soc. Am.*, **43**:376 (1953).
126. C. Kooy and J. M. Nieuwenhuizen, in "Basic Problems in Thin Film Physics" (R. Niedermayer and H. Mayer, eds.), p. 181, Vandenhoeck and Ruprecht, Göttingen, 1966.
127. R. H. Wade and J. Silcox, *Appl. Phys. Letters,* **8**:7 (1966).
128. K. L. Chopra, *Appl. Phys. Letters,* **7**:140 (1965).
129. D. I. Kennedy, R. E. Hayes, and R. W. Alsford, *J. Appl. Phys.*, **38**:1985 (1967).
130. R. M. Hill, *Elec. Res. Assoc. Rept.* El.SC/20-3, 1966.
131. W. R. Sinclair and C. J. Calbick, *Appl. Phys. Letters,* **10**:214 (1966).
132. H. W. Larson and G. A. Walker, *J. Appl. Phys.*, **38**:4513 (1967).
133. C. J. Paparoditis, in "Single Crystal Films" (M. H. Francombe and H. Sato, eds.), p. 79, Pergamon Press, New York, 1964.
134. K. L. Chopra, A. N. Mariano, and S. K. Bahl, *14th Natl. Vacuum Symp. Abstracts*, 1967, p. 33.
135. For example, R. B. Belser, *J. Appl. Phys.*, **28**:109 (1957); K. L. Chopra and L. C. Bobb, *Acta Met.*, **12**:807 (1964).
136. S. G. Fleet, *Mullard Res. Lab. Rept.* 466, 1963.
137. L. S. Palatnik, M. J. Fuks, B. T. Boiko, and V. B. Pariiskii, *Phys. Metals Metallog.*, **11**(6):44 (1961).
138. L. S. Palatnik, B. T. Boiko, and M. J. Fuks, in "Basic Problems in Thin Film Physics" (R. Niedermayer and H. Mayer, eds.), p. 110, Vandenhoeck and Ruprecht, Göttingen, 1966.
139. J. P. Borel, *Compt. Rend.*, **233**:296 (1951).
140. M. L. Gimpl, A. D. McMaster, and N. Fuschillo, *J. Appl. Phys.*, **35**:3572 (1964).
141. L. Bachman, D. L. Sawyer, and B. M. Siegel, *J. Appl. Phys.*, **36**:304 (1965).
142. J. G. Skofronick and W. B. Phillips, *J. Appl. Phys.*, **38**:4791 (1967).
143. J. M. Blakely, *J. Appl. Phys.*, **35**:1756 (1964).
144. K. L. Chopra (unpublished).
145. R. W. Vook and F. Witt, *J. Vacuum Sci. Tech.*, **2**:49, 243 (1965).
146. H. D. Keith, *Proc. Phys. Soc. London,* **B69**:180 (1956).
147. For example, R. W. Vook, F. R. L. Schoening, and F. Witt, in "Single Crystal Films" (Francombe and Sato, eds.), p. 69, Pergamon Press, New York, 1964.
148. For example, I. H. Khan and R. N. Summergrad, *Appl. Phys. Letters,* **11**:12 (1967).
149. R. W. Vook, T. Parker, and D. Wright, in "Surfaces and Interfaces—Chemical and Physical Characteristics" (Burke, Reed, and Weiss, eds.), p. 347, Syracuse University Press, Syracuse, N.Y., 1967; R. W. Vook, private communication.
150. O. Beeck, A. E. Smith, and A. Wheeler, *Proc. Roy. Soc. London,* **177**:62 (1940).
151. O. Beeck, *Advan. Catalysis,* **2**:151 (1950).
152. A. S. Potter and F. C. Tompkins, *Proc. Roy. Soc. London,* **A217**:544 (1953).

153. D. Brennan, D. O. Hayward, and B. M. W. Trapnell, *Proc. Roy. Soc. London,* **A256**:81 (1960), and other references cited.
154. T. L. O'Connor and H. H. Uhlig, *J. Phys. Chem.,* **61**:402 (1957).
155. J. A. Allen, C. C. Evan, and J. W. Mitchell, in "Structure and Properties of Thin Films" (Neugebauer et al., eds.), p. 46, John Wiley & Sons, Inc., New York, 1959.
156. M. S. Blois and L. M. Rieser, *J. Appl. Phys.,* **25**:338 (1964).
157. N. Wainfan and L. G. Paratt, *J. Appl. Phys.,* **31**:1331 (1960).
158. A. R. Wolter, *J. Appl. Phys.,* **36**:2377 (1965).
159. J. Edgecumbe, *J. Vacuum Sci. Tech.,* **3**:28 (1966).
160. T. E. Hartman, *J. Vacuum Sci. Tech.,* **2**:239 (1965).
161. D. W. Pashley, *Advan. Phys.,* **5**:173 (1956).
162. D. W. Pashley, in "Thin Films," p. 59, American Society for Metals, Metals Park, 1963.
163. C. A. Neugebauer, in "Physics of Thin Films" (G. Hass and R. E. Thun, eds.), vol. 2, p. 1, Academic Press Inc., New York, 1964.
164. M. H. Francombe in "The Use of Thin Films for Physical Investigations" (J. C. Anderson, ed.), p. 29, Academic Press Inc., New York, 1966.
165. R. E. Thun, in "Physics of Thin Films" (G. Hass, ed.), vol. 1, p. 187, Academic Press Inc., New York, 1963.
166. G. I. Finch and S. Fordham, *Proc. Phys. Soc. London,* **48**:85 (1936).
167. F. W. C. Boswell, *Proc. Phys. Soc. London,* **A64**:465 (1951).
168. J. S. Halliday, T. B. Rymer, and K. H. R. Wright, *Proc. Roy. Soc. London,* **A225**:548 (1954).
169. R. Shuttleworth, *Proc. Phys. Soc. London,* **A63**:444 (1950).
170. T. B. Rymer, *Nuovo Cimento,* **6**:294 (1957).
171. A. Cimino, P. Porta, and M. Valigi, *J. Am. Ceram. Soc.,* **49**:1952 (1966).
172. G. I. Finch and A. G. Quarrell, *Proc. Roy. Soc. London,* **A141**:398 (1933); *Proc. Phys. Soc. London,* **46**:148 (1934).
173. W. Cochrane, *Proc. Phys. Soc. London,* **48**:723 (1936).
174. H. Raether, *J. Phys. Radium,* **11**:11 (1950); L. N. D. Lucas, *Proc. Phys. Soc. London,* **A64**:943 (1951).
175. R. C. Newman, *Proc. Phys. Soc. London,* **B69**:432 (1956).
176. N. A. Shishakov, *Zh. Eksperim. i Teor. Fiz.,* **22**:241 (1952).
177. J. P. Jones, *Proc. Roy. Soc. London,* **A284**:469 (1965).
178. R. E. Schlier and H. E. Farnsworth, *J. Phys. Chem. Solids,* **6**:271 (1958).
179. N. J. Taylor, *Surface Sci.,* **4**:161 (1966).
180. C. A. Haque and H. E. Farnsworth, *Surface Sci.,* **4**:195 (1966).
181. A. J. Melmed, *J. Chem. Phys.,* **38**:1444 (1963).
182. A. J. Melmed and R. F. McCarthy, *J. Chem. Phys.,* **42**:1466 (1965).
183. A. J. Melmed, *J. Chem. Phys.,* **42**:3332 (1965).
184. R. D. Gretz, to be published.
185. See, for example, the list compiled in Ref. 1, p. 90.
186. R. B. Kehoe, *Phil. Mag.,* **2**:455 (1957).
187. C. W. B. Grigson and D. B. Dove, *J. Vacuum Sci. Tech.,* **3**:120 (1966).
188. B. Abeles, R. W. Cohen, and G. W. Cullen, *Phys. Rev. Letters,* **9**:402 (1966); *Phys. Rev. Letters,* **18**:902 (1967).
189. K. L. Chopra, M. R. Randlett, and R. H. Duff, *Appl. Phys. Letters,* **9**:402 (1966).

190. K. L. Chopra, M. R. Randlett, and R. H. Duff, *Phil. Mag.*, **16**:261 (1967).
191. T. E. Hutchinson, *J. Appl. Phys.*, **36**:270 (1965); *Appl. Phys. Letters*, **3**:51 (1963).
192. P. Duwez and R. H. Willens, *Trans. AIME*, **227**:362 (1963).
193. M. Moss, D. L. Smith, and R. A. Lefever, *Appl. Phys. Letters*, **5**:120 (1964).
194. R. Hilsch, in "Non-crystalline Solids" (V. D. Frechette, ed.), chap. 15, John Wiley & Sons, Inc., New York, 1960.
195. W. Buckel, *Proc. Conf. Elec. Magnetic Properties Thin Metallic Layers, Louvain*, 1961, p. 264.
196. W. Buckel, *Z. Physik*, **138**:136 (1954); W. Buckel and R. Hilsch, *Z. Physik*, **138**:109 (1954).
197. H. Bülow and W. Buckel, *Z. Physik*, **145**:141 (1956).
198. H. Richter, H. Berckhemer, and G. Breitling, *Z. Naturforsch.*, **9**:236 (1954) (for Sb); B. G. Lazarev et al., *Zh. Eksperim. i Teor. Fiz.*, **37**:1461 (1959) (for Be).
199. E. Feldtkeller, *Z. Physik*, **157**:64 (1959).
200. N. Barth, *Z. Physik*, **142**:58 (1955).
201. W. Rühl, *Z. Physik*, **138**:121 (1954).
202. K. L. Chopra, *Ledgemont Laboratory Rept.* TR-179, August, 1968; to be published.
203. S. Mader, *J. Vacuum Sci. Tech.*, **2**:35 (1965).
204. S. Mader, A. S. Nowick, and H. Widmer, *Acta Met.*, **15**:203 (1967); S. Mader and A. S. Nowick, *Acta Met.*, **15**:215 (1967).
205. S. Mader, in "The Use of Thin Films for Physical Investigations" (J. C. Anderson, ed.), p. 433, Academic Press Inc., New York, 1960.
206. H. Richter and S. Steeb, *Naturwiss.*, **45**:461 (1958).
207. S. Fujime, *Japan. J. Appl. Phys.*, **5**:59, 764 (1966).
208. W. Mönch and W. Sander, *Z. Physik*, **157**:149 (1959).
209. W. Mönch, *Z. Physik*, **164**:229 (1961).
210. W. Ostwald, *Z. Physik. Chem.*, **22**:289 (1897).
211. R. W. Cohen, B. Abeles, and G. S. Weisbarth, *Phys. Rev. Letters*, **18**:336 (1967).
212. R. C. Newman, unpublished (referred to in Ref. 161).
213. T. Honma and C. M. Wayman, *J. Appl. Phys.*, **36**:2791 (1965).
214. L. S. Palatnik, A. G. Ravlik, and A. N. Stetsenko, *Phys. Metals Metallog.*, **18**(4):147 (1964).
215. A. I. Bublik and B. I. A. Pines, *Dokl. Akad. Nauk SSSR*, **87**:215 (1952).
216. M. Shiojiri, *J. Phys. Soc. Japan*, **21**:335 (1965).
217. K. L. Chopra and S. K. Bahl, *J. Appl. Phys.*, **40** (1969).
218. E. Kneller, *J. Appl. Phys.*, **33**:1355 (1962); **35**:2210 (1964).
219. A. S. Nowick and S. Mader, *IBM J. Res. Develop.*, **9**:358 (1965).
220. P. N. Denbigh and R. B. Marcus, *J. Appl. Phys.*, **37**:4325 (1966).
221. M. H. Read and C. Altman, *Appl. Phys. Letters*, **7**:51 (1965).
222. H. C. Cook, *J. Vacuum Sci. Tech.*, **4**:80 (1967).
223. T. E. Hutchinson and K. H. Olsen, *J. Appl. Phys.*, **38**:4933 (1967).
224. K. Bahadur and P. V. Shastry, *Proc. Phys. Soc. London*, **78**:594 (1961).
225. J. E. Davey and R. H. Deiter, *J. Appl. Phys.*, **36**:284 (1965) and other references on hexagonal gold cited here.

226. N. Takahashi, *Compt. Rend.*, **234**:1619 (1952).
227. J. G. Wright and J. Goddard, *Phil Mag*, **11**:485 (1965).
228. D. W. Pashley, M. J. Stowell, and T. J. Law, *Phys. Stat. Solidi*, **10**:153 (1965).
229. L. G. Schulz, *Acta Cryst.*, **4**:487 (1951); D. W. Pashley, *Proc. Phys. Soc. London*, **A65**:33 (1952).
230. K. Mayerhoff and J. Ungelenk, *Acta Cryst.*, **12**:32 (1959).
231. M. Blackman and I. H. Khan, *Proc. Phys. Soc. London*, **77**:471 (1961).
232. S. A. Semiletov, *Kristallografiya*, **1**:306 (1956).
233. C. A. Escoffery, *J. Appl. Phys.*, **35**:2273 (1964).
234. M. Shiojiri and E. Suito, *Japan. J. Appl. Phys.*, **3**:314 (1964).
235. R. Ludeka and W. Paul, *Phys. Stat. Solidi*, **23**:413 (1967).
236. K. V. Shalimova A. F. Andrushko, and I. Dima, *Soviet Phys. Crys.*, **10**:414 (1966).
237. M. Weinstein, G. A. Wolff, and B. N. Das, *Appl. Phys. Letters*, **6**:73 (1965).
238. J. E. Davey and T. Pankey, *J. Appl. Phys.*, **39**:1941 (1968).
239. J. E. Davey and T. Pankey, *Appl. Phys. Letters*, **12**:38 (1968).
240. A. N. Mariano and K. L. Chopra, *Appl. Phys. Letters*, **10**:282 (1967).
241. L. Valenta and A. Sukiennicki, *Phys. Stat. Solidi*, **17**:903 (1966).
242. J. Burton and G. Jura, *Phys. Rev. Letters*, **18**:740 (1967).
243. J. J. Lander, in "Progress in Solid State Chemistry" (H. Reiss, ed.), vol. 2, p. 26, Pergamon Press, New York, 1965.
244. R. E. Schlier and H. E. Farnsworth, *Advan. Catalysis*, **9**:434 (1957); H. E. Farnsworth and J. Tuul, *J. Phys. Chem. Solids*, **9**:48 (1959); H. E. Farnsworth and H. H. Madden, *J. Appl. Phys.*, **32**:1933 (1961).
245. A. U. MacRae, *Science*, **139**:379 (1963).
246. C. W. Tucker, *J. Appl. Phys.*, **35**:1897 (1964); **37**:528, 3013 (1966).
247. W. Peria, Electronics Conference, MIT, Boston, 1963.
248. S. Goldsztaub and B. Lang, *Compt. Rend.*, **257**:1908 (1963); **258**:117 (1964).
249. J. M. Charig, *Appl. Phys. Letters*, **10**:139 (1967).
250. R. E. Schlier and H. E. Farnsworth, *J. Chem. Phys.*, **30**:917 (1959).
251. F. Jona, *Appl. Phys. Letters*, **9**:235 (1966).
252. R. N. Thomas and M. H. Francombe, *Appl. Phys. Letters*, **11**:108 (1967); 134 (1967).
253. S. Hagstrom, H. B. Lyon, and G. A. Somorjai, *Phys. Rev. Letters*, **15**:491 (1965); H. B. Lyon and G. A. Somorjai, *J. Chem. Phys.*, **46**:2539 (1967).
254. A. M. Mattera, R. M. Goodman, and G. A. Somorjai, *Surface Sci.*, **7**:26 (1967).
255. D. G. Fedak and N. A. Gjostein, *Acta Met.*, **15**:827 (1967).
256. P. W. Palmberg and T. N. Rhodin, *Phys. Rev.*, **161**:586 (1967).
257. E. Bauer, in "Single Crystal Films" (M. H. Francombe and H. Sato, eds.), p. 42, Pergamon Press, New York, 1964; *Trans. 9th Natl. Vacuum Symp.*, p.35, The Macmillan Company, New York, 1962.
258. A. Wassermann and J. Grewen, "Texturen metallischer Werkstoffe," pp. 127-147, Springer-Verlag OHG, Berlin, 1962.
259. N. J. Mel'nikov, V. D. Shchukov, and M. M. Umanskii, *Zh. Eksperim. i Teor. Fiz.*, **22**:775 (1952).
260. A. V. Shubnikov, *Dokl. Akad. Nauk SSSR*, **51**:679 (1946); A. N. Kolmogorov, *Dokl. Akad. Nauk SSSR*, **65**:681 (1949); G. I. Finch, H. Wilman, and L. Yang,

Discussions Faraday Soc., **1**:144 (1947); D. M. Evans and H. Wilman, *Acta Cryst.*, **5**:731 (1952); K. R. Dixit, *Phil. Mag.*, **16**:1049 (1933).
261. H. Sato, in "Single Crystal Films" (M. H. Francombe and H. Sato, eds.), p. 341, Pergamon Press, New York, 1964.
262. L. Royer, *Bull. Soc. Franc. Mineral.*, **51**:7 (1928); *Compt. Rend.*, **194**:1088, 1932.
263. M. L. Frankenheim, *Ann. Physik*, **37**:516 (1836).
264. T. V. Barker, *J. Chem. Soc. Trans.*, **89**:1120 (1906), *Z. Kris.*, **45**:1 (1908).
265. G. L. Bailey, S. Fordham, and J. T. Tyson, *Proc. Roy. Soc. London*, **50**:63 (1938) (for Si); E. Rasmanis, *J. Electrochem. Soc.*, **110**:57c (1963).
266. B. W. Sloope and C. O. Tiller, *Appl. Phys. Letters*, **8**:223 (1966).
267. J. A. Koutsky, A. G. Walton, and E. Baer, *J. Polymer Sci.*, **4**:611 (1966); *Polymer Letters*, **5**:177 (1967); for earlier references on epitaxy of polymers, see E. W. Fisher in "Newer Methods of Polymer Characterization" (B. Ke, ed.), vol. 6, p. 279, Interscience Publishers, Inc., New York, 1964.
268. L. Brück, *Ann. Physik*, **26**:233 (1936).
269. P. W. Palmberg, T. N. Rhodin, and C. J. Todd, *Appl. Phys. Letters*, **11**:33 (1967).
270. A. Yelon and R. W. Hoffman, *J. Appl. Phys.*, **31**:1672 (1960); R. W. Vook, *J. Appl. Phys.*, **32**:1557 (1961).
271. B. W. Sloope and C. O. Tiller, *J. Appl. Phys.*, **36**:3174 (1965); *J. Appl. Phys.*, **33**:3458 (1962).
272. B. W. Sloope and C. O. Tiller, *J. Appl. Phys.*, **32**:1331 (1961).
273. E. Krikorian and R. J. Sneed, *J. Appl. Phys.*, **37**:3665 (1966) and earlier references by the authors cited here.
274. H. Jahrreiss and H. J. Isken, *Phys. Stat. Solidi*, **17**:619 (1966).
275. For example, J. E. Davey, *J. Appl. Phys.*, **32**:877 (1961); S. P. Wolsky, T. R. Piwkowski, and G. Wallis, *J. Vacuum Sci. Tech.*, **2**:97 (1965); J. D. Williams and L. E. Terry, *J. Electrochem. Soc.*, **114**:158 (1967).
276. S. Ino, D. Watanabe, and A. Ogawa, *J. Phys. Soc. Japan*, **17**:1074 (1962); **19**:881 (1964).
277. J. W. Matthews and E. Grünbaum, *Appl. Phys. Letters*, **5**:106 (1964); *Phil. Mag.*, **11**:1233 (1965); E. Grünbaum and J. W. Matthews, *Phys. Stat. Solidi*, **9**:731 (1965); also, see S. Shinozaki and H. Sato, *J. Appl. Phys.*, **36**:2320 (1965).
278. M. Harsdorff, *Solid State Commun.*, **1**:218 (1963); **2**:133 (1964); M. Harsdorff and H. Raether, *Z. Naturforsch.*, **19a**:1497 (1964).
279. M. Harsdorff, *Fortschr. Mineral.*, **42**:250 (1966) (review); R. W. Adam and M. Harsdorff, *Z. Naturforsch.*, **20a**:489 (1965).
280. E. Bauer, A. K. Green, K. M. Kunz, and H. Poppa, in "Basic Problems in Thin Film Physics" (R. Niedermayer and H. Mayer, eds.), p. 135, Vandenhoeck and Ruprecht, Göttingen, 1966.
281. J. W. Matthews, *Phil. Mag.*, **12**:1143 (1965).
282. M. H. Jacobs, D. W. Pashley, and M. J. Stowell, *Phil. Mag.*, **13**:129 (1966).
283. J. W. Matthews, *Appl. Phys. Letters*, **7**:131, 255 (1965).
284. E. Bauer, A. K. Green, and K. M. Kunz, *Appl. Phys. Letters*, **8**:248 (1966).
285. Y. Tarui, H. Teshima, K. Okura, and A. Minamiya, *J. Electrochem. Soc.*, **110**:1167 (1963).

286. P. W. Palmberg, T. N. Rhodin, and C. J. Todd, *Appl. Phys. Letters*, **10**:122 (1967).
287. K. M. Kunz, A. K. Green, and E. Bauer, *Phys. Stat. Solidi*, **18**:441 (1966).
288. G. Menzer, *Naturwiss.*, **26**:385 (1938); *Z. Krist.*, **99**:378 (1938).
289. F. C. Frank and J. H. van der Merwe, *Proc. Roy. Soc. London*, **A198**:205, 216 (1949).
290. J. H. van der Merwe, reviews of a series of publications are given in "Single Crystal Films" (Francombe and Sato, eds.), p. 139, Pergamon Press, New York, 1964; and in "Basic Problems in Thin Film Physics" (R. Niedermayer and H. Mayer, eds.), p. 122, Vandenhoeck and Ruprecht, Göttingen, 1966.
291. J. W. Matthews, *Phil. Mag.*, **6**:1347 (1961); P. Delavignette, T. Tournier, and S. Amelinckx, *Phil. Mag.*, **6**:1419 (1961); E. Grunbaum and J. W. Mitchell in "Single Crystal Films," p. 221, Pergamon Press, New York, 1964; H. Gradmann, *Ann. Physik*, **13**:213 (1964); E. R. Thompson and K. R. Lawless, *Appl. Phys. Letters*, **9**:138 (1966); W. A. Jesser, J. W. Matthews, and D. Kuhlmann-Wilsdorf, *Appl. Phys. Letters*, **9**:176 (1966); a comprehensive review is given by J. W. Matthews in "Physics of Thin Films" (G. Hass and R. E. Thun, eds.), vol. 4, Academic Press Inc., New York, p. 137, 1967.
292. P. D. Dankov, *Zh. Eksperim. i Teor. Fiz.*, **20**:853 (1946).
293. G. P. Thomson and W. Cochrane, "The Theory and Practice of Electron Diffraction," The Macmillan Company, New York, 1939.
294. O. G. Engel, *J. Chem. Phys.*, **20**:1174 (1952); *J. Res. Natl. Bur. Std.*, **50**:249 (1953).
295. H. Göttsche, *Z. Naturforsch.*, **11a**:55 (1956).
296. I. I. Papirov and L. S. Palatnik, *Kristallografiya*, **7**:286 (1962); D. A. Brine and R. A. Young, *Phil. Mag.*, **8**:651 (1963).
297. K. R. Dixit, *Phil. Mag.*, **16**:1049 (1933); *Indian J. Phys.*, **40**:117 (1957); *Proc. Indian Acad. Sci.*, **A48**:330 (1958).
298. M. J. Stowell, in "The Use of Thin Films in Physical Investigations" (J. C. Anderson, ed.), p. 131, Academic Press Inc., New York, 1966.
299. J. W. Matthews, *Phil. Mag.*, **4**:1017 (1959); **7**:915 (1962); J. W. Matthews and D. L. Allinson, *Phil. Mag.*, **8**:1283 (1963); a review by J. W. Matthews in "Physics of Thin Films" (G. Hass and R. E. Thun, eds.), vol. 4, Academic Press Inc., New York, 1967.
300. V. A. Phillips, *Phil. Mag.*, **5**:571 (1960); *J. Appl. Phys.*, **33**:712 (1962).
301. S. Ino, *J. Phys. Soc. Japan*, **21**:346 (1966).
302. L. O. Brockway and R. B. Marcus, *J. Appl. Phys.*, **34**:921 (1963); A. P. Rowe and L. O. Brockway, *J. Appl. Phys.*, **37**:2703 (1966).
303. S. Mendelson, reviews in "Single Crystal Films" (Francombe and Sato, eds.), p. 251, Pergamon Press, New York, 1964; and in *Mater. Sci. Eng.*, **1**:42 (1966); *Surface Sci.*, **6**:233 (1967) bring up to date the reference to the works of this author.
304. D. P. Miller, S. B. Watelski, and C. R. Moore, *J. Appl. Phys.*, **34**:2813 (1963).
305. T. L. Chu and J. R. Gavaler, *Phil. Mag.*, **9**:993 (1964).
306. G. R. Booker, *Discussions Faraday Soc.*, **38**:298 (1964); *Phil. Mag.*, **11**:1007 (1965).
307. H. Holloway, a review in "The Use of Thin Films for Physical Investigations" (J. C. Anderson, ed.), p. 111, Academic Press Inc., New York, 1966; H.

Holloway and L. C. Bobb, *J. Appl. Phys.*, **38**:2893 (1967). This paper gives reference to the rest of the series of papers by Holloway et al.
308. B. W. Sloope and C. O. Tiller, *J. Appl. Phys.*, **37**:887 (1966).
309. T. B. Light, *AIME Conf. Met. Semiconductor Mater.*, 1962, p. 137.
310. R. Reizman and H. Basseches, *AIME Conf. Met. Semiconductor Mater.*, 1962, p. 87.
311. J. A. Amick, in "Single Crystal Films" (Francombe and Sato, eds.), p. 283, Pergamon Press, New York, 1964; R. E. Ewing and P. E. Greene, *J. Electrochem. Soc.*, **111**:885 (1964).
312. I. Teramoto, *Phil. Mag.*, **8**:357 (1963).
313. Review by R. C. Newman, *Microelectron. Reliability*, **3**:121 (1964).
314. A. Brown and B. Lewis, *J. Phys. Chem. Solids*, **23**:1597 (1962).
315. M. H. Francombe, *Phil. Mag.*, **10**:989 (1964).
316. Y. Takeishi, I. Sasaki, and K. Hirabayashi, *Appl. Phys. Letters*, **11**:330 (1967).
317. G. A. Kurov and Z. G. Pinsker, *Zh. Tekhn. Fiz.*, **28**:2130 (1958).
318. S. Shirai and Y. Fukuda, *J. Phys. Soc. Japan*, **17**:319 (1962).
319. R. L. Schwoebel, *J. Appl. Phys.*, **34**:2784 (1963).
320. P. E. Doherty and R. S. Davis, *J. Appl. Phys.*, **34**:619 (1963).
321. J. E. Mee and G. R. Pulliam, *Proc. Intern. Conf. Crystal Growth, Boston,* 1966, p. 333.
322. P. S. Schaffer, *J. Am. Ceram. Soc.*, **48**:508 (1965).
323. R. E. Cech and E. I. Alessandrini, *Trans. Am. Soc. Metals*, **51**:150 (1959).
324. H. Takei and S. Takasu, *Japan. J. Appl. Phys.*, **3**:175 (1964); J. L. Archer, G. R. Pulliam, R. G. Warren, and J. E. Mee, *Proc. Intern. Conf. Crystal Growth, Boston,* 1966, p. 337.
325. J. E. Mee, J. L. Archer, R. H. Meade, and T. N. Hamilton, *Appl. Phys. Letters*, **10**:289 (1967).
326. E. K. Muller, B. J. Nicholson, and G. L'. E. Turner, *J. Electrochem. Soc.*, **110**:969 (1963).
327. R. F. Tramposch, *Appl. Phys. Letters*, **9**:83 (1966).
328. L. R. Weisberg, *J. Appl. Phys.*, **38**:4537 (1967), and other references cited here.
329. M. Tamura and M. Nomura, *Appl. Phys. Letters*, **11**:196 (1967).
330. J. L. Richards, in "The Use of Thin Films in Physical Investigations" (J. C. Anderson, ed.), p. 419, Academic Press Inc., New York, 1966.
331. G. S. Kamath and D. Bowman, *J. Electrochem. Soc.*, **114**:192 (1967).
332. C. Weaver and R. M. Hill, *Advan. Phys.*, **8**:375 (1959).
333. C. A. O. Henning, *Surface Sci.*, **9**:277, 296 (1968).
334. J. P. Hirth and K. L. Moazed, in "Physics of Thin Films" (G. Hass and R. Thun, eds.), p. 97, Academic Press Inc., New York, 1967.
335. M. J. Knight and K. N. Jha, *Thin Solid Films*, **2**:131 (1968).
336. W. A. Jesser, J. W. Matthews, and D. Kuhlmann-Wilsdorf, *Appl. Phys. Letters*, **9**:176 (1968).
337. N. Cabrera, *Surface Sci.*, **2**:320 (1965); W. A. Jesser and D. Kuhlmann-Wilsdorf, *Phys. Stat. Solidi*, **21**:533 (1967).
338. J. W. Matthews, *Phil. Mag.*, **13**:1207 (1966).
339. J. W. Matthews and W. A. Jesser, *Acta Met.*, **15**:595 (1967).

340. W. A. Jesser and J. W. Matthews, *Phil. Mag.*, **16**:475 (1968).
341. W. A. Jesser and J. W. Matthews, *Phil. Mag.*, **16**:595 (1968).
342. W. A. Jesser and J. W. Matthews, *Phil. Mag.*, **16**:461 (1968).
343. J. Goddard and J. G. Wright, *Brit. J. Appl. Phys.*, **15**:807 (1964).
344. J. G. Wright and J. Goddard, *Phil. Mag.*, **11**:485 (1965).
345. A. Cisman et al., *Rev. Roum. Phys.*, **12**:273 (1967).
346. W. Nowak, *J. Vacuum Sci. Tech.*, **2**:276 (1965).
347. A. P. Clark and R. G. Alibozek, *J. Appl. Phys.*, **39**:2156 (1968).
348. M. Braunstein, R. R. Henderson, and A. I. Braunstein, *Appl. Phys. Letters*, **12**:66 (1968).
349. R. W. Vook and M. A. Otooni, *J. Appl. Phys.*, **39**:2471 (1968).
350. G. I. Distler and B. B. Zvyagin, *Nature*, **212**:807 (1966); G. I. Distler, S. A. Kobzareva, and Y. M. Gerasimov, *J. Crystal Growth*, **2**:45 (1968); Y. M. Gerasimov and G. I. Distler, *Naturwiss.*, **3**:132 (1968).
351. C. N. J. Wagner, T. B. Light, N. C. Handler, and W. E. Luckens, *J. Appl. Phys.*, **39**:3690 (1968).
352. K. Mihana and M. Tanaka, *J. Crystal Growth*, **2**:51 (1968).
353. T. Inuzuka and R. Ueda, *Appl. Phys. Letters*, **13**:3 (1968).
354. K. L. Chopra, *J. Appl. Phys.*, **40** (February, 1969).
355. C. Weaver and L. C. Brown, *Phil. Mag.*, **17**:881 (1968).

V MECHANICAL EFFECTS IN THIN FILMS

1. INTRODUCTION

Whatever the applications of thin films may be, their mechanical stability and strong adhesion to the substrate are essential qualities. Large internal stresses ($\sim 10^9$ to 10^{10} dynes/cm^2) are known to develop in thin films during their growth. A stress of this magnitude is comparable to the yield strength of most bulk metals. It is also sufficient to overcome the film-substrate interfacial adhesion and thus result in peeling of the film if adhesion is due only to physical adsorption. The presence of large stresses is expected to have considerable effect on the mechanical, electron-transport, magnetic, superconducting, and optical properties of films.

Thin films exhibit unusually high tensile strength, which is related to their internal microstructure, growth, and in particular to the high density of defects frozen in during the atomistic deposition process. For instance, a dislocation density of $\sim 10^{10}$ to 10^{12} lines/cm^2 is character-istic of epitaxially grown (heterogeneous) single-crystal metal films, which is up to two orders of magnitude higher than that produced by the severest cold-work treatment of a bulk material. Thus, thin films are novel specimens for studies of mechanical effects and their relation to the microstructure and to structural defects.

This chapter contains a critical review of the mechanical measure-ments and the results obtained on stresses in films and on tensile strength and adhesion of films. Various models proposed to explain the observed unusual mechanical behavior are discussed in their respective sections. A substantial part of this chapter is based on a comprehensive review on the subject of stresses in films and the tensile strength of films by Hoffman [1].

2. INTERNAL STRESSES

2.1 Experimental Techniques

The study of internal stresses in thin films has attracted much attention since the early work of Stoney [2] on electrodeposited films in 1909. Most of the recent work has been carried out on evaporated films. A wide variety of techniques have been employed to measure internal stresses. Bending-plate methods are, however, the most commonly used ones. If a film adheres strongly to the substrate, the film-substrate composite will bend because of stresses in the film. If the film length tends to contract and is restrained from doing so by the substrate, the film is in a state of tension. If, on the other hand, the film tends to expand, it will be placed under a compressive stress by the substrate.

(1) Bending-plate or -beam Methods. The bending of a film-substrate composite caused by the deposition of a stressed film is the basis of this commonly employed method to determine stresses in films. In order for the bending or deflection to be easily measurable, the thickness of the substrate must be small, and thus thin ($\sim$ 0.1 mm) glass or mica substrates are used. The substrate, in a rectangular or long, thin beam form, is either clamped (cantilever) at one side for an observation of the deflection of the free end, or held on knife-edges for measuring the center deflection. The deflection itself can be measured in a variety of ways: optically [3-5], through a capacitance change [6] using the film as one plate of the condenser, mechanically [7] by using a stylus-type probe, or by electromechanical [8, 9] or magnetic [10, 11] devices to restore the null position.

Finegan and Hoffman [12] used the deflection of a circular plate to measure stress. The deflection is related to the radius of curvature of the circular plate, which can be measured interferometrically by observing changes in the optical-fringe system between the plate and an optical flat forming the Newton's-rings apparatus. The departure from circularity of the fringes offers the possibility of observing stress anisotropy. The radius of curvature of the plate can also be measured [13, 14] by determining the surface profile along a diameter using an optical microscope. One needs an instrument to project a horizontal, slit-shaped beam of light on the sample at $45°$; the reflected image is observed in the eyepiece. Deviations of the reflecting surface from planarity cause a deflection of the beam, the micrometer measurement of which determines the deflection and hence the curvature. The deflection of a plate may also be measured by means of a laser interferometer technique developed by Ennos [15].

The stress in the film can be related to the end deflection of the composite cantilever, or the edge or center deflection of the circular-plate substrate, by applying the standard theory of elasticity. The relation in the particular case of thin films may be obtained in a simple form only after several assumptions are made. In view of the uncertainty of the validity of some of these assumptions, the results obtained must be accepted with caution. The assumptions are: (1) The film strains the substrate, which bends until equilibrium is reached. (2) The film-substrate bond is strong enough to suppress slippage. (3) The substrate is linearly elastic, homogeneous, and uniformly thick. (4) The bending displacement is small compared with the thickness of the substrate. (5) The width of the substrate is less than half the length. (6) The stress is uniform throughout the film thickness. (7) No stress relief or changes in elastic constants take place as the film is built up.

Some of the assumptions are hard to justify. For instance, stress in a film is not expected to be uniform because of the presence of a free surface. The frequent curling away of the films on peeling is, of course, the result of a nonuniform stress. Stress relief is known to occur in many systems, caused in some cases by plastic flow and in others by the substrate curvature. The measured displacements often are not much smaller than the substrate thickness (~0.005 cm). Timoshenko [16] solved the problem for large deflections when a bending moment is applied only to two opposite edges of the plate, a condition which is similar to the film-substrate case. The substitution of displacement data for films in Timoshenko's equation yields results which are essentially the same as those obtained with the simple plate theory, indicating the latter is reasonably accurate for thin films.

With the assumptions outlined above, the theory [17] of a bending plate yields the following expression for an isotropic stress S in equilibrium with the resultant strain (of the substrate):

$$S = \frac{\delta E_s D^2}{3L^2 t(1 - \nu)}\left(1 + \frac{E_F t}{E_s D}\right) \tag{1}$$

where E_s and E_F are the Young's moduli of the substrate and the deposit, respectively, D and t are the thicknesses of the substrate and the deposit, respectively, L is the free length of the substrate, δ the deflection of the substrate, and ν Poisson's ratio for the substrate. The factor $1/1-\nu$ is not needed for a cantilever geometry. Most stress values reported in literature on geometries which correspond more closely to the case of a plate neglect this factor of $1/1-\nu$ ~1.5. For a very thin deposit $(t \ll D)$, Eq. (1) can be written in the form originally derived by Stoney [2] (taking Poisson's factor into account):

$$S = \frac{E_s D^2}{6rt(1 - \nu)} \tag{2}$$

where r is the radius of curvature of the bent substrate. If stresses in the film are anisotropic, the displacement δ_z perpendicular to the isotropic substrate plane at a point (x, y) is given [12] by

$$\delta_z = \frac{3(S_x - \nu S_y)x^2}{E_s t^2} + \frac{3(S_y - \nu S_x)y^2}{E_s t^2} \tag{3}$$

where S_x and S_y are the stress components along x and y directions.

The bending-plate or cantilever methods measure directly the product $St = F$, called the force per unit width from which the average stress S is calculated. The so-called "instantaneous" stress is given by the derivative of F with respect to the film thickness. The calculation of S from F becomes dubious as $t \to 0$, except in the ideal case of a constant stress independent of film thickness. It must be emphasized that, if stress distribution at the film-substrate interface is different from that in the bulk of the film, then increasing or decreasing thickness dependence of the average stress may be obtained, depending on the assumed stress distribution.

(2) X-ray and Electron-diffraction Methods. Stress may be determined from the measurement of changes in lattice parameters and line broadening, using x-ray [18-27] and electron-diffraction [28-35] techniques. A detailed treatment of the method of determination of stress by electron-diffraction analysis is given by Halliday et al. [35]. A higher accuracy is obtained by x-ray diffraction techniques because of the larger Bragg angles available. The stress (assumed isotropic) is computed from the relation

$$S = \frac{E_F}{2\nu_F} \frac{a_0 - a}{a_0} \qquad (4)$$

where a_0 and a are the lattice constants of the bulk material and the strained film, respectively, and ν_F is the Poisson ratio. Here, a refers to the lattice constant perpendicular to the film plane. If a is the lattice constant in the plane of the film, the stress is given by

$$S = \frac{E}{1 - \nu_F} \frac{a - a_0}{a_0} \qquad (5)$$

If broadening of the diffraction lines is used for the determination of stress, stress gradient and microstrain, the contributions to broadening due to the small crystallite size must be taken into account.

(3) Other Techniques. Any physical property of the film affected by stress can in principle be used to measure the stress. For instance, the existence of stress will produce an anisotropy in ferromagnetic films as a result of the magnetoelastic coupling (Chap. X, Sec. 4). Since the ferromagnetic resonance frequency depends on the anisotropy as well as the magnetization, a shift in the resonance peak will occur. With a proper film geometry, this shift can be separated from other anisotropy shifts. This method [36] is obviously limited to ferromagnetic films. Stresses in films may also be deduced from changes in the band gap of semi-conducting films [37, 38], electrical resistance [39], and super-conducting transition temperature [40-42]. Table I lists the various commonly used stress-measuring techniques and their sensitivities, as compiled by Hoffman [1] from the published literature.

2.2 Experimental Results

Stress in a film consists of two major components, a "thermal" component arising from the difference in the thermal-expansion coefficients of the film and the substrate, and an "intrinsic" one resulting from the structure and growth of films. The former is well understood

Table I. Comparison of Various Stress-measuring Techniques (Compiled by Hoffman [1])

Method of observing deflection	Type of plate*	Detectable force per unit width, dynes/cm	Ref.
Optical	B	800	44
Capacitance	C	500	6
Optical	C	250	3
Magnetic restoration	C	250	10
Electromechanical restoration	C	150	8
Mechanical	C	1	7
Electromechanical	C	1	9
Interferometric	C	0.5	
Interferometric	P	15	12
Ferromagnetic resonance		1,000†	36
X-ray		500†	18

*B, beam supported on both ends; C, cantilever beam; and P, circular plate.

† Approximate equivalent; a force is not measured.

and behaves in a predictable fashion. The latter component is generally the dominant one in most thin film cases, but its understanding still presents a problem.

(1) Thermal Stress. The thermal stress due to the constraint imposed by the film-substrate bonding is given by

$$S_{Th} = (\alpha_F - \alpha_s) \, \Delta T E_F \qquad (6)$$

where α_F and α_s are the average coefficients of expansion for film and substrate, ΔT is the temperature of the substrate during film deposition minus its temperature at measurement, and E_F is the Young's modulus of the film. The typical value of α is ~ 8 for glass, ~ 10 to 20 for many metals, and 30 to 40 for the alkali halides, all in units of $10^{-6}/°C$. For a positive ΔT, therefore, a tensile stress (positive) is found for metals on glass, but a compressive stress (negative) on alkali-halide substrates. Hence S_{Th} can be of either sign and can be controlled by a judicious choice of substrate material and deposition temperature. Thermal stress is evidently significant when intrinsic stress is negligible. Films of low-melting metals as well as those with a high state of structural order show little intrinsic stress, and thermal stress is therefore the main contribution to stress in these cases.

For Ni films deposited on soft glass at $75°C$ and measured at $25°C$, S_{Th} is about 5 percent of the total stress. At higher substrate temperatures, S_{Th} increases, whereas intrinsic stress decreases, and at about $250°C$, S_{Th} is the dominant contribution. Nickel films deposited on NaCl substrates at a temperature of $275°C$ were found by Freedman [20] to be under a compressive stress of $\sim 10^{10}$ dynes/cm^2 when cooled to room temperature. The observed tetragonal distortion of the structure of Ni films was ascribed to this stress. The stress is entirely elastic and is relieved when the film is floated free of the substrate, thereby restoring the normal structure.

Films of low-melting metals such as In, Sn, and Pb deposited at room temperature show little or no intrinsic stress ($< 5 \times 10^7$ dynes/cm^2), presumably because of their rapid annealing at room temperature. Upon cooling, a biaxial thermal stress is generated. The thickness dependence of the thermal stresses generated in films of Sn [40], In [41], and Pb [42] on cooling from room temperature to liquid-helium temperatures is shown in Fig. 1. The data on Pb films are the measured values, whereas those on Sn and In films are calculated from the observed increase of their superconducting transition temperature with decreasing film thickness (Chap. IX, Sec. 3.2). The calculations assume that the change in transition temperature is due to a stress which has a maximum observable value limited by the critical shear stress necessary to move dislocations. If one further assumes that dislocations are pinned at the surface so that their length is nearly equal to the film thickness, the maximum stress is given by

$$S_m = S_0 + \frac{0.5G}{t} \tag{7}$$

where S_0 is the friction stress in the bulk and G is the bulk shear modulus. Blumberg and Seraphim [40] and Toxen [41] used Eq. (7) to explain their observations of the thickness dependence of the transition temperature. The values of G so obtained are in good accord with the known bulk values; those of S_0 are higher than the bulk values. The data on Pb films show a deviation from the theoretical curve and can be

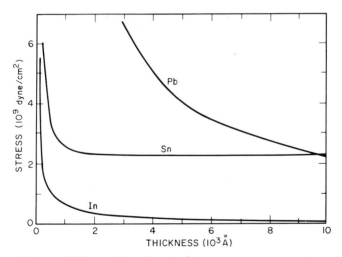

FIG. 1. Thickness dependence of thermal strain, induced by cooling from room temperature down to liquid helium, in superconducting films of In (Toxen [41]), Sn (Blumberg and Seraphim [40]), and Pb (Caswell et al. [42]). (*Compiled by Hoffman* [1])

explained by an additional dislocation locking and, therefore, a stress-raising mechanism.

An extensive, in situ x-ray study of thermal strains in Cu films deposited on glass at $\sim 80°$K and subsequently annealed to room temperature was made by Vook and Witt [22, 23], who studied line shifts and line broadening of as-deposited and annealed films and attempted to separate the thermal contribution. The observed tendency to form an [hkl] annealing texture was found to increase with decreasing Young's modulus and hence the strain energy for different $<hkl>$ directions (Sec. 3.2.1).

Films of PbS grown on NaCl at 275°C and cooled to 25°C are expected to develop a 0.5 percent thermal strain, corresponding to a stress of ~ 4 kbars. This stress should shift the band gap of PbS by 32×10^{-3} eV. However, such a shift was not observed by Palik et al. [37, 38] and was attributed to the fact that stress relief may be taking place. The degradation of the Hall mobility of these films on annealing after growth and subsequent cooling to room temperature was cited [38] in support of the possible occurrence of stress relief. On the other hand, a significant shift in the band gap of PbS films was observed by Palik et al. on subsequent cooling down to 77 and 4°K. The shift ascribed to the differential thermal strain is in agreement with the bulk strain data for PbS but there are significant differences in the cases of PbSe and PbTe films. That the observed shift is entirely a thermal-stress effect therefore remains questionable.

Thermal stress in epitaxial Si films [14] vapor-grown on sapphire at 1100°C is $\sim 10^9$ to 10^{10} dynes/cm^2. A 5×10^9 dynes/cm^2 stress in (111) p-type Si is expected to cause a 30 percent rise in the longitudinal resistance, a 20 percent drop in the transverse resistance, and a 10 percent loss in the mobility. None of these predictions has been verified experimentally, so far.

(2) Intrinsic Stress. This second contribution to the total stress reflects film structure in a way not well understood at present. Since the stress appears to mirror the film structure, we would expect to observe dependences on thickness, condensation rate, deposition temperature, ambient pressure, and type of substrate used. The wide variety of inconsistent experimental data on intrinsic stress published by numerous authors is certainly due to the difficulty of obtaining the same deposition parameters.

Hoffman [1] has collected published data for average stress for ~1,000-Å-thick metal and ~5,000-Å-thick dielectric films of a number of materials. Most of this collection is reproduced in Tables II and III. The values given in these tables should be considered representative rather than precise. In some cases, the Poisson factor is neglected and in others only total stress is listed for lack of more detailed information.

Table II. Intrinsic Stresses in Metal Films Approximately 1,000 Å Thick (Compiled by Hoffman [1])

Metal	Substrate temp,°C^a	Substrate material	Stress, 10^9 dynes/cm^2	Sign^b	Method^c	Ref.
Ag	90	Glass	0.75	T	CL	6
	A	Copper	1.0	T	CL	51
	A	Mica	0.2	T	CL	48
	A	Copper	0.75	T	CL	3
Al	A	Cellulose	1.2	T	ED	28
	A	Copper	0.1	T	CL	3
Au	A	Cellulose	4.6	T	ED	28
	A	Quartz	2.9	T	P	18
	A	Copper	0.85	T	CL	3
Bi	A	Copper	0		CL	3
Co	200	Glass	3.4	T	B	44
Cu	A	Copper	0.9	T	CL	3
	25	Mica	0.2	T	B	44
	A	Cellulose	4.4	C	ED	35
	A	Copper	1.5	T	CL	43
	75	Mica	0.1	T	CL	8
	−150	Mica	3.6	T	CL	8
Fe	75	Mica	10.5^d	T	B	45
	175	Mica	5.9^d	T	B	45
	75+A	Glass	9.6^d	T	P	12
	A	Glass, silica	8.5^e	T	P	50
	A	Copper	3.1	T	CL	3
Ga	−180	Aluminum	2	C	CL	63
	27	Aluminum	Small	T	CL	63
In	A	Silica	0		CL	41
Mg	A	Copper	0		CL	3
Ni	A	Glass	5–8	T	P	46
	A	Copper	3.5	T	CL	3
	A	Glass	7.7	T	CL	49
	75	Mica	6.4	T	B	44

Table II. Intrinsic Stresses in Metal Films Approximately 1,000 Å Thick (Compiled by Hoffman [1]) (Continued)

Metal	Substrate temp. °C[a]	Substrate material	Stress, 10^9 dynes/cm^2	Sign[b]	Method[c]	Ref.
	175	Mica	2.6	T	B	44
	A	Mica	5–8	T	FR	36
Pb	A	Nickel	0		CL	42
Pd	A	Copper	1.4	T	CL	3
Permalloy	75	Glass	9	T	P	46
	75	Glass, mica	9	T	CL	47
Sb	A	Copper	0.8	T	CL	93
	A	Copper	0.25	T	CL	3
Sn	A	Glass	0		CL	40
Zn	A	Copper	0		CL	3

[a] A, ambient temperature.

[b] T, refers to tension, C to compression.

[c] B, beam supported on both ends; CL, cantilever beam; ED, electron diffraction; and FR, ferromagnetic resonance.

[d] Poisson-corrected.

[e] Poisson-corrected, 36.5° angle of incidence.

Despite the diversity of results in the literature, it is possible to note some general and perhaps the most significant observations as follows: (1) The average stress in thick films is relatively *independent* of the film thickness. (2) Its magnitude is ~10^{10} dynes/cm^2, and it is tensile in films of metals and most dielectrics deposited at room temperature. Some dielectric films exhibit compressive stresses at low deposition temperatures, and at relatively large film thicknesses. (3) The intrinsic stress *does not* seem to depend strongly on the substrate material.

(3) Substrate Temperature Dependence. At lower substrate temperatures, metal films commonly exhibit tension which decreases with increasing substrate temperature, often linearly, and finally goes to zero or changes to compression. The changeover to compression takes place at relatively lower temperatures for lower-melting-point metals. Figure 2 shows typical data for the substrate-temperature dependence of the intrinsic stresses in Cu [43], Ni [44], Fe [45], and Permalloy [46, 47] films. The Permalloy curves indicate some influence of the deposition

Table III. Intrinsic Stresses in Dielectric Films Approximately 5,000 Å Thick
(Compiled by Hoffman [1])

Material	Substrate temp,°C[a]	Substrate material	Stress, 10^9 dynes/cm²[b]	Sign[c]	Method[d]	Ref.
Al_2O_3 (anodic)	A	Aluminum	~1	C	Tensile	59
B_2O_3	90	Glass	0.1	T	CL	6
BaO	50	Glass	0.15	C	CL	6
C	A	Glass	4.0	C	CL	7
CaF_2	110	Glass	0.2	T	CL	6
	A	Mica	(<0.0003)	T	CL	67
CdS	110	Glass	0.8	C	CL	6
CeF_3	40	Glass	2.8	T	CL	6
Ce_2O_3	50	Glass	1.6	C	CL	6
Chiolite	A	Glass	(0.029)	T	CL	4
Cryolite	A	Glass	(0.061)	T	CL	4
	A	Glass	(0.06)	T	CL	67
LiF	110	Glass	0.4	T	CL	6
	A	Cellulose	2.0	T	ED	35
	A	Mica	(0.023)	T	CL	67
	A	Glass	(0.023)	T	CL	4
MgF_2	110	Glass	2.0	T	CL	6
	75	Mica	2.2	T	B	38
	A	Glass	(0.11)	T	CL	4
	A	Mica	(0.11)	T	CL	67
MoO_3	A	Glass	(0.013)	T	CL	4
$PbCl_2$	50	Glass	0.18	T	CL	6
	A	Glass	(0.014)	T	CL	4
PbF_2	110	Glass	0.8	T	CL	6
Sb_2S_3	A	Glass	(0.007)	T	CL	4
Si	1100	Sapphire	~5	C	CL	14
SiO	110	Glass	1.2	C	CL	6
	A	Nickel	4	T	CL	10
SnO_2	A	Glass	(0.008)	T	CL	4
TiO_2 (thermal)	585	Glass	~0.5	C	CL	60
ZnS	110	Glass	1.0	C	CL	6
	A	Glass	(0.022)	C	CL	4
	A	Mica		C	CL	67

[a] A, thermally floating at ambient temperature.
[b] Values in parentheses are relative.
[c] C and T, compression or tension.
[d] B, end-supported beam; CL, cantilever beam; and ED, electron-diffraction technique.

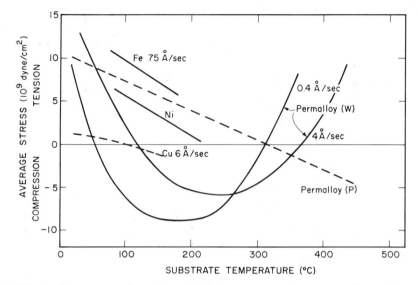

FIG. 2. Temperature dependence of the average intrinsic stress in films of Permalloy (W) (Weiss and Smith [47]), Cu (Horikoshi et al. [43]), Fe Finegan [45]), Ni (Hoffman et al. [44]). The Permalloy (P) curve is the total stress measured at room temperature (Prutton [46]). (*Compiled by Hoffman* [1])

rate on the average stress. A similar dependence reported by Horikoshi et al. [43] for Cu films indicated that stress increased with increasing rate, but showed some saturation at the highest rate studied (70 Å/sec). On the other hand, Hoffman [1] observed no rate dependence for films of several transition metals for a 20 percent rate change about a mean of 60 Å/sec, and for Ni [44] in the 5 to 10 Å/sec range.

Compressional stress developed in ZnS [6, 15, 48] films decreases with increasing substrate temperature, becoming small at about 160°C. By contrast, the tensile stress in SiO [6, 10, 11] films deposited at a 45° angle of incidence at a source temperature of 1350°C shows little dependence on substrate temperature between −125 and +140°C. The behavior of SiO films is known to be sensitive to the source temperature and the oxygen partial pressure during deposition and therefore cannot be compared with other materials.

(4) Thickness Dependence. The thickness dependence of stress in films has been studied extensively. The average stress for Permalloy [47] and Ni [49] films is almost constant, while Fe [50] exhibits a constant stress superimposed on a $1/t$-dependent (t=film thickness) term for thicknesses greater than 400 Å. Thus in Fe films, the average stress decreases with increasing thickness.

The thickness dependence of the average stress in Cu [43] and Ag [51] films obtained by continuous and sequential observations employing a cantilever deflection is shown in Fig. 3. Similar results were obtained by Blackburn and Campbell [6] on Ag films, and by Horikoshi and Tamura [43] on Sb films. The films exhibited no stress until they reached thicknesses of a few hundred angstroms, at which point a sudden rise to a maximum followed by an almost constant level was observed. These threshold thicknesses coincide roughly with the average thickness at which the growing three-dimensional nuclei coalesce with each other and then form a network stage. The hyperbolic dependence observed in Ag films led Kinosita et al. [48] to suggest that the stress is concentrated in a 10-Å layer near the free surface. The same authors, however, later failed to confirm their own observation.

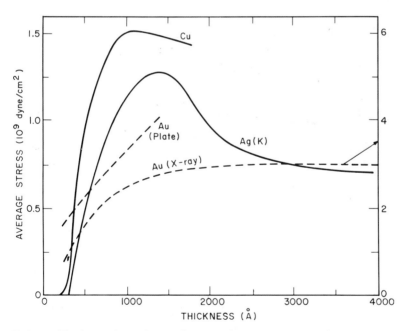

FIG. 3. Thickness dependence of the average intrinsic stress (measured by the cantilever method) in films of Cu (Horikoshi et al. [43]), Ag (Kato et al. [51]), and Au (Kinbara and Haraki [19]). Both bending-plate and x-ray data for Au films are shown for comparison.

In sharp contrast to the results of Cu and Ag films, Au films do not show a thickness-dependent maximum. The results of Kinbara and Haraki [19] obtained by bending-plate as well as x-ray methods are also shown in Fig. 3 for comparison. The x-ray data yield thickness-independent stress for thick films ($>$1,000 Å), while the bending-plate method yields a linear dependence up to 1,500 Å. This difference is ascribed to the fact that highly disordered grain-boundary areas contribute to stress in the bending-plate method, but not to the x-ray line intensities (due to incoherent scattering). An interpretation of the data on Au films is possible by assuming a superficial layer of

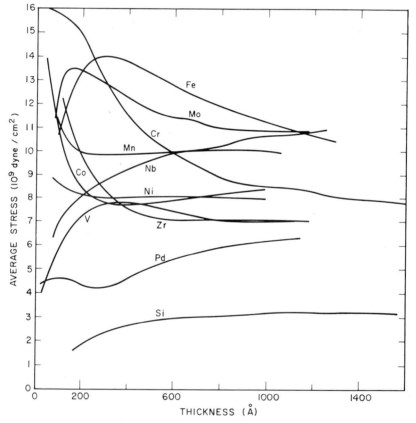

FIG. 4. Thickness dependence of the average intrinsic stress in films of several metals deposited on glass at room temperature measured in situ by cantilever technique. (*Klokholm and Berry* [49])

inhomogeneous stress, superimposed on the rest of the film under uniform stress.

Extensive investigations of stresses in films of a large number of metals were undertaken by Klokholm and Berry [49]. The results are shown in Fig. 4. Two significant conclusions may be deduced: (1) The stresses approach a constant value in thick continuous films. In ultrathin films (presumably not completely continuous), a marked thickness dependence is observed; both an increase and a decrease in stress, depending on the metal, are observed. This suggests that the interfacial region between the film and the substrate is strained differently from the rest of the film. Assuming a suitable stress gradient, the observed behavior can be qualitatively understood. These results cast doubt on the explanation of the maximum shown in Fig. 3 in terms of the structure of the films. (2) The magnitude of stress in continuous films is usually about 0.1 percent of the shear modulus of the bulk metal, and it increases with increasing melting point of the metals. Since the frozen-in structural defects during the deposition process are expected to increase with increasing value of the melting point of the metals relative to the deposition temperature, the intrinsic stresses may be related to the defects incorporated into the film as a result of the quenching process (that is, a thermal effect).

The tensile stress vs. thickness for LiF [7] films deposited on mica at different deposition rates is shown in Fig. 5. These results indicate that stress does not exhibit a thickness threshold and that it rises to a peak as the discrete crystallites coalesce with each other to form a continuous film with a subsequent decrease in stress. The stress increases with increasing crystallite size. The recent studies of Carpenter and Campbell [52] on stresses in several alkali-halide films, however, showed that in clean conditions the stress increased rapidly to reach a saturation value without exhibiting a maximum. Further, a plot of the stress in island-like ($\sim$50-Å) alkali-halide films deposited on glass showed a linear decrease with increasing lattice constant a, being tensile for $a > 5.3$ Å (for LiF and NaF) and becoming compressive for $a < 5.3$ Å (for NaCl, NaBr, KCl, KBr). This result has been explained by a questionable model according to which adsorption takes place at negative oxygen sites on the surface of pyrex. These sites, which are separated by an average distance of $\sim$5.3 Å, are assumed to determine the sign and magnitude of the net stress in the islands as a result of the lattice mismatch.

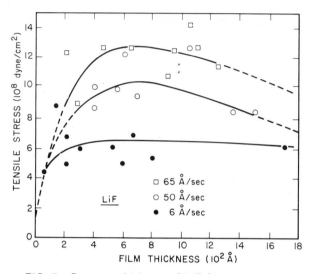

FIG. 5. Stress vs. thickness of LiF films deposited by thermal evaporation onto mica at different deposition rates. The crystallite size increases with the increasing rate and film thickness. (*Campbell* [68])

(5) Deposition Rate and Angle-of-incidence Dependence. The observed dependence of stress on deposition rate is not systematic enough to allow meaningful conclusions to be drawn. Increases as well as decreases in stress occurring with increasing deposition rate have been reported. Dielectric films, e.g., of SiO, show large variations because of their sensitivity to contamination [53]. As the rate of deposition of a film is increased, the number N of SiO molecules for every molecule of the residual gas impinging on the substrate increases, and thereby affects stress. The sensitivity to deposition conditions is further enhanced by increasing angle of incidence of the SiO vapors depositing on the substrates. The oblique incidence yields a more porous structure which is highly susceptible to adsorption effects (Chap. IV, Sec. 3). In Fig. 6 is shown the stress variation in obliquely deposited SiO films as a function of the angle of incidence of the vapor atoms for differenct values of N At high deposition rates or at lower pressures during deposition of the film, the stress is compressional at normal incidence, and it changes to tensile at oblique deposition. With a value of N exceeding unity, only tensile stress is observed.

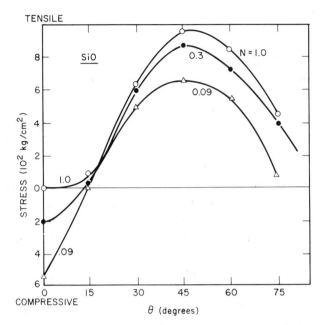

FIG. 6. Stress in SiO films in vacuum as a function of the angle of incidence (away from the normal) of the vapor beam. N is the ratio of the number of vapor atoms to the residual gas atoms striking the glass substrate. (*Hill and Hoffman* [53])

Priest et al. [11] observed that for a 45° angle of incidence, the stress in SiO films changed from 1.0×10^9 dynes/cm² tension to 1.0×10^9 dynes/cm² compression within 3 min after exposure to air. The resulting strain was often accompanied by buckling of the film from the substrate. The observed aging effects were attributed to oxidation and hygro-scopicity of the films. These authors also found the stress to be independent of the deposition rate (in contrast to the data in Fig. 6), but a strong function of the source temperature. Thus the situation about the rate dependence of stress in SiO films remains confused.

In this connection, it is significant to note that stresses in metal films deposited in high-vacuum systems with a low water-vapor content are not very sensitive to moderate changes in the residual gas pressure. The oxidation does, however, change the stress. Murbach and Wilman [3] found compression in Al films whenever the chamber pressure exceeded 10^{-4} Torr during deposition, presumably because of oxidation of the film. Similarly, an increase in compressive stress was observed by Halliday et al. [35] when Cu films were aged in air at room temperature, which resulted in the formation of Cu_2O. The stress could be reduced considerably by etching away the oxide layer.

(6) Annealing Effects. Since the internal stress depends much on the structural order of a film, it is useful to observe its irreversible changes due to an extended annealing and relate it to the elementary lattice defects and recrystallization process. The intrinsic stress for 1,500-Å Fe and Ni films as a function of annealing temperature using a pulse-anneal technique showed [8] the expected stress decrease with increase of the annealing temperature. At sufficiently high temperatures, the stress increased because of the occurrence of agglomeration of the film. The x-ray technique [19] showed a decrease of stress in an Au film with increase of annealing temperature. The bending-plate method, on the other hand, exhibited a marked increase of stress, again probably because of the highly disordered grain-boundary areas which contribute to bending but not the x-ray line intensities.

The stress changes in Cu films [8] as a function of the annealing temperature are shown in Fig. 7. The stress decreases with temperature both above and below room temperature. At sufficiently high temperatures, the stress increases. The concurrent variations in the electrical

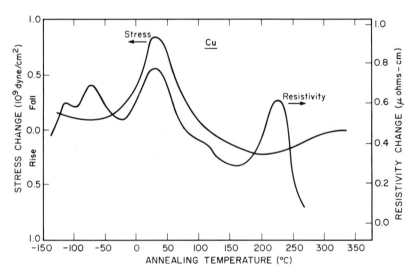

FIG. 7. Correlation of stress and electrical resistivity changes in Cu films during "pulse-anneal" treatment. (*Story and Hoffman* [8])

resistivity are also displayed in the same figure. A discussion of the correlation of the two curves is given later.

(7) Anisotropic Stress. When a vapor beam arrives at the condensing substrate at an angle other than normal, stress becomes anisotropic in the plane of the film. Anisotropy in stress is also expected owing to thermal mismatch of a film deposited on a low-symmetry single-crystal substrate of anisotropic expansion coefficient.

If we denote the plane containing the vapor beam and the substrate normal as the plane of incidence and x and y directions perpendicular and parallel to it in the film plane, respectively, a measure of the anisotropy is obtained by the ratio S_y/S_x of stresses along y and x directions. The data for this stress ratio as a function of the angle of incidence for Fe [12] and SiO [11] films are presented in Fig. 8. The stress is evidently isotropic for normal incidence in both cases. The stress anisotropy ratio in Fe films decreases roughly as the cosine of the angle of incidence, but SiO films show a more complex behavior. When pulse-annealed [54], the stress ratio for Fe films decreases monotonically with temperature. The results on the annealing behavior of anisotropic stresses are sketchy and not well understood at present.

Cohen et al. [55] observed large stress anisotropy in Permalloy films deposited on a circular plate and attributed it mostly to nonuniformities in the substrate. Further, Smith et al. [56] established the existence of strain anisotropy in obliquely deposited Permalloy films from the observed change in magnetic anisotropy (Chap. X, Sec. 4) which occurred upon stripping the films from their substrates. The results on the sense and magnitude of anisotropy were similar to those for Fe films (Fig. 8).

(8) Stresses in Chemically Prepared Films. The earliest studies [2] on stresses were made on electrodeposited films. The magnitude and the sign of stress in electrodeposited films are known to be influenced strongly by the conditions prevailing during electrolysis, such as temperature, current density, composition, and acidity of the solution. In an attempt to formulate a generalized theory of stresses in electrolytically prepared films, Popereka [57] observed (primarily from the Russian literature) that such films can be classified into three groups: (1) films of Ni, Co, Fe, Cr, and Rh exhibit tensile stress; (2) films of Zn, Cd, Sn, Pb, and Bi are under compressive stress; and (3) films of Cu, Ag, and Sb have an equal tendency to stresses of either kind depending on

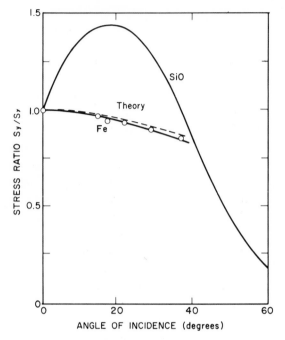

FIG. 8.   Stress anisotropy (S_y/S_x) for films of Fe
(Finegan and Hoffman [12]) and SiO (Priest et al.
[11]) as a function of the angle of oblique (with
respect to the normal) deposition. The dashed curve
is calculated for the stress ratio varying as the cosine
of the angle of incidence. (*Hoffman* [1])

the conditions of electrolysis. This classification is certainly oversimplified, but it does point out the complex and chaotic state of this
subject at present.

Vermilyea [58] studied stresses in anodic oxide films of Ta, Nb, Al,
Ti, W, and Zr using a bending-beam technique. The observed stresses
were always tensile at small thicknesses, and their magnitude depended
strongly on the anodizing conditions. A tendency toward compressive
stress was found when a large amount of gas evolved. Bradhurst and
Leach [59] (using a hard-beam tensometer to measure stress in the oxide
film formed on an aluminum foil by anodizing) observed that the oxide
films developed compressive stress at a low current density (that is, a low
rate of formation), but a tensile stress at a high current density. As the
current density increased, both the tensile stress and the cationic

transport increased to a saturation value. The crossover from compressive to tensile stress also depended on the pH value of the anodizing solution as seen by the dotted curve in Fig. 9. Data on thickness dependence of the stress showed no consistent behavior since an increase as well as a decrease were observed. The measured stress values varied over a wide range, as indicated by the spread of the data in Fig. 9.

Schröder and Spiller [60] studied thickness dependence of stress in thin titanium oxide films prepared by pyrolysis and thermal oxidation. Pyrolytically prepared films exhibited tension ($\sim 10^9$ dynes/cm^2), whereas thermally oxidized films were in tension for small thicknesses but in compression above a thickness of about 150 Å.

2.3 Origin of Intrinsic Stress

As already mentioned, thermal stress is easily understood. Here we are concerned with intrinsic stress only. A number of phenomenological models of limited applicability have been proposed to explain various features of the observed intrinsic stress. The wide variation in the experimental results, which are invariably limited in scope and information on the structural details of the film, have made it difficult to

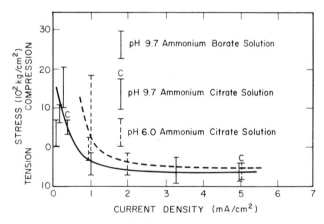

FIG. 9. Stress in anodic oxide films on Al as a function of the current density of formation in ammonium borate and ammonium citrate solutions. The vertical lines represent the total range of stress values observed in each experiment for 150- to 3,000-Å-thick oxide films. Comparison with the dotted curve shows the effect of the pH value of the electrolyte. (*Bradhurst and Leach* [59])

formulate an acceptable theory of the intrinsic stress. This section contains a discussion of some of the more plausible models.

(1) Thermal Effect. Murbach and Wilman [3] suggested that the surface of a film during deposition is at a substantially higher temperature than the substrate so that a thermal stress proportional to this temperature differential will be generated. This hypothesis accounts for the observed nearly linear decrease in stress as the substrate temperature is raised. But Hoffman et al. [1, 44] questioned the hypothesis on the basis that (1) thermal-relaxation processes at the surface are expected to be rapid enough to maintain the surface at approximately the substrate temperature and (2) the observation of irreversible stress changes in a typical annealing process subsequent to the deposition of a film appears to be not in accord with the thesis of an effective high temperature of deposition.

Although the steady-state temperature of a thick, continuous film may not be greatly affected by the heat liberated during the condensation of a reasonable flux of vapor atoms, a marked temperature rise is, however, expected and indeed observed (see Chap. IV, Sec. 1.4) for discontinuous films. As pointed out by Gafner [61], this rise would be in the form of a "thermal spike," the duration of which depends naturally on various heat-loss mechanisms. Assuming some realistic values of thermal conductivity of an island-like film, the thermal spike may have a relaxation time $\sim 10^{-6}$ to 10^{-9} sec. Thus, cooling rates ($\gtrsim 10^{6}°C/sec$) resembling those produced in the "splat-cooling" technique [62] are obtained. The two consequences of this rapid-quenching process which are relevant to our present discussion are: (1) Stresses must be generated in the lattice because of the freezing in of a large number of structural defects. Since these defects lock their own movements, subsequent annealing is essential to increase the mobility of the vacancies and other point defects, and consequently irreversible changes in the stress behavior result. (2) Quenching is also known to freeze in high-temperature and/or metastable phases of some materials (see Chap. IV, Secs. 4.2 and 4.3, for further discussion). In these cases, stress changes should take place on annealing because of structural transformations attended by changes in the atomic volume. For example, Günther and Buckel [63] observed compressive stresses in $\sim 1,000$-Å-thick Ga films deposited on Al foils at low temperatures. Subsequent warming of the film reduced the compressive stress to a negligible value, which was attributed to the fact that the frozen-in liquid phase of Ga is denser than its solid phase.

(2) Volume Changes. Processes such as phase transitions, polymorphic transformations, and changes in stoichiometry, which result in a change in the atomic volume, lead to stresses if constraint is present. This is illustrated in the case of Sb films which show a rapid increase in stress as the film thickness increases above ∼1,000 Å where an amorphous-to-crystalline transformation takes place. Similarly, the compressive stress in Ga films [63] deposited at low temperatures is also due to the frozen-in denser liquid phase. Many other cases of polymorphic transformations in thin films are known, but no stress measurements have been made in these cases. Face-centered-cubic (fcc) structures with larger atomic volumes than the normal structures are observed [64] in the initial growth of sputtered films of a number of transition metals such as Ta, Mo, W, Nb, Zr, Hf, and Re. The continuous transformation of the metastable fcc crystallites to the normal structure results in localized coexistence of large compressive and tensile stresses which produce fracture and buckle patterns (Sec. 3.5) in these films.

(3) Surface Layer. Oxidation has a profound effect on the intrinsic stress. For example, tension is converted to compression when films of Cu [35], Fe [54], and Al [59] are oxidized. The compressive stress increases with the thickening of a surface oxide layer and decreases when the oxide is chemically thinned. The observed hyperbolic relation between stress and oxide film thickness in Fe films has been explained phenomenologically by Kinosita et al. [48] on the assumption that the total stress is localized to a superficial layer of about 10 Å thickness.

Highly strained or pseudomorphic layers at the substrate interface would similarly give rise to stress. This consideration led Gimpl et al. [65] to suggest that the observed high stress and high strength in metal films may be due to a thin strained amorphous layer at the surface of the film. The amorphous layer may be an oxide, which would be responsible for most of the observed stress, since the body of the film should behave like a bulk metal which would relieve high stress by plastic flow. Stress would similarly arise if the lattice constant of the interface layer is different (pseudomorphic, see Chap. IV, Sec. 4) from that of the bulk film. A compressive stress is expected in an oxide film if the volume ratio for oxide is greater than 1. For α-alumina, this ratio is 1.28 and corresponds to a compressive stress of ∼3.6 x 10^{10} dynes/cm^2 at the metal-oxide interface, assuming the growth proceeds by the inward diffusion of oxygen. The observed [59] compressive stress in anodic oxide of Al is $\sim 2 \times 10^9$ dynes/cm^2, which is at least an order of magnitude smaller than the calculated one. This discrepancy is explained on the basis that movement of both oxygen and metal ions may take place in thicker films so that the volume ratio is not the stress-determining factor as the thickness of the oxide increases. The presence of stresses at the oxide-metal interface would modify the conclusions considerably.

Jaccodine and Schlegal [66] measured stresses at Si-SiO$_2$ interfaces induced by thermal oxidation of (111) and (100) surfaces of Si. The cantilever and bulge techniques yielded a compressive stress of 3×10^9 dynes/cm^2 for oxidation at 1200°C. This stress is attributed to thermal mismatch of Si and SiO$_2$ interfaces.

(4) Surface Tension. If the surface tension is σ_1 at the film-vacuum interface and σ_2 at the film-substrate interface, the film of thickness t is under a total apparent stress given by $S = (\sigma_1 + \sigma_2)/t$. A positive surface energy would manifest itself as tension in a bending-plate experiment. Several workers [35, 42, 67] have suggested that surface tension can provide an important contribution to stress in thin films. With a typical value of surface energy of $\sim 1,000$ ergs/cm^2, a stress of $\sim 10^9$ dynes/cm^2 is expected, however, only in films of thickness less than 100 Å.

Stresses due to surface tension could have a significant effect [68] on the lattice constants of a film consisting of small isolated crystallites (see Chap. IV, Sec. 3.1). Accurate measurements by Rymer [30] from x-ray line broadening showed that the lattice constant of LiF films increases with decreasing crystallite size, in agreement with the negative surface energy in thin LiF films predicted theoretically by Shuttleworth [69]. As the crystallites grow and their lattice constant tends to become normal, one would expect an increasing tensile stress, which agrees with the observation of Blackburn and Campbell [7, 68] on LiF films.

Smith et al. [56] proposed that the coalescence of two crystallites gives rise to tension partly because they are prevented from doing so by the bonding to the substrate and partly because of the reduction of the surface area due to decreased surface energy on coalescence. In addition to the tensile stress, a compressive stress may be caused by the surface tension of the individual grains. Hence the sign of the film stress should depend on the balance of the two effects. Weiss and Smith [47] observed a transition from tension to compression in Permalloy films with increasing temperature which shifted toward higher temperatures when faster deposition rates were employed. They attributed this result to the domination of the surface-tension term with the reduction in grain size.

(5) Electrostatic Effects. The presence of electrostatic charges on small crystallites enhances their total energy [70]. For example, 100 unit charges present on a 100-Å Ag crystallite increase the free energy by about 10^{-9} ergs, which is comparable with the surface-tension energy and can cause appreciable geometrical distortion, thereby producing intrinsic stress. Electrostatic charges present on the vapor beam or on the substrate can play an important role in the growth of island-like films (Chap. IV, Sec. 2.3). In addition to the free-energy contributions, the net attractive and repulsive forces [70, 71] between charged crystallites, which depend on the sign of the charges and the size and spacing between the crystallites, are of sufficient magnitude to produce large shear forces. The shear stress between a 100-Å island having 100 unit charges and a 20-Å island having a unit charge separated by a distance of 200 Å from each other is $\sim 10^{10}$ dynes/cm^2.

(6) Lattice-misfit Accommodation Model. A theory of accommodation of lattice misfit between substrate and monolayer type of overgrowth by elastic strain was suggested by Frank and van der Merwe [72] and later extended by van der Merwe [73-76]. According to this theory, the equilibrium strain of an epitaxial film of thickness t is given by

$$\eta = -\frac{(1 - 2\nu)(2 - f)(1 + f)G_i a_f}{8\pi^2(1 - \nu)(2 + f)fG_f t}\beta \ln\left[2\beta(1 + \beta^2)^{1/2} - 2\beta^2\right]$$

where $f = (a_f - a_s)/a_f$ and $\beta = 8\pi G_f f/(1 - \nu)(1 + G_f/G_s)(2 + f^2)G_i$. Here, a_f and a_s are the lattice parameters of the overgrowth and substrate and G_f, G_s, and G_i are the shear moduli of the film, substrate, and interface between film and substrate. The Poisson ratio of the film and substrate are assumed equal and represented by ν. It is further assumed that the misfit is only along one axis in the interface; that the misfit in excess of that accommodated by strain is taken up by an array of identical edge dislocations whose Burgers vector lies in the interface and is parallel to the direction of misfit; and that the substrate is infinitely thick and unstrained.

The above expression for η predicts that for strong interfacial bonding ($G_i \sim G_f \sim G_s$) and small misfits (less than ~ 4 percent), the misfit is accommodated by an elastic strain up to a critical thickness (~ 5 Å for 4 percent and 400 Å for 0.2 percent misfit). The strain is then inversely proportional to thickness. Above this critical thickness, the misfit is accommodated by misfit dislocations. For large misfits (>10 percent), much of the misfit is accommodated by misfit dislocations even for a monolayer deposit.

The constraining of the film lattice to the substrate lattice (called pseudomorphism; for further discussion see Chap. IV, Sec. 4.1), and the occurrence [77, 78] of misfit dislocations with both increasing film thickness and increasing misfit (see the review by Matthews [78]) are now well established in several systems where a monolayer-by-monolayer type of deposition is obtained. The separation of moiré-fringe patterns (Chap. III, Sec. 2.4) has been utilized for measuring the elastic strain. The separation becomes smaller with decreasing fraction of the misfit accommodated by elastic strain.

The qualitative variation of the elastic strain with film thickness and misfit is in agreement with the van der Merwe theory. Jesser et al. [151] measured elastic strain in individual Pt islands grown on Au substrate and found a good quantitative agreement with the theory. Generally, however, the observed decrease in elastic strain with film thickness is considerably smaller than given by the inverse thickness dependence. The observed strain may be ~ 10 to 100 times higher than the predicted value. This disagreement may be partly due to interface alloying if it reduces the misfit to be accommodated. The more important factor probably is the result of the fact that misfit dislocations require certain activation energy for their generation. If they cannot be generated so easily, the film growth will be accompanied by a smaller decrease in elastic strain than is predicted by the theory.

Just how the aforementioned theory can be used to explain the observed stresses in thick polycrystalline films is not clear. The general observation that neither the magnitude nor the sign of the stress observed in most cases is related to either the sign or the absolute value of the misfit sharply contradicts the van der Merwe theory.

(7) Structural-defect Hypothesis. The existence of correlation [8, 79, 80] between the changes in stress and electrical resistivity during pulse annealing of Cu and Ni films led Story and Hoffman [8] to suggest that the intrinsic stress might be related to the point defects and defect clusters in films. This simple model can be used to explain qualitatively the temperature-dependent features of intrinsic stress. The model was later criticized by Hoffman [1] himself on the ground that the recovery of resistance and stress on annealing corresponds [80-83] to a loss of about 1 percent vacancies, which is many orders of magnitude higher than the equilibrium vacancy-concentration value expected at about room temperature. It was further asserted that the observed annealing peak in the resistance and stress curves in Cu films is not due to simple point vacancy migration but is more probably a recrystallization phenomenon. These objections are not completely valid. It is reasonable to believe that point defects play an important role in determining intrinsic stress in films. The occurrence of ~0.1 percent vacancies is not uncommon in nonstoichiometric, severely cold-worked, or rapidly quenched materials. The process of deposition of a film at or near room temperature essentially consists of vapor quenching, which freezes in an enormous number of structural defects. It is likely that vacancies collapse at defect centers, but their mobility at or near room temperature is very low because of the high concentration of locked-in structural defects. Further, the recrystallization process and the removal of structural defects in a film due to annealing are intimately related to each other. By postulating suitable relaxation processes of migration of vacancies and other point defects, the observed tensile behavior may be explained qualitatively. A quantitative analysis may, however, be difficult.

(8) Crystallite-boundary Mismatch Model. Finegan and Hoffman [84] discussed a model in which stresses are due to a combined effect of the surface tension discussed already and the growth processes at the grain boundaries. Accordingly, each nucleus (island) in the discrete-nuclei stage is strained by its surface tension. The nuclei are considered mobile enough so that a large shear force is not transmitted across their interface to the substrate. As the nuclei grow, the surface-tension-induced strains help to freeze the bonds of the atoms in the nucleus to the surface. This anchoring of the crystallites should induce the first measurable compressive strain at the interface. For example, the contraction of a 10-Å Ag crystallite due to surface tension is expected to be about 1 percent. As the size of the nuclei increases with further deposition, the lattice expands because of the relaxation of surface-tension forces, and thus compression results. As the nuclei become large enough to coalesce and form a nearly continuous film, the interatomic forces at the boundaries of the islands tend to close any gap, with the result that the neighboring crystallites are strained in tension. A sharply increasing tensile stress is expected at this stage which will increase to approach a constant value as the film becomes continuous. The tensile strain is given by the ratio of the thickness of the region of interatomic influence at the grain boundary to the size of the crystallite.

This model therefore gives zero stress followed by a compression and a rapid crossover to tension as the film grows in thickness. The model yields little or no stress for single-crystal films and indicates a small intrinsic stress for polycrystalline films with large grains. The model fails to provide a mechanism for compressive stress in thick dielectric films and Permalloy films in the low-temperature range (see Fig. 2). While the predicted compressive stress in ultrathin films also lacks experimental support so far, the model seems to fit qualitatively the observed behavior of most thick metal films. A quantitative fit to the observed data may be obtained when suitable assumptions are made as to the size of the crystallites and the stage at which they coalesce. Since the growth stages and the crystallite size depend on film thickness, substrate temperature, annealing, vapor temperature, deposition rate, etc., it is reasonable to expect that a generalized version of this model could explain the effects of these deposition variables on the stress in films.

It is interesting to note that Popereka [57] proposed a somewhat similar model to explain the variety of stress behavior observed in electrodeposited films. He considered the total stress to be the sum of a tensile stress existing inside the crystallites, and a compressive stress at the boundaries of the crystallites, the latter arising because of defects at the mismatched boundary. Thus, depending on the relative values of the two contributions, tensile or compressive stress may be obtained in films of different materials.

(9) Anisotropic Growth. Stress anisotropy observed in obliquely deposited films is, essentially, the result of a geometrical growth effect.

Smith et al. [56] suggested that anisotropy results from the self--shadowing effect in the film as it grows from random nucleation centers by adding material from the obliquely directed vapor beam. Thus chains of crystallites grow preferentially along the direction of incidence of vapor as illustrated by the electron micrograph of an Al film [Fig. 9(b), Chap. IV]. The anisotropic growth produces anisotropic stress.

Finegan and Hoffman [12] suggested that a cosine distribution of the crystallite size can explain the observed anisotropy in Fe films (Fig. 8). Further, they proposed that anisotropic nucleation, if present, would also contribute to anisotropic stress since nuclei might be slightly farther apart in a direction parallel to the plane of incidence than perpendicular to it. This process will result in larger crystallites and, hence, a smaller number of grain boundaries along the perpendicular direction. If grain boundaries contribute to stresses, the stress will be smaller. The structural studies of Chopra and Randlett [85] on obliquely deposited Ag films show no evidence of anisotropic nucleation for oblique angles up to 80°. With the increasing angle of incidence, the size of the discrete nuclei and the average distance between them increase isotropically in a manner similar to that effected by an elevated substrate temperature. Geometrical shadowing effects, which should lead to stress anisotropy, occur only when the nuclei are large and close to each other prior to the formation of the network structure in the film.

3. MECHANICAL PROPERTIES

Interest in the deformation behavior of a thin film and its relation to the interaction of dislocations with the free surfaces, the frozen-in structural defects, and the microstructure of a film has attracted numerous investigations on the subject of mechanical properties of films. Unfortunately, the wide variety of experimental results and interpretations obtained have clouded the general picture. This subject has been reviewed by Hoffman [1], by Menter and Pashley [86], and by Neugebauer [87]. We shall summarize and discuss some of the significant results. Since in many cases, the results depend on the technique of measurement, we shall first describe some of the techniques employed to measure the tensile properties of films.

3.1 Experimental Techniques

(1) High-speed Rotor. Beams et al. [88] measured both the tensile strength and adhesion of films electroplated onto the cylindrical surface of a rotor by spinning it at high speeds in vacuo with a rotating magnetic field. Assuming the centrifugal force is balanced by the hoop stress in the film and the adhesion of the film to the rotor, then for a hollow, thin-wall, cylindrical rotor unsupported at the ends,

$$4\pi^2 N^2 R^2 \tilde{\rho} = \Gamma + \frac{AR}{t} \tag{8}$$

where N is the rotor speed in rps, R the rotor radius, $\tilde{\rho}$ the density of the deposited film, t the film thickness, Γ the tensile strength of the film, and A the adhesion. Depending on the choice of substrate and deposition conditions, either the adhesion or the tensile-strength term can be made negligible, allowing Eq. (8) to be simplified. This method is obviously specialized and, further, does not yield a stress-strain curve.

(2) Bulge Test. This test, used by Beams [89] and others [90, 91], is based on the principle that, if a film mounted at the end of a hollow cylindrical tube is allowed to bulge as a result of differential pressure, the biaxial stress of the film can be calculated from the pressure P and the strain η from the height or deflection D of the center of the deformed film. The relations for the hemispherical bulge are

$$S = \frac{PR^2}{4tD} \qquad \eta = \frac{2}{3}\frac{D^2}{R^2} \tag{9}$$

where P is the pressure difference between the two sides of the film and R is the internal radius of the tube.

Papirno [92] suggested that the bulge surface is better described as a general quadratic surface of revolution, in which case the strain is given by $0.30 (D/R)^{1.9}$, which is smaller by a factor of 2 than that given by Eq. (9).

If an initial stress S_0 is present, the stress in Eq. (9) should be replaced by

$$S = S_0 + \frac{2}{3}\frac{E_F}{1 - \nu_F}\frac{D^2}{R^2} \tag{10}$$

where ν_F is the Poisson ratio for the film and E_F is its modulus. For the bulge test, the films are either mounted or deposited directly onto a plastic film mounted at the end of the tube, and the plastic film is dissolved. In another approach, a hole is drilled into the substrate (for example, rock salt) to provide direct pressure access to the film. Both circular and rectangular bulges have been used with similar results.

(3) Tensile Test. A number of microtensile testers have been built for the observation of small loads and small elongations typical of thin film work. "Soft" testers, which are common in thin film studies, are those in which the load is given and the elongation is measured, whereas "hard" types prescribe the strain rate and measure the load.

A number [93-98] of soft and hard microtensile machines have been designed and used for thin film studies. A uniaxial load may be applied magnetically, electromagnetically [94, 95, 97, 98], or gravimetrically [96]. The elongation is measured by direct observation through an optical microscope, interferometrically, or with a differential transducer. Marsh [93] has designed a null torsion balance to apply the force. The extension in this case is measured from a mechanical linkage with a sensitive optical null detector, permitting the determination of elongations greater than 5 Å. The hard machines described in the literature use a strain-gauge-proof ring to measure the load while the strain is observed using a differential transducer or an optical arrangement. The various tensile-test apparatuses are summarized by Hoffman [1]. The original papers should be consulted for more details.

Common to all the methods is the difficulty of mounting the film so that a uniform loading can be obtained without tearing the film at the edges. Flat, corrugated, and Orowan [99] types of grip have been employed for loading the specimens, which are glued or cemented onto the grip or simply held by friction. The experimental techniques vary from observer to observer and are primarily determined by the equipment available. There is, therefore no standard method which can be relied upon to yield reproducible results. This difficulty is further aggravated by the fact that the details of the stress-strain curves may depend on the shape of the specimen, its flaws and nonuniformities, residual stress, and the damage caused by the mounting process and any creep resulting therefrom.

(4) Electron-microscope Devices. Fixtures which permit straining of the specimen at a constant rate while observing it in the microscope have been built by Wilsdorf [100] and Pashley [101]. A detailed study of dislocation movements is made possible by the use of such techniques. Accurate values of stress and strain are generally difficult to measure and are complicated by the fact that the strain is nonuniform.

(5) Direct Measurement of Strain. The most direct method of measuring elastic strain is to determine the change in lattice parameters under the influence of an applied stress. While the electron-diffraction method has been used by several workers [28-35], accuracy is poor. Transmission electron diffraction is limited to very thin films (<1,000 Å) since the diffraction spots become diffuse in thicker films. A high accuracy may be obtained by x-ray diffraction techniques such as the divergent-beam technique of Kossel [102] and Lansdale [103].

The x-ray and electron-diffraction lines are broadened because of microstrain as well as particle size. By separating the two contributions, the line broadening may be used for determining microstrain, strain gradient, and strain anisotropy. This technique has been successfully used for a number of metal films, although several authors have placed insufficient emphasis on separation of the particle-size contribution.

3.2 Experimental Results

Numerous studies have been reported on the tensile properties of polycrystalline films of Au [86, 89, 97], Ag [24, 88, 89, 108], Cu [24, 25, 108], Ni [95], Al [23, 108], and cold-rolled foils of Ni [91, 95] and Cu [109]. Epitaxially grown Au films with (100) and (111) orientations obtained by deposition onto heated (~300 to 350°C) substrates of cleaved rock salt and mica substrates, respectively, have also been extensively investigated [86, 90, 94, 97, 104-108]. The significant results may be summarized as follows.

(1) Stress-Strain Curves. 1. A stress-strain curve of a metal film is generally irreversible on initial loading but becomes nearly reversible on unloading and then reloading, provided that the previous load level is not exceeded. The slope of the initial loading curve is smaller than the slope of the unloading curve, which may be due to the misalignment of the sample, or creep in the sample or adhesive.

2. Stress-strain curves for two 1,050- and 2,790-Å-thick epitaxial (100) Au films [90] are shown in Fig. 10. The data were obtained by the bulge test with stress increased at a rate of about 5×10^8 dynes/(cm^2)-(min). The small value of stress at zero strain is due to the initial pressure

difference required to remove wrinkles at the center of the film. The nonlinearity of the initial-load curves may also be due to the film wrinkles. Stress-relaxation steps (Fig. 10) are reproducible straight lines and do not exhibit significant hysteresis. The separation at the top of these steps is due to a creep of $\sim 5 \times 10^{-3}$ percent over a period of 15 sec. Creep has been observed in most stress studies of films. A creep rate varying from 10^{-7} to 10^{-4}/min, depending on load, dimensions, and amount of prestraining, has been observed in Au films.

3. Values of the elastic modulus are obtained from the slope of the stress-strain curves. With a few exceptions, the calculated modulus of films has the normal bulk value. Bulge-test studies [88-90] on (100) Au films indicate a thickness-dependent increase in the modulus for films thinner than 4,000 Å. Conflicting results [91] are reported for Ni films. Both high and low values have been obtained, depending on the technique for preparation of films. Large changes [110] in Young's modulus are obtained in Ni films electrodeposited on Cu substrates when the pH value of the plating bath is varied. On the other hand, the stress-strain curves of the unsupported films of thermally grown SiO_2

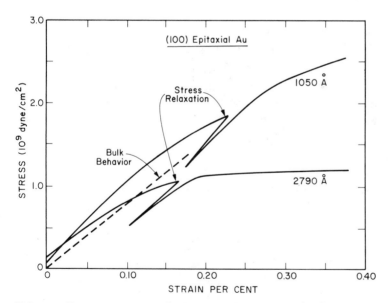

FIG. 10. Stress-strain curves (with stress-relaxation steps) of epitaxially grown (100) Au films for two thicknesses. The dotted line corresponds to the bulk elastic behavior. (*Catlin and Walker* [90])

[66] yield a value of 6.6×10^{11} dynes/cm^2 for Young's modulus, which compares favorably with the value for silicate glasses.

4. Large amounts of plastic deformation and work hardening at higher strains are commonly observed, as illustrated by the 1,050-Å curve in Fig. 10. Catlin and Walker [90] observed mechanical twinning in (100) Au films under stress. A few twin bands were formed at low stresses, but a large number of small bands appeared suddenly just below the ultimate tensile strength. In thicker films, these small bands grouped into one or more large bands with a laminated fine structure. The twin bands were oriented in the [110] direction. They are probably formed because of the junction of twins formed on different planes. Once formed, some unknown mechanisms impede their propagation.

The electron-microscopic investigations of Pashley [111] showed a slow motion of screw dislocations in (111) Au films of ~200 Å thickness. A faster motion confined to one slip plane, and single and multiple cross slips were observed. In addition, a dislocation was found to split into two partials separated by a very wide ribbon of stacking faults, apparently caused by the pinning of the partials at the surface by the stress due to the surface contamination layer. In annealed films, he noted low-angle boundaries and dislocation arrays. Pashley also showed that the extension is completely elastic at fracture in (111) Au films thinner than 500 Å. Thicker films exhibited plastic behavior which was confined to thin regions in the specimen.

5. The deformation processes show no significant dependence on the crystallite size, as illustrated by Blakely [97] in a schematic representation in Fig. 11 of his stress-strain data on polycrystalline and single-crystal Au films. The epitaxial films are expected to consist of large crystallites (subgrains) about 1 μ in size, which is at least an order of magnitude greater than the crystallite size commonly obtained in polycrystalline Au films. The slope (Young's modulus) of the stress-strain curves in Fig. 11 is, however, significantly lower than that of the bulk behavior.

6. The total elongation at fracture in polycrystalline metal films is usually 1 to 2 percent and exhibits a large elastic component. The relative elastic and plastic extension at fracture is estimated to vary from completely elastic to about three-quarters elastic for Au. For Ni, elastic strains of about 0.85 percent and plastic strains of 0.67 percent have been noted at fracture, and total strains of 2 percent have been found in

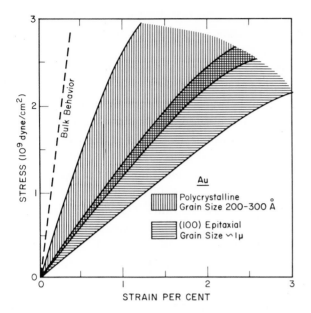

FIG. 11. Schematic comparison of tensile behavior of polycrystalline and epitaxially grown (100) Au films of estimated grain size of ~1 μ. The graph indicates representative values only. The dotted line corresponds to the bulk elastic behavior. (*Blakely* [97])

Cu films. Electron-diffraction studies of Palatnik et al. [25] yielded a maximum elastic deformation of 0.5 and 0.3 percent for Al and Ag films, respectively.

7. Unconstrained single-crystal Au films annealed in air for 3 hr at 1000°C were found by Blakely [97] to be softer and more ductile and exhibit a total strain at fracture of about 10 percent. Such specimens showed a decided "necking" near the center.

8. The fracture mechanism in thin films, as proposed by Menter and Pashley [86], is based on localized plastic deformation with a resultant thinning of the film and a rise in the stress level. Eventually, the smaller cracks formed in this manner join and the film fails. The dislocations responsible for the deformation are not the grown-in dislocations, but those which nucleate and multiply in holes and discontinuous regions of a film. They move on their glide planes until sufficient shear takes place to open the small cracks. Blakely [97] obtained some evidence of necking just prior to fracture, although it is generally not observed. The

maximum stress seems to correspond to that needed to propagate cracks from flaws existing in the specimen. For polycrystalline Ni films, a "clean-cleavage" type of fracture results.

9. The elastic anisotropy of a single crystal persists to a certain extent in a polycrystalline aggregate and is determined by such factors as grain size, grain boundaries, coupling between crystallites, orientation effects, and intrinsic strains. X-ray and electron-diffraction methods have been used by Palatnik et al. [25] to determine the fractional change $\Delta d/d$ in a lattice spacing for various (hkl) reflections of 400- to 500-Å-thick vacuum-deposited films of Al, Ag, and Ni. The films were detached from the substrate and were strained inside the diffraction camera. Large elastic anisotropy was found in Ag and Ni films. The variation of $\Delta d/d$ for various (hkl) reflections as a function of macrodeformation is shown in Fig. 12. A negligible effect was observed in Al films. It is clear from the data of Fig. 12 that, for Ag and Ni films, an inequality relation

$$\left(\frac{\Delta d}{d}\right)_{111} \sim \left(\frac{\Delta d}{d}\right)_{220} < \left(\frac{\Delta d}{d}\right)_{311} < \left(\frac{\Delta d}{d}\right)_{200} \tag{11}$$

holds. Since $(\Delta d/d) \rightarrow 0$ after fracture, no new defects are produced by the deformation. The lattice thus undergoes only elastic deformation. The high $(\Delta d/d)$ for (200) reflection cannot be ascribed to packing defects because effects from these are completely eliminated experimentally by the use of the relative change in the diffraction ring diameter.

The anisotropy of Young's modulus has also been deduced by Vook and Witt [23] from their studies on the growth of textures on deposition at low temperatures and subsequent annealing of films of Cu and Au. The [111] deposition texture for Cu films is attributed to the tendency of impinging Cu atoms to form close-packed layers. The [100] annealing texture is ascribed to the fact that Young's modulus is lowest in ⟨100⟩ directions so that [100]-oriented grains, which thus contain the least strain energy U, should grow preferentially. The inequality [Eq. (11)] in reverse gives the tendency for the textural growth in fcc metals provided that other factors such as anisotropy of structural imperfections are not significant. In films of bcc Fe and Ta, [100] texture grows because $U_{100} < U_{111}$, but [111] texture is observed in films of bcc Cr, Mo, Nb, and V since here $U_{111} < U_{100}$.

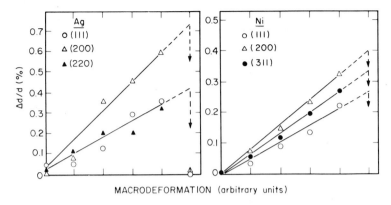

FIG. 12. Relative percentage change ($\Delta d/d$) in the lattice spacing corresponding to various (*hkl*) planes vs. macrodeformation for films of Ag and Ni. (*Palatnik et al.* [107])

(2) Tensile Strength. 1. Tensile strength is the maximum value of the breaking force divided by the original cross-sectional area. The tensile strength of metal films has been found to be up to 200 times larger than that of annealed bulk samples and exceeds the handbook values for hard-drawn materials by a factor of 3 to 10. Table IV lists the available data as compiled by Hoffman [1]; tensile strength is quoted numerically and as a fraction of the shear modulus G. The values are only representative since a large scatter is obtained experimentally.

2. Whether or not the ultimate strength of a film depends on the thickness still remains unresolved, since both a dependence and no dependence have been reported by different observers on the same materials and with similar tests. The bulge test on Au films and tensile-machine tests on Cu, Ag, and Au [96], and Ni [95] films show a large increase in strength for thicknesses below about 2,000 Å (Fig. 13). Pashley [111] as well as Neugebauer [87] found no thickness dependence of the tensile strength for Au films in the thickness range of 500 to 12,500 Å. The average strength observed was $\sim 4 \times 10^9$ dynes/cm^2, but values from 2 to 8×10^9 dynes/cm^2 were measured. It is interesting to note that the strength level at large thicknesses for cases where thickness dependence is observed is about half that found in samples which show no thickness dependence. A $1/\sqrt{t}$ dependence has also been reported by Lawley and Schuster [109] in Cu foils less than 26 μ thick.

Table IV. Tensile Properties of Metal Films and Foils (Compiled by Hoffman [1])

Material	Structure	Test	Modulus	Strength properties			Ref.
				Max tensile strength, 10^9 dynes/cm^2	Strain at fracture, %	Thickness dependence	
Au	Bulk hard-drawn			25* = (G†/114)			
	(111) film	Tensile	Normal	79 = (G/36)	1.2	No	86
	(100) film	Bulge	Thickness dependent	26 = (G/110)	0.5	Yes	90
	(100), (111) and polycrystalline film	Tensile		48 = (G/59)	1	No	94
	Polycrystalline film	Bulge	High	53 = (G/54)	0.7	Yes	89
	(100) film	Tensile	Normal	25 = (G/115)	3.5	No	97
	Polycrystalline film	Tensile	Normal	31 = (G/92)	2.3	No	97
Ag	Bulk hard-drawn			36* = (G/75)			
	Polycrystalline film	Bulge	High	57 = (G/47)	0.7	Yes	89
	Polycrystalline film	El. diff.	Assumed normal	40 = (G/68)	0.3-0.4		106
Cu	Bulk hard-drawn			47* = (G/98)			
	Polycrystalline film	Tensile	Normal	90 = (G/51)	1.8	Yes	96
	Polycrystalline film	Tensile	Normal	85 = (G/54)		No	23
	Rolled foil	Tensile		18 = (G/256)	10-15	Below 26μ	109
Ni	Bulk cold-rolled			120* = (G/67)			
	Polycrystalline film	Tensile	Normal	200 = (G/40)	1.8	Yes	95
Al	Bulk cold-rolled			15* = (G/171)			
	Polycrystalline film	El. diff.	Assumed normal	40 = (G/66)	0.5-0.8		106

*Bulk tensile strength from "Handbook of Chemistry and Physics," 42d ed., 1961.
†Shear moduli G from "AIP Handbook," 1957.

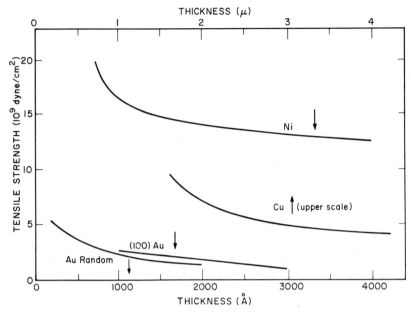

FIG. 13. Thickness dependence (arrow indicates the corresponding thickness scale) of tensile strength of films of Ni (D'Antonio et al. [95]), randomly oriented and epitaxial (100) Au (Beams et al. [88], Catlin and Walker [90]), and Cu (Oding and Aleksanyan [96]). (*Compiled by Hoffman* [1])

3. A constant, high tensile strength continues in very thick deposited films. Palatnik and Il'yinskii [24] found no thickness-dependent change in the 20- to 150-μ thickness of Cu films.

4. The tensile strength is higher for (111) than for (100) orientations of Au films. But polycrystalline films are only slightly stronger than (100) epitaxial Au films (Fig. 11). A large spread in data precludes any conclusions of a systematic dependence of the tensile strength on the microstructure of thin films. However, since the epitaxial films used in the various mechanical studies probably have structural defects comparable in magnitude with those found in polycrystalline films, a significantly different behavior may not be expected anyway.

(3) Microhardness. Microhardness has been measured in several-microns-thick vacuum-evaporated Cu and Al films by Palatnik et al. [24, 25]. A strong correlation of microhardness with tensile strength was found. Both decrease as the deposition temperature is increased; the decrease is monotonic for Cu films but shows two broad peaks for Al films (Fig. 14). The significance of these peaks is not clear.

Palatnik et al. [112] studied the effect on the microhardness of Al films due to the presence of aluminum oxide, and also due to alloying with 9 percent Cu, 0.6 percent each Mg and Mn, and 0.7 percent each Si and Fe. The aluminum oxide doping was provided by evaporation from Al_2O_3 crucibles. Microhardness results as a function of the deposition temperature are shown in Fig. 14. Aluminum oxide as well as multicomponent alloying of Al films greatly enhance their micro-hardness. Annealing decreases the microhardness, but specimens condensed at high substrate temperatures do not soften on annealing. The hardness of the alloy films is also increased with aluminum oxide. The alloy films are more ductile than pure Al films produced under analogous conditions.

3.3 Origin of the Tensile-strength Effects

(1) Structural-defect Hypothesis. The high strength observed in thin metal films not supported on substrates is evidently determined by the atomistic film-deposition process since foils [86, 109] thinned down from bulk specimens do not show such a high strength. The slope of the stress-strain curves for thin films is slightly smaller than Young's modulus

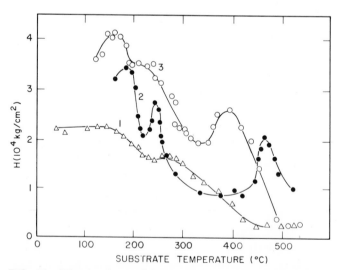

FIG. 14. Microhardness of films of (1) pure Al, (2) Al doped with Al_2O_3, and (3) Al + 9 percent Cu + 0.6 percent each Mg and Mn + 0.7 percent each Si and Fe, as a function of the substrate deposition temperature. (*Palatnik et al.* [112])

of the bulk material and decreases slowly at higher stresses, indicating that only very few dislocations are generated or mobilized, even at the highest stresses observed. This observation suggests [87, 94] a certain similarity between the condensation process of a film, and severe cold working, since the latter also leads to a high concentration of dislocations and other defects accompanied by a strong reduction of the dislocation mobility. The dislocation density in films is $\sim 10^{10}$ to $10^{12}/cm^2$, while it is about two orders of magnitude less in cold-worked bulk materials. Cold working also increases the strength; the fracture strength of cold-worked Au has been measured as high as $\sim 1.2 \times 10^{10}$ dynes/cm^2.

Although the tensile strength of bulk material can be increased by precipitation hardening, internal oxidation, and impurities, these factors are not important in most thin film cases that have been studied. The yield stress of a bulk material may be affected by the grain size, but its effect in thin films cannot be too significant in view of the observed small difference between the polycrystalline and single-crystal films, unless the grain size is so small (<100 Å) that the surface-tension contributions become pronounced. In that case, since the grain size varies with film thickness, this mechanism would lead to a thickness-dependent tensile strength of the film.

As suggested by Menter and Pashley [86], the high strength of metal films appears to result from the absence of operating dislocation sources and the restrained motion of grown-in dislocations. Since no necking has been observed in films at high stresses, the mechanisms must act uniformly throughout the film. Let us now examine various factors that may contribute to dislocation generation and/or locking mechanisms.

As for the sources of dislocations, Bilby [113] estimated that stresses of the order of $G/25$ are necessary for the nucleation of dislocations at a perfect surface. However, this stress level may be lowered at the tapered edges of a film. Dislocation networks have been observed in films thicker than 1,000 Å, but no sources have ever been observed to be directly operating in the films. Menter and Pashley discussed the operation of both single- and double-ended sources; the main factor in preventing their operation is the thickness itself. An extreme bowing would have to take place giving rise to a high stress needed for a source to continue to operate. Because of the surface pinning of dislocations, Frank-Read sources are not expected to occur in thin films. However, Wilsdorf [114] found such sources in thick stainless-steel foils. He also observed that grain boundaries, twin boundaries, and precipitates are good generators of dislocations.

(2) Surface Effects. The surface can increase the strength of bulk materials in a number of ways [115].

1. It may form a barrier* to the escape of dislocations so that the dislocations pile up and increase in density near the surface. This process is not expected to operate in very thin films since dislocations are not expected to travel in a direction perpendicular to the plane of the film. Dislocations moving parallel or at a small angle to the film plane, however, have been observed to pile up in thinned foils.

2. Dislocations may be pinned at the surface because of surface defects, surface oxidation, or a surface drag effect caused by the motion of a screw dislocation which intersects the surface. The stress necessary to move the pinned dislocations can be calculated from the observed curvature of the dislocations. In the case of bulk Al and stainless steel, these stresses correspond to flow stresses in the work-hardened material and so would not be large enough to account for the high strength of films or their large elastic strain.

Considerable influence on the mechanical properties of bulk Al is known to be exerted by the presence of a thin oxide film which appears to increase the elastic strain by pinning surface dislocation sources. Thick oxide films, on the other hand, act in pinning as well as in preventing the escape of dislocations having internal sources. Moverover, if changes in the elastic modulus, lattice parameters, etc., occur at the surface because of surface oxidation or the formation of a new interface, the strength of films will be greatly affected.

Thus, although various surface contributions are not large enough individually to account for the observed tensile strength, their effective total contribution may be significant enough to give rise to characteristic thickness-dependent effects.

(3) Volume Effects. If surface effects are not significant, one must conclude that the pinned length of dislocations must be much smaller than the film thickness. The dislocations must then be locked in the

---

*In the case of a low interfacial misfit, such as that of epitaxial overgrowth of Ni on a Cu substrate, activation of surface dislocation sources may be promoted. This results in a substantial reduction in the mechanical strength of the Cu substrates [150].

body of the film by defects such as stacking faults, twins, point defects, or impurities. Dislocations could also be pinned by severe tangling or jog formation after vacancy condensation. Pashley [101] observed in Au films that the dislocations formed during the growth tend to be as short as possible and are randomly distributed throughout the film area. No tangles or regions of extremely high dislocation concentration were observed.

Although few details are known about the actual dislocation configurations and pinning modes in thin films, it is commonly assumed that the point defects are the main contribution to the locking of dislocations. The existence of a high density of point defects is also indicated by the annealing behavior of thin metal films. Blakely [97] estimated that in the case of Au films an obstacle spacing of the order of 1,500 Å would be sufficient to account for the observed strength. The number of pinning centers thus required would be $\sim 10^{10}/cm^2$, not an unreasonable figure for thin films. Annealing increases the ductility of films, presumably as a result of reduction in the number of pinning centers.

(4) Phenomenological Approach. It seems reasonable that several of the mechanisms mentioned above may operate simultaneously in a thin film. To incorporate them all into a generalized theory, Tarshis [116] and Grunes et al. [117], following the analogy with the electrical resistivity of films, proposed that the film strength Γ_F may be described as the sum of contributions from bulk behavior, imperfections, and a surface or thickness effect. Therefore,

$$\Gamma_F = \Gamma_B + \Gamma_I + \Gamma_S \tag{12}$$

where Γ_B represents the value of tensile strength for annealed material, Γ_I is the contribution due to pinning of dislocations at impurities and imperfections, and Γ_S accounts for the effects due to the limited film thickness and the grain size. In analogy with the treatment of the electrical resistivity, one may write

$$\Gamma_F = \Gamma_B \left(1 + \frac{L}{t}\right) \tag{13}$$

where Γ_B is the strength of a bulk material with the same defect structure as that exhibited by the film, and L is a constant (corresponding to the mean free path) which takes into account changes in dislocation motion with film thickness t.

Equation (13) can be used to explain the observed thickness variation in Γ_F. As mentioned already, a similar relation for stress [Eq. (7)] has been used to explain the thickness dependence of the superconducting transition temperature of Pb, Sn, and In films. Tarshis found a good fit to Eq. (13) for his data on Ni films (Fig. 13) in the thickness range of 750 to 4,000 Å. However, the same data, if plotted against $1/\sqrt{t}$, as in a Cracknell-Petch analysis [118] (taking t as the crystallite size), yields an equally good fit [117]. This procedure has, however, been questioned by Tarshis and Wilshaw [119]. More refined data on single-crystal films with a low density of structural defects are required to test the validity of Eq. (13) and the significance of L. This is particularly desirable since Lawley and Schuster [109] also obtained a $1/\sqrt{t}$ dependence for the tensile strength of Cu foils thinner than 26 μ.

3.4 Films vs. Whiskers

The high tensile strength and the lack of movement of dislocations observed in Au films under stress led Menter and Pashley [86] to suggest that these properties were analogous to those found in whiskers. Neugebauer [87, 94] compared the tensile properties of thin films and whiskers with following observations:

The high strength in whiskers is due to the abnormally high stress required to nucleate dislocations, probably because of the high degree of surface smoothness. The high strength of evaporated films, however, appears to be the result of a high concentration of uniformly quenched-in defects which greatly impede the motion of dislocations and their multiplication. The dislocation density in whiskers is several orders of magnitude lower than that found in metal films. In relatively perfect iron whiskers, a dislocation density of $\sim 10^6/\text{cm}^2$ has been deduced by Gorsuch [120] from the x-ray rocking-curve data.

Metal whiskers display an entirely elastic stress-strain behavior up to very high stresses. The stress at the yield point is generally sharply defined and coincides with the ultimate tensile strength. Creep is observed only after yield has taken place. Only thin and perfect whiskers exhibit high strength, but thickness dependence of the strength is observed if the surface of the whiskers is imperfect or damaged.

By contrast, thin metal films exhibit plastic deformation, creep, and an ill-defined yield point. No systematic thickness dependence of the strength has been established so far. The high strength continues even in films as thick as 50 to 100 μ without an apparent upper limit to the thickness.

3.5 Stress Relief

We have already seen that large stresses of the order of 10^9 to 10^{10} dynes/cm^2 exist in thin films as a result of the process of deposition. A stress of this magnitude corresponds to about 1 percent strain in many materials and is comparable to the yield strength of most bulk materials. The shear stress at the film-substrate interface increases with increasing film thickness until the adhesion of the film to the substrate is not sufficient to prevent lifting or peeling of the film. Thus, peeling is commonly observed with thick deposits. If the film does not peel off, mechanical failure may take place and stress relief occurs by means of a fracture, or buckling of the film, or by plastic flow if the stress exceeds the yield strength of the film. Fracture or buckling will take place depending on whether the net stress is tensile, or compressive, respectively.

Vacuum-evaporated films of SiO [6, 121, 122], Ni-Fe [123], and ZnS [124], and ion-beam-sputtered films of transition metals [125] show crazing and buckling during and/or after deposition in vacuo as well as on exposure to air. Priest et al. [121] observed that obliquely deposited SiO films showed a strong tendency toward buckling and thus suggested that buckling resulted from anisotropic growth arising out of oblique deposition. Behrndt [124] examined ZnS films and established that buckling occurs even for normal incidence, and that oblique deposition simply enhances the tendency for buckling. Yelon and Voegeli [123] observed sinusoidal buckle patterns in 75 percent Fe-25 percent Ni films grown epitaxially on rock-salt substrates. The patterns were shown to run along $\langle 110 \rangle$ directions of the substrate. They attributed buckling to the high-volume magnetostriction of the oriented γ (fcc) phase which is found to coexist with an unoriented α (bcc) phase.

Chopra [125] studied in detail the kinetics of stress-relief patterns in sputtered films of W, Mo, Ta, Hf, Zr, and Re having mixed normal and metastable fcc structures. The following results were obtained. Fracture lines commonly observed in SiO and ZnS films were only occasionally seen in these transition-metal films. When fracture lines were observed [Fig. 15(a)], they were always accompanied by perpendicular sinusoidal buckling waves which propagated initially at a speed of $\sim$1 to 10 cm/sec

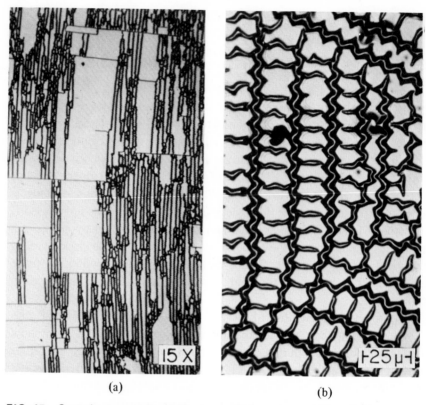

(a)   (b)

FIG. 15. Optical micrographs of the stress-relief patterns in 2,000-Å-thick Ar ion-beam-sputtered films of (a) W on glass at 200°C, showing fracture lines perpendicular to the sinusoidal buckle waves; (b) W on mica at 300°C, showing sinusoidal buckle waves running along ⟨100⟩ directions of the substrate. (*Chopra* [125])

until they intersected another fracture line or a buckle wave and came to a stop.

A large number of parallel sine waves [Fig. 15(b)], sometimes running parallel to particular directions of the substrate, appeared in various regions of a specimen. At the extrema of the sine waves, new sine waves were initiated in perpendicular directions which linked up the parallel-wave pattern. Where buckle waves crossed, the film wrinkled and lifted off the substrate. The amplitude of the sine waves in the plane of the film was nearly half the wavelength and was typically $\sim 10^{-4}$ cm.

Buckling occurred at or beyond a certain critical thickness which was higher for lower substrate temperatures during deposition and for greater adhesion of the film. For example, the critical thickness for Mo and W films deposited at $300°C$ was about a thousand angstroms for a glass substrate, as compared with less than a few hundred angstroms for a rock-salt substrate for which the film adhesion is also relatively weaker. Buckling was not found in films deposited on metallic substrates, presumably because of stronger adhesion. Furthermore, the tendency to buckle was reduced markedly for films sputtered in a glow-discharge environment rather than by low-pressure ion-beam sputtering. It was also possible to regenerate buckling by a thermal shock (e.g., from a point source of light) or by a mechanical fracture and thereby change or even completely reverse the direction of propagation of the relief pattern. Once initiated, the buckle wave propagated by itself.

Fracture lines can be explained as due to the intrinsic tensional stress present in the films. The simultaneous occurrence of fracture and buckle patterns requires the presence of localized tension and compression regions in the films. This may arise in SiO films as a result of selective oxidation of, and/or absorption of water by different crystallites. On complete oxidation, tension changes over to compression. The co-existence of crystallites of two structures with different atomic volumes can yield localized tension and compression regions. This is probably the case with epitaxial Permalloy films and the ion-beam sputtered transition-metal films. In the latter case, Chopra et al. [64] established that the films in which fracture and buckle patterns occurred always consisted of mixed normal and metastable fcc phases; the latter having larger atomic volumes (ranging from about 5 percent for Ta to 29 percent for Hf) than the normal structures. When the fcc phases reverted gradually to the normal structure with increasing film thickness and/or with increasing substrate temperature, stress-relief patterns occurred. The coexistence of the two structures in different crystallites having random distribution could explain the existence of tensile- and compressive-stress regions in the same film. The sinusoidal buckling behavior is not simple to understand. The phenomenon can, however, be compared with the sinusoidal buckling of a column under an eccentric load [126].

4. ADHESION OF FILMS

The durability of a relatively fragile film is largely dependent on the adhesion between the film and the substrate. Since adhesion is related to the nature and the strength of the binding forces at interfaces between the two materials in contact with each other, a study of thin film adhesion is of both fundamental and practical interest. Although no reliable and reproducible technique to obtain quantitative information on adhesion exists at present, a rather primitive scratch test (described later) yields interesting results. We shall review whatever little is known in this field and hope this will stimulate further work in this important area.

4.1 Measurement of Adhesion

Early attempts to measure adhesion used either the "scotch tape" method suggested by Strong [127], or some method of abrasion testing [128]. The tape method employs an adhesive tape to lift the film off the substrate and gives only qualitative results. Abrasion testing gives results which depend on both the hardness and the adhesion of the films and are affected by the burnishing action of the abrasive head.

The technique of x-ray topography may be employed to obtain qualitative information on adhesion of epitaxial films deposited on single-crystal substrates (Si, for example). The strains at the film-substrate interface as well as poor adherence of a film modify the diffraction contrast of the surface and thus provide qualitative but nondestructive information on strain and adhesion. The method is obviously limited to epitaxial films.

A direct measure of adhesion may be obtained by applying a force normal to the interface between film and substrate. Tensile tester [129], ultracentrifuge [88], and ultrasonic vibrations [130] have been employed to apply the required force but with inconsistent results. Weaver [131] reviewed these techniques as they apply to thin films and concluded that the "scratch test" used by Heavens [132] may yield reasonable results. This method uses a smoothly rounded chrome-steel point which is drawn across the film surface. A vertical load applied to the point is gradually increased until a critical value is reached, at which time the film is stripped from the substrate, leaving a clear channel. The critical load is taken as a measure of the adhesion of the film.

The scratch test was used by Benjamin and Weaver [133, 134] and others [135, 136]. A commercial hardness tester with a stylus similar to the Knoop hardness indentor with replaceable tungsten carbide or diamond tips of spherical radius of about 0.05 mm is commonly used.

The critical loads using this size of tip radius are several grams for weakly adherent films and increase to hundreds of grams for stongly adherent films. The stripping of the film is observed by means of a microscope which has both a reflection and a transmission stage. For dielectric films, the establishment of an electrical contact between the stylus and the metallic substrate may be conveniently used to indicate the occurrence of stripping.

Benjamin and Weaver [133] (B-W) analyzed the scratch test experimentally and theoretically in detail. They measured the critical load for Ag films evaporated onto various substrates, and for various metal films deposited on glass substrates. The load was found to be independent of the film thickness except for thin films (<800 Å), which probably have agglomerated structures. They concluded from the data that the load required depends upon the nature of both the film and the substrate without being directly dependent upon the mechanical properties of either. This conclusion is valid only if the critical load refers to the condition for peeling of the film. In actual practice, the stylus scratches the film so that the critical load is determined by both adhesion and the mechanical strength of the film. Consequently, the critical load shows a strong film-thickness dependence (see Fig. 18). For thin films (<2,000 Å), this load is probably a reasonable measure of the adhesion.

An analysis of the forces involved in the stripping action suggested by B-W is as follows: At the loads normally used, the substrate surface is drastically deformed at the loaded point. If a film is deposited on the surface, it is deformed to contour to the shape of the indentation, causing stretching of the film and the development of a shearing force between the film and the substrate surface. This shearing force would have a maximum value at the tip of the indentation, and the tensile forces in the film would be small in comparison. At the critical load, the shearing force at the tip would be sufficient to break the adhesive bonds between film and substrate. Thereafter, the horizontal motion of the point would just push the sheared metal aside. If W is the critical load on the point of radius r, F the shearing force per unit area due to the deformation of the surface, a the radius of the circle of contact, and P the indentation hardness of the substrate material, then

$$F = \frac{aP}{\sqrt{r^2 - a^2}} \qquad a = \sqrt{\frac{W}{\pi P}} \qquad (14)$$

This shear force is assumed to move an atom of one layer from one equilibrium position to the next and is a direct measure of adhesion. If the distance between the symmetrical equilibrium positions for an atom is x, then $1/2$ Fx corresponds to the height of the potential barrier which, according to B-W, is the energy of adhesion.

4.2 Experimental Results

Benjamin and Weaver [134] made extensive investigations of the adherence of metal films to alkali-halide substrates. Films of Ag, Au, Al, Cr, Cu, and Cd deposited on the (100) faces of NaCl, KCl, and KBr were found to adhere poorly, but no aging effects were noticed. Both polycrystalline and epitaxially grown films showed comparable adhesion. The magnitude of the observed shearing force suggests strongly that the initial bonding could be attributed to the van der Waals forces. These authors verified that a linear relation (Fig. 16) exists between the experimentally measured shearing force and the theoretically calculated van der Waals energy per unit area of crystal face for each metal-substrate pair. Since the condensation energies of Ag and Al films deposited on NaCl and KBr have also been found to bear a correlation with the van der Waals energy, the condensation energy is a good measure of the energy of adhesion.

The adhesion of films to the most commonly used microscope-glass-slide substrates has interesting features [133]. The magnitude of the critical load is such that the adhesion could be attributed to the van der Waals forces. In the case of oxygen-active metals, however, the initial adhesion increases markedly with aging (Fig. 17) for Fe, Al, and Au films. The oxidation behavior is dependent on the structure of the film and the availability of oxygen to the film during or after deposition. Fine-grained structures adhere more strongly with aging because of enhanced oxidation activity. The ultimate adhesion appears to be correlated with the free energy of formation of the oxides of the film materials [133, 135]. Table V lists the critical loads of some metals as measured by Karnowsky and Estill [135]. It is clear that the load increases with increasing negative free energy of oxide formation.

Gold films do not adhere strongly to glass substrates, which is attributed to the low chemical activity of gold with oxygen. By providing a diffusion type of interface or an intermediate layer of an oxygen-active metal which has solid solubility with Au, strong adherence

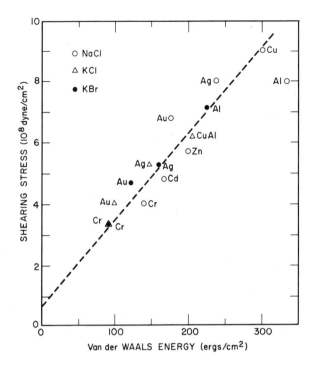

FIG. 16. The measured shearing force in a scratch test (proportional to energy of adhesion) plotted against the calculated van der Waals energy per unit area of the substrate plane for films of various metals deposited on NaCl(O), KCl (Δ), and KBr (●). (*Benjamin and Weaver* [134])

of Au films is obtained. For example, Au does not ordinarily adhere well to silica surfaces, but does so when melted on silica surface in the presence of appreciable oxygen. This is accompanied by diffusion into and etching of the silica surface. Mattox [137] sputtered Au films onto fused-silica substrates with argon in the presence of a partial pressure of oxygen, and obtained strong adherence of films when more than 20 percent oxygen was present. Similar results were obtained by glow discharge, triode, and rf sputtering. Heat treatment of films subsequent to deposition deteriorated adhesion, indicating that the strong adherence was due to the effect of oxygen. The chemical reaction of gold with silica is not understood, although the presence of oxygen ions and atomic oxygen in the sputtering process may be responsible for the formation of stable gold oxides.

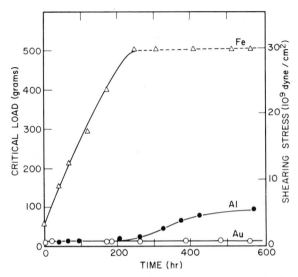

FIG. 17. The variation of the scratch-test critical load (proportional to adhesion) with time for various metal films deposited on glass at a residual vacuum of ~10^{-5} Torr. (*Benjamin and Weaver* [133])

It is generally known [129, 138] that sputtered films adhere more strongly than evaporated films. This is probably because the sputtered atoms have considerably higher energies (see Chap. II, Sec. 3.1) than the evaporated atoms. The condensation of energetic atoms may modify the substrate surface in a number of ways: (1) by removing adsorbed gases

Table V. Scratch-test Critical Loads for Several Metal Films Deposited on Glass and Their Relation to the Free Energy of Oxide Formation, ΔE_f (Karnowsky and Estill [135])

Metal film	Failure load, g	Oxide	ΔE_f, kcal/mole
Au	Less than 25	Au_2O_3	+39.0
Ag	Less than 25	Ag_2O	−2.59
Cu	50-100	CuO	−30.4
Cd	100-200	CdO	−53.79
Ti	600-700	TiO_2	−203.8
Cr	Chipped substrate	Cr_2O_3	−250.2

and impurities, (2) by sputtering of the substrate, (3) by limited penetration into the substrate, and finally (4) by the creation of point defects on the substrate surface. Unfortunately, all these factors are expected to enhance the adhesion of a film to the substrate, and thus an understanding of the individual contributions is difficult, if not impossible.

Chopra [139] measured the scratch-test critical load for Au films deposited on glass and other substrates by evaporation at 10^{-6} Torr, by glow-discharge sputtering in Ar at 5×10^{-2} Torr, and by Ar ion-beam sputtering at $\sim 10^{-4}$ Torr. The results of the comparative study for glass substrates are shown in Fig. 18. The critical load in all cases was found to increase rapidly with the film thickness above about 2,000 Å.

A comparison of the results for the variously prepared films of the same thickness yields the following conclusions: (1) The critical load is generally higher for sputtered than for evaporated films, at sputtering voltages above 1.5 kV. (2) The load increases more rapidly with film thickness for increasing accelerating voltages for the sputtering process. (3) The loads for low-pressure ion-beam-sputtered and glow-discharge-sputtered films are comparable and show little dependence on the deposition rate ranging between 1 and 10 Å/sec. (4) The load is minimum for an amorphous, inert glass substrate and maximum for a metallic substrate. The relative approximate values for 2,000-Å-thick Au films sputtered (at 1.5 kV) onto various substrates are:

Substrate	Glass	Quartz	NaCl	Mica	Pt
Load, g	5	8	20	50	300

Chopra's studies show the enhancement of adhesion and mechanical strength of the sputtered films due to the increasing kinetic energy of sputtered atoms. It has also been found that the adhesion of glow-discharge-sputtered films depends considerably on the position of the substrate relative to the cathode and the plasma current density employed. If the film prepared by sputtering is not held at ground potential, it builds up a high potential because of the collection of electrostatic charges. The high-voltage arcing results in the deterioration of adhesion and the physical appearance of the film. On the other hand, the adhesion of an evaporated film can be considerably improved by irradiating the film with electrons during the deposition process. Whether

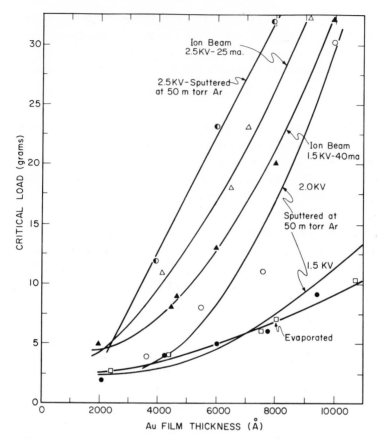

FIG. 18. The critical load in a scratch test as a function of the thickness of polycrystalline Au films deposited on glass at an average rate of ~ 1 Å/sec by evaporation at 10^{-6} Torr, and glow-discharge sputtering at 50 mTorr and argon ion-beam sputtering at 10^{-4} Torr with different accelerating voltages. (*Chopra* [139])

this effect is due to the desorption of contaminants by the energetic electrons or is a consequence of the high-atomic-mobility growth process of a film (leading to more perfect films) due to the presence of electrostatic charges (Chap. IV, Sec. 2.3) is not clear.

Adhesion of a film to its substrate is strongly dependent on the chemical nature, cleanliness, and the microstructure of the substrate. Generally, the adhesion is greater, the higher the adsorption energy of the deposit, and/or the higher the number of nucleation centers in the early growth stage of a film. Thus, surface contaminants may *increase* or *decrease* the film adhesion depending on whether the adsorption energy is increased or decreased, respectively. Generally, however, the adhesion can be enhanced by providing more nucleation centers as with a fine-grained substrate or a substrate precoated with suitable materials.

4.3 Origin of Adhesion

Benjamin and Weaver [134] concluded from their studies that both the energy of film adhesion and the heat of condensation of the vapor atoms could be related to and explained by the van der Waals forces at the film-substrate interface.

In case of oxygen-active evaporated metals, the film adhesion is generally attributed to a bond with an oxide layer. The energy of adhesion is proposed [140-142] to be closely related to the change in the free energy of oxide formation. Thus, the film adhesion is high in those cases where a transition region of suitable oxides and suboxides of the film or the substrate at the interface exists. The formation mechanism of an oxide layer at the glass-metal interface is by no means clear. Bateson [140] suggested that an oxide bond may be formed by the reaction of the impinging atoms with the hydroxyl groups present at the glass surface. More studies are needed to establish the nature of this reaction.

In certain cases, adhesion at the metal-insulator interfaces has been attributed to an electric double layer produced by charge transfer [143]. The existence of opposite charges on the two surfaces when a polymer film is stripped off a metal surface has been observed [144, 145]. The electric forces [144, 146] are expected to be equal to or greater than van der Waals forces. Benjamin and Weaver [134] argued that since electron transfer would depend on the work function of the metal and since they found no obvious relation between the measured adhesion and the work function, an electronic mechanism is probably not operative for metal-alkali halide interfaces. On the other hand, the role of charges in determining adhesion is indicated by the observed marked dependence of adhesion of electrodeposited films on the current density for electrolysis. One can obtain strongly adherent or porous films depending on the prevailing electrolysis conditions. Further, the presence of the electrostatic charges during the deposition of evaporated and sputtered films also appears to influence their adhesion.

Derjaguin and Smilga [147] developed an electronic theory of adhesion and showed that the electrostatic forces generated by the double electric layer formed on the interface are the chief factors of adhesion between two surfaces. This mechanism yields value of adhesion energy $\sim 10^8$ to 10^9 dynes/cm^2. The theory suggests that by modifying the electronic nature of the surfaces, both an increase and decrease in adhesion can be achieved.

5. CONCLUDING REMARKS

1. Despite a multitude of experimental studies, no satisfactory understanding of the stresses introduced in films during their growth exists at present. Whereas tensile stress may be understood qualitatively on the basis of several models, the occurrence of compressive stress in some dielectric films and crossover of tensile to compressive stress at high substrate temperatures is still puzzling. Nevertheless, it is now abundantly clear that structural defects in films play a dominant role in determining stresses as well as the high tensile strength of films. An unambiguous understanding of the role of defects will be possible only if careful experiments are undertaken on epitaxially grown films with concentrations of structural defects significantly lower than in poly-crystalline films. No study reported in literature so far satisfies this condition.

2. Profound effects on the electrical, magnetic, superconducting, and semiconducting properties are expected because of the enormous intrinsic stresses in films. The observed effect of stresses on the superconducting transition temperature (Chap. IX, Sec. 3.5) and the band gap of lead-salt films [37, 148] indicates the intrinsic value of such studies. Externally [148] applied stresses can further aid such investigations.

3. The dependence of the tensile properties and elastic constants on film thickness remains questionable because of the lack of reliable data, although such dependence is expected from elementary considerations of the effect of a surface discontinuity. Improved and new experimental techniques are necessary for such studies. A promising technique developed by Gelles [149] uses a micromanipulator to propagate ultrasonic stress waves in films as thin as $\sim 1,000$ Å to measure the stress-wave velocity (and hence elastic constants) in the films.

4. Mechanical failure of films and the nature of the film adhesion to substrates are areas of much practical importance and need systematic studies.

REFERENCES

1. R. W. Hoffman, in "Physics of Thin Films" (G. Hass and R. E. Thun, eds.), vol. 3, p. 211, Academic Press Inc., New York, 1966. This review encompasses an earlier one by the same author in "Thin Films," p. 99, American Society for Metals, Metals Park, 1964.
2. G. G. Stoney, *Proc. Roy. Soc. London,* **A82**:172 (1909).
3. H. P. Murbach and H. Wilman, *Proc. Phys. Soc. London,* **B66**:905 (1953).
4. A. F. Turner, Thick Thin Films, *Bausch and Lomb Tech. Rept.,* Rochester, N. Y., 1951.
5. R. W. Hoffman and E. C. Crittenden, Jr., *Phys. Rev.,* **78**:349 (1950).
6. H. Blackburn and D. S. Campbell, *Trans. 8th Natl. Vacuum Symp.,* p. 943, Pergamon Press, New York, 1961.
7. H. Blackburn and D. S. Campbell, *Phil. Mag.,* **8**:823 (1963).
8. H. S. Story and R. W. Hoffman, *Proc. Phys. Soc. London,* **B70**:950 (1957).
9. E. Klokholm, *IBM Res. Rept.* RC 1352, 1965.
10. J. R. Priest and H. L. Caswell, *Trans. 8th Natl. Vacuum Symp.,* p. 947, Pergamon Press, New York, 1961.
11. J. Priest, H. L. Caswell, and Y. Budo, *J. Appl. Phys.,* **34**:347 (1963).
12. J. D. Finegan and R. W. Hoffman, *Trans. 8th Natl. Vacuum Symp.,* p. 935, Pergamon Press, 1961.
13. R. Glang, R. A. Holmwood, and R. L. Rosenfeld, *Rev. Sci. Instr.,* **36**:7 (1965).
14. D. J. Dumin, *J. Appl. Phys.,* **36**:2700 (1965).
15. A. E. Ennos, *Appl. Opt.,* **5**:51 (1966).
16. S. P. Timoshenko, *Mech. Eng.,* **259** (1923).
17. A. Brenner and S. Senderoff, *J. Res. Natl. Bur. Std.,* **42**:105 (1949).
18. A. Kinbara, *Oyo Butsuri,* **30**:391, 496 (1961).
19. A. Kinbara and H. Haraki, *Japan. J. Appl. Phys.,* **4**:423 (1965).
20. J. F. Freedman, *IBM J. Res. Develop.,* **6**:449 (1962).
21. H. D. Keith, *Proc. Phys. Soc. London,* **B69**:180 (1956).
22. R. W. Vook and F. Witt, *J. Appl. Phys.,* **36**:2169 (1965).
23. R. W. Vook and F. Witt, *J. Vacuum Sci. Tech.,* **2**:49, 243 (1965).
24. L. S. Palatnik and A. I. Il'yinskii, *Soviet Phys.-Solid State English Transl.,* **3**:2053 (1962).
25. L. S. Palatnik, G. V. Fedorov, and A. I. Il'yinskii, *Phys. Metals Metallog.,* **11**(5):159 (1961).
26. J. E. Davey and R. H. Deiter, *J. Appl. Phys.,* **36**:284 (1965).
27. K. Haruta and W. J. Spencer, *J. Appl. Phys.,* **37**:2232 (1966).
28. T. B. Rymer, *Proc. Roy. Soc. London,* **A235**:274 (1956).
29. T. B. Rymer, F. J. Fayers, and S. J. Hewitt, *Proc. Phys. Soc. London,* **B69**:1059 (1956).
30. T. B. Rymer, *Nuovo Cimento,* **6**:294 (1957).

31. E. K. Halteman, *J. Appl. Phys.*, **23**:150 (1952).
32. R. S. Smith, *IBM J. Res. Develop.*, **4**:205 (1960).
33. A. Segmüller, *IBM J. Res. Develop.*, **6**:464 (1963).
34. A. Segmüller, *Z. Metallk.*, **54**:248 (1963).
35. J. S. Halliday, T. B. Rymer, and K. H. R. Wright, *Proc. Roy. Soc. London,* **A225**:548 (1954).
36. J. R. MacDonald, *Phys. Rev.*, **106**:890 (1957).
37. E. D. Palik, D. L. Mitchell, and J. N. Zemel, *Phys. Rev.*, **135**:A763 (1964).
38. J. N. Zemel, in "The Use of Thin Films in Physical Investigations" (J. C. Anderson, ed.), p. 332, Academic Press Inc., New York, 1966.
39. R. L. Parker and A. Krinsky, *J. Appl. Phys.*, **34**:2700 (1963).
40. R. H. Blumberg and D. P. Seraphim, *J. Appl. Phys.*, **33**:163 (1962).
41. A. M. Toxen, *Phys. Rev.*, **123**:442 (1961).
42. H. L. Caswell, J. R. Priest, and Y. Budo, *J. Appl. Phys.*, **34**:3261 (1963).
43. H. Horikoshi, Y. Ozawa, and H. Hasunuma, *Japan. J. Appl. Phys.*, **1**:304 (1962); H. Horikoshi and M. Tamura, *Japan. J. Appl. Phys.*, **2**:328 (1963).
44. R. W. Hoffman, R. D. Daniels, and E. C. Crittenden, Jr., *Proc. Phys. Soc. London,* **B67**:497 (1954).
45. J. D. Finegan, *Case Inst. Tech., AEC Tech. Rept.* 15, Cleveland, Ohio, 1961.
46. M. Prutton, *Nature*, **193**:565 (1962).
47. G. P. Weiss and D. O. Smith, *J. Appl. Phys.*, **33**:1166S (1962).
48. K. Kinosita, H. Kondo, and I. Sawamura, *J. Phys. Soc. Japan*, **15**:942 (1960).
49. E. Klokholm and B. S. Berry, *J. Electrochem. Soc.*, **115**:823 (1968).
50. J. Riesenfeld and R. W. Hoffman, *Case Inst. Tech., AEC Tech. Rept.* 39, Cleveland, Ohio, 1965.
51. H. Kato, K. Nagasima, and H. Hasunuma, *Oyo Butsuri*, **30**:700 (1961).
52. R. Carpenter and D. S. Campbell, *J. Mater. Sci.*, **2**:173 (1967).
53. A. E. Hill and G. R. Hoffman, *Brit. J. Appl. Phys.*, **18**:13 (1967).
54. J. Riesenfeld and R. W. Hoffman, *Case Inst. Tech., AEC Tech. Rept.* 32, Cleveland, Ohio, 1964.
55. M. S. Cohen, E. E. Huber, Jr., G. P. Weiss, and D. O. Smith, *J. Appl. Phys.*, **31**:291S (1960).
56. D. O. Smith, M. S. Cohen, and G. P. Weiss, *J. Appl. Phys.*, **31**:1755 (1960).
57. M. Ya. Popereka, *Phys. Metals Metallog.*, **20**(5):114 (1965).
58. D. A. Vermilyea, *J. Electrochem. Soc.*, **110**:345 (1963).
59. D. H. Bradhurst and J. S. L. Leach, *J. Electrochem. Soc.*, **113**:1245 (1966).
60. H. Schröder and M. Spiller, in "Basic Problems in Thin Film Physics" (R. Niedermayer and H. Mayer, eds.), p. 233, Vandenhoeck and Ruprecht, Göttingen, 1966.
61. G. Gafner, *Phil. Mag.*, **5**:1041 (1960).
62. P. Duwez, R. H. Willens, and W. Klement, Jr., *J. Appl. Phys.*, **31**:1137 (1962).
63. G. Günther and W. Buckel, in "Basic Problems in Thin Film Physics" (R. Niedermayer and H. Mayer, eds.), p. 231, Vandenhoeck and Ruprecht, Göttingen, 1966.
64. K. L. Chopra, M. R. Randlett, and R. H. Duff, *Phil. Mag.*, **16**:261 (1967).
65. M. L. Gimpl, A. D. McMaster, and N. Fuschillo, *J. Appl. Phys.*, **35**:3572 (1964).
66. R. J. Jaccodine and W. A. Schlegal, *J. Appl. Phys.*, **37**:2429 (1966).

67. O. S. Heavens and S. D. Smith, *J. Opt. Soc. Am.,* **47**:469 (1957).
68. See the general review by D. S. Campbell in "Basic Problems in Thin Film Physics" (R. Niedermayer and H. Mayer, eds.), p. 113, Vandenhoeck and Ruprecht, Göttingen, 1966.
69. R. Shuttleworth, *Proc. Phys. Soc. London,* **A63**:444 (1950).
70. K. L. Chopra, *J. Appl. Phys.,* **37**:2249 (1966).
71. D. B. Dove, *J. Appl. Phys.,* **35**:2785 (1964).
72. F. C. Frank and J. H. van der Merwe, *Proc. Roy. Soc. London,* **A198**:205 (1949).
73. J. H. van der Merwe, *Phil. Mag.,* **7**:1433 (1962).
74. J. H. van der Merwe, *J. Appl. Phys.,* **34**:117,123 (1963).
75. J. H. van der Merwe, *J. Appl. Phys.,* **34**:3420 (1963).
76. J. H. van der Merwe, in "Single-crystal Films" (M. H. Francombe and S. Sato, eds.), p. 139, Pergamon Press, New York, 1964.
77. J. W. Matthews, in "Physics of Thin Films" (G. Hass and R. Thun, eds.), vol. 4, p. 137, Academic Press Inc., New York, 1967.
78. J. W. Matthews, Ref. 76, p. 165.
79. R. W. Hoffman, F. J. Anders, and E. C. Crittenden, Jr., *J. Appl. Phys.,* **24**:231 (1953).
80. R. W. Hoffman and H. S. Story, *J. Appl. Phys.,* **27**:193 (1956).
81. V. V. Shah and Y. G. Naik, *Indian J. Pure Appl. Phys.,* **3**:20 (1964).
82. W. Sander and E. Strieder, *Z. Physik,* **188**:99 (1965).
83. H. G. van Bueren, "Imperfections in Crystals," North Holland Publishing Company, Amsterdam, 1961.
84. J. D. Finegan and R. W. Hoffman, *Case Inst. Tech., AEC Tech. Rept.* 18, Cleveland, Ohio, 1961.
85. K. L. Chopra and M. R. Randlett, *J. Appl. Phys.,* **39**:1874 (1968).
86. J. W. Menter and D. W. Pashley, in "Structure and Properties of Thin Films" (C. A. Neugebauer, J. D. Newkirk, and D. A. Vermilyea, eds.), p. 111, John Wiley & Sons, Inc., New York, 1959.
87. C. A. Neugebauer, in "Physics of Thin Films" (G. Hass and R. E. Thun, eds.), vol. 2, p. 1, Academic Press Inc., New York, 1964.
88. J. W. Beams, J. B. Breazeale, and W. L. Bart, *Phys. Rev.,* **100**:1657 (1955).
89. J. W. Beams, in "Structure and Properties of Thin Films" (C. A. Neugebauer, J. D. Newkirk, and D. A. Vermilyea, eds.), p. 183, John Wiley & Sons, Inc., New York, 1959.
90. A. Catlin and W. P. Walker, *J. Appl. Phys.,* **31**:2135 (1960).
91. S. Jovanovic and Cyril S. Smith, *J. Appl. Phys.,* **32**:121 (1961).
92. R. Papirno, *J. Appl. Phys.,* **32**:1175 (1962).
93. D. M. Marsh, *J. Sci. Instr.,* **38**:229 (1961).
94. C. A. Neugebauer, *J. Appl. Phys.,* **31**:1096 (1960).
95. C. D'Antonio, J. S. Hirschhorn, and L. A. Tarshis, *Trans. AIME,* **227**:1346 (1963).
96. A. Oding and I. T. Aleksanyan, *Soviet Phys. "Doklady" English Transl.,* **8**:818 (1964).
97. J. M. Blakely, *J. Appl. Phys.,* **35**:1756 (1964).
98. A. Lawley and S. Schuster, *Rev. Sci. Instr.,* **33**:1178 (1962).
99. E. Orowan, *Z. Physik,* **82**:235 (1933).

100. H. G. F. Wilsdorf, *Rev. Sci. Instr.*, **29**:323 (1958).
101. D. W. Pashley, *Proc. Roy. Soc. London*, **A255**:218 (1960).
102. W. Kossel, *Ann. Physik*, **26**:533 (1936).
103. K. Lansdale, *Phil. Trans. Roy. Soc. London*, **240**:219 (1947).
104. G. A. Bassett and D. W. Pashley, *J. Inst. Metals*, **87**:449 (1959).
105. D. M. Marsh, *J. Sci. Instr.*, **36**:165 (1959).
106. J. E. Gordon, in "Growth and Perfection of Crystals" (R. H. Doremus, B. W. Roberts, and D. W. Turnbull, eds.), p. 219, John Wiley & Sons, Inc., New York, 1958.
107. L. S. Palatnik, M. Ya. Fuks, B. T. Boiko, and A. T. Pugachev, *Soviet Phys. "Doklady" English Transl.*, **8**:818 (1964).
108. A. Oding and I. T. Aleksanyan, *Soviet Phys. "Doklady" English Transl.*, **8**:818 (1964).
109. A. Lawley and S. Schuster, *Trans. AIME*, **230**:27 (1964).
110. Y. Yamamoto, *Kanazawa Univ. Japan, Tech. Rept.* 2, 1954.
111. D. W. Pashley, *Phil. Mag.*, **4**:324 (1959).
112. L. S. Palatnik, G. V. Fedorov, A. I. Prokhvatilov, and A. I. Fedorenko, *Phys. Metals Metallog.*, **20**:574 (1965) [Russian].
113. B. A. Bilby, *Nature*, **182**:296 (1958).
114. H. G. F. Wilsdorf, in "Structure and Properties of Thin Films" (C. A. Neugebauer, J. D. Newkirk, and D. A. Vermilyea, eds.), p. 151, John Wiley & Sons, Inc., New York, 1959.
115. E. S. Machlin, in "Strengthening Mechanisms in Solids," p. 375, American Society for Metals, Cleveland, Ohio, 1960.
116. L. A. Tarshis, M. S. Thesis, Brooklyn Polytechnic Institute, June, 1965.
117. R. L. Grunes, C. D'Antonio, and F. Kies, *J. Appl. Phys.*, **36**:2735 (1965).
118. A. Cracknell and M. J. Petch, *Acta Met.*, **3**:186 (1955).
119. L. A. Tarshis and T. R. Wilshaw, *J. Appl. Phys.*, **37**:3322 (1966).
120. P. D. Gorsuch, *J. Appl. Phys.*, **30**:837 (1959).
121. J. R. Priest, H. L. Caswell, and Y. Budo, *Trans. 9th Natl. Vacuum Symp.*, p. 121, The Macmillan Company, New York, 1962.
122. G. Siddall, *Vacuum*, **9**:274 (1960).
123. A. Yelon and G. Voegeli, in "Single Crystal Films," p. 321, Pergamon Press, New York, 1964.
124. K. H. Behrndt, *J. Vacuum Sci. Tech.*, **2**:63 (1965).
125. K. L. Chopra, *Ledgemont Lab. Rept.* TR 134, May, 1967.
126. S. P. Timoshenko, "Strength of Materials," p. 258, D. Van Nostrand Company, Inc., Princeton, N.J., 1955.
127. J. Strong, *Rev. Sci. Instr.*, **6**:97 (1935).
128. L. Holland, "Vacuum Deposition of Thin Films," p. 102, John Wiley & Sons, Inc., New York, 1956.
129. R. B. Belser and W. H. Hicklin, *Rev. Sci. Instr.*, **27**:293 (1956).
130. S. Moses and R. K. Witt, *Ind. Eng. Chem.*, **41**:2334 (1949).
131. C. Weaver, *Proc. 1st Intern. Conf. Vacuum Tech.*, Pergamon Press, 1958.
132. O. S. Heavens, *J. Phys. Radium*, **11**:355 (1950).
133. P. Benjamin and C. Weaver, *Proc. Roy. Soc. London*, **A254**:163, 177 (1960).
134. P. Benjamin and C. Weaver, *Proc. Roy. Soc. London*, **A274**:267 (1963).

135. M. M. Karnowsky and W. B. Estill, *Rev. Sci. Instr.*, **35**:1324 (1964); A. J. Raffalovich, *Rev. Sci. Instr.*, **37**:368 (1966).
136. J. R. R. Frederick and K. C. Ludema, *J. Appl. Phys.*, **35**:256 (1964).
137. D. M. Mattox, *J. Appl. Phys.*, **37**:3613 (1966).
138. D. M. Mattox and J. E. McDonald, *J. Appl. Phys.*, **34**:2493 (1963).
139. K. L. Chopra, *Ledgemont Lab. Rept.*, June, 1968; to be published.
140. S. Bateson, *Vacuum,* **2**:365 (1952).
141. W. A. Weyl. *Research,* **3**:230, 1950; also, *Proc. ASTM,* **46**:1506, 1946.
142. R. C. Williams and R. C. Backus, *J. Appl. Phys.*, **20**:98 (1949).
143. S. M. Skinner, R. L. Savage, and J. E. Rutzler, *J. Appl. Phys.*, **24**:438 (1953).
144. S. M. Skinner, *J. Appl. Phys.*, **26**:498 (1955).
145. B. V. Derjaguin and V. P. Smilga, *Proc. 3d Intern. Congr. Surface Activity,* **3**:349 (1960).
146. D. C. West, *J. Appl. Phys.*, **25**:1054 (1954).
147. B. V. Derjaguin and V. P. Smilga, *J. Appl. Phys.*, **38**:4609 (1967).
148. V. Prakash, *Harvard Univ. Tech. Rpt.* HP-13, Cambridge, Mass., June, 1967.
149. I. L. Gelles, *J. Acoust. Soc. Am.*, **40**:138 (1966).
150. G. E. Ruddle and H. G. F. Wilsdorf, *Appl. Phys. Letters,* **12**:271 (1968).
151. W. A. Jesser, J. W. Matthews, and D. Kuhlmann-Wilsdorf, *Appl. Phys. Letters,* **9**:176 (1968).

VI ELECTRON-TRANSPORT PHENOMENA IN METAL FILMS

1. INTRODUCTION

Electron-transport properties of two-dimensional solids have inspired much curiosity and motivated extensive investigations ever since the first known work on thin films of Ag and Pt by Moser [1] in 1891. The voluminous literature on the subject is full of inconsistencies and violent disagreements among the results of early workers. Much of it may be attributed to the poor thin film technology and the lack of understanding of the microstructure (internal structure) of thin films until about a decade ago. Although some anomalous results continue to appear in literature, it is now generally possible to obtain reliable and reproducible data on films of well-defined structures and deduce fundamental information therefrom with confidence.

A comprehensive account of the work done before 1955 is given by Mayer [2]. Brief topical reviews on the developments in the last 15 years have also appeared and are referred to in the appropriate sections of this

328

chapter. We propose here to review critically the well-accepted experimental and theoretical results on electrical conduction, galvanomagnetic effects, and thermal transport in thin films. The emphasis is primarily on the size effects due to the thickness (size) and microstructure of films and their interpretation in terms of the fundamental transport parameters and the electron-scattering processes. Only metal films are discussed in this chapter; the following chapters deal with semiconducting and dielectric films. Wherever data on thin films are not available, thickness or size-effect data on bulk materials that are of dimensions comparable with the mean free path of conduction electrons (and thus are electronically equivalent to thin films) are presented.

2. ELECTRICAL CONDUCTION IN DISCONTINUOUS FILMS

Depending on the growth stages (Chap. IV, Sec. 2) a film may be granular or island-like (that is, consisting of discrete particles), porous (network), or continuous. Each stage has its characteristic electrical properties and will be considered separately. This section deals with physically discontinuous films.

2.1 Conduction Mechanisms

Electrical conductivity of a granular film is many orders of magnitude smaller than that of the bulk material and is generally characterized by a negative temperature coefficient of resistivity (TCR). The conductivity is found to vary exponentially with the inverse of temperature, suggesting that the conduction mechanism is thermally activated. It is ohmic at low applied fields, but nonlinear at high fields. Experimental results have been interpreted in terms of various conduction mechanisms, such as thermionic and Schottky emission, and tunneling through the vacuum gap and/or via traps in the dielectric substrate. Brief reviews of the subject have been published in the literature [3-5]. A discussion of these conduction mechanisms and the experimental results cited in their support follows.

(1) Thermionic (Schottky) Emission. Nifontoff [6] calculated the effective resistance of the gap between two identical metal particles separated by a distance d for a negligible applied potential at $300°K$. His calculations show that thermionic and Schottky emission predominate only if particles have a large separation ($d > 100$ Å). At $d \sim 20$ to 50 Å, and with a typical metal work function ~4 to 5 eV, tunneling is the dominant mechanism of electron transport unless the temperature is higher than $300°K$. Similar results were obtained by Simmons [7] (see Fig. 6, Chap. VIII), who calculated the relative contributions of thermionic and tunneling currents between two plane metal electrodes separated by a thin dielectric film.

Thermionic emission depends exponentially on the height of the potential barrier between particles. When two particles are very close together, the height of the barrier may decrease significantly because of the overlap of the image-force potentials, thus giving rise to higher conductivity. On this basis, Minn [8] derived the following expression for the conductivity:

$$\sigma = \frac{AeT}{k} d \exp\left(-\frac{\psi - Be^2/d}{kT}\right) \qquad (1)$$

where A is a constant characteristic of each film, T the temperature, e the electronic charge, k the Boltzmann constant, d the distance between particles, and ψ the bulk work function of the metal.

The term Be^2/d (B = constant) represents the contribution of the image forces. If d is sufficiently small (~ several angstroms), the effective work function $\psi_{eff} = (\psi - Be^2/d)$ can become quite small. Under an applied field E between particles (Schottky emission), the effective work function will be further reduced to

$$\psi_{eff} = \psi - \frac{Be^2}{d} - e^{3/2}\sqrt{E} \qquad (2)$$

Thus, Schottky emission leads to an $\exp(-1/T)$ dependence of the conductivity, an ohmic behavior at low fields, and an $\exp\sqrt{E}$ dependence at high fields.

(2) **Quantum-mechanical Tunneling.** Charge transfer by tunneling (discussed in Chap. VIII) was considered by several workers [6, 9, 10] to explain conduction in granular films. This mechanism predicts the transfer current to be nearly independent of temperature [11] and exponentially proportional to the inverse of the applied field.

(3) **Activated Tunneling.** Gorter [12] and Darmois [13] introduced a thermal-activation term in the tunneling mechanism by using the

activated process involving transfer of charge from one *initially* neutral particle to another some distance removed. The activation energy $\approx e^2/\epsilon r = 14.4/\epsilon r$ electron volts (r is the linear dimension of the particle in angstroms, and ϵ is the dielectric constant) is thus the electrostatic-potential barrier. Only electrons or holes excited to states of at least this energy from the Fermi level will be able to tunnel from one neutral particle to another. Since $e^2/r \approx kT$, electrons can be created by thermal activation over the electrostatic-potential barrier. Charged particles would thus exist because of thermal fluctuations, and tunneling between a charged and a neutral particle could occur. This, however, is not an activated process. Neugebauer and Webb [14] (N-W) proposed an activated transfer mechanism in which charge carriers created by thermal activation over the electrostatic-potential barrier are transported by tunneling from one neutral particle to another.

Conductivity is the product of the equilibrium concentration of charge carriers n, the charge of the carriers e, and the mobility μ, *provided* an equilibrium concentration of charges is maintained and interaction between carriers is negligible. The thermally activated equilibrium charge-carrier density (one electron per particle) is given by

$$n = \frac{n_0}{n_0 r^3} e^{-e^2/\epsilon r k T} \tag{3}$$

where n_0 is the total number of particles assumed to form a linear array in the film. The net probability of a transition between particles in the field direction due to an applied potential drop V between particles is

$$P_{net} = P_{+V} - P_{-V} \tag{4a}$$

The probability of transition of the "extra" charge from i to j particle in the field direction is given by

$$P \propto \int_{-\infty}^{\infty} Df_i(1 - f_j)\, dE \tag{4b}$$

where f_i and f_j are the Fermi distribution functions for energy E and $E + V$ (or $-V$), respectively. D is the transmission coefficient, which at low applied field between two electrodes is approximated by

$$D \propto \frac{\sqrt{2m\phi}}{h^2 d} \exp\left(-\frac{4\pi d}{h}\sqrt{2m\phi}\right) \tag{5}$$

Here, m is the electronic mass, h is Planck's constant, and ϕ is the potential barrier between particles, which can be approximated by the work function of the metal. By substituting Eq. (4b) and the expressions for the Fermi functions in Eq. (4a), and further assuming that the equilibrium carrier concentration is not disturbed (which implies $kT \gg eV$), we obtain

$$P_{net} \propto D\left(\frac{eV}{1 - e^{-eV/kT}} + \frac{eV}{1 - e^{eV/kT}}\right) \tag{6}$$

At vanishingly small fields, Eq. (6) reduces to

$$P_{net} \propto DeV \quad \text{transitions/sec} \tag{7}$$

The transition probability from one particle to another of cross section $\sim r^2$ is

$$P_r \propto Dr^2 eV \quad \text{transitions/sec} \tag{8}$$

Since the velocity of the charge carriers moving a distance d in a time $t = 1/P_r$ is $v = d/t = dP_r$, the mobility in an applied field V/d is given by

$$\mu = \frac{d^2 P_r}{V} \propto Der^2 d^2 \quad \text{cm}^2/(\text{V})(\text{sec}) \tag{9}$$

Therefore, using Eqs. (3) and (9), conductivity is given by

$$\sigma = ne\mu \propto \frac{d^2}{r} e^2 D \exp\frac{-e^2/\epsilon r}{kT} \propto \frac{d\phi^{1/2}}{r} \exp\left(\frac{-e^2}{\epsilon rkT} - \frac{4\pi d}{h}\sqrt{2m\phi}\right) \tag{10}$$

Thus, an ohmic behavior (at low fields) of conductivity, an exponential dependence on reciprocal temperature, i.e., Arrhenius-type behavior, and a marked dependence on the island size and interisland distance are predicted. The activation energy $e^2/\epsilon r$ is the energy required to remove an electron to infinity. But to transfer the electron to a distance d, the energy required is less and is given by

$$\mathscr{E} \approx \frac{e^2}{\epsilon r} - \frac{e^2}{\epsilon(d - r)} \tag{11}$$

At low fields, the conductivity [Eq. (10)] obeys Ohm's law. Deviations from Ohm's law are expected at high fields because of an appreciable reduction of the activation energy as a result of the energy picked up by the charge during its flow along the field direction, and also because of enhanced field emission through the modified potential barrier. If E is the field between the particles (clearly, E is always greater than the field applied across the whole film), the effective activation energy $\mathscr{E}_{\text{eff}}$ can be shown to be

$$\mathscr{E}_{\text{eff}} \approx \frac{e^2}{\epsilon r} - \frac{2e^{3/2} E^{1/2}}{\epsilon} + reE \tag{12}$$

Thus, at high fields, the second term will give strong temperature-dependent deviations from Ohm's law and an $\exp(\sqrt{E})$ behavior. The field at which deviations become important is $\sim 10^3$ to 10^4 V/cm for $r \approx$ 20 Å. The last term, reE, in Eq. (12) is negligible for small r and intermediate fields, while at high fields its effect will be to saturate the conductivity. Physically, this means that at very high fields the electrons are transferred faster than they are thermally created.

The N-W model has essentially considered tunneling of the electron after it is thermionically emitted over the electrostatic-potential barrier. The electron may penetrate through this barrier (tunnel emission) rather than pass over it, in which case the electrostatic barrier becomes a part of the total tunneling barrier. The number of charge carriers is then determined by the Fermi statistics. This treatment, however, results in nearly the same expression as that obtained by using the N-W model.

(4) Tunneling between Allowed States. The electronic-energy levels of a metallic particle of finite size are expected to be quantized. Hartman [15] suggested as a result of this effect that the transfer of electrons between particles takes place by tunneling between their overlapped or "crossed" energy levels. Ground levels are not expected to be crossed because of their small width and even if they are crossed, the application of an electric field will uncross them. According to Hartman, the width of the first excited level is sufficient ($\sim 10^{-3}$ eV for $\sim$100-Å-size potential box) to allow crossing at fields of up to several hundred volts per centimeter. Thus, activated tunneling can occur only between excited levels of which the first excited level is the most likely. The activation energy then corresponds to the difference in energy between the nth and $(n + 1)$th level and is

$$\mathcal{E} = \frac{h^2}{8mrL} \tag{13}$$

for large particles $(r \gg L)$. Here, h is Planck's constant, m the mass of the electron, and L the nearest-neighbor distance in the bulk material. For zero applied field, and temperatures which are not too low [i.e., $kT \gg \delta E_{n+1}$ = the spread of $(n + 1)$th level], the conductivity is given by

$$\sigma \propto \exp - \frac{\mathcal{E}}{kT} \tag{14}$$

The Hartman model predicts an ohmic behavior and an exponential dependence on the reciprocal of temperature. The activation energy for the conduction process depends on the particle size and not on the dielectric constant. Although the theory does not predict any definite field dependence, a reduced conductivity at high fields is indicated. This model has been questioned by Neugebauer and Wilson [4] on the basis that level widths are actually much narrower than indicated by Hartman and therefore no appreciable tunneling is expected. Moreover, because of a variety of shapes and sizes of particles present in granular films, the exact matching of bands may not be possible for interband tunneling to occur.

(5) **Tunneling via Substrate and Traps.** So far, we have considered tunneling through the free space between the particles. The barrier height for free tunneling is of the order of the work function of the metal, and that for tunneling through a dielectric is much less, being the difference between the work function of the metal and the electron

affinity of the dielectric. The reduced barrier height enhances the transmission coefficient. Moreover, because of the higher dielectric constant of the substrate, the activation energy $\mathcal{E}$ is reduced as compared with that for free tunneling. Thus, conduction by tunneling through the substrate is expected to be significant. Using subscripts D and S for dielectric and free-space tunneling, one can write the conductivity ratio [neglecting the preexponential dependence in Eq. (10)] as

$$\frac{\sigma_D}{\sigma_S} = \exp\left[\frac{4\pi}{h}\sqrt{2m}\left(d_S\phi_S^{1/2} - d_D\phi_D^{1/2}\right) + \frac{e^2}{rkT}\left(\frac{1}{\epsilon_S} - \frac{1}{\epsilon_D}\right)\right] \quad (15)$$

The conduction path through the substrate may be direct tunneling or via the stable energy states created by impurities and/or traps. Conductivity in such a case depends on the details of the potential barrier, which in turn is determined by the level, number, and distribution of these states. An impurity-conduction mechanism [16] would be similar to that in semiconductors which exhibit $e^{-\mathcal{E}/kT}$ dependence of the conductivity.

In order to explain the anomalously high current densities observed in granular films, Milgram and Lu [17] proposed that carriers are created by thermal excitation as well as by field injection and are transported by a hopping-process type of tunneling between traps in the substrate. The postulated injection process should yield a space-charge-limited type of current-voltage behavior given by

$$I = AV + BV^n \quad (16)$$

where A and B are constants and the value of $n (\geq 2)$ depends on temperature and energy distribution of the trapping states.

2.2 Experimental Results

An exponential dependence of the conductivity of granular films on the inverse temperature has been reported by several workers [8, 14, 17-21]. Minn [8] carried out a comprehensive work on Au films at temperatures down to 1.4°K. Neugebauer and Webb [14] studied films of Pt, Au, and Ni, and their results for Pt films are shown in Fig. 1. An exponential dependence of conductivity and an activation energy

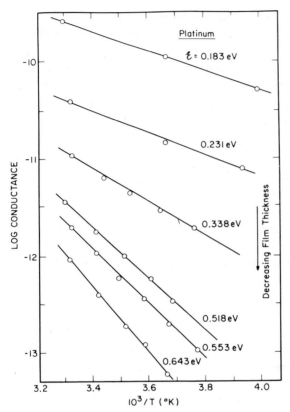

FIG. 1. Log conductance vs. reciprocal temperature for discontinuous Pt films of various thicknesses. The calculated activation energies are given with each plot. (*Neugebauer and Webb* [14])

increasing with the decreasing film thickness were found. Using a dielectric constant of 3, halfway between those of vacuum and glass, values of $r = 8$ Å for the thinnest and $r = 26$ Å for the thickest film in Fig. 1 were obtained. These deduced values agree with the fact that the particle size increases with the film thickness. The smaller the size of the particles, the greater is the activation energy. The larger the distance between particles, the lower is the tunneling probability. In both cases, the conductivity is lowered. The size of the particles and the interparticle separation are generally relatively greater for a lower melting point of the deposit material, higher substrate temperature, and smoother substrates. Thus, given the deposition conditions, one can predict the conduction behavior of granular films.

At high applied fields, the observed conductivity exhibits nonohmic behavior which becomes more marked at low temperatures and follows an $\exp\sqrt{E}$ dependence [Fig. 2(a)]. A decrease in the conductance at very high fields observed by Bashara and Weitzenkamp [20] is shown in Fig. 2(b). A more complicated field dependence was observed by Minn [8] and Milgram and Lu [17]. The latter authors found their data on Cr films at low temperatures to fit the space-charge-limited-current V^2 dependence [Eq. (16)].

The observed temperature and field dependence of the conductance of granular films can be explained qualitatively by both the thermionic emission and the N-W model. The observed low activation energies and the high conductance have generally formed the basis for rejecting the thermionic model. But this is not completely justified since both these observations may be explained if significant image-force lowering of the barrier potential occurs. On the other hand, although the N-W model can

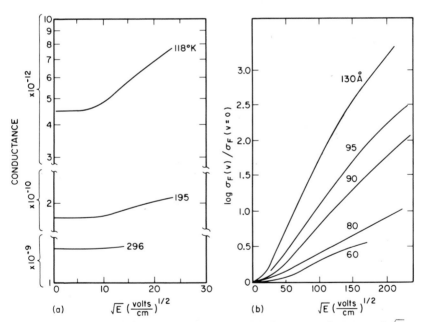

FIG. 2. (a) Relative conductance at different temperatures as a function of $\sqrt{E}$ (E is the applied field) for a discontinuous Ni film, illustrating $\sqrt{E}$ dependence and increasingly nonohmic behavior at lower temperatures. (*Neugebauer and Webb* [14]) (b) Normalized conductance (with respect to that measured at a negligibly small applied voltage) of discontinuous Au films vs. $\sqrt{E}$, showing a decrease of conductivity at high fields. (*Bashara and Weitzenkamp* [20])

explain qualitatively most of the general features of conductivity of granular films, the magnitude of conductivity calculated by using the typical geometrical configuration of particles is much lower than observed. This discrepancy suggests [17] that the tunneling distance is considerably smaller than the physical interparticle distance and may be indicative of the role of substrate impurities and/or traps.

Wei [22] observed an $e^{-\mathscr{E}/kT}$ dependence of conductivity of potassium films deposited on KCl crystals and interpreted the data on the basis of an impurity-conduction mechanism. He assumed that metallic islands acted as donor impurities and the ionic substrate contributed acceptor impurities. The activation energy is then related to the density of the impurities. Wei asserted that, since various metal films deposited on different substrates yield similar dependence on temperature, conduction must take place through impurity levels. But these observations, on the other hand, are consistent with the fact that the electrostatic barrier (activation energy) in the tunnel model depends on the particle size only and not on the metal. Of course, the difference in work functions of different metals would affect the transmission coefficient in Eq. (10) and hence the absolute value of the conductivity. Because of the poorly defined structure of thin films, it has not yet been possible to calculate the absolute value of conductivity and thus throw light on the influence of the substrate on conduction.

Herman and Rhodin [23] explained their results on granular films of Pt and Au on the basis of tunneling between the filled and unfilled surface states or traps. A high density of empty traps was assumed to exist between the metal particles. Some of the traps under the metallic region of the surface may be filled, the degree of filling depending on the position of the Fermi level. The activation energy was taken as the electrostatic energy required to move a charge carrier out of the metallic region to an unfilled trap in the surface region midway between the two metallic particles. As expected, this model yielded an activation-energy expression as well as a conductivity equation very similar to that obtained on the basis of the N-W model. A dielectric constant of 5.6 (static dielectric constant of NaCl substrate) was required to make the experimentally determined activation energy agree with that determined from the average size and separation of islands measured from electron micrographs. On the other hand, for low-temperature deposits annealed and measured at room temperature, only a value of $\epsilon = 1.9$ was required.

This low value of ϵ was attributed to the rearrangement or surface movement of islands as the film was annealed.

It is clear that the above-mentioned results do not provide a convincing evidence of the influence of the substrate. Any conclusion about a conduction mechanism based on the effective value of the activation energy calculated from the conductivity data is open to criticism. The statistical distribution of the particle sizes, the inter-particle-separation distances, and the complicated image-force corrections make an accurate theoretical estimate of the activation energy extremely difficult and are apt to belittle the importance of the value of the dielectric constant used to fit the data.

The influence of substrate on conduction was shown by Hill's [24] observation that the conductivity of granular Au films deposited on borosilicate glass could be modified reversibly by an applied transverse field of $\sim$250 to 1,000 V/cm. The resistance was found to increase slightly for a negative voltage applied to the Au film, but a marked initial increase followed by a rapid decrease occurred for the positive voltage. The maximum change decreased with increasing temperature. Fehlner and Irving [25] confirmed this effect and further found evidence for the formation of an intermetallic compound of Na in the substrate with the Au film. The Na ions present in this particular glass apparently move readily under the influence of the applied field and thus the film-substrate interface is altered. If traps can be associated with the intermetallic compound, then the applied field provides a mechanism for altering the density (and hence the activation energy) of the filled and the unfilled traps. But if the formation of Na-Au alloy at the interface is responsible for surface states, the observed reversible changes in conductance with the applied field are hard to understand. The absence of a similar effect for substrates of other dielectric or amorphous materials with well-known electron-trapping properties is indeed puzzling.

The results of detailed studies of van Steensel [26] on Au and Pb films showed better agreement with the thermionic-emission model rather than the tunneling theories. Van Steensel also established that the conductivity did not change as the dielectric constant of the $BaTiO_3$ substrate was varied by heating it to various temperatures. This significant result shows conclusively that the low-frequency dielectric constant has no influence on the conduction process in this particular case. Further, only small ($<$50 percent) changes in conductance were observed due to the deposition of an SiO overlay. This result suggests that either the conduction takes place through the substrate and is thus not influenced by the lowering of the potential barrier between the islands due to the SiO deposit, or the overlay has no appreciable effect on the initial potential barrier.

Further conductivity studies as a function of the work function of the particle material, dielectric constant of the substrate, frequency of the measuring current, photoionization of the electrons from the particles, transverse-field effect, etc., should obviously provide more information in identifying the tunneling medium. Hirsch and Bazian [27] studied the frequency dependence of the conductivity at frequencies up to 2×10^6 cps and at temperatures between 4.2 and 297°K. The resistance was found to decrease with increasing frequency, the decrease being more marked for films having a large dc resistance. The results were interpreted in terms of the impedance of the intergranular capacitance. The effects of surface states, if any, were not considered.

If tunneling takes place from one energy level of the particle to another, then it should be possible to raise the electrons to higher energy levels by the application of a sufficiently high field. This acceleration process is possible because the mean free path of electrons is much greater than the particle dimension. The transition of the excited electrons to the ground level would yield photoemission (electro-luminescence) and electron emission. Borzjak et al. [28] observed such effects and found electron emission to depend strongly on the work-function changes brought about by a BaO overlay on 40- to 60-Å-thick Au films. These observations, however, remain suspect in view of the abnormally low voltages (~10 V) used to obtain electro-luminescence.

2.3 Temperature Coefficient of Resistivity (TCR)

Temperature coefficients of resistivity of both signs and of variable magnitudes have been observed in discontinuous metal films. The exact behavior of TCR, like that of conductivity, depends on the details of the film microstructure. Whatever the details of the conduction mechanism, the significant temperature-dependent function in the conductivity expression is $e^{-\mathcal{E}/kT}$ (provided that the interparticle spacing does not change significantly), which yields an expression for the TCR (α) as*

*Neglecting the temperature coefficient of expansion of the film (see Sec. 3.3).

$$\alpha \equiv -\frac{1}{\sigma}\frac{d\sigma}{dT} = -\frac{d\ln\sigma}{dT} = \frac{-\mathcal{E}}{kT^2} = -\frac{e^2/\epsilon r}{kT^2} \qquad (17)$$

since $\mathcal{E} = e^2/\epsilon r$.

The temperature variation of the particle spacing can also be taken into account [29]. If the particles are strongly bound to the substrate, the expansion of the substrate increases the spacing by $\Delta d = a_s d\Delta T$, where a_s is the substrate expansion coefficient. On the other hand, if the particles are not constrained by the substrate, the separation distance is decreased by $\Delta d = (a_m - a_s)r\,\Delta T$. Here, ΔT is the temperature change, r is the linear dimension of the island, and $(a_m - a_s)$ is the difference of the thermal-expansion coefficients of the film material and the substrate. The variation in d primarily affects the tunneling-transmission coefficient D [Eq. (5)]. The ratio of the tunneling-transmission probabilities at two different temperatures is given by

$$\frac{D(T)}{D(0)} = 1 - \frac{4\pi\sqrt{2m\phi}}{h}(\Delta d) \qquad (18)$$

For a typical film with $r = 100$ Å, $d = 10$ Å, $\phi = 5$ eV, $a_m - a_s = 5 \times 10^{-6}/°C$ and $\Delta T = 100°C$, the change is 12 percent for the constrained case, and 1.2 percent for the nonconstrained case.

Neglecting the small temperature variation of the preexponential terms in the conductivity expression [Eq. (10)], one may write the general expression for the TCR as

$$\alpha = -\frac{d\ln\sigma}{dT} = \frac{4\pi}{h}\sqrt{2m\phi}\left(\frac{\Delta d}{\Delta T}\right) - \frac{e^2}{kr\epsilon T^2} \qquad (19)$$

provided that

$$a_s \gg a_m \quad \text{and} \quad \Delta d = a_s d\,\Delta T$$

Thus, depending on the predominance of one or the other of the opposing terms, we expect to observe positive or negative TCR of varying magnitudes. For a Teflon substrate with a high value of a_s ($\sim 10 \times 10^{-5}/°C$), which is about 10 times larger than that of a typical metal and 20 to 50 times that of glass, anomalously large positive TCR, exceeding that of the bulk metal, is expected and is in agreement with the observations [29]. Conversely, if r is very small the second term will give a predominently negative TCR. Figure 3 shows data for Au films deposited on Pyrex and Bi_2O_3 substrates, demonstrating the occurrence of both positive and negative TCR. Further, with increasing porosity and hence film resistance, the negative TCR increases rapidly in agreement with the prediction of Eq. (19). One can vary the particle size and the separation distance by controlling deposition parameters (Chap. IV, Sec. 2.2) and thus obtain a desired TCR. For example, films deposited at very low substrate temperatures, and also those of high-melting-point materials deposited at ambient temperatures, exhibit small particle sizes. Subsequent annealing of such films results in an increased particle size because of recrystallization and coalescence brought about by surface diffusion in the case of physically touching particles.

2.4 Network (Porous) Films

After the stage at which coalescence of the particles (islands) takes place, the physical structure of the film resembles a network. Electrical

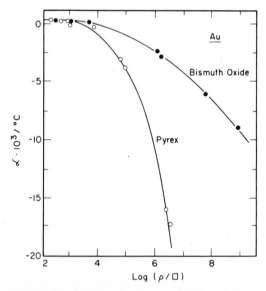

FIG. 3. The TCR vs. the sheet resistivity of ultra-thin Au films deposited on pyrex and Bi_2O_3. The TCR was measured between 77.4 and 293°K. (*Minn* [8])

conduction in such films is due to the particles, and bridges and gaps between the particles, and is very sensitive to the physical and electrical changes in the filamentary bridges brought about by aging, annealing, and adsorption (Sec. 3.6). The TCR of such films is the sum of the metallic, positive (particles) and the activated, negative (gaps) contributions. Thus a variety of behavior, including a temperature-dependent minimum in the resistance observed by Feldman [30], is expected.

As the network structure fills up, the porosity decreases and the amount of porosity depends on the deposition conditions (Chap. IV, Sec. 3). Conductivity of a porous film is determined largely by grain-boundary scattering, diffuse scattering at grain surfaces, and intergranular tunneling. Electrical conduction in continuous, fine-grained films of various materials such as refractory metals [31-33] is also dominated by these mechanisms. A summary of the experimental results on such films is given by Siddall [32].

Ion-beam-sputtered films of W, Mo, and Re with grain size of ~20 Å have very small (<50 ppm) TCR over an extended temperature range of 700 to 4.2°K. The refractory-metal films are highly susceptible to oxidation and contamination at grain boundaries. The resistance of the widely studied Ta films is known [34] to increase on aging because of oxidation of grains, and is attended by a large negative TCR. If an overlayer of Au is allowed to diffuse into the intergrain spacing, the resistance is stabilized, presumably becuase of the prevention of further oxidation. Further, a considerable reduction and reversal to positive values occurs in the TCR as a result of the metallic behavior of the grain boundaries.

The conduction mechanism in porous films is complex and difficult to analyze. Drumheller [35] explained his results for Bi films on the basis of tunneling between grains across the potential barrier imposed by the grain interfaces. The transport of charges was considered to occur because of a charge-density gradient in the presence of the field. This tunneling process yields zero TCR. A more complicated and rather artificial approach is due to Grigorovici and Ciobanu [36], who attempted to explain their results [37] on Be and Pb films by considering the tunneling of "three" types of carriers through barriers between crystallites. The impurities in the grain boundaries are postulated to supply the third type of carriers thermionically in addition to the electrons and holes in the bulklike crystallites. This phenomenological approach is equivalent to adding thermally activated tunneling to a direct temperature-independent tunneling process. Thus the negative TCR may be explained.

2.5   Elastoresistance

If an electrical conductor is deformed elastically in tension or compression, its resistance changes because of changes in its dimensions. The ratio χ of the fractional change in the resistance to the fractional change in the length of the conductor is called the strain coefficient or the gauge factor. If a film is ultrathin, consisting of discrete islands, and tunneling is the dominant conduction mechanism, the resistance [Eq. (10)] is proportional to $\exp(-d\sqrt{\phi_{eff}})$. Here d is the spacing between islands and ϕ_{eff} is the effective barrier height between islands. If the size of the islands does not change during elastic strain, one can show [38] that $\chi_F \approx \sqrt{\phi_{eff}}\, d$, where ϕ_{eff} is expressed in eV and d in angstrom units. For a typical value of $\phi_{eff} \sim 4$ eV and $d \sim 50$ Å a value of $\chi_F \sim 100$ is expected. The observed film values are generally less than 10 and are also considerably lower than the bulk values, contrary to what is expected. A possible explanation is that the total strain is not transmitted to the film grains owing to the poor adherence of the films to the substrate.

A number of studies [38, 39] on strain-induced changes in the electrical resistance of films of metals, semimetals, and semiconductors have reported diverse results which make it difficult to draw meaningful conclusions. Linear changes in the resistance with both tensional and compressional strain were observed by Parker and Krinsky [38] in films of Au, Co, Pd, Pt, and Te. The coefficient χ was found to be nearly constant in thick films, but it increased rapidly and nonlinearly in thin discontinuous films. Bismuth and antimony films have been found [39] to show large and strongly anisotropic elastoresistance behavior. These films lend themselves to applications as pressure transducers.

3.   ELECTRICAL CONDUCTION IN CONTINUOUS FILMS

Electrical conductivity of a metal in the free-electron-gas theory of Drude-Lorentz-Sommerfeld is treated in standard textbooks [40] and is given by

$$\sigma = \frac{ne^2 l}{mv} \tag{20}$$

where n is the number of free electrons per unit volume

$$n = \frac{8\pi}{3} \left(\frac{mv}{h}\right)^3 \tag{21}$$

e is the electronic charge, m is the effective mass of the electron, v is the average velocity of the electron at the surface of the Fermi distribution, l is the mean free path (mfp) of the conduction electrons, and h is Planck's constant.

A detailed quantum-mechanical treatment [40] of electrical conductivity of metals concludes that the simple relation of Eq. (20) is correct for an isotropic, monovalent metal with a single conduction band. Its application to the complicated cases of multivalent metals is expected to give at least a semiquantitative description of the conduction phenomenon. The generalized conductivity expression deduced from the free-electron model in the modern theory consists of an ideal and residual resistance terms. The ideal resistance term is given by Eq. (20) and is proportional to T at high temperatures and to T^5 at very low temperatures.

Electrical conductivity of a metal is directly proportional to l, which is defined by the relaxation time of electron-phonon interaction $\tau = l/v$. One can rearrange Eq. (20) to obtain

$$\frac{\sigma}{l} = \frac{1}{\rho l} = \frac{e^2 n}{mv} = \left(\frac{8\pi}{3}\right)^{1/3} \frac{e^2 n^{2/3}}{h} = 7.1 \times 10^7 n^{2/3} \tag{22}$$

and

$$\frac{\sigma}{l} = \frac{e^2}{6\pi^2} \frac{A}{h} \qquad A = 4\pi(3\pi^2 n)^{2/3} \tag{23}$$

where A is the Fermi surface area. The ratio σ/l therefore determines directly the number of conduction electrons per unit volume and the area of the Fermi surface.

3.1 Theories of Size Effect

As the thickness of a metal film becomes comparable in magnitude with the mfp, the film boundaries impose a geometrical limitation on the movement of the conduction electrons and thus the effective value of the mfp. Physical effects arising because of this geometrical limitation of the mfp are termed "mean-free-path" or "size" effects.

Thomson [41] was the first to propose a size-effect theory to explain the observed high electrical resistivity of thin specimens as compared with that of the same metal in bulk. The size-effect theory for a free-electron model was worked out by Fuchs [42] for a spherical Fermi surface and extended by Sondheimer [43, 44] to include galvanomagnetic effects. Lucas [45] generalized the Fuchs calculations to the case of scattering from the two film surfaces with different specularity parameters.

Fuchs' treatment is a statistical analysis based on the Boltzmann equation for the distribution function of conduction electrons. Chambers [46] formulated the same problem by using the solution of the Boltzmann equation in a form suggested by simple kinetic-theory considerations. Chambers' approach was used to extend the size-effect theory to spheroidal energy surfaces by Ham and Mattis [47] and to ellipsoidal energy surfaces (such as may describe semimetals) by Price [48]. Kaganov and Azbel [49] obtained generalized expressions for conductivity and galvanomagnetic effects. Soffer [50] used the Chambers-Price approach to derive a generalized expression for conductivity for any arbitrary band structure, anisotropic relaxation time, and specularity parameter of an arbitrary magnitude. Here, as well as in the work of Kaganov and Azbel, the generalized expressions as such are too complicated for direct comparison with the experimental results. The reader is referred to the original works for mathematical details of the treatments. The Fuchs treatment for a free-electron model is simple and instructive and therefore is presented here.

The Boltzmann equation for the distribution function f of the conduction electrons is formed by equating the rate of change of f due to external fields to the rate of change due to the collision mechanism. In the presence of an electric field E and a magnetic field H, the Boltzmann equation for quasi-free electrons takes the form

$$\frac{-e}{m}(E + v \times H) \cdot \text{grad}_v f + v \cdot \text{grad}_r f = \left(\frac{\partial f}{\partial t}\right)_{\text{coll.}} \tag{24}$$

which is purely classical except that the mass m is to be regarded as an effective mass. Here, f is the Fermi-Dirac function of the velocity vector v and the space vector r.

The analysis of size effects depends essentially on the appearance in Eq. (24) of the terms involving the space derivative of f which take into account the nonuniform distribution in space of the conduction electrons. The term $(\partial f/\partial t)_{coll.}$ can be determined in a general way by invoking a relaxation time τ which is defined by the time taken for the distribution function f to relax to the steady undisturbed state f_0 when the external constraint is removed. The approach to equilibrium is then supposedly given by

$$\left(f - f_1\right)_t = \left(f - f_0\right)_{t=0} e^{-t/\tau} \tag{25}$$

Now, since $\partial f_0/\partial t = 0$,

$$\left(\frac{\partial f}{\partial t}\right)_{coll.} = -\frac{f - f_0}{\tau} \tag{26}$$

If v is the mean velocity of these electrons to which τ refers, the corresponding free path l is defined by $l = v\tau$. The detailed theory of conduction shows that a free path can be unambiguously defined for scattering by randomly distributed impurity atoms as well as by lattice vibrations if the temperature is above the Debye temperature of the lattice.

We may solve Eq. (24) for the one-dimensional case of a metal film of thickness t with the z axis perpendicular to the film and an electric field E applied in the x direction. The distribution function may be written in the form

$$f = f_0 + f_1(v, z) \tag{27}$$

Neglecting the product of E with f (that is, neglecting deviations from Ohm's law), the Boltzmann equation reduces to

$$\frac{\partial f_1}{\partial z} + \frac{f_1}{\tau v_z} = \frac{eE}{mv_z} \frac{\partial f_0}{\partial v_x} \tag{28}$$

The general solution of this equation is

$$f_1(v, z) = \frac{eE\tau}{m} \frac{\partial f_0}{\partial v_x} \left[1 + F(v) \exp\left(\frac{-z}{\tau v_z}\right) \right] \tag{29}$$

where $F(v)$ is an arbitrary function of v determined by the boundary conditions.

Let us assume the boundary conditions that every free path is terminated by collision at the surface (diffuse scattering) so that the distribution function of the electrons leaving each surface must then be independent of direction, and further that the relaxation process for surface scattering is the same as for the bulk. This first condition is satisfied if we choose $F(v)$ so that $f_1(v, 0) = 0$ for all v such that $v_z > 0$ and $f_1(v, t) = 0$ for all v such that $v_z < 0$. Thus, there are two values for f_1, depending on whether electrons are moving away from ($v_z > 0$), or moving toward ($v_z < 0$) the surface. These may be written as

$$f_1^+(v, z) = \frac{eE\tau}{m} \frac{\partial f_0}{\partial v_x} \left[1 - \exp\left(-\frac{z}{\tau v_z}\right) \right] \quad (v_z > 0)$$

and

$$f_1^-(v, z) = \frac{eE\tau}{m} \frac{\partial f_0}{\partial v_x} \left[1 - \exp\left(\frac{t - z}{\tau v_z}\right) \right] \quad (v_z < 0) \tag{30}$$

The current density for a position z is given by

$$j(z) = -2e\left(\frac{m}{h}\right)^3 \iiint v_x f_1 \, dv_x \, dv_y \, dv_z \tag{31}$$

which can be simplified by introducing polar coordinates (v, θ, ϕ) in the v space (with $v_z = v \cos \theta \equiv v/a$), and remembering that f_0 depends only on the magnitude of v. The total current density is obtained by integrating Eq. (31) with respect to z. The current density J_0 in bulk metal may be obtained by putting $z = \infty$ in the integration of Eq. (31). Since the effective conductivity is $\sigma \equiv J/E$, the ratio of the conductivities is given by an expression

$$\frac{\sigma_B}{\sigma_F} \equiv \frac{\rho_F}{\rho_B} = \frac{\phi(\gamma)}{\gamma} \tag{32}$$

where $\gamma = t/l$, and

$$\frac{1}{\phi(\gamma)} = \frac{1}{\gamma} - \frac{3}{8\gamma^2} + \frac{3}{2\gamma^2} \int_1^\infty \left(\frac{1}{a^3} - \frac{1}{a^5}\right) e^{-\gamma a} \, da \tag{33}$$

The limiting form for large γ (which is generally considered valid for $\gamma \gg 1$ but is in fact reasonably accurate down to $\gamma = 1$) is given by

$$\frac{\sigma_B}{\sigma_F} = \frac{\rho_F}{\rho_B} \approx 1 + \frac{3}{8\gamma} \quad (\gamma \gg 1) \tag{34}$$

Equation (34) is referred to in the early literature as the Planck-Weale or Nordheim relation, since it was originally obtained empirically by these authors.

For very thin films,

$$\frac{\sigma_B}{\sigma_F} \approx \frac{4}{3\gamma[\ln(1/\gamma) + 0.4228]} \quad (\gamma \ll 1) \tag{35a}$$

which may be written as

$$\frac{\sigma_B}{\sigma_F} \approx \frac{4}{3\gamma \ln(1/\gamma)} \tag{35b}$$

We assumed in the above discussion that the scattering at the surfaces is entirely diffuse. We can generalize the treatment for the case where a fraction p of the electrons is scattered elastically (specular) from both surfaces of the film with reversal of the velocity component z, while the rest are scattered diffusely with complete loss of their drift velocity. For simplicity of calculations, p is considered here to be a constant (which is a highly artificial model, see Sec. 3.4). The distribution functions of the electrons are then given by

$$f_0 + f_1^+(v, z) = p\left[f_0 + f_1^-(-v_z, 0)\right] + (1 - p)f_0 \quad \text{at} \quad z = 0$$

and

$$f_0 + f_1^-(v, z = t) = p\left[f_0 + f_1^+(-v_z, t)\right] + (1 - p)f_0 \quad \text{at} \quad z = t \tag{36}$$

These equations determine $F(\mathbf{v})$, and we obtain from Eq. (29) two values for f_1

$$f_1^+(\mathbf{v}, z) = \frac{eE\tau}{m} \frac{\partial f_0}{\partial v_x} \left[1 - \frac{1 - p}{1 - p \exp(-t/\tau v_z)} \exp\left(\frac{-z}{\tau v_z}\right) \right] \quad (v_z > 0)$$

and

$$f_1^-(\mathbf{v}, z) = \frac{eE\tau}{m} \frac{\partial f_0}{\partial v_x} \left[1 - \frac{1 - p}{1 - p \exp(t/\tau v_z)} \exp\left(\frac{t - z}{\tau v_z}\right) \right] \quad (v_z < 0)$$

The effective conductivity is calculated as before. The function $\phi(\gamma)$ defined by Eq. (33) is now replaced by

$$\frac{1}{\phi_p(\gamma)} = \frac{1}{\gamma} - \frac{3}{2\gamma^2} (1 - p) \int_1^\infty \left(\frac{1}{a^3} - \frac{1}{a^5} \right) \frac{1 - e^{-\gamma a}}{1 - pe^{-\gamma a}} \, da \tag{37}$$

This form [Eq. (37)] reduces to Eq. (33) when $p = 0$, and to bulk-metal value $1/\gamma$ when $p = 1$. This expression can be further rearranged in a form convenient for numerical calculations. Such calculations have been published in the literature [51].

The limiting forms of Eq. (37) for thick and thin films yield, respectively,

$$\frac{\sigma_B}{\sigma_F} \equiv \frac{\rho_F}{\rho_B} \approx 1 + \frac{3}{8\gamma} (1 - p) \quad (\gamma > 1) \tag{38}$$

and

$$\frac{\rho_F}{\rho_B} \approx \frac{4}{3} \frac{1 - p}{1 + p} \frac{1}{\gamma [\ln(1/\gamma) + 0.4228]} \approx \frac{4}{3} \frac{1}{\gamma (1 + 2p)} \frac{1}{\ln(1/\gamma)}$$

$$(\gamma \ll 1; p < 1) \tag{39}$$

Note that Eq. (39) is valid *only* for small p and $\gamma < 0.1$. Theoretical variation of ρ_F/ρ_B given by Eqs. (32) and (38) as a function of γ is shown in Fig. 4 for several values of p. The dotted curves are calculated from the limiting equation (38). A comparison of the two sets of curves shows that the approximate equation departs from the exact equation by a maximum of 7 percent for values of γ down to ~0.1. The deviation

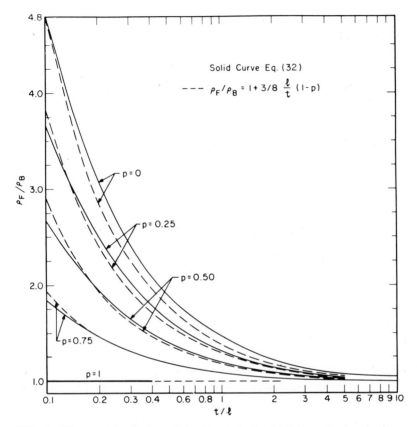

FIG. 4. Theoretical variation of the film-to-bulk resistivity ratio (ρ_F/ρ_B) with normalized thickness t/ℓ (ℓ is the mfp) for several values of the scattering coefficient p. The dotted curves represent the approximate form [Eq. (38)] of the exact equation in the limiting case of $\gamma \gg 1$.

becomes much smaller for $p > 0$. Thus Eq. (38) is a sufficiently accurate approximation for values of γ as low as 0.1 and therefore represents the practical range of most thin film experiments.

Size effects for wire geometry were calculated by MacDonald and Sarginson [52] for a square cross section, and by Dingle [53] for a circular cross section. For the latter case (taking t = diameter)

$$\frac{\rho_{wire}}{\rho_B} = 1 + \frac{3}{4\gamma}(1-p) \qquad (\gamma \gg 1)$$

(40)

and

$$\frac{\rho_{wire}}{\rho_B} = \frac{1-p}{1+p}\frac{1}{\gamma} \qquad (\gamma \ll 1; p < 1)$$

(41)

The calculated resistivity-ratio values for thin wires as a function of γ are given in Table I.

Fuchs' theory is based on the assumption that the value of p is the same for both surfaces. Lucas [45] obtained a general expression for the conductivity ratio, assuming different scattering coefficients p and q for the two surfaces, which is

$$\frac{\sigma_F}{\sigma_B} = 1 - \frac{3}{4\gamma} \int_1^\infty \left(\frac{1}{a^3} - \frac{1}{a^5}\right) \frac{1 - e^{-\gamma a}}{1 - pqe^{-2a}}$$
$$\times [2 - p - q + (p + q - 2pq)e^{-\gamma a}] \, da \tag{42}$$

The asymptotic forms of this equation are given by [54]

$$\frac{\sigma_F}{\sigma_B} = 1 - \frac{3}{8\gamma}\left(1 - \frac{p + q}{2}\right) \qquad (\gamma \gg 1) \tag{43}$$

and

$$\frac{\sigma_F}{\sigma_B} = \frac{3}{4}\frac{(1 + p)(1 + q)}{(1 - pq)} \gamma \ln \frac{1}{\gamma} \qquad (\gamma \ll 1) \tag{44}$$

Comparison of these equations with the corresponding Fuchs' equations indicates that p in the Fuchs theory is an effective parameter lying within

$$p_{\text{eff}} \approx \frac{p + q}{2} \text{ for } \gamma \gg 1 \text{ and } p_{\text{eff}} \approx \frac{p + q}{2} - \frac{[(p - q)/2]^2}{1 + (p + q)/2}$$

for $\gamma \ll 1$. Values of ρ_F/ρ_B computed from Eq. (42) for several values of p and q are given in Table I.

Several comments of experimental interest may now be made.

1. We can rewrite Eq. (38) for $\gamma \gg 1$ in the following form:

$$\rho_F = p\rho_{\text{Sp}} + (1 - p)\rho_{\text{Diff}} \tag{45}$$

where $\rho_{\text{Sp}} \equiv \rho_B$ is the specular-scattering resistivity and $\rho_{\text{Diff}} = \rho_B (1 + 3/8\gamma) = \rho_B + \rho_D$ is the resistivity in the diffuse-scattering case. Here, $\rho_D = (3/8\gamma)\rho_B$ is the surface-scattering contributions for $p = 0$. This form suggests a method for determining p by adding the known ρ_{Sp} and ρ_{Diff} values to obtain a fit to the observed ρ_F data.

Table I. Resistivity Ratio ρ_F/ρ_B and ρ_{wire}/ρ_B as a Function of Thickness/Mean Free Path (t/l) for Several Values of p and q

| | ρ_F/ρ_B | | | | | | | | | | ρ_{wire}/ρ_B | |
| | | | | | | | | | | | | |
t/l	$p\to0$ $q\to0.0$	0 0.5	0 1.0	0.25 0.25	0.25 0.5	0.25 1.0	0.5 0.5	0.5 1.0	0.75 0.75	0.75 1.0	$p=0$	$p=0.5$
0.001	179.0	128.0	99.5	118.0	96.2	66.3	73.1	41.4	36.7	21.2	1,000	337
0.002	99.5	71.5	55.8	66.3	54.1	37.6	41.4	23.8	21.2	12.5	503	170
0.005	46.6	33.8	26.5	31.5	25.9	18.2	20.0	11.8	10.6	6.51	202	69.3
0.01	26.5	19.4	15.3	18.2	15.1	10.8	11.8	7.18	6.51	4.16	102	35.7
0.02	15.3	11.4	9.09	10.8	9.01	6.55	7.18	4.53	4.16	2.80	52.1	18.7
0.05	7.73	5.91	4.78	5.63	4.81	3.62	3.95	2.67	2.50	1.83	21.6	8.31
0.1	4.78	3.76	3.10	3.62	3.15	2.46	2.67	1.93	1.83	1.45	11.45	4.88
0.2	3.10	2.52	2.13	2.46	2.20	1.79	1.93	1.50	1.45	1.24	6.33	3.02
0.5	1.92	1.66	1.46	1.64	1.52	1.33	1.40	1.21	1.19	1.10	3.14	1.84
1.0	1.462	1.331	1.221	1.327	1.266	1.159	1.206	1.101	1.098	1.049	2.04	1.422
2.0	1.221	1.159	1.103	1.159	1.130	1.075	1.101	1.049	1.049	1.024	1.475	1.208
5.0	1.081	1.060	1.039	1.060	1.049	1.029	1.039	1.019	1.019	1.010	1.172	1.080
10.0	1.0390	1.028	1.0191	1.0289	1.0240	1.0143	1.0191	1.0095	1.0095	1.0047	1.081	1.038
20.0	1.0191	1.014	1.0095	1.0143	1.0119	1.0071	1.0095	1.0047	1.0047	1.0023	1.039	1.0191
50.0	1.0076	1.005	1.0038	1.0057	1.0047	1.0028	1.0038	1.0019	1.0019	1.0009	1.0152	1.0076
100.0	1.0038	1.002	1.0019	1.0028	1.0023	1.0014	1.0019	1.0009	1.0009	1.0005	1.0076	1.0038

2. We may define an effective free path l_{eff} for the size effect so that the film conductivity may be written in analogy with Eq. (20) as

$$\sigma_F = \frac{1}{\rho_F} = \frac{ne^2 l_{eff}}{mv}$$

Thus, from Eqs. (38) and (39), we get

$$l_{eff} = \frac{l}{1 + (3/8\gamma)(1 - p)} \qquad (\gamma \gg 1) \tag{46}$$

and

$$l_{eff} = \frac{3}{4} \frac{1 + p}{1 - p} t \ln \frac{1}{\gamma} \qquad (\gamma \ll 1) \tag{47}$$

Note that for very thin wires ($t \ll l$), on the other hand, the free path is limited by the dimension of the wire diameter, that is, $l_{eff} = t$. This condition arises because of the more severe geometrical restrictions of free path in the case of a wire than in thin films. In films, the contribution to conduction is from electrons which move in directions nearly parallel to the surface so that their free paths remain of the order of the bulk free path.

3. In addition to film surfaces and the lattice, impurities and the enormous number of frozen-in structural defects in films will also scatter conduction electrons. According to Matthiessen's rule, various electron-scattering processes (and hence resistivity contributions) are additive, provided that the lattice scattering remains predominant. We can therefore write

$$\rho_F = \rho_B + \rho_S + \rho_I \tag{48}$$

where the contributions ρ_B, ρ_S, and ρ_I are due to the ideal lattice, surface scattering, and imperfections including impurities, respectively, Additional contributions to resistivity may arise because of scattering (a) by the pinned spins at a ferromagnetic surface [55], and (b) by electrostatically charged surfaces (Sec. 3.5).

Depending on the deposition process, the contribution ρ_I can be quite significant as compared with ρ_B and ρ_S. Since the frozen-in imperfections vary with the growth stages of a polycrystalline film, ρ_I is expected to show thickness dependence. If a film is susceptible to oxidation and contamination, further thickness-dependent effects result. This situation complicates the analysis and interpretation of the film resistivity vs. thickness data. It needs to be emphasized that Fuchs theory is applicable *only* to the thickness dependence arising out of the limitation of the free path by the geometrical boundaries

4. Conductivity size-effect expressions for films involve three transport parameters, e.g., p, l, and σ_B. Here the free path l pertains to an infinitely thick film with defect structure very similar to that of the thin film. A direct, simultaneous determination of these three parameters from the experimental data on the conductivity (or resistivity) vs. film thickness alone is clearly not possible. One normally attempts to fit the observed data with the theoretical curves for various combinations of the values of ρ_B, l, and p. One must assume, of course, the constancy of these parameters for all films, which is certainly difficult to justify since each film is expected to have its own characteristic structure and, therefore, characteristic values of these parameters.

The interpretation of the data is simpler in the two limiting cases. For $\gamma > 1$, a plot of ρ_F vs. $1/t$ should yield a straight line with the intercept as ρ_B and the slope = $\tfrac{3}{8}\rho_B l(1 - p)$ Thus, values of ρ_B and $l(1 - p)$ can be determined. Similarly for $\gamma \ll 1$, a plot of $1/\rho_F t$ vs. $\ln t$ gives a straight line with the intercept = $(3/4\rho_B)[(1 + p)/(1 - p)](\ln l/l)$, and the slope = $\tfrac{3}{4}(1/l\rho_B)[(1 + p)/(1 - p)]$. Thus, ρ_B, l, and $(1 + p)/(1 - p)$ can be determined. Since the region $\gamma \geq 1$ is not very sensitive to the choice of the value of p (see Fig. 4), no conclusion on the value of p may be derived from the data in this region. It should be noted that if the number of electrons per atom is known, l can be calculated from ρ_B [Eq. (22)], thus allowing a determination of p.

3.2 Experimental Results

The size effects in electrical conduction have been studied primarily to determine the mfp and surface-scattering parameters. Most studies have been carried out on monovalent metal films to which the free-electron theory is expected to be best applicable.

(1) **Thickness Dependence.** The pioneering works of Appleyard and Lovell [56] on alkali-metal film at 90°K were repeated and verified by Nossek [57]. A review of these results is given by Mayer [58]. A plot of $\rho_F t$ vs. t, shown in Fig. 5, yields straight lines for Na, K, Rb, and Cs

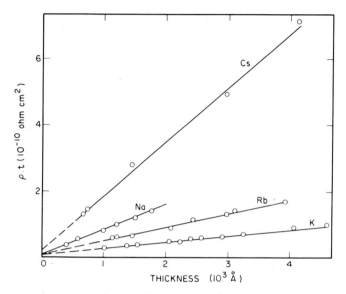

FIG. 5. A plot of ρt vs. t (film thickness) for Na, K, Rb, and Cs films condensed and measured at 90°K. Assuming $p = 0$, the straight lines yield 0.99, 0.81, 0.87, and 0.39 electrons per atom for the respective metals. (*Nossek* [57])

films, in agreement with Eq. (34). The values of $\rho_B l$ deduced from the intercept of the straight lines agree well with the theoretical values for all except Cs films. The departure observed for Cs films may be due to possible structural changes in these films. These results suggest that $p = 0$ is valid. By contrast, however, Cirkler's [59] data on Na films agree with the theory only if $p = 0.5$ is assumed.

Reynolds and Stilwell [60] studied fast-evaporated Cu and Ag films and obtained verification of the Eq. (34) and also of Matthiessen's rule. These results, as well as those of Nossek, did not consider the possibility of p being different from zero, which is actually not precluded by their data. Gillham et al. [61] observed little thickness dependence of the conductivity of continuous Au films sputtered onto bismuth oxide substrates and thus concluded the existence of specular reflection. This conclusion is hardly justified (as is also true of the results of Learn and Spriggs [62] on Pb and Sn films) because the condition for the size-effect regime, e.g., $\gamma < 1$, was not well satisfied to allow unambiguous conclusions to be drawn. Ennos [63] evaporated Au films onto bismuth oxide substrates and observed nearly bulk resistivity for film thicknesses for which $\gamma < 1$, thereby indicating the existence of specular reflection. On the other hand, the observed TCR of such films was only half of the bulk value, which is not consistent with the interpretation of specular reflection.

Detailed investigations of the conduction mechanism in the polycrystalline and epitaxially grown single-crystal films of Au and Ag were made by Chopra and coworkers [64–66]. Sputtered and evaporated films were grown epitaxially by deposition onto freshly cleaved mica substrates heated to about 300°C. The (111) crystallites of many microns in linear dimensions showed Kikuchi lines in the electron-reflection diffraction patterns, indicating surface smoothness on an atomic scale. The resistance of the films was measured in situ during film growth. The thickness variation of the resistivity of sputtered (S) and evaporated (E) Au films is shown in Fig. 6. As the films become physically continuous, the resistivity approaches to within about 10

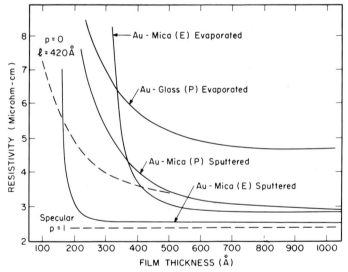

FIG. 6. Thickness variation of the electrical resistivity of evaporated and sputtered, polycrystalline (P) Au films deposited on glass and of epitaxial (E) Au films deposited on cleaved mica. Epitaxial films were obtained by depositing onto mica at ~300°C. Polycrystalline films were deposited at 25°C and were unannealed. Measurements were made in situ with continuous deposition. Experimental points have been eliminated for the sake of clarity. Dotted curves represent theoretical specular ($p = 1$) and diffuse ($p = 0$ for $\ell = 420$ Å) scattering behavior. (*Chopra et al.* [64])

percent of the bulk value, and is independent of the film thickness for thicknesses greater than 250 and 300 Å for S and E films, respectively. Since these values are smaller than the mfp of 420 Å obtained from the resistivity of the thick Au films, the condition of nearly perfect specular scattering must hold. Sputtered films, which become physically continuous [64, 67] at an earlier stage, exhibit a slightly lower resistivity than the evaporated films under epitaxial conditions.

The polycrystalline Au films deposited at low substrate temperatures, on the other hand, show size effects which follow the dotted theoretical curve for $l = 420$ Å. The departure of the data from the theoretical curve at thicknesses below 400 Å is due to the imperfections and voids in the films. Further annealing of such films was found to extend the agreement with the dotted curve down to 250 Å. Polycrystalline films deposited on glass substrates have resistivities much higher than the bulk value. The mfp thus being very short, no conclusions on the size effects are possible. Since the observed thickness variation in this case can be changed considerably by factors such as annealing and prenucleation of the substrate [64], the thickness dependence is certainly not due to a size effect. A similar situation holds for films where contamination continues with increasing deposition. The Fuchs theory is not expected to be valid for such cases. The observed deviations from the theory can be interpreted in terms of scattering by imperfections and impurities, as has been done by Young and Lewis [68] for Cr films, Cirkler [59] for Na films, and Marcus [69] for epitaxially grown Ta films.

Hirschhorn's [70] results on polycrystalline Ni films agreed with Eq. (34) for $p = 0$, but marked deviations from the Fuchs theory were reported by Crittenden and Hoffman [71]. The deviations for Fe films observed by Rosette and Hoffman [72] suggested a nonzero value of p. The results of Grigorovici et al. [73] on Al films agreed with the Fuchs theory.

(2) Low-temperature Results. Since the mfp increases at lower temperatures, relatively thicker specimens can be used to observe size effects. Andrew [74] studied the resistance of 3- to 2,000-μ-thick rolled Sn foils at 3.8°K. His results on Sn foils as well as Hg wires showed satisfactory agreement with the theory for diffuse scattering. Moreover, the number of conduction electrons per atom calculated from his data (0.43 and 0.15 for Sn and Hg, respectively) are entirely reasonable values even in this case of complicated multivalent metals. Similar studies by

MacDonald and Sarginson [52] on thin Na wires concluded that the condition of entirely diffuse scattering was not fulfilled.

Niebuhr [75] obtained interesting results on the thickness dependence of the resistivity of Sn films deposited on quartz at temperatures of 20, 90, 130, and 200°K. The results verified Matthiessen's rule and indicated that $p = 0$ for films deposited at 200°K but that $p > 0$ for films deposited at lower temperatures. An extensive study of the size effects in high-purity (and hence long mfp) bulk materials of Sn, Zn, Cd, Bi, Al, In, and Pb at 1.65 to 4.2°K was made by Aleksandrov [76]. His results suggested a diffuse scattering condition. But the value of γ obtained was not low enough to justify a definitive conclusion.

Chopra and Bobb [65] studied the ratio $\rho_{300°K}/\rho_{78°K}$ as a function of the thickness of the Au films of Fig. 6 which exhibit specular and diffuse scattering behavior. Clearly, the size effects are enhanced at low temperatures and thus allow a more accurate determination of p. A $p = 0.8$ was found by fitting the experimental data to a theoretical curve for epitaxial Au films. Larson and Boiko [77] studied the resistivity ratio $(RR) = \rho_{300°K}/\rho_{4.2°K}$ for thick epitaxial Ag films. At 300°K, $\gamma \gg 1$, and at 4.2°K, $\gamma \ll 1$ is satisfied for these films. Thus, from Eqs. (38) and (39);

$$\frac{(RR)_F}{t} = \frac{3}{4l} (RR)_B \frac{1 + p}{1 - p} \ln \frac{1}{\gamma} \tag{49}$$

By fitting the data to this relation, a value of $l_{eff} = 31.2\mu$ and $p = 0.5$ was deduced.

Specular reflection in high-purity Cu whiskers is shown by Isaeva's [78] data (Fig. 7). Here, $\rho_F \approx \rho_B$ at 300°K and $\gamma < 1$ at 4.2°K, so that instead of Eq. (49), we get $(RR)_F = (RR)_B /[1 + (3/8)(l/t)(1 - p)]$. A comparison of the experimental and theoretical plots of $(RR)_F$ vs. $1/t$ (Fig. 7) yields $p = 0.6$. Similarly Skove and Stillwell [79] obtained evidence of specular reflection in Zn whiskers.

Size effects in a film can also be studied by varying the mfp instead of the film thickness. This is achieved by a study of the temperature dependence of the film resistivity. At low temperatures ($T < \theta_D$), however, the mfp of conduction electrons in thin films may not be the same as that in bulk. Further, because of the longer free path available at low temperatures, small-angle electron-phonon scattering is very effective

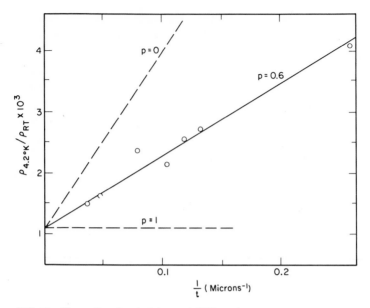

FIG. 7. The ratio of resistivity at 4.2°K to that at room temperature for Cu whiskers as a function of the inverse thickness. The $p = 0$, and 0.6 curves are calculated from Eq. (38) using $\rho_F \approx \rho_B$ at 300°K and $\rho_B \ell_B = 0.65 \times 10^{-11}$ ohm-cm^2. (*Isaeva* [78])

in deflecting the electron toward the geometrical surface for further diffuse scattering. This electron-phonon-assisted diffuse boundary scattering produces an additional size effect and was first suggested by Olsen [80] to explain his data on In wires. The additional scattering occurs as a deviation from Matthiessen's rule and is characterized by both a thickness and a temperature dependence of the resistivity in the temperature range where normally constant residual resistivity is obtained.

A T^3 dependence is expected from elementary considerations of the density of phonons. Blatt and Satz [81] treated this problem for a wire geometry and obtained a$(T^{2.3} t^{-2/3})$ dependence for the conductivity ($t =$ wire diameter). The thin film case [82-84] shows a complex behavior. Azbel and Gurzhi [83] suggested that at low temperatures l_B in the size-effect equations should be replaced by an "effective" thin film value l_F given by $1/l_F = 1/l_I + (1/l_L)(\theta_D/T)^2$, where I and L refer to impurity and lattice scattering and θ_D is the Debye temperature. This substitution results in a complex log T (or a power series of T) dependence of the low-temperature film resistivity. Gurzhi [84] further showed that the resistance of a film should show a shallow minimum at sufficiently low temperatures, followed by a rise to the saturation value.

The results on Hg [77] and In [80] wires show fair agreement with the theory of Blatt and Satz. Results of Gaide and Wyder [85] on temperature-dependent size-effect variation for In foils are shown in Fig. 8 and are claimed to be in agreement with the Monte-Carlo calculations based on Olsen's kinetic scattering process described above. Chopra [86] compared the temperature variation of the resistance of a 1.5-μ-thick polycrystalline film with an epitaxial Ag film of the same thickness. The linear temperature dependence was followed by a T^n variation in both cases below about 40°K. The value of n was between 2 and 3 for the polycrystalline film and 4 and 5 for the epitaxial film. The reported rapid decrease of resistivity below 10°K was not observed in more

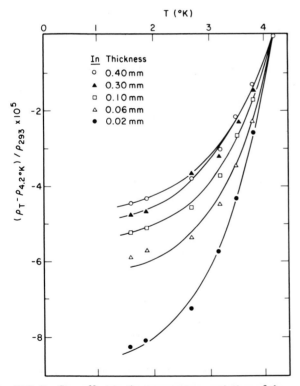

FIG. 8. Size effect in the temperature variation of the relative change in the resistivity below 4.2°K for In foils of different thicknesses. The data points fit the ballistic Monte Carlo calculations of electron-phonon-assisted diffuse boundary scattering. (*Gaide and Wyder* [85])

careful experiments. Since the electron-phonon-assisted enhanced temperature dependence should disappear for specular scattering, and further a T^5 dependence should be observable in pure bulk (specular), the observed higher temperature dependence in the epitaxial films may be attributed to the existence of partial specular reflection. Because the power index in the temperature dependence is not well defined over a reasonable range of temperature, a quantitative determination of the value of p is difficult.

(3) Size-effect Anisotropy. Since the main contribution to the film conductivity comes from electrons traveling nearly parallel to the plane, within the angle $\sin^{-1}(t/l)$, the conductivity should depend not only on the direction of current flow but also on the orientation of the surface relative to the crystal. Kaganov and Azbel [87], and Engleman and Sondheimer [88] treated this problem theoretically and showed that a study of the anisotropy would yield information about the Fermi surface of the metal. Clearly, a determination of the anisotropy would be possible only with thick single-crystal films of high purity. The evidence of large anisotropy in the size effects has been found [76] in high-purity bulk Sn, Zn, and Cd specimens but there are at present no data on films.

(4) Magnetic Boundary Scattering. A contribution to the electrical resistance of ferromagnetic metals may be provided by the scattering of electrons by magnetic domains and disordered spins. Since, in thin ferromagnetic films, the ratio of the disordered spins pinned at the surface to the free spins in the bulk is a function of thickness (see Chap. X, Sec. 2), the conductivity should exhibit a size effect. Results of Schüler [55] on Gd films above and below the Curie temperature verified the spin contribution; however, owing to the lack of systematic thickness-dependence studies, the existence of a size effect has not been established.

(5) Superimposed Films. Lucas [89] observed an increase ($\sim$10 percent) in the resistance of specularly scattering, annealed Au films when an unannealed overlay film of Au, or another material, was deposited onto the base film. The observation was attributed to the roughening of the surface of the specular film and hence increased diffuse surface scattering of electrons. This conclusion may be questioned since the statistical process of condensation onto a continuous film should not result in any unusual increase in roughness with the deposit of an additional few atomic layers of the same or another material. The observed resistance increase is too large to be caused by the small interfacial diffusion of the overlay at ~25° C. The more likely explanations of the observation are: (1) microstructural changes in the partially continuous 80-Å-thick Au films used on further deposition, and (2) changes in the amount of scattering at the new surface.

Because of the importance of the electronic behavior of interfaces in thin film studies, Chopra and Randlett [90] investigated the influence of a superimposed overlay on the resistance of the base films of Au, Ag, Cu, and Al. The results can be summarized as follows:

1. Little change is observed in the resistance of a continuous base film (>100 Å thick) with further deposit of the *same* material. The effect observed by Lucas occurs only in discontinuous films.

2. The resistance of *both* specular and diffuse scattering base films changes because of an overlay of semiconducting, insulating, or a higher-resistivity metal film. The magnitude and the sign of the change depend on the base-overlay combination but not noticeably on the specularity parameter. The maximum change occurs in most cases for an overlay deposit of only a few atomic layers. Figure 9 shows typical curves for changes in the resistance of several underlayer films as a function of the thickness of the overlay materials. The resistance of Au films increases with an overlay of a film of Permalloy as well as SiO. In Ag films, Ge overlay increases the resistance but SiO has negligible effect. The resistance of Cu films first increases and then decreases to a constant value with the deposition of Ge and Cr films. As in the case of Ag films, SiO overlay has a negligible effect on Cu films. When the effects are small (<2 percent), the sign of the effect may change with increasing thickness of the base film. Since some annealing may take place with further deposition, no significance can be attached to this reversal.

The increase in the base film resistance shows a thickness dependence of the type shown in Fig. 10 for Au/SiO and Au/Permalloy super-imposed films. The thickness dependence of the resistivity of Ag/Ge films with and without overlays fits the Fuchs equation [Eq. (38)], from which a value $l(1 - p) = 580$ Å is calculated. Since the theoretical bulk mfp for Ag = 550 Å, $p \sim 0$ is suggested. If diffuse scattering already exists, the observed thickness-dependent increase in the resistance of superimposed films cannot be explained by a further increase in the value of l or $(1 - p)$. An explanation was proposed by Chopra and Randlett in terms of an additional surface-scattering mechanism caused

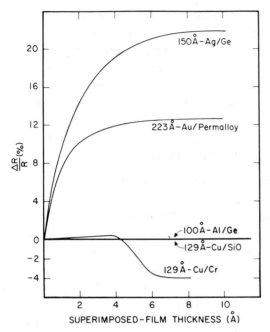

FIG. 9. Percentage change in the resistance of thin polycrystalline Au, Ag, Cu, and Al films as a function of the thickness of a superimposed layer of Permalloy on Au, Ge on Ag and Al, and Cr and SiO on Cu. (*Chopra and Randlett* [90])

by the fact that deposition of the overlay material modifies the electrical nature of the surface of the base film. It should be pointed out that a similar model has also been proposed for resistance changes in a film due to gas adsorption (Sec. 3.6). Depending on the combination of the base and overlay materials, the surface may be left unchanged or may be modified electronically so that the amount of scattering of electrons is enhanced or reduced. This additional scattering will naturally be a function of t/l. The results on Ag/Ge films may be explained by a 25 percent increase in the surface-scattering contribution. Note that this procedure is mathematically equivalent to increasing l or using a negative value of p.

It must be pointed out that the above-mentioned effects were observed for both diffuse and specular films, and the results therefore cannot be explained in terms of changes in p. Further, since these effects are observed at room temperature or below, interfacial alloying is not

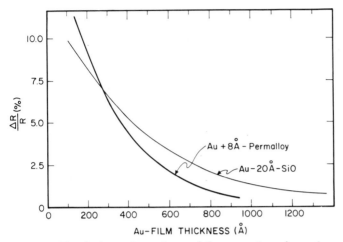

FIG. 10. The thickness dependence of the percentage change in the resistance of polycrystalline Au films with superimposed 8- and 20-Å layers of Permalloy and SiO, respectively. (*Chopra and Randlett* [90])

expected to affect seriously the conductivity of the relatively thicker base layer. The influence of alloying is further discounted by the fact that subsequent annealing at ~200°C of the composite film tends to reduce rather than increase the resistivity in most cases.

3.3 Temperature Coefficient of Resistivity (TCR)

(1) Theory. If the temperature variation of the free path is the same as that of the conductivity (which is strictly valid for $T > \theta_D$), then

$$\frac{1}{\sigma_B} \frac{d\sigma_B}{dT} = \frac{1}{l} \frac{dl}{dT}$$

Using this relation, the general expression [91] for the TCR of a film,

$$\alpha_F = -\frac{1}{\sigma_F} \frac{d\sigma_F}{dT}$$

may be derived from Eq. (32), as

$$\frac{\alpha_F}{\alpha_B} = 1 - \phi(\gamma) \frac{d[\gamma/\phi(\gamma)]}{d\gamma} \tag{50}$$

The calculated values of α_F/α_B vs. γ for several values of p are shown in Fig. 11.

Let us note that, in calculating temperature coefficient of resistivity α_ρ we generally assume that it is equal to the temperature coefficient of resistance α_R. This is valid only if the thermal-expansion coefficient α_T is negligible, since $\alpha_\rho = \alpha_R + \alpha_T$. For pure bulk metals, $\alpha_T < 10^{-2}\alpha_\rho$ so that $\alpha_\rho = \alpha_R$ holds. The expansion-coefficient correction can be significant [280] for thin granular and high-resistivity films having small values of α_R. The magnitude of the correction depends on the differential thermal expansion of the film with respect to the substrate material and the strain coefficient (Sec. 2.5).

By using the conductivity expressions for the limiting cases of very thick and thin films, we can conveniently derive the following expressions for the TCR:

$$\frac{\alpha_F}{\alpha_B} \approx 1 - \frac{3}{8}\frac{1-p}{\gamma} \quad (\gamma \gg 1) \tag{51}$$

As noted already, this form is reasonably accurate down to $\gamma = 0.1$.

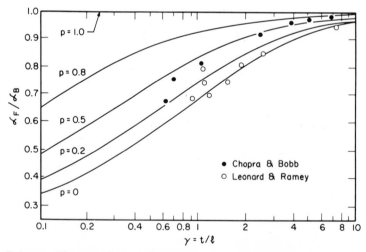

FIG. 11. Theoretical ratio of the film to the bulk TCR as a function of the reduced thickness $\gamma = t/\ell$. The open circles are the data of Leonard and Ramey [93] on polycrystalline Au films, and the full circles refer to the data of Chopra and Bobb [65] on sputtered, epitaxial Au films.

Further,

$$\frac{\alpha_F}{\alpha_B} \approx \frac{1}{\ln(1/\gamma) + 0.4228} \approx \frac{1}{\ln(1/\gamma)} \qquad (\gamma \ll 1; p < 1) \tag{52}$$

Note that Eq. (52) is independent of p *only* if p is small.

If the temperature-independent contribution ρ_I is significant as compared with ρ_S and ρ_B in Eq. (48), then the calculated value of the TCR = $[1/(\rho_S + \rho_B + \rho_I)](d\rho_F/dT)$ is considerably lower than that expected on the basis of the size-effect theory. The interpretation of the data on the basis of the size-effect theory is thus complicated, as was the case in the conductivity expression.

From Eq. (48), we can write

$$\frac{d\rho_F}{dT} = \frac{d\rho_B}{dT} + \frac{d\rho_S}{dT}$$

i.e.,

$$\alpha_F \rho_F = \alpha_B \rho_B + \alpha_S \rho_S \tag{53}$$

which in the limiting case of $\gamma \gg 1$ reduces to

$$\alpha_F \rho_F = \alpha_B \rho_B \tag{54}$$

This is an alternate expression for Matthiessen's rule, and note that it is independent of p. For $\gamma \ll 1$, deviations from Matthiessen's rule are expected for all values of $p < 1$ because of the size effects.

Equation (54) suggests a convenient way of determining ρ_B from the measured values of α_F, ρ_F, and α_B, for the $\gamma \gg 1$ case. But the value of p cannot be deduced in this range since size effect is insensitive to the value of p. Similarly, the size dependence of the product $\alpha_F \rho_F$ is considerably reduced as compared with that of either α_F or ρ_F and is thus not suitable for determining p.

(2) Experimental Results. Crittenden and Hoffman [71] measured the TCR of ultrathin Ni films and found marked deviations from the theoretical relation [Eq. (50)]. This result is probably a consequence of the errors in the thickness of the extremely thin films employed. Savornin's [92] results on Co films were essentially in agreement with

Eq. (51) for $p = 0$. Reynolds and Stillwell [60] verified Matthiessen's rule for polycrystalline Cu and Ag films. The results of Rosette and Hoffman [72] on Fe films showed disagreement with the diffuse-scattering theory although a better fit of the data could be obtained if $p > 0$ is assumed.

Nossek [57] made extensive measurements of the TCR of alkali films (Na, K, Cs, and Rb) at 90 and $60°K$. The general variation of the TCR in the range $\gamma < 1$ is in qualitative agreement with the diffuse-scattering theory. Quantitatively, the observed values fall off faster than predicted. One possible explanation for this behavior is the enhanced temperature dependence of the film conductivity at temperatures below θ_D discussed already.

Chopra and Bobb [65] measured the TCR of films for which the resistivity data are already given in Fig. 6. The results on polycrystalline Au films agree with the theory for $p = 0$. The comparison of the data on epitaxially grown Au films with the theoretical curves (Fig. 11), on the other hand, suggests a partial specular behavior. The value of p for the best fit is 0.5, and for $\gamma > 2$, this value increases to 0.8. Leonard and Ramey [93] showed that both their data on polycrystalline Au, Ag, and Co films, and those of others available in the literature fit Eq. (51) for $p = 0$. Their results for Au films are also shown in Fig. 11 for comparison with epitaxial Au films.

3.4 Specular Scattering

The specularity parameter p introduced by Fuchs is a measure of the size-effect *deviation* from the bulk behavior, which ranges from zero for perfect specular scattering ($p = 1$) to the maximum value for perfect diffuse scattering ($p = 0$). For ellipsoidal energy surfaces [48], in contrast to spherical surfaces, size effects are expected owing to the anisotropy of the mfp even when $p = 1$.

Although most early studies indicated $p = 0$, the existence of partial specular ($p > 0$) scattering at surfaces is now well established for films and whiskers of several bulk metals (Table II), Bi [94], Ge, and Si [95]. However, the origin of nonspecular scattering is still not clear. One would expect [96] diffuse scattering due to (1) any physical deviations from an ideal surface, (2) the presence of atomic disorder and impurities on the surface, (3) phonons associated with surface waves, (4) termination of the charge carrier-wave function at the surface, (5) the charge gradient due to the presence of surface states and the fact that

Table II. Some Electrical Data for Bulk and Thin Specimens of Several Metals at about 20°C (Unless given Otherwise in Parentheses) The bulk mfp is calculated from the bulk resistivity data (see Ref. 251). The measured mfp and p (scattering coefficient) are calculated from the size-dependent resistivity data on specimens listed in the last column

Metal	Assumed free electron per atom	Bulk resistivity, $\mu\Omega$ – cm	Bulk TCR, °C^{-1}	Bulk mfp, Å	Measured mfp, Å	p	Specimens (refs. in brackets)
Na	1	4.8 / 0.276 (90°K)		305	820 (90°K)	0.45	Films [59]
K	1	7.3 / 2 (90°K)		346	1,220 (90°K)	?	Films [57]
Cs	1	22 / 17 (90°K)		145	400 (90°K)	?	Films [57]
Cu	1	1.67	0.0068	387	450	0	Polycrystalline films [60]
					390	>0	Epitaxial films [223]
					34,000 (4.2°K)	0.6	Whiskers [78]
Au	1	2.44	0.0034	380	400	0	Polycrystalline films [64]
					420	0.8	Epitaxial
					360	0.5	Polycrystalline films [106]
Ag	1	1.59	0.0041	523	31,000 (4.2°K)	0.5	Epitaxial films [77]
					26,800 (4.2°K)	~1	Epitaxial films [105]
Al	3	2.65	0.00429	380 (?)	169 (273°K)	?	Films [73]
Co	0.7	7.7	0.006	117	100	?	Films [92]
Sn	0.4	11.5	0.0047		550 (200°K)	0.69	Films [75]
					(90°K)	0.38	
					9.5×10^5 (3.8°K)	0	Wire [74]
Fe	0.2	9.7	0.0065	192		?	Films [72]
Ni	0.6	6.8	0.0069	110		0	Films [70]
In	3	~9.3	0.0047		202 (4.2°K)	>0	Wire [211]

more electrons at the surface move parallel to it than in the bulk, and (6) the variation of the number of scattering events with the angle of incidence of the carriers at the surface.

As introduced by Fuchs, the specularity parameter does not depend on electron wavelength and the angle of incidence (at the surface) of carriers and is essentially an empirical constant. The assumption of a bulk relaxation time right up to the surface in Fuchs' formalism is undoubtedly a grave oversimplification. Electron states at the surface are perturbed by any one of the factors listed above, and perturbations by the corresponding scattering mechanisms should therefore be considered in determining the surface scattering. Greene [97] considered some aspects of this problem and obtained an expression for the so-called "kinetic specularity parameter" in terms of a general scattering process. Greene and O'Donnell [98] calculated p for scattering by localized surface charges and showed that p can have a strong angular dependence because of the anisotropy of the scattering amplitude. This effect should be negligible for metal surfaces, owing to the large carrier concentration. Stern [99] showed that the forced movements of the carriers parallel to the surface due to the physical boundary may be an important factor in determining the surface scattering.

By analogy with the reflection of light at an interface, Parrott [100], and Brändli and Cotti [101] considered a model based on a certain angular dependence for p. In simple terms, they assumed that if the angle of incidence of carriers with respect to the surface normal is below a critical value, say θ_0, then $p = 1$; otherwise $p = 0$. This model would obviously lead to saturation of the size-effect conductivity for small values of t/l.

Smith [102] suggested that a diffuse scattering may be the result of the surface being rough as compared with the wavelength of the carriers. The observed specular behavior in thin Bi platelets was attributed to the fact that the surface smoothness in this case was comparable with the long wavelength ($\sim 100 \text{ Å}$) of carriers in Bi. Since the wavelength of electrons in metals is only of atomic dimensions ($\sim 5 \text{ Å}$ for Au) and further, specimen surface smoothness is expected to approach this state only under ideal conditions, metal surfaces should not be expected to exhibit specular-scattering behavior according to this model.

A surface-roughness model was analyzed by Ziman [40] to obtain theoretical dependence of the specularity parameter on the height h of the surface asperities relative to a wavelength λ of the carriers. Assuming a Gaussian height probability distribution and tangential correlation, the value of p was found to decrease rapidly when $h \sim \lambda$ was satisfied. This result is what one would expect from simple physical considerations. Soffer [103] extended Ziman's treatment to include flux conservation and oblique-incidence conditions. The analysis showed that diffuse scattering is anisotropic and that p varies with the angle of incidence θ of carriers as $p = \exp[-16\pi^2(h^2/\lambda^2)\cos^2\theta]$.

A quantitative comparison of the size-effect resistivity as derived from the theories of Fuchs, Parrott-Cotti, and Soffer (for zero correlation) is shown in Fig. 12. In contrast to the Fuchs model, other models indicate a saturation of the size-effect resistivity for $t/l \ll 1$, that is, with increasing value of p. All models agree with each other in both the extreme rough and smooth limits.

The observed saturation behavior of size-effect conductivity in specular Bi platelets has been previously explained on the basis of an anisotropic Fermi surface. An alternative explanation is provided [104] by the Parrott-Cotti model. A direct support for the angular dependence of the p model was suggested by Chopra's [105] results on the longitudinal magnetoresistance of Ag films (Sec. 4.2). It was found that for $t/l < 0.5$ (that is, angles of incidence less than $60°$), Ag films exhibited partial specular scattering ($p > 0$). Abeles and Theye [106] combined resistivity and optical data to determine [107] (discussed later) the l and p of annealed polycrystalline Au films deposited on quartz. Their results (Fig. 13) show an increasing value of p with increasing l. These data also suggest an angular dependence of p. However, the functional dependence cannot be determined since the film thicknesses are not given.

The occurrence of specular or partial specular scattering in films of several metals (see Table II) is now a well-established fact, and it implies that surface roughness cannot be the dominating factor in determining p. Bennett and Bennett [108] deduced $p = 1$ for polycrystalline Au and Ag films deposited on supersmooth quartz from the optical-reflectivity data (Fig. 15, Chap. XI). The electrical resistivity of these films, however, was considerably higher than the bulk value, and the thickness dependence corresponded to $p = 0$. Bennett et al. [109] studied changes in the optical reflectivity with increasing film surface roughness. The latter was obtained by depositing Ag films on different thicknesses of CaF_2 films (having platelet-like microstructure) which, in turn, were deposited on

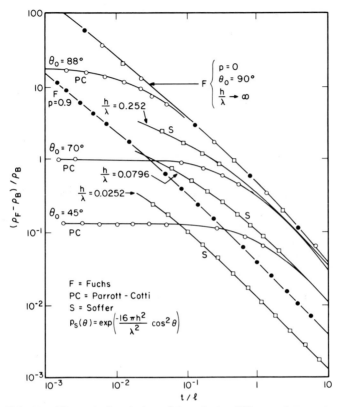

FIG. 12. Theoretical variation of the relative difference between the film and the bulk resistivities as a function of t/ℓ for: Fuchs' diffuse-boundary-scattering model; Parrott-Cotti angular-dependence model with various critical scattering angles (below which scattering is specular and above which it is diffuse); and Soffer's surface-roughness model for zero correlation. Note the trend toward saturation of the size-effect variation for both the angular-dependence model and the surface-roughness model. (*Data compiled by Soffer* [103])

supersmooth quartz. After the theoretically expected diffuse scattering of light caused by surface roughness was taken into account, the remaining decrease in the reflectivity was attributed to changes in the value of p. A very rapid decrease from $p = 1$ to $p = 0$ was obtained for surface roughness $\gtrsim 45$ Å rms. It was necessary to assign a negative value to p for rougher films in order to account for the observed decrease, which was larger than that expected theoretically for a total change in p of unity. The inconsistency between the conclusions on surface

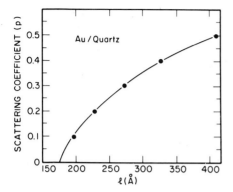

FIG. 13. Dependence of the surface scattering coefficient on the mean free path in annealed polycrystalline Au films deposited on amorphous quartz. (*Abeles and Theye* [106])

scattering of electrons provided by the resistivity data on the one hand and optical data on the other hand suggests that *either* p has a different physical meaning for these two cases, *or* the interpretation of the data by Bennett et al. lacks considerations of other unknown factors responsible for the decrease in the reflectivity. The former is a more likely explanation in view of the differences in the transport kinetics of the carriers relative to the film surface in the two cases, and the possible frequency dependence [110] of the carrier relaxation time.

Methods of Determining p **and** l**.** The size-effect conductivity equations involve three transport parameters ρ_B, l, and p. It is clear that data on the thickness dependence of either ρ_F or α_F alone cannot yield unambiguous values of these parameters. Generally, one assumes the constancy of these parameters for films of different thicknesses and then compares the observed variation with the theoretical one for different values of these parameters to obtain the best fit. Each specimen has a characteristic microstructure and, therefore, its own value of these parameters. A meaningful determination of these parameters is therefore possible only by measuring, *on the same specimen,* three or more size-dependent physical quantities, such as resistivity, TCR, temperature dependence of resistivity, magnetoresistance, optical constants, and anomalous skin effect. No such studies have been reported in the literature so far. Some attempts to measure two physical quantities on the same films form the basis of the following methods.

1. For thick films ($\gamma \gg 1$), the product $\rho_F \alpha_F$ provides a good method for determining l. Young and Lewis [68] applied it to Au films.

2. If a film is thick enough so that $\gamma \gg 1$ at room temperature but $\gamma \ll 1$ holds at helium temperatures, then the resistivity ratio (RR) is given by Eq. (49). The ratio of the intercept to the slope of the $(1/t)(RR)_F$ vs. $\ln t$ straight line determines the value of l, and hence p can be calculated, provided ρ_B is known from the thickness dependence of the resistivity. This method was used by Larson and Boiko [77] and Isaeva [78].

3. The variation of p with γ is given by Eq. (38) as

$$p = 1 + \frac{8}{3}\gamma\left(\frac{\rho_B}{\rho_F} - 1\right) \quad (\gamma \gg 1) \tag{55}$$

Another relation between p and l, derived by Dingle [107] in the anomalous skin-effect region, is

$$\frac{1}{l} = 2\pi\left(\frac{c}{v}\frac{1}{\lambda_\tau} - \frac{3}{8}\frac{1-p}{\lambda_0}\right) \tag{56}$$

where v and c are the velocity of electrons at the Fermi surface and the speed of light, respectively. The constants λ_0 and λ_τ are related to the optical constants n and k by

$$n^2 - k^2 = A - \frac{(\lambda/\lambda_0)^2}{1 + (\lambda/\lambda_\tau)^2} \qquad 2nk = \frac{(\lambda/\lambda_0)^2(\lambda/\lambda_\tau)}{1 + (\lambda/\lambda_\tau)^2} \tag{57}$$

Here, A is a constant. Thus, by measuring ρ_F and the optical constants on the same films (assuming p and ρ_B are the same for all films and, further, that p is independent of the optical frequency), one can determine [106, 107] p and l from the simultaneous graphical solution of Eqs. (55) and (56). Abeles and Theye [106] used this method to determine p and l in Au films (Fig. 13).

4. The eddy-current decay time in thin films may be used to determine the film resistivity and hence the mfp. This method (see Sec.

4.7) is convenient only for $\gamma > 1$ and becomes complicated for analysis if $p > 0$.

5. Galvanomagnetomorphic effects in materials with small bulk contributions offer the best promise for an accurate determination of p. The lack of a generalized theory for an arbitrary Fermi surface, anisotropic relaxation time, and an arbitrary value of p makes it difficult, however, to analyze the data. Chopra [105] used a phenomenological approach to analyze the longitudinal magnetoresistance size-effect data on Ag films to determine p (Sec. 4.2). Hall-effect studies (Sec. 4.5) can also be used to estimate l and p.

6. Among other specialized measurements which allow a determination of p are those of the anomalous skin effect [44] (Sec. 4.6) and optical reflectivity in far infrared [108-110] (Chap. XI, Sec. 3.3).

Table II lists values of bulk α and l (experimental), and l [calculated from Eq. (20)], for several metals. These bulk values are clearly the highest values that an ideal film with perfect specular scattering ($p = 1$) would yield. Some values of l and p obtained from the data on thin specimens are listed for comparison.

3.5 Field Effect

The term "field effect" describes the change in the sample conductance due to the electrostatic charges induced by an electric field applied normal to the sample surface. For a semiconductor surface, this charge is distributed among the surface states and the space-charge region, and the large resulting changes provide a powerful tool for surface studies.

If induced electrostatic charges δQ are distributed uniformly, the change in the electrical conductance of a metal sample is given by

$$\delta\sigma = \mu\,\delta Q \tag{58}$$

where μ is the bulk mobility. Since the screening distance [111] of the electric field in metals is of the order of atomic dimensions, the field effect should essentially correspond to a field-induced modification of the interaction of the conduction electrons with the surface-potential barrier. The effect would be significant only if the surface-scattering contribution to the conductivity is considerable as in the case of a film of thickness comparable in magnitude with the free path. Thus, the field effect would appear as a modification of the already discussed conductivity size effect.

Juretschke [112, 113] analyzed the metallic field effect and later extended the theory to include angular dependence of the specularity parameter. He calculated the field-induced changes and their temperature variation with the density of the carriers, effective film thickness, and scattering coefficient.

The mathematical treatment followed that of Schrieffer [114] for nonuniform electron density as applied to semiconductors. The calculated field effect appears as the sum total of the various contributions. The separation of the individual contributions from the data may be difficult, if not impossible.

Juretschke's theory predicts the effective mobility for the added charge carriers to be higher than the mobility of the charge-carrier population already present, and to increase with increased surface diffuse scattering. This prediction goes against intuitive expectations as pointed out by McIrvine [115], who solved the Boltzmann transport equation to obtain the general expression for the effective conductivity in the case of partially diffuse boundary scattering. He obtained the following simplified expressions in the limiting cases:

$$\frac{\delta\sigma_F}{\sigma_B} = \frac{\delta n}{n} \qquad (p = 1) \tag{59}$$

$$\frac{\delta\sigma_F}{\sigma_B} = \frac{\delta n}{n} \frac{1}{2}\left[1 + \frac{1}{ql}\log(1 + ql)\right] \qquad (p = 0, \gamma \gg 1) \tag{60a}$$

and

$$\frac{\delta\sigma_F}{\sigma_B} = \frac{\delta n}{n} \frac{\gamma}{2}\ln\frac{1}{\gamma} \qquad (p = 0, \gamma \ll 1) \tag{60b}$$

where $\delta\sigma_F$ is the change in the conductivity of the film due to an increase in the electron density δn, and $q = (6\pi ne^2/\zeta)^{1/2}$ is the screening length for a metal with Fermi energy ζ.

Attempts to observe the field effect date back to 1906 with inconclusive results [116]. The effect was later observed by Deubner and Rambke [117], Bonfigioli and Malvano [118], and Glover and Sherrill [119]. The magnitude of the effect is small, as expected. With a breakdown strength of 10^7 V/cm for the insulation of a capacitor, the maximum possible surface charge density is 4×10^{13} electrons/cm^2.

Assuming one conduction electron per atom, this charge density will result in a relative change of $\sim 10^{-3}$ in the conductivity of 35 atom layers (~ 100 Å thick). This being a small change, much care is required in such measurements. These measurements of field effect are carried out by applying a transverse field to the film which forms the electrode of a capacitor with a thin mica substrate acting as the dielectric medium.

Normal effect, i.e., negative charging resulting in a linear increase in conductivity with the amount of charge added, has been observed for Au [119] and Ag [120] films. Since no temperature dependence was observed for Au films over a wide range (300 to 4.2°K), thermally activated surface effects cannot be important. The observed linear relationship suggests the dominance of the effective thickness term in Juretschke's theory, which means that the field effect primarily measures the change in the surface barrier under an applied transverse field. The contributions from other parameters cannot be ruled out until detailed measurements on fully oriented interfaces for both the metal deposit and the substrate are made over a wide temperature range.

In contrast to Au and Ag films, Sn films exhibit a reverse effect which is strongly temperature-dependent. Indium films show (Fig. 14) a slight temperature dependence, but the magnitude and the sign of the effect are erratic and vary from film to film. This variation is probably due to structural changes arising from the annealing process used.

An interesting observation of the charge effect on Au films deposited on ferroelectric $BaTiO_3$ and TGS is due to Stadler [121]. Since the

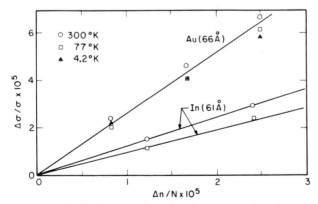

FIG. 14. Field-effect conductivity changes as a function of negative charging of 66-Å-thick Au and 61-Å-thick In films at various temperatures. (*Glover and Sherrill* [119])

polarization field for $BaTiO_3$ is about 3×10^8 V/cm, which is up to two orders of magnitude higher than the breakdown strength of common dielectrics, much higher charges can be induced by reversing the polarization of a $BaTiO_3$ capacitor than by applying the maximum field to a dielectric. Thus, charges of $\sim 3 \times 10^{14}$ electrons/cm^2 may be induced. This method of charging also allows field effect to be measured when the polarization is reversed with no electric field applied to the dielectric. A maximum effect of about 2 percent was found for a 100-Å Au film deposited on $BaTiO_3$. The effect is of right polarity but is considerably higher than that predicted by a uniform charge distribution. That the observed conductivity change varies inversely with film thickness suggests nonuniform charge distribution, which contradicts the conclusions of Glover and Sherrill [119].

3.6   Influence of Absorption and Adsorption on Conductivity

The electrical resistance of metal films changes because of adsorption and absorption of gases and other impurities. A study of such changes may throw light on the activity of a clean and smooth surface of a vapor-deposited film, and also on the nature of the adsorption phenomenon.

Beeck et al. [122, 123] compared the resistivity of vacuum-deposited Ni films with those evaporated and sputtered in Ar and N_2 ambients. The resistance of films deposited in an Ar atmosphere was generally higher than that of the films deposited in vacuum. Similar results for Hg films deposited in the presence of H_2, O_2, and Ar were obtained by Johnson and Starkey [124]. Ingersoll and Hanewalt [125] deposited Ni-Fe films in He, Ar, N_2, and H_2 and showed that the films contained considerable gas which was occluded on annealing. Films prepared by evaporation and sputtering in a 0.5-mm pressure of Ar exhibited irreversible decrease of the resistance on annealing because of the occlusion of the gas. The changes were more marked for sputtered than for evaporated films. Similar observations were also made by Itterbeek et al. [126] on Sn films.

Since the sputtering process enhances absorption and adsorption of impurities present in the system, much care is needed to obtain reliable data for sputtered films. With improved techniques, it is now possible to prepare Ar sputtered metal films at elevated temperatures with resistivities comparable with if not better than those of evaporated films, thereby suggesting that absorption and adsorption effects of Ar are not significant under suitable conditions.

Adsorption of neutral Ar atoms during the sputtering process is expected to be negligible. Owing to their high kinetic energy [127], however, Ar ions may be strongly adsorbed on film surfaces. Winters and Kay [128] studied Ar trapped during sputtering by using a non-Q-spoiled ruby laser to induce flash evaporation of the gas from the film and then analyzed it by standard mass-spectrometry technique. They found that increased trapping of Ar ions takes place only for kinetic energy greater than 100 eV. At lower energies (30 to 70 eV), sorbed Ar is released during sputtering. Thus, sorption is significant only at energies above 100 eV.

Argon-ion bombardment is found to change the electrical resistance of films. Ivanovskii and Radzhabov [129] observed a 1.3 percent increase in the resistance of 1,000-Å Ti films sputtered by a 0.1 to 4-keV Ar ion beam and ascribed it to a purely mechanical effect of the adsorbed Ar atoms. An initial decrease followed by an increase in the resistance of 100- to 1,000-Å-thick films of Ti, Ag, and Au due to bombardment by Ar ions of energy 0.25 to 4 keV was observed by Navinsek and Carter [130]. The initial decrease was explained by the removal of surface-contaminant gas, and the following increase by the introduction of defects. Similar radiation-induced initial decrease in resistance of Au and Ag films observed by others [131] has been attributed to an increase in the size-effect specularity parameter, an explanation for which there is little justification since a number of other factors may also cause small resistance changes.

The effect of adsorption of various solute impurities such as metaphosphoric acid, glycol, tartar, and citrate on the resistivity of electrolytically prepared Cu and Ag films was studied by Raub [132]. He found that annealing of these films to a certain temperature resulted in a sharp fall of the resistance. Similar observations were made by Kramer and Zahn [133], and by Fery [134] on sputtered Pt films.

Quantitative studies of the conductivity changes due to adsorption were made by Suhrmann and coworkers [135] on Ni and Pd films, by Zwietering et al. [136, 137] on Ni, Ti, and Fe films, by Bliznakov and Lazarow [138] on Ag films, and by Wortman and Canady [139] on Au films. These changes are small and are generally of the order of a few percent at room temperature.

The decrease in conductance was found by van Heerden and Zwietering [137] to depend upon the number of molecules n adsorbed on the films (itself made of N atom) and is given by

$$-\frac{\Delta\sigma}{\sigma} = x\frac{n}{N} \tag{61}$$

where x is a small number and is of the order of 2 for H_2, O_2, and CO. These values of x are rationalized on the assumption that each adsorbed molecule influences x conduction electrons of the film. If the adsorbed atoms have more affinity for electrons than for the metal, then the adsorption is expected to decrease the conductivity. The reverse would hold if the metal had more affinity for electrons than the adsorbed atoms. No effect is expected if the adsorbed atoms have no affinity for the electrons.

The observed decrease of conductivity with O_2 adsorbed on Ni, and an increase of conductivity with H_2 adsorbed on Ni, can be explained on the basis of the aforementioned model. The conductivity of Ag films is found to decrease with the adsorption of oxygen, methanol, and water vapors, which therefore act as acceptors. In sharp contrast to the behavior of Ag films, the conductivity of Au films increases with adsorption of oxygen in a *reversible* manner but remains unaffected by the presence of inert gases [139].

Suhrmann et al. [135] observed that the adsorption of H_2 on Fe, Ni, and Pd films first increases and then decreases the resistance at higher coverages. Up to a certain coverage of the adsorbate, chemisorption is partially reversible. At higher coverages, the resistance change is proportional to the amount of occluded H_2. The effect is much larger for Pd than for Ni. This behavior was explained by assuming that the adsorbed H_2 before the maximum forms a negative atomic layer, whereas behind the maximum it is dissociated in protons and electrons.

Adsorption studies on thin films are clearly an important research tool if an unambiguous interpretation can be obtained. It is necessary to emphasize that the electrical conductivity of discontinuous, or nearly continuous films is very sensitive to slight changes in the structural rearrangement due to aging, annealing, and adsorption. This marked sensitivity is due to the dependence of electrical conduction on the fragile interconnections between the discrete islands. Changes in the electronic structure of the surface of a film due to adsorption can also alter surface scattering of electrons and thereby the conductivity. Interpretation of the data involving only a few percent change in the conductivity is obviously fraught with uncertainties.

3.7 Conductivity Changes Due to Annealing

Electrical resistivity of thick $(t \gg l)$ films is generally appreciably higher than the bulk value. The excessive resistivity may be ascribed to the scattering of electrons by structural defects and impurities. Although impurities have a profound effect on the resistivity, their role in ultrahigh-vacuum prepared films of pure materials must be negligible. In some cases, such as Au and Ag, the evaporation process is even beneficial in reducing the oxygen contamination so that films exhibit resistivity ratios (a good measure of purity) which may be an order of magnitude higher than those obtained with highest-purity starting bulk materials.

Many structural defects are frozen in during deposition of a film (see Chap. IV, Sec. 6). The most abundant of these defects are dislocations, typically $\sim 10^{10}$ to 10^{12} lines/cm^2. Their contribution to resistivity is, however, negligible ($\sim 10^{-19} \Omega$-cm per dislocation). Similarly, the small density of stacking faults ($\sim 10^5$/cm) and interstitials has little effect. Large contributions (~ 1 to 6 $\mu\Omega$-cm/atomic percent) to resistivity may arise from vacancies. The observed excess resistivity in noble-metal films requires a vacancy concentration of ~ 0.1 to 1.0 atomic percent. This is considerably larger than the equilibrium value at 300°K but is entirely feasible in a rapidly quenched system (Chap. V, Sec. 2.3.7).

Another significant source of electron scattering is the grain boundaries in the film. If the grain size t is smaller than the mfp l, internal diffuse scattering at surfaces will increase the resistivity considerably. If l does not change appreciably with temperature so that $t \ll l$ is satisfied, this contribution will be temperature-independent. On the other hand, the transport of electrons across a grain-boundary potential may be a thermally activated tunneling process characterized by a negative TCR (Sec. 2.4).

Some understanding of the influence of structural defects on electron transport may be obtained from annealing studies. An extensive review of the early work on this subject is given by Mayer [2]. The principal change on annealing encountered in partially continuous films deposited at low temperatures (that is, vapor-quenched) is grain growth which results from the surface diffusion between the physically touching islands. Electron micrographs [140] of Ag films support this picture. This growth produces changes in the size and the separation of islands in the film and thereby rapid changes in the electrical conductivity. Consequently, a variety of time and temperature variations of conductivity are observed in such cases.

The observed behavior of Ag [141, 142], Pb [143], and Au [144] films deposited at low temperatures points out a general feature on warming, namely, an initial reversible decrease frequently with hysteresis followed by an $\exp(-1/T)$ increase. These changes can be interpreted, at least qualitatively, in terms of changes in the film structure by surface-diffusion-controlled sintering process and by relative movements of the metal islands.

The annealing behavior of continuous thin films deposited at low temperatures was studied by Suhrmann and coworkers [141, 145] on Au, Ag, Bi, and Ni, Buckel and Hilsch [146] and Buckel [147] on Cu, Al, Ti, Pb, Sn, In, and Hg, by Armi [143] on Pb, by Mönch [148] on Sn, and by Sander and Strieder [149] on Cu and In films. Typical cyclic annealing curves for a 500-Å-thick Sn film condensed at 4°K are shown in Fig. 15. Above 5°K, the resistance decreases irreversibly with reproducible cooling cycles which depend on the maximum annealing temperature attained in the previous cycle. Annealing at or about 140°K produces nearly bulklike temperature variation of the resistance. The resistance of the film, however, continues to decrease with further annealing at higher temperatures. Note that the film is initially superconducting at 4.6°K, which is 0.9°K higher than the bulk value. The superconducting transition temperature continues to decrease with successively increasing annealing temperatures, being 4.1°K for annealing at 90°K and nearly bulk value at higher temperatures.

The observed annealing characteristic at low temperatures shows a similarity to the resistance decrease of bulk metals first irradiated at 4.2°K with heavy particles and then heated to temperatures up to 78°K. The annealing mechanism consists, in this case, of a rearrangement of microcrystals within displacement spikes or, perhaps, the annihilation of Frenkel defects. On the other hand, mere deformation and cold working of bulk materials at 4.2°K does not introduce defects which anneal between 4.2 and 78°K, indicating that a wider spectrum of defects occurs in condensed films than in high cold-worked materials.

At temperatures above about 100°K, annealing begins with the migration of interstitials, and eventually, at sufficiently high tempera-

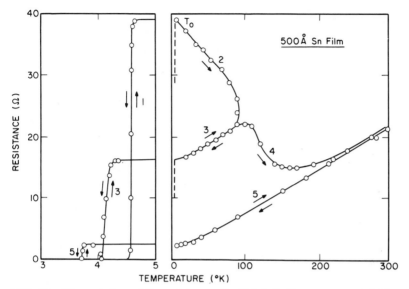

FIG. 15. The variation of the resistance of a 500-Å Sn film vapor-quenched onto quartz at 4.2°K with cyclic annealing at higher temperatures. The expanded scale on the left shows the decrease of superconducting transition temperature with increasing annealing temperature. Cycles 1, 2, and 4 go up to 5, 90, and 300°K annealing temperatures, respectively. (*Buckel and Hilsch* [146])

tures, includes the migration of single vacancies. Grain growth may also occur, and the disappearance of much of the grain-boundary area is probably the main cause of the decrease in resistance.

At temperatures above 140°K, the change in resistance is determined by the superimposition of the continuing irreversible decrease due to further annealing of defects and the normal metallic increase in resistance. This causes a minimum in the resistance vs. temperature curve. Suhrmann and Schnackenberg [145], and Buckel and Hilsch [146] observed that the temperature at which the minimum in resistance occurs coincides with the Debye temperature of that metal, so that the largest part of the lattice defects is removed at a temperature at which the lattice vibrations have the maximum frequency.

Sander and Strieder [149] studied the resistance of vapor-quenched Cu and In films as a function of time and temperature when the films were warmed up to any temperature between 20 and 300°K within 1 min. The resistance variations were analyzed by the Meechan-Brinkham [150] method to derive the activation energies. The observed percentage change in the resistance of Cu films as a function of the calculated activation energies is shown in Fig. 16. For comparison, the observed [151] recovery changes after deuteron irradiation, cold work, and splat cooling are also shown. Clearly, the recovery processes in thin films condensed at low temperatures have similarities to those found with other defect-inducing agencies. Deuteron bombardment produces several types of defects which may anneal by processes such as migration of interstitials, vacancies, vacancy pairs, and formation of clusters. Seeger et al. [152] showed that the recovery stage III was due to the migration of interstitials corresponding to an activation energy of 0.5 eV, which compares well with 0.68 eV obtained from recovery after deuteron bombardment. Studies of Marx et al. [153] showed that the electrical-resistivity changes induced by 12-MeV deuteron bombardment of Cu,

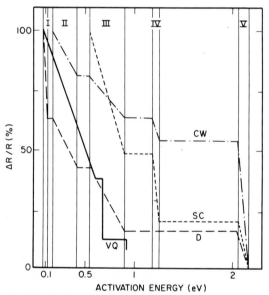

FIG. 16.   Percentage change in the recovery of resistance due to annealing (VQ, solid curve) of vapor-quenched Cu films deposited on quartz at 4°K as a function of the activation energy. For comparison, the observed resistance recoveries in bulk Cu after deuteron irradiation (D), cold work (CW), and splat cooling (SC) are shown. The recovery takes place in definite activation-energy regions depicted by roman numerals. (*Sander and Strieder* [149])

Ag, Au, Ni, and Ta foils at about $120°K$ recovered with a slight increase of the specimen temperature. The activation energy of 0.2 to 0.3 eV deduced in these cases may be ascribed to the migration of vacancy pairs.

Suhrmann and Schnackenberg [145] observed that the rate of variation of the resistance R of low-temperature condensed films on warm-up followed approximately a parabolic relation, that is, proportional to R^2. They interpreted this relation to imply a bimolecular reaction in which the collision of the two disturbed atoms results in the removal of the defect. The activation energies calculated from the data were very small and corresponded closely to the limiting Debye frequency, which indicated that the defects were removed on annealing by transient lattice vibrations rather than by a diffusion process. On the other hand, the annealing results of Kinbara and Sawaturi [154] on Au films showed the rate to vary as R^6, indicating that the recovery processes must be more complex in nature.

In contrast to the electrical-resistivity behavior of annealed vapor-quenched films of ordinary metals, those of Bi, Ga [146], and Sb [155] exhibit additional features in the conductivity behavior resulting from the occurrence of new, metastable structures. The amorphous (or liquid-like) structures obtained in these materials at low temperatures change to the normal structures via other metastable crystalline phases, which are normally observed only at high temperatures and/or high pressures (see Chap. IV, Sec. 4.3).

The annealing behavior of metastable-alloy films [155-160] formed by vapor quenching (Chap. IV, Sec. 4.2) is of fundamental interest [161] and is illustrated for a Cu + 50 percent Ag alloy film in Fig. 17. The resistivity of amorphous films [158-160] is about 10 to 100 times higher than the bulk value. This value of the resistivity is independent of the alloy composition and is nearly the same as that of a vapor-quenched film of one of the pure components. The resistance of metal films decreases rapidly on annealing, but that of amorphous films exhibits nearly zero (or slightly positive) TCR up to an annealing temperature of about one-third of the average melting point of the components. At this point, a rapid decrease occurs as the film assumes single-phase crystalline structure and the grain size increases from $\gtrsim 10$ to ~ 50 Å. Further annealing produces a two-phase structure with a rapid fall in the resistance. The resistivity thereafter is still about 1 to 2 $\mu\Omega$-cm higher than bulk but has nearly the same TCR as bulk. Comparison with the extrapolated resistivity data of liquid alloy suggests that the amorphous

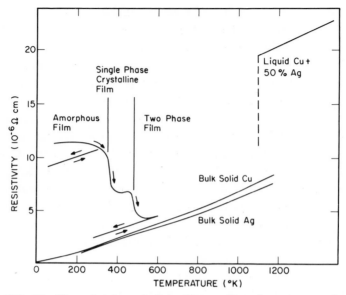

FIG. 17. The resistivity variation with annealing of a vapor-quenched amorphous Cu + 50 percent Ag alloy film deposited on glass at 77°K. With annealing, the amorphous structure changes from a single-phase to a two-phase crystalline structure. The temperature variations of the resistivity of bulk Ag and Cu and their liquid alloy are shown for comparison. (*Mader et al.* [158])

phase resembles a frozen-in liquid structure. The ~1.0-eV activation energy obtained from resistivity data for the first stage may be associated with a diffusion process, whereas the second stage, with activation energy of ~1.4 eV, may be interpreted as a spinodal decomposition process. The combined-resistivity and Hall-effect studies [160] suggest that the conductivity changes during the amorphous-to-crystalline transformation are largely due to changes in the mobility of carriers.

The natural aging and annealing behavior of films deposited at ambient temperatures depends on the deposition conditions. The situation is best summarized by the extensive work of Belser [162] on films of 24 metals prepared by evaporation, sputtering, and electrolysis. A reduction in the resistance of 20 to 50 percent as a result of annealing was commonly observed. The resistance of ultrathin films may, however, increase markedly, primarily because of increasing physical discontinuities on annealing. A preferred annealing temperature was found to depend on film thickness but was close to the recrystallization temperature of the metal. Agglomeration of the film occurred at high temperatures and was attended by a sharp increase in resistance.

The effect of annealing is also dependent on the substrate [163]. The resistance decrease is larger for substrates with smaller grain size and high adsorption energy. For example, Bi_2O_3 substrate yields much larger resistance changes in metal films on annealing than mica.

Structural defects distort the lattice and contribute significantly to the high resistivity of films. Vand [164] formulated a theory for understanding the decay of the resistance of films in terms of decay of lattice distortions. The theory, supposedly valid for continuous films deposited at ambient temperatures, enables calculation of the decay energies and the relative number of distortions from measurements of time- and temperature-dependent decrease in the resistance. This method makes it possible to construct an energy spectrum of the decay processes. This analysis was utilized by Wilkinson [165] to interpret his data on Au films evaporated at various partial pressures of N_2, and by Shah and Naik [166] on vacuum-evaporated Cu and Au films. The activation energies calculated from the data range from 0.95 to 1.03 eV for Au films which correlate with the theoretical and experimental values [151] of the activation energy for migration of vacancies in the bulk. The recovery of electrical conductivity on annealing the films thus appears to be due to the migration of vacancies to the surface or to sinks such as dislocations or stacking faults. Using the contribution to the resistivity by 1 atomic percent of vacancies as 1.5 $\mu\Omega$-cm for Au, the concentration of vacancies for Au films is found to range from 0.06×10^{-2} to 2.58×10^{-2}. Similar high values were also deduced by Story and Hoffman [167], and Fumi [168] from their annealing studies.

3.8 High-resistivity Films

Investigations on the electrical behavior of low- and high-resistivity film materials have been stimulated largely by the industrial demand for microminiaturized electronic components. It is therefore appropriate to present here a brief report on the passive resistor components.

The chief requirements for such a component are small size, low cost, high thermal, mechanical, and aging stability, good control on the value of resistance, low TCR, low high-frequency losses, and compatibility with the substrate used. These requirements can be met to a large extent by thin films of suitable materials. Commercial success and pilot-plant production of some types of film resistors are already a reality and will no doubt have a profound influence on future integrated microcircuitry. Both evaporation and sputtering techniques have been employed for deposition of resistors. A typical density of 10^4 units (resistors) per cm^2 is conveniently obtainable with present-day technology. Advanced photoetching techniques have replaced mechanical masks and make it possible to obtain film line widths and spacings of $\sim 10\,\mu$.

Several excellent summaries [32, 169-172] on various aspects of resistor films have been published. The references given in this section are only representative. Thin film materials of interest for resistor and interconnection applications, and their properties are listed in Table III.

Film resistors are commonly characterized by the normalized parameter called sheet resistivity R (Ω/sq), which is related to resistivity by

$$R = \frac{\rho}{t} = \frac{rW}{L} \tag{62}$$

Here r is the resistance of a film of thickness t, length L, and width W.

Three types of film materials have been considered for resistor applications.

1. Pure and continuous metal [31, 162] films prepared under clean conditions have low ($\sim 10\,\mu\Omega$-cm) resistivities and are primarily of interest for contacts and interconnections. Aluminum, which is commonly used

Table III. Resistance per Square of Films of Various Materials of Interest for Resistor Applications

Film material	Preparation technique	Resistance, Ω/sq	TCR, % per °C	Ref.
Cr-Ni (20:80)	Evap.	10–400	+0.01 to +0.02	181
Cr-Si (24:76)	Flash evap.	100–4,000	±0.02	182,183
Cr-Ti (35:65)	Flash evap.	250–650	±0.015	182
Cr-SiO (70:30)	Flash evap.	Up to 600	−0.005 to −0.02	190-192
Cr	React. sputt.	Up to 10^4	+0.25 to 0.3	179
SnO$_2$	React. sputt.	Up to 30,000	±0.015	187
Ta	Sputt.	Up to 100	±0.01	176,188
TaN	React. sputt.	10–100	−0.0085	179
W, Mo, Re	Sputt.	10–500	−0.002 to −0.01	33

for connections, is fairly reactive with many systems. It also shows electromigration [173] under an applied electric field, leading to failure of devices. A Pt film sandwiched between the substrate and the Al film is highly resistant to solid-state diffusion of the latter. Gold-molybdenum films are gaining popularity for use in interconnections [174].

Resistivities of refractory and semirefractory metal films can be increased 10 to 100 times by decreasing their grain size, by depositing them in the presence of a partial pressure of oxygen, by vapor quenching, or by ion-beam sputtering. Small-grain-size films show considerable aging and sensitivity to further oxidation. Tantalum films can be protected from further oxidation by anodization [169], thermal oxidation, or diffusion [34] of Au in the grain boundaries. The grain-boundary oxidation results in a large negative TCR, presumably because of an activated tunneling process of conduction across the thin oxide layers between grains. Amorphous ($\sim$10-Å grain size) films [33, 175] of W, Mo, and Re produced by ion-beam sputtering have resistivities of $\sim$100 $\mu\Omega$-cm, and a very small TCR ($<$50 ppm) over 4 to 700°K range and thus are very promising for resistor application. Their resistivities can be increased with further oxidation by a factor of up to 10 with some increase in the TCR.

Among the various refractory-metal films, those of Ta have received the most attention [34, 176-179] because of their compatibility with thin-film capacitor applications using tantalum oxide. Surface oxidation, thermally or by anodization, may be used to protect the films as well as to help trim the resistors to the design value. Tantalum-film resistors exhibit negligible losses [34] from dc to 1 kMc/sec. The same result should hold for resistors of other refractory metals. Chromium-film [180] resistors have not found acceptance, but Cr films may be used as underlays to increase the adherence of other conductive films.

2. This class consists of alloys [181-186], oxides [187], nitrides [182, 188, 189], and silicides [182, 183] of refractory metals. The most widely studied are the Ni-Cr alloys produced by flash evaporation [181], filament evaporation [184], sublimation [186], and sputtering [184]. Nonstoichiometry and poor stability of films are the serious problems for reproducibility. The amount of nonstoichiometry depends largely on deposition conditions and occurs [184] in both sputtered and evaporated films. The most promising material for production purposes appears to be reactively sputtered TaN [188, 189] films which, like Ta

films, can be protected from aging by anodization and at the same time be trimmed accurately.

3. Metal-dielectric (ceramic) mixtures (called "cermets"), such as those of Au and Cr with SiO, yield films of high resistivities ($\sim 10^3$ to 10^5 $\mu\Omega$-cm). Films of such materials have been prepared by flash evaporation [190], by rf sputtering [191], and by diode sputtering [192]. As expected, the resistivity greatly depends on the composition of the cermet and the amount of disproportionation. The films are normally amorphous [193] and probably conduct by thermally activated tunneling which would be modified considerably by the presence of segregated metallic particles resulting from disproportionation. A review of the technology of cermet resistors is given by Lessor et al. [170] and by Layer [171].

4. GALVANOMAGNETIC SIZE EFFECTS IN THIN FILMS

4.1 Introduction

The electrical resistance of a metal increases [40] when placed in a magnetic field. This ordinary bulk magnetoresistance effect depends in a complicated way on the binding of the electrons in the lattice. Since this effect is theoretically zero for quasi-free electrons on spherical Fermi surface, the very existence of bulk magnetoresistance implies that the Fermi surface is not spherical. Therefore, the validity of the free-electron concept to the galvanomagnetic size effects is questionable.

An applied magnetic field forces the conduction electrons to move in helical orbits whose axes lie parallel to the magnetic field. The velocity of the electrons in a direction parallel to the field remains unchanged. The radius of the helical orbit of a free electron is given by

$$r = \frac{mvc}{He} \tag{63}$$

where mv is the average momentum of the electrons at the Fermi surface and is in a plane perpendicular to the field H.

If conduction electrons are scattered by the boundaries of a finite specimen, the alteration of the electron trajectories in a magnetic field in general would lead to a nonzero resistance change even if the electrons are regarded as free [43, 44]. This is essentially a geometrical effect (galvanomagnetic size effect) and its details therefore vary with the shape of the specimen and with the relative configurations of specimen, current, and magnetic field. Such galvanomagnetic size effects depend on the orbital radius given by Eq. (63) and therefore should yield a direct estimate of the average momentum and hence the number of free electrons [from Eq. (22)]. This method is, however, severely restricted in practice by the disturbing effect of bulk magnetoresistance, which is small for alkali metals and, for other metals, only in the longitudinal geometry.

Clearly, the galvanomagnetic size effects would be observable only if the condition $r > l > t$ is satisfied. The value of l in normal metals at room temperatures is $\sim 10^{-5}$ to 10^{-6} cm. Since v is about 10^8 cm/sec, the magnetic field required to obtain $r = 10^{-5}$ cm is $\approx 10^3$ kG, which is beyond present-day technology. Thus, it is essential to work with several-micron-thick films at low temperatures so that l is large and magnetic fields of reasonable magnitude are required. Thin single-crystal films of noble metals can now be grown which have mean free paths of several microns at helium temperatures, enabling the study of galvano-magnetic size effects in thin films of noble metals for the first time to supplement the results on high-purity bulk specimens of alkali metals.

Theories of galvanomagnetic size effects in different configurations are complicated. The analyses are generally based on the unrealistic assumptions of a free-electron model, diffuse surface scattering, and isotropic relaxation time which lead to zero bulk effects contrary to the experimental facts. Kaner [194] obtained a generalized theory of the galvanomagnetic and thermomagnetic effects for a metallic film located in a magnetic field directed at an arbitrary angle with respect to the surface of the film. The kinetic coefficients were calculated for an arbitrary electron-dispersion law and relaxation time dependent on the quasi-momentum of the electron. The end result of these calculations is a complicated set of equations which can be simplified only by assuming a constant relaxation time and an isotropic, square dispersion law (free-electron model). According to Kaner, however, it can be shown that qualitative results, in particular the dependence on the magnetic field, do not depend significantly on the form of the dispersion law and the collision integral. This conclusion lends confidence to the results based on the simplified theories.

Experimental data on the magnetoresistance of bulk metals are often analyzed in terms of Kohler's rule, according to which

$$\rho_H - \rho_0 = \rho_0 f\left(\frac{H}{\rho_0}\right) \tag{64a}$$

where f is a universal function and ρ_0 and ρ_H are the resistivities at zero and H applied field. In the presence of surface scattering (the film resistivity is then $\rho_F = \rho_B + \rho_S$, where F, B, and S stand for film, bulk, and surface contributions), Olsen [195] suggested an empirical modification of Kohler's rule as

$$\rho_{F-H} - \rho_{S-H} = \rho_{S-H} f\left(\frac{H}{\rho_{S-H}}\right) \tag{64b}$$

Here, ρ_{S-H} corresponds to the resistivity of an ideal bulk metal (zero bulk magnetoresistance) at field H. This suggested modification has not received any strong support from the existing experimental results [196].

4.2 Longitudinal Magnetoresistance

An exact analysis with a solution by graphical integration was given by Chambers [197] in the case of a wire. The problem of a film geometry was solved by Kaner [194], Koenigsberg [198], and Kao [199] following Chambers' method and assuming a free-electron model. The results should obviously depend on the two dimensionless parameters $\beta = t/r$ and $\gamma = t/l$ which are measures of the strength of the magnetic field and of the effective thickness, respectively. Theoretical values of the resistivity ratio ρ_F/ρ_B vs. β for the case of a wire (Chambers), and a thin film (Kao) are shown in Fig. 18(a) and (b), respectively. The values for $\beta = 0$ are those given by the Fuchs theory [Eq. (32)]. As expected from the considerations of the confinement of electron trajectories inside the film, the resistance decreases steadily as β increases, and tends to the bulk value as β tends to infinity when surface scattering is completely eliminated. A comparison of these curves with the experimental results should determine the mfp and momentum of the electrons.

Note that the calculations given by Koenigsberg and Kaner for thin films yield no magnetoresistance for $\beta < 1$. For $\beta \gg 1$, however,

$$\frac{\rho_F}{\rho_B} = 1 + \frac{3\pi}{8\beta} \tag{65}$$

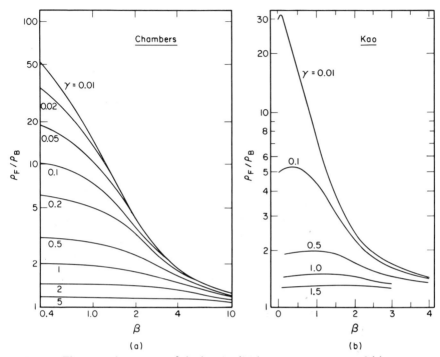

FIG. 18. Theoretical variation of the longitudinal magnetoresistance of (a) a wire based on Chambers' [197] calculations, (b) a film based on Kao's [199] calculations, as a function of $\beta = t/r$ (r is the orbital radius and t the specimen thickness). Both calculations assume $p = 0$ and *zero* bulk magnetoresistance.

In contrast to the monotonic decrease of ρ_F/ρ_B predicted for films by Eq. (65) (and also for wires), Kao's theory predicts an initial enhancement [Fig. 18(b)] of the size-effect magnetoresistance. Kao appears to have considered a more detailed treatment of the trajectories, and the initial enhancement is the result of additional scattering due to the curving of some electrons which move nearly parallel to the surface in zero magnetic field. The theory gives an approximate relation between the position of the peak and the value of γ (lying between 0.01 and 1) as $\beta_{\text{peak}} = 1.26\gamma^{0.57}$. If the theory is valid, this formula may be used conveniently to estimate l provided mv is known, or vice versa.

Chambers [197] measured the longitudinal magnetoresistance of a 30-μ-diameter Na wire at various temperatures between 15.5 and 4.2°K. The results agree well with the theoretical curves and yield $\sigma/l = 8.1 \times 10^{22}$ and $mv = 9.1 \times 10^{-20}$ for Na. One obtains values of 6.3 × 10^{22} and

9.7 × 10^{-27}, respectively, from Eq. (22), assuming one conduction electron per atom. MacDonald and Sarginson [52] also obtained similar results on Na wires.

Chopra [105] studied longitudinal magnetoresistance of thin films of Au, Ag, Al, and In at 4.2°K using dc and pulsed fields up to 150 kG. Size-effect departures from the bulk value [199] were observed in all cases, but because of the higher resistivity ratio (RR) obtained for Ag films, the effects were more marked in Ag films. The observed variation generally showed an initial rise to a maximum and then a slow decrease toward a saturation value for $\beta > 1$. Polycrystalline films, having a smaller RR, and thus a shorter mfp, exhibited behavior similar to that predicted by Kao's theory for diffuse scattering and showed an initial enhancement effect over and above the bulk value. It may be noted that a maximum in the magnetoresistance curve can also occur at a field when the monotonically increasing bulk contribution falls below that due to the monotonically decreasing size effect. The enhancement predicted by Kao's theory is, however, distinguishable by the fact that the magneto-resistance value at the maximum should exceed the bulk value. Since the bulk values are not known, the observed maximum at low fields in case of platelets of Bi [200] and Sb [201] may be due to either of these effects.

For films with $\gamma < 0.5$, the observed results (Fig. 19) for an annealed 1.35-μ-thick polycrystalline Ag film represent a typical variation, which is clearly a characteristic of neither perfect specular nor perfect diffuse scattering behavior. By contrast, the data on a 2.6-μ-thick epitaxial Ag film ($\gamma = 0.05$) exhibits a perfect specular behavior. The results for the partial specular case can be analyzed to determine p. Since no theory is available for such a case, only a phenomenological approach is possible. If the specularity parameter does not vary with the applied field, then Eq. (45) suggests that the film magnetoresistance can be written as

$$\left(\frac{\Delta\rho}{\rho}\right)_F = p\left(\frac{\Delta\rho}{\rho}\right)_B + (1 - p)\left(\frac{\Delta\rho}{\rho}\right)_{\text{Diff}} \tag{66}$$

The observed magnetoresistance curve can therefore be reconstructed by adding the theoretical diffuse and the experimentally known bulk contributions, and thus p may be determined. The diffuse contributions $\Delta\rho_{\text{Diff}}/\rho_B$ and $\Delta\rho_{\text{Diff}}/\rho_F$ as calculated from Fig. 18(b) are shown by the

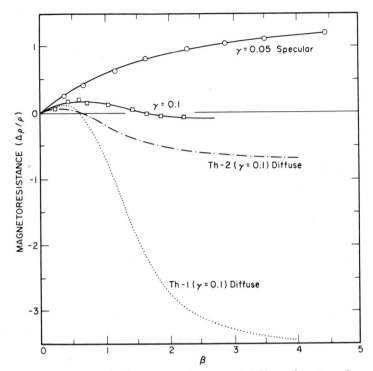

FIG. 19. Longitudinal magnetoresistance at 4.2°K as a function of $\beta = t/r$ for a 2.68-μ epitaxially grown (111) Ag/mica film (○○○); and for a 1.35-μ polycrystalline Ag film (□□□). The dashed Th-1 and Th-2 curves are the variations of $\Delta\rho_D/\rho_B$ and $\Delta\rho_D/\rho_F$, respectively, for diffuse scattering as calculated from Fig. 18(b) for $\gamma = 0.1$. (*Chopra* [105])

dotted curves in Fig. 19. The "100 percent diffuse" curve given in the original publication [165] is the sum of the bulk and $\Delta\rho_{\mathrm{Diff}}/\rho_B$ terms. The choice of the latter term is questionable. It is more appropriate to use the $\Delta\rho_{\mathrm{Diff}}/\rho_F$ term in Eq. (66) to fit the observed data. This procedure yields a value of $p \sim 0.5$ for the 1.2-μ-thick Ag film.

At sufficiently high fields ($\beta \gg 1$), another method for estimating p may be conveniently used. Under this condition,

$$\Delta\rho_F \rightarrow \rho_S \qquad \rho_F - \rho_B$$

$$\approx \frac{3}{8}\,\rho_B\,\frac{1-p}{\gamma} \qquad (\gamma > 1) \qquad\qquad (67a)$$

$$\approx \frac{4}{3}\,\rho_B\!\left(\frac{1-p}{1+p}\frac{1}{\gamma}\ln\frac{1}{\gamma}-1\right) \qquad (\gamma \ll 1) \qquad (67b)$$

The observed variation $\Delta\rho_F$ appears to have saturated at $\beta = 2.25$ and can therefore be extrapolated to higher values of β without serious error. The value of p so derived is 0.7, which is higher than that obtained by using Eq. (66).   But the difference is not significant in view of the approximate nature of both the analyses.

4.3 Transverse Magnetoresistance (Field Perpendicular to the Film Surface— $H\perp$)

This transverse case was analyzed by Sondheimer [43] for a free-electron model. The theory leads to an interesting result that the magnetoresistance (electrical as well as thermal) in this geometry is an oscillatory function of the strength of the magnetic field (Fig. 20). Chambers [46] explained the resistance oscillations as being essentially due to the oscillations in the speed of an electron moving in perpendicular electric and magnetic fields. The speed at time t is a trigonometric function of $eHt/mc = vt/r$, and this causes the distribution function $f(v, z)$ to be a fluctuating function of z/r. This effect occurs only in thin films because the presence of the metal surface is required to provide a finite limit to the distance from which electrons can come to contribute to the current at z. For large thickness, the elementary

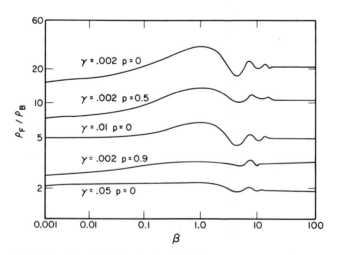

FIG. 20. Theoretical variation of the transverse magnetoresistance of a thin-film specimen with $\beta = t/r$ (proportional to the magnetic field) for various values of p and γ. Bulk effects are assumed to be zero. (*Sondheimer* [43])

oscillating contributions add up to give a nonfluctuating total, and f is then independent of z. The resistance increases initially with β and reaches its first maximum when $\beta \sim 1$. The oscillations are of roughly constant spacing in H (unlike the characteristic $1/H$ oscillations of the de Haas-van Alphen type), and the maxima occur at $\beta = 1, 7, 13, \ldots$. The amplitude of higher oscillations decreases rapidly, and in very strong fields the resistance tends to the constant asymptotic, zero-field value. The oscillations die out when γ becomes large, or the scattering is specular (see Fig. 20). This oscillatory effect suggests a method for determining β and also for mapping the topography of the Fermi surface if single-crystal metal films are used.

Gurevich [202] showed that the oscillating component of the current in Sondheimer oscillations arises because of the cutoff of an electron trajectory at some fraction of a turn. Accordingly, oscillations ought to exist for an arbitrary dependence of the electron energy on the quasi-momentum and for an arbitrary direction of the field, except parallel to the surface.

Sondheimer oscillations were first observed by Babiskin and Siebenmann [203] for a thin Bi wire at low temperatures. Even though the case of a wire is complicated because the magnetic field is both parallel and perpendicular to the surface of the wire, the oscillations were found to occur at expected values of β. The observed first maximum, however, occurred at $r = l$ rather than at $r = t$ as predicted in Sondheimer's theory, suggesting that the first maximum might be due to an unexplained bulk rather than a surface-scattering effect.

The Sondheimer oscillations in thin mechanically prepared foils of In and Al were observed and studied in great detail by Førsvoll and Holwech [204]. Amundsen and Olsen [205] studied such oscillations in both the electrical and thermal resistivities of Al foils. Figure 21 shows their experimental results of magnetoresistance (electrical and thermal) for an Al-foil specimen. The theoretical curves [206] were calculated by adding to Sondheimer's formula the bulk magnetoresistance derived from Kohler's rule [Eq. (64a)]. The observed amplitudes of oscillations depart considerably from the theoretical values for thinner specimens and may be attributed to the nonapplicability of Kohler's rule to this case where surface scattering must introduce asymmetry in the relaxation time.

The amplitude of oscillations was found to be smaller for In than for Al foils. The amplitude observed by Cotti [207] for folded In foils was

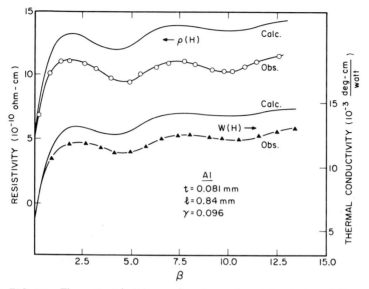

FIG. 21. Theoretical (solid curves) and experimental variation of the electrical and thermal resistivities as a function of $\beta = t/r$ (proportional to magnetic field) for 0.081-mm-thick Al foil with $\gamma = t/\ell = 0.096$ at 4.2°K. The specimen is placed in the Sondheimer position. (*Amundsen and Olsen* [205])

also about an order of magnitude smaller than that predicted by the Sondheimer theory. The value 1.3×10^{-19} g-cm/sec of the electron momentum for In is in agreement with the second-zone Fermi surface. For Al, the value of 1.33×10^{-19} g-cm/sec gives a good fit to the theory, at least for low fields. This value indicates that the oscillations are caused by the electrons in the second Brillouin zone.

4.4 Transverse Magnetoresistance (Field Parallel to the Film Surface—$H \parallel$)

MacDonald and Sarginson [52] treated the problem of a thin film with a magnetic field in the plane of the film but perpendicular to the current direction. This problem is mathematically complicated because the Hall field will not be constant across the film thickness. Nevertheless, the authors assumed that the Hall field was constant and obtained complicated theoretical expression of the effect which showed that the resistivity should pass through a single maximum and then decrease toward the bulk value. This conclusion is just what one would expect from elementary considerations of the electron trajectories.

Experiments by MacDonald and Sarginson [52] on Na wires, by Cotti et al. [208] on In wires, by Førsvoll and Holwech [204] on foils of In and Al, and by Chopra [209] on Ag films, confirmed qualitatively the general variation. Figure 22 shows the magnetic-field variation of the electrical resistivity of an Al foil for different field configurations. Førsvoll and Holwech replotted their data for the MacDonald-Sarginson position after subtracting the normal magnetoresistance which was calculated by using the modified Kohler's rule [Eq. (64b)]. These curves exhibited a single maximum, but the maximum occurred at a value of β which is several times higher than that given by the MacDonald-Sarginson theory. The curves calculated from the experimental data also showed a small dip at low fields, which is not expected from the theory. At high values of β, marked departures from the modified Kohler's rule were observed. The authors suggested that the deviations may be due to the presence of normal open orbits or open orbits established by mangetic breakthrough phenomena. One does expect the surface collisions to influence the conductivity of both the open and closed orbits in a region of $\gamma < 1$ even in the high-field limit.

Førsvoll and Holwech also showed that if an effective free path of about one-quarter of the zero-field value is assumed, the observed

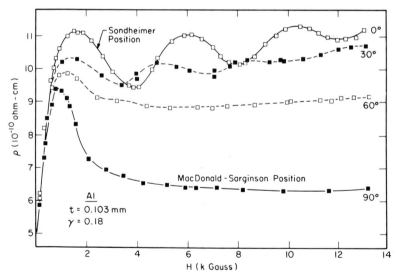

FIG. 22. The resistivity at 4.2°K as a function of the magnetic-field strength for different directions of the field relative to the Al foil. (*Førsvoll and Holwech* [204])

thickness dependence of magnetoresistance follows the same variation as that given by the Fuchs theory for zero field. Consequently, the high-field size effect may be predicted with a fair accuracy.

Size-effect anisotropy in the transverse magnetoresistance of epitaxial Ag, Au, Sn, and In films at 4.2°K was studied by Larson and Coleman [210]. Marked effects were observed in high-resistivity-ratio Ag films. The rotational anisotropy (Fig. 23) of magnetoresistance in a (111) Ag film with current along ⟨321⟩ shows several interesting features. A strong geometric anisotropy, with $\Delta\rho_\perp/\Delta\rho_\parallel$ ratio up to 10, was observed. The anisotropy increased with increasing film thickness and magnetic field, and decreasing temperature, and was observed in Ag and Au films, but

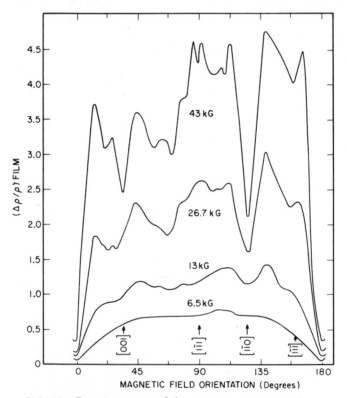

FIG. 23. Rotation curves of the transverse magnetoresistance for a 3.2-μ epitaxial Ag film in various applied magnetic fields. Resistivity ratio of the film is 275. Current direction ⟨321⟩. Zero degree corresponds to $H\|$ configuration. The maxima and minima for different crystallographic directions are indicated. (*Larson and Coleman* [210])

not in In films. Such an anisotropy, first observed by MacDonald [211] on Na wires and Rb platelets, and by Førsvoll and Hollwech [212] on Al and In foils, occurred when the condition $l > t > r$ was obtained. In Larson and Coleman's experiment, although $l > t$ was satisfied, the applied field was small so that $r \gg t$ was obeyed. The anisotropy was observed whenever $\omega_c \tau < 1/2$, where $\omega_c = eH/mc$ is the cyclotron frequency and τ is the relaxation time. Moreover, this observation does not appear to be related to another geometrical effect called "zigzag" effect [213, 214] which arises from the variation of the Hall voltage near the current leads when the width-to-length ratio of the film changes. Thus, the large geometric anisotropy effect remains puzzling. It is probably related to the nonuniform distribution of Hall voltage over the cross section in the $H \parallel$ geometry, which suggests that the MacDonald-Sarginson calculations for the $H \parallel$ case are in serious error.

With increasing field and for $\omega_c \tau > 1/2$, orientation-dependent maxima and minima characteristics of the Fermi surface topology were observed by Larson and Coleman, thereby demonstrating the usefulness of thin films for studies of the Fermi surface. At very high fields when $r \ll t$, the single-crystal anisotropy due to open orbits in Ag becomes dominant, producing an additional maximum, and an adjacent maximum at $35°$ orientation. Quantitative analysis of the results has not, however, been done.

4.5 Hall Effect in Thin Films

When a conductor is placed in a magnetic field perpendicular to the direction of current flow, a field E_Y is developed across the specimen in the direction perpendicular to both the current and the magnetic field H_z. The field, called the "Hall field," is given by

$$E_Y = \frac{J_x H_z}{ne} \tag{68}$$

where J_x is the current density and n is the number of carriers per unit volume. The ratio

$$R_H = \frac{E_Y}{J_x H_z} = \frac{1}{ne} \tag{69}$$

is called the "Hall coefficient." A measurement of R_H thus determines n. Since $\rho = 1/ne\mu$ for a free electron (μ = mobility),

$$R_H = \mu\rho \tag{70}$$

Equations (69) and (70) are valid for a free-electron model of the bulk materials with only one type of current carriers.

The Hall effect should be modified by the geometrical scattering of conduction electrons in thin specimens. As in the case of transverse magnetoresistance, there are two possible geometries for the Hall effect in a thin film, the Sondheimer case $(H\perp)$, and the MacDonald-Sarginson case $(H\parallel)$. At small applied magnetic fields, the Hall coefficient in both cases is given [43] by complicated expressions which in the limiting cases approximate to

$$R_{H-F} \approx R_{H-B} \qquad\qquad (\gamma > 1) \tag{71}$$

$$R_{H-F} \approx R_{H-B} \frac{4}{3}\frac{1-p}{1+p}\frac{1}{\gamma[\ln(1/\gamma)]^2} \quad (\gamma \ll 1;\ p \ll 1) \tag{72}$$

The expressions for mobility follow from Eq. (70) as

$$\mu_F \approx \frac{\mu_B}{1 + \dfrac{3}{8\gamma}(1-p)} \qquad (\gamma > 1) \tag{73}$$

and

$$\mu_F \approx \mu_B \frac{1}{\ln(1/\gamma)} \qquad (\gamma \ll 1) \tag{74}$$

The limiting case $\gamma > 1$ is sufficiently accurate down to $\gamma \sim 1$. Plots of the theoretical variation of R_H and μ_F vs. γ are shown in Fig. 24 for various values of p. Size effects are evidently more pronounced in μ_F than in R_H.

At high fields ($\beta > 1$), R_H is predicted [206, 215] to oscillate around the bulk value. Sondheimer's calculations do not predict an oscillatory behavior, which is the result of a truncation in the numerical calculations. In the MacDonald-Sarginson geometry, R_H is expected to show only one maximum, as in the magnetoresistance curve.

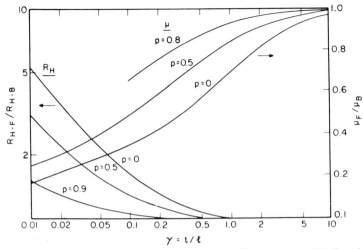

FIG. 24. Theoretical variation of the Hall coefficient R_H and Hall mobility μ (normalized against bulk values) of a thin film at low magnetic field strength as a function of $\gamma = t/\ell$ for various values of p.

The above discussion suggests that a study of the size effects in the Hall effect would provide information on n, l, and p. A number of anomalous results on R_H of thin films have been reported in the literature (see, for example, Ref. 2). The very first studies of the Hall effect were done by Hall [216] on chemically prepared Ag films. The results of Leverton et al. [217] on Bi and Sb films, Wait [218] on chemically prepared Ag films, Steinberg [219] on evaporated Ag and Au films, and Peacock [220] on sputtered Fe, Pt, and Pd films showed no size effects. Cirkler [59] observed size effects in K films, but the increase of R_H at small thicknesses was found to be slower than predicted by Eq. (72). Below 30 Å, R_H decreased rapidly. While no explanation of this behavior was given, it may be due to errors in the film thickness. Actually, in this thickness range, films become increasingly porous with a reduction in the conductivity so that a rapid rise of R_H for thinner films may be expected. For thick K films, a constant value of R_H corresponding to 1.07 electrons per atom was obtained.

Barker and Caldwell [221] observed an increase in R_H with decreasing thickness of evaporated Ag films, but the increase did not occur in the size-effect regime. Jeppesen's [222] results on evaporated polycrystalline Au films showed no variation of R_H down to a thickness of about 30 Å. The value of $R_H = -9.7$ cm^3/C is 30 percent higher, and that of $\mu = 18$

cm^2/volt-sec for thick films is 30 percent lower than the bulk value. The mobility decreased with the decrease of thickness. There is some inconsistency in these results since the constancy of R_H indicates a bulklike behavior whereas the thickness variation of μ suggests a size effect.

Chopra and Bahl [223] made simultaneous measurements of ρ and R_H (and hence μ) on polycrystalline and epitaxial Au, Ag, and Cu films at 296 and 77°K using van der Pauw's method (Chap. III, Sec. 1.1). Below 600 Å, size-effect thickness dependence was observed. The fact that this dependence was weak for epitaxial films indicated a predominantly specular reflection of conduction electrons. Below 300 Å, a large scatter in the data occurs and is greatly influenced by the temperature. It is apparently due to physical changes in the porous structure of the films.

The thickness dependence of R_H and μ in polycrystalline Cu films is shown in Fig. 25. The thickness variation of R_H for epitaxial Cu films and the theoretical curves of R_H and μ for diffuse scattering with $l = 390$ and 1,650 Å at 296 and 77°K, respectively, are also shown. A qualitative

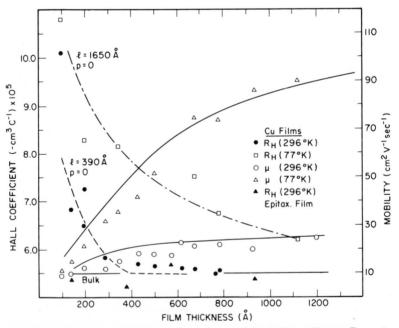

FIG. 25. The observed thickness dependence of the Hall coefficient R_H and mobility μ for Cu films at 296 and 77°K. Theoretical curves shown are for $p = 0$ with $l_{296°K} = 390$ Å and $l_{77°K} = 1,650$ Å. (*Chopra and Bahl* [223])

agreement of the polycrystalline film data with the diffuse-scattering theory is indicated. The mobility, however, falls more rapidly for $t < l$ than expected, which may be attributed to additional scattering at the crystallite boundaries. A linear relation between μ_F and μ_F/t predicted by Eq. (73) was verified for Cu and Au films and was used to determine μ_B. Values of R_H μ, and n for thick films of several metals agree closely with the bulk values, as seen in the data listed in Table IV.

Contrary to the theory, R_H shows size effects even for $\gamma > 1$. The theory, of course, is oversimplified and the discrepancy may result from the assumed (1) isotropic relaxation time, even in the presence of a free surface, and an applied magnetic field, and (2) the uniform current distribution throughout the film. The observed departure of n of monovalent noble metals from one free electron per atom value is indeed strong evidence for the existence of considerable anisotropy of the relaxation time of electrons in these metals.

The high-field oscillatory behavior of R_H has been observed in pure, thin Cd platelets at helium temperatures by Zebouni et al. [215] and Mackey and coworkers [224], and in Al foils by Førsvoll and Holwech [212]. The latter authors studied both the Sondheimer and the MacDonald-Sarginson geometry and found the general behavior in accordance with the theoretical predictions. Figure 26 shows their typical results for an Al foil. The comparison with the theoretical curves, using $mv = 1.33 \times 10^{-19}$ g-cm/sec, gives a good qualitative agreement. The amplitude of the oscillations is larger than predicted. The phase difference between the theoretical and experimental curves suggests that, as also in the case of transverse magnetoresistance size effect, it is impossible to get perfect agreement for all values of β using a single value of the electron momentum. This discrepancy further stresses the shortcoming of the modified Kohler's rule.

4.6 The Anomalous Skin Effect

The size effects which we have discussed so far have been caused by the restriction on the mfp of the electrons imposed by the geometry of very thin specimens. Electrical conduction in thick specimens at high frequencies can also exhibit similar size effects. As is well known, a high-frequency electric field does not penetrate very far into a metal and all the current flows in a skin at the surface. London [225] first suggested that if the frequency could be increased to a value sufficient

Table IV. The Observed Hall Coefficient and Mobility of Some Metals in Bulk and in Thin Films (Thickness $\gg$ mfp) at Room Temperature

The value of Hall coefficient calculated in free-electron theory is given in the third column

Metal	Valency	Hall coefficient $\times 10^5$ cm^3/C			Mobility, cm^2/V-sec,		Ref.
		Calculated $R_H = \dfrac{1}{ne}$	Observed		Observed		
			$R_{H\text{-}B}$	$R_{H\text{-}F}$	μ_B	μ_F	
K	1	-47	-42	-42.3 (90°K)	60	11.6 (90°K)	59
Cu	1	- 7.4	- 5.5	- 5.6	34.8	34	223
Ag	1	-10.4	- 8.4	- 8.6	56.3	45	223
Au	1	-10.5	- 7.2	- 9.7	29.7	18	222
				- 7.5		30	223
Al	3	- 3.4	- 3.0	- 2.6	10.1	8.7	223

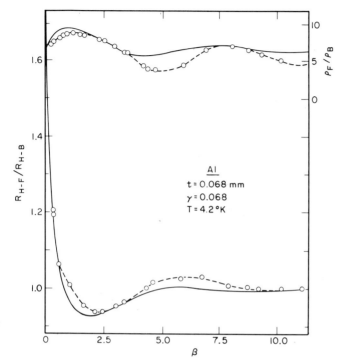

FIG. 26.   The observed and theoretical (solid curve) oscillatory variation of the normalized Hall coefficient and the resistivity of a 0.068-mm-thick Al foil at 4.2°K as a function of $\beta = t/r$ (proportional to the applied magnetic field). *(Førsvoll and Holwech [212])*

for the penetration layer to be thinner than the mfp, there should be a change in the behavior of the skin resistivity. This change or departure from the classical behavior is called the anomalous skin effect and was first investigated in detail by Pippard [226]. Detailed calculations of the effect were made by Reuter and Sondheimer [227] and Dingle [228].

Classical electromagnetic theory shows that the penetration (the classical skin depth) of an electric field of angular frequency ω into a conductor of dc conductivity σ_B is approximately given by

$$d \approx (2\pi\sigma_B \omega)^{-1/2} c \tag{75}$$

The surface resistance of the metal $R(\omega)$ is then

$$R(\omega) = \left(\frac{2\pi\omega}{\sigma_B}\right)^{1/2} \frac{1}{c} \tag{76}$$

If the mfp is greater than d, then the current is carried by only those electrons which are moving in directions which make an angle of less than $\sin^{-1}(d_F/l)$ with the surface (d_F being the effective skin depth in this case) and therefore travel nearly parallel to it. Very roughly, then, these so-called "effective" electrons will comprise a fraction of the order of d_F/l of the total number of conduction electrons. Using this concept of effective electrons and the associated effective conductivity $\sigma_{eff} \sim \sigma_B d_F/l$, the penetration depth of the field in the anomalous skin region is given by

$$d_F = \left(\frac{2\pi\sigma_B \omega d_F}{l}\right)^{-1/2} c = \left(2\pi\sigma_B \frac{\omega}{l}\right)^{-1/3} c^{2/3} \qquad (77)$$

The surface resistance in the anomalous skin region $R(\omega)$ will be given by

$$R_F(\omega) = \left(\frac{2\pi\omega l}{d_F \sigma_B}\right)^{1/2} \frac{1}{c} = \left(\frac{4\pi^2\omega^2 l}{\sigma_B c^4}\right)^{1/3} \qquad (78)$$

A rigorous mathematical treatment gives

$$R_F(\omega) = \left(\sqrt{3}\pi \frac{\omega^2 l}{\sigma_B}\right)^{1/3} \frac{1}{c^{4/3}} \qquad \text{when } p = 0$$

$$= \frac{8}{9}\left(\sqrt{3}\pi \frac{\omega^2 l}{\sigma_B}\right)^{1/3} \frac{1}{c^{4/3}} \qquad \text{when } p = 1 \qquad (79)$$

Note that R_F is dependent on $(l/\sigma_B)^{1/3}$ in contrast to the $(1/\sigma_B)^{1/2}$ dependence of the classical effect, and since σ_B is proportional to l, R_F is independent of the conductivity. The measurement of R_F gives the value of σ_B/l, which according to Eq. (22) gives us an estimate of the number of conduction electrons per unit volume and thereby the area of the Fermi sphere. By making measurements of R_F on single crystals of several orientations (by measuring the Q of a resonant cavity of which the specimen forms a part), it is possible to calculate the shape of the Fermi surface.

The value of p plays only a subordinate role in determining the surface resistance. Extensive experiments on the surface impedance of Cu, Ag, and Sn at 1.2 Gc/sec were carried out by Chambers [229] at temperatures between 2 and 90°K. His results fit the theoretical curves for $p = 0$ rather than $p = 1$. The values of n determined from these data were rather low for Au and Ag and were found to be very sensitive to the surface conditions, indicating that the observed diffuse scattering may be due to the surfaces not being ideally clean and smooth.

The extreme anomalous skin region ($\omega\tau \gg 1$) is, however, quite sensitive to the value of p. The surface-resistance studies of Smith [102] on Bi single crystals at 2°K and 23.5 Gc/sec established that the reflection of the carriers from the surface of the sample is predominantly specular. Similar studies on evaporated single-crystal films would be of much interest.

Cyclotron-resonance Size Effects   The cyclotron-resonance phenomenon in metals and semiconductors under the anomalous-skin-effect condition with magnetic field applied parallel to the surface of the specimen has been treated by Azbel and Kaner [230] and by Jones and Sondheimer [231]. Both longitudinal and transverse surface resistance for $p = 0$ are predicted to show strong oscillatory behavior for $\omega/\omega_c > 1$, where ω_c is the cyclotron frequency. The phenomenon offers a method for determining the effective mass and the nature of surface scattering of the carriers. The experimental conditions required to observe this effect may be difficult to achieve. For example, at 20 μm wavelength in the extreme anomalous relaxation region at low temperatures for a good metal of high purity, $\omega = \omega_c$ is achieved at a magnetic field of ~600 kG.

4.7   Eddy-current Size Effects

Eddy currents induced by changes in the magnetic field parallel to the surface of a plate specimen of resistivity ρ_B decay with a time constant

$$\tau = \frac{4}{\pi} \times 10^{-9} \frac{t^2}{\rho_B} \tag{80}$$

(t = plate thickness)

In a specimen thin enough so that the resistivity is size-dependent, the decay time will exhibit size effects. Following the derivation for anomalous skin effect, and utilizing the concept of effective electrons, Cotti [232] showed that the decay time for a film is

$$\tau_F = \frac{4}{\pi} \times 10^{-9} \frac{t^2}{\rho_{\tau - F}} \tag{81}$$

where the effective resistivity is

$$\rho_{\tau - F} = \rho_B \frac{2\pi^2 l}{3t} \frac{1}{C(l/t)} \tag{82}$$

The function $C(l/t)$ becomes constant in the limit $l/t \ll 1$, yielding

$$\frac{\rho_{\tau - F}}{\rho_B} \approx 1.725 \frac{l}{t} \tag{83}$$

Thus, by measuring the decay time in thin films we can calculate $\rho_{\tau - F}$ and by using Eqs. (83) and (34) and the measured value of ρ_F we can determine the value of l/t. The need for the knowledge of the value of ρ_B is thus eliminated. Cotti used this method to calculate the value of ρ_B and l for In foils at low temperatures and obtained reasonable values. The method also offers the advantage that no electrical contacts are necessary for making measurements.

In the presence of a strong magnetic field perpendicular to the plane of the plate, the decaying eddy currents rotate around the direction of the field. This phenomenon is called "helicons." In the size-effect region, deviations from the ordinary behavior may occur in this case too, giving an anomalous dispersion relation for the helicons. One could use the Sondheimer treatment for this configuration and show [232] that the time constant of the decay of the helicon wave can be expressed by a size-dependent resistivity which is an oscillatory function of the applied field. Such a treatment also leads to an oscillatory field dependence of the frequency of the helicon waves for sufficiently thin plates. Such oscillations were observed by Bowers et al. [233] on Na wires. The frequency of the oscillations was found to be independent of the conductivity in the high-conductivity limit and proportional to the magnetic field. However, Bowers et al. explained the oscillations as due to the excitations of standing modes of free electrons.

5. TRANSPORT OF HOT ELECTRONS

So far we have discussed the ordinary transport processes which are determined by the scattering of conduction electrons. These electrons are in thermal equilibrium with the lattice and therefore have energies no greater than kT above the Fermi level. The Pauli principle inhibits electron-electron collisions in this case. Carriers which have energies well above the Fermi level, and are thus not in thermal equilibrium with the lattice, are called "hot." The transport of such carriers is determined by electron-electron interaction in addition to the electron-phonon and electron-defect interactions.

The attenuation length (also called range) and its relation to the mean free paths l_p and l_e, respectively, for electron-phonon and electron-electron interactions of hot carriers in solids, is of considerable interest for hot-electron devices (Chap. VIII, Sec. 4.7). The attenuation length L is defined in terms of the probability that electrons, excited at a distance x from a barrier and escaping over that barrier, are proportional to $e^{-x/L}$. The barrier may be that at a metal-insulator or semiconductor junction, or a solid-vacuum interface.

The transport of hot electrons in metal films (particularly Au) has been studied by many workers. Several methods have been used to determine the electron range and its energy dependence. In the photoelectric method [234-240] the yield (*photoresponse* or number of electrons (holes) that escape over a barrier $h\nu_0$ at the *front* surface per photon of a particular energy $h\nu$ absorbed on the opposite surface of the film) is measured. The front yield is given by [236]

$$Y = A \frac{KL}{KL - 1} \frac{e^{-t/L} - e^{-Kt}}{1 - e^{-Kt}} \approx A e^{-t/L} \quad (Kt > 1; KL > 1) \tag{84}$$

where A is a constant for given $h\nu$, L is the range of electrons of energy $E = h\nu_0 + (2/3) h(\nu - \nu_0)$ which exceeds the work function of the metal (typically ~4 to 5 eV) and is measured from the Fermi level, t is the film thickness, and K is the optical absorption coefficient assumed to be energy-independent and defined by the exp $(-Kt)$ absorption law, neglecting multiple reflections.

For $t > L$, the front yield is found [238] both theoretically and experimentally to be approximately given by

$$Y \approx \left(\frac{L_0\, h\nu_0}{t}\right)^2 e^{-t/L_0} \quad (L = L_0 \text{ at threshold energy } h\nu_0) \qquad (85)$$

For relatively thick films, Y should be independent of the excitation photon energy.

The yield may be measured by collecting photoexcited carriers in vacuum or by an internal collection measurement across a Schottky barrier of devices such as Au-GaP and Au-Si. A linear plot of the square root of the photoresponse vs. photon energy (Fowler plot) is used to determine the barrier height (threshold energy).

The tunnel-emission method [241-244] measures the transfer ratio in a thin film, sandwich structure, or tunnel diode of the type metal-oxide-metal. An exponential dependence of α on thickness of the counter-electrode film (see Fig. 16(b), Chap. VIII) determines L. Sze et al. [244] used the low-frequency, common-base, current transfer ratio for hot-electron transport in semiconductor-metal-semiconductor point-contact transistor structures to determine L in the metal electrode.

A direct method [245], which is not subject to the uncertainties of momentum and energy of the injected electrons and also allows an easy control over the initial energy, measures the attenuation of slow-electron beams through unsupported thin films of different thickness. The use of a plane-parallel retarding field in the collector region permits measurements of only those electrons whose energies and momenta have not been changed appreciably by passing through the film.

Thomas [234] used the photoelectric method to obtain $L \sim 1,000 \text{ Å}$ in K films for photoenergies in excess of 2.2 eV. A rapidly decreasing L at higher energies was observed and was attributed to the excitation of the plasma oscillations in potassium. The large value of L was questioned by Katrich and Sarbej [235], who obtained values of 70 Å for Au and 35 Å for Cr at work functions of 3.2 eV for Au and 2.8 eV for Cr. The electron range has also been measured in films of Au, Ag, Cu, Al, and Pd. The energy dependence of L for Au films obtained by Sze et al. [238] is shown in Fig. 27 (a) and compared with data from other sources. A marked decrease of L with energy is the prominent feature of the results.

The values of L for Au (Table V) obtained by different authors vary considerably. Soshea and Lucas [240] showed that, by taking the optical absorption of Au on Si into account, the data of Crowell et al. [236] reduce the value of L to 332 Å, which is in agreement with their own

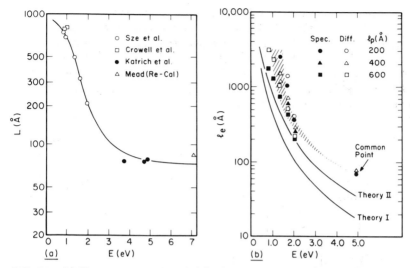

FIG. 27. (*a*) The attenuation length (*L*) of hot electrons in Au films as a function of their energy above the Fermi level. (*compiled by Sze et al.* [238]) (*b*) The energy dependence of the mean free path ℓ_e for electron-electron interaction. The points in the shaded region are calculated from the data of Fig. 27(*a*) by Monte Carlo calculations for different values of ℓ_p, and for specular and diffuse boundary-scattering conditions. The curves marked Theory I and Theory II refer to the calculations of Quinn [247] without, and with, *d*-electron-shielding correction, respectively. (*Stuart et al.* [246])

value. The high value of Sze et al. [238] is attributed to the Si surface condition. These authors also pointed out that values derived from hot-electron triode devices are not the same as those obtained photoelectrically. The latter method imposes a stringent electron-collection requirement, since the electrons injected into the metal perpendicular to the plane of the emitter barrier must not be deflected if they are to surmount the collector barrier. A single collision with a phonon or an electron can thus prevent collection and hence decrease the calculated value of *L*.

The interpretation of the measured range in terms of electron-phonon and electron-electron interaction is difficult since neither a diffusion nor the Fermi age theory is adequate for problems in which the initial source is concentrated within a few mean free paths of the surface, as is the case here. Consequently, an exact functional dependence of *L* on l_e and l_p which would allow a convenient decoupling of the influence of l_e and l_p is not known at present. Stuart et al. [246] applied Monte Carlo calculations to deduce l_e from the published data [236, 237] on the

Table V. Attenuation Lengths and Mean Free Paths for Electrons

The energy of electrons above the Fermi level (in eV) is given in parentheses below the attenuation length L. The mfp l_p is calculated from the conductivity [251]. ΔE_p is the energy loss per electron-phonon collision at $300°$K. l_e for diffuse and specular boundary conditions is derived from the Monte Carlo calculations in Ref. 246. l_e (calculated) is given by Quinn's theory; the values in parentheses are those corrected for the shielding effect 247

Material	L, Å	Ref.	l_p, Å	ΔE_p, meV	Electron-electron mfp, l_e, Å		
					Diffuse	Specular	Calc.
Au	740(332)* (0.8-1.0)	236,237	406	3.7	> 4,000	3,000	415 (910)
	220 (1.0)	244					
	350 (0.95)	240					
	45-80 (5.5)	241					
	34 (6.0)	245					
Ag	440 (0.65-0.95)	236,237	570	7.0	900	650	560 (1,000)
	265 (1.0)	244					
Cu	50-200 (0.55-0.95)		420	15	<300	<200	720 (1,700)
	70-110 (0.93)	239					
Pd	170 (0.7-0.8)	236	110	9.6	~900	~900	
	90 (1.0)	244					
Al (holes)	<50 (0.93)	239					
	100 (1.0)	244					

*Corrected value; see Ref. 240.

range of electrons in Au by assuming certain values of electron-phonon interaction energy loss with various values of l_p and for diffuse and specular boundary condition. The results for Au are shown in Fig. 27(*b*) and are discussed later.

Sze et al. [238] showed that in the limiting case of weak absorption $(l_e \gg l_p)$, the Fermi age theory may be applied to obtain

$$L^2 = \frac{1}{3} \frac{l_e^2 l_p}{l_e + l_p} \approx \frac{1}{3} l_e l_p \qquad (86)$$

On the other hand, when $l_e < l_p$, the electron motion is ballistic (that is, elastic phonon interactions which change directions are negligible), and the measured range should correspond to the electron-electron mfp, i.e.,

$$L = l_e \qquad (87)$$

These two limiting forms agree with the results of the Monte Carlo calculations.

Quinn [247] analyzed the problem of electron-electron interaction theoretically on the basis of a self-consistent dielectric-constant approach in a free-electron model, and obtained an explicit expression for the energy dependence of l_e. His results of the original theory as well as those of the modified theory, which takes into account the effect of shielding by d-band electrons, are presented in Fig. 27(b).

Figure 27(b) shows that, as expected, the Monte Carlo values of l_e are insensitive to l_p at high energies but are sensitive to l_p (that is, elastic collisions with phonons and imperfections) for large values of L at low values of electron energy. Despite any uncertainty in l_p, it is clear that l_e becomes much larger than 1,000 Å for energies less than 1.0 eV above the Fermi level. As the electron energy increases, l_e decreases, reaching values of about 70 Å for 5 eV. The attenuation length is affected somewhat by the type of electron surface scattering but to a lesser extent by the average energy loss per lattice scattering. The comparison of the Monte Carlo values with the theoretical values of Quinn shows only a fair qualitative agreement. The Coulombic-shielding correction seems to be an important factor in producing the long mean free paths. The qualitative agreement is only the result of high experimental values of L used for calculation. The more recent data from careful experiments yield substantially lower values of L, making the discrepancy between Quinn's theory and the derived values of l_e even more marked. The low values of L for Cu, for example [239], are now well established to be a real rather than an impurity effect. It may be necessary to take into account [248] the anisotropy of the Fermi surface, which is more pronounced in Cu than in Au.

If the observed values of L are small and comparable with l_p and the estimated values of l_e* are large, then phonon scattering must be significant. Sze et al. [244] observed that, for Au, Ag, and Pd films, the measured range was proportional to $1/E^2$. Since for low-energy excitation (E much less than the Fermi energy), a theoretical consideration of electron-electron scattering also yields

$$l_e \propto \frac{1}{E^2} \tag{88}$$

one may conclude that L is determined only by electron-electron interaction. But on the other hand, the influence of phonon interaction is definitely indicated by the observation that L of Au films near 1 eV increases with decrease of temperature from 300 to 90°K. The observed 25 percent increase is, however, 7 percent smaller than that predicted by Eq. (86).

Some confusion exists as to the energy dependence of l_p. According to Wilson [249], l_p increases with energy, i.e., $l_p \propto (E + \zeta)^2$, which would yield l_p exceeding 1,000 Å for an energy of about 4 to 5 eV above the

---

*An estimate of l_e at the Fermi level can be obtained as follows. The screening length $(6\pi n e^2/\zeta)^{1/2}$ yields cross section for collisions between electrons in a typical metal as $\sim 10^{-15}$ cm^2. As a consequence of the Fermi-Dirac statistics, the scattering of one electron by another takes place only if both initial electrons lie in or near the energy range $\approx kT$ about the Fermi energy ζ. Thus, the rate at which electron-electron collisions can take place is reduced from the classical value by a factor of $(kT/\zeta)^2$ which is about 10^{-4} at room temperature in a typical metal. The effective cross section A is then $\sim 10^{-19}$ cm^2. Therefore

$$l_e \sim \frac{1}{nA} \sim 10^{-4} \text{ cm}$$

which is about 10 times the value of l_p so that electron-electron collisions near the Fermi level at room temperature or below are negligible as compared with the electron-phonon collisions. The situation is reversed, however, for electrons with energies above the Fermi level.

Fermi level of Au. Consequently, electron-phonon interaction should not be significant for range values below a few hundred angstroms. On the other hand, Koshiga and Sugano [250], using Bardeen's self-consistent field approach, showed that l_p decreases with energy and thus phonon interaction should be significant unless L is very small (< 100 Å).

Observed values of L for several metal films and the calculated l_p and l_e are collated in Table V [251].

6. THERMAL TRANSPORT

6.1 Thermal Conductivity

The kinetic theory yields the following expression for the thermal conductivity of a classical electron gas:

$$K = \tfrac{1}{3} Cvl \tag{89}$$

where C is the heat capacity per unit volume, v is the average carrier velocity, and l is the mfp of the carrier. For Fermi-Dirac statistics, this result reduces to

$$K = \frac{\pi^2}{3} \frac{nk^2 T\tau}{m} \tag{90}$$

The relaxation time $\tau = l/v$ refers to the carrier responsible for thermal conduction. In insulators, phonons are primarily responsible for thermal conduction. The mfp l is then determined principally by geometrical scattering and scattering by other phonons. The latter mechanism gives the mfp an $\exp(1/T)$ dependence at high temperatures and $1/T$ dependence at lower temperatures where Umklapp processes are dominant. Size effects occur in thin samples or at low temperatures where the mfp is long. Since C, the heat capacity, varies as T^3 at low temperatures, the size-effect thermal conductivity should show a T^3 variation, which is verified in a number of cases (see the review by Berman [252]).

In normal pure metals, the electrons usually carry almost all the heat current, whereas in very impure metals or in disordered alloys, the phonon contribution may be comparable with the electron contribution since collisions between electrons and phonons become important. Thermal conductivity for a metal is therefore generally the sum of both electron and phonon contributions, i.e.,

$$K = K_e + K_p \tag{91}$$

The electron term follows a $(A/T + BT^n)$ variation and the phonon term shows CT^n dependence, where A, B, and C are constants. From Eqs. (90) and (20), the following relation between σ and K is obtained:

$$\frac{K}{\sigma T} = \frac{\pi^2 k^2}{3e^2} \frac{l_{th}}{l_{el}} \tag{92}$$

If the free paths l_{th} and l_{el} for thermal and electronic conduction are equal (i.e., similar scattering mechanisms are operative in both types of transport processes), then

$$\frac{K}{\sigma T} = \frac{\pi^2 k^2}{3e^2} \equiv \mathcal{L} = 2 \cdot 45 \times 10^{-8} \text{ watt-ohm/deg}^2 \tag{93}$$

This is the Wiedemann-Franz law, and the constant $\mathcal{L}$ is the Lorentz constant. In the regime of the validity of this relation, $\mathcal{L}$ is independent of temperature. This relation is obeyed in impurity-dominated conduction at very high and very low temperatures.

Thermal conductivity should exhibit size effects of the same type, as discussed in Sec. 3 for electrical conduction. Indeed, we can borrow the same formulation for calculating the effective values of the free paths to account for the size effects. In addition, galvanomagnetomorphic effects analogous to those described in Sec. 4 should occur. A study of these size effects should yield information on the various scattering processes and the corresponding free path lengths. The data on thermal conductivity of thin films are also of much value in understanding the role of heat transfer during nucleation and growth of films, the thermal failure of microelectronic circuitry, thermal aging effects, etc.

Despite the fundamental and technical interest in this area, very little experimental work has been done. Scanty data exist on thin wires and foils at low temperatures in the size-effect regime. Size-dependent thermal conductivity was observed by White and Woods [253] in Na wires, by Wyder [254, 255] in In wires, and by Amundsen and Olsen [205] in Al foils. Of particular interest are the electrical- and thermal-conductivity size-effect studies by Wyder on the same In wire specimens to determine l_{el} and l_{th}. An approximate agreement with Eq. (92) was verified. Wyder also measured [255] the thermal conductivity of various In wire specimens, above and below the superconducting transition temperature of In, and established that the mfp of thermal electrons is the same in the normal and in the superconducting state.

Amundsen and Olsen [205] studied the thermal and electrical resistivities of rolled Al foils of thickness ranging from 0.20 to 0.066 mm in a transverse magnetic field at 4.2°K. The high resistivity ratio of the pure foils ensured the size-effect regime. The results for the thermal resistivity showed size effects similar to those predicted by Sondheimer and by MacDonald and Sarginson for the electrical conductivity. The oscillations in the thermal resistivity were observed in the Sondheimer position (see Fig. 21). These observations are in qualitative agreement with the theory. That the observed amplitude is lower than theoretically predicted is probably due to the unsatisfactory treatment of the normal magnetoresistance effect. Since both electrical and thermal resistivities exhibit oscillatory behavior, the Lorentz constant also shows an oscillatory variation.

The thermal magnetoresistance of Al foils is found to obey Kohler's rule provided that the normal resistance is taken to include the zero-field size-effect value. It is also found that Kohler's rule for thermal magnetoresistance is the same as that for electrical magnetoresistance. Discrepancies in the MacDonald-Sarginson position are observed at high fields and are similar to those observed in electrical resistance (Sec. 4.4). These discrepancies may be explained as due to additional scattering by open electron orbits.

6.2 Thermoelectric Power

Thermoelectric power is the emf produced per unit temperature difference between the two junctions of materials. Among the various transport properties, thermopower is perhaps the most sensitive to distortions of the Fermi surface. The free-electron theory describes the conductivity and galvanomagnetic behavior of noble metals reasonably well. However, it predicts a negative thermopower in noble metals in sharp contrast to the positive one observed above about 40°K. Although no theory [256] presently explains this anomalous behavior of thermo-power of noble metals, it is generally assumed to be the result of their Fermi surface touching the zone boundary. The positive thermopower must be explained by the energy-dependent terms of the mfp and/or the Fermi surface area in the electronic contribution to thermopower (the phonon drag component is negligible at and above room temperature). The electronic thermopower of a pure metal (bulk) as given by the free-electron model is

$$S_B = -\frac{\pi^2}{3e}\frac{k^2 T}{\zeta}(U + V) \tag{94}$$

where

$$U = \left(\frac{\partial \ln l}{\partial \ln E}\right)_{E = \zeta}, \quad V = \left(\frac{\partial \ln A}{\partial \ln E}\right)_{E = \zeta},$$

A is the area of the Fermi surface, ζ the Fermi energy, T the temperature, and e the electronic charge. In the free-electron approxi-mation, $V = 1$. According to the Bloch quantum theory of electrical conduction in metals, $U = 2$ (i.e., $l \propto E^2$). It is obvious that these values of U and V cannot explain the observed positive sign of S_B. The free-electron size-effect theory yields the following asymptotic expres-sions for the thermopower of thin films [2]:

$$S_F = S_B\left[1 - \frac{3}{8}\frac{(1 - p)}{\gamma}\frac{U}{1 + U}\right] \quad (\gamma \gg 1) \tag{95}$$

and

$$S_F = S_B\left[1 + \frac{U}{1 + U}\frac{\ln\gamma - 1.42}{\ln\gamma - 0.42}\right] \quad (\gamma \ll 1; p \sim 0) \tag{96}$$

Note that in this theory, it is assumed that scattering of carriers at the surface and in the bulk are additive processes (Matthiessen's rule), and the mean free paths for electronic and thermal relaxation processes are the same (Wiedman-Franz law).

From Eqs. (94) and (95), we obtain

$$\Delta S_F \equiv S_B - S_F = \frac{-\pi^2}{8e} \frac{k^2 T}{\zeta} \frac{(1-p)U}{\gamma} = -9.2 \times 10^{-3} \frac{T(1-p)U}{\zeta \gamma}$$

$$\mu V/\text{deg}$$

(97)

Thus, a plot of ΔS_F vs. $1/t$ should yield a straight line through the origin, the slope of which determines $Ul(1-p)$. Since the value of p is nearly zero for polycrystalline films, and the effective mfp l_{eff} can be calculated from Eq. (22), the value of U can thus be deduced.

Results on the thickness dependence of the size-effect contribution for alkali-metal films (K, Rb, and Cs) [58] agree well with Eq. (95), and a linear relation between $(S_B - S_F)$ and $1/t$ is verified. Results for Co films are shown in Fig. 28. Comparison with the calculated curves

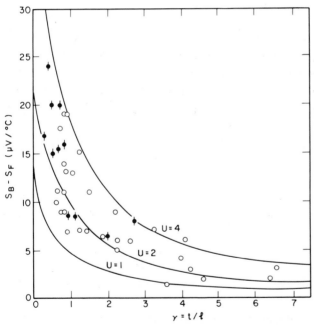

FIG. 28. The size-effect contribution to the thermoelectric power (the difference between the bulk and the observed thin film values) as a function of the normalized thickness of evaporated Co films. Measurements were made against bulk Ag. The solid curves represent Eq. (95) for $U = 1, 2,$ and 4 with a value of $l = 100$ Å (obtained from the TCR data, on the same films). (*Savornin* [257])

(assuming $l = 100$ Å and $p = 0$), indicates fair fit for $U = 2$. This means $l \propto E^2$, in agreement with Bloch's theory. Similar results have been obtained on films of Al and Co [257] and Bi and Pb [258]. Reimer [259], however, found no size effects in Cu and Ni films down to a thickness of 30 Å but his results are questionable because of the poor quality of his films.

Huebener [260] studied thermopower of thin Au foils against Au wires between 77 and 296°K and obtained $U = -0.53 + 0.19$. Worobey et al. [261, 262] studied evaporated Au films from 1.4 to 300°K and obtained $U = -0.61 + 0.20$ above 200°K. The magnitude and sign of low-temperature thermopower were found [262] to be very sensitive to oxidation. An anomalously large value of $U = -18.7$ was deduced by Chopra et al. [263] for Cu films from thin film Cu-constantan thermocouple data which showed an excellent qualitative fit with Eq. (95). This anomaly is not understood and may be due to the properties of a transition region formed between Cu and constantan films by diffusion.

The aforementioned results clearly indicate a very unsatisfactory understanding of the phenomenon of thermopower. The observed sign of thermopower in noble metals definitely requires a negative value of U and, therefore, the free-electron theory cannot apply here. We must take into account the distortions in the Fermi surface. It is also necessary to perform more careful experiments on films of noble metals so that consistent values of U may be obtained to help guide theoretical approaches.

Large contributions to thermopower in metals are provided at low temperatures by the existence of an electron drag induced by the flux of phonons. This generally arises at low temperatures when the actual mfp of phonons is much longer than the mfp for phonon-electron interaction. Size-effect reduction in the phonon-drag component of the thermopower should occur when phonons are appreciably scattered at the surface of a finite specimen. This effect can yield information on the phonon mean free path.

Huebener [264] compared the thermoelectric power data on a Pt foil with that of a thick Pt wire at low temperatures and thus determined the size-effect phonon drag component. The mfp was calculated by using a Fuchs type of size-effect equation [Eq. (38)] and was found to depend exponentially on temperature, indicating that phonon-phonon Umklapp processes are the main source of the lattice thermal resistivity in this case. If phonons interact only with electrons, a T^3 dependence of the phonon drag component is expected. Studies [261] on Au films showed an approximate agreement with this relation.

Since the thermopower of films approaches bulk value for small thicknesses, the use of films in low-thermal-capacity thermocouples for measurement of surface temperatures is suggested. The preparation and use of thin-film thermocouples has been discussed by a number of workers [263, 265-270]. The possible use of suitable alloy films as thermogenerators has been discussed by Abowitz et al. [268].

Marshall et al. [269] studied thermocouples made from a combination of evaporated films of Ni, Fe, Cu, constantan, Chromel, and Alumel, and found that a maximum thermal emf is obtained when the film thickness of each element is greater than 2,500 Å. Thermal emf's of 0.027 and 0.035 mV/°C for Ni-Fe and Cu-constantan thermocouples, respectively, were obtained. Typical thickness dependence of the thermo emf of Cu-constantan thermocouples obtained by Chopra et al. [263] is shown in Fig. 29. These authors also studied the response time of such thermocouples deposited on glass slides by simultaneously displaying the heating pulse from a neodymium doped laser and the thermal emf generated. A response time of better than 10^{-6} sec was measured

6.3 Heat Transport across Film-Insulator Interface

Thin films are generally deposited on insulating substrates. The heat transfer from the film into the substrate determines the equilibrium

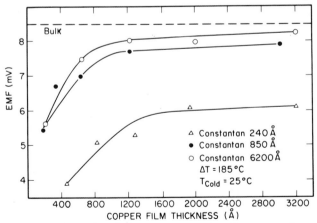

FIG. 29. Thermal emf of copper-constantan thermocouples as a function of the thickness of Cu films for various constantan film thicknesses. (*Chopra et al.* [263])

temperature of a microelectronic device. Heat flow may become a limiting factor [271] in the performance of certain thin-film devices and is therefore a subject of considerable interest [272-273].

Heat transport across a metal-film insulator interface is also of interest for an understanding of the nature of the thermal barrier [274] at the interface. This interest has been further stimulated by the possible use of thin films as high-frequency thermometers [275] for studying the existence of second sound in solids and for other technical applications. If phonons carry heat, then the thermal barrier at the interface is due to the limited number of phonons available. The barrier is most apparent at low temperatures where the mfp is long and the number of excited phonons is small. The source of the barrier can be considered to be localized within a distance of one phonon mfp. An acoustic mismatch [274] between the materials on the two sides of the interface can also be used to explain the thermal barrier. Since heat in metals is carried predominantly by the electrons, the application of either the phonon-blackbody-radiation or acoustic-mismatch concept to heat transfer across a metal-insulator interface should be accepted with caution.

The heat-transport problem has been subjected to several investigations [276-279]. Gutfeld et al. [276,277] studied the thermal relaxation times at 4 to 300°K in In films, and at 4°K in Pb films evaporated onto cylindrical sapphire and quartz single-crystal substrates. The evaporated film was heated with a short (~4 nsec for GaAs laser and ~25 nsec for ruby laser) light pulse from a laser. The subsequent rate of temperature decrease was monitored by the observation of the resistance of the film. At low temperatures, a high sensitivity of resistance change was achieved by operating the film very near to its superconducting transition temperature.

The decay times characterizing the initial part of the thermal decay of a 2,200-Å In film on a sapphire substrate were measured to be as fast as 2×10^{-9} sec near 100°K, increasing to about 8×10^{-9} sec at room temperature [277]. These results are in agreement with a thermal-diffusion theory. Below 10°K, decay times as long as 30×10^{-9} sec were found, which correspond to ballistic phonon flow, i.e., perfect black-body phonon radiation from the film into the insulator. The observed values of the relaxation speeds are, however, smaller than predicted by any of the proposed models.

The short thermal decay time (~10^{-8} sec) observed on quartz and sapphire substrates suggests the possibility of using thin films as high-speed thermometers over a wide temperature range. Films of Au and Ag which show a high TCR and a linear dependence of the resistance on temperature down to about $40°K$, and also films of semiconducting materials (e.g., Ge) with a large negative TCR, have been shown in the author's laboratory to be useful for such applications.

7. CONCLUDING REMARKS

Two significant conclusions have emerged from the experimental studies of the various electron-transport phenomena in metal films. First, it is possible to prepare suitable metal films which exhibit bulklike transport properties at film thicknesses comparable with or smaller than the carrier mean free path. Secondly, if metal films exhibit size effects, their studies yield reasonable values of such bulk-transport parameters as the number of free electrons per atom, mean free path and its energy dependence, mobility, and scattering coefficient. These conclusions thus establish the usefulness of conveniently available thin-film specimens to obtain information on the transport processes.

Several areas of technical and fundamental importance which should attract future theoretical and experimental studies are (1) the physical origin and the amount of diffuse scattering of carriers at a free surface or an interface; (2) conduction mechanisms in structures with a high concentration of lattice defects, and in metastable, amorphous metal and alloy films; (3) more refined theories and experimental studies of size effects in the Hall-coefficient, magnetoresistance, and magnetic-field orientation-dependent effects; (4) a clearer understanding of the energy-loss mechanisms during hot carrier transport and the respective mean free paths; and (5) the energy dependence of the mean free path of conduction electrons in relation to the poorly understood phenomenon of the sign of the thermoelectric power.

REFERENCES

1. J. Moser, *Wied. Ann.*, **42**:639 (1891).
2. H. Mayer, "Physik dünner Schichten," vol. 2, Wissenschaftliche Verlag, Stuttgart, 1955.
3. C. A. Neugebauer, in "Physics of Thin Films" (G. Hass and R. E. Thun, eds.), vol. 2, p. 1, Academic Press Inc., New York, 1964.

4. C. A. Neugebauer and R. H. Wilson, in "Basic Problems in Thin Film Physics" (R. Niedermayer and H. Mayer, eds.), p. 579, Vandenhoeck and Ruprecht, Göttingen, 1966.
5. N. M. Bashara, *IEEE Trans. Component Pts.*, **CP-11** (1):4 (1964).
6. N. Nifontoff, *Compt. Rend.*, **236**:1538, 2486 (1953); **237**:24 (1953).
7. J. G. Simmons, *J. Appl. Phys.*, **35**:2472 (1964).
8. S. S. Minn, *J. Rech. Centre Natl. Rech. Sci. Lab. Bellevue Paris* **51**:131 (1960).
9. H. Mayer, R. Nossek, and H. Thomas, *J. Phys. Radium*, **17**:204 (1956).
10. A. Blank-Lapierre and N. Nifontoff, *J. Phys. Radium*, **17**:230 (1956).
11. R. Stratton, *J. Phys. Chem. Solids*, **23**:1177 (1962).
12. C. J. Gorter, *Physica*, **17**:777 (1951).
13. G. Darmois, *J. Phys. Radium*, **17**:211 (1956).
14. C. A. Neugebauer and M. B. Webb, *J. Appl. Phys.*, **33**:74 (1962).
15. T. E. Hartman, *J. Appl. Phys.*, **34**:943 (1963).
16. N. F. Mott and W. D. Twose, *Advan. Phys.*, **10**:107 (1961).
17. A. A. Milgram and C. Lu. *J. Appl. Phys.*, **37**:4773 (1966).
18. N. Mostovetch and B. Vodar, "Semiconducting Materials," Proceedings of Conference University of Reading, p. 260, Butterworth & Co. (Publishers), Ltd., London, 1951.
19. B. Vodar, *J. Phys. Radium*, **17**:213, 298 (1956).
20. N. M. Bashara and L. A. Weitzenkamp, *J. Appl. Phys.*, **35**:1983 (1964).
21. J. G. Skofronick and W. B. Phillips, in "Basic Problems in Thin Film Physics" (R. Niedermayer and H. Mayer, eds.), p. 591, Vandenhoeck and Ruprecht, Göttingen, 1966.
22. L. Y. Wei, *J. Chem. Phys.*, **39**:2709 (1963).
23. D. S. Herman and T. N. Rhodin, *J. Appl. Phys.*, **37**:1594 (1966).
24. R. M. Hill, *Nature*, **204**:35 (1964).
25. F. P. Fehlner and S. M. Irving, *J. Appl. Phys.*, **37**:3313 (1966).
26. K. van Steensel, *Philips Res. Rept.*, **22**:246 (1967).
27. A. A. Hirsch and S. Bazian, *Physica*, **30**:258 (1964).
28. P. G. Borzjak, O. G. Sarbej, and R. D. Fedorowitsch, *Phys. Stat. Solidi*, **8**:55 (1965).
29. C. A. Neugebauer, *Trans. 9th Natl. Vacuum Symp.*, p. 45, The Macmillan Company, New York, 1962.
30. C. Feldman, *J. Appl. Phys.*, **34**:1710 (1963).
31. R. B. Belser and W. H. Hicklin, *J. Appl. Phys.*, **30**:313 (1959).
32. See, for example, G. Siddall in "Thin Film Microelectronics" (L. Holland, ed.), p. 1, John Wiley & Sons, Inc., New York, 1966.
33. K. L. Chopra, unpublished, Ledgemont Laboratory Invention Disclosure, 1966.
34. L. I. Maissel, *Trans. 9th Natl. Vacuum Symp.*, p. 169, The Macmillan Company, New York, 1962; P. M. Schaible and L. I. Maissel, *Trans. 9th Natl. Vacuum Symp.*, p. 190, The Macmillan Company, New York, 1962.
35. C. E. Drumheller, *J. Phys.*, **25**:198 (1964).
36. R. Grigorovici and G. Ciobanu, in "Basic Problems in Thin Film Physics" (R. Niedermayer and H. Mayer, eds.), p. 596, Vandenhoeck and Ruprecht, Göttingen, 1966.
37. R. Grigorovici et al., *J. Phys.*, **27**:133 (1966); *Z. Physik*, **160**:277 (1960).

38. R. L. Parker and A. Krinsky, *J. Appl. Phys.*, **34**:2700 (1963); J. LeBas, *Compt. Rend.*, **258**:4952 (1964).
39. R. Koike and H. Kurokawa, *Japan. J. Appl. Phys.*, **6**:546 (1967).
40. J. M. Ziman, "Electrons and Phonons," Oxford University Press, Fair Lawn, N.J., 1962.
41. J. J. Thomson, *Proc. Cambridge Phil. Soc.*, **11**:120 (1901).
42. K. Fuchs, *Proc. Cambridge Phil. Soc.*, **34**:100 (1938).
43. F. H. Sondheimer, *Phys. Rev.*, **80**:401 (1950).
44. E. H. Sondheimer, *Advan. Phys.*, **1**:1 (1952).
45. M. S. P. Lucas, *J. Appl. Phys.*, **36**:1632 (1965).
46. R. G. Chambers, *Proc. Roy. Soc. London*, **A202**:378 (1950).
47. F. S. Ham and D. C. Mattis, *IBM J. Res. Develop.*, **4**:143 (1960).
48. P. J. Price, *IBM J. Res. Develop.*, **4**:152 (1960).
49. M. I. Kaganov and M. Ya. Azbel, *Soviet Phys. JETP English Transl.*, **27**:762 (1954).
50. S. B. Soffer (unpublished), 1966.
51. S. B. Soffer, *J. Appl. Phys.*, **36**:3947 (1965).
52. D. K. C. MacDonald and K. Sarginson, *Proc. Roy. Soc. London*, **A203**:223 (1950).
53. R. B. Dingle, *Proc. Roy. Soc. London*, **A201**:545 (1950).
54. H. Juretschke, *J. Appl. Phys.*, **37**:435 (1966).
55. C. Schüler, *Symp. Elec. Magnetic Properties Thin Metallic Layers*, Louvain, 1961, p. 30.
56. E. T. S. Appleyard and A. C. B. Lovell, *Proc. Roy. Soc. London*, **A158**:718 (1937).
57. R. Nossek, *Z. Physik*, **142**:321 (1955); *Z. Naturforsch.*, **16a**:1162 (1961).
58. A review by H. Mayer in "Structure and Properties of Thin Films" (C. A. Neugebauer, J. B. Newkirk, and D. A. Vermilyea, eds.), p. 225, John Wiley & Sons, Inc., New York, 1959.
59. W. Cirkler, *Z. Physik*, **147**:481 (1957).
60. F. W. Reynolds and G. R. Stilwell, *Phys. Rev.*, **88**:418 (1952).
61. E. J. Gillham, J. S. Preston, and B. E. Williams, *Phil. Mag.*, **46**:1051 (1955).
62. A. J. Learn and R. S. Spriggs, *J. Appl. Phys.*, **34**:3012 (1963).
63. A. E. Ennos, *Brit. J. Appl. Phys.*, **8**:113 (1957).
64. K. L. Chopra, L. C. Bobb, and M. H. Francombe, *J. Appl. Phys.*, **34**:1699 (1963).
65. K. L. Chopra and L. C. Bobb, *Acta Met.*, **12**:807 (1964).
66. K. L. Chopra and L. C. Bobb, in "Single Crystal Films" (M. H. Francombe and H. Sato, eds.), p. 373, Pergamon Press, New York, 1964.
67. C. E. Ellis and G. D. Scott, *J. Appl. Phys.*, **23**:31 (1952).
68. I. G. Young and C. W. Lewis, *Trans. 10th Natl. Vacuum Symp.*, p. 428, The Macmillan Company, New York, 1963.
69. R. B. Marcus, *J. Appl. Phys.*, **37**:3121 (1966).
70. J. S. Hirschhorn, *J. Phys. Chem. Solids*, **23**:1821 (1962).
71. E. C. Crittenden, Jr., and R. W. Hoffman, *Rev. Mod. Phys.*, **25**:310 (1953).
72. K. H. Rosette and R. W. Hoffman, in "Structure and Properties of Thin Films" (C. A. Neugebauer, J. B. Newkirk, and D. A. Vermilyea, eds.), p. 370, John Wiley & Sons, Inc., New York, 1959.

73. R. Grigorovici, A. Dévényi, and T. Botilă, *J. Phys. Chem. Solids,* **23**:428 (1962).
74. E. R. Andrew, *Proc. Phys. Soc. London,* **A62**:77 (1949).
75. J. Niebuhr, *Z. Physik,* **132**:468 (1952).
76. B. N. Aleksandrov, *Soviet Phys. JETP English Transl.,* **16**:286 (1963).
77. D. C. Larson and B. T. Boiko, *Appl. Phys. Letters,* **5**:155 (1964).
78. R. V. Isaeva, *Soviet Phys. JETP Letters English Transl.,* **4**:209 (1966).
79. M. J. Skove and E. P. Stillwell, *Appl. Phys. Letters,* **7**:241 (1965).
80. J. L. Olsen, *Helv. Phys. Acta,* **31**:713 (1958).
81. F. J. Blatt and H. G. Satz, *Helv. Phys. Acta,* **33**:1007 (1960).
82. M. Ya. Azbel and E. A. Kaner, *Soviet Phys. JETP English Transl.,* **5**:730 (1957).
83. M. Ya. Azbel and R. N. Gurzhi, *Soviet Phys. JETP English Transl.,* **15**:1133 (1962).
84. R. N. Gurzhi, *Soviet Phys. JETP English Transl.,* **20**:953 (1965).
85. A. Gaide and P. Wyder, in *Symp. Elec. Magnetic Properties Thin Metallic Layers,* Louvain, 1961, p. 411.
86. K. L. Chopra, *Phys. Letters,* **15**:21 (1965).
87. M. I. Kaganov and M. Ya Azbel, *Zh. Eksperim. i Teor. Fiz.,* **27**:762 (1954).
88. R. Engleman and E. H. Sondheimer, *Proc. Soc. London,* **B69**:449 (1956).
89. M. S. P. Lucas, *Appl. Phys. Letters,* **4**:73 (1964).
90. K. L. Chopra and M. R. Randlett, *J. Appl. Phys.,* **38**:3144 (1967).
91. For example, a review by D. S. Campbell, in "The Use of Thin Films in Physical Investigations" (J. C. Anderson, ed.), p. 299, Academic Press Inc., New York, 1966.
92. F. Savornin, *Ann. Phys.,* **5**:1355 (1960); *Compt. Rend.,* **248**:2458 (1959).
93. W. F. Leonard and R. L. Ramey, *J. Appl. Phys.,* **37**:3634 (1966).
94. A. N. Friedman and S. H. Koenig, *IBM J. Res. Develop.,* **4**:158 (1960).
95. See, for example, A. Many, Y. Goldstein, and N. B. Grover, "Semi-conductor Surfaces," North Holland Publishing Company, Amsterdam, 1965.
96. R. F. Greene, *Surface Sci.,* **2**:101 (1964).
97. R. F. Greene, *Phys. Rev.,* **141**:687 (1966).
98. R. F. Greene and R. W. O'Donnell, *Phys. Rev.,* **147**:599 (1966).
99. E. A. Stern, *Phys. Rev.,* **162**:565 (1967).
100. J. E. Parrott, *Proc. Phys. Soc. London,* **85**:1143 (1965).
101. G. Brändli and P. Cotti, *Helv. Phys. Acta,* **38**:801 (1965).
102. G. E. Smith, *Phys. Rev.,* **115**:1561 (1959).
103. S. B. Soffer, *J. Appl. Phys.,* **38**:1710 (1967).
104. J. E. Aubrey, C. James, and J. E. Parrott, *Proc. Intern. Conf. Semicond., Paris,* 1964, p. 689.
105. K. L. Chopra, *Phys. Rev.,* **155**(3):660 (1967).
106. F. Abeles and M. L. Theye, *Phys. Letters,* **4**:348 (1963).
107. R. B. Dingle, *Physica,* **19**:729 (1953).
108. H. E. Bennett and J. M. Bennett, in "Optical Properties and Electronic Structure of Metals and Alloys" (F. Abeles, ed.), p. 175, North Holland Publishing Company, Amsterdam, 1966.
109. H. E. Bennett, J. M. Bennett, E. J. Ashley, and R. J. Motyka, *Phys. Rev.,* **165**:755 (1967).
110. A. Lonke and A. Ron, *Phys. Rev.,* **160**(3):577 (1967).

111. R. F. Furth and E. Morris, *Proc. Phys. Soc. London,* **73**:869 (1959).
112. H. Juretschke, *Surface Sci.,* **2**:40 (1964).
113. H. Juretschke, *Surface Sci.,* **5**:111 (1966).
114. J. R. Schrieffer, *Phys. Rev.,* **97**:641 (1955).
115. E. C. McIrvine, *Surface Sci.,* **5**:171 (1966).
116. E. Bose, *Z. Physik,* **7**:373, 462 (1906).
117. A. Deubner and K. Rambke, *Ann. Physik,* **17**:317 (1956).
118. G. Bonfigioli and R. Malvano, *Phys. Rev.,* **115**:330 (1959).
119. R. E. Glover, III, and M. D. Sherrill, in *Symp. Elec. Magnetic Properties Thin Metallic Layers,* Louvain, 1961, p. 316.
120. A. Berman (unpublished), cited in Ref. 112.
121. H. L. Stadler, *Phys. Rev. Letters,* **14**:979 (1965).
122. O. Beeck, A. E. Smith, and A. Wheeler, *Proc. Roy. Soc. London,* **A177**:62 (1941).
123. O. Beeck, *Advan. Catalysis,* **2**:151 (1950).
124. M. C. Johnson and T. V. Starkey, *Proc. Roy. Soc. London,* **A140**:126 (1933).
125. L. R. Ingersoll and J. D. Hanewalt, *Phys. Rev.,* **33**:1094 (1929); **34**:972 (1929).
126. A. van Itterbeek and L. deGreve, *Physica,* **15**:80 (1949); A. van Itterbeek, L. deGreve, and F. Heremans, *Appl. Sci. Res.,* **B2**:352 (1952).
127. C. Y. Bartholomew and A. A. LaPadula, *J. Appl. Phys.,* **31**:445 (1960).
128. H. F. Winters and E. Kay, *J. Appl. Phys.,* **38**:3928 (1967); H. F. Winters, *J. Chem. Phys.,* **14**:550 (1963).
129. G. F. Ivanovskii and T. D. Radzhabov, *Soviet Phys.-Solid State English Transl.,* **8**:1013 (1966).
130. B. Navinsek and G. Carter, *Appl. Phys. Letters,* **10**:91 (1967).
131. V. Teodosic, *Appl. Phys. Letters,* **9**:209 (1966); K. L. Merkle and L. R. Singer, *Appl. Phys. Letters,* **11**:35 (1967).
132. E. Raub, *Z. Metallk.,* **39**:33 (1948).
133. J. Kramer and H. Zahn, *Naturwiss.,* **20**:792 (1932).
134. A. Fery, *Ann Phys. Paris,* **19**:421 (1933).
135. See, for example, R. Suhrmann, *Advan. Catalysis,* **7**:303 (1955); R. Suhrmann and G. Wedler, *Advan. Catalysis,* **9**:258 (1960).
136. P. Zwietering, H. L. T. Koks, and C. van Heerden, *J. Phys. Chem. Solids,* **11**:18 (1959).
137. C. van Heerden and P. Zwietering, *Koninkl. Ned. Akad. Wetenschap.,* **B60**:160 (1957).
138. G. Bliznakov and D. Lazarow, *Z. Phys. Chem. Leipzig,* **223**:33 (1963).
139. J. J. Wortman and K. S. Canady, *Appl. Phys. Letters,* **9**:75 (1966).
140. J. P. Borel, *Compt. Rend.,* **233**:296 (1951); K. L. Chopra (unpublished).
141. R. Suhrmann and W. Berndt, *Z. Physik,* **115**:17 (1940).
142. L. Hamburger and W. Reinders, *Res. Trans. Chim. Pays-Bas,* **50**:441 (1931).
143. E. Armi, *Phys. Rev.,* **63**:451 (1943).
144. J. G. Skofronick and W. B. Phillips, *Appl. Phys. Letters,* **7**:249 (1965).
145. R. Suhrmann and H. Schnackenberg, *Z. Physik,* **119**:287 (1942).
146. W. Buckel and R. Hilsch, *Z. Physik,* **131**:420 (1952); **138**:109 (1954); see also R. Hilsch, in "Non-crystalline Solids" (V. D. Frechette, ed.), p. 348, John Wiley & Sons, Inc., New York, 1958.
147. W. Buckel, *Z. Physik,* **138**:136 (1954).

148. W. Mönch, *Z. Physik,* **170**:93 (1962); W. Monch and W. Sander, *Z. Physik,* **157**:149 (1959).
149. W. Sander and E. Strieder, *Z. Physik,* **188**:99 (1965).
150. C. J. Meechan and J. A. Brinkham, *Phys. Rev.,* **103**:1193 (1956).
151. H. G. van Bueren, "Imperfections in Crystals," Interscience Publishers, Inc., New York, 1960.
152. A. Seeger, P. Schiller, and H. Kronmuller, *Phil. Mag.,* **5**:853 (1960).
153. J. W. Marx, H. G. Cooper, and J. W. Henderson, *Phys. Rev.,* **88**:106 (1952).
154. A. Kinbara and Y. Sawaturi, *Japan. J. Appl. Phys.,* **4**:161 (1965).
155. W. Rühl, *Z. Physik,* **138**:121 (1954).
156. H. Richter, H. Berckhemer, and G. Breitling, *Z. Naturforsch.,* **9a**:236 (1954).
157. E. Feldtkeller, *Z. Physik,* **157**:65 (1959).
158. S. Mader, A. S. Nowick, and H. Widmer, *Acta Met.,* **15**:203 (1967).
159. S. Mader, *J. Vacuum Sci. Tech.,* **2**:35 (1965).
160. K. L. Chopra, to be published in *J. Appl. Phys.* (1969).
161. See, for example, N. F. Mott, *Advan. Phys.,* **16**:49 (1967).
162. R. B. Belser, *J. Appl. Phys.,* **28**:109 (1957).
163. See, for example, Refs. 61 and 65.
164. V. Vand, *Proc. Phys. Soc. London,* **55**:222 (1943).
165. P. G. Wilkinson, *J. Appl. Phys.,* **22**:419 (1951).
166. V. V. Shah and Y. G. Naik, *Indian J. Pure Appl. Phys.,* **3**:20 (1965).
167. H. S. Story and R. W. Hoffman, *Proc. Phys. Soc. London,* **B70**:950 (1957).
168. F. G. Fumi, *Phil. Mag.,* **46**:1007 (1955).
169. N. Schwartz and R. W. Berry in "Physics of Thin Films" (G. Hass and R. E. Thun, eds.) vol. 2, p. 363, Academic Press Inc., New York, 1964.
170. A. E. Lessor, L. I. Maissel, and R. E. Thun, *Semicond. Prod. Solid State Tech.,* March-April, 1964.
171. E. H. Layer, *Proc. Natl. Electron. Conf.,* **20**:191 (1964).
172. R. Glang, *J. Vacuum Sci. Tech.,* **3**:37 (1966).
173. J. K. Howard and R. F. Ross, *Appl. Phys. Letters,* **11**:85 (1967); P. B. Ghate, *Appl. Phys. Letters,* **11**:14 (1967).
174. J. A. Cunningham, *Solid-State Electron.,* **8**:735 (1965).
175. K. L. Chopra, M. R. Randlett, and R. H. Duff, *Phil. Mag.,* **16**:261 (1967).
176. D. A. McLean, *Proc. Natl. Electron. Conf.,* **16**:206 (1960).
177. R. W. Berry and N. Schwartz, *Proc. 4th Natl. Conv. Military Electron.,* p. 214, 1960.
178. C. Altman, *Trans. 9th Natl. Vacuum Symp.,* p. 174, The Macmillan Company, New York, 1962.
179. D. Gerstenberg and W. H. Mayer, *Electron. Reliability Micromin.,* **1**:353 (1962).
180. K. B. Scow and R. E. Thun, *Trans. 9th Natl. Vacuum Symp.,* p. 151, The Macmillan Company, New York, 1962.
181. W. Himes, B. F. Stout, and R. E. Thun, *Trans. 9th Natl. Vacuum Symp.,* p. 144, The Macmillan Company, New York, 1962.
182. E. H. Layer and H. R. Olson, *Elec. Mfg.,* **58**:104 (1956).
183. E. H. Layer, *Trans. 6th Natl. Vacuum Symp.,* p. 210, Pergamon Press, New York, 1959.
184. I. H. Pratt, *Proc. Natl. Electron. Conf.,* **20**:215 (1964).
185. E. R. Dean, *J. Appl. Phys.,* **35**:2930 (1964).

186. T. K. Lakshmanan, *Trans. 8th Natl. Vacuum Symp.*, p. 868, Pergamon Press, New York, 1961.
187. L. Holland and G. Siddall, *Vacuum,* **3**:375 (1953).
188. R. W. Berry, W. H. Jackson, G. I. Parisi, and A. H. Schafer, *IEEE Trans. Component Pts.*, **CP-11**(2):86 (1964).
189. For example, R. E. Scott, *J. Appl. Phys.*, **38**:2652 (1967).
190. M. Beckerman and R. E. Thun, *Trans. 8th Natl. Vacuum Symp.*, p. 905, Pergamon Press, New York, 1961.
191. N. C. Miller and G. A. Shirn, *Appl. Phys. Letters,* **10**:86 (1967).
192. W. O. Freitag and V. R. Weiss, *Res. Develop.*, August, 1967, p. 44.
193. W. J. Ostrander and C. W. Lewis, *Trans. 8th Natl. Vacuum Symp.*, p. 881, Pergamon Press, New York, 1961.
194. E. A. Kaner, *Soviet Phys. JETP English Transl.* **7**:454 (1958).
195. J. L. Olsen, *Helv. Phys. Acta,* **31**:713 (1958).
196. O. S. Lutes and D. A. Clayton, *Phys. Rev.*, **138**:A1448 (1965); F. De La Cruz, M. E. De La Cruz, and J. M. Cotignola, *Phys. Rev.*, **163**:575 (1967).
197. R. G. Chambers, a review given in Ph.D. Dissertation, Cambridge, 1951.
198. E. Koenigsberg, *Phys. Rev.*, **91**:8 (1953).
199. Yi-Han Kao, *Phys. Rev.,* **138**:A1412 (1965).
200. J. Babiskin, *Phys. Rev.,* **107**:981 (1957).
201. M. C. Steele, *Phys. Rev.,* **97**:1720 (1955).
202. V. L. Gurevich, *Soviet Phys. JETP English Transl.* **8**:464 (1959).
203. J. Babiskin and P. G. Siebenmann, *Phys. Rev.,* **107**:1249 (1957).
204. K. Førsvoll and I. Holwech, *Phil. Mag.,* **9**:435 (1964).
205. T. Amundsen and T. Olsen, *Phil. Mag.,* **11**:561 (1965).
206. J. Feder and T. Jøssang, *Phys. Norvegica,* **1**:217 (1964).
207. P. Cotti, *Proc. Intern. Conf. High Magnetic Fields (MIT)*, 1961, p. 539.
208. P. Cotti, J. L. Olsen, J. G. Daunt, and M. Kreitman, *Cryogenics,* **4**:45 (1964).
209. K. L. Chopra, *Bull. Am. Phys. Soc.,* **10**(1):126 (1965).
210. D. C. Larson and R. V. Coleman, in "Basic Problems in Thin Film Physics" (R. Niedermayer and H. Mayer, eds.), p. 574, Vandenhoeck and Ruprecht, Göttingen, 1966.
211. D. K. C. MacDonald, *Phil. Mag.,* **2**:97 (1957).
212. K. Førsvoll and I. Holwech, *Phil. Mag.,* **10**:921 (1964); I. Holwech, *Phil. Mag.,* **12**:117 (1965).
213. P. Cotti, *Helv. Phys. Acta,* **34**:777 (1961).
214. H. J. Lippmann and F. Kuhrt, *Z. Naturforsch.,* **13a**:462 (1958).
215. N. H. Zebouni, R. E. Hamburg, and H. J. Mackey, *Phys. Rev. Letters,* **11**:260 (1963).
216. E. H. Hall, *Phil. Mag.,* **11**:157 (1881).
217. W. F. Leverton and A. J. Dekker, *Phys. Rev.,* **80**:732 (1950); **81**:156 (1951).
218. G. R. Wait, *Phys. Rev.,* **19**:615 (1922).
219. J. C. Steinberg, *Phys. Rev.,* **21**:22 (1923).
220. H. B. Peacock, *Phys. Rev.,* **27**:474 (1926).
221. D. B. Barker and W. C. Caldwell, *USA Rept.* AEC-JSC, 1952, p. 215.
222. M. A. Jeppesen, *J. Appl. Phys.,* **37**:1940 (1966).
223. K. L. Chopra and S. K. Bahl, *J. Appl. Phys.,* **38**:3607 (1967).
224. C. G. Grenier, K. R. Efferson, and J. M. Reynolds, *Phys. Rev.,* **143**:406 (1966); Mackey et al., *Phys. Rev.,* **157**:578 (1967); **158**:658 (1967); **161**:611 (1967).

225. H. London, *Proc. Roy. Soc. London,* **A176**:522 (1940).
226. A. B. Pippard, *Proc. Roy. Soc. London,* **A191**:385 (1947).
227. G. E. H. Reuter and E. H. Sondheimer, *Proc. Roy. Soc. London,* **A195**:336 (1948); E. H. Sondheimer, *Proc. Roy. Soc. London,* **A224**:260 (1954).
228. R. B. Dingle, *Physica,* **19**:311 (1953); E. H. Sondheimer, *Proc. Roy. Soc. London,* **A224**:260 (1954).
229. R. G. Chambers, *Proc. Roy. Soc. London,* **A215**:481 (1952).
230. M. Ya. Azbel and E. A. Kaner, *Zh. Eksperim. i Teor. Fiz.,* **30**:811 (1956).
231. M. C. Jones and E. H. Sondheimer, *Proc. Roy. Soc. London,* **A278**:256 (1964).
232. P. Cotti, *Phys. Letters,* **4**:114 (1963).
233. R. Bowers, C. Legendy, and F. Rose, *Phys. Rev. Letters,* **7**:339 (1961).
234. H. Thomas, *Z. Physik,* **147**:395 (1957).
235. G. A. Katrich and O. G. Sarbej, *Soviet Phys.-Solid State English Transl.,* **3**:1181 (1961).
236. C. R. Crowell, W. G. Spitzer, L. E. Howarth, and E. E. LaBate, *Phys. Rev.,* **127**:2006 (1962); W. G. Spitzer, C. R. Crowell, and M. M. Atalla, *Phys. Rev. Letters,* **8**:57 (1962).
237. J. L. Moll and S. M. Sze (unpublished).
238. S. M. Sze, J. L. Moll, and T. Sugano, *Solid-State Electron.,* **7**:509 (1964).
239. L. B. Leder, M. E. Lasser, and D. C. Rudolph, *Appl. Phys. Letters,* **5**:215 (1964).
240. R. W. Soshea and R. C. Lucas, *Phys. Rev.,* **138**:A1182 (1965).
241. C. A. Mead, *Phys. Rev. Letters,* **8**:56 (1962).
242. H. Kanter, *J. Appl. Phys.,* **34**:3629 (1963).
243. W. M. Feist, *IEEE Spectrum,* **1**:57 (1964).
244. S. M. Sze, C. R. Crowell, G. P. Carey, and E. E. LaBate, *J. Appl. Phys.,* **37**:2690 (1966).
245. H. Kanter, *Appl. Phys. Letters,* **10**:73 (1967).
246. R. N. Stuart, F. Wooten, and W. E. Spicer, *Phys. Rev.,* **135**:A495 (1964).
247. J. J. Quinn, *Phys. Rev.,* **126**:1453 (1962); *Appl. Phys. Letters,* **2**:167 (1963).
248. S. L. Adler, *Phys. Rev.,* **130**:1654 (1963).
249. A. H. Wilson, "Theory of Metals," p. 264, Cambridge University Press, New York, 1955.
250. F. Koshiga and T. Sugano, *Japan. J. Appl. Phys.,* **5**:1036 (1966).
251. N. F. Mott and H. Jones, "The Theory of the Properties of Metals and Alloys," p. 268, Dover Publications, Inc., New York, 1958.
252. R. Berman, *Advan. Phys.,* **2**:103 (1953).
253. G. K. White and S. B. Woods, *Phil. Mag.,* **1**:846 (1956).
254. P. Wyder, *Proc. 7th Low Temperature Conf.,* Toronto, 1960, p. 266.
255. P. Wyder, *Rev. Mod. Phys.,* **36**:116 (1964).
256. J. M. Ziman, *Advan. Phys.,* **10**:1 (1961).
257. F. Savornin, in *Symp. Elect. Magnetic Properties of Thin Metallic Layers,* Louvain, 1961, p. 69.
258. E. Justi, M. Kohler, and G. Lautz, *Z. Naturforsch.,* **8a**:544 (1953).
259. L. Reimer, *Z. Naturforsch.,* **12A**:525 (1957).
260. R. P. Huebener, *Phys. Rev.,* **136**:A1740 (1964).

261. W. Worobey, P. Lindenfeld, and B. Serin, in "Basic Problems in Thin Film Physics" (R. Neidermayer and H. Mayer, eds.), p. 601, Vandenhoeck and Ruprecht, Göttingen, 1966.
262. W. Worobey, P. Lindenfeld, and B. Serin, *Phys. Letters,* **16**:15 (1965).
263. K. L. Chopra, S. K. Bahl, and M. R. Randlett, *J. Appl. Phys.,* **39**: 1525 (1968).
264. R. P. Huebener, *Phys. Letters,* **15**:105 (1965).
265. L. Harris and E. Johnson, *Rev. Sci. Instr.,* **5**:153 (1934).
266. E. Justi, K. Kohler, and G. Lautz, *Z. Naturforsch.,* **6a**:456 (1951).
267. R. D. Morrison and R. R. Lachenmayer, *Rev. Sci. Instr.,* **34**:106 (1963).
268. G. Abowitz, V. Klints, M. Levy, E. Lancaster, and A. Mountuala, *Semicond. Prod.,* **8**:18 (1965).
269. R. Marshall, L. Atlas, and T. Putner, *J. Sci. Instr.,* **43**:144 (1966).
270. M. G. Cooper and A. J. P. Lloyd, *J. Sci. Instr.,* **42**:791 (1965).
271. D. Abraham and T. O. Poehler, *J. Appl. Phys.,* **36**:2013 (1965).
272. H. Seki and I. Ames, *J. Appl. Phys.,* **35**:2069 (1964).
273. D. Griffiths and R. Watton, *Brit. J. Appl. Phys.,* **17**:535 (1966).
274. W. A. Little, *Can. J. Phys.,* **37**:334 (1959).
275. M. Chester, *Phys. Rev.,* **145**:76 (1966).
276. R. J. von Gutfeld and A. H. Nethercot, Jr., *J. Appl. Phys.,* **37**:3767 (1966).
277. R. J. von Gutfeld, A. H. Nethercot, Jr., and J. A. Armstrong, *Phys. Rev.,* **142**:436 (1966).
278. D. A. Neeper and J. R. Dillinger, *Phys. Rev.,* **135**:A1028 (1964).
279. V. E. Holt, *J. Appl. Phys.,* **37**:798 (1966).
280. P. M. Hall, *Appl. Phys. Letters,* **12**:212 (1968).

VII TRANSPORT PHENOMENA IN SEMICONDUCTING FILMS

1. INTRODUCTION

Surface transport phenomena are well known to have a strong influence on the electronic properties of bulk semiconductors. These phenomena play an important role in the transport properties of semiconducting films of about 1 μ thickness and having a carrier concentration up to $\sim 10^{18}/cm^3$. This role results from the fact that, when transport takes place through thin specimens, the carriers are subject to considerable scattering by the boundary surfaces in addition to the normal bulk scattering. This additional scattering will reduce the effective carrier mobility below the bulk value and will thus give rise to conductivity size effects of the type we have already discussed for metal films in the preceding chapter. A study of these size effects can yield information on the electronic structure of a surface and is therefore of considerable fundamental and practical importance.

Surface transport in bulk semiconductors has received much attention in recent years. An excellent review of the subject is given by Many et al.

[1]. By contrast, however, very little work has been done on thin semiconducting films. This chapter presents a review of the brief theoretical and experimental work related to thin films only.

2. THEORETICAL CONSIDERATIONS

(1) Mobility. Analysis of size effects in semiconductors is modified from that of metals owing to the introduction of an additional feature of surface space charge and hence surface-potential barrier. Schrieffer [2] considered this factor and extended the Fuchs-Sondheimer treatment [3] for metal films (Chap. VI, Sec. 3.1). Because of the assumptions involved, Schrieffer's theory is valid only for strong accumulation and inversion surface layers. The theory has been improved by several workers and extended to include galvanomagnetic effects. A review of the theory has been given by Greene [4]. The influence of the quantization of the transverse motion of the carriers on the transport equations in thin-film semiconductors has been studied by Tavger [5]. We shall not discuss these theoretical treatments but rather present elementary considerations to estimate the surface mobility and surface conductivity. It must be recognized that, for mathematical convenience, rigorous theories [1] generally deal with a nondegenerate semiconductor characterized by spherical energy surfaces and a constant relaxation time. The more general but complicated case of nonspherical energy surfaces has also been treated in the literature [6,7].

We consider a one-carrier system, corresponding to an extrinsic n-type sample in the form of a thin film of thickness t comparable with the mean free path (mfp) of the carriers. The band edges are assumed to continue flat up to the surfaces, the electron density n being uniform throughout the sample and equal to bulk value n_B. The effect of surface scattering will be introduced in the form of some average collision time τ_S, just as bulk scattering is characterized by the relaxation time τ_B (generally $\sim 10^{-12}$ sec). If we further assume that the bulk and the surface-scattering processes are additive, then the average relaxation time τ_F for electrons in thin films is given by

$$\frac{1}{\tau_F} = \frac{1}{\tau_S} + \frac{1}{\tau_B} \tag{1}$$

As an estimate of τ_S, we take the mean distance t of a carrier from the surface divided by the unilateral mean velocity v_z, so that

$$\tau_S \approx \frac{t}{v_z} = \frac{t}{l}\tau_B = \gamma\tau_B \tag{2}$$

where l is the mfp defined by $l \equiv \tau_B v_z$ and $\gamma \equiv t/l$. The mfp is given by

$$l = \mu_B \frac{h}{e}\left(\frac{3}{8\pi}n_B\right)^{1/3} \tag{3}$$

where μ_B is the mobility of the carriers in the bulk, e is a unit electronic charge, and h is Planck's constant. For $\mu_B = 1,000$ cm^2/V-sec and $n_B = 10^{18}$/cm^3, the value of l is ~200 Å and is typical of films of degenerate semiconductors.

The average electron mobility μ_F is taken as $\mu_F = e\tau_F/m^*$ in analogy with the corresponding expression for the bulk mobility $\mu_B = e\tau_B/m^*$. By using Eqs. (1) and (2), we obtain

$$\mu_F = \frac{\mu_B}{1 + 1/\gamma} \tag{4}$$

As expected also on the basis of the physical considerations, the average mobility decreases with decreasing film thickness, while it approaches bulk value for $\gamma \gg 1$. Equation (4) is valid for (1) thick films in the flat-band approximation only, and (2) sufficiently thin films such that t is small compared with the *effective* Debye length L_D, which for an intrinsic semiconductor is given by

$$L_D = \left[\frac{4\pi\epsilon kT}{e^2(n_B + p_B)}\right]^{1/2} \tag{5}$$

Here ϵ is the static dielectric constant, and n_B and p_B are the densities of negative and positive carriers, respectively. The value of L_D for $n_B = 10^{16}$/cm^3, $p_B = 0$, and $\epsilon = 10$ is calculated to be about 350 Å. Case (2) arises because the potential in ultrathin film ($t \ll L_D$) will be essentially constant regardless of the value of the surface potential V_s. However, the uniform electron density which assumes a value n_B for $V_s = 0$ (flat-band approximation) will be different from n_B for $V_s \neq 0$ (bending of bands).

We have assumed diffuse scattering at the boundary (see Chap. VI, Sec. 3.1) in the above discussion. The reciprocal of τ_S expresses the probability that an electron will be scattered by the surface in unit time. Since p, the scattering coefficient, is the probability that an electron is specularly reflected, only a fraction $(1 - p)$ of the electrons will be scattered diffusely, i.e., with a loss of memory, so that $1/\tau_S$ must be replaced by $(1 - p)/\tau_S$. This is equivalent to replacing t in Eq. (2) by $t/(1 - p)$. The average mobility [Eq. (4)] then becomes

$$\mu_F = \frac{\mu_B}{1 + (1 - p)/\gamma} \tag{6}$$

In thick films $(t \gg L_D)$, the carrier transport is dominated by the size effect because of the surface space charge. One should therefore expect the surface mobility to be determined by the amount of band bending (surface-potential drop) and the ratio l/L_c, where L_c is the effective charge distance from the surface to the center of the space charge. The parameter L_c is related to L_D by $L_c = (|v_s|/F_s)L_D$, where $v_s = eV_s/2kT$ is the normalized value of the surface-potential drop V_s, k is the Boltzmann constant, and F_s is a complicated space-charge-distribution function plotted in Ref. 1 (p. 144). The function F_s decreases rapidly with increasing value of v_s. For small v_s (flat band), L_c approaches L_D.

With an accumulation layer and symmetrical potential barriers at the two surfaces, the carriers can be considered as moving in a thin film with one surface, $z = 0$, a diffuse scatterer, and the other, $z = L_c$, a specular reflector. The kinetic energy of the electrons is increased because of the surface potential V_s through which they drop. Taking this into account, the electron surface mobility is given [1] by

$$\mu_S = \frac{\mu_B}{1 + (l/L_c)(1 + v_s)^{1/2}} \tag{7}$$

The same relation holds for holes provided the sign of v_s is changed. If the reflection of electrons is partially specular, L_c should be replaced by $L_c/1 - p$, so that

$$\mu_S = \frac{\mu_B}{1 + (l/L_c)(1 - p)(1 + v_s)^{1/2}} \tag{8}$$

Thus, μ_S is a decreasing function of barrier height. Note that $\mu_S/\mu_B \to 1/\{1 + [l(1 - p)]/L_c\}$ as $v_s \to 0$ (i.e., no band bending). The expression for the surface mobility in a depletion layer is somewhat complicated by the fact that only those electrons having sufficient energy to surmount the potential barrier are able to reach the surface. The expression derived in Ref. 1 shows again that μ_S approaches μ_B for large negative values of v_s, as expected. For flat bands ($v_s \to 0$), $\mu_S/\mu_B \to 1 - [l(1 - p)]/L_c$. For $l \ll L_c$, the size effect is absent in both accumulation and depletion layers.

The expression for the conductivity of thin films in the flat-band approximation is similar to that for metals (Chap. VI, Sec. 3.1). Further, if we introduce a boundary condition such that all carriers incident at angles less than θ_0 to the surface normal are diffusely scattered but that all incident at greater angles are specularly reflected, then the conductivity can be shown [8] to be given by

$$\sigma_F = e n_B \mu_B \left(1 - \frac{3}{8}\frac{1}{\gamma} \sin^4 \theta_0\right) \qquad (\gamma \gg 1) \qquad (9)$$

and

$$\sigma_F = e n_B \mu_B \left[\frac{3}{2}\cos\theta_0 \left(1 - \frac{\cos^2\theta_0}{3}\right)\right] \qquad (\gamma \ll 1) \qquad (10)$$

(2) Galvanomagnetic Surface Effects. The contribution of surface scattering to such galvanomagnetic effects as Hall effect and magneto-resistance can be derived by solving the Boltzmann equation. Zemel [9], Amith [10], and Petritz [11] have made calculations for semiconductors along lines similar to those used by Sondheimer for metals (Chap. VI, Sec. 4). The results are similar in the two cases.

In small magnetic fields, the Hall coefficient R_{H-F} of a thin film ($t \ll L_D$) is given by

$$R_{H-F} = R_{H-B}\,\eta\!\left(\frac{1}{\gamma}\right) \qquad (11)$$

where R_{H-B} is the bulk value. The correction function $\eta(1/\gamma)$ is a complicated expression [1] which has been calculated numerically by

Amith [10]; it approaches unity when $\gamma \gg 1$. The transverse magnetoresistance expressed by the ratio $\Delta\rho/\rho H^2$ is given by

$$\frac{\Delta\rho}{\rho H^2} = \mu_B^2 - \eta^2 \left(\frac{1}{\gamma}\right)\mu_F^2 \qquad (12)$$

Amith has extended this treatment to include two-carrier systems and the presence of potential barriers at the surface.

(3) Anisotropy Effects. So far we have concerned ourselves with the simple case of a spherical energy surface. Following the treatment of Engleman and Sondheimer [12] for ellipsoidal energy surfaces in the band structure of a metal film, Ham and Mattis [6] derived expressions for the conductivity and Hall coefficient of a thin-film nondegenerate semiconductor. These transport parameters were shown to vary with the orientation of the normal to the film with respect to the crystallographic axes, even for materials of cubic symmetry such as Ge or Si. This contrasts with the isotropic behavior of these quantities in the bulk material. This anisotropy in films should permit an experimental determination of the ratio of the components of the effective mass tensor.

Further, for ellipsoidal energy surfaces, even for specular reflection, the conductivity and the effective mobility decrease to a finite limiting value for very thin films and exhibit crystallographic anisotropy. These features are also present in a general theory for nonspherical energy surfaces given by Price [7]. Ham and Mattis [6], however, pointed out that the use of ellipsoidal energy surfaces leads to differences that are small compared with probable experimental uncertainties, so that the use of the spherical model with a suitable effective mass should in practice be quite satisfactory, except, of course, for the case of total specular reflection.

(4) Quantum Size Effects. Quantum size effects [13-15] appear in semiconductor and semimetal films when their thickness is comparable with or smaller than both the mfp and the effective deBroglie wavelength of the carriers. Because of the finite thickness of a film, the transverse component of the quasi-momentum is quantized. Consequently, the electron states assume quasi-discrete energy values in a thin film. The discreteness of these levels will be evident only if the spacing ΔE between the subbands is larger than the broadening of the subbands $\sim kT$ due to thermal scattering, and $\sim h/\tau_B$ due to miscellaneous scattering mechanisms. The latter broadening is small for films of thickness t ~1,000 Å and an effective carrier mass m^* ~0.01 m_e, when the bulk mobility exceeds 10^3 cm^2/V-sec.

As a result of the quantization, the bottom of the conduction and the top of the valence band are separated by an additional amount ΔE. An estimate of ΔE is given by the uncertainty principle as

$$\Delta E \sim \frac{h^2}{8m^* t^2} \tag{13}$$

We have assumed a degenerate semiconductor obeying a square-dispersion law $(E = p^2/2m^*)$, and further a free motion of the carriers parallel to the film surface. The upward shift ΔE may have a profound effect on the electrical and optical properties of a material which in bulk form has a comparable overlap between the conduction and the valence bands. This condition may be satisfied by semimetals such as Bi and Sb. For example, the overlap in bulk Bi is known to be about 0.0184 eV. Using $m^* = 0.01$ m_e for the carriers along the trigonal axis of Bi, a shift $\Delta E = 0.0173$ eV should be obtained for a 500-Å-thick film.

When the conditions for the occurrence of the quantum size effects are satisfied, it can be shown [14, 15] that all transport properties will exhibit oscillatory behavior as a function of the film thickness with a period

$$\Delta t = \frac{h}{\sqrt{8m^* \zeta}} \tag{14}$$

where $1/m^* = 1/m_e^* + 1/m_h^*$, and $\zeta = (\zeta)_e + (\zeta)_h$. The e and h subscripts stand for electrons and holes and the ζ's correspond to the respective Fermi energies.

3. EXPERIMENTAL RESULTS

3.1 Size Effects

An enormous amount of effort has been devoted to the study of preparative techniques, and structure and growth of epitaxial semiconducting films. But the size effects and, in particular, the surface transport phenomena have received little attention despite their obvious practical and fundamental importance.

The only systematic measurement of the thickness dependence of the film conductivity has been made by Davey et al. [16] on evaporated polycrystalline Ge films. Their data on the resistivity of Ge films deposited at different rates are shown in Fig. 1 and are compared with the theoretical curve calculated for l = 500 Å, p = 0, and v_s = 0. The agreement is surprisingly good in spite of the fact that large thickness variation of the resistivity is also expected because of changes in the structural defects and the microstructure of the films. The effect of the defects in reducing the mfp is indicated by its deduced value (required for agreement) being substantially lower than the expected value of 1,500 Å for pure bulk Ge at 300°K, assuming a mobility of 3,900 cm²/V-sec and m^* = 0.3 m_e. These studies certainly demonstrate the presence of the size effects although more careful studies on epitaxial films are, of course, necessary for quantitative analysis.

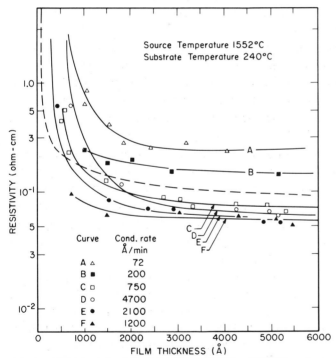

FIG. 1. Thickness dependence of the resistivity of Ge films deposited on glass at different rates, with source and substrate temperature held constant. The dotted curve is calculated for v_s = 0 (Fuchs' theory), ℓ = 500 Å, and p = 0 (diffuse scattering). (*Davey et al.* [16])

Waxman et al. [17] measured the Hall and the field-effect surface mobilities in polycrystalline CdS films of thickness 1,000 to 4,000 Å. They observed the mobility to increase with small values of increasing surface potential, a result which is in direct contradiction to the theory already outlined. This behavior was explained by assuming that the film consisted of blocks of crystallites and the mobility was primarily determined by the scattering at the crystallite boundaries. Because of the resulting small mfp of electrons, the size effect was indicated only at high surface potentials when the $l \sim L_c$ condition was realized.

Neugebauer and Joynson [18] studied field-effect changes on the surfaces of CdS films by applying a field in a thin-film metal-oxide-semiconductor configuration and then measuring the capacitance changes in this arrangement. These changes can be related to the carrier density, interface surface-state density, and surface potential. Their results indicated that the thickness of the depletion layer increased with decreasing temperature, which was ascribed to the thermionic nature of the Au-CdS contact barrier.

Detailed and significant analysis of the surface transport properties of epitaxial PbSe films was undertaken by Brodsky and Zemel [19]. These films are degenerate semiconductors due to high ($\sim 10^{19}/cm^3$) carrier concentration. But owing to the lack of a proper theory, the results were interpreted in terms of the theory for nondegenerate semiconductors. The Hall mobility of $\sim$1,000-Å-thick films was found to be *independent* of the resistivity over a range of a decade, and it increased rapidly at low temperatures. The observed *linear* relation between the Hall coefficient and the resistivity implies the *absence* of surface scattering and the surface potential. The fact that mobility continues to rise at low temperatures for films as thin as 300 Å (so that the mfp is comparable with or larger than the film thickness) supports the existence of specular scattering in thin epitaxial PbSe films.

We have already mentioned the conditions for observing quantum size effects in thin films. Because of the long wavelength and long mfp of carriers in Bi, Sb, As, etc., reasonably thick films of these materials are expected to exhibit size effects. The effects will be enhanced at helium temperatures where the mfp is long and is about 1 mm for pure bulk Bi. As already mentioned, a reduction and even complete elimination of the overlap between the conduction and valence bands may occur in thin Bi films. Thus, one would expect a metallic behavior for thick films and a semiconductor-like behavior for thinner films. Duggal et al. [20]

observed such a behavior in thin Bi films grown epitaxially on mica with their trigonal axis normal to the substrate. The resistivity was found to fit $\rho_F = \rho_B \exp(\Delta E/kT)$ relation characteristic of a semiconducting transport process. The value of ΔE was reasonably consistent with Eq. (13). These authors, however, reported no oscillatory behavior which Ogrin et al. [21] observed in similar films. The thickness-dependent oscillatory variation of the resistivity, Hall coefficient, Hall mobility, and magnetoresistance at various temperatures is shown in Fig. 2. The amplitude of the oscillations increases at lower temperatures. The period of oscillation $\Delta t = 400$ Å yields from Eq. (14) a value of $m^* = 0.01\ m_e$,

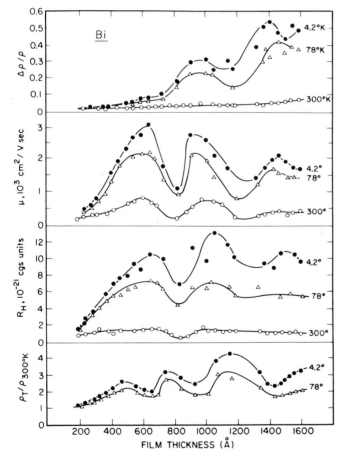

FIG. 2.   Thickness dependence of transverse magnetoresistance $\Delta\rho/\rho$, Hall mobility μ, Hall coefficient R_H, and resistivity ratio $\rho_T/\rho_{300°\,K}$ of epitaxial Bi films. (*Ogrin et al.* [21])

assuming $\zeta = 0.18$ eV. This value of the effective mass is in reasonable accord with the known bulk value.

Antimony films have also been found [22] to exhibit a similar behavior with $\Delta t = 29$ Å. If we take $\zeta = 0.2$ eV, then $m_e^* = m_h^* = 0.5 \, m_e$ is obtained. Thus it appears that Eqs. (13) and (14), and hence the quantization hypothesis, have been substantially verified. (See Chap. XI, Sec. 3.3 for the effect of the quantization on the optical properties of thin films.) But an independent verification of the quantization effect is still lacking.

3.2 Transport Properties of Thick Films

Thick epitaxial semiconducting films with bulk or nearly bulklike properties are interesting for device applications and have therefore received much attention. The electron-transport properties of these films may be characterized by the resistivity, Hall coefficient, and Hall mobility. A study of the temperature dependence of these parameters is generally essential to an understanding of the carrier-scattering mechanisms responsible for these transport properties. This remark may be best illustrated by the following brief discussion of the bulk properties.

Electrical conduction, in general, is determined by various overlapping carrier-scattering mechanisms [23]. Provided that one of the scattering mechanisms is dominant, we can use the analog of Matthiessen's rule and add the inverse of the various mobilities, so that

$$\frac{1}{\mu_F} = \frac{1}{\mu_L} + \frac{1}{\mu_I} + \frac{1}{\mu_S} \tag{15}$$

where μ_L, μ_I, and μ_S are contributions due to lattice (intrinsic), impurities and imperfections (extrinsic), and surface scattering including crystallite boundary scattering, respectively.

In a bulk semiconductor [23] the parameters l, R_H, and μ_L increase with the decrease of temperature in the intrinsic range (lattice-dominated); μ_L has a characteristic $T^{-3/2}$ dependence (or $T^{-5/2}$, if the optical phonons contribute significantly). In the extrinsic range, where the ionized impurities are the dominant scatterers, ρ_I and μ_I decrease with the decrease of temperature; μ_I has a characteristic $T^{+3/2}$ dependence, the inverse of the lattice contribution. If the density of impurity states is so high that they overlap into the conduction band (degeneracy thus arises), no thermal excitation is necessary to produce carriers, and therefore ρ, μ, and R_H are independent of temperature. Temperature-independent behavior is also predicted for neutral impurity scattering. In general, however, the various scattering mechanisms will overlap and their additive effect will give rise to a complicated temperature dependence. An investigation over a large range of temperatures and carrier concentrations is thus essential for the interpretation of the data.

The preparation, structure, and growth of epitaxial films of a large number of semiconducting materials obtained by various deposition techniques have been discussed in Chaps. II and IV. The materials on which transport measurements have been reported are Ge [13, 24-31], Si [32, 33], Bi_2Te_3 [34], Sn Te [35], PbS [35], PbSe [16, 35, 36], PbTe [35, 37-39], CdS [40-42], CdSe [43], CdTe [44, 49], GaAs [45], GaP [46], Cd_3As_2 [47], Ag_2Se [48], Ag_2Te [49], InSb [50-52], HgTe [53], InAs [53], GeTe [54], and GaSb [55]. In Table I, we have compiled the best available data on various transport parameters of epitaxial films of several common semiconductors. The values given are only representative rather than precise since these vary considerably with the deposition conditions.

The observed temperature dependence of the film-transport parameters is, in general, complex. Because of the limited scope of most of the reported measurements, an unambiguous explanation of the wide variety of experimental results is impossible at present. There is, however, no doubt that the microstructure, impurities, and structural imperfections dominate the transport behavior in most cases. We shall illustrate this point by way of a few examples.

The temperature dependence of ρ, μ, and R_H for a thick pyrolytically prepared epitaxial GaAs film is shown in Fig. 3. The variation of μ and R_H with temperature is not as pronounced as expected, although it is characteristic of an intrinsic semiconductor behavior. On the other hand, the resistivity variation is metallic, suggesting ionized impurity conduction.

The mobility in polycrystalline CdS films varies as exp $(1/T)$ for high-resistivity films, and as $T^{3/2}$ for low-resistivity films, which is characteristic of impurity scattering.

Amorphous films of Cd_3As_2 [47] prepared by evaporation are n-type with a carrier concentration of $\sim 10^{17}/cm^3$. Here, ρ and R_H increase with increasing temperature and saturate at high temperatures. Mobility follows a nearly $T^{-3/2}$ dependence and saturates at low temperatures.

Table I. Transport Parameters of Thin Epitaxial Films of Some Semiconducting Materials at Room Temperature

Material	Thickness, μ	Conductivity type	Carrier concentration, cm^{-3}	Mobility, cm^2/V-sec	Deposition method*	Ref.
CdS	0.1–2.5	n	$\sim 10^{15}-10^{11}$	1–30 (dark) $\sim$300 (photo)	Ev	41
Cd$_3$As$_2$	50–100	n	10^{18}	500–3,000	Ev	47
GaAs	>100	n	2.5×10^{15}	>7,000	CVD	45
GaP	>0.4	p	10^{14}	150	CVD	46
Ge	0.2	p	$10^{16}-10^{18}$	100–1,100	Ev	30
GeTe	0.25–0.5	p	$10^{19}-10^{20}$	$\sim$15–50	Ev	54
InSb		n	10^{17}	15,000	Ev	51
		p	10^{17}	$\sim$500	Ev	50
PbS	$\sim$0.3	n	2×10^{18}	$\sim$500	(polycrys.) Ev	35
		p	2×10^{18}	10	Ev	
PbSe	$\sim$0.3	n	7×10^{17}	1,000	(polycrys.) Ev	35
	0.3–0.6	p	10^{18}	300–1,000	Ev	36
PbTe	$\sim$0.3	n	5×10^{17}	1,100	Ev	35
	1.8	n	$10^{17}-10^{18}$	2,000		37
SnTe		p	2.5×10^{20}	200	Ev	35
Si	3.5	p (<1000°C) n (annealed above 1000°C)		$\sim$50 $\sim$120	Ev Ev	35 33

*Ev stands for evaporation and CVD for chemical-vapor deposition.

446

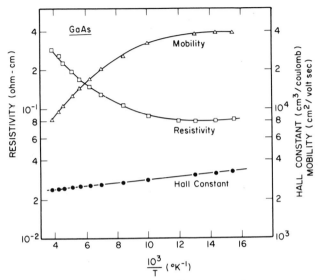

FIG. 3. Resistivity, mobility, and Hall constant of a pyrolytically prepared epitaxial GaAs film (25 μ thick) as a function of the reciprocal temperature. (*Knight et al.* [45])

Indium antimonide [51] films grown epitaxially by flash evaporation onto mica substrates held at 330 to 350°C show anomalous electrical behavior. R_H is independent of temperature between 300 and 77°K. The films are extrinsic n-type with carrier concentration $\sim 10^{17}/\text{cm}^3$. The resistivity increases whereas the mobility decreases with the decrease of temperature.

Evaporated and sputtered films of Ge, as well as evaporated films of Si, when deposited at low substrate temperatures (below about 130°C for Ge, and 200°C for Si) are amorphous, and their conductivity type is at best uncertain. The polycrystalline and epitaxially grown films, on the other hand, are *invariably* p-type, independent of the type and the purity of the starting material. A large carrier concentration of $10^{18}/\text{cm}^3$ is generally observed. Conflicting results on the transport properties of Ge films have been reported by many workers. The most extensive studies are due to Sloope and Tiller [30], who investigated the effect of the film thickness, temperature, deposition rate, substrate temperature, and annealing on the transport properties of Ge films deposited epitaxially on (111) CaF$_2$. Thinner films and lower deposition rates yielded lower values of R_H. The mobility decreased with decreasing thickness in the 20,000- to 4,000-Å range. The decrease may be partly ascribed to size effects and partly to the increase in the density of structural defects.

Hole carrier concentration of 10^{16} to 10^{18} per cm^3 and room-temperature mobility of 100 to 1,100 cm^2/V-sec were observed for thicknesses greater than 4,000 Å. The Hall coefficient decreased slightly at higher temperatures and approached a nearly constant value at low temperatures. The mobility exhibited a maximum which occurred at lower temperatures for higher-mobility films, and a $T^{-2.3}$ decrease above 300°K (lattice scattering).

What causes a p-type conductivity in Ge and Si films is a subject of some discussion. Bylander et al. [28] suggested chemical impurities as a source of acceptors in their films. The observations of the p-type behavior in thick films prepared under ultraclean conditions rules out this explanation. Davey et al. [16] attributed the p-type conduction in textured Ge films deposited on quartz to surface states and the associated space charge. This conclusion was supported by Humphris and Catlin [27] on the basis that they observed neither a change in the conductivity on annealing of films to 600°C nor a change in the carrier concentration due to changes in the defect density by two orders of magnitude. The effect of surface states, if any, should obviously decrease with increasing film thickness. Davey [29] measured the carrier concentration in epitaxial Ge films deposited on GaAs substrates as a function of film thickness up to 10^6 Å. He observed a monotonic decrease of carrier concentration from $\sim 10^{18}$ to 2.7×10^{17}/cm^3 with increasing film thickness which is clearly not compatible with a simple surface-state model.

Several authors [24-26, 30] have suggested that the structural defects in Ge films are responsible for the p-type conductivity. The studies of Sloope and Tiller [30] on epitaxial Ge films support this hypothesis. Further support for the structural-defect hypothesis is provided by the observations [24, 25] that an initially p-type film obtained by the thermal evaporation of n-type bulk Ge can be converted to n-type by annealing or remelting the film. Similarly, Si films [32] can be transformed from p- to n-type on annealing at or above 900°C. Naber [56] observed that the transformation from n- to p-type occurs in electron-beam-heated Ge films as the crystallite size starts to increase.

Which structural defects lead to acceptor states in Ge films is not clear at present. The high concentration of broken bonds may yield p-type behavior in the amorphous structure. Considerations [1] similar to those employed with regard to impurities indicate that vacancies and interstitials in group IV crystalline semiconductors should act as acceptors and donors, respectively. Further, dislocations in Ge should introduce [57] deep-lying acceptor states. Sloope and Tiller [30] noted that the results of their transport studies can be understood in terms of the defects in the film microstructure. Wolsky et al. [26] estimated the acceptor levels in Ge films from their transport data to be about 0.1 eV above the valence band. This level could be attributed to vacancies or vacancy complexes.

The resistivity of amorphous Ge, Te, Se, and GeTe films is significantly higher than that of the crystalline films of these materials. For example [118, 119], Ge films obtained by evaporation from a 40 Ω-cm source have resistivities $\sim 10^{-1} \Omega$-cm in the crystalline form as compared with $\sim 10^{2}$-$10^{3} \Omega$-cm for the amorphous structure. In contrast to the small temperature dependence of the resistivity of the former, that of the latter increases by several orders of magnitude from room temperature down to 77°K. The log resistivity versus $1/T$ plot shows a continuously variable slope, yielding an activation energy ranging from ~ 0.6 to ~ 0.05 eV. Measurements of the Hall effect in the amorphous materials are very difficult and uncertain. However, crude measurements [118] on Ge films suggest that the density of carriers is the same ($\sim 10^{18}$-10^{19}/cm^3) in both structures and that the variation in the mobility dominates the transport properties of the amorphous structure. The results on GeTe films [120] are even more dramatic. The crystalline (rhombohedral structure) GeTe films [120, 121] behave like a p-type degenerate semiconductor ($\sim 10^{20}$/cm^3 carriers, 20-50 cm^2/V-sec mobility) with the Fermi level well inside the valence band (~ 0.4 eV). The resistivity ($\sim 10^{-4} \Omega$-cm) changes slightly with temperature, showing a metallic behavior (positive TCR). On the other hand, the resistivity of the amorphous structure [120] is $\sim 10^{3} \Omega$-cm at room temperature and it increases by a factor of $\sim 10^{6}$ at 77°K, yielding an activation energy of ~ 0.4 eV. The observation that the optical absorption edge of GeTe is nearly the same for the crystalline and the amorphous structures down to 77°K indicates [120] that the Fermi level remains unchanged and thus the resistivity changes must be interpreted in terms of the variation in the mobility of the carriers.

The band structure and electrical conduction in amorphous conductors has been treated theoretically by Gubanov [122] and reviewed by Cusack [123] and Mott [58]. It is generally concluded that even though long-range order is not preserved in an amorphous material, the features of the energy-band model are preserved, provided the molecular bonds are not significantly disturbed. Further, the superimposition of random changes on the normal interatomic potential gives rise to "fluctuation" energy levels. The position and density of these levels depend on the magnitude of the fluctuation potential. If a continuum of levels is formed at or near the band edges, the levels are "nonlocalized" resulting in smeared-out band edges. The levels formed well inside the band become "localized" states. The optical behavior of amorphous Ge and GeTe films suggests (Chap. X, Sec. 3.2) that the band structure is sufficiently well defined in the amorphous films. This conclusion is also supported by rather dubious tunneling spectroscopy data [124] on Ge films. On the other hand, the transport behavior can only be understood by considering the role of the fluctuation levels in dominating the carrier mobility. As compared with the intrinsic behavior of dc conductivity, the thermal activation energy for ac conduction decreases continuously to zero at high frequencies. The ac conductivity increases nearly linearly with frequency to a saturation value at high frequencies. These observations on GeTe films [120] can be understood in terms of conduction by trap-controlled hopping of carriers in the localized levels above the Fermi level.

Zemel et al. [35] carried out a detailed study of epitaxial films of PbS, PbSe, and PbTe. These films were found to possess structure, crystal perfection, dielectric, transport, and mechanical properties comparable with the best available single crystals. In the temperature range of 77 to 300°K, R_H was found to be nearly constant for all cases; ρ decreased with decrease of temperature; μ followed an approximately $T^{-5/2}$ dependence at higher temperatures and tended to approach saturation at low temperatures. This variation of ρ and R_H indicates the dominant role of conduction by ionized impurities and/or defects.

Since the inverses of the mobilities due to different scattering mechanisms are additive [Eq. (15)], we can subtract the $T^{+5/2}$ dependence of $1/\mu_L$ (the lattice contribution) from the observed temperature dependence of $1/\mu_F$ to obtain the extra, unknown contribution. The result is shown in Fig. 4 for the data on two PbTe films. The $(1/\mu_F - 1/\mu_L) = 1/\mu_S$ contribution is thus independent of temperature and has been ascribed by Zemel et al. [35] to a boundary-scattering mechanism which arises when the mfp of the carriers is comparable with the average dimension of the crystallites.

If the mfp is assumed to be limited by the crystallite size D (i.e., $l = D$), then since the mobility is defined as $\mu = e\tau/m$, we obtain $\mu = eD/mv$, where mv is the average momentum of the carriers. Using the measured value of μ, a reasonable value of $D \sim 250$ Å is obtained. Contrary to the suggestion of Zemel et al., this value of D should bear no relation to the observed voids in thin epitaxial films. The crystallite-boundary-scattering argument may also be criticized since the assumption that $l = D$ is valid only for a wire geometry, while for a platelet-like specimen, $l_{eff} \sim {}^3\!/_4 t \ln(l_B/t)$ holds for diffuse scattering [Eq. (47), Chap. VI]. Furthermore, the contribution μ_I from impurities and structural defects in Eq. (15) has been completely neglected.

The results of Zemel et al. on PbTe films are similar to those of Makino [38, 39]. On the other hand, Gobrecht et al. [36] found that polycrystalline PbSe films exhibited a T^x ($x = 1.5$ to 2.0) variation for

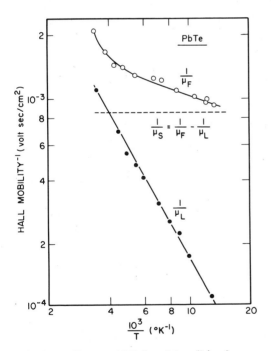

FIG. 4. Reciprocal Hall mobility $(1/\mu_F)$ vs. reciprocal temperature for PbTe films (1,600 and 3,200 Å thick); μ_L refers to the intrinsic or lattice contribution in an ideal bulk material. (*Zemel et al.* [35])

the hole mobility. But the hole mobility in epitaxial films followed a T^{-x} variation at higher temperatures and approached saturation at low temperatures. The deviation from the bulk behavior depended on annealing, which enhanced the mobility and its temperature dependence at low temperatures. This behavior was interpreted as due to scattering by dislocations rather than by point defects. Voronina and Semiletov [37] reported electrical properties of 1- to 2-μ-thick PbTe films grown epitaxially on KCl crystals. The resistivity followed a metallic variation with temperature. The mobility showed a $T^{-0.8}$ dependence which was attributed to defects in the films.

Ramey and McLennan [59] analyzed their Hall-effect data on polycrystalline Ge films on the basis of Eq. (15). By subtracting the known lattice and impurity contributions from the measured film thickness and temperature dependence of the mobility, they attributed the remaining temperature-dependent part to scattering by dislocations. The remaining temperature-independent part was found to depend on the grain size and thickness, and varied inversely with log t, suggesting its origin in the surface and grain-boundary-scattering processes.

4. PHOTOCONDUCTION IN SEMICONDUCTOR FILMS

Photoconductivity in several activated II-VI and IV-VI compounds in thin-film form is of much practical interest. The large uniform surface area of a film is useful in the preparation of photovoltaic and photoconductive devices, infrared detectors, image intensifiers, high-resolution cathode-ray screens, scintillation counters, electroluminescent devices, etc. The device interest has led to extensive investigations aimed at producing photosensitive films of the various sulfide, selenide, and telluride compounds of Cd, Zn, and Pb. Lead-salt photoconductors have been reviewed by several workers [60, 61]. Activation and recrystallization processes in Zn and Cd salts have been reviewed by Vecht [62].

4.1 Activation Process

Vacuum-evaporated films of II-VI and IV-VI compounds deposited at room temperature generally have fine-grained structures and low electrical resistivities (owing to the excess of the metal component), and exhibit little or no photoconductivity. By depositing the films at suitable elevated temperatures, or by annealing subsequent to deposition, the grain size and the resistivity are increased. The films then show weak photoconductivity. If sensitized or activated by incorporating certain impurities, these films become efficient photoconductors. The activator may be evaporated simultaneously with or sequentially to the host film. A postevaporation annealing is, however, essential to achieve homogeneity and uniformity and is also accompanied by the necessary recrystallization. The independent roles of the activation and recrystallization processes in obtaining photoconduction are not well understood.

Lead-salt films are commonly sensitized by incorporating oxygen during or after deposition. The sensitization process is at present an art rather than a science, and it varies from material to material. For instance [63], the addition of oxygen converts the PbTe films from n- to p-type, and the optimum degree of sensitization is obtained when a maximum in the resistance is obtained. The changes in the electrical characteristics of PbS films during sensitization have been studied by several workers [64-66]. The oxygenation process is reversible to a limited extent when oxygen is pumped out. The nature of the sensitizing centers is not well established. Since photoconductivity has been observed [67] in both polycrystalline and single-crystal n- and p-type PbS films, it is clear that the sensitization centers are not simply related to the microstructure of these films.

The II-VI compound films show photoconductivity only when they are activated and are crystalline. The activation process introduces certain levels between the conduction and valence bands so that the minority carriers are readily trapped to prevent recombination with the majority carriers and thus enhance photocurrent. The activator is usually a group I metal (e.g., Ag, Cu) and its effect, in most cases, can be regarded as the introduction of acceptors. The process of incorporation of a nonmetal from group VII (e.g., Cl), or a metal from group III, is termed "coactivation" and is believed to introduce donors.

Controlled recrystallization of CdS [68, 69] and ZnSe [70] films has been obtained by a method due to Gilles and van Cakenberghe [68]. According to this method, CdS films are covered with films of metals, such as Ag, Al, Cu, Bi, Pb, In, and Zn, and recrystallized by heating to a threshold temperature of ~500°C. The induced recrystallization front grows at a rate of about 0.1 mm/sec. During the process, only a small fraction of the metal diffuses into the crystallites; most of it precipitates at the grain boundaries. Another promising method to obtain recrystallization is by the application of a lateral electric field during deposition.

This method has been demonstrated by Chopra and Khan [71] for CdS films and by Paparoditis [72] for CdTe films.

The photoconducting properties of the films are strongly dependent on the deposition techniques and the history of activation and recrystallization. There is unfortunately a noticeable lack of systematic work on structurally well-defined films. Generally speaking, thin films are less photosensitive than the corresponding bulk materials but have shorter response times. The fundamental response time constants [63, 73] are in the microsecond range but are generally limited to milliseconds by trapping and recombination mechanisms. An ingenious technique to decrease the response time by a factor of up to 100 in a Te photodetector has been employed by Cohen [74]. The electrons photogenerated in the Te-film electrode of the Al-Al$_2$O$_3$-Te tunnel structure are collected at the Al electrode after tunneling, thereby avoiding their trapping and recombination in the Te film.

4.2 Photoconductivity Mechanisms

Photoconduction may be explained by an increase in one or a combination of the parameters of concentration, mobility, and lifetime of the majority carriers. Both carrier-modulation [63, 75, 76] and mobility-modulation [77] theories have been proposed to account for photoconductivity in the films discussed above. Slater [77] proposed (for Pb salts) that *p*-type crystallites have *n*-type space-charge regions. The modulation of the barrier height of these *p-n* junctions by photoinjected carriers should yield large changes in the mobility. According to Volger's [78] analysis of inhomogeneous films, the electronic behavior of such films would be dominated by the barrier region $\approx L_D (E/kT)^{1/2}$ (which is ~100 Å for PbS, assuming the barrier height $E = 0.4$ eV, carrier concentration ~10^{19}/cm^3, and L_D is the Debye length). Petritz [79] combined the carrier- and mobility-modulation theories in a complicated model consisting of the space-charge barriers and intercrystalline oxide-film barriers.

Woods [80] measured the Hall effect in chemically prepared PbS films and ascribed photoconductivity to changes in the carrier concentration. Snowden and Portis [81] supported this conclusion on the basis of their observation that the mobility changed little from dc to 10 Gc/sec measuring frequencies. On the other hand, Levy [82] observed changes in the mobility rather than carrier concentration in PbTe films on illumination. Similar results have been obtained [61] on PbSe films. The significant role of the mobility variations and the trapping of the minority carriers is indicated by the temperature dependence of the photoconductive decay times studied by Klaassen et al. [83] on PbS films, and by Khan and Chopra [84] on CdS films.

Photoconducting properties of semiconducting films must be profoundly influenced by the trapping centers. Information on the depth and density of traps in thin films is scanty at present. Dresner and Shallcross [41, 42] used the well-known technique of thermally stimulated currents (TSC) to study CdS films. They obtained a total trap density of $\sim 10^{19}$ to $10^{21}/cm^3$ in films as compared with 10^{14} to $10^{16}/cm^3$ in single crystals of bulk CdS. The depth (level) of traps depended on the heat treatment and the impurities present. Shimizu [43], using the TSC technique on CdSe films, found trap depths at 0.13 and 0.4 eV and a trap density of $\sim 10^{20}/cm^3$. The 0.13-eV level apparently corresponds to Se vacancies; and 0.4 eV may be due to the different ionization states of some crystal imperfections.

Klaassen et al. [83] used the temperature dependence of photocurrent in PbS films to determine the depth of traps. A value of 0.165 eV was obtained. Similar studies by Khan and Chopra [84] on CdS films indicated that the depth of the traps increased with increasing number of sequential depositions from the same source material (increasing non-stoichiometry), or with increasing annealing temperatures. Several discrete activation energies of about 0.008, 0.015, 0.03, 0.060, and 0.11 eV were observed. The level 0.034 eV corresponds to the Cd vacancies, while others are probably due to unknown impurities present in the starting material.

4.3 High-voltage Photovoltaic Effect

Photovoltages larger than the band gap have been observed in thin films of PbS [85-87], ZnS [88-92], Ge and Si [93], and antimony and bismuth chalcogenides [94]. Anomalously high photovoltages of up to a few hundred volts in obliquely evaporated CdTe films were reported by Goldstein and Pensak [95, 96] and later confirmed by others [97-99]. High photovoltages were also observed in obliquely evaporated Ge and Si [93, 100, 101], and GaAs and CdS [102] films, and nonuniform films of Si and SiC [103].

A wide range of values of the photovoltages may be obtained for any one material by varying deposition conditions. It is generally agreed by different workers that this phenomenon is related to the inhomogeneity of the films. The photovoltage is enhanced by oblique deposition, due probably to the increased inhomogeneity and porosity (Chap. IV, Sec. 2.3) of films and their higher sensitivity to adsorption effects.

Goldstein and Pensak [96] explained the occurrence of high photovoltages on the basis of a series combination of a large number of *p-n* junctions, or other photovoltaic elements which may result from the warping of the energy bands at the interfacial surfaces of the oriented crystallites. Ellis et al. [88] suggested that photoelements in ZnS films formed because of repetitive stacking faults. Observation of a mixed hexogonal structure with the normal cubic structure in CdTe films led Semiletov [97] to suggest that stratified layers of different structures may yield photoelements in series. Novik [98] and Palatnik and Sorokin [99] cited their results on epitaxial CdTe films in support of the Semiletov hypothesis. These authors observed that the enrichment of the hexagonal phase by depositing epitaxially on Al_2O_3 increased the photovoltage. These conclusions are questionable since the phenomenon is common to a variety of films and is definitely dependent on the microstructure rather than the crystallographic structure.

Brandhorst and Potter [103] concluded from their detailed study of the kinetics of the growth and decay behavior of photovoltage in Si and SiC films that slow traps must play an important role in this phenomenon. They proposed that the anomalously high photovoltage arises from the space charge produced by a nonuniform distribution of minority carriers trapped at the structural defects in the films.

5. FIELD EFFECT—THIN-FILM TRANSISTOR (TFT)

The surface conductivity of a semiconductor is very sensitive to surface contaminants, preparation techniques, contact effects, and induced localized charges. This sensitivity arises from the low carrier density of the semiconductor, since as few as 10^{11} to 10^{12} positive surface charges, if equalized by a surface density of electrons, give an equivalent density of 10^{16} to 10^{17} carriers/cm^3 when spread to a depth of 1,000 Å. It is just this effect that we can exploit advantageously to obtain significant changes in the surface conductance by varying the surface potential by means of an applied transverse electric field. An electrode structure based on this concept is called the "insulated-gate thin-film transistor" (TFT).

Figure 5(*a*) and (*b*) shows two of the several possible electrode configurations for a typical TFT structure. The electrode terminology is self-explanatory. We may point out that the metal-oxide-semiconductor (MOS) transistor [104] is also based on the same field-effect principle.

An excellent review of the fabrication technology, principle of operation, historical and theoretical background, and characteristics and applications of the TFT is given by Weimer [105]. While detailed discussion is beyond our scope, an elementary account of the perform-ance of a TFT may be instructive.

In common with the conventional field-effect transistors, the TFT is a high-input-impedance device with well-saturated pentode-like character-istics. The gain is flat from dc up to several megacycles, where the response drops off at the 6-dB octave rate expected from the output impedance. Typical plots of the drain current vs. drain voltage for a TFT operating with the source grounded are illustrated in Fig. 5(*c*). These drain characteristics differ from those of the conventional *p-n* junction field-effect device in that the gate may be biased either positively or negatively without drawing gate current.

If the energy bands at the surface of an *n*-type semiconductor are bent down at the surface (i.e., excess electrons at the surface), a negative bias

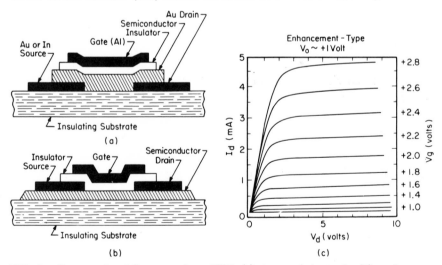

FIG. 5. Cross-sectional diagrams of two TFTs (*a*) staggered-electrode; (*b*) coplanar-electrode structure. The diagrams are self-explanatory. (*c*) The "enhancement" mode characteristics of a coplanar-electrode *n*-type CdS TFT. This unit has $G_m = 4,000$ μmhos at V_g (gate bias) = +2.6 V and $V_0 = 1$V. The drain-gate spacing (*L*) is ~0.001 cm and CdS-film thickness = 4,000 Å. (*Weimer* [105])

is applied to the gate to decrease the source-drain current (depletion mode). Conversely, if the bands are bent up (i.e., surface depleted of electrons), a positive potential is needed to increase the current (enhancement mode). In either case, current saturation occurs when the conduction channel is pinched off in the neighborhood of the drain.

A significant parameter of a TFT is its transconductance defined by

$$G_m = \frac{\Delta I_d}{\Delta V_g} \tag{16}$$

i.e., a change in the drain-current flow ΔI_d per unit change ΔV_g in the gate potential. The intrinsic conductivity of the drain is $\sigma = ne\mu$, where n is the density of carriers and μ is the effective drift mobility. The change in conductivity due to an excess charge q in an effective channel area A and length L of the drain is given by

$$\Delta\sigma = \frac{q\mu}{AL} \tag{17}$$

The change of current for a voltage V_d across L is

$$\Delta I_d = \frac{AV_d \, \Delta\sigma}{L} \tag{18}$$

If the conductivity is modulated by an increment ΔV_g in the voltage applied to the capacitance C across the insulator, then

$$q = C \, \Delta V_g \tag{19}$$

By combining Eqs. (17), (18), and (19), we get

$$G_m \equiv \frac{\Delta I_d}{\Delta V_g} = \frac{\mu V_d}{L^2} C \tag{20}$$

A figure of merit which characterizes the high-frequency performance of a three-terminal active device is its gain-bandwidth product, which is given by

$$GBw \approx \frac{G_m}{2\pi C} \approx \frac{\mu V_d}{2\pi L^2} \tag{21}$$

(below the knee of the $I-V$ curves). At or above the knee, this expression changes to $\mu(V_g - V_0)/2\pi L^2$, where V_g is the gate voltage and V_0 is the gate voltage required for the onset of drain current.

Typical values for a CdS TFT are G_m = 4,000 to 25,000 μ mhos, which is independent of the frequency up to 30 Mc/sec in the best devices, and an effective μ ~200 cm^2/V-sec. The most encouraging results have been obtained by Weimer and coworkers, and others, with CdS and CdTe [105-107] and CdSe [107, 108] TFTs. Other materials such as Te(p-type) [109], InSb [110], SnO$_2$ [111], Si [112], PbS [113], PbTe [114], and InAs [115] have also been successfully used. A G_m of 10,000 μ mhos, effective field-effect mobility of 1,800 cm^2/V-sec, and a gain-bandwidth of 8 Mc have been obtained [115] in an InAs-TFT with a 100-μ source-drain spacing. Performance in the Gc/sec region is predicted for TFT devices of proper source-drain geometry. The actual frequency response of a TFT may, however, depart considerably from Eq. (21) because of the presence of trapping centers [116, 117] and surface states. It should be possible to operate these devices over a wide frequency range provided the semiconducting and insulating materials employed are those which do not tend to form a high density of slow surface states. A wide variety of materials apparently satisfy this condition.

Thin-film high-frequency transistors show a great promise for applications in microelectronic circuitry. Their fabrication process by means of mechanical masks is straightforward. The small gaps (5 to 50 μ) between source and drain may be obtained by using a wire-grill mask. The material problems are particularly simple because evaporated polycrystalline semiconductor films obtained by conventional deposition techniques can be employed. The choice of the substrate material also appears to be not critical. Tellurium TFTs are simpler to fabricate than those using compound semiconductors since the stoichiometry problems do not arise. The ability to build both p- and n-type TFTs should make it possible to employ the principle of complementary symmetry in designing thin-film circuits.

6. CONCLUDING REMARKS

Studies of the transport phenomena in semiconducting films have so far been very limited in scope. It has therefore not been possible to shed significant light on the various scattering mechanisms operating in films. Despite their suitability, the clean and smooth surfaces of epitaxial semiconductor films have not been exploited fully for the study of surface transport phenomena. Further, mechanisms of photoconductivity in certain semiconducting films remain obscure.

It is clear that, in order to understand the variety of behavior exhibited by semiconducting films, extensive measurements of the conductivity, galvanomagnetic properties, and field effect must be undertaken on polycrystalline and epitaxial films. The information so obtained will be valuable not only to the understanding of the basic scattering processes but also to the successful utilization of semiconducting films for practical applications such as thin-film transistors, heterojunctions, solar cells, and photoconductive devices.

Amorphous semiconductors prepared in thin film form are of much interest for basic research and future technological applications. Extensive electrical and optical investigations of the subject are needed to understand the transport behavior of these materials.

REFERENCES

1. A. Many, Y. Goldstein, and N. B. Grover, in "Semiconductor Surfaces," chap. 8, North Holland Publishing Company, Amsterdam, 1965.
2. J. R. Schrieffer, *Phys. Rev.*, **97**:641 (1955); also, in "Semiconductor Surface Physics" (R. H. Kingston, ed.), p. 55, University of Pennsylvania Press, Philadelphia, 1957.
3. K. Fuchs, *Proc. Cambridge Phil. Soc.*, **34**:100 (1938); E. H. Sondheimer, *Advan. Phys.*, **1**:1 (1952).
4. R. F. Greene, *J. Phys. Chem. Solids*, **14**:291 (1960) (a review).
5. B. Tavger, *Phys. Stat. Solidi*, **22**:31 (1967).
6. F. S. Ham and D. C. Mattis, *IBM J. Res. Develop.*, **4**:143 (1960).
7. P. J. Price, *IBM J. Res. Develop.*, **4**:152 (1960).
8. J. E. Parrott, *Proc. Phys. Soc. London*, **85**:1143 (1965).
9. J. N. Zemel, *Phys. Rev.*, **112**:762 (1958).
10. A. Amith, *J. Phys. Chem. Solids*, **14**:271 (1960).
11. R. L. Petritz, *Phys. Rev.*, **110**:1254 (1958).
12. R. Engleman and E. H. Sondheimer, *Proc. Phys. Soc. London*, **B69**:449 (1956).
13. F. J. Blatt, in "Solid State Physics," vol. 4, p. 199, Academic Press Inc., New York, 1957.
14. I. M. Lifshitz and A. M. Kosevich, *Izv. Akad. Nauk SSSR, Ser. Fiz.*, **19**:395 (1955); V. B. Sandomirskii, *Soviet Phys. JETP English Transl.*, **16**:1630 (1963); **25**:101 (1967).

15. B. A. Tavger and V. Ya. Demikhovskii, *Soviet Phys.-Solid State English Transl.*, 5:469 (1963); *Soviet Phys. JETP English Transl.*, 21:494 (1965).
16. J. E. Davey, R. G. Turner, T. Pankey, and M. D. Montgomery, *Solid-State Electron.*, 6:205 (1963).
17. A. Waxman, V. E. Henrich, F. V. Shallcross, H. Borkan, and P. K. Weimer, *J. Appl. Phys.*, 36:168 (1963).
18. C. A. Neugebauer and R. E. Joynson, *Abstracts 13th Natl. Vacuum Symp.*, 1967, p. 105; C. A. Neugebauer, *J. Appl. Phys.*, 39:3177 (1968).
19. M. H. Brodsky and J. N. Zemel, *Phys. Rev.*, 155(3):780 (1967).
20. V. P. Duggal, R. Rup, and P. Tripathi, *Appl. Phys. Letters*, 9:293 (1966).
21. Yu. F. Ogrin, V. N. Lutskii, and M. I. Elinson, *Soviet Phys. JETP Letters*, 3:71 (1966).
22. Yu. F. Komnik and E. I. Bukhshtab, *Soviet Phys. JETP Letters*, 6:58 (1967).
23. See J. M. Ziman, "Phonons and Electrons," Cambridge University Press, New York, 1964.
24. G. A. Kurov, S. A. Semiletov, and Z. G. Pinsker, *Kristallografiya*, 2:59 (1957).
25. O. A. Weinreich and G. Dermit, *J. Appl. Phys.*, 34:225 (1963); O. A. Weinreich et al., *J. Appl. Phys.*, 32:1170 (1961).
26. S. P. Wolsky, T. R. Piwkowski, and G. Wallis, *J. Vacuum Sci. Tech.*, 2:97 J (1965).
27. R. R. Humphris and A. Catlin, *Solid-State Electron.*, 8:957 (1965).
28. E. G. Bylander, J. R. Piedmont, L. D. Shubin, and R. C. Smith, *J. Appl. Phys.*, 34:3407 (1963).
29. J. E. Davey, *Appl. Phys. Letters*, 8:164 (1966).
30. B. W. Sloope and C. O. Tiller, *J. Appl. Phys.*, 38:140 (1967). This paper gives references to a series of other papers on the subject by the same authors.
31. R. L. Schalla, L. H. Thaller, and A. E. Potter, Jr., *J. Appl. Phys.*, 33:2554 (1962).
32. J. Marucchi, *Compt. Rend.*, 258:3848 (1964); 260:6580 (1965).
33. Y. Kataoka, *J. Phys. Soc. Japan*, 17:967 (1962); A. J. Mountuala and G. Abowitz, *Vacuum*, 15:359 (1965).
34. D. J. Dumin, *J. Appl. Phys.*, 38:1909 (1967).
35. J. N. Zemel, J. D. Jensen, and R. B. Schoolar, *Phys. Rev.*, 140:A330 (1965).
36. H. Gobrecht, K. E. Boeters, and H. J. Fleischer, *Z. Physik*, 187:232 (1965).
37. I. P. Voronina and S. A. Semiletov, *Soviet Phys.-Solid State English Transl.*, 6:1494 (1964).
38. Y. Makino, *Japan. J. Phys. Soc.*, 19:580 (1964).
39. Y. Makino and T. Hoshina, *Japan. J. Phys. Soc.*, 19:1242 (1964).
40. R. G. Mankarious, *Solid-State Electron.*, 7:702 (1964).
41. J. Dresner and F. V. Shallcross, *Solid-State Electron.*, 5:205 (1962).
42. J. Dresner and F. V. Shallcross, *J. Appl. Phys.*, 34:2390 (1963).
43. K. Shimizu, *Japan. J. Appl. Phys.*, 4:627 (1965).
44. R. Glang, J. G. Kren, and W. J. Patrik, *J. Electrochem. Soc.*, 110:407 (1963).
45. J. R. Knight, D. Effer, and P. R. Evans, *Solid-State Electron.*, 8:178 (1965).
46. G. S. Kamath and D. Bowman, *J. Electrochem. Soc.*, 114:192 (1967).
47. L. Zdanowicz, *Phys. Stat. Solidi*, 6:K153 (1964).
48. G. Kienel, *Ann. Physik*, 5:229 (1960).
49. C. J. Paparoditis, *J. Phys. Radium*, 23:411 (1962).
50. R. F. Potter and H. H. Wieder, *Solid-State Electron.*, 7:153 (1964).

51. C. Juhasz and J. C. Anderson, *Phys. Letters,* **12**:163 (1964).
52. W. J. Williamson, *Solid-State Electron.,* **9**:213 (1966).
53. C. Paparoditis, *J. Phys.,* **25**:226 (1964).
54. K. L. Chopra and S. K. Bahl, *Abstracts 13th Natl. Vacuum Symp.,* 1967, p. 33.
55. S. A. Aitkhozhin and S. A. Semiletov, *Soviet Phys. Cryst. English Transl.,* **10**:409 (1966).
56. C. T. Naber, Electrochemical Society Meeting, Dallas, Tex., *Abstract* 6, 1967.
57. R. M. Broudy, *Advan. Phys.,* **12**:135 (1963); W. Shockley, *Phys. Rev.,* **91**:228 (1953).
58. A review by N. F. Mott, *Advan. Phys.,* **16**:49 (1967).
59. R. L. Ramey and W. D. McLennan, *J. Appl. Phys.,* **38**:3491 (1967).
60. R. J. Cashman, *Proc. IRE,* **47**:1471 (1959); T. S. Moss, *Proc. IRE,* **43**:1869 (1955).
61. D. E. Bode, in "Physics of Thin Films" (G. Hass and R. E. Thun, eds.), vol. 3, p. 275, Academic Press Inc., New York, 1966.
62. A. Vecht, in "Physics of Thin Films" (G. Hass and R. E. Thun, eds.), vol. 3, p. 165, Academic Press Inc., New York, 1966.
63. D. E. Bode and H. Levinstein, *Phys. Rev.,* **96**:259 (1954).
64. H. Hintenberger, *Z. Physik,* **119**:1 (1942).
65. R. T. Harada and H. T. Minden, *Phys. Rev.,* **102**:1258 (1956).
66. G. W. Mahlman, *Phys. Rev.,* **103**:1619 (1956).
67. J. L. Davis, H. R. Riedl, and R. B. Schoolar, *Appl. Phys. Letters,* **10**:155 (1967).
68. J. M. Gilles and J. van Cakenberghe, *Nature,* **182**:862 (1958). Also, in "Solid State Electronics and Telecommunications" (M. Desirant and J. L. Michiels, eds.), vol. 2, p. 900, Academic Press Inc., New York, 1960.
69. R. R. Addiss, Jr., *Trans. 10th Natl. Vacuum Symp.,* p. 1963, Pergamon Press, New York, 1963.
70. Te Velde, Electrochemical Society Meeting, Toronto, *Abstract* 13, p. 70, 1964.
71. K. L. Chopra and I. H. Khan, *Surface Sci.,* **6**:33 (1967).
72. C. J. Paparoditis in "Single Crystal Films" (M. H. Francombe and H. Sato, eds.), p. 79, Pergamon Press, New York, 1964.
73. H. Levinstein, "Photoconductivity Conference," John Wiley & Sons, Inc., New York, 1956.
74. J. Cohen, *Appl. Phys. Letters,* **10**:118 (1967).
75. T. S. Moss, *Research (London),* **6**:285 (1953).
76. O. Simpson, *Phil. Trans. Roy. Soc. London,* A243:564 (1951).
77. J. C. Slater, *Phys. Rev.,* **103**:1631 (1956).
78. J. Volger, *Phys. Rev.,* **79**:1023 (1950); H. Berger, *Phys. Stat. Solidi,* **5**:739 (1961).
79. R. L. Petritz, *Phys. Rev.,* **104**:1508 (1956).
80. J. F. Woods, *Phys. Rev.,* **106**:235 (1957).
81. D. P. Snowden and A. M. Portis, *Phys. Rev.,* **120**:1983 (1960).
82. J. L. Levy, *Phys. Rev.,* **92**:215 (1953).
83. F. M. Klaassen, J. Blok, H. C. Booy, and F. J. DeHoog, *Physica,* **26**:623 (1960).

84. I. H. Khan and K. L. Chopra, Electrochemical Society Meeting, Washington, D. C., 1964.
85. J. Starkiewicz, L. Sosnowski, and O. Simpson, *Nature,* **158**:28 (1946).
86. G. Schwabe, *Z. Naturforsch.,* **10a**:78 (1955).
87. T. Piwkowski, *Acta Phys. Polon.,* **15**:271 (1956).
88. S. G. Ellis, F. Herman, E. E. Loebner, W. J. Merz, C. W. Struck, and J. G. White, *Phys. Rev.,* **109**:1860 (1958).
89. G. Cheroff and S. P. Keller, *Phys. Rev.,* **111**:98 (1958); Cheroff et al., *Phys. Rev.,* **116**:1091 (1959).
90. W. J. Merz, *Helv. Phys. Acta,* **31**:625 (1958).
91. A. Lempicki, *Phys. Rev.,* **113**:1204 (1959).
92. G. F. Neumark, *Phys. Rev.,* **125**: 838 (1962).
93. H. Kallman, et al., *J. Electrochem. Soc.,* **108**:247 (1961); *J. Phys. Chem. Solids,* **28**:279 (1967).
94. V. M. Lyubin and G. A. Fedorova, *Soviet Phys. "Doklady" English Transl.,* 5: 1343 (1960); *Soviet Phys. - Solid State* 4:1486 (1963).
95. L. Pensak, *Phys. Rev.,* **109**:601 (1958); B. Goldstein, *Phys. Rev.,* **109**:601 (1958).
96. B. Goldstein and L. Pensak, *J. Appl. Phys.,* **30**:155 (1959).
97. S. A. Semiletov, *Soviet Phys.-Solid State English Transl.,* 4:909 (1962).
98. F. T. Novik, *Soviet Phys.-Solid State English Transl.,* 4:2440 (1962).
99. L. S. Palatnik and V. K. Sorokin, *Soviet Phys.-Solid State English Transl.,* 8:2233 (1967).
100. P. P. Konorov and K. Liubits, *Soviet Phys.-Solid State English Transl.,* 6:55 (1964).
101. M. Takahashi and J. Nakai, *Japan. J. Appl. Phys.,* 3:364 (1964).
102. S. Martinuzzi, M. Perrot, and J. Fourny, *J. Phys. Radium,* 25:203 (1964).
103. H. W. Brandhorst, Jr., and A. E. Potter, Jr., *J. Appl. Phys.,* 35:1997 (1964).
104. M. M. Atalla, *NEREM Record,* p. 162, 1962.
105. P. K. Weimer, in "Physics of Thin Films" (G. Hass and R. E. Thun, eds.), vol. 2, p. 147, Academic Press Inc., New York, 1964.
106. P. K. Weimer, *Proc. IRE,* **50**:1462 (1962).
107. F. V. Shallcross, *Proc. IEEE,* **51**:851 (1963).
108. R. Zuleeg, *Solid-State Electron.,* **6**:645 (1963).
109. P. K. Weimer, *Proc. IEEE,* **52**:608 (1964).
110. V. L. Frantz, *Proc. IEEE,* **53**:760 (1965).
111. H. A. Klasens and H. Koelmans, *Solid-State Electron.,* **7**:701 (1964).
112. C. A. T. Salama and L. Young, *Proc. IEEE,* **53**:2156 (1965).
113. W. B. Pennebaker, *Solid-State Electron.,* **8**:509 (1965).
114. J. F. Skalski, *Proc. IEEE,* **53**:1792 (1965).
115. T. P. Brody and H. E. Kunig, *Appl. Phys. Letters,* **9**:259 (1966).
116. R. R. Haering, *Solid-State Electron.,* **7**:31 (1964).
117. M. G. Miksik, E. S. Schlig, and R. R. Haering, *Solid-State Electron.,* **7**:39 (1964).
118. A. H. Clark, *Phys. Rev.,* **154**:750 (1967).
119. P. A. Walley and A. K. Jonscher, *Thin Solid Films,* **1**:367 (1967).
120. S. K. Bahl and K. L. Chopra, a series of three papers on the structural, electrical, and optical properties of GeTe films which will be published in *J. Appl. Phys.* (1969).

121. R. Tsu, W. E. Howard, and L. Esaki, *Phys. Rev.,* **172**:779 (1968).
122. A. Gubanov, "Quantum Electron Theory of Amorphous Conductors," Consultants Bureau, New York, 1965.
123. N. Cusack, *Rept. Progr. Phys.,* **26**:361 (1963).
124. A. Nwachuku and M. Kuhn, *Appl. Phys. Letters,* **12**:163 (1968).

VIII TRANSPORT PHENOMENA IN INSULATOR FILMS

1. INTRODUCTION

The mechanisms of transport of carriers through thin insulator films have been the subject of intensive theoretical and experimental investigations for the last several years. These studies have been stimulated by the attractive possibilities of development of a variety of miniaturized solid-state devices such as diodes, hot-electron triodes, switching devices, fixed and variable capacitors, piezoelectric transducers, photocells, and electroluminescent devices. The quest for practical applications has yielded much information on such properties of thin insulator films as their dielectric behavior, dielectric breakdown, band structure, potential barriers at the interface electrodes, and current transport at high ($\sim 10^6$ to 10^7V/cm) fields. The available information is, however, by no means complete or even totally consistent. Consequently, an authoritative review of the electronic behavior of thin insulators is not possible at present. Brief reviews of some aspects of this subject have appeared in the literature and are referred to in the text. We shall present in this chapter the salient features of the transport process in thin insulators and the associated electronic phenomena of basic and practical interest.

Most of the work on thin insulators has been carried out on amorphous or fine-grained films, primarily metal oxides, of thickness less than 100 Å. The techniques for preparing insulator films are described in Chap. II.

An amorphous, stoichiometric dielectric material is an ideal specimen for investigations. If, however, deviations from the stoichiometry exist in these materials, their dielectric properties and hence electronic behavior are profoundly affected. It must be pointed out that technical advances have now made it possible to prepare epitaxial single-crystal dielectric films of a number of materials (see Table VIII, Chap. IV). But the electronic properties of these films have received little attention so far.

2. DIELECTRIC PROPERTIES

The dielectric behavior of thin insulating films is of direct interest to both the basic studies of electrical conduction through such films and their applications in devices. This section deals with the dielectric-constant and the dielectric-loss properties of thin films.

2.1 Thin Films

It has been shown theoretically [1] that the dielectric constant of an insulator should be preserved down to a few atomic layers so that no thickness dependence is expected. The only experimental verification [2] of this conclusion has been reported in structurally perfect organic films of cadmium stearate down to a monolayer thickness which is equal to 24.6 Å.

If, however, insulator films are obtained by vapor-deposition techniques, the growth processes (Chap. IV, Sec. 2) produce agglomeration and hence porosity in ultrathin films (generally less than 200 Å). The dielectric constant falls rapidly with decreasing thickness in the porous films, as shown by the data [3] on insulating ZnS films in Fig. 1. The thickness below which the effect of the structural porosity becomes significant depends on the deposition technique and the deposition conditions. In the case of ZnS films (Fig. 1), this thickness is lower for lower deposition temperatures and certain substrates such as mica as

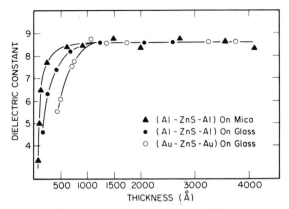

FIG. 1. The thickness variation of the dielectric constant of ZnS films: (Al-ZnS-Al) on mica at 23°C; (Al-ZnS-Al) on glass at 23°C; and (Au-ZnS-Au) on glass at 300°C. (*Chopra* [3])

compared with glass. The commonly used amorphous oxide films are completely continuous at thicknesses of ~20 to 50 Å as judged from the electron micrographs. The anodic oxide films appear to be more perfect than those grown thermally. Thus, although the dielectric constant of ultrathin oxide films has not been measured, bulklike values are expected for thicknesses above 50 Å. Below 50 Å, depending on the deposition conditions, marked departure from the bulk values may be expected.

Large intrinsic stresses generally present in atomistically deposited films (Chap. V, Sec. 2.2) appear to have little effect on their dielectric properties. When mechanical crazing or cracking of films occurs because of the stresses, the effective dielectric constant decreases. In the case of piezoelectric films, the effect of the intrinsic stresses, if any, is not well established at present. Hacskaylo and Feldman [4] observed a thickness-dependent maximum in the dielectric constant of thin, unoriented ZnS films. At the position (~900 Å) of the maximum, a considerable increase in the loss factor was observed. The authors suggested that the dielectric-constant maximum may be due to the intrinsic strain-induced piezoelectricity in films. Chopra [3] showed that this explanation was inconsistent with other experimental observations and further that these results could not be reproduced under careful experimental conditions.

Thickness dependence of the dielectric constant in ultrathin (< 50 Å) films may also arise because of changes in the "effective" thickness of the insulator. The anomalous capacitance of tunnel junctions (discussed later) observed by Mead [5] was ascribed to a reduced effective insulator thickness resulting from the electric-field penetration of the electrodes. The estimated [6] field-penetration depth is, however, too small (~ 1 Å for typical metals) to explain the observations. The effective insulator thickness may be reduced because of (1) the penetration of energetic metal-vapor atoms during deposition of the metal electrodes for measurements, and (2) the presence of ionized impurities and/or traps in the insulator. It should be pointed out that small changes in the effective thickness have a very large effect on the conductance of thin insulators (see Sec. 4.3).

The fields applied across thin-film insulators for device applications are generally high ($\sim 10^6$V/cm) and approach breakdown values for typical bulk dielectrics. The poorly understood phenomenon of electrical breakdown of bulk insulators has been the subject of numerous theoretical treatments which have been reviewed elsewhere [7]. For our present purpose, the various types of "excess" current breakdowns in solids may only be mentioned. (1) Thermal breakdown, in which the high temperature due to Joule heating leads to cumulative thermal ionization of the insulator. (2) Avalanche breakdown, in which avalanche multiplication of carriers takes place because of ionization of the lattice by collisions with energetic electrons. (3) Defect breakdown, induced by ionization of impurities or structural defects. (4) Intrinsic breakdown due to field emission in thin films.

Breakdown type (4) is of interest in thin-film tunnel structures and was analyzed by Forlani and Minnaja [8]. They predicted the breakdown voltage to depend on the work function of the negative electrode and the square root of the film thickness. If, however, the avalanche type of breakdown occurs, the breakdown voltage should increase when the film thickness is comparable with or less than the mean free path (mfp) of the electrons owing to the decrease in the possible number of inelastic collisions with the lattice. Since the mfp's in insulators are estimated [9] to be of the order of atomic dimensions (~ 10 Å for ZnS), the size-dependent breakdown voltage may be observable in thicknesses below 50 Å.

Chopra [3] measured the breakdown voltage of ZnS films and obtained a linear thickness dependence down to 200 Å (Fig. 2). Below 200 Å, the breakdown voltage increased. A similar linear dependence was observed by Pakswer and Pratinidhi [10] for Al_2O_3 films down to

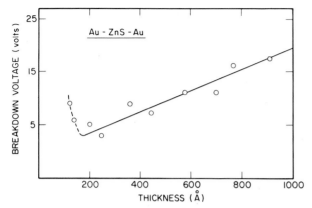

FIG. 2. The dc breakdown strength of Au-ZnS-Au structures at 23°C as a function of the thickness of ZnS films. (*Chopra* [3])

~30 Å. The observed breakdown fields of 1.7×10^6 and 2.3×10^7 V/cm for ZnS and Al_2O_3 films, respectively, approach the maximum intrinsic dielectric strength of inorganic dielectrics.

An electrical and electron-microscope study of the breakdown phenomenon in thin-film Al-SiO-Al capacitors (SiO film thickness 1,200 to 5,500 Å) was reported by Budenstein and Hayes [11]. Breakdown was found to originate at some unknown inhomogeneities or impurities and propagate symmetrically in all directions in about 10^{-6} sec. The breakdown resulted in a change of conductivity by a factor of 10^{10} in less than 10^{-7} sec. The voltage threshold for the onset of the breakdown varied approximately as $t^{1/2}$ (t = dielectric-film thickness), while the voltage for the cessation of breakdown was found to be independent of the dielectric thickness. Both these voltages were found to be nearly independent of temperature between 80 to 380°K. The breakdown mechanism was proposed to be an electrochemical solid reaction in the presence of the high electric field (~10^6 V/cm), resulting in the decomposition of SiO. The authors compared the observed thickness and temperature variation with the various breakdown theories and concluded that their results were consistent with the Forlani-Minnaja theory. This conclusion may be questioned since the insensitivity of the breakdown to the ambient temperature could also be due to the poor heat exchange between the hot film and the ambient during the short ($<10^{-6}$ sec) time of the breakdown. Further, a $\sqrt{t}$ dependence of the breakdown voltage has also been observed [12] in thick (~0.5 μ) anodic Al_2O_3 films for which the Forlani-Minnaja theory is not expected to hold.

There is no definite evidence for a dependence of the breakdown voltage on the polarity of the electrodes. However, Worthing [13] observed the dc breakdown of steam-grown SiO_2 films (~6,000 Å) on Si to depend on the length of time for which the negative voltage is applied to silicon. The breakdown followed a $1/\tau^4$ dependence (τ = time after which breakdown occurs at a particular voltage). For the positive polarity of Si, the breakdown occurred abruptly at 5×10^6V/cm.

An increase in the breakdown voltage with decreasing film thickness has been observed in films of ZnS below 200 Å (Fig. 2), NaCl [14] below 1.6 μ, and $BaTiO_3$ [15] below ~0.5 μ. These thicknesses are much larger than the estimated values of the mfp's in these materials, so that these observations cannot be ascribed to size effects. An explanation in terms of the size effect is, however, possible if the size of the individual crystallites rather than the film thickness is considered to dominate the dynamics of the avalanche breakdown.

2.2 Thick Films

Most of the studies of the dielectric properties reported in the literature have been concerned with thick insulator films for applications in capacitors for microelectronics. The capacitance C of a parallel-plate condenser of surface area A with a dielectric constant ϵ and thickness t in centimeters is given by

$$C = \frac{\epsilon A}{4\pi t} = 8.85 \times 10^{-14} \frac{\epsilon A}{t} \quad \text{farads} \tag{1}$$

Evidently, a high dielectric constant and small thickness of the dielectric are necessary to obtain high values of the capacitance. Because of the difficulty of obtaining structurally continuous and stable ultrathin films, capacitor applications are generally limited to thick films (>100 Å for anodic, and >500 Å for thermal oxides).

The evaluation of the dielectric properties of insulator films is generally carried out by measuring simultaneously the capacitance and dissipation factor at a particular frequency. The measured dissipation factor is the sum of the various dielectric-loss processes and the loss due to the series resistance and capacitance of the electrodes. The latter contribution yields an apparent frequency-dependent loss which must be taken into account. Thus, in order to obtain a low dissipation factor at reasonably high frequencies, low-resistance metal electrodes should be used.

A number of reports on the dielectric properties of films of oxides [16-20], alkali halides [20-22], polymer [20, 23, 24], and glass [25] have been published. A summary of the representative results on dielectric constant, loss factor, and breakdown strength of thick films of various dielectric materials of interest is given in Table I. The given references [26-38] should be consulted for further details.

Anodic tantalum oxide and evaporated SiO films are best suited for capacitor applications. A brief summary [18] of the characteristics of these materials will familiarize the reader with some magnitudes involved in such applications. Tantalum oxide film capacitors have a high dielectric constant ($\sim$25), high working field ($\sim10^6$V/cm), low dissipation factor ($\sim$0.01 for up to 10 kc/sec), reasonable temperature coefficient (+250 ppm/°C over a range of −196 to +100°C), and good yield and life performance. Although variability still exists to some extent in processing, the fabrication process of tantalum oxide capacitors is compatible with the simultaneous production of sputtered Ta and tantalum nitride resistors and is therefore attractive for manufacturing microcircuits.

The various properties of evaporated SiO films depend critically on the deposition conditions. Nonstoichiometry, continued oxidation, and crazing or fracture phenomena (Chap. V, Sec. 3.5) due to the presence of large intrinsic stresses are some of the undesirable properties of SiO films. Despite these disadvantages, evaporated SiO is at present a frequently used dielectric for capacitors and insulation in thin-film circuits. Much research has led to a definition of the deposition parameters to minimize the deleterious effects. Best results are obtained with films of a composition near SiO, deposited at high rates, low pressures, and source temperatures around 1300°C. Their dielectric constant ranges from 4.8 to 6.8; the dissipation factor in the audio-frequency ranges from 0.01 to 0.1, the dielectric strength is 1 to 3 × 10^6V/cm, and the temperature coefficient of capacitance is $\sim$100 to 400 ppm/°C.

2.3 Dielectric Losses

If the electric polarization in a dielectric is unable to follow the varying electric field, dielectric losses occur. Electronic, ionic, dipole, or

Table I. Dielectric Properties of Thin Insulating Films of Various Materials Prepared by Different Techniques
If not mentioned, thickness is ~1μ. Aluminum (or gold) films are generally used as electrodes for measurements

Material	Dielectric constant	Dissipation factor	Frequency, kc	Break-down V/cm	TC,* % per °C	Thickness, μ	Deposition technique†	Ref.
Al oxide	8.8	0.008	1.0		0.03	1.5	Ev	27
BaSrTiO$_3$	$1-10 \times 10^3$		1.0				Flash Ev	11
BaTiO$_3$	~200	0.05	1.0				Ev	30
CaF$_2$	3.2	0.05	0.1				Ev	21
LiF	5.2	0.031	0.1					21
MgF$_2$	4.9	0.016	0.1					21
NaCl	5.1	0.35	0.1					
Nb oxide	39	0.07	1.0	6.0×10^5			RS	29
PbTiO$_3$	33–50	Variable	1.0					20
Polymer styrene	2.6–2.7	0.001–0.002	0.1–100				GD	38
Polymer butadiene	2.6–2.75	0.002–0.01	0.1–100				uv	33
SiO	3.3–5.2	0.004–0.02	0.1–1,000	3.0×10^5	0.03	~2	Ev	27
	6	0.015–0.02	1–1,000				Ev	26
SiO$_2$	3.4	0.005	1.0	3.0×10^6	0.0012	0.1	RS	29
	3–4	0.001		~10^6			RS	29
Si$_2$N$_4$	5.5			10^7		0.03–0.3	CVD	
Ta oxide	22	<0.01	1–30	7.0×10^5		0.18	RS	37
	25	0.008–0.01	0.1–50	6.0×10^6	0.025		An	32

Ti oxide	55	0.04–0.09	1–100		0.021–0.044	0.16	RS	38
	40	0.02–0.05	0.1–400		0.03		An	36
	50	0.01	1.0				An	34
W oxide	40	0.6					An	33
ZnS	8.5	<0.01	1.0	1.7×10^6			Ev	3
Zr oxide	25	0.045					RS	29

*TC = temperature coefficient of capacitance.

†Ev stands for evaporation, An for anodization, RS for reactive sputtering, GD for glow discharge, uv for ultraviolet, and CVD for chemical-vapor deposition.

interfacial polarizations would contribute their characteristic relaxation processes. A study of the frequency and temperature dependence of the dielectric losses should provide information on these relaxation processes.

Dielectric loss is essentially a bulk property and should not depend on film thickness *unless* the defect structure of the film takes part in a particular relaxation process, or else the film is thin enough so that the surfaces influence the internal reorientation of dipoles. Dielectric losses may be modified by deviations from the stoichiometry and thus exhibit strong dependence on deposition conditions. This is amply supported by widely divergent results in the published literature.

The dielectric losses in anodic oxide films have been studied extensively and are reviewed by Young [16]. These films generally display a constant loss factor at audio frequencies. This observation suggests the occurrence of a range of relaxation times, which is not unreasonable for amorphous and highly disordered films. The losses increase at low frequencies and may be ascribed to interfacial polarization involving ionic movement. A mathematical treatment of such a system is, however, lacking.

Detailed studies of the dielectric loss in thin films as a function of frequency and temperature [39, 40] are scarce. At higher temperatures above ~200 to 400°C, depending on the capacitor materials, the diffusion of the film electrodes into the insulator film is significant and produces a rapidly rising dielectric loss [39]. The frequency dependence [41] of the loss factor for 1,000-pF evaporated capacitors of SiO, MgF_2, and ZnS is shown in Fig. 3. The apparent high loss at increasing frequencies is due to the uncorrected electrode resistance. As discussed later in this section, the dielectric loss in these films can be explained on the basis of excess vacancy concentration. The dielectric loss in ZnS films exhibits peaks at about 210 and 475°K which correspond to activation energies of 0.16 and 1.18 eV, respectively. The higher value suggests a vacancy-migration process. The origin of the lower-energy activation process is unknown.

Weaver studied the dielectric properties of thin films of alkali halides (CaF_2, LiF, NaF, NaCl, RbBr, and MgF_2) and has reviewed his work in several articles [21, 22]. These ionic materials were chosen to study the dielectric-loss mechanisms primarily due to their simple structures. Capacitance and dissipation-factor measurements over a frequency range from 0.01 cps to 100 kc/sec showed the following characteristics for

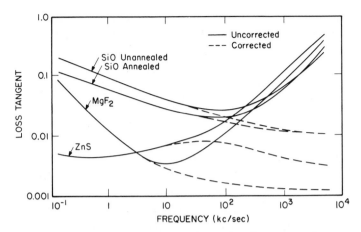

FIG. 3. Loss tangent vs. frequency for 1,000-pF evaporated capacitors of SiO, MgF$_2$, and ZnS. The solid lines are the measured values, the dotted are corrected for electrode series resistance. (*Maddocks and Thun* [41])

evaporated alkali-halide films at room temperature: (1) The dielectric losses decreased steadily with aging time after deposition. (2) Both the dissipation factor and the capacitance increased toward the low-frequency range in a manner that suggests a relaxation mechanism operating at some frequencies below 100 cycles. (3) The loss factor showed a plateau at about 1 cps, suggesting a broad loss peak at this frequency resulting from contributions from at least two different loss mechanisms. The second, more pronounced dispersion region appeared to be centered on a very low frequency, certainly well below 0.01 cps at room temperature. (4) The magnitude of the losses increased as the anion size increased and/or the cation size decreased. The rate of aging, however, decreased for the same cases. (5) Marked increase in the capacitance and loss factor was observed as a result of adsorbed moisture, even for water-insoluble MgF$_2$. This increase masks the effects due to normal aging and basic loss mechanisms.

Dielectric losses of the type and magnitude described above are not observed in bulk single crystals [42]. Experiments show that these losses cannot be attributed to adsorption or moisture effect or the dc conductivity. Weaver suggested that the basic loss mechanisms in alkali-halide films are due to the excess vacancy concentration (which is $\sim 10^{19}$/cm^3 in KCl films as deduced from F-center measurements [43]). Loss peaks due to a vacancy-pair orientation in an ionic crystal as proposed by Breckenridge [42] have not been observed. Observation (4) indicates that the losses are produced by the motion of cation vacancies. The rate at which the Schottky type of defects with an equal number of positive- and negative-ion vacancies disappear (aging thus takes place) is, however, determined by the mobility of the slower anion vacancies. If vacancies are assumed to condense at the intercrystallite boundaries, their observed rate of condensation is in good agreement with the calculated rate for anion vacancies, using the activation energies obtained from independent measurements.

With a high vacancy concentration, the absence of any appreciable dc conductivity is surprising, and it was explained by Weaver on the basis that the polycrystalline films consist of individual platelet-like crystallites. Weaver proposed that the vacancies are blocked at intercrystallite boundaries and cause the high-frequency loss peak. Adsorbed moisture increases the size of the crystallites by recrystallization, thereby yielding a longer relaxation time and increased loss due to the crystallite-polarization mechanism. The blocking of vacancies at intercrystallite boundaries may not be complete. If some leakage takes place which is further blocked at the electrodes, a very low frequency loss peak, as exhibited by NaBr films (Fig. 4), will be observed. As the deposition temperature increases, the position of the peak shifts to higher

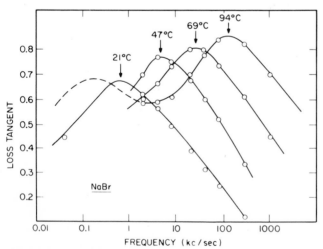

FIG. 4. Loss tangent vs. frequency for a NaBr film at 21, 47, 69, and 94°C. A low- as well as a high-frequency loss peak appear at temperatures above 94°C. (*Weaver* [22])

frequencies, presumably because of the increased structural uniformity of films. At temperatures above 90°C, the low-frequency peak becomes quite pronounced. The activation energies of the two peaks are about 1 eV (blocking effect) and 0.7 eV (crystallite polarization).

In the light of the above-mentioned mechanism for the loss peaks, significantly lower losses are expected in single-crystal epitaxial films. The results [44] for (100) epitaxial LiF films deposited on an epitaxial silver film deposited in turn on a rock-salt substrate confirm this conclusion. Within the range of the frequencies used, the loss factor for single-crystal films is at least an order of magnitude less than for the polycrystalline material under the same conditions, and none of the curves shows any evidence of the high-frequency dispersion which appears in polycrystalline films. The low-frequency losses in epitaxial and poly-crystalline films are comparable since they are determined by the polarization effects at the electrodes. This discussion on alkali-halide films may also be applied to films of other ionic materials.

Hirose and Wada [45] measured the dielectric losses in SiO films from 10^{-2} to 10^7 cps and observed a loss peak at 0.1 cps. This peak corresponds to an activation energy of ~0.4 eV, a value also obtained from the dc conductivity data. This observation indicates that the same ions (probably free silicon ions) contribute to losses at both low and high frequencies.

3. PIEZOELECTRIC FILMS

Hypersonic waves, 10^8 to 10^{10} cps, in dielectric materials have been generated by direct surface excitation of quartz, conventional quartz transducers at high harmonics, or magnetostrictive films. A more convenient and efficient technique for generation of either compres-sional or shear waves in the gigacycle range is provided by the use of thin-film piezoelectric transducers. The small thickness of a film makes it possible to obtain a high fundamental resonant frequency. Active films of CdS as thick as 8μ and as thin as 300 Å have been used to provide fundamental resonant frequencies of about 250 Mc/sec and 75 Gc/sec, respectively. A frequency of ~10^{12} cps, which is the upper limit of the phonon spectrum, may indeed be possible with the use of thinner films.

Piezoelectric films of vacuum-evaporated hexagonal CdS [46-50] (cubic structure is not active), hexagonal and cubic ZnS [47-49], and reactively sputtered [51] as well as evaporated [52] hexagonal ZnO have been investigated as transducers. Piezoelectric activity is exhibited only by stoichiometric and high-resistance films. The deposition techniques, structure, and growth of such films have been studied by various investigators. In particular, the orientation effects in the epitaxial growth of films have received considerable attention since the mode of generation of sound depends on the crystallographic orientation of the active film with respect to the applied electric field for excitation. If, for example, a hexagonal CdS film has its c axis normal to the substrate and an electric field is applied across the film thickness, a longitudinal ultrasonic wave will be generated and launched into the substrate medium. For a film in which the c axis is in the film plane, a shear ultrasonic wave polarized in the direction of the c axis will be propagated into the substrate medium. For a film in which the c axis lies at some intermediate angle, both types of waves will, in general, be generated. If the c axis is oriented at an angle of about $40°$ with respect to the film normal, theoretical calculations [53] show that *only* a shear mode should then be generated.

Interestingly, however, it is possible to control the crystallographic orientation of the active films by epitaxial deposition onto suitable substrates and under certain deposition conditions (see Chap. IV). Active CdS and ZnS films have been grown [47, 49] epitaxially on single-crystal substrates of quartz Al_2O_3, MgO, and TiO_2. Foster [50] studied and exploited the orientation effects due to oblique deposition of CdS films (Chap. IV, Sec. 2.3.3). A correlation exists between the oblique angle of deposition and the inclination of the c axis to the film surface which makes it possible to produce $40°$ films capable of generating only the shear mode. The extension of this useful oblique-deposition technique to other piezoelectric materials should also be possible. An ingenious method of flipping the c axes of consecutive active CdS or ZnS layers was utilized by deKlerk [184]. By placing an epitaxial sandwich of (111) PbS film, with either Pb or S atoms on both of its surfaces depending on the polarity of the active base film, the polarity of the overgrowing epitaxial film of CdS or ZnS is fixed to yield a reversal of the c axis.

Thin-film transducers have so far been used as microwave phonon generators for studies of phonon-phonon interactions in dielectric materials [49], and as active elements in microwave acoustic-delay lines [48]. For the former application, transducers should be capable of independent generation of each of the pure acoustic modes in the material and should exhibit a low insertion loss so as to reduce the piezoelectric (electromechanical conversion) losses to a level below the lowest attenuation encountered in the material. The low loss requirement is obviously met by a transducer which has the maximum possible conversion efficiency and a minimum possible insertion loss. Thus, ZnO, with a higher coupling constant and a higher sound velocity (which provides a higher electric input impedance), is a more attractive transducer material than CdS.

The maximum power-handling capability of a single thin-film transducer will be determined by the maximum electric-field gradient which can be applied to it without electrical breakdown. A gain in the total power delivered can be obtained by employing multilayer film transducers. DeKlerk [49] deposited multilayer structures consisting of active CdS $\lambda/2$ layers separated by passive SiO $\lambda/2$ layers. By maintaining a constant electric-field gradient through the multilayer structure, deKlerk realized a gain in power proportional to the square of the number of the active $\lambda/2$ layers employed at an operating frequency of 1 Gc/sec. These studies also showed that bulk excitation, as opposed to surface excitation, of a piezoelectric material can be achieved by introducing periodic discontinuities in the appropriate piezoelectric modulus. Multilayer structures result in greatly reduced capacitance and hence lower impedance mismatch between the electric input and the transducer. Thus, they are very useful for high-frequency devices.

Sliker and Roberts [54] developed a thin-film CdS-quartz composite resonator for the 100 to 1,000 Mc/sec range. The device consists of a $\lambda/2$ thick piezoelectric CdS transducer evaporated onto a single-crystal quartz wafer of thickness equal to an integral number of half-waves (typically, $9\lambda/2$). The principle of energy trapping was applied to localize the acoustic energy in the quartz wafer to the volume under and immediately surrounding the electrically driven CdS film. Both shear and longitudinal mode composite resonators may be fabricated. Further, by choosing suitable composite materials, the temperature coefficient for the shear mode can be made negligibly small.

4. ELECTRICAL CONDUCTION IN INSULATOR FILMS

4.1 Conduction Mechanisms

According to the band structure of solids, an insulator is characterized by a full valence band separated from an empty conduction band by a forbidden energy gap of a few electron volts. Evidently, conduction cannot take place in either the filled or the empty band unless additional carriers are introduced. Carriers may be generated or modulated inside the insulator (bulk-limited processes), or injected from the metal electrode (injection-limited processes). The various mechanisms affecting current transfer through a thin insulator sandwiched between two metal electrodes (henceforth called a tunnel junction or a sandwich structure) are illustrated schematically in the energy-level diagram shown in Fig. 5(A).

The simplest mechanism is the direct, quantum-mechanical tunneling of electrons from one metal electrode to the other (a). The carriers can be injected into the conduction (or valence) band of the insulator by thermionic or Schottky emission over the metal-insulator interface barrier (b). The carriers may tunnel through the insulator barrier gap (c) at high applied fields (field or cold emission). The (b) and (c) processes are analogous to the corresponding emission from a metal into vacuum.

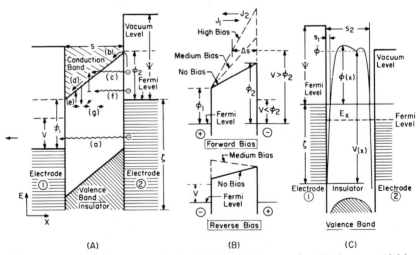

(A)   (B)   (C)

FIG. 5. (A) Schematic representation of energy diagram of an ideal trapezoidal barrier in a metal ① –insulator-metal ② structure with a positive voltage V applied to electrode 1 (forward bias); ϕ_1 and ϕ_2 ($\phi_1 < \phi_2$) are the barrier heights at interfaces; Ψ and ζ refer to the work function and the Fermi energy of the metal electrodes. Conduction from ① to ② electrode may take place via any one, or a combination of the (a) to (g) conduction mechanisms explained in the text. (B) A symmetrical rectangular barrier in the intermediate- and high-bias-voltage conditions. (C) Schematic representation of a general practical barrier with image-force correction included. Note that J_1 is the reverse current from ① to ② electrodes.

The transport of carriers in the conduction band is modulated by the scattering processes (d). The lattice scattering is probably not significant because of the small insulator thickness. On the other hand, since amorphous films must contain a large number of traps, trapping of carriers (e) can be expected to have a profound effect on the conduction processes. In the presence of traps, the conduction can also take place by tunneling via traps (f) or, depending on the type of traps, by a hopping process from trap to trap (g). If the number of traps is small and the contacts are ohmic (i.e., the Fermi level of the contact material is very close to the conduction band of the insulator so that an unlimited supply of electrons is available for conduction), the current flow may be regulated by the prevailing space-charge conditions.

A quantitative description of the various mechanisms mentioned above is possible under certain simplifying conditions. The analysis of the experimental data is, however, complicated and fraught with uncertainties because a number of these mechanisms may operate simultaneously in a practical case of a thin insulator. We shall analyze in this section the important conduction mechanisms and present experimental evidence in support of these mechanisms. Brief reviews of the subject have been given by Seraphim [55], Eckertová [56], and Hill [57].

4.2 Thermionic (Schottky) Emission

Thermal evaporation of electrons from a heated metal is called thermionic emission. The emission of electrons into vacuum or into the conduction band of a dielectric from the metal-contact electrode by thermal activation over the interface barrier and under an applied electric field is called Schottky emission. The well-known Richardson-Schottky formula for Schottky emission has been applied to tunnel structures with well-defined rectangular energy barriers (Fig. 5) by Simmons [58], Gundlach [59], and Mann [60]. Simmons extended the theoretical treatment to the case of a trapezoidal energy barrier [Fig. 5(B) and (C)] separating two dissimilar electrodes and took into account the corrections for the image-potential lowering of the two interface barriers.

In a sandwich structure, it is necessary to add the thermionic current in both directions. Thus, since thermionic emission over a barrier ϕ is given by $AT^2 e^{-\phi/kT}$, the net current density is given by

$$J = AT^2(e^{-\phi_1/kT} - e^{-\phi_2/kT}) \qquad (2)$$

where ϕ_1 and ϕ_2 are the maximum barrier heights above the Fermi level of the negatively biased and positively biased electrodes, respectively. When a potential V exists between electrodes so that $\phi_2 = \phi_1 + eV$ (e = electric charge), then

$$J = AT^2 e^{-\phi_1/kT}(1 - e^{-eV/kT}) \tag{3}$$

where $A = 120 \text{ A/cm}^2 {}^\circ\text{K}^2$ is the Richardson constant. If the barrier is symmetrical, then by applying a parabolic image-force correction for one barrier, we obtain the Richardson-Schottky formula*

$$J = AT^2 \exp\left[-\frac{\phi - (14.4eV/\epsilon s)^{1/2}}{kT}\right] \tag{4}$$

Here, ϵ is the dielectric constant and s is the effective thickness of the insulator barrier in angstroms.

Schottky emission was experimentally identified by Emtage and Tantraporn [61] in thin-film sandwich structures of Au-polymerized silicone oil-Au, Al-Al$_2$O$_3$-Al, and Al-GeO$_2$-Al; by Mead [62] in Zn-ZnO-Au; by Standley and Maissel [63] in Ta-Ta$_2$O$_5$-metal; by Flannery and Pollack [64] in Ta-Ta$_2$O$_5$-Au; and by Hartman et al. [65] in Al-SiO-Al. By fitting the observed temperature dependence of the current density in Eq. (4), one can calculate the constant A and the barrier height. The temperature and field dependence observed in most cases is only approximately in accord with the theory. The value of A is found to be lower by an order of magnitude for polymer films and a factor of 2.2 for Ta$_2$O$_5$ films. However, the barrier heights determined from the data on Ta-Ta$_2$O$_5$-Au and Zn-ZnO-Au systems agree with those obtained from other experiments (see Table II, Sec. 4.3).

At low temperatures, high applied fields, and small effective barrier widths, the dominant mechanism of conduction is expected to be tunneling rather than thermionic emission. Simmons [58] calculated the J-V characteristics of a tunnel structure on the basis of both thermionic emission and tunneling mechanism for several values of the barrier parameters. His results for a symmetric structure are shown in Fig. 6. A comparison of the curves shows that at a temperature of 300°K, and for

*Using $e^2 = 14.4$ eV-Å as a convenient unit.

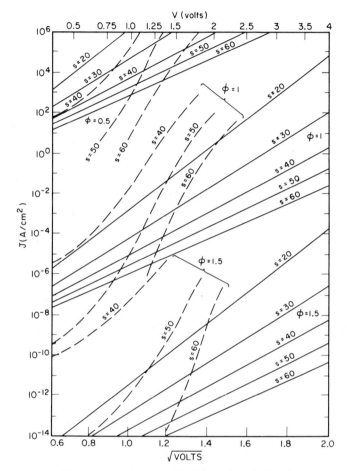

FIG. 6. Comparison of thermal (solid lines) and tunnel current density (dashed lines) vs. $V^{1/2}$ for symmetric tunnel structures at $300°K$; insulator dielectric constant $\epsilon = 3$; s is barrier thickness expressed in angstroms. (*Simmons* [58])

a barrier thickness s less than 40 Å, conduction is predominantly by the tunneling mechanism. However, for $s > 40$ Å, either thermionic or tunnel mechanism can predominate, depending upon the barrier height and the applied voltage.

4.3 Quantum-mechanical Tunneling

(1) Theories. According to the quantum theory, the wave function of an electron has finite values within the classically forbidden barrier of the insulator separated by two metal electrodes. The wave function decays exponentially with the thickness of the barrier. Therefore, if the barrier is very thin, the electron has a finite penetration probability which depends on the size and shape (roughly, the area) of the potential barrier. Tunneling (field emission) may also take place through barriers whose effective thickness is reduced by a strong applied field [see Fig. 5(B)].

In order to analyze mathematically the problem of current transport through a thin-film sandwich structure, the following assumptions need to be made: (1) That the band theory applies to the thin insulator and further a bulklike band-energy diagram with sharp and well-defined interfaces as shown in Fig. 5 can be constructed. Thus, the insulator barrier is in general a trapezoid, its shape being determined by the interface barrier potentials, which in turn depend on factors such as the work functions of the electrodes, electron affinity of the insulator, dielectric constant, applied field, and image forces. Geppert [66] calculated the shape of the potential barrier for a thin insulator and showed that a trapezoid is a good approximation; however, significant modification due to the image-force correction is required in the case of ultrathin films. A practical, generalized barrier with image-force correction is shown in Fig. 5 (C). (2) That the energy-momentum relation is parabolic (free electrons) and the carrier mass in both the metals and the insulator is that of a free electron. (3) That all carriers incident at the metal-insulator interface with energies exceeding the barrier height are transmitted through the barrier.   These assumptions cannot be completely justified and are criticized later in Sec. 4.4.5.

When two parallel electrodes are separated by a thin insulator, the tunnel current density from electrode 2 to 1 is given [67] by the basic integral

$$J_{2-1} = \frac{4\pi m e}{h^3} \int_0^{E_m} D(E_x)\,dE_x \int_0^\infty f(E + eV)\,dE \tag{5}$$

where m is the electron mass, e is the electron charge, h Planck's constant, E_m the maximum energy of the electrons in electrode 2, E_x the energy associated with the x-directed momentum (x being in the direction of tunneling), V the positive voltage applied to electrode 2 relative to 1, $f(E + eV)$ the source Fermi-Dirac function, and $D(E_x)$

the tunneling probability. The tunneling probability is given by the well-known WKB approximation as

$$D(E_x) = \exp\left\{-\frac{4\pi}{h}(2m)^{1/2}\int_{s_1}^{s_2}[\zeta + \phi(x) - E_x]^{1/2}\,dx\right\}\qquad(6)$$

where s_1 and s_2 are the classical turning points at energy E_x and $\phi(x)$ is the barrier-potential energy measured with respect to the Fermi level ζ of the negatively biased electrode [see Fig. 5 (C)].

Several authors [67-69] have discussed the formulation of tunnel equations (5) and (6). Frenkel [70] was the first to obtain an approximate solution of these equations. Since the integral in Eq. (6) cannot, in general, be evaluated in a closed form, approximations of one form or another must be made. Sommerfeld and Bethe [71], Holm and Kirschstein [72], and others [73, 74] simplified $\phi(x)$ by substituting a rectangular barrier of height equal to the average of $\phi(x, V)$. Simmons [75-77] generalized Holm's concept of the average-height approximation and extended it to an arbitrary barrier shape, and symmetric as well as asymmetric potential barriers.

An alternative approach to evaluate Eq. (6) due to Murphy and Good [78] (see the review by Good and Müller [79]) is to expand the integral in a Taylor series with respect to $(\zeta - E_x)$. This expansion is justified if the current flow is predominantly due to electrons whose E_x values are close to the Fermi energy. This approach was used by Stratton [80] and Tantraporn [81] to derive general tunnel formulas. In particular, Stratton's analysis encloses a broader class of arbitrary barrier shapes and temperatures than Holm's or Simmons', and is thus necessarily more complicated.

Chow [82] exploited the separate merits of Stratton's and Holm's approximations and evaluated the basic integral [Eq. (6)] by first replacing the true barrier $\phi(x)$ by a rectangular barrier $\phi(r)$ which is the square of the average of $\sqrt{\phi(x)}$. Then, with the help of the binomial expansion, the tunnel equation can be integrated. Chow's treatment of symmetric and asymmetric barriers runs parallel to that of Simmons, and substantially similar results have been obtained.

Whatever the approximations and errors involved, the functional dependence of the tunnel current density on field and temperature as obtained by various approaches is the same in all cases. At high fields and low temperatures, the basic tunnel integral yields the Fowler-Nordheim equation [73], while at low fields and high temperatues the Richardson-Schottky equation is obtained. Wherever comparison is possible, Chow [82] showed that there is little difference between Holm's and Stratton's equations. Hartman and Chivian [83, 84] compared the Simmons and Stratton theories and pointed out that the former takes into account the reverse current more accurately than the latter while the latter holds for nonzero temperatures and is more easily tractable to experimental analysis. The inaccuracy in the reverse current is small (~3 percent at 0.1 V bias). Both theories are equivalent at $T = 0$ and $V \to 0$. Simmons [77] introduced a temperature-dependent distribution function in his generalized theory and obtained temperature dependence identical to that given by the Stratton theory.

Only the Stratton theory has considered the effect of the nonparabolicity of the energy-momentum relation on tunnel characteristics of an insulator. When the metal-insulator work function is an appreciable fraction of the band gap of the insulator, nonparabolicity should have a significant effect in increasing the current density. The theory of Stratton et al. [85] suggests a method for determining the energy-wave number relationship from the tunnel data. Stratton [80] showed that if the energy bands are parabolic and the effective mass in the metal electrodes is not much less than that in the insulator, then the free-electron mass can be replaced by an insulator effective mass in the insulator. Further, the tunnel current density was found to be insensitive to the choice (elastic or inelastic) of surface scattering of the carriers at the barrier interface, which is physically a somewhat surprising result.

An outline of Simmons' method [75] for obtaining a generalized formula for the tunnel current density through a barrier of width $\Delta s = s_2 - s_1$ will now be presented. By replacing $\phi(x)$ in Eq. (6) with a mean barrier height $\bar{\phi}$ defined by

$$\bar{\phi} = \frac{1}{\Delta s} \int_{s_1}^{s_2} \phi(x)\,dx \tag{7}$$

Eq. (6) may be integrated to yield

$$D(E_x) \simeq \exp\left[-C(\zeta + \bar{\phi} - E_x)^{1/2}\right] \tag{8}$$

Here $C = [4\pi\beta(\Delta s/h)](2m)^{1/2}$, and β is a function of the barrier shape and is approximately equal to unity. The tunnel current is obtained by adding the current contributions [Eq. (5)] from 1 to 2 and 2 to 1 electrodes. If electrode 2 is at a positive potential V, with respect to electrode 1, then the net current density is given by (*at $T = 0°K$*)

$$
J = \frac{4\pi m e}{h^3}\left\{ eV \int_0^{\zeta-eV} \exp\left[-C(\zeta + \overline{\phi} - E_x)^{1/2}\right] dE_x \right.
$$

(9)

$$
\left. + \int_{\zeta-eV}^{\zeta} (\zeta - E_x) \exp\left[-C(\zeta + \overline{\phi} - E_x)^{1/2}\right] dE_x \right\}
$$

Integration of Eq. (9) yields

$$
J = J_0\{\overline{\phi}\exp-(C\overline{\phi}^{1/2}) - (\overline{\phi} + eV)\exp[-C(\overline{\phi} + eV)^{1/2}]\} \tag{10}
$$

where $J_0 = e/2\pi h(\beta\,\Delta s)^2$. If $\overline{\phi}$ is in volts and s_1 and s_2 are in Å units, then J in A/cm^2, *for all values of V*, is given by

$$
J = \frac{6.2 \times 10^{10}}{(\Delta s)^2}\{\overline{\phi}\exp(-1.025\,\Delta s\overline{\phi}^{1/2})
$$

(11)

$$
-(\overline{\phi} + V)\exp[-1.025\,\Delta s(\overline{\phi} + V)^{1/2}]\}
$$

Equation (10) is applicable to *any* shape of potential barrier provided the mean barrier height is known. Let us now apply it to some simple cases. For the low- and intermediate-voltage range $0 \le V \le \phi$, the tunnel current density for a rectangular barrier of height ϕ_0 with similar electrodes [Fig. 5(A)] is obtained by substituting $\Delta s = s$ (physical thickness of the insulator) and $\overline{\phi} = \phi_0 - eV/2$ in Eqs. (10) or (11). In the limit of high voltages $V > \phi_0/e$, $\Delta s = s\phi_0/eV$ and $\overline{\phi} = \phi_0/2$, so that the tunnel current density as given by Eq. (10) is (setting $F = V/s$)

$$
J = 3.38 \times 10^{10}\frac{F^2}{\phi_0}\left\{\exp\left(-0.689\frac{\phi_0^{3/2}}{F}\right)\right.
$$

(12)

$$
\left. - \left(1 + \frac{2V}{\phi_0}\right)\exp\left[-0.689\frac{\phi_0^{3/2}}{F}\left(1 + \frac{2V}{\phi_0}\right)^{1/2}\right]\right\}
$$

When V is greater than $(\phi_0 + \zeta)/e$, the last term in Eq. (12) is negligible compared with the first, and one obtains the familiar Fowler-Nordheim equation $\left[\text{i.e., } J \propto F^2 \exp(-0.689\, \phi_0^{3/2}/F)\right]$.

If the junction is asymmetric, we can still apply Eq. (10). Denoting the reverse-direction current by J_1 [with the electrode of lower work function positively biased as in Fig. 5(B)], and the forward-direction current by J_2, we obtain for the voltage range $0 \le V \le \phi/e$, $J_1 = J_2$ by substituting $\bar{\phi} = (\phi_1 + \phi_2 - eV)/2$ and $\Delta s = s$ in Eq. (10). For $V > \phi_2, \phi_1, J_2$ is given by substituting $\bar{\phi} = \phi_2/2$ and $\Delta s = s\phi_2/(eV - \Delta\phi)$ where $\Delta\phi = (\phi_1 - \phi_2)$, and J_1 is given by using $\bar{\phi} = \phi_1/2$ and $\Delta s = s\phi_1/(eV - \Delta\phi)$. Thus $J_1 \ne J_2$; that is, the junction is rectifying. Hartman [84], and Chow [82] also calculated $J-V$ characteristics for asymmetric junctions and pointed out some errors in Simmons' treatment. Chow showed that refinements of the theory predict an asymmetry (rectification) even at low voltages for trapezoidal barriers.

An interesting property of the tunneling equations of various theories was pointed out by Gundlach [86]. The function $Q = (1/\Delta V) \ln[J(V + \Delta V)/J(V)]$ corresponding to J_1 and J_2 shows sharp cusplike maxima at voltages close to ϕ_1 and ϕ_2. This variation thus affords a convenient method for determining the barrier heights.

(2) Image-force Correction. The image force arises because of the polarization of the metal-electrode interfaces by the electrons within the barrier. It reduces the area of the potential barrier by rounding off the corners and by reducing the thickness of the barrier [see Fig. 5(C)]. As a result of the image-force correction, the optical dielectric constant ϵ is introduced into the tunnel equation. The image potential is a hyperbolic function which may be approximated by a symmetric parabola at low voltages. Simmons corrected the general tunnel equation with a simple hyperbolic image-force term. The tunnel current density [in A/cm^2, with ϕ_0 in V and s_1, s_2 (classical turning points) and s (barrier physical thickness) in Å units] for a rectangular barrier for *all values of V* is given by (at $T = 0°$K)

$$J = \frac{6.2 \times 10^{10}}{\Delta s^2}\left\{\phi_1 \exp\left(-1.025\,\Delta s\phi_1^{1/2}\right)\right.$$

$$\left. - (\phi_1 + V)\exp\left[-1.025\,\Delta s(\phi_1 + V)^{1/2}\right]\right\} \tag{13}$$

where

$$\phi_1 = \phi_0 - \frac{V}{2s}(s_1 + s_2) - \frac{5.75}{\epsilon(s_2 - s_1)} \ln \left[\frac{s_2(s - s_1)}{s_1(s - s_2)} \right]$$

and

$$\left. \begin{array}{l} s_1 = \dfrac{6}{\epsilon\phi_0} \\[2em] s_2 = s\left(1 - \dfrac{46}{3\phi_0\epsilon s + 20 - 2V\epsilon s}\right) + \dfrac{6}{\epsilon\phi_0} \end{array} \right\} \, V < \phi_0$$

$$\left. \begin{array}{l} s_1 = \dfrac{6}{\epsilon\phi_0} \\[2em] s_2 = \dfrac{\phi_0\epsilon s - 28}{\epsilon V} \end{array} \right\} \, V > \phi_0$$

Figure 7 shows the theoretical voltage dependence of the resistance of a symmetric tunnel junction for some values of s, ϵ, and ϕ. A comparison of $\phi = 2$ eV curves with the corresponding curves without the image-force correction shows the significant effect of the dielectric constant on the magnitude of the current density although the qualitative variation is not appreciably altered. Numerical calculations of the tunnel equations of Simmons' and Chow's theories for various ranges of the parameters involved have been published in the literature [75-77, 82, 87].

(3) **Temperature-field (TF) Emission.** The maximum field that can be applied to most thick-film tunnel structures generally corresponds to neither the low-field limit nor the high-field Fowler-Nordheim region. Thus, the tunnel current corresponding to the intermediate case, called the temperature-field emission, is of practical interest. The basic integral in the tunnel equation [Eq. (5)] can be evaluated for this case by a point-by-point integration technique due to Dolan and Dyke [88]. Mann [60] employed this technique to evaluate the tunnel current density for

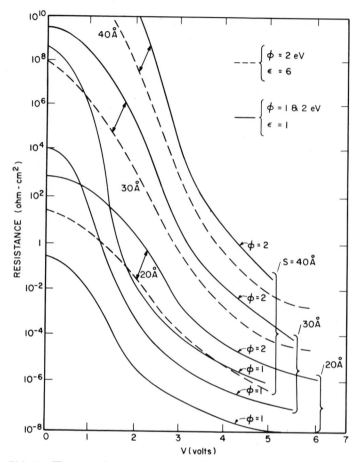

FIG. 7. Theoretical resistance-voltage characteristics of a symmetric tunnel structure having an ideal rectangular barrier (solid curves); dotted curves include image correction for ϵ = 6. (*Simmons* [75])

a trapezoidal barrier with ϵ = 1. The range of the variables used, $10^6 \leq F \leq 10^7$V/cm, $100 \leq T \leq 300°$K, and $1.0 \leq \phi \leq 2.5$ eV, represents practical values in the intermediate case between the pure thermionic emission and pure field emission. Some of the theoretical plots of current density vs. the inverse of the field for ϕ = 1.5 and 2.5 with ϵ = 1 and for different temperatures are shown in Fig. 8. These results are valid for thick films with a bias $V > \phi/e$ and show the transition from temperature-dependent to temperature-independent regions. Although the current is primarily due to tunneling, the ratio of

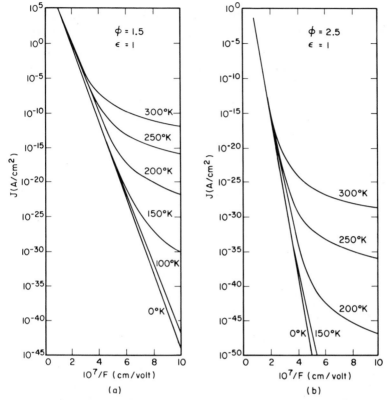

FIG. 8. Theoretical current density vs. inverse field ($1/F = s/V$) obtained by numerical integration of the tunnel equation [Eq. (5)] for a symmetric structure at various temperatures. (*a*) $\phi = 1.5$ eV; (*b*) $\phi = 2$ eV; $\epsilon = 1$ for both cases. (*Mann* [60])

tunnel to thermionic current is both temperature- and field-dependent. At high temperatures, the tunnel current density shows a better fit to $\exp (F)^{1/2}$ than to the expected $\exp (-1/F)$ Fowler-Nordheim dependence.

(4) Temperature Dependence. The expression for tunnel current density $J(V, T)$ at a temperature $T°K$ as deduced by Stratton and Simmons from the temperature dependence of the distribution function is

$$\frac{J(V, T)}{J(V, 0)} = \frac{\pi BkT}{\sin \pi BkT}$$

where $B = (P/2)\bar{\phi}^{1/2}$, or

$$\frac{J(V, T)}{J(V, 0)} \approx 1 + \frac{3 \times 10^{-9}(\Delta s T)^2}{\bar{\phi}} + QT^4 + \cdots \tag{14}$$

Here P and Q are constants and Δs is expressed in angstroms and $\bar{\phi}$ in volts. Thus, for $T = 300°K$, $\bar{\phi} = 4.5$ eV, $\Delta s = 20$ Å, $J(V, T) = 1.02\, J(V, 0)$. At $1000°K$, $J(V, T) = 1.5\, J(V, 0)$;

By substituting the values of Δs and $\bar{\phi}$ for their respective cases (discussed already), Eq. (14) can be rewritten for symmetric and asymmetric junctions. Denoting $\hat{J}$ as $[J(V, T) - J(V, 0)]/J(V, 0)$, we obtain for the reverse-biased asymmetric junction in Simmons' theory,

$$
\hat{J}_1 =
\begin{cases}
\dfrac{6 \times 10^{-9}(sT)^2}{(\phi_1 + \phi_2 - V)} & V \leq \phi_2 \\[4mm]
6 \times 10^{-9}\,\phi_1 \left(\dfrac{sT}{V - \Delta\phi}\right)^2 & V \geq \phi_2
\end{cases}
\tag{15a}
$$

For forward bias,

$$
\hat{J}_2 =
\begin{cases}
\dfrac{6 \times 10^{-9}(sT)^2}{(\phi_1 + \phi_2 - V)} & V \leq \phi_1 \\[4mm]
6 \times 10^{-9}\,\phi_2 \left(\dfrac{sT}{V + \Delta\phi}\right)^2 & V \geq \phi_1
\end{cases}
\tag{15b}
$$

Thus, the $\hat{J}$'s are identical for a given T and intermediate voltages. At higher voltages, $\hat{J}_1$ exceeds $\hat{J}_2$, but they approach each other at sufficiently high voltages. The variations of $\hat{J}$'s exhibit cusplike maxima at voltages equal to the barrier heights ϕ_1 and ϕ_2 (or ϕ_0, in the case of symmetric barrier). The image-force correction has only a moderating effect on the sharpness of these cusps. The positions of these cusps offer a method for determining the barrier heights.

We should point out that tunnel current can also vary with temperature because of any temperature dependence of the barrier heights. Assuming the temperature-dependent decrease of the barrier to be the same as that of the band gap, Mann's [60] calculations showed that the current density increases markedly with temperature at high fields although the slope of the tunnel curves remains approximately unchanged.

(5) **Experimental Results.** Numerous studies [83, 84, 89-96] of the current transport through thin insulators have been published in the literature. Fisher and Giaever [89] measured the conductance of

Al-Al$_2$O$_3$-Al structures with an oxide thickness less than 100 Å. They observed an ohmic behavior at low voltages and an exponential rise of current at higher voltages, in qualitative agreement with Holm's calculations. The observed current was, however, several orders of magnitude lower than calculated and required an effective mass $m^* = \frac{1}{9} m_e$ (m_e = electron mass) to explain this discrepancy. Stratton's [80] attempt to fit these data to his calculations yielded $m^* = 0.3 m_e$—a value critically dependent on the estimate of the thickness of the oxide film which could easily be in error.

Tantraporn's [81] comparison of his calculations with the results of Advani et al. [90] suggested the necessity of including the image force for an insulator thickness below 150 Å. An effective mass of 1.78 m_e was deduced for Al$_2$O$_3$ films.

Hartman and Chivian [83] studied Al-Al$_2$O$_3$-Al for oxide thicknesses less than 35 Å, applied voltages between zero and 10 V, and temperatures between 79 to 300°K. A comparison of their results with Stratton's theory yielded qualitative agreement. The WKB approximation used by Stratton was found to be valid only for $0 \leq V \leq 0.3$ V. Assuming a trapezoidal barrier shape, approximate values of $m^* = m$ and $\phi_0 = 0.78$ eV were deduced.

Meyerhofer and Ochs [91] obtained a good agreement over several decades for their data on the Al-Al$_2$O$_3$-Al system with the numerical solution of the tunneling equation of the type used by Stratton. The data compared well with the theoretical curves provided an effective thickness of 22 Å (instead of 52 Å, determined from capacitance measurement) was used. Barrier heights $\phi_1 = 2.3$ and $\phi_2 = 2.1$ eV were deduced. Simmons [77] questioned these results. He reanalyzed the data by multiplying the current density with an arbitrary factor and obtained $\phi_1 = 1.6$ eV and $\phi_2 = 2.5$ eV, again using an effective thickness. These values are in accord with those of Pollack and Morris [92, 93], who obtained an excellent agreement with Simmons' equations over five decades of tunnel current at 300°K as illustrated in Fig. 9. However, in order to obtain the fit, the effective active area of various samples had to be adjusted to values between 10 and 0.1 percent of the geometrical area and the tunnel thickness of 1.5 to 2 times smaller than that given by the capacitance data. Note that J_1 (Fig. 9) exhibits a marked departure from the theoretical curve, which is attributed to the existence [94] of a small barrier in the transition region between the parent-metal electrode and the oxide layer which becomes effective at low temperatures. A similar

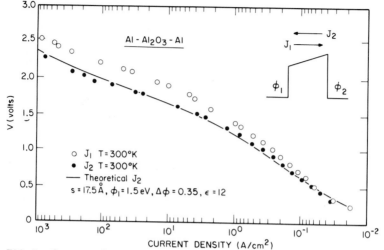

FIG. 9. Current-voltage characteristics for an asymmetric Al-Al$_2$O$_3$-Al tunnel structure at 300°K. The solid curve is calculated from Simmons' theory using s = 17.5 Å, ϕ_1 = 1.5 and ϕ_2 = 1.85 eV, and ϵ = 12. (*Pollack and Morris* [93])

behavior of J_1 was observed with other counterelectrode materials, suggesting its origin at the electrode interface.

Electron transfer through ideal, single-crystal films of vacuum-cleaved mica was studied by McColl and Mead [95]. The $J-V$ data on 30- and

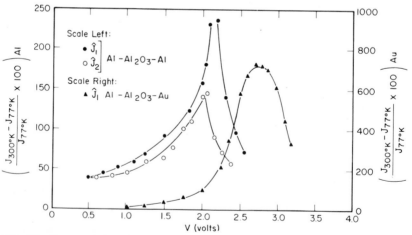

FIG. 10. Percent change $(\Delta J/J) \times 100 = [(J_{300°K} - J_{77°K})/J_{77°K}] \times 100$ in J_1 and J_2 between 77 and 300°K as a function of voltage for an Al-Al$_2$O$_3$-Al tunnel structure of Fig. 9. The change in J_1 for an Al-Al$_2$O$_3$-Au sample is also shown. (*Pollack and Morris* [93])

40-Å-thick films were in good agreement with the Stratton theory. At applied voltages greater than the barrier height, the magnitude and functional dependence of tunnel current departed markedly from the theory. The departure may be a characteristic of injection into the conduction band of the muscovite mica.

The temperature dependence of the tunnel current was studied by several authors [84, 91, 92]. Although a greater temperature dependence than that predicted by the theory is observed, the general features of the variations of $\hat{J}_1$ and $\hat{J}_2$ (Fig. 10) for Al-Al$_2$O$_3$-Al are, however, in accord with the theory [Eq. (14)]. The position of the peaks of $\hat{J}_1$ and $\hat{J}_2$ yield barrier heights of ϕ_1 = 1.5 eV and ϕ_2 = 1.8 eV for the Pollack and Morris data, and 1.6 eV and 2.4 eV for the Hartman data for a gaseous anodized oxide. There is no apparent explanation for the difference in the two sets of results.

The voltage dependence of Q_1 and $Q_2 [Q = d \ln J(V)/dV]$ functions derived from the tunnel data on an Al-Al$_2$O$_3$-Mg structure by Gundlach [86] is shown in Fig. 11. The position of the cusps yields $\phi_1 \sim 1.7$ eV and $\phi_2 \sim 2$ eV. For an Al-Al$_2$O$_3$-Al structure, $\phi_1 \sim 1.7$ eV and $\phi_2 \sim 2.5$ eV are obtained.

Asymmetric sandwich structures with dissimilar electrode materials have been studied by a number of workers [84, 92, 94, 95]. For the same electrode, the tunnel resistance increases with increasing work function (and hence barrier height) of different counterelectrode materials. The studies of Pollack and Morris [92] on Al-Al$_2$O$_3$-M (M = Al, Au, Bi, Mg, Pb, and Cu) structures suggest that the interface barriers are not sharp, possibly because of the penetration of the vapor atoms of the counterelectrode material into the oxide film [94].

Barrier heights at interfaces are important parameters which dominate the transport of carriers through tunnel structures. Their values may be determined by comparing the experimental and theoretical $J-V$ curves, the position of cusps in the temperature and voltage dependence of tunnel current, and photoemission measurements (Sec. 5.1). Values of barrier heights for some interfaces measured by these techniques are listed in Table II.

(6) Conclusions. Let us now examine critically the discrepancies between the theoretical and experimental results and discuss their possible origin. (1) Qualitative, and, in some cases, quantitative correlation has been obtained between the observed current densities through thin (<50 Å) insulators and quantum-mechanical tunneling theories. But

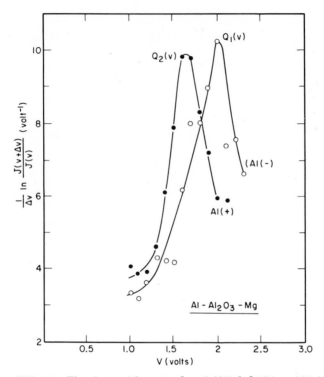

FIG. 11. The observed function $Q = 1/\Delta V \, \ln[J(V + \Delta V)/J(V)]$ for normalized tunnel currents J_1 and J_2 in an Al-Al$_2$O$_3$-Mg structure at 77°K. (*Gundlach* [86])

this correlation is generally possible *only* if "effective" values for tunnel area, tunnel barrier thickness, or dielectric constant are used. For example, tunnel areas $\sim 10^{-2}$ times (up to 10^{-5} times in some cases) the geometrical area, and tunnel thicknesses about 2 to 5 times smaller than those determined from the capacitance data are required for agreement with the theory. (2) Analyses of the tunnel data yield different values of the effective mass of the electron. The values for Al$_2$O$_3$, for example, range from 0.01 to 1.78. (3) The observed temperature dependence of the tunnel current is too high to be explained by the present theories, even with the inclusion of the temperature dependence ($\sim 10^{-4}$ eV/°C) of the interface barrier heights, and also that of the dielectric constant.

The aforementioned discrepancies can be qualitatively understood on the basis that a practical barrier departs considerably from the ideal barrier assumed in the theoretical treatments. The following points may

Table II. (A) Interfacial Barrier Heights for Some Thin-film Metal-oxide Structures at Room Temperature. Metal Films are Deposited by Vacuum Evaporation

Interface	Barrier height, eV		Method	Remarks	Ref.
	(ϕ_1)	(ϕ_2)			
Al-Al$_2$O$_3$-Al	1.49	1.92	Photo	Evaporated oxide	173
	1.0	2.9	Photo	Thermal oxide	172
	0.31	0.74	Schottky	Thermal oxide	90
	1.5	1.85	Tunnel	Evaporated oxide	92
	2.0	2.2	$\Delta J(T)/J(0)$ cusps (300-77°K)	Gas anodized	93
	1.6	2.4	$\Delta J(T)/J(0)$ cusps	Gas anodized	83
Al-Al$_2$O$_3$-Al	1.7	2.5	$d \ln J/dV$ cusps (77°K)	Thermal oxide	86
Al-Al$_2$O$_3$-Mg	1.7	2.0			
Al$_2$O$_3$-Au	1.6		Schottky	Thermal	179
Al$_2$O$_3$-In	0.55		Tunnel	Thermal	179
Be-BeO	2.0-2.5		Tunnel	Thermal	91
Si-SiO$_2$	0.65		Tunnel-Schottky	Thermal oxide	178
SiO$_2$-Au	3.94		Photo	Evaporated	177
-Al	3.19		Photo	Evaporated	177
-Pt	4.18		Photo	Evaporated	177
-Cr	3.27		Photo	Evaporated	177
Ta-Ta$_2$O$_5$-Au	1.1	1.6	Schottky	Gas anodized	64

Table II. (B) Metal-Semiconductor (Bulk) Barrier Height for Some Cases. Metal Films are Deposited by Vacuum Evaporation

Interface	Barrier height, eV (ϕ_1)	(ϕ_2)	Method	Remarks	Ref.
n CdS-Au	0.78		Photo	VC*-CdS	Taken from the collection in Ref. 173
-Al	0.68		Photo		
n CdSe-Au	0.49		Photo	VC-CdSe	
-Ag	0.43		Photo		
n GaAs-Au	0.90		Photo		
p GaAs-Au	0.42		Photo		
-Al	0.50		Photo		
n Ge-Au	0.45		Capacitance	VC-Ge	
n Si-Au	0.78		Photo	Chem. Si	
n ZnS-Au	2.0		Photo	VC-ZnS	
-In	1.50		Photo		
-Al	0.8		Photo		

*VC= vacuum-cleaved.

498

be noted: (1) Because of the finite penetration of the energetic vapor atoms of the electrode metal into the insulator film, the interface and hence the barrier is not expected to be sharp and well defined. (2) Owing to the statistical process of deposition, the insulator film thickness must show fluctuations which will necessitate the use of an effective thickness value. (3) The assumptions of the parabolic energy-momentum relation and the free-electron mass will certainly not be valid when the forbidden energy gap is small and the barrier height is an appreciable fraction thereof. (4) The barrier height, instead of being constant, may depend significantly on the insulator film thickness and the applied field (Sec. 5.2). (5) If ionic defects and trapping centers are present in the insulator (which is very likely), the tunneling current will be appreciably modified [97]. If the tunnel time is less than the response time of the ions in the insulator ($\sim 10^{-12}$ sec), the optical dielectric constant should be used. If, however, the time is increased markedly by the trapping centers, one must use a low-frequency or static dielectric constant in the tunnel equation. (6) The transmission coefficient is a continuous rather than discontinuous function of the electron energy and is nonzero for energies equal to the maximum barrier. (7) Finally, there is overwhelming evidence [98-100] that the tunnel electrons in practical structures interact strongly with the insulator material. Several authors have discussed the hot-electron interaction with the optical phonons of the insulator lattice. Emtage [99] showed that this interaction would yield a highly temperature-dependent tunneling current even at room temperature. Savoye and Anderson [100] have discussed their experimental results on tunnel emission in terms of a model based on this interaction (Sec. 4.7). Further discussion of inelastic scattering in insulator films is given in Sec. 4.8.

4.4 Bulk-limited Conduction

(1) **Space-charge-limited Current (SCLC) Flow.** When injection into the conduction band, or tunneling, is not the rate-limiting process for conduction in insulators, a space-charge buildup of the electrons in the conduction band or at trapping centers may occur which will oppose the applied voltage and impede the electron flow. At low applied biases, if the injected carrier density is lower than the thermally generated free carrier density, Ohm's law is obeyed. When the injected carrier density is greater than the free carrier density, the current becomes space-charge-

limited. Two requirements must be satisfied if SCLC flow is to be observed: (1) at least one electrode must make ohmic contacts to the insulator, (2) the insulator must be relatively free from trapping effects.

For the simple case of one-carrier trap-free SCLC, the well-known Mott-Gurney relation gives [101]

$$J = 10^{-13} \mu\epsilon \frac{V^2}{t^3} \quad A/cm^2 \qquad (16)$$

where V is the applied voltage, t the thickness of the insulator, and μ the drift mobility of the charge carriers and ϵ is the dielectric constant. Equation (16) is similar to the Child law for electron emission in a vacuum diode. More general treatments of the SCLC which include the diffusion of charge carriers and effects of traps in the insulator have been published [101]. The case of two-carrier SCLC was treated by Lindmayer and Slobodskoy [102].

In insulators, as in semiconductors, a significant number of lattice defects capable of accepting one or more charge carriers may be present. These charge-trapping centers must obviously modify the equilibrium concentration of the carriers and thus the SCLC flow in an imperfect insulator. Lampert and Rose [101] have given a detailed discussion of the influence of traps on SCLC. If only shallow traps are present, the current density given by Eq. (16) for SCLC flow is simply multiplied by θ, which is the ratio of the free to the trapped charges. Discrete trap levels are expected in single-crystal materials of high purity, while distributed trap levels may be expected in amorphous and polycrystalline materials, and may be related to intrinsic disorder of the lattice. For a trap distribution whose density decreases exponentially as the energy from the band edge increases, the current density for the voltage V is given by

$$J \propto \frac{V^{n+1}}{t^{2n+1}} \qquad (17)$$

where n is a parameter characteristic of the distribution of traps and t is the thickness of the specimen. A V^n dependence is expected if the Fermi level lies in a uniform trap distribution. After the traps are filled, the current density will increase rapidly with the bias and then follow a square law again.

Space-charge-limited flow has been observed in nearly single crystal CdS films [103] sandwiched between Au and Te electrodes. Oxide films are generally amorphous and are expected to have a large number of traps, making the observation of SCLC flow difficult, if not impossible. Surprisingly, a V^2 dependence has been observed in thick films ($\sim$500 Å or more) of SiO_2 [104] and Nb_2O_5 [105, 106]. The thickness dependence remains uncertain, however. Hickmott [106] observed a V^2/t^6 dependence for Nb-Nb_2O_5-In structures. He also reported a V^3/t^5 dependence for Nb-Nb_2O_5-Au structures. A V^n/t^3 dependence was observed by Chopra [105] in Nb-Nb_2O_5-Au, Al structures at 300 and 78°K. The value of n was found to be 2 before a sudden rise in the current takes place preceding the occurrence of a negative-resistance region (Sec. 4.6). The occurrence of the SCLC flow in amorphous films suggests that the traps are possibly made ineffective by filling or by other compensating mechanisms.

The effect of traps on conduction through *thin* insulators has been treated theoretically by several workers [107-111]. Frank and Simmons [107] discussed the effect of the SCLC flow on Schottky emission. Geppert [108] showed that space charge can severely limit the tunnel current density, even in the case of very thin films. A comparison of the theoretical curves of current density through a thin insulator with and without trapped charges is presented in Fig. 12. The physical parameters used in the calculations are given in the figure caption. The drastic reduction of the tunnel current density due to the filling of traps is clearly seen. After the traps are filled, the current should rise rapidly with the increase of the bias and merge into the "space-charge with no traps" curve. The observed "slow" $J-V$ characteristics with hysteresis effects in some thin oxide film structures in BeO [91] and SiO [65] are most likely due to the trapping centers. Since the slow characteristics are observed even at low temperatures, only deep traps must be present.

Despite the importance of the role of traps in influencing the transport processes, very little work has been done on the study of the density and level of traps in thin insulators. Ullman [112] studied photocurrents and thermally stimulated currents [113] in reactively sputtered films of tantalum oxide. The results surprisingly indicated the presence of *only* a single set of traps situated about 0.25 eV from the nearest band edge. These empty traps were associated with oxygen vacancies occupied by one electron. The small value of mobility made it difficult to determine the trap population.

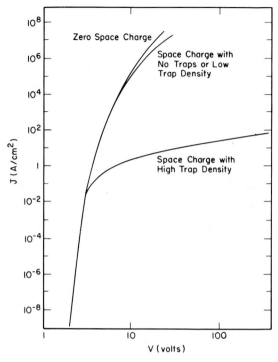

FIG. 12. Theoretical curves for field emission through a thin-film symmetric tunnel structure with zero space charge, space charge with no traps, and space-charge conditions. The following barrier and insulator parameters are assumed for calculations: $\phi = 1$ eV, $\zeta = 7$ eV, $s = 100$ Å, $\epsilon = 9$, θ (ratio of trapped to free charges) $= 10^7$, v_m (the maximum drift velocity of the carriers) $= 10^6$ cm/sec. (*Geppert* [108])

(2) Trap and Impurity Effects. We have already mentioned the influence of traps on the SCLC flow. Depending on the number and level of traps, thermal and/or field ionization of traps can contribute significantly to the current transfer in thin insulator films. A field-enhanced thermal excitation of trapped electrons into the conduction band, that is, internal Schottky effect (known as Poole-Frenkel effect), gives a current-voltage relation of the form [114]

$$J \propto \exp \frac{1}{kT} \left[\left(57.7 \times \frac{eV}{\epsilon s} \right)^{1/2} - E_t \right] \qquad (18)$$

where E_t is the depth of the trap-potential well and s is the electrode spacing in angstroms. The same relation holds if, instead of traps, impurities are ionized, in which case E_t is the energy required to ionize an impurity under zero field conditions. By comparing Eq. (18) with Eq. (4), we note that the slope of the log J vs. $V^{1/2}$ plot for the Poole-Frenkel effect is twice that for the Schottky emission. Mark and Hartman [182] pointed out that this is valid only if the donor levels are partially ionized and uncompensated. If conductivity arises from complicated defect structures with partial compensation, the slope relationship is altered.

Mead [115] studied conduction through thick (up to 870 Å) oxide films in Ta-Ta$_2$O$_5$-Au structures. The results were interpreted in terms of Eq. (18). On the other hand, Simmons [114] found the $J-V$ data on thick SiO and Ta$_2$O$_5$ films to fit the Schottky emission rather than the Poole-Frenkel mechanism. Therefore, the evidence for trap and impurity ionization effects is not conclusive.

4.5 Voltage-controlled Negative Resistance (VCNR)

Thin (>100 Å) anodized as well as thermally oxidized films of valve metals are generally good insulators with high resistivities (10^{14} to $10^{18}\,\Omega$-cm) under low applied electric fields. As the voltage across such films sandwiched between the parent-metal electrode and a thin-film counterelectrode is increased, a critical voltage is reached at which a "forming" process sets in and is attended by an increase in the transfer current by several orders of magnitude. On further increase of the voltage, the $J-V$ characteristic shows an N-shaped voltage-controlled negative resistance (VCNR) for both increasing and decreasing voltages (Fig. 13). The initially noisy and erratic currents become more stable on successive tracings of the $J-V$ characteristics.

Hickmott [116-119] reported the first observation of the VCNR and made an extensive investigation of the phenomenon in anodic oxide films. The VCNR has been observed and investigated by a number of other workers [120-129]. Eckertová [56] has given a detailed review of the subject. The salient features of the phenomenon are as follows:

1. The VCNR has been observed in "formed" films of SiO, Al$_2$O$_3$, Ta$_2$O$_5$, ZrO$_2$, and TiO$_2$ at 3.1, 2.9, 2.2, 2.1, and 1.7 V, respectively. A similar effect has been found in evaporated oxide and alkali-halide films.

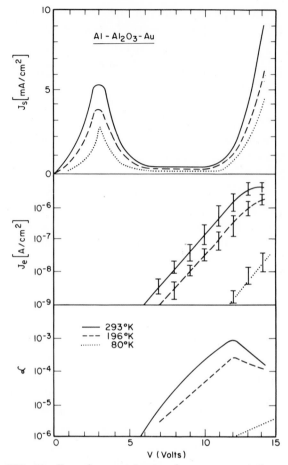

FIG. 13. Tunnel current density J_s, vacuum-emission current density J_e and transfer ratio α at 80, 196, and 293°K as a function of the voltage across an Al-Al$_2$O$_3$-Au structure which exhibits voltage-controlled negative-resistance behavior. (*Eckertová* [56])

Multiple VCNR regions are observed in some cases as the applied voltage is increased beyond the first negative-resistance region.

2. The voltage for maximum current shows [116] a $1/\sqrt{\epsilon}$ dependence. It is independent of the film thickness between 100 and 20,000 Å, i.e., the phenomenon is apparently not field-dependent.

3. The polarity of the "formed" film can be reversed symmetrically by reversing the bias. The forming voltage [121, 122] for the evaporated films is generally higher than that required for the anodic films.

4. The voltage at which the VCNR occurs is independent of the ambient temperature, although the VCNR disappears at sufficiently low temperatures. A rapid cooling gives a temporary memory effect.

5. The VCNR cannot be traced at frequencies above 60 cps for anodic films and above 1 kc/sec for evaporated films.

6. Following the onset of the VCNR, electron emission into vacuum (Fig. 13) and electroluminescence are observed [117-119].

Typical $J-V$ characteristics of an Al-Al$_2$O$_3$-Au structure exhibiting the VCNR at different temperatures are shown in Fig. 13. Included in the same figure are the variations of vacuum-emission current and the transfer ratio (ratio of the emitted to the tunnel or sandwich current).

The above observations suggest that the "formation" process consists of a buildup of positive space charge due to ions in the vicinity (within a few angstroms) of the cathode setting up a high field at the cathode, thereby reducing the effective metal-insulator barrier, and thus enhancing conductivity as well as electron emission through the counterelectrode. If sufficient electron emission takes place at the cathode, a recombination process with the emission of light and reduction of the conductivity would occur. With further increase of applied voltage, conductivity may again increase because of increased ionization by energetic electrons. This explanation is, however, not consistent with Hickmott's [119] results obtained from his studies of the potential distribution in a thin-film triode structure (Al-SiO-Al-SiO-Au). These studies indicated that at the onset of the VCNR the applied potential occurred across a thin (<120 Å) region of the insulator film. This region may not shift readily with the reversal of the field polarity but could be anywhere within the insulator. Hickmott proposed tentatively that the field ionization of impurities from an impurity band, located somewhere between the valence and conduction bands of the insulator, increases the conductivity in the high-field acceptor region. As the field is increased further, a competitive recombination process at an impurity acceptor level above the valence band sets in. This process would decrease the conductivity again and may be accompanied by the emission of the recombination light and would thus explain the observed electroluminescence. How the electron emission is also increased is not clear, although it may be due to the tunneling of carriers (obtained by field ionization) via impurity levels. This explanation may also account for the observed electron emission through rather thick insulator films at low electron energies.

Rectifying $J-V$ characteristics with well-defined asymmetric VCNR have been observed in 100-Å polyethylene single crystals between Pt whiskers and Cu substrates. Van Roggen [129] attributed his observation to a double-injection mechanism [101] but offered no supporting evidence.

4.6 Current-controlled Negative Resistance (CCNR)

A "formation" process which takes place at a critical current, instead of a critical voltage as with the VCNR, was observed by Chopra [105, 130] in thin (50 to 500 Å) anodic and thermal oxide films of Nb, Ta, and Ti, and thick (~1,000 Å) films of CdS [131] sandwiched between metal electrodes. If the voltage applied across a rectifying virgin Nb-Nb oxide-Au structure is increased until a current density of ~1 mA/cm^2 is obtained in the reverse direction, the rectifying $I-V$ curves [Fig. 14(a)]

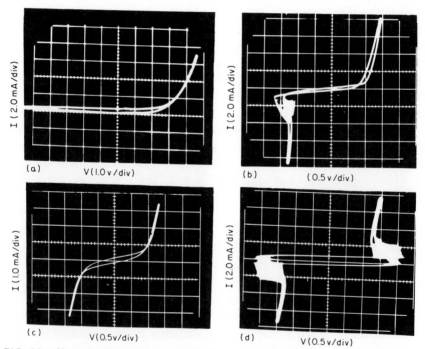

FIG. 14. 60 cps current-voltage characteristics of an Nb-Nb$_2$O$_5$ (200 Å)-Au structure. (a) Undeformed (virgin) state. (b) Slight deformation yields current-controlled negative resistance (CCNR) in reverse direction. (c) Complete deformation produces symmetrical nonlinear characteristics. (d) Symmetrical CCNR. (*Chopra* [105])

transform to exhibit an S-shaped current-controlled negative resistance (CCNR) in the reverse direction (*b*). If the current density is further increased to about 100 mA/cm^2, the $I-V$ curves transform *irreversibly* from rectifying to the nonrectifying (symmetrical) one (*c*). The current rises very rapidly with voltage at this stage, and one observes symmetrical CCNR for both directions of current flow (*d*). The irreversible deformation process is often fast so that a direct transition from (*a*) to (*d*) is observed.

CCNR has been observed by other workers [132-136, 185-187]. It is not clear what conditions produce a VCNR or a CCNR in the same structures, although the two never occur simultaneously. Niobium oxide in general yields the latter rather than the former. The salient features of the CCNR are:

1. The formation process and the CCNR condition are irreversible. In most cases, the complete formation process produces a symmetrical CCNR.

2. The voltage required to sustain the negative resistance is independent of the electrode materials, apparent area, and thickness of the oxide films. It is about 0.4, 1.3, and 3.5 V for Ti, Nb, and Ta oxide films, respectively. A $1/\sqrt{\epsilon}$ dependence of the voltage on the dielectric constant suggests the role of an impurity or trap ionization process.

3. Hysteresis effects, sometimes attended by multiple CCNR regions, are observed in some samples.

4. The short relaxation time ($<10^{-6}$ sec) determined from the self-sustained oscillations of more than 60 Mc/sec suggests that the CCNR is of electronic origin. This is further supported by the observation of only very weak dependence of the CCNR region on temperature down to 77° K.

5. Illumination of the sandwich increases the current but does not alter the sustaining voltage for the CCNR.

6. The current following the CCNR increases rapidly as a high power of the voltage.

The observed CCNR may be tentatively understood as follows: The effective insulator thickness decreases with increasing bias voltage (as indicated by the capacitance data) until at the onset of CCNR (just as in VCNR), the sustaining voltage seems to appear across a thin critical (<50 Å) region of the insulator. The electrons would thus be accelerated and produce cumulative ionization of traps and/or impurities in the vicinity of this critical region. The attenuation length of the energetic electrons in the insulator region being of atomic dimensions, avalanche ionization is a reasonable possibility. The ionization process would start at a certain minimum current density. Once avalanche is initiated, the voltage will drop to a minimum value to sustain the process. This explanation is similar to that suggested by Gunn [135] to explain the CCNR phenomenon in point-contact diodes of Ge and Si heterogeneous junctions. The explanation is not unambiguous since the marked increase of current following the CCNR can be fitted into a number of field-sensitive injection mechanisms. But the mechanism appears to be definitely determined by the insulator and not the electrode materials. Braunstein et al. [136] attributed the CCNR in Si films to a double-injection mechanism but presented no supporting evidence.

Some CCNR exhibiting films also show bistable switching and memory phenomena. Such effects have been observed in polycrystalline (CdS [131], Si [136], GaAs [185]) and amorphous (Nb_2O_5 [105, 134], NiO [186], TiO [187]) films of wide band gap materials.

4.7 Tunnel Emission (Hot-electron Transport)

We have already discussed the transfer of electrons from metal electrode 1 to metal electrode 2 (Fig. 5A) by tunneling through a thin insulating film under the influence of a strong electric field of $\sim 10^6$ to 10^7V/cm. In metal 2, the tunnel electrons are "hot," i.e., they are not in equilibrium with other electrons in the metal. If the potential drop between 1 and 2 is greater than the work function of 2, some of the tunnel electrons could pass through a thin layer of 2 (assumed smaller than or comparable with the mean free path) without being attenuated significantly, surmount the barrier to the electron emission, and then appear in the vacuum or some suitable collector as *tunnel emission.*

The early works of Mahl [137] and Vudynskii [138] on cold or tunnel emission have been revived recently by Mead's attempt [139, 140] to utilize the tunnel effect in sandwich structures to obtain a cold cathode of a controlled source of majority carriers. Such a device may in principle be operated at very high current densities which would be insensitive to temperature changes. The tunnel-emitted electrons could be collected in vacuum or by another insulated metal electrode or a semiconductor blocking contact with the base. The resulting triode configuration and its energy-level diagram are shown schematically in Fig. 15. The device resembles a transistor and is called a tunnel-emission

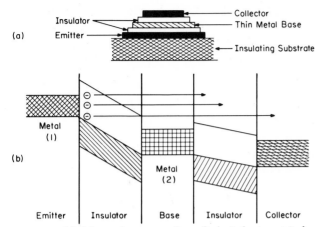

(a)

(b)

Insulator

Emitter

Collector

Thin Metal Base

Insulating Substrate

Metal
(1)

Metal
(2)

Emitter   Insulator   Base   Insulator   Collector

FIG. 15. (*a*) Schematic cross section of a hot-electron triode;
(*b*) an idealized energy-level diagram of the hot-electron-
triode structure.

or hot-electron triode or a metal-base transistor. These devices are
expected to be very fast, although switching times less than 10^{-8} sec may
be limited by the space-charge conditions.

The critical parameter for such a device is the transfer ratio α, the
fraction of the current which leaves the second electrode, which should
clearly be as large as possible. Rather stringent conditions must be met in
order to obtain a high transfer ratio. The thickness of the insulator used
for tunneling should be comparable with the mfp of the electrons so that
the electrons are not lost by collisions. Moreover, the thin insulator
should be able to withstand voltages in excess of the work function of
the base-metal electrode ($\sim$4V) which would be required for significant
tunnel emission. Further, the thickness of the base film should be smaller
than or comparable with the attenuation length to allow a reasonable
probability of transmission of the hot electrons. We may note that the
attenuation length of hot electrons (Chap. VI, Sec. 5) decreases rapidly
with increasing energy and ranges between 170 and 740 Å for Au films.

Triodes of the type shown in Fig. 15 have been studied by Mead
[140] and others [141-145]. The value of α varies from specimen to
specimen and is very small ($\sim 10^{-4}$ to 10^{-5}), although in some cases values
as high as 0.1 have been obtained. An unusually high value of $\alpha = 0.8$ for
a triode with a semiconductor collector, reported by Spratt et al. [144]
was questioned [145] and later shown to be due to a geometrical
pinhole effect.

Tunnel emission into vacuum has been studied by a number of workers [116-120, 123-128, 139-148] using structures such as Al-Al₂O₃-Al or Au. Typical tunnel, vacuum-emission, and transfer-ratio characteristics of an Al-Al₂O₃(30 Å)-Al structure are shown in Fig. 16(a). The fact that the transfer ratio α decreases monotonically with the sandwich voltage V clearly suggests the absorption of hot electrons in the insulator and/or counterelectrode. The value of α increases with the increasing counterelectrode film thickness in an exponential manner [Fig. 16(b)]. An increase in the insulating film thickness was found by Savoye and Anderson [100] to shift the log α – V plot to higher V values. These authors also established that the electron emission through holes in the Au film counterelectrode *cannot* be a significant fraction of the total emission current.

The images formed on a fluorescent screen by the emitted hot electrons (with postacceleration) show [100] uniform brightness, indicating an isotropic spatial distribution of the emitted electrons. The

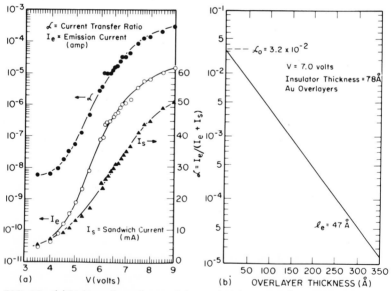

FIG. 16. (a) Voltage dependence of the sandwich current I_s, vacuum-emission current I_e, and current-transfer ratio α of an Al-Al₂O₃-Al tunnel structure. Thickness of the oxide film is 30 Å. (*Cohen* [147]) (b) The transfer ratio α in an Al-Al₂O₃-Au tunnel structure as a function of the thickness of the Au overlayers. (*Savoye and Anderson* [100])

absence of a geometrical edge effect in the emission pattern of carefully prepared crossed sandwiches is noteworthy and implies the uniformity of the field despite the presence of edges. Simmons et al. [149] observed symmetrical optical images of hot electrons and attributed them to coherent scattering (electron diffraction) close to pinhole edges in the counterelectrode.

Using a retarding-potential method, Collins and Davies [146] measured the energy distribution of the emitted electrons. An exponential distribution in energy, ranging from 0.1 to 0.2 eV, depending on the bias field and the distance from the oxide-metal interface, was found. Transfer ratios of $\sim 10^{-3}$ were obtained. An average value of 5 Å was deduced for the mfp of hot electrons in amorphous Al_2O_3 films. These authors discussed the possible energy-loss mechanisms during transport through a thin insulator and attributed the short mfp to the inelastic collisions of electrons with the lattice.

Kanter and Feibelman [120] studied vacuum emission from Al-Al_2O_3-Au structures with oxide film thickness ranging from 67 to 151 Å, and Au film thickness between 200 and 300 Å, and at applied voltages of up to 10 V. Transfer ratios of $\sim 10^{-4}$ were observed which increased to $\sim 10^{-2}$ when the work function of the Au films was decreased because of a superimposed Ba film. Assuming an attenuation length of 100 Å in Au, an attenuation length of about 24 Å in Al_2O_3 was deduced for electrons accelerated from the conduction band to more than 4.7 eV above the Fermi level. The average energy of the emitted electrons was found to increase linearly with the applied diode voltage. The very low transfer ratios observed clearly suggest the absorption of a large fraction of the hot electrons. A typical energy distribution of the emitted electrons is shown in Fig. 17. The energy shows a maximum value much below the peak energy of the injected distribution, and thus a considerable loss of the energy of electrons must take place during transport through the tunnel structure.

Savoye and Anderson [100] made a comprehensive and elegant in situ (during sequential deposition) study of tunnel emission in Al-Al_2O_3-Au (Al) structures. They analyzed the data to decouple the energy loss in the insulator film and the overlayer film. Since the losses due to scattering by impurities are not expected to be significant, and further since the diode voltages used are too low for impact ionization to occur, these authors (as well as others [97, 98]) have proposed that significant losses in the insulator occur because of the creation of optical phonons.

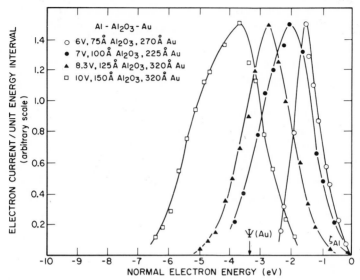

FIG. 17. Normal energy distributions of vacuum-emitted electrons from Al-Al$_2$O$_3$-Au tunnel structures of different oxide, and Au film thicknesses. (*Kanter, quoted by Handy* [98])

This process would require an energy loss of ~0.1 eV per interaction, which is consistent with the fact that $k\theta_D$ ~0.1 eV, where θ_D = 1200°K is the Debye temperature for Al$_2$O$_3$.

Savoye and Anderson performed Monte Carlo calculations based on a model in which the electrons are treated as being free between the inelastic interactions which result in the creation of optical phonons. The free electrons follow a ballistic trajectory determined by the magnitude of the applied field and their initial velocity, following the last interaction. The initial direction subsequent to a given scattering event, and the path length traveled between the interactions were chosen using a random-number generator in such a way that the scattering is isotropic and the mfp l_e corresponds to the electron-electron interaction.

Calculation of the transfer ratio for 1,000-electron samples showed a qualitative agreement with the α vs. V data for l_e = 6 Å. It must be emphasized that the exact value of l_e is not meaningful here since the oversimplified analysis is not too sensitive to the choice of l_e. Quantitatively, the fraction of electrons escaping the insulator calculated from this model is much larger than that deduced from the observed [Fig. 16(*b*)] thickness dependence of α,

$$\alpha = \alpha_0 \exp\left(-\frac{t}{l_e}\right) \tag{19}$$

where t is the overlayer thickness, l_e is the mfp for electron-electron interaction in the metal electrode (Chap. VI, Sec. 5), and α_0 is the zero overlayer-thickness value of α. The slope of the straight line in Fig. 16(b) yields a value of l_e = 47 Å for a 7.0-V bias. This value agrees well with those estimated by other workers.

That the observed vacuum emission in the tunnel structures is indeed due to tunnel electrons can be easily verified by negatively biasing the top electrode to prevent emission. The verification has apparently been effected by some workers [98, 147]. But the observations of Delord et al. [150] have cast *very serious* doubts on the observations of tunnel emission in some structures. These authors studied emission of electrons and light from Be-BeO (40 Å)-Au, and Al-Al$_2$O$_3$ (100 to 150 Å)-Ag thin-film sandwich structures. The structures were held close to liquid-nitrogen temperature to minimize thermionic contribution, and the measurements were carried out under pulse conditions. The most significant observations were: (1) electron emission occurred for *both* polarities of the top electrode but *only* when the luminescence of the sandwich was excited; and (2) the emitted current was always proportional to the intensity of the light generated, whatever the polarity of the driving pulse. These observations, therefore, show that the observed tunnel emission is actually due to photoemission from the counterelectrode and is caused by the luminescence of the sandwich.

Kanter [183] repeated the experiments of Delord et al. using Al-Al oxide (~100 Å thick)-Au sandwich structures. He found that *only* at bias voltages larger than about 8 V the observed electron emission was predominantly caused by photoemission due to easily observable electroluminescence. At lower bias voltages in the forward direction (Au film positive with respect to Al film), the increase in current with voltage was found to be much slower, and further the magnitude of current at 7.5-V bias was about three orders of magnitude smaller than that expected on the basis of photoemission.

Vacuum emission of electrons has been observed [117-119] in tunnel structures using insulator films which are too thick (> 100 Å) to allow any appreciable tunneling current. Moreover, in some cases, the emission occurs at bias voltages which are substantially lower than the work function of the counterelectrode. These puzzling observations may find explanations in terms of the photoemission phenomenon mentioned above.

4.8 Tunnel Spectroscopy

As discussed already, the conductance of a tunnel junction is determined by the supply functions, the barrier potentials, and the barrier profile. Changes in this conductance may occur at certain applied voltages in the following cases: (1) if the traps and/or impurities at a certain energy level in the insulator are ionized by inelastic collisions with the tunneling electrons; (2) if the vibrational and possibly electronic spectra of the impurity atoms present in the insulator are excited by the inelastic collisions; (3) if one of the metal electrodes in the metal-insulator-metal (MIM) junction is replaced by a semiconductor (MIS junction). Case (1) involves changes in the barrier profile and the supply function. Case (2) may be thought of in terms of perturbations in the barrier height. In these cases, the conductance should change when the energy of the tunneling electrons exceeds the energy of ionization or excitation, as the case may be, of the impurity atoms.

In the case (3), the conductance behavior of the MIS junction may be markedly changed by the presence of an energy gap and relatively small Fermi energy of the semiconductor. (The situation here is somewhat similar to tunneling in superconducting tunnel junctions discussed in Chap. IX, Sec. 7.1.) On the basis of a simple energy-level diagram of the junction, it is easy to show that, with increasing tunnel voltage, the current tunneling from the metal to the semiconductor electrode would first decrease (intrinsic conduction) or increase (degenerate conduction) followed by a decrease with a negative resistance when the metal Fermi level moves across the semiconductor energy gap. Thereafter, the current will rise and exhibit sharp increases when the negative bias is equal to the respective interface potentials. However, the current from the semi-conductor to the metal electrode will exhibit behavior similar to that of an MIM junction. Thus, the *I-V* characteristics will be highly asymmetric with respect to the voltage polarity. Since the voltages at which the tunnel current shows changes are directly related to the parameters of the energy-level diagram of the junction, a powerful technique is thus offered for determining these parameters.

It is clear from the foregoing discussion that tunneling characteristics may be employed as a spectrometric tool to measure some energy levels of adsorbed or absorbed insulating impurities in an MIM junction, and the semiconductor parameters in an MIS junction. This tool, called "tunneling spectroscopy," is still in its infancy. A few examples of such studies will now be mentioned.

The size quantization of the energy levels in ultrathin Bi films (Chap. VII, Sec. 2.4) may be studied [151] by measuring the tunneling characteristics between a thin Bi film and a thick bulklike film. Lutskii et al. [152] studied tunnel junctions made of 800- to 1,300-Å Bi film and a thick Bi or Ag film separated by an estimated air gap of ~50 Å produced by bringing two films deposited on mica together in a way (crude) shown in Fig. 18. The observed $I-V$ characteristics showed nonmonotonic behavior at certain values of the bias voltage. By plotting the first derivative of the current, the positions of sharp changes in the $I-V$ characteristics become marked singularities (Fig. 18). When the Fermi level has shifted below the first quantized subband of the thin film, the $I-V$ characteristics should become monotonic again. The voltage at

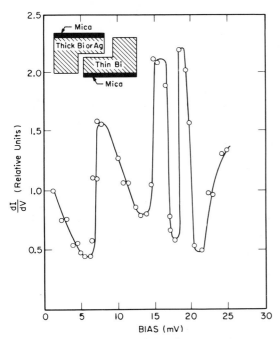

FIG. 18. The voltage derivative of the tunnel current vs. applied voltage for a Bi (thin film)–vacuum–Bi (thick film) tunnel structure. The cross section of the structure is shown in the inset. (*Lutskii et al.* [152])

which this condition is met yields the value of the Fermi level and is ~0.02 to 0.027 eV. The voltage separation between the singularities allows a determination of the effective mass. The value so obtained agrees well with the known bulk value.

The $I-V$ characteristics of an MIS junction have been analyzed theoretically by Chang et al. [153]. These authors obtained a good agreement between the theoretical and the experimental curves on Al-Al$_2$O$_3$-SnTe (or GeTe) junctions. Comparison of the two curves yielded reasonable values of the barrier heights, and the Fermi level, energy gap, work function, and electron affinity for the semiconducting electrode. Similar studies have also been carried out by Hauser and Testardi [153] on Al-Al$_2$O$_3$-Bi(Sb) tunnel junctions. The observed positions of the tunneling spectrum were found to be in accord with the optical data on Bi and Sb films.

Jaklevic and Lambe [154] observed changes (~1 percent) in the conductance of Al-Al oxide-Pb tunnel junctions at definite voltages. These voltages were identified with the vibrational frequencies (bending and stretching modes) of the OH molecule, which may be present in the oxide film as an impurity. These authors then deliberately doped the oxide films by exposing them to the vapors of a variety of high-molecular-weight organic materials (e.g., acetic and propionic acid, methyl and ethyl alcohols) which form strong absorbing bonds to the oxide. The voltage positions of the observed sharp peaks in the second derivative of the I-V curves could be associated with the vibrational spectra of the molecular species. Following the explanation of Scalapino and Marcus [155] in terms of inelastic tunneling, these authors showed that both infrared and Raman modes could be excited. The experimental results could not, however, distinguish between the two modes.

The application of the tunneling technique to study surface adsorption and catalysis is clearly demonstrated in the above experiment. If sufficient energy were available to the tunneling electrons, inelastic interactions leading to electronic transitions could occur and thus enhance the versatility of this technique.

5. PHOTOEFFECTS IN TUNNEL STRUCTURES

5.1 Electroluminescence

When a hole in the valence band combines with an electron in the conduction band, a photon may be given off having a wavelength characteristic of the energy gap of the semiconductor. Jaklevic et al. [156] injected holes into single crystals of n-type CdS by biasing an MIS tunnel junction. When the metal electrode is made positive, the Fermi level of the metal can be shifted to the same level as the valence band of the insulator. Electrons can then tunnel from the valence band of the semiconductor through the insulator into the metal. Holes are thus formed in the valence band of the semiconductor which are filled by electrons from the conduction band, resulting in the emission of a photon of energy equal to the energy gap. Using a polymer thin film as the tunneling medium at $77°K$ and a semiconductor as the receiver electrode, nearly monochromatic photon emission with a spectral width of ~50 Å has been reported [56]. The details of the recombination process in this case however, remain obscure.

Low-voltage electroluminescence of the type mentioned above is also observed [157, 158] in MIM structures with insulating films of SiO, Al_2O_3, Nb_2O_5, TiO_2, and Ta_2O_5. When the electrical conductivity is "developed" in these structures preceding the occurrence of the VCNR, light is emitted and is accompanied by emission of electrons into vacuum. The light appears to be emitted uniformly from the whole or at least large parts of the structure. The intensity of light increases rapidly with the applied sandwich voltage. The low-voltage electroluminescence phenomenon is puzzling. We may note that a similar effect is also observed [159] in the electroluminescence of bulk CdS and is not understood at present.

A crude spectral-distribution study of the emitted light emitted from the tunnel junctions suggests that the energy of photons at the maximum intensity is nearly the same as that observed for photoconduction in the same structures, a point illustrated by the data compiled in Table III.

Table III. Threshold of Photoresponse, Maximum Photovoltage, and Peak Energy of Electroluminescence Observed in Some Metal-Oxide (> 100 Å)-Metal Sandwich Structures

Structure	Insulator band gap*, eV	Photo-response threshold, eV	Max photo-voltage, V	Electro-luminescence energy, eV	Ref.
Al-Al$_2$O$_3$-Al	~8 [119]	1.1	±4 mV[†]	1.2–1.3	167
Nb-Nb$_2$O$_5$ (thermal)-Au	~3.3 [180]	1.2	+0.35–0.45	1.2–1.3	105, 106

Table III. Threshold of Photoresponse, Maximum Photovoltage, and Peak Energy of Electroluminescence Observed in Some Metal-Oxide (>100 Å)-Metal Sandwich Structures *(Continued)*

Structure	Insulator band gap*, eV	Photo-response threshold, eV	Max photo-voltage, V	Electro-luminescence energy, eV	Ref.
Ta-Ta$_2$O$_5$ (anodic)-Au	~4.3 [181]	1.3	+0.94	1.5–1.7	158
Ti-TiO$_2$ (anodic)-Au	~3 [113]	1.3	+0.39	1.2–1.4	158

*The references for this column are given in brackets.

†Polarity, with respect to upper electrode, is positive or negative depending on whether the upper or lower electrode is illuminated.

The fact, that the energy of both the peak electroluminescence and threshold for photoresponse is nearly the same for several "formed" anodic films, and further is considerably smaller than the band gap of the insulator film, suggests that a recombination level below the impurity-conduction band must exist. Such a level is also consistent with the phenomenological explanation of the VCNR phenomenon given already.

5.2 Photoconduction and Photoemission

The occurrence of photoconduction and photovoltaic effect in a metal-anodic oxide-electrolyte system under ultraviolet illumination has been known since 1839 and investigated extensively [160-163]. Similar photoeffects in the visible region are exhibited [164-172] by thin-film metal-oxide-metal sandwich structures. The short-circuit photocurrent in either case increases linearly with the intensity of light, and the open-circuit photovoltage increases exponentially to approach a saturation value. These characteristics are illustrated in Fig. 19 for thermally oxidized niobium oxide films of 50 to 500 Å thickness sandwiched between Nb and Au film electrodes. The load characteristics of the niobium oxide photocells for different levels of illumination are similar to those of a *p–n* junction photocell. Although the quantum efficiency is low (~10^{-2} to 10^{-3}) and is up to 100 times smaller than that of the silicon solar cells, the oxide-film photocells offer the advantages of microminiaturization and less susceptibility to radiation damage.

The explanation of the photoeffect and the origin of photoelectrons in the sandwich structures continues to be a subject of discussion. Van

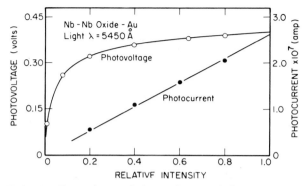

FIG. 19. Dependence of photovoltage and photocurrent in an Nb-Nb$_2$O$_5$-Au structure on the intensity of light of wavelength 5,450 Å. (*Chopra and Bobb* [166])

Geel [162] suggested the bulk or volume excitation of electrons from the valence to the conduction band of the insulator. Sasaki [163] proposed that the oxide is nonstoichiometric, resembling a *p-i-n* structure, and that the photoelectrons are created in the *i* region. A volume-excitation hypothesis must be ruled out since the absorption of light in a thin transparent oxide film is expected to be negligible, and further since it is not consistent with the observed dependence of the polarity and the magnitude of the photocurrent on the electrode material and the photon energy.

A satisfactory understanding of the various photoeffects is obtained [167] by considering that electrons are generated in the metal electrodes. Since the attenuation length for the hot electrons is small, the useful electrons must originate near the metal-oxide interfaces. The photoelectrons are then emitted over the trapezoidal barrier (see Fig. 5) of a metal-insulator-metal structure. The rectifying property of such an asymmetric structure explains the occurrence of the photovoltaic effect. The polarity of the observed photovoltage is opposite to the rectification direction, as one would expect from the elementary considerations of the photoelectrons which move from the higher barrier side to the lower one in order to equalize the potential difference. The transport of photoelectrons is complicated by the presence of trapping centers in the insulator, as indicated by the observed long time constants of the order of seconds for the photocurrent to reach equilibrium values.

If we assume the trapezoidal barrier model, then a bias V may be applied externally to modify the intrinsic field present because of the two unequal barrier heights. For $V \geq (\phi_2 - \phi_1)$, the direction of the photocurrent should obviously reverse, which offers a convenient way to determine $(\phi_2 - \phi_1)$.

Interesting photovoltage polarity-reversal effects [167] are observed for slightly asymmetric structures (i.e., $\phi_1 \sim \phi_2$) such as "formed" Al-Al$_2$O$_3$-Al. These effects arise since the density of photoelectrons reaching the two interfaces is not the same because of the different optical-absorption paths and the hot-electron-transport losses. If more photoelectrons are created by direct illumination on the low barrier side, a direction-dependent reversal of the intrinsic field and hence the polarity of the photovoltage is found. A time-dependent reversal of polarity occurs if the unequal electron-density distributions on the two sides of the barrier equilibrate slowly.

A quantitative study of the spectral dependence of photoresponse is a valuable tool for measuring the barrier heights and for determining the barrier profile. The intercept of a plot of the square root of the photoresponse vs. photon energy (called "Fowler plot") gives the barrier height. This technique has been used by several investigators [173, 174]. The data so obtained for some tunnel structures are listed in Table II (Sec. 4.3). This method of analysis is illustrated in Fig. 20 by the results obtained from an Al-Al$_2$O$_3$-Al tunnel structure (oxide thickness $\sim$20 to 40 Å). With no voltage applied to the sample and for photon energies $h\nu$ less than $\phi_1 (\phi_1 < \phi_2)$, equal and opposite photon-induced tunnel currents flow so that there is no net photocurrent. For $\phi_2 > h\nu > \phi_1$, the electrons from electrode 2 contribute *more* to the photocurrent because the flow of electrons in the conduction band of the insulator is in the direction of the intrinsic field rather than against it. For $h\nu > \phi_2$, *most* of the photocurrent is contributed by electrons from electrode 2. A Fowler plot yields a straight line, the intercept of which gives ϕ_2. To obtain predominant photocurrent contributions from electrode 1, the intrinsic field must first be removed by a bias voltage. The Fowler plot obtained under the bias condition yields a threshold value which is equal to ϕ_1. Values obtained in this case are $\phi_1 = 1.42$ and $\phi_2 = 1.77$ eV.

The barrier heights obtained from photoemission data need to be corrected for the Schottky-effect lowering of the barriers and electric-field penetration of the electrodes. These corrections are, however, uncertain in view of the anomalously high field dependence of the barrier height observed by Braunstein et al. [175] on Al$_2$O$_3$ and by

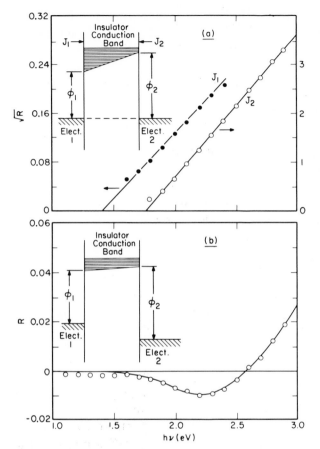

FIG. 20. (*a*) Square root of response per incident photon vs. photon energy in an Al-Al$_2$O$_3$-Al structure. J_2 curve (right-hand scale) is for zero bias; J_1 curve (left-hand scale) is for a bias of 0.53 V with electrode 2 positive. (*b*) Relative photo-response vs. photon energy for a bias of 0.4 V with electrode 2 positive. Positive response is defined as the flow of electrons from electrode 2 to electrode 1. This is the case of almost zero internal field in the Al$_2$O$_3$, as illustrated in the inset. (*Braunstein et al.* [170])

Lewicki and Mead [176] on AlN tunnel junctions. The latter authors also reported a linear twofold increase of ϕ with the thickness of the AlN film from 35 to 90 Å. The interpretation of these results is not clear at present, but it is probably related to the electronic properties of the particular metal-insulator interfaces.

Nelson and Anderson [177] studied in detail the photoemission from $Al\text{-}Al_2O_3$-metal tunnel junctions and analyzed the results to obtain the barrier profile. An image-force-corrected trapezoidal barrier determined by the work functions of the metal electrodes minus the electron affinity of the oxide was found to be in good agreement with the observations.

6. CONCLUDING REMARKS

1. A thin-film insulator departs considerably from the idealized model assumed in the theoretical treatments of the transport of carriers through thin insulators. While studies of electrical conduction have attracted much attention, the basic electronic parameters of thin ($\sim$50 Å) insulators remain largely unknown. The dielectric constant, dielectric breakdown, effective mass of the carriers, energy-momentum relation, distribution of traps, and the band structure and inelastic interactions of the electrons with the insulator lattice are some of the parameters of major importance to the transport phenomenon in thin insulators. Reliable and quantitative information on these parameters is certainly difficult to obtain. But it is clear that systematic transport studies must be made on various thicknesses of only reproducible and structurally well-defined thin insulators such as cleaved mica, organic films, and epitaxial oxides.

2. Because of the occurrence of electroluminescence and thus the photoemission of electrons in metal-insulator-metal structures, the results obtained hitherto on tunnel emission are open to question. Clearly, careful experiments are needed to understand both electroluminescence and the transport of hot electrons in MIS structures. The present experimental evidence offers little optimism for useful hot-electron-tunnel devices because of the very high transport losses of hot electrons.

3. Tunneling studies offer a sensitive and powerful spectrometric tool for investigating the band structure of semiconductors and suitable insulators. This tool will undoubtedly be exploited extensively in the future.

REFERENCES

1. W. Kohn, *Phys. Rev.,* **110**:857 (1958).
2. K. H. Drexhage and H. Kuhn, in "Basic Problems in Thin Film Physics" (R. Niedermayer and H. Mayer, eds.), p. 339, Vandenhoeck and Ruprecht, Göttingen, 1966.

3. K. L. Chopra, *J. Appl. Phys.*, **36**:655 (1965).
4. M. Hacskaylo and C. Feldman, *J. Appl. Phys.*, **33**:3042 (1962).
5. C. A. Mead., *Phys. Rev. Letters*, **6**:545 (1961).
6. N. F. Mott and H. Jones, "Metals and Alloys," p. 87, Dover Publications, Inc., New York, 1958.
7. S. Whitehead, "Dielectric Breakdown of Solids," Oxford University Press, Fair Lawn, N. J., 1951; also, J. J. O'Dwyer, *Advan. Phys.*, **7**:349 (1958); also, *J. Phys. Chem. Solids*, **28**:1137 (1967).
8. F. Forlani and N. Minnaja, *Phys. Stat. Solidi.* **4**:311 (1964).
9. F. Seitz, 'The Modern Theory of Solids," p. 563, McGraw-Hill Book Company, New York, 1940.
10. S. Pakswer and K. Pratinidhi, *J. Appl. Phys.*, **34**:711 (1963).
11. P. P. Budenstein and P. J. Hayes, *J. Appl. Phys.*, **38**:2837 (1967).
12. R. C. Merrill and R. A. West, *Electrochem. Soc. Meeting Abstracts*, Pittsburgh, 1963.
13. F. L. Worthing, *J. Electrochem. Soc.*, **115**:88 (1968).
14. I. S. Pikalova, *Soviet Phys.-Solid State, English Transl.*, **8**:1784 (1967).
15. A. Moll, *Z. Angew. Phys.*, **10**:410 (1958).
16. L. Young, "Anodic Oxide Films," Academic Press Inc., New York, 1961.
17. F. J. Burger and L. Young, *Progr. Dielectrics*, **5**:1 (1963).
18. N. Schwartz and R. W. Berry, in "Physics of Thin Films" (G. Hass and R. E. Thun, eds.), vol. 2, p. 363, Academic Press Inc., New York, 1964.
19. G. Siddall, *Vacuum*, **9**:174 (1959).
20. G. Siddall, in "Thin Film Microelectronics" (L. Holland, ed.), p. 1, John Wiley & Sons, Inc., New York, 1965.
21. C. Weaver, *Advan. Phys.*, **11**:85 (1962).
22. C. Weaver, in "The Use of Thin Films in Physical Investigations" (J. C. Anderson, ed.), p. 283, Academic Press Inc., New York, 1966; also, *Vacuum*, **15**:171 (1965).
23. P. White, *Insulation*, September, 1965, p. 57.
24. L. V. Gregor, in "Physics of Thin Films" (G. Hass and R. E. Thun, eds.), vol. 3, p. 133, Academic Press Inc., New York, 1966.
25. W. A. Pliskin, in "Measurement Techniques for Thin Films," p. 280, Electrochemical Society, New York, 1967.
26. H. G. Manfield, *Microelec. Reliability*, **3**:5 (1964).
27. B. Lewis, *Microelec. Reliability*, **3**:109 (1964).
28. E. M. DaSilva and P. White, *J. Electrochem. Soc.*, **109**:12 (1962).
29. E. E. Smith and S. G. Ayling, *Proc. Electron. Components Conf.*, 1962, p. 82; also, *Proc. IEE London*, **109**:504 (1962).
30. P. Bickley and D. S. Campbell, *Vide*, **99**:214 (1962).
31. T. K. Lakshmanan, *Trans. 8th Natl. Vacuum Symp.*, p. 868, Pergamon Press, New York, 1961.
32. D. A. McLean, *Proc. Natl. Electron. Conf.*, **16**:206 (1960).
33. E. R. Bowerman, *IEEE Trans. Component Pts.*, **CP-10**:86 (1963).

34. H. G. Rudenberg, J. R. Johnson, and L. C. White, *Proc. Electron. Components Conf.*, 1962, p. 90.
35. M. E. Sibert, *J. Electrochem. Soc.*, **110**:65 (1963).
36. F. Huber, *IEEE Trans. Component Pts.*, **CP-11**(2):38 (1964).
37. P. Lloyd, *Solid-State Electron.*, 3:74 (1961).
38. M. Stuart, *Nature*, **199**:59 (1963).
39. C. Feldman and M. Hacskaylo, *Rev. Sci. Instr.*, **33**:1459 (1962).
40. C. Feldman, *Rev. Sci. Instr.*, **26**:463 (1955).
41. F. S. Maddocks and R. E. Thun, *J. Electrochem. Soc.*, **109**:99 (1962).
42. R. G. Breckenridge, in "Imperfections in Nearly Perfect Crystals," John Wiley & Sons, Inc., New York, 1952.
43. F. Seitz, *Rev. Mod. Phys.*, **18**:384 (1946).
44. D. C. Mullen, cited by Weaver in Ref. 22.
45. H. Hirose and Y. Wada, *Japan. J. Appl. Phys.*, 3:179 (1964).
46. J. deKlerk and E. F. Kelly, *Appl. Phys. Letters*, **5**:2 (1964).
47. J. deKlerk and E. F. Kelly, *Rev. Sci. Instr.*, **36**:506 (1965).
48. N. F. Foster, *Proc. IEEE*, **53**:1400 (1965).
49. J. deKlerk, *J. Appl. Phys.*, **37**:4522 (1966).
50. N. F. Foster, *J. Appl. Phys.*, **38**:149 (1967).
51. N. F. Foster and G. A. Rozgonyi, *Appl. Phys. Letters*, **8**:221 (1966).
52. R. M. Malbon, D. J. Walsh, and D. K. Winslow, *Appl. Phys. Letters*, **10**:9 (1967).
53. R. W. Gibson, *Electron. Letters*, **2**:213 (1966).
54. T. R. Sliker and D. A. Roberts, *J. Appl. Phys.*, **38**:2350 (1967).
55. D. P. Seraphim, in "Thin Films," p. 135, American Society for Metals, Metals Park, Ohio, 1964.
56. L. Eckertová, *Phys. Stat. Solidi*, **18**:3 (1966).
57. R. M. Hill, *Thin Solid Films*, **1**:39 (1967).
58. J. G. Simmons, *J. Appl. Phys.*, **35**:2472 (1964).
59. K. H. Gundlach, *Phys. Stat. Solidi*, **4**:527 (1964).
60. H. T. Mann, in "Basic Problems in Thin Film Physics" (R. Niedermayer and H. Mayer, eds.), p. 691, Vandenhoeck and Ruprecht, Göttingen, 1966.
61. P. R. Emtage and W. Tantraporn, *Phys. Rev. Letters*, **8**:167 (1962).
62. C. A. Mead, in "Basic Problems in Thin Film Physics" (R. Niedermayer and H. Mayer, eds.), p. 674, Vandenhoeck and Ruprecht, Göttingen, 1966.
63. C. L. Standley and L. I. Maissel, *J. Appl. Phys.*, **35**:1530 (1964).
64. W. E. Flannery and S. R. Pollack, *J. Appl. Phys.*, **37**:4417 (1966).
65. T. E. Hartman, J. C. Blair, and R. Bauer, *J. Appl. Phys.*, **37**:2468 (1966).
66. D. V. Geppert, *J. Appl. Phys.*, **34**:490 (1963).
67. P. J. Price and J. M. Radcliffe, *IBM J. Res. Develop.*, 3:364 (1959).
68. W. A. Harrison, *Phys. Rev.*, **123**:85 (1961).
69. J. Bardeen, *Phys. Rev. Letters*, **6**:57 (1961).
70. J. Frenkel, *Phys. Rev.*, **36**:1604 (1930).
71. A. Sommerfeld and H. Bethe, in "Handbuch der Physik" (H. Geiger and K. Scheel, eds.), vol. 24/2, p. 450, Springer-Verlag OHG, Berlin, 1933.

72. R. Holm and B. Kirschstein, *Z. Tech. Phys.*, **15**:488 (1935).
73. R. H. Fowler and L. Nordheim, *Proc. Roy. Soc. London,* **A119**:173 (1928).
74. R. Holm, *J. Appl. Phys.*, **22**:569 (1951).
75. J. G. Simmons, *J. Appl. Phys.*, **34**:1793 (1963).
76. J. G. Simmons, *J. Appl. Phys.*, **34**:2581 (1963).
77. J. G. Simmons, *J. App. Phys.*, **35**:2655 (1964); *Trans. Met. Soc. AIME,* **233**:485 (1965).
78. E. L. Murphy and R. H. Good, Jr., *Phys. Rev.*, **102**:1464 (1956).
79. R. H. Good, Jr., and E. W. Müller, in "Handbuch der Physik" (S. Flügge, ed.), vol. XXI, p. 176, Springer-Verlag OHG, Berlin, 1956.
80. R. Stratton, *J. Phys. Chem. Solids,* **23**:1177 (1962); the role of the boundary conditions is discussed in *Phys. Rev.,* **136**:837 (1964).
81. W. Tantraporn, *Solid-State Electron.*, **7**:81 (1964).
82. C. K. Chow, *J. Appl. Phys.*, **34**:2490, 2599, 2918 (1963); **36**:559 (1965).
83. T. E. Hartman and J. S. Chivian, *Phys. Rev.*, **134**:A1094 (1964).
84. T. E. Hartman, *J. Appl. Phys.*, **35**:3283 (1964).
85. R. Stratton, G. Lewicki, and C. A. Mead, *J. Phys. Chem. Solids,* **27**:1599 (1966); also, see J. Antula, *Phys. Stat. Solidi,* **24**:89 (1967).
86. K. H. Gundlach, in "Basic Problems in Thin Film Physics" (R. Niedermayer and H. Mayer, eds.), p. 696, Vandenhoeck and Ruprecht, Göttingen, 1966.
87. G. J. Unterkofler, *J. Appl. Phys.*, **34**:3143 (1963); J. G. Simmons and G. J. Unterkofler, *J. Appl. Phys.*, **34**:1828 (1963).
88. W. W. Dolan and W. P. Dyke, *Phys. Rev.*, **95**:327 (1954).
89. J. C. Fisher and I. Giaever, *J. Appl. Phys.*, **32**:172 (1961).
90. G. Advani, N. Gottling, and T. Osman, *Proc. IRE,* **50**:1130 (1962).
91. D. Meyerhofer and S. A. Ochs, *J. Appl. Phys.*, **34**:2535 (1963).
92. S. R. Pollack and C. E. Morris, *J. Appl. Phys.*, **35**:1503 (1964).
93. S. R. Pollack and C. E. Morris, *Trans. Met. Soc. AIME,* **233**:497 (1965).
94. R. M. Handy, *Phys. Rev.,* **126**:1968 (1962).
95. M. McColl and C. A. Mead, *Trans. Met. Soc. AIME,* **233**:502 (1965).
96. J. G. Simmons and G. J. Unterkofler, *J. Appl. Phys.*, **34**:1828 (1963); *Appl. Phys. Letters,* **2**:78 (1963).
97. F. W. Schmidlin, *J. Appl. Phys.*, **37**:2823 (1966).
98. R. M. Handy, *J. Appl. Phys.*, **37**:4620 (1966).
99. P. R. Emtage, *J. Appl. Phys.*, **38**:1820 (1967).
100. E. D. Savoye and D. E. Anderson, *J. Appl. Phys.*, **38**:3245 (1967); also, see O. L. Nelson and D. E. Anderson, *J. Appl. Phys.*, **37**:66 (1966).
101. M. A. Lampert and A. Rose, *Phys. Rev.,* **121**:26 (1961); M. A. Lampert, *Rept. Progr. Phys.*, **27**:329 (1964) (a review).
102. J. Lindmayer and A. Slobodskoy, *Solid-State Electron.*, **6**:495 (1963).
103. J. Dresner and F. V. Shallcross, *Solid-State Electron.*, **5**:205 (1962).
104. R. W. Brander, D. R. Lamb, and P. C. Rundle, *Brit. J. Appl. Phys.*, **18**:23 (1967).
105. K. L. Chopra, *J. Appl. Phys.*, **36**:184 (1965).
106. T. W. Hickmott, *J. Appl. Phys.*, **37**:4380, 4588 (1966).

107. R. I. Frank and J. G. Simmons, *J. Appl. Phys.*, **38**:832 (1967).
108. D. V. Geppert, *J. Appl. Phys.*, **33**:2993 (1962).
109. J. C. Penley, *J. Appl. Phys.*, **33**:1901, 2906 (1962).
110. J. C. Penley, *Phys. Rev.*, **128**:596 (1962).
111. E. Pittelli, *Solid-State Electron.*, **6**:667 (1963).
112. F. G. Ullman, *J. Phys. Chem. Solids*, **28**:279 (1967).
113. R. H. Bube, "Photoconductivity of Solids," chap. 9, John Wiley & Sons, Inc., New York, 1960; G. A. Dussel and R. H. Bube, *Phys. Rev.*, **155**(3):764 (1967).
114. J. Frenkel, *Phys. Rev.*, **54**:647 (1938); J. G. Simmons, *Phys. Rev.*, **155**:(3):657 (1967).
115. C. A. Mead, *Phys. Rev.*, **128**:2088 (1962).
116. T. W. Hickmott, *J. Appl. Phys.*, **33**:2669 (1962).
117. T. W. Hickmott, *J. Appl. Phys.*, **34**:1569 (1963).
118. T. W. Hickmott, *J. Appl. Phys.*, **35**:2118 (1964).
119. T. W. Hickmott, *J. Appl. Phys.*, **35**:2679 (1964).
120. H. Kanter and W. A. Feibelman, *J. Appl. Phys.*, **33**:3580 (1962); also, **34**:3629 (1964).
121. S. R. Pollack, *J. Appl. Phys.*, **34**:877 (1963).
122. S. R. Pollack, W. O. Freitag, and C. E. Morris, *Electrochem. Tech.*, **1**:96 (1963).
123. G. S. Kreinina, L. N. Selivanov, and T. I. Schumskaya, *Radiotekhn. i Elektron.*, **5**:1338 (1960).
124. G. S. Kreinina, *Radiotekhn. i Elektron.*, **7**:182, 2096 (1962).
125. G. S. Kreinina, *Radiotekhn. i Elektron.*, **9**:2051 (1964).
126. L. Eckertová, *Czech. J. Phys.*, **B15**:111 (1965).
127. Z. Hajek, *Czech. J. Phys.*, **12A**:573 (1962).
128. T. Lewowski, S. Sedecki, and B. Sujak, *Acta Phys. Polon.*, **28**:343 (1965).
129. A. van Roggen, *Phys. Rev. Letters*, **9**:368 (1962).
130. K. L. Chopra, *Proc. IEEE*, **51**:941 (1963).
131. K. L. Chopra, *Proc. IEEE*, **51**:1242 (1963).
132. D. V. Geppert, *Proc. IEEE*, **51**:223 (1963).
133. W. R. Beam and A. L. Armstrong, *Proc. IEEE*, **52**:300 (1964).
134. W. R. Hiatt and T. W. Hickmott, *Appl. Phys. Letters*, **6**:106 (1965).
135. J. B. Gunn, *Proc. Phys. Soc. London*, **69B**:781 (1956).
136. M. Braunstein, A. I. Braunstein, and R. Zuleeg, *Appl. Phys. Letters*, **10**:313 (1967).
137. H. Mahl, *Phys. Z.*, **12**:985 (1937).
138. M. M. Vudynskii, *Zh. Tekhn. Fiz.*, **20**:1306 (1950).
139. C. A. Mead, *Proc. IRE*, **48**:1478 (1960).
140. C. A. Mead, *J. Appl. Phys.*, **32**:646 (1961).
141. D. V. Geppert, *Proc. IRE*, **48**:1644 (1960).
142. M. I. Elinson and A. G. Zhdan, *Radiotekhn. i Elektron.*, **5**:1862 (1960).
143. L. Eckertová, *Czech. J. Phys.*, **10**:412 (1962); *Naturwiss.*, **9**:201 (1962).
144. J. P. Spratt, R. F. Schwarz, and W. Kane, *Phys. Rev. Letters*, **6**:341 (1963).

145. R. N. Hall, *Solid-State Electron.*, 3:320 (1961).

146. R. E. Collins and L. W. Davies, *Solid-State Electron.*, 7:445 (1964).

147. J. Cohen, *J. Appl. Phys.*, 33:1999 (1962); *Appl. Phys. Letters*, 1:61 (1962).

148. J. Cohen, *J. Appl. Phys.*, 36:1504 (1965).

149. J. G. Simmons, R. R. Verderber, J. Lytollis, and R. Lomax, *Phys. Rev. Letters*, 17:675 (1966); J. G. Simmons and R. R. Verderber, *Appl. Phys. Letters*, 10:197 (1967).

150. J. F. Delord et al., *Appl. Phys. Letters*, 11:287 (1967).

151. L. V. Iogansen, *Soviet Phys. Usp. English Transl.*, 8:413 (1965); G. A. Gogadze and I. O. Kulik, *Soviet Phys.-Solid State English Transl.*, 7:345 (1965).

152. V. N. Lutskii, D. N. Korneev, and M. I. Elinson, *Soviet Phys. JETP Letters English Transl.*, 4:179 (1966).

153. L. L. Chang, P. J. Stiles, and L. Esaki, *J. Appl. Phys.*, 38:4440 (1967); J. J. Hauser and L. R. Testardi, *Phys. Rev. Letters*, 20:12 (1968).

154. R. C. Jaklevic and J. Lambe, *Phys. Rev. Letters*, 17:1139 (1966); J. Lambe and R. C. Jaklevic, *Phys. Rev.*, 165:821 (1968).

155. D. J. Scalapino and S. M. Marcus, *Phys. Rev. Letters*, 18:459 (1967).

156. R. C. Jaklevic, D. K. Donald, J. Lambe, and W. C. Vassell, *Appl. Phys. Letters*, 2:7 (1963).

157. T. W. Hickmott, *J. Appl. Phys.*, 36:1885 (1965).

158. T. W. Hickmott, *J. Electrochem. Soc.*, 113:1223 (1966).

159. W. A. Thornton, *Phys. Rev.*, 116:893 (1959).

160. E. Becquerel, *Compt. Rend.*, 9:561 (1839).

161. L. Young, *Trans. Faraday Soc.*, 50:164 (1954).

162. W. C. van Geel, C. A. Pistorius, and P. Winkel, *Philips Res. Rept.*, 13:265 (1958).

163. Y. Sasaki, *J. Phys. Chem. Solids*, 13:177 (1960).

164. F. Huber, *J. Electrochem. Soc.*, 110:846 (1963).

165. F. Huber and M. Rottersman, *J. Appl. Phys.*, 33:3385 (1962).

166. K. L. Chopra and L. C. Bobb, *Proc. IEEE*, 51:1784 (1963).

167. K. L. Chopra, *Solid-State Electron.*, 8:715 (1965).

168. G. Lucovsky, C. J. Repper, and M. E. Lasser, *Bull. Am. Phys. Soc.*, 7:399 (1962).

169. K. W. Shepard, *J. Appl. Phys.*, 36:796 (1965).

170. A. I. Braunstein, M. Braunstein, G. S. Picus, and C. A. Mead, *Phys. Rev. Letters*, 14:219 (1965).

171. A. I. Braunstein, M. Braunstein, and G. S. Picus, *Phys. Rev. Letters*, 15:956 (1965).

172. T. M. Lifshitz and A. L. Musatov, *Soviet Phys. JETP Letters English Transl.*, 4:199 (1966).

173. For example, C. A. Mead, *Solid-State Electron.*, 9:1023 (1966).

174. For example, A. M. Goodman and J. J. O'Neill, Jr., *J. Appl. Phys.*, 37:3580 (1966).

175. A. I. Braunstein, M. Braunstein, and G. S. Picus, *Appl. Phys. Letters*, 8:95 (1966).

176. G. Lewicki and C. A. Mead, *Appl. Phys. Letters,* 8:98 (1966).
177. O. L. Nelson and D. E. Anderson, *J. Appl. Phys.,* 37:77 (1966).
178. T. Hayashi, *Japan. J. Appl. Phys.,* 3:230 (1963).
179. H. G. Hurst and W. Ruppel, *Z. Naturforsch.,* 19a:573 (1964).
180. R. A. Schell and R. E. Salomon, *J. Electrochem. Soc.,* 113:24 (1966).
181. L. Apker and E. A. Taft, *Phys. Rev.,* 88:58 (1952).
182. P. Mark and T. E. Hartman, *J. Appl. Phys.,* 39:2163 (1968).
183. H. Kanter, *Appl. Phys. Letters,* 12:243 (1968).
184. J. deKlerk, *Appl. Phys. Letters,* 13:102 (1968).
185. J. L. Richardson, *General Electric Co. Rept.* AD66645 (1968).
186. J. F. Gibbons and W. E. Beadle, *Solid-State Electron.,* 7:785 (1964).
187. F. Argall, *Solid-State Electron.,* 11:535 (1968).

IX SUPERCONDUCTIVITY IN THIN FILMS

1. INTRODUCTION

The disappearance of the electrical resistance of mercury at helium temperatures was first observed by Kammerlingh Onnes in 1911. He called this phenomenon "superconductivity." Since then, a large number of metals, alloys, and compounds have been found to be superconductive. This remarkable phenomenon of superconductivity is now well established as a macroscopic manifestation of the quantum effects. Much theoretical and experimental effort undertaken to understand the nature and origin of superconductivity has culminated during this decade in a successful microscopic theory. In these works, studies of the superconducting behavior of thin film specimens have made profound contributions in providing significant conclusions in support of the various models and theories of superconductivity. Furthermore, thin film studies have opened many new and fascinating areas of solid-state research, such as superconductive tunneling and enhanced superconductivity. Some attractive possibilities for future electronic-device

529

applications have further stimulated research on the superconducting behavior of thin films.

The literature on superconductivity of films is growing at an enormous rate. A number of books and reviews on superconductivity of bulk [1-11] and/or thin film [10-14] materials have been published. However, the subject related to thin films is not covered adequately by any single review. This chapter purports to describe the salient features of the superconducting behavior of thin films. Since the theoretical treatments in this field are specialized and are also dealt with in standard books, only the results in the form of final expressions are stated. Their discussion in the light of the available experimental data is emphasized.

For the sake of the uninitiated reader, it is necessary first to introduce some basic concepts and notations pertinent to the phenomenon of superconductivity. No references are given for this section, since the subject is discussed in standard textbooks.

2. BASIC CONCEPTS

1. The temperature T_c at or below which electrical resistance vanishes, is called the superconducting critical or transition temperature. An upper limit of $10^{-23}\Omega$-cm for the resistivity in the superconducting state has been estimated from the time decay of a current circulating in a superconducting thin-film tube. The superconducting transition of an ideal superconductor is sharp and is of the order of a millidegree.

2. In nonideal cases, the transition is broad; here T_c is the temperature at which the resistance falls to half the normal value. Normal conductivity is restored by the application of a critical magnetic field H_c, which varies with temperature T approximately as

$$H_c \approx H(0)\left[1 - \left(\frac{T}{T_c}\right)^2\right]$$

(1a)

where $H(0) = H_c$ at $T = 0°K$. Using the reduced coordinate $T_R \equiv T/T_c$, Eq (1a) reduces at $T \rightarrow T_c$ to

$$H_c \approx 2H(0)(1 - T_R)$$

(1b)

3. In addition to being a perfect conductor, an ideal superconductor is also perfectly diamagnetic; i.e., it expels the magnetic flux reversibly. This is called the "Meissner effect" (discovered by Meissner and Ochsenfeld). However, in the case of a multiply connected, or a simply connected but nonhomogeneous superconductor, flux is trapped by an indefinitely persisting current (incomplete Meissner effect) and cannot change unless the superconductivity of the specimen is quenched.

4. Superconductivity can be quenched by passing current through a specimen. The current is critical when the surface magnetic field exceeds H_c (Silsbee rule).

5. The brothers F. and H. London incorporated the perfect-conductivity and the perfect-diamagnetic characteristics into a phenomenological theory of superconductivity. The spatial penetration of an external magnetic field H into a superconductor is given by this theory as

$$H(x) = He^{-x/\lambda_L} \tag{2}$$

where

$$\lambda_L = \left(\frac{mc^2}{4\pi n_s e^2} \right)^{1/2} \tag{3}$$

is called the London penetration depth λ_L. With the free-electron values, the value of λ_L for the usual range of electron densities in superconductors is $\sim 10^{-6}$ cm. The observed values of the penetration depth, called the empirical λ, are generally about 5×10^{-6} cm, several times larger than the London value. Utilizing the Gorter-Casimir two-fluid model of a superconductor, the temperature dependence of the empirical λ is given by

$$\lambda(T_R) = \lambda(0) \left(1 - T_R^4 \right)^{-1/2} \tag{4a}$$

where $\lambda = \lambda(0)$ at $T = 0°$K. As $T \to T_c$,

$$\lambda(T_R) \approx \frac{\lambda(0)}{2} \left(1 - T_R \right)^{-1/2} \tag{4b}$$

6. The sharpness of the superconducting transition in the absence of a magnetic field, and the marked dependence of the observed λ on the

impurities in a superconductor led Pippard to suggest that the superconducting state should be characterized by a finite (rather than infinite as
in the London theory) range of momentum coherence such that the
order parameter changes gradually over a distance ξ called the "coherence length." He estimated the value of $\xi \sim 10^{-4}$ cm from the data on the
field dependence of λ. The coherence range is essentially the spatial
definition of the superconducting electron and may be estimated from
the uncertainty principle as

$$\xi \sim \frac{\hbar}{\Delta p} \sim \frac{\hbar v_F}{kT_c} \qquad (v_F \text{ is the Fermi velocity}) \tag{5}$$

Pippard's "nonlocal" modification of the London theory is based on
the finiteness of the coherence distance. To explain the dependence of λ
on the electron mfp l, he assumed an effective coherence length $\xi(l)$
related to ξ_0 of a pure metal by

$$\frac{1}{\xi(l)} = \frac{1}{\xi_0} + \frac{1}{Al} \tag{6}$$

where A is a constant of the order unity. The nonlocal theory reduces to
simpler local form in two limiting cases and yields explicit expressions
for λ as

$$\lambda = \lambda_L \left(\frac{\xi_0}{\xi(l)} \right)^{1/2} \qquad \text{for } \xi \ll \lambda \text{ (London limit)} \tag{7a}$$

where λ_L is the London value of the penetration depth. Using Eq. (6),
Eq. 7(a) reduces to

$$\lambda \approx \lambda_L \sqrt{1 + \frac{\xi_0}{l}} \tag{7b}$$

Tinkham pointed out that experimental data fit Eq. 7(b) better if the
theoretical λ_L is replaced by an empirically determined bulk value λ_B for
a *pure* bulk material. One finds the use of both λ_B and λ_L in the
literature, however.

The condition $\lambda \gg \xi$ holds for pure metals close to T_c (where $\lambda \to \infty$) and also for alloys and impure thin films where l and ξ are reduced or limited by electron scattering at imperfections, impurities, or the film boundary so that as $l \to 0$, $\xi \to l$. The opposite limit $\xi \gg \lambda$, which holds for most pure bulk superconductors ($l \to \infty$) at temperatures not too close to T_c, also yields a simpler local form:

$$\lambda_{l \to \infty} = \left(\frac{\sqrt{3}}{2\pi} \xi_0 \lambda_L^2 \right)^{1/3} \qquad \text{for } \xi \gg \lambda \text{ (Pippard limit)} \qquad (8)$$

7. The total exclusion of an external field does not lead to a state of lowest energy for a superconductor unless an appreciable surface energy of bonding between the superconducting and normal phases exists. It is therefore energetically more favorable for a suitably shaped supercon-ductor (any shape other than an infinitely long cylinder in a longitudinal field) to divide into a large number of alternately normal and superconducting layers such that the width of the latter is less than λ, and that of the former very much smaller. This is called the intermediate or mixed state. The exclusion of magnetic flux (Meissner effect) occurs when the interphase boundary energy is positive.

8. Ginzburg and Landau (G-L) introduced an effective wave function or order parameter to incorporate the interphase boundary energy in their theory. They predicted a positive interphase boundary energy when $\kappa > 1/\sqrt{2}$, where the dimensionless parameter κ in the low-field limit (London limit) is given by

$$\kappa = \frac{\sqrt{2}\,\overset{*}{e} H_c \lambda^2}{\hbar c} \qquad (9)$$

Here $\overset{*}{e}$ is the charge of the superelectron. Gorkov showed that the G-L theory may be derived from the BCS theory provided that $\overset{*}{e} = 2e$ so that the superelectron is a Cooper pair. For a pure superconductor very close to T_c (i.e., $\lambda > \xi_0$),

$$\kappa \approx 0.96 \frac{\lambda}{\xi_0} \qquad (10)$$

The G-L theory predicts that a superconductor with $\kappa > 1/\sqrt{2}$ remains superconducting up to a field

$$H_{c2} = \sqrt{2}\kappa H_c \tag{11}$$

without exhibiting the Meissner effect. Abrikosov has called it a type II or hard superconductor as compared with the type I or soft superconductor for which $\kappa < 1/\sqrt{2}$. The state below H_c is called the mixed state and it extends down to

$$H_{c1} = \frac{H_c}{\kappa}(\ln\kappa + 0.08) \qquad \text{for } \kappa \gg 1 \tag{12}$$

9. Abrikosov's analysis of type II superconductors based on the G-L theory predicts flux penetration in filaments parallel to the external field, and arranged in a regular pattern. These filaments are believed to consist of individual quanta of flux. The G-L theory as applied to type II superconductors is denoted as Ginzburg-Landau-Abrikosov-Gorkov (GLAG) theory.

Saint-James and deGennes have shown that even above H_{c2} and up to $H_{c3} = 1.69\,H_{c2}$, a superconducting sheath persists on surfaces parallel to the applied field. This phenomenon is called "surface superconductivity."

Pure and mechanically soft materials which exhibit low critical fields, first-order superconducting-normal transition, and $\xi \gg \lambda$ belong to type I (Pippard type). Materials with a large number of defects and/or impurities which exhibit high transition temperatures, high critical fields, broad second order transition, and $\xi \ll \lambda$ belong to type II (London type). The G-L thoery is applicable to both types at $T \to T_c$ and reduces to the London theory in the low-field limit.

10. The microscopic theory of Bardeen, Cooper, and Schrieffer (BCS), which has successfully accounted for the unique properties of superconductors, is based on the condensation into a lower energy of Cooper pairs of electrons of opposite spins, and equal and opposite momentum due to an attractive electron-phonon interaction. The low-energy state is the superconducting state separated from the normal state by an energy gap 2Δ which decreases from $3.52\,kT_c$ at $T = 0$ to zero at T_c. The temperature dependence of $\Delta(T)$ is given by the BCS theory as

$$\frac{\Delta(T)}{\Delta(0)} = \tanh\left[\frac{T_c}{T}\frac{\Delta(T)}{\Delta(0)}\right] \tag{13}$$

The size of the Cooper pair or the correlation distance of the BCS theory corresponds to Pippard's coherence distance and is

$$\xi_0 = 0.18 \frac{\hbar v_F}{kT_c} \tag{14}$$

where v_F is the electron Fermi velocity.

11. F. London pointed out that the flux passing through a hole in a superconducting ring must be quantized if the wave function describing a superconducting electron in the presence of the flux is to be single-valued with respect to the circulation around the hole. Byers and Yang [15], Onsager [16], and Bardeen [17] modified the London argument to take into account the coherent electron-pair states of the BCS theory, and showed that the flux threading a superconducting ring must be quantized in units of

$$\phi_0 = \frac{hc}{2e} = 2 \times 10^{-7} \text{ G-cm}^2 \tag{15a}$$

This value is half that predicted by London. Using Eqs. (9) and (10), we can express ϕ_0 as

$$\phi_0 = \sqrt{8}\,\pi\lambda\xi_0 H_{cB} \tag{15b}$$

For a cylinder 10 μ in radius, the calculated value of the trapped flux corresponding to one quantum of flux is 0.063 Oe, a value that is large enough to be detectable. The measurement of this flux was reported simultaneously by Deaver and Fairbank [18] and by Doll and Näbauer [19], using superconducting cylinders made, respectively, by depositing Sn films on Cu wires, and Pb films on quartz fibers. These experiments employed different detection methods and showed the existence of flux quanta of magnitude ϕ_0 given by Eq. (15a). This exciting result provided the most striking verification of the existence of the Cooper pair.

3. TRANSITION TEMPERATURE OF THIN-FILM SUPERCONDUCTORS

3.1 Introduction

The superconducting properties of thin films are extremely sensitive to the deposition parameters. Thick, annealed films of pure, soft (type I) superconductors prepared by evaporation in reasonably clean vacuum conditions display properties similar to the corresponding bulk materials. Values of some useful parameters such as T_c, H_c, λ, ξ, and Δ of several common bulk superconductors are listed in Table I for general reference.

The characteristic superconducting behavior of thin films has been thoroughly investigated primarily in type I materials such as In, Tl, Pb, Sn, and Al. While the superconducting behavior of hard or type II superconductors is much more complicated, the interesting and challenging problems in understanding this behavior in bulk and in thin films, coupled with their great potential in possible device applications, are now attracting increasing attention. With the availability of ultrahigh-vacuum and clean conditions and reproducible methods for preparing films by thermal evaporation and sputtering (see Chap II), thin-film hard superconductors such as Nb, Ta, V, W, Mo, and Re can now be readily prepared [20-23] in polycrystalline and epitaxial forms with T_c's very close to those of the bulk materials.

Thin film deposition technology is well suited to the preparation of hard superconductors in the form of homogeneous and inhomogeneous alloys and compounds. For example, Gerstenberg and Hall [23] prepared superconducting carbides and nitrides of Ta and Nb by reactive sputtering in the presence of methane and nitrogen, respectively. Films of Nb_3Sn have been prepared by codeposition [24] of individually

Table I. Some Parameters for Several Common Bulk Superconductors

Material	T_c,°K	$H_c(0)$, G	$\lambda(0)$, Å		ξ, Å	$2\Delta(0)/kT_c$	
			obs.	calc.*	calc.*	obs.†	calc.*
Al	1.19	99	~500	530	16,000	3.43	3.5
Ga	1.09	51					3.5
Hg-a	4.153	411	380–450			4.6	3.7
In	3.407	283	640		4,400	3.63	3.5
Nb	9.46	1,944				3.8	3.7
Pb	7.18	803	390	480	830	4.33	3.9
Sn	3.722	306	510	560	2,300	3.46	3.6
Ta	4.482	830				≤ 3.6	3.6
V	5.30	1,310					3.6

*Calculated from the BCS theory (see Ref. 5).

†Thin film tunneling data (Ref. 7). Values obtained from other methods are also listed in the same reference.

evaporated components followed by annealing, as well as by pyrolytical methods [25]. Theuerer and Hauser [26], and Edgecumbe et al. [27] prepared and studied the properties of thin films of Nb-Zr, Nb-Ti, Nb-Sn, V-Ti, V-Si, V-Ge, and V-Ga systems. These materials are of much interest because of their high T_c, high H_c, and high current-carrying capacity in the presence of a field. Edgecumbe et al. employed sequential sputtering from two targets onto planar and cylindrical glass substrates held at 300 to 450°C. The observed transition temperature, critical magnetic field, and current-carrying capability (see Fig. 14) of as-prepared thin films ($\sim$1,000 Å) of several Nb and V systems compared well with those of the corresponding bulk materials.

Various aspects of the superconducting behavior of type I and type II films are discussed in the rest of this chapter. It may be pointed out that, although the criterion for the type classification of a superconductor is the same for both bulk and thin films, a thin film prepared from a type I superconductor may behave as type I or type II, depending on the "effective" value of the electron mean free path, which is strongly influenced by its thickness, impurities, imperfections, and grain size. Thus, for example, by increasing the mean free path by annealing a vapor-quenched film, its type II behavior may be changed to type I.

The critical temperature of a film is influenced by a number of film parameters, which we shall now discuss.

3.2 Thickness Dependence

The question of a lower limit of a specimen thickness to which superconductivity can persist has been examined by several workers [28-33]. Bulklike superconductivity was observed by Shalnikov [30] in Pb films as thin as 50 Å. Armi [31] reported normal conductivity only at or below $\sim$7 Å for Pb films. A similar lower limit at 11 and 40 Å, respectively, was found by Zavaritskii [32] for Sn and Tl films. These observations suggest that superconductivity persists down to the thinnest continuous films. It is, of course, meaningless to talk of a lower limit in vapor-deposited ultrathin films since they generally consist of three-dimensional discrete nuclei so that only an average thickness is known (Chap. IV, Sec. 2.1).

The transition temperature T_c increases with decreasing thickness of In [34] and Al [35-37] films. A typical variation is shown in Fig. 1 for In films. A reverse behavior is exhibited by Pb [33] films; an initial

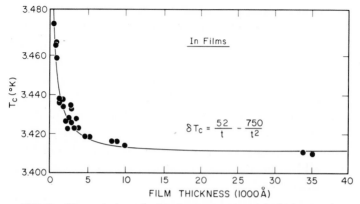

FIG. 1. The variation of critical temperature with thickness of evaporated In films. The solid curve represents the calculated variation arising from the differential thermal stress induced on cooling. (*Toxen* [34])

increase followed by a decrease of T_c is observed for Sn and Tl [32] films. A depression in T_c inversely proportional to the product of the width and thickness (and hence the film volume) of very narrow Sn films was reported by Hunt and Mercereau [38]. For example, a maximum depression of $0.047°K$ in T_c was observed for a 495-Å-thick, 1-μ-wide, and 300-μ-long film. This observation was explained on the basis of the possible existence of size-dependent uncorrelated differences in the quantum phase of different microscopic regions in a small-volume specimen. This explanation further implied that below a critical cross-sectional area ($\sim 10^{-12}$ cm^2) of a link connecting two superconductors, the quantum phase of the two superconductors should remain uncorrelated at all temperatures so that the link will not be superconductive. Both the observed size effect and the explanation offered need further verification.

Toxen [34] explained his results on the size-dependent transition temperature of In films (Fig. 1) on the basis of the hydrostatic stress resulting from the differential thermal contraction on cooling (to helium temperatures) of the vitreous-silica substrate relative to the film material. Thermal stress is expected to yield a change in the T_c of about $0.13°K$. The observed decrease was, however, smaller and was explained by assuming that the maximum strain in the film was limited by the occurrence of a plastic flow. Assuming that the maximum stress is given by Eq. (7) of Chap. V, the change in T_c can be shown to be given by the relation $T_c = (A/t - B/t^2)$, which is plotted in Fig. 1. Here A and B are constants and t is the film thickness. Blumberg and Seraphim [39] also used the same explanation to account for their similar results on Sn films.

3.3 Superconductivity-enhancement Phenomenon

Thermal stress is known to increase the T_c of Al [37] films deposited on glass (tension) but it decreases T_c of films deposited on Teflon (compression). The observed thickness-dependent increase of up to ~2°K in the T_c of Al films is, however, much too large to be explained solely by stress effects. This type of enhancement of superconductivity, i.e., superconductivity at a T_c higher than that of the corresponding bulk material, is obtained in fine-grained and in amorphous-like films of a variety of materials. Data on a number of such films and the corresponding enhancement factors, the ratio of the transition temperatures of the film to that of the bulk (T_{cF}/T_{cB}), are listed in Table II.

Bismuth [40, 41] and beryllium [42], which are not superconducting in bulk down to ~10^{-2}°K, are superconducting at 6 and 8°K, respectively, in amorphous-like vapor-quenched (VQ) films obtained by vapor deposition onto substrates at helium or low temperatures. On annealing, superconductivity is irreversibly quenched in Bi and Be films. Gallium [41] is normally superconducting at 1.08°K, but its amorphous films deposited at 4°K have $T_c = 8.4$°K. On warming to about 15°K, Ga films crystallize to a γ structure with $T_c = 6.4$°K. The claim [43] of superconductivity in Fe films remains questionable [44].

A thorough study of the enhancement of superconductivity in VQ films of Al, Zn, In, Tl, Pb, Hg, Sn, Sn + 10 percent Cu (Au or Ag), Ga, Bi, and Be was undertaken by Buckel and Hilsch [41, 45] and complemented by electron-diffraction investigations by Buckel [46] and Rühl [47]. They made the following observations: (1) All VQ films yielded very small crystallite size (as judged from diffuse electron-diffraction patterns). (2) The T_c increased linearly with increasing Debye temperature of the material. (3) The T_c of Sn films decreased with the increasing deposition temperature (Fig. 2), as also with the annealing temperature (see Chap. VI, Fig. 15). (4) The enhancement *did not depend* on the film thickness (Fig. 2), or the metallic or dielectric nature of a variety of substrates used. (5) The increase in the T_c

Table II. Superconductivity Enhancement (Ratio of the Transition Temperature of the Film T_{cF} to That of the Bulk T_{cB}) and Other Physical Properties of Films of Several Materials

Material	T_{cF}/T_{cB}	Thickness, Å	Grain size, Å	Deposition* technique	θ_{Debye}, °K	Ref.
Al/SiO/Al/- (multilayer)	4	60		VQ		57
Al + TCM†	4.5			VQ		248
Al	2.27	1,000	40	VQ	370–400	54
	2.6			RE		
Mo	~5	100–4,000	<50	SP, EB	375	52
	~7	200–4,000	50–1,000	ISP		
W	~400	100–4,000	<50	SP, EB	315	51, 52
	~600	200–4,000	50–1,000	ISP		
Zn	1.55			VQ	200–250	45
Cd	1.4			VQ	164	45
Sn	1.26	500	90	VQ	110–210	45
	1.1	>2,000	110	RE		
Sn + 10% Cu	~2		~40	VQ	100–150	45
In	1.22	>2,000	110	VQ		45
				RE		
Tl	1.1				75–135	45
	1.1			VQ		45
Be	>800			VQ	925	45
Pb	1.0	>2,000		RE, VQ	~90	54,45
Bi	>600			VQ	111	45
Ga	6.5	>2,000		VQ		54

*VQ = vapor-quenched by deposition at 4.2°K; RE = reactive evaporation in presence of oxygen partial pressure; EB = electron beam; SP = glowdischarge sputtering, and ISP = ion-beam sputtered.

†Tetracyanoquinodimethan.

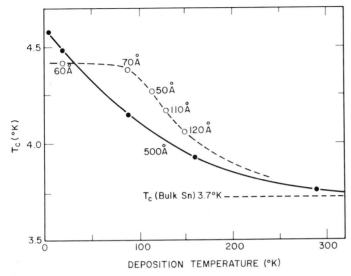

FIG. 2. The dependence of the critical temperature (resistive transition) of Sn films of various thicknesses on the temperature of deposition on quartz substrates. (*Buckel and Hilsch* [45])

correlated with a decrease in the crystallite size, and an increase in the number of structural defects (and hence the normal state electrical resistivity). For example, by codepositing of 10 percent Cu (or Au, Ag) with Sn, the crystallite size was decreased from 90 to 30 Å and the T_c rose from the bulk value of 3.7 to above $7°$K.

Richter and Steeb [48] and Fujime [49] analyzed the diffraction patterns of the VQ films of Bi, Ga, and Be and established that these highly disordered structures were essentially amorphous. As discussed in detail in Chap. IV, Sec. 4.2, on annealing of these amorphous structures at certain temperatures, nucleation and growth of some metastable (generally, high-temperature and/or high-pressure polymorphs) phases occur. These phases have their characteristic transition temperatures. For example, the T_c's of some unidentified high-pressure phases of Ga and Bi are about 7.5 and $6°$K, respectively. These values are very close to those observed for films of these materials on suitable annealing of the amorphous structures. It should also be pointed out that, although no systematic structural studies on VQ films of pure metals have been reported in literature, there is definite evidence [47] that the Sn lattice is considerably distorted. Similar distortion effects should occur in other cases and thus influence the transition temperature.

The role of a fine-grained or amorphous structure in yielding anomalously high enhancements of T_c's is indicated by the studies on vacuum-evaporated and sputtered films of W [50-52] and Mo and Re [50, 52]. Although subnitrides and subcarbides of these transition metals are also known to be high T_c superconductors, the observed enhancement is certainly a property of the structure of the films. Chopra et al. [53] have established that sputtered films of several transition elements can be prepared in amorphous, metastable fcc, or the normal structure, depending on the deposition conditions. The fcc phases of W and Mo have been found to be superconducting. The respective T_c's are 4.6 and 6.4°K, which do not strongly depend on the crystallite size (Fig. 3). On the other hand, the T_c of the amorphous structure decreases rapidly with increasing crystallite size because of annealing. Electron-microscope studies [53] of W films showed that annealing converted the amorphous to the normal (bcc) phase via the metastable fcc phase. The crystallite size dependence of T_c may be understood if we assume the existence of traces of superconducting fcc phase in the normal state

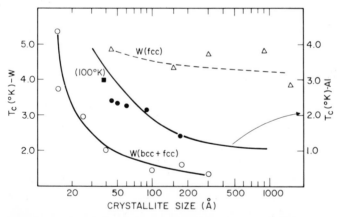

FIG. 3. The crystallite-size (estimated from electron-diffraction patterns) dependence of the critical temperature of films of Al (*Abeles et al.* [54]), W (fcc), and W (mixed fcc and bcc). (*Chopra* [52]) Aluminum films were deposited on glass at 300°K (except for one film at 100°K) by evaporation in a partial pressure of ~10^{-5} Torr of oxygen. Tungsten films were sputtered by an argon ion beam. The solid curve through the Al data is calculated on the basis of the deGennes theory assuming a layer of 5 Å thickness with a BCS coupling constant = 0.27 superimposed on the ordinary Al films.

matrix. A detailed discussion of the metastable structures is given in Sec. 4.3 of Chap. IV.

Abeles et al. [54] observed enhancement in the T_c of several metal films (see Table II) with fine-grained structures obtained by evaporation in the presence of $\sim 10^{-5}$ Torr oxygen partial pressure (reactive evaporation). The variation of T_c with the average crystallite size of as-prepared Al films is shown in Fig. 3 and is compared with a theoretical curve based on Ginzburg's surface-superconductivity model (discussed later). Further studies by Abeles et al. [55] established that the transverse critical fields of fine-grained (~ 40 Å) films of Al, Sn, and Ga were up to two orders of magnitude larger than the thermodynamic critical fields. These results indicated that the superconducting grains were strongly coupled to each other, and that the enhancement in T_c was independent of the coupling over a wide range of the coupling strength.

Cohen et al. [56] obtained fine-grained Ga films at room temperature by depositing in the presence of 10^{-4} Torr oxygen pressure. Using these films as one electrode of several tunnel junctions (Sec. 7), the energy gap of Ga films was measured. Three well-defined values of energy gap of Ga films were measured. Three well-defined values of energy gap corresponding to T_c = 6.4, 7.9, and 8.4°K were obtained which may be assigned to the crystalline β and γ, and amorphous structures of Ga, respectively. These results strongly suggest that these Ga polymorphs are nucleated and stabilized in the "amorphous phase," most likely by the small grain size (see page 196 for a definition of an amorphous phase).

Following the theoretical suggestion of Cohen and Douglass [57], Strongin et al. [57] studied the T_c enhancement of several VQ films separated by dielectric barrier layers. They observed enhancement in a structure consisting of alternate 60-Å layers of superconductor S and barrier material B, whenever the sequence SBS was formed. A T_c of 5.7°K was obtained for four layers of Al separated by SiO films. The effect was observed for Al, Sn, Zn, In, and Pb and for a variety of barrier materials. Although the enhancement was attributed by these authors to the presence of the dielectric layer, the experimental evidence was not unambiguous. Since the deposition process employed provides an efficient means of producing fine-grained structures, and further, since similar results have been obtained by various observers using fine-grained films deposited on different metal and dielectric substrates, the most likely explanation of these results must be sought in the structural disorder* and the finite crystallite size [249] of these films.

We can summarize the experimental results of the enhancement phenomenon by saying that a small crystallite size (diffuse electron-diffraction patterns) is a necessary condition for enhancement in T_c to occur. Whether it is achieved by the VQ method, codeposition of incompatible materials, reactive evaporation, or ion-beam sputtering is immaterial. The small crystallite size is known to introduce enormous lattice defects and intrinsic stresses and, in some cases, appears to be responsible for stabilizing high-temperature and/or high-pressure meta-stable phases of the material. The influence of a dielectric barrier or that of a surface layer on the enhancement mechanism is at best doubtful at present.

3.4 Mechanisms of Enhanced Superconductivity

Several theoretical mechanisms have been proposed which predict the occurrence of high-temperature superconductivity. Some of these interesting predictions of high-temperature superconductivity rest on the application of the following generalized BCS expression for the transition temperature:

$$kT_c = 1.14\hbar\omega \exp\left[-\frac{1}{N(0)U}\right] \quad \text{for } N(0)U \ll 1 \qquad (16)$$

where ω is the characteristic cutoff frequency and corresponds to the Debye frequency [$\omega \sim (k\theta_D/\hbar)$] in the case of electron-phonon inter-action mediated superconductivity. The BCS coupling constant $N(0)U$ is the product of the electron density of states N of one spin at the Fermi surface, and the average, net attractive electron-lattice interaction U. Normally, $N(0)U$ is weakly sensitive to the number of lattice dimensions and its value lies between 0.1 and 0.5. It is clear from Eq. (16) that by increasing either $\hbar\omega$ or $N(0)U$, we can obtain higher values of the T_c.

Let us now examine the proposed enhancement mechanisms in light of the scanty experimental data available at present.

---

*This conclusion is further supported by the observed [248] enhancement of T_c of Al to 5.24°K by codepositing Al and an organic material (tetracyanoquinodi-methan) at 4°K.

1. Little [58] proposed the possibility of obtaining high-temperature superconductivity in suitable one-dimensional organic chains. The proposed interaction for superconductivity has been shown to be theoretically [59] impossible. Note, however, that although superconductivity is usually considered to be a three-dimensional-order effect, its existence in one or two dimensions is justified by a formal application of the BCS theory.

2. The effect of the quantization of the transverse motion of electrons in samples of finite size on the BCS coupling constant has been discussed by several workers [60-65]. It has been shown [63, 64] that for phonon-mediated superconductivity with Cooper pairs formed by electrons possessing opposite two-dimensional momenta and identical quantum numbers, the T_c should increase for thicknesses below 100 Å, the increase becoming exponential for $t < 10$ Å.

The observed [55] strong coupling between very small grains of Al, Sn, and Ga films does not favor an enhancement mechanism based on the quantum quantization of electronic motion. Similarly, enhancement of T_c by Josephson tunneling between grains, as suggested by Parmenter [61], is not in accord with the observed absence of correlation between enhancement and the intergrain coupling.

3. Several phononless (not mediated through a phonon) superconductivity mechanisms have been proposed [65-69]. Ginzburg's [66] suggestion of the occurrence of surface superconductivity at $T_c \sim 10^2$ to $10^{4\circ}$ K is based on the possibility of a suitable attractive electron-electron interaction at the surface, or between electrons in the surface states. In the latter case, the bulk of the material may be nonmetallic.

According to Ginzburg and coworkers, it may be possible to change the surface interaction energy by superimposing a monolayer of a suitable material onto the surface of another material so that $\hbar\omega$ may reach a value of several electron volts. The effective value of $N(0)U$ may, however, still remain very small, and thus T_c will not be increased drastically. Silvert [68, 69] suggested that, in some cases, a form of a high-temperature localized superconductivity may occur because of short-range correlation effects at the surface. But he also pointed out that, at high T_c, the correlation length $\hbar v_F/2\pi k T_c$ given by Eq. (14) becomes comparable with the interatomic spacing, and thus the very concept of a coherent many-electron state may be open to question [68].

Clearly, the exponential factor in Eq. (16) is the most likely variable to provide high values of T_c. Since this factor contains the volume of the specimen, a thickness dependence of T_c can be incorporated by assuming Ginzburg's superconducting surface layer with an enhanced interaction superimposed on the ordinary superconductor below. By volume averaging the surface interaction over the whole volume of the dual sample, a suitable thickness or size dependence may be obtained. This procedure was used by Strongin et al. [37] to explain the thickness dependence and by Abeles et al. [54] to explain the crystallite-size dependence of the T_c of Al films. Toxen's results on In films (Fig. 1) may also be explained by the same procedure. This method is strictly phenomenological, and further the underlying principle lacks justification, particularly in view of the fact that the observed enhancement is most likely a property of the film structure and not of the deposition technique, oxidation, dielectric shielding, or impurity effects.

4. An electronic mechanism based on a suitable attractive interaction between different electron groups in the volume of the film may result in the enhancement of superconductivity. For example, intervalley electronic transitions giving rise to phonon-induced electron-electron interaction was proposed by Cohen [70] to explain superconductivity in semiconducting GeTe and ferroelectric $SrTiO_3$. Appel [71], however, showed that superconductivity in ferroelectric $SrTiO_3$ can also be explained by the contemporary electron-phonon interaction theory.

Kresin and Tavger [65] proposed that the Cooper pairing may arise from the Coulomb interaction of the electrons from different groups which in a thin film are the overlapping subbands. In addition, owing to the decrease in the symmetry, the degeneracy in a thin semiconducting film may be lifted [72] resulting in the formation of two or more groups of electrons differing in their effective masses and wave functions. This electronic mechanism should yield an increase in T_c with decreasing thickness.

In this connection of an electronic mechanism, the Cohen-Douglass [57] enhancement mechanism is of interest. They suggested that a thin insulating barrier between two superconductors weakens the attractive electron-phonon interaction and may greatly diminish and even change the sign of the effective Coulomb interaction, thereby making it possible for electrons across the barrier to form Cooper pairs with each other. The observed enhancement in alternate layers of a superconductor separated by a dielectric was cited in support of this theory. This conclusion is, however, questionable since, as already discussed, the experimental results on the sandwich structures are most likely attributable to the structural disorder in the films produced by the specific deposition method rather than to some fundamental phenomenon.

5. Parmenter [73] proposed an enhancement process to utilize the operation of a tunnel junction to extract and thus relatively reduce the number of normal electrons in a superconductor. This process has not been subjected to direct experimental verification. It is, however, consistent with the observed change in T_c due to electrostatic charging of the surface of a superconductor by a transverse electric field (Sec. 3.7).

In summary, none of the models proposed so far has yet received an unambiguous supporting evidence. On the other hand, there is overwhelming evidence that enhancement in most cases is due to structural disorder in particular films. As pointed out earlier, the only necessary condition to obtain enhancement is a highly disordered structure with small crystallite size, which gives rise to a high density of lattice defects, large intrinsic stresses ($\sim 10^{10}$ dynes/cm^2), and significant distortion of the lattice (see Chap. V). These factors are known to stabilize the high-temperature and high-pressure polymorphic phases of some materials (Chap. IV, Sec. 4.3). These polymorphic phases in Bi, Ga, Be, W, and Mo are also high T_c superconducting phases. The enhancement of T_c can occur in a highly distorted lattice, possibly as a result of changes in the Debye frequency and the BCS-coupling-constant factor in Eq. (16).

3.5 Influence of Stress

Stress affects T_c of a bulk specimen. As discussed in Chap. V, high stresses of about 10^9 to 10^{10} dynes/cm^2 are developed in a thin film because of the resultant effect of the frozen-in structural defects (intrinsic), and the differential thermal contraction between the substrate and the deposit during the process of condensation of a thin film and also during subsequent cooling for measurement purposes. The observed increase in the T_c of In and Sn films with decreasing film thickness and its explanation in terms of the thickness-dependent critical shear stress due to thermal strains has been mentioned in the preceding section.

Tensile stress increases whereas compressive stress decreases the T_c of Sn [39] films. In addition to the dependence on film thickness, the changes in the T_c of Sn [39] films are markedly dependent on the crystallographic orientation of the films relative to the substrate because of the strongly anisotropic stress behavior and the electrical resistivity of tin. The change in T_c with stress along the diad axis is small. A Sn film oriented with its c axis normal to the plane of a glass substrate is stressed only slightly because of a relatively smaller thermal contraction for this combination. Values of the T_c ranging from 3.53 to 4.15°K (bulk T_c = 3.73°K) depending on the film orientation are observed. The change in T_c with uniaxial stress for single-crystal Sn may increase by a factor of up to 20 in going from the diad to the tetrad axis.

The effect of a thermal stress and of an externally imposed stress by stretching or bending of the substrate has been studied by a number of investigators [74-76]. Notarys [75] reported a linear increase in the T_c of 7°K per unit tensile stress in Al films deposited on Mylar plastic. No hysteresis was observed up to a strain of 3 percent. Hall [76] observed a linear increase in the T_c of Sn films with stress. Values of $d \ln T_c/d \ln V$ (V is the volume of the film) equal to +8.6 from differential thermal-contraction data and +8.1 from the substrate bending measurements were obtained. These values compare well with the +7.5 value for bulk tin. These data suggest that the elastic limit of Sn films is at least an order of magnitude higher than that of the bulk.

3.6 Influence of Impurities

Impurities influence [77-82] the normal-superconducting transition. The magnitude of the effect depends on the film material and the nature of the impurity or the alloying materials. The transition temperature of Ta and Nb films, for example, decreases markedly with the addition of interstitial O_2 and N_2. A decrease of T_c about 1°K per atomic percent of dissolved O_2 is observed [24] in Nb films. Some of the contribution to the change in T_c may well be the result of expansion of the Nb lattice.

Caswell [80] studied the effect of residual gases on the supercon-ducting properties of In and Pb films; N_2, H_2, CH_4, Ar, CO, and CO_2 were found to have no measurable effect on In films when present in partial pressures of 10^{-5} Torr. For O_2, pronounced effects were observed when the ratio of O_2 molecules to In atoms striking the substrate exceeded 3 percent. It was shown that In films could be deposited at 100 Å/sec on clean, baked substrates at 10^{-5} Torr with characteristics indistinguishable from those of films deposited in ultrahigh vacuum provided the partial pressures of O_2 and H_2O were below 10^{-7} and 10^{-5} Torr, respectively.

Suitable deposition procedures may be used to prepare films of homogeneous and inhomogeneous alloys. By coevaporation of two materials onto a cold substrate, a matrix of the more abundant material containing the other as an homogeneously distributed impurity is obtained. This process has been used to prepare specimens and study the depression of the T_c of Pb films due to small impurities of transition metals (Mn, Fe, Gd, and Co) and their oxides. The T_c of Pb films was found to decrease approximately linearly with the composition of the impurities, and the energy gap Δ tended to zero. Superconductivity was quenched at impurity concentrations of 0.26, 1.2, 2.8, and 6.3 atomic percent of Mn, Fe, Gd, and Co, respectively. These results were explained theoretically by Abrikosov and Gorkov [83] on the basis of a spin-dependent interaction between the conduction electrons and the impurity atoms.

3.7 Electrostatic-charge (Field) Effect

The possibility of influencing the T_c of a superconductor by changing its normal-electron density was realized by Glover and Sherrill [84]. They used a high transverse electric field ($\sim 10^6$V/cm) (Chap. VI, Sec. 3.5) to charge the surface of thin films of In, Tl, and Sn. A T_c change of $\sim 10^{-4}$°K due to a charge density of $\sim 1.5 \times 10^{-7}$ C/cm^2 was observed. Negative charging was found to increase the normal conductivity but decrease the T_c of Tl and annealed In films. However, the reverse was observed for Sn and unannealed In films. The effect in all cases is reversed by positive charging.

It may be noted that the normal electron density can also be varied and hence the T_c changed by alloying a material with a different valency element as is the case in bulk Sn and In alloys [85]. Such a study [86] with superconducting, amorphous Bi films containing small amounts of Te or Pb impurities, however, showed no simple relation with the valence or atomic size of the impurity atoms. This result is probably the consequence of the fact that, in an amorphous material, the resistivity is not simply related to the valency of an impurity atom.

In order to obtain a higher charge density (and hence a larger change in the T_c), Rühl [87,88] exploited the charge-transfer process during the oxidation of a metal. According to the Mott model of oxidation, electrons from the metal move to the acceptor levels of the adsorbed oxygen on the metal and convert oxygen atoms to ions. The electric field so built up forces positive metal ions to move to the surface to form oxide molecules. When the oxide film is thick enough to prevent further tunneling, oxidation stops. A high ($\sim 10^8$V/cm) field appears across the metal-oxide interface to charge the metal surface positively and thus cause a change of T_c by several tenths of a degree. These values are about 10^3 times greater than those obtained by Glover and Sherrill with applied transverse field, and correspond to their data extrapolated to a field of $\sim 10^8$V/cm. These observations were later confirmed by Glover et al. [89].

In Rühl's experiment, thin metal films were deposited at low temperatures. Oxygen (~ 50 Å thick) was then condensed onto these films and the film was allowed to warm up in order for oxidation reaction to take place. Oxidation was found to occur at temperatures as low as $20°$K. Below $50°$K, a self-limited oxide thickness of ~ 20 Å was generally obtained. An increase in the T_c of trivalent metals Al, In, and Tl, and a decrease in T_c of the tetravalent Pb and Sn was observed. The sign of the T_c shift corresponds to an excess positive-charge effect. Since the change in T_c was found to be proportional to the reciprocal of the thickness of the metal film, the excess charge may be interpreted as being distributed uniformly throughout the film. By depositing another thin layer of the metal on the oxidized surface, the normal value of T_c was restored. This result was attributed to the neutralization of the acceptor levels of oxygen by reaction with the superimposed film.

3.8 Proximity Effects in Superimposed Films

Misener and coworkers [90] observed that the superconducting transition temperature of Pb films electroplated onto constantan wires decreased below the bulk value with decreasing film thickness. Similar observations made by Meissner [91] on Sn films electroplated onto Au and Cu wires suggested that this lowering of the T_c was caused by the proximity or contact of a normal metal. Extensive investigations [91-111] have since been made on this so-called "proximity" effect using superimposed thin metal films. These studies have employed a variety of analytical methods such as the measurement of T_c [91-98], H_c [99-100], persistent currents [98], flux penetration, and tunneling [101-111]. Although the quantitative significance of the results in many cases can be criticized, the existence of the proximity effect appears well established. A detailed review of the subject is given by Hilsch and Minningerode [92].

Several theoretical [112-123] treatments of the various aspects of the proximity effect have appeared in the literature. An instructive physical argument due to Cooper [112] is as follows: In the BCS theory, the range of the attractive electron interaction between electrons is very short ($\sim 10^{-8}$ cm); however, the size of the wave packet or the correlation distance of the attractively bound Cooper pairs is of the order of the coherence length ($\sim 10^{-4}$ cm). Because of this longer coherence length, the Cooper pairs can extend a considerable distance into a region in which the interaction between electrons is not attractive. Thus when a thin layer of superconducting material is in contact with a layer of normal metal, the zero-momentum pairs formed in the superconductor extend into both layers. As a result, the ground-state energy of this thin bimetallic layer is characterized by some average of the interaction parameter $N(0)U$ over both metals, which in turn determines the energy gap of the layer and its transition temperature from Eq. (16). Implicit solutions for the T_c of a superimposed structure are given by the de Gennes-Werthamer theory [114-117]. This theory suggests that quantitative studies of the proximity effect should yield information on the electron-electron interaction in the normal metal.

Maki [119-121] and Fulde [121] analyzed the proximity effect and showed that the effect of a magnetic or a nonmagnetic metal contact with a superconductor is to break Cooper pairs and hence suppress superconductivity. This process is equivalent to that of paramagnetic impurities, or a critical current in a superconductor.

Qualitatively, one can surmise that the proximity effect must depend on the nature of the boundary between the two metals and the relative thickness of the two layers. The thicker the normal layer, the smaller the average interaction and thus the smaller the influence on the properties of the superconducting film. If normal metal is magnetic, it should have a more pronounced effect because of the additional interaction of spins with the conduction electrons.

Smith et al. [96] studied the thickness dependence of T_c of Pb in Ag/Pb superimposed films. The induced superconductivity in a normal 5,700-Å Ag film in contact with a superconducting 1,500-Å-thick Pb film was verified by establishing the existence of persistent currents through the couple. The superconductivity gap induced in the Ag film was measured by tunneling (Sec. 7.1) through Pb-PbO-Ag-Pb thin-film structures and found to be 1.6×10^{-4} eV at 4.2°K. This value is ten times smaller than that of pure Pb and indicates that the observed effect is not due to pinholes in the Ag film. Marcus [102] studied the energy gap in similar tunnel structures and found an exponentially decreasing energy gap with increasing Ag film thickness. The gap was found to be insensitive to temperatures far below the T_c of Pb, which is consistent with the proximity mechanism of de Gennes. The gap was independent of the applied longitudinal field up to a certain value above which it decreased linearly. This behavior, also observed by Burger et al. [109] in contact effect with dirty superconductors, has been explained (tentatively) by de Gennes as due to domains in Ag film where Cooper pairing is decreased above a critical field.

P. and R. Hilsch [93-95] carried out extensive measurements of the T_c of superimposed films deposited onto crystalline quartz substrates held at 10°K. Deposition at low temperatures was considered essential to minimize diffusion and alloying problems. Typical curves for the thickness dependence of the transition temperature of a Pb film in contact with Cu films of various thicknesses, and of a Cu film in contact with Pb films of various thicknesses are shown in Fig. 4. Note that the depression of the T_c of Pb/Cu couples *depends* on whether Pb or Cu film

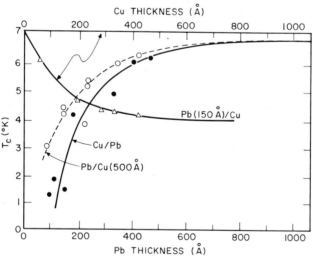

FIG. 4. Critical temperatures of superimposed films: of Pb/Cu and Cu/Pb for various thicknesses of Pb and Cu films. Both systems were deposited at 10.5°K on quartz substrates. The solid and dotted curves for the variable Pb thickness data refer to Cu/Pb/quartz and Pb/Cu/quartz sequence of preparation, respectively. (*Hilsch* [94])

is deposited first, *suggesting* that interfaces are different for the two cases despite the careful experimentation.

The reduced transition temperature T_s/T_c (T_s is the transition temperature of the couple and T_c that of single pure film) for couples of several metals, fits an empirical relation:

$$\left(\frac{T_s}{T_c}\right)^2 = 1 - \frac{1}{0.2 + 0.8t_s} \tag{17}$$

where t_s is the ratio of the thickness of the superconducting film and the critical thickness below which no superconductivity is observed for a constant thickness of the normal metal film. A plot of T_s/T_c vs. t_s is shown in Fig. 5. The data of these authors also show that the critical thickness, which is always smaller than the coherence length of the ideal superconductor, decreases with the decrease of the effective mean free path.

When applied to the Pb/Cu couple data of P. and R. Hilsch, the de Gennes-Werthamer theory yields a positive value of the BCS interaction for Cu. This value, $N(0)U \approx +0.05$, predicts a T_c of a few millidegrees

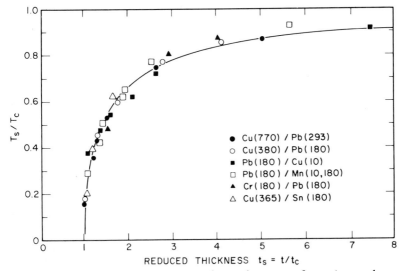

FIG. 5. Reduced critical temperatures of several systems of superimposed films as a function of the reduced thickness (see text). The thickness of the normal films is kept constant. The condensation temperatures (°K) of films are given inside the parentheses. (*Hilsch and Hilsch* [93])

for Cu. The results of Hauser et al. [104] on Cu/Pb and Pt/Pb films also predict a T_c ~6 millidegree for Cu. Based on the available data, the interaction for Pt is indicated to lie between −0.25 (repulsive) and a slightly positive (attractive) interaction, the latter corresponding to a T_c of ~2 millidegree for Pt. No verification of these predictions exists so far.

Strong ferromagnetism and superconductivity being apparently mutually exclusive, proximity effects with magnetic materials are of interest. P. and R. Hilsch [95] observed no special effects with overlays of antiferromagnetic Cr and Mn. Reif and Woolf [107, 108], on the other hand, found that only 15 Å of Fe on 500 Å of In was sufficient to eliminate the superconducting energy gap, and that 15 Å of Mn on 3,000 Å of Sn smeared out the density of states considerably. Hauser et al. [105] studied proximity effects in Pb films due to overlays of ferromagnetic Fe, Ni, and Gd, antiferromagnetic Cr, and dilute magnetic alloys 1 percent Fe in Mo, and 2.9 percent Gd in Pb. The proximity effect with a ferromagnetic film is very large compared with nonmagnetic material with the same coherence length. The results on the thickness dependence of T_c of Pb films with overlays of Pt, Cr, and Ni agree well with the de Gennes-Werthamer theory of the proximity effect together with the spin-scattering theory of Abrikosov and Gorkov.

Duffy and Meissner [99] studied the critical field of double films and established that their observation of the proximity effects could be explained by assuming a modified (effective) mean free path for the couple and a total film thickness which corresponds to that of the superconducting film only. This explanation is somewhat similar to that offered for the normal conductivity of superimposed films (see Chap. VI, Sec. 3.2).

If both metal films are superconductors of different T_c's, the T_c of the couple varies from one extreme to another as the relative thickness of the two films is varied [110,111]. Such a behavior is intuitively expected and is illustrated by the data on Pb/Al film couples in Fig. 6. The quantitative results are in good agreement with the de Gennes theory applicable to such a case.

It must be remarked in conclusion that the reasonable agreement between theories and experimental results of the proximity effect is somewhat surprising. This is so because the practical interface between the superimposed films is far from being a sharp discontinuity as is assumed in the theoretical treatments. The following significant factors

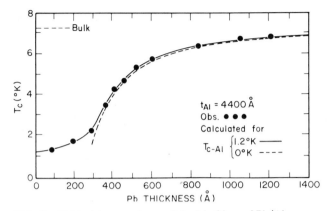

FIG. 6. Critical temperatures of double films of Pb/Al as a function of the thickness of Pb film. Aluminum was deposited at 673°K followed by Pb film at 85°K. The curves are calculated from a refined version of the deGennes theory for two assumed values of T_c for Al. (*Hauser and Theuerer* [110])

must clearly influence the results and should be taken into account in any interpretation of the data. (1) Owing to the kinetic energy of the impinging vapor atoms, penetration into the underlying film surface and consequent alloying [124-126] at the interface (see Chap. IV, Sec. 2.3) is unavoidable even when deposition is carried out at low temperatures. This is a particularly serious problem in sputtered films. Further, the structure of the interface will depend much on the deposition parameters, so that each interface is essentially unique. (2) The interface and hence surface scattering of conduction electrons in a thin metal film may be profoundly modified by a superimposed film of another material (Chap. VI, Sec. 3.2). The resulting changes (up to ~25 percent) in the normal electrical conductivity must influence the superconducting behavior of the couple. (3) Electrochemical cell reaction such as anodic oxidation at the interface can take place [105] when dissimilar metals form a couple. As discussed already, interface oxidation will set up a high transverse electric field which will change the T_c of the films.

4. CRITICAL MAGNETIC FIELD

Superconductivity is destroyed if the excess magnetostatic energy per unit volume $H^2/8\pi$ arising because of the exclusion (diamagnetic behavior) of the magnetic field H equals the free-energy difference between the superconducting and the normal states. This thermo-dynamical argument forms the basis of the method of theoretical calculation of the critical field H_c. The expressions for the critical field are derived in standard textbooks. The pertinent expressions will be presented in the following sections without derivation.

Let us, however, note on the basis of a physical reasoning that since an applied magnetic field is only very imperfectly excluded in a thin film, the excess magnetostatic energy per unit volume is smaller than that for a bulk superconductor. Hence the field causing a superconducting-to-normal transition in a film should be correspondingly larger than the bulk value. On the basis of this physical reasoning, the enhancement in the critical field is expected to be larger the smaller the t/λ ratio and/or the mean free path.

The superconducting-to-normal transition in films shows a variety of behavior. A sharp transition, where the order parameter (or the number of superconducting electrons) and the energy gap drop discontinuously to zero, is called a first-order transition. If, on the other hand, these parameters are reduced continuously to zero by the applied magnetic field, the transition is of second order. The former is characteristic of thick $(t \gg \lambda)$, annealed films of type I superconductors while the latter is a property of type II superconductors to which thin $(t < \sqrt{5}\lambda)$, dirty, or highly disordered and fine-grained films belong. We shall now examine the experimental and theoretical results on both types of films.

4.1 Type I Films

Both the London and the G-L theories yield [127-129] similar expressions (but with slightly different numerical constants) for the parallel critical field H_{cF} for films. The dependence of H_{cF} on film thickness, the mean free path [130], and the temperature [131] has been discussed thoroughly by Toxen on the basis of the G-L theory. Bardeen [129] obtained expressions for the H_{cF} which are applicable over a wide temperature range. The limiting cases of interest obtained from the G-L theory are: For thick films with $t \gg \lambda$,

$$H_{cF}(T) = H_{cB}(T)\left[1 + \frac{\lambda(T)}{t}\right] \tag{18}$$

where $H_{cB}(T)$ is the bulk critical field and $\lambda(T)$ is the effective penetration depth. Using Eqs. (1b) and (4b), Eq. (18) reduces at $T \rightarrow T_c$ to

$$H_{cF}(T) = H_{cB}(0) \left[2(1 - T_R) + \frac{\lambda(0)}{t}(1 - T_R)^{1/2} \right] \tag{19}$$

One can further rewrite Eq. (19) by replacing λ by $\lambda_B \sqrt{1 + \xi_0/l}$ for $T = 0$.

In the limit of thin films with $t < \sqrt{5}\lambda(T)$, the expression* for H_{cF} is given by

$$H_{cF}(T) \approx \sqrt{24} \frac{\lambda(T)}{t} H_{cB}(T) \tag{20}$$

For $T_R \to 1$,

$$H_{cF}(T) \approx \sqrt{24} H_{cB}(0) \frac{\lambda_B(0)}{t} \sqrt{1 - T_R} \sqrt{1 + \frac{\xi_0}{l}} \tag{21}$$

If the mfp is small and is limited by the film thickness, i.e., $l \sim t$ and $l \ll \xi_0$, then $\xi \sim l$ follows from Eq. (6). In this case, Eq. (21) reduces to

$$H_{cF}(T) = \sqrt{24} H_{cB}(0) \sqrt{1 - T_R} \left(\frac{\xi_0 \lambda_B^2(0)}{t^3} \right)^{1/2} \tag{22}$$

Note that the assumption $l = t$ is correct only for the wire geometry. For a thin film with perfect diffuse surface scattering of electrons, the effective value of the size-limited mfp for $t \ll l_B$ (l_B is the mfp of a bulklike thick film) given by $l_{eff} \sim \frac{3}{4} t \ln(l_B/t)$ [Chap. VI, Eq. (47)] should be substituted for ξ in Eq. (21).

The foregoing discussion suggests that a study of the thickness and temperature dependence of H_{cF} would enable us to verify the theory underlying Eqs. (19) and (20), and also to determine the values of $\lambda(0)$ and ξ_0.

Numerous studies of the H_{cF} of films of Hg [28, 29], In [74, 130-132], Pb [74, 133], Sn [74, 133-139], and Tl [137] have been made. The critical field is determined by methods such as by detecting the transition by resistance measurement, by measuring the penetration of the magnetic flux ballistically or by means of static forces, or by tunneling. Since H_{cF} varies inversely with film thickness, the thin edges

---

*The London theory gives $H_{cF} = H_{cB}[1 - (2\lambda/t) \tanh (t/2\lambda)]^{-1/2} \approx \sqrt{2}(\lambda/t) H_{cB}$. This expression is the same as Eq. (20) except for a factor of $\sqrt{12}$.

of a flat film have a much higher critical field than the thicker central region. As a consequence of the thickness gradient, the transition (both field- as well as current-induced) will be very broad. It is therefore essential to eliminate the edge effect either by mechanically removing the edges or by placing the current and voltage contacts in the central region of a relatively wide film.

The results of various investigations generally agree with the theoretically predicted $\sqrt{T_c - T}$ and $\lambda(T)/t$ dependencies of the H_{cF}. Typical results of the dependence of H_{cF} on $(T_c - T)$ for Sn films deposited on glass held at 4.2°K and subsequently annealed at room temperature are shown in Fig. 7. Thick films exhibit two regions, a bulklike region for large ΔT, and a thin-film-like region for $T \to T_c$. The former shows a $(T_c - T)$ dependence [Eq. (19)]. The expected $(T_c - T)^{1/2}$ dependence is observed for thinner films throughout the entire measured temperature range of about 2°K, and for thicker films sufficiently close to T_c where

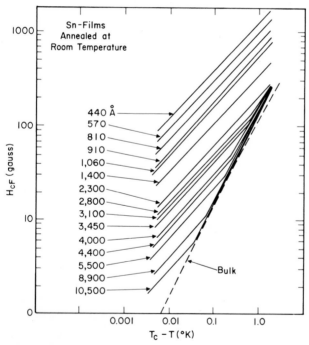

FIG. 7.   Critical magnetic field H_{cF} necessary to destroy superconductivity in annealed Sn films of various thicknesses as a function of the change in temperature from the critical-temperature value T_c at zero field. The superconducting transition was measured resistively. (*Zavaritskii* [137])

$\lambda \to \infty$. The observed thickness dependence for the films studied in Fig. 7 follows $H_{cF} \propto 1/t^{5/4}$. Similar results were reported by Douglass and Blumberg [138] on Sn films deposited on glass substrates at 77°K, and by Toxen [130] on In films. These authors, however, found a better fit of the data with a $1/t^{3/2}$ dependence [Eq. (22)] and interpreted it to imply the validity of the assumption $\xi \sim l \sim t$.

Figure 8 displays the thickness dependence of H_{cF} for Sn films as obtained from the Zavaritskii [136, 137] and the Douglass-Blumberg data for $T_R = 0.99$. By fitting the data to Eq. (19), values of $H_{cB}(0) = 304$ G, $\lambda_B(0) = 355$ Å, and $\xi_0 = 2,300$ Å are obtained. The Zavaritskii data appear to show a 30 percent disagreement with the theoretical curve and can be explained [13] by a systematic error in the film-thickness

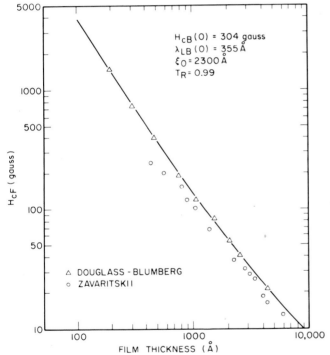

FIG. 8. Critical magnetic field $H_{cF} \equiv H_{c2}$ for Sn films as a function of thickness at a reduced temperature $T_R = 0.99$. Films were condensed on glass and annealed at room temperature. The solid curve is calculated from Eq. (19) with parameters given inside the figure. (*Data from Zavaritskii* [136] *and Douglass and Blumberg* [138], *compiled by Glover* [13])

data. The results on thin In films yield a $(T_c - T)^{1/2}$ dependence of H_{cF}/H_{cB} as $T_R \to 1$. Values of $\xi_0 = 2,600 \pm 400$ Å and $\lambda_L(0) = 350 \pm 30$ Å obtained from the data are in reasonable agreement with the bulk values determined by other techniques.

4.2 Type II Films

The behavior of type II superconductors has been discussed by Abrikosov [140], Goodman [141], and others [142-144], on the basis of the G-L theory. Assuming zero demagnetizing factor and magnetic field parallel to the film surface, the magnetic flux penetrates a thick sample in quantum units of $\phi_0 = hc/2e = 2 \times 10^{-7}$ G-cm^2 at a field (lower critical limit) H_{c1} which is smaller than the critical field H_{cB}. As the field is increased further, more flux penetrates until an upper critical field $H_{c2} \equiv H_{cF} = \sqrt{2}\kappa H_{cB}$ for volume superconductivity is approached, resulting in a second-order transition.

For thick films $(t \gg \lambda)$, the theory yields

$$H_{cF}(T) = \sqrt{2}\kappa H_{cB} = \sqrt{2}\kappa H(0)\left(1 - T_R^2\right) \tag{23}$$

Using the Gorkov relation $\kappa \approx \lambda/\xi_0$, and $\sqrt{8}\,\pi\lambda\xi_0 H_{cB} = \phi_0$, we get for the limit $T \to T_c$,

$$H_{cF}(T) \approx \frac{8\pi H_{cB}^2(0)}{\phi_0}\lambda_B^2(0)\left(1 + \frac{\xi_0}{l}\right)(1 - T_R) \tag{24}$$

Saint-James and de Gennes [142] have calculated universal curves for the critical field of thick and thin films as a function of t and ξ. They also showed that, as H_{cF} is approached, volume superconductivity is destroyed. However, surface superconductivity in a sheath parallel to the applied field continues to exist up to a critical field $H_{c3} = 1.69\,H_{c2} = 2.4\,\kappa H_{cB}$.

The parallel critical field for thin $(t \ll \lambda)$ films is given by $H_{cF}(T) = \sqrt{24}\,(\lambda/t)H_{cB}(T)$, which is the same as Eq. (20) for a thin film of type I superconductor. At $T_R \to 1$, this relation reduces to Eq. (21).

Thus, we expect near T_c, a $(T_c - T)$ dependence of the size-independent critical field for thick films of type II superconductivity.

Size-dependent enhancement of the critical field, however, occurs in thin films. The film thickness at which the changeover of the thick- to thin-film behavior takes place is obtained by comparing Eqs. (22) and (23). A simple relation for this critical thickness given by the numerical solution [143, 144] of the G-L equations is

$$t \approx \frac{4.5\lambda(T)}{\kappa} \qquad (25a)$$

Physically, this value corresponds to the distance between two Abrikosov flux vortices. One can further show that

$$t \approx \left(\frac{\phi_0}{H_{cF}}\right)^{1/2} \qquad (25b)$$

which implies that the changeover to thick-film behavior occurs when the Abrikosov vortices containing one quantum of flux ϕ_0 would just fit the cross section of the film.

Measurement of the critical fields of 100 to 200,000-Å-thick films of Nb and Ta prepared by getter sputtering, and of V_3 Ge prepared by pyrolysis were reported by Hauser and Theuerer [145]. The data on V_3Ge films obtained from the broad resistive transitions are shown in Fig. 9. A linear $(T - T_c)$ dependence is indicated in a limited temperature range. Although the data show size-dependent enhancement of the critical field, the enhancement occurs for film thicknesses which are considerably larger than given by Eq. (25a). Similar enhancements of critical field by a factor of about 10 have been observed in electron-beam evaporated films of Nb (6,000 Å thick) and Ta [21] (3,800 Å thick), W [37] (2,000 Å thick), and evaporated and sputtered films of Nb-Zr, Nb-Ti, Ti-V, V-Si, and Nb-Sn (~1,000 Å thick). These observations of the enhancement phenomenon may be explained if we assume that these films consist of discrete superconducting grains which limit the mfp so that the grain size should replace the film thickness t in Eq. (22). The grain size of the unannealed films of the materials cited being generally very small (~100 Å or less), a considerable enhancement of H_{cF} is possible. A discussion of the mechanism of enhancement of H_{cF} in granular superconductors is given by Abeles et al. [55].

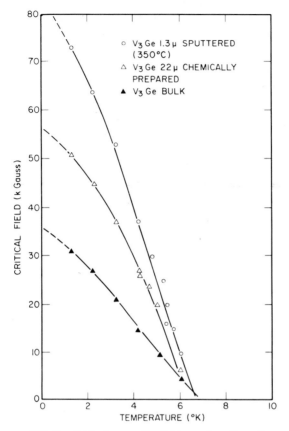

FIG. 9. Critical magnetic field as a function of temperature for bulk V_3Ge, for a 22-μ chloride-reduced film, and for a 1.3-μ sputtered film. (*Hauser and Theuerer* [145])

4.3 Mean-free-path Dependence of H_{cF}

The mean-free-path dependence of the Ginzburg-Landau parameter κ is given by

$$\kappa = \kappa_0 + \frac{0.73\lambda_B(0)}{l} \tag{26}$$

where κ_0 is the value for a pure $(l \to \infty)$ material. For $t \gg \lambda$, the parallel (or transverse) critical field H_{cF} [Eq. (23)] may be rewritten as

$$H_{cF} = H_{cB}\sqrt{2}\left[\kappa_0 + \frac{0.73\lambda_B(0)}{l}\right] \tag{27}$$

For $t \gg l$, ρl_{eff} = constant [Chap. VI, Eq. (22)], then

$$H_{cF} = H_{cB}\sqrt{2}(\kappa_0 + A\lambda_B\rho) \tag{28}$$

where A is a constant and ρ is the normal-state resistivity. An alternative approximate expression for the transverse critical field due to Goodman [141] is

$$H_{cF} = H_{cB}\left(\sqrt{2}\kappa_0 + 1.06 \times 10^4 \beta^{1/2}\rho\right) \tag{29}$$

where β is the coefficient of normal-state electronic specific heat per unit volume in ergs/$^\circ K^2$-cm^3.

A determination of κ_0 can therefore be made by measuring the variation of the critical field with the mfp. The mfp can be either increased by gradual annealing out of the structural defects in a film condensed at low temperatures, or decreased by adding impurities. A linear dependence of H_{cF} on the resistivity ($\rho \propto 1/l$) was observed by Glover [146] on Sn films condensed at 4.2°K and subsequently annealed at successively higher temperatures. The results are shown in Fig. 10. A value of $\kappa_0 = 0.158$ for Sn films was deduced from the data. Similar results were obtained by Toxen [130], and by Chang and Serin [147] on Sn-In alloy films. The latter authors found the value of κ_0 to increase rapidly to 0.7 with a few percent of Sn in In.

In the size-effect regime ($t \ll l$), the effective mfp is related to the film thickness [Chap VI, Eq. (47)]. This relationship may also be utilized to determine κ_0.

4.4 Transverse Critical Field

So far we have assumed a zero demagnetizing factor which is true for an infinitely long cylinder in a parallel magnetic field. For any other geometry, flux penetration should occur at fields determined by the demagnetizing factor. For example, the demagnetizing factor for a flat circular film in a transverse field is $F \sim (1 - t/w)$, where w is the diameter of the circular film. Thus, an applied field H_{cF} produces an

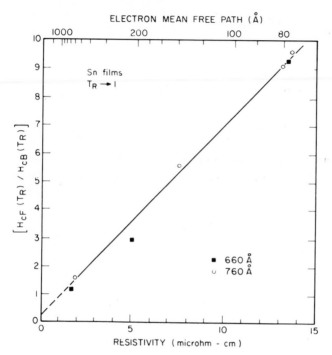

FIG. 10. Dependence of the critical magnetic field of Sn films near T_c relative to that of well-annealed bulk material as a function of the normal-state resistivity and electron mean free path. Films were condensed on crystalline-quartz substrates at 4.2°K and subsequently annealed at successively higher temperatures to vary the resistivity. (*Glover* [146])

internal field $H_c \sim H_{cF}/1 - F = H_{cF}(t/w)$, so that flux penetration occurs at fields much smaller than H_{cF}.

The behavior of a thin film in a transverse magnetic field has been discussed by Tinkham [143]. A detailed solution of the G-L equations is given by Maki [148]. Tinkham showed that for *both* thin ($t < \lambda$) and thick ($t > \lambda$) films, the critical transverse field is given by

$$H_{cF}\perp = \sqrt{2}\,\kappa H_{cB}$$

which is the same as Eq. (23). *This result implies that in a transverse field all films behave as type II superconductors.*

The resistive transition of Sn films in transverse fields was studied extensively by Rhoderick and by Broom and Rhoderick [149].

Relatively sharp transitions with no hysteresis but marked dependence on the measuring current were observed. Penetration of the field occurred at fields far lower than those at which resistance appeared. The penetration of a transverse field through Pb films was investigated by DeSorbo and Newhouse [150] using the magneto-optic technique for detection of the trapped flux. Domain structure similar to that in bulk type II superconductors was observed in these films.

Abeles et al. [55] studied the transverse critical field of granular superconductors obtained by vacuum evaporation of films of Al, Sn, and Ga in the presence of oxygen partial pressure. Aluminum films of 300 to 800 Å thickness with an average grain size of ~40 Å yielded good agreement with $\left(1 - T_R^2\right)$ dependence of $H_{cF}\perp$ over an appreciable temperature range. The critical field increased with increasing resistance of the films. These results were interpreted in terms of the limitation of the mfp by the individual grains.

For orientations other than parallel or transverse, the critical field of a thin film [143, 151] ($t \ll \xi$) depends on the angle of orientation θ of its plane with respect to the applied field. The angular variation of H_{cF} for a thin film shows a second-order transition and is described by

$$\left(\frac{H_{cF}\cos\theta}{H_{cF}\|}\right)^2 + \frac{H_{cF}\sin\theta}{H_{cF}\perp} = 1 \tag{30}$$

where $\|$ and $\perp$ signs denote parallel ($\theta = 0$) and perpendicular ($\theta = \pi/2$) orientation. The ratio $H_{cF}\|/H_{cF}\perp$ decreases from infinity for thin films to a constant value 1.69 when t/ξ increases to infinity. Equation (30) is in excellent agreement with experimental results of various investigators [152].

Note that we can invert the various critical field expressions and obtain

$$\xi = \left(\frac{\phi_0}{2\pi H_{cF}\perp}\right)^{1/2} \tag{31}$$

$$t = \left(\frac{6\phi_0 H_{cF}\perp}{\pi H_{cF}^2\|}\right)^{1/2} \tag{32}$$

$$\lambda = \left(\frac{H_{cF} l \, \phi_0}{4\pi H_{cB}^2} \right)^{1/2} \tag{33}$$

These relations show that the critical-field data can be used to determine t, ξ, and l for thin films.

5. CRITICAL CURRENT

If current carried by a superconducting film exceeds a critical value, a transition to the normal state is induced. The critical current can be conveniently calculated theoretically only for geometries in which the current is distributed uniformly over the whole film so that edge effects are negligible. For a thick cylindrical film ($t \gg \lambda$), the current flows in an outer region of approximate depth 3λ. The critical current in this case is given by Silsbee's rule, and is that which produces a surface field equal to H_{cF} outside the cylinder. On the other hand, if the film is thin ($t \ll \lambda$), the current is distributed approximately uniformly throughout the film thickness and the surface field H_I corresponding to the critical current density J_c is smaller than H_{cF}. For a cylindrical film, or a film deposited on a shield plane (Sec. 9.1), H_I is given by [11]

$$H_I = \frac{4\pi J_c t}{c} \tag{34}$$

where c is the velocity of light. The expression for J_c is easily obtained by equating the kinetic energy of the critical current in a thin film with the difference $H_{cB}^2/8\pi$ in the free energies of the normal and supercon-ducting states. Thus, one obtains

$$J_c(T) = \frac{c H_{cB}(T)}{4\pi\lambda(T)} \tag{35}$$

Substituting for $\lambda(T)$ from Eq. (7a) we have the nonlocal value

$$J_c(T) \approx \frac{c H_{cB}}{4\pi\lambda_B} \left(\frac{\xi}{\xi_0} \right)^{1/2} \tag{36}$$

A more exact calculation by Bardeen [129], which takes into account the slight variation of the energy gap with current, modifies this result by a multiplying factor $(2/3)^{3/2}$. One can rewrite Eq. (35) as

$$\frac{J_c(T)}{J_c(0)} = \left(1 - T_R^2\right)^{3/2}\left(1 + T_R^2\right)^{1/2} \tag{37}$$

Near T_c, H_{cB} varies as $(T_c - T)$ and λ_B varies as $(T_c - T)^{-1/2}$, so that $J_c(T)$ varies as $(T_c - T)^{3/2}$.

Note that an expression similar to Eq. (37) is obtained from the G-L theory for type I material which in the limit $t \gg \lambda(T)$ and $T \to T_c$ is given by

$$J_c(T) = \frac{\sqrt{2}c \, H_{cB}(T)}{6\sqrt{3} \, \lambda_B(T)} \approx \frac{2c}{\sqrt{27}} \frac{H_{cB}(0) \, (1 - T_R)^{3/2}}{\lambda_B(0) \, (1 + \xi_0/l)^{1/2}} \tag{38}$$

The dependence of J_c on film thickness does not appear explicitly, but it enters through the mfp, so that J_c decreases slowly with decreasing film thickness. The ratio $J_c(T)/J_c(0)$ is, however, *independent* of ξ_0 and l [Eq. (37)].

Experimentally, the critical current is measured by passing current from an external source or by inducing it in a ring-shaped sample with the aid of a magnetic field. The current distribution over the whole film must be uniform since an intense magnetic field created near the film edges has a complex effect on the transition. The two examples of film geometries that possess uniform current distribution are: (1) a film deposited on a cylinder much longer than its radius, which itself is much larger than the film thickness, and (2) a flat film adjacent and parallel to a much larger insulated superconducting plane of higher critical current and temperature (shielded plane).

In the case of the commonly used geometry of a flat film, it can be shown [153] that the surface field and the current density vary from point to point of the surface, and that if the film is thin compared with the penetration depth, fields normal to the surface are present. Consequently, the measured critical-current density of a flat film must be much less than that calculated by assuming a uniform current distribution. This results from the fact that the requirement that the surface field generated by the current be parallel to the film surface leads to very uneven current distribution which approaches infinity at the edges.

The critical-current (dc) density of Sn films has been extensively studied [154-158]. A $\left(1 - T_R^2\right)^{3/2}$ dependence is observed near T_c. The measured current density is, however, at least an order of magnitude smaller than that predicted by Eq. (37). For Sn films, assuming $H_{cB}(0) = 304$ G, and $\lambda_B(0) = 355$ Å, $J_c(0) \sim 7 \times 10^7$ A/cm^2 is predicted. The observed values for Sn films a few thousand angstroms thick typically give values of 2×10^6 A/cm^2 at temperatures well below T_c. Edwards and Newhouse [159] observed an enhancement of this value by a factor of 2.5 by using a shielded plane geometry. The ideal long cylindrical film [154, 155] geometry also yields comparable values of $J_c(0)$. The measurements on flat as well as ring-shaped films are expected to be complicated by the nonuniform distribution of current unless the ring diameter is very large relative to the film width. The critical current in ring-shaped films has been found [156, 157] to vary as $\left(1 - T_R^2\right)^{3/2}$. The magnitudes are, however, considerably smaller than predicted.

By assuming a certain distribution along the width, correction for the nonuniform current distribution in planar films was made by Glover and Coffey [158]. Thus, they obtained a good agreement of their data on Sn films with the theory. These results yielded $J_c(0) \sim 2 \times 10^7$ A/cm^2, which agrees in order of magnitude with the theoretically predicted value.

If small normal regions are present, the Joule heating due to the high current density would lead to an abrupt transition [160] to the normal state at current densities substantially lower than predicted theoretically. Such a transition is irreversible and exhibits hysteresis (Fig. 11). To avoid Joule heating, it is essential to make pulse measurements. The transition measured in this manner is found to be very broad (Fig. 11). Pulse measurements have been performed by a number of workers [160-168]. Hagedorn [163] measured the J_c of a 1,700-Å cylindrical Sn film by using current pulses of 1 nsec width and 0.1 nsec rise time. He claims to have observed a bulklike Silsbee transition. Gittleman and Bozowski [162] observed a finite relaxation time associated with the first onset of resistance. Both investigations placed the intrinsic superconducting-to-normal switching times at less than 10^{-10} sec.

Extensive dc and pulse measurements of J_c have been reported by Mydosh and Meissner [165] on planar and cylindrical films of Sn of 650 to 2,000 Å thickness. Both the magnitude and the temperature

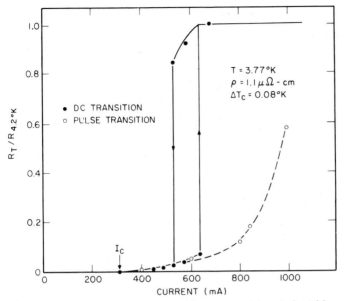

FIG. 11. Superconducting-to-normal-state transitions induced by dc and pulsed currents in a 3,000-Å-thick 4-mm-wide Sn film deposited on single-crystal sapphire. (*Bremer and Newhouse* [160])

dependence of J_c obtained by dc measurements are in close agreement with Bardeen's calculations. The observed three-halves-power dependence on temperature is illustrated in Fig. 12. The pulse data deviate from theory and from the dc data for $T_R < 0.9$. The discrepancy appears to be a characteristic of fast-pulsed critical-current-measurement technique. Its origin may have to be ascribed to a new mechanism rather than Joule heating. The critical-current density at $T = 0$ obtained by extrapolation is 6.1×10^6 A/cm² from dc, and 8.2×10^6 A/cm² from pulse data as compared with the theoretical value of 1 to 2×10^7 A/cm².

These authors have also studied the critical-current density in the presence of an external magnetic field H. The data on a 940-Å cylindrical Sn film are shown in Fig. 13 and are compared with the solid curves computed from the theoretical expression derived from the G-L theory

$$J_c(T) = 0.544 \frac{c}{4\pi} \frac{H_{cB}(0)}{\lambda(0)} \left(1 + T_R^2\right)^{1/2} \left[1 - T_R^2 - \frac{H^2 t^2 \left(1 + T_R^2\right)}{24 H_{cB}^2(0) \lambda^2(0)}\right]^{3/2}$$

(39)

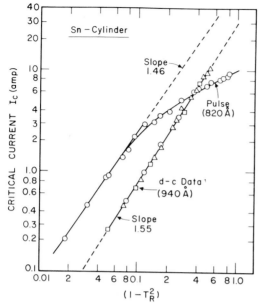

FIG. 12. Temperature $(T_R \equiv T/T_c)$ dependence of the dc and pulsed (80 nsec) critical current of cylindrical Sn films deposited on 2-mm o.d. glass tubes. (*Mydosh and Meissner* [165])

The constant is adjusted to match the observed $J_c(T)$ for $H = 0$. A reasonable agreement is seen with only slight departures from the theoretical curve which may be attributed to local variations in the film thickness.

A temperature dependence of the critical-current density which varies with the alloy concentration has been reported by Meiklejohn [167] in In-Sn alloy films with the compensated geometry arrangement. On the other hand, Burton [168] observed the expected three-halves dependence in In-Bi and Sn-Bi alloy films.

The foregoing discussion was confined to type I superconductors. Unfortunately, no satisfactory theory exists for the critical-current density of type II superconductors since the critical current is limited by the occurrence of a complicated mixed state consisting of flux threads or vortices. Marked degradation effects due to the Lorentz forces are characteristic of such materials. These materials are, however, of technical interest. Some pertinent experimental results are given below.

A $\left(1 - T_R^2\right)^{3/2}$ dependence of the critical current was observed by London and Clark [169] in an evaporated 1,050-Å Nb film. Meiklejohn

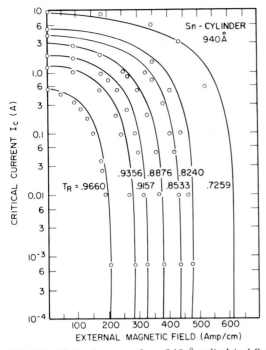

FIG. 13. Critical current for a 940-Å cylindrical Sn film at various temperatures as a function of the external longitudinal magnetic field. (*Mydosh and Meissner* [165])

[167] obtained 2.3 instead of the three-halves power for a 1,000-Å Ta film. Fowler [21] reported $J_c \sim 10^6$ A/cm^2 for an 800-Å Nb film at 4.2°K with and without an applied field. At 30 kG, the value of J_c reduces to $\sim 10^4$ A/cm^2. Edgecumbe et al. [27] made extensive investigations of the magnetic-field dependence of sputtered as well as vapor-deposited 1,000-Å films of Nb-Zr, Nb-Ti, Ti-V, V-Si , Nb-Sn, and Nb systems. The characteristics for Nb-Sn and V-Si systems shown in Fig. 14 are similar to those of the corresponding bulk superconductors. At an applied field of 135 kG, $J_c \sim 3 \times 10^5$ A/cm^2 is obtained at 4.2°K in Nb-Sn films (T_c = 15°K). The high current-carrying capabilities of thin-film hard superconductors in high magnetic fields are of much interest for compact high-field solenoids (Sec. 8.3) and high-Q high-inductance coils.

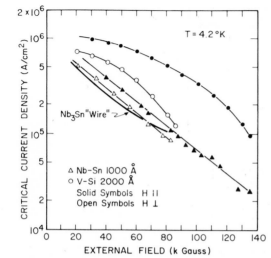

FIG. 14. Critical-current density vs. magnetic field (longitudinal and transverse) at 4.2°K for Nb-Sn (T_c = 15°K) and V-Si (T_c = 13.5°K) vapor-deposited and homogenized planar thin films of width 2.5 mm on glass substrates. Measurements were made by keeping the field constant and applying a short current pulse and observing the voltage across the specimen. Data for bulk Nb_3Sn wire are shown for comparison. (*Edgecumbe et al.* [27])

6. PROPAGATION OF NORMAL AND SUPERCONDUCTING PHASES

The propagation of normal phase into a superconducting slab under the influence of a magnetic field has been discussed by Pippard [170] on the basis that as the normal phase moves forward, the magnetic field penetrating the material sets up eddy currents which impede further progress of the boundary between the two phases. According to this model, the characteristic time constant τ for propagation of the superconducting-normal phase boundary is given by

$$\tau = \frac{2\pi x^2 H_c}{\rho \, \Delta H} \tag{40}$$

where x is the distance the boundary moves in time τ, ρ is the normal-state resistivity of the material, and ΔH is the magnetic field in excess of the critical field H_c. For typical values of $x \sim 6{,}000$ Å, $\rho \sim 1.5 \times 10^{-6}\,\Omega$-cm, and $\Delta H/H_c$ of a few percent, the time constant is about 1×10^{-9} sec.

A detailed discussion of the subject of nucleation and propagation of a normal phase in a bulk material is given by Faber and Pippard [171].

A current-induced transition from superconducting to normal phase is very sensitive to the existence of any normal nuclei in the specimen. The Joule heat generated at the normal region causes its further growth by heat conduction. This phenomenon, called "thermal propagation," takes place at velocities of up to 10^6 cm/sec.

Experimentally, the velocity is measured by passing a steady or broad pulse current below the critical value through the film. The normal region is initiated by passing a narrow nucleating pulse. The time variation of the voltage across the film is measured with an oscilloscope, and thus the velocity is determined.

The velocity of thermal propagation clearly depends on the Joule heating, the thermal properties of the substrates, and the efficiency of heat transfer from the film to the substrate. The velocity has been calculated for highly simplified models. The Broom and Rhoderick theory [172] predicts a positive, zero, or negative velocity depending on the magnitude of the current density and the ambient temperature. Although the qualitative features of the theory have been verified [172-174], the measured velocities in 1-μ-thick Sn films are at least two orders of magnitude smaller than predicted.

7. SUPERCONDUCTIVE TUNNELING

The transport of electrons between two metal electrodes separated by a thin (~ 10 to 50 Å) insulator by means of a quantum-mechanical tunneling process is discussed in Chap. VIII. The current-voltage (I-V) relation of a metal-insulator-metal structure (called a sandwich structure or a tunnel junction) is ohmic at low applied bias and nonlinear at high bias. The resistance of such a junction is nearly independent of temperature and increases exponentially with the insulator film thickness. In studying tunneling through such a junction, Giaever [175] discovered that if one of the metals of the junction became superconducting, the I-V characteristics became nonlinear. Nicol et al. [176], and later Giaever [177], observed that a negative-resistance region developed in the *I-V* characteristics when both metal electrodes became superconducting. Typical *I-V* characteristics of an Al-Al_2O_3-In junction are shown in Fig. 15.

These *I-V* characteristics can be understood on the basis of a one-particle approximation of a BCS superconductor with its characteristic energy gap and the density-of-states function. The density-of-state functions of junctions with various combinations of normal and superconducting metal electrodes form semiconductor-type energy-band diagrams* as shown in Fig. 16. The expected *I-V* characteristics are also indicated in the same figure. A negative voltage applied to an electrode raises its Fermi level with respect to the other electrode, and thus current

---

*It may be noted that the BCS ground state contains admixtures of one-electron states from above as well as below the Fermi energy. That is, the states in a superconductor are filled somewhat like a Fermi-Dirac distribution for some finite temperature. The one-particle states in the BCS formulation are of course occupied in pairs.

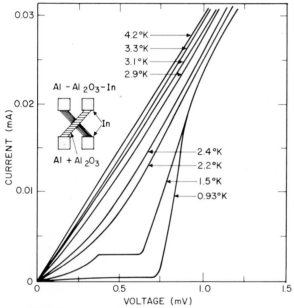

FIG. 15. Current-voltage characteristics of an Al-Al$_2$O$_3$-In crossed-strip thin-film tunnel structure at various temperatures. (*Giaever and Megerle* [178])

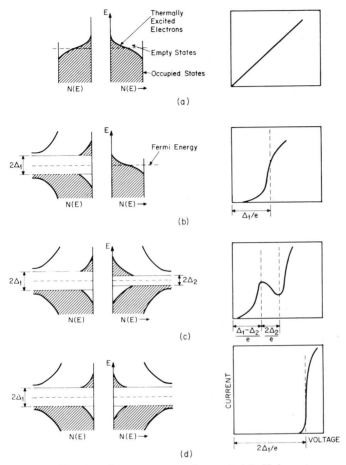

FIG. 16.   Density-of-states curves and expected I–V characteristics of various tunneling junctions near absolute zero. (*a*) Both
metals normal. (*b*) One metal superconducting, one metal normal.
(*c*) Both metals superconducting. (*d*) Two identical superconducting metals. (*Giaever and Megerle* [178])

flows. The tunnel current is proportional to the product of the density
of electrons at a particular energy in the emitter metal and the density of
unoccupied states at the same energy level in the absorber metal.
Assuming the transition probability is proportional to the density of
states in a superconductor, the tunnel current from electrode 1 to 2 is
given by

$$I_{1-2} = C_{n_1 n_2} \int_{-\infty}^{\infty} \rho_1(E)\,\rho_2(E + eV)\,[f(E) - f(E + eV)]\,dE \qquad (41)$$

where $C_{n_1 n_2}$ is the conductance when both metals are normal, $f(E)$ the Fermi function, E the energy measured from the Fermi level, V the applied voltage, and ρ_1 and ρ_2 are the ratios between the superconducting and normal densities of states in the two metal films. The density of states of a superconductor in the BCS theory is given by

$$N_s = N_n \frac{E}{(E^2 - \Delta^2)^{1/2}} \tag{42}$$

where 2Δ is the energy gap and subscripts n and s refer to the normal and superconducting states of the electrodes, respectively. Giaever and Megerle [178] have discussed the evaluation of the integral in Eq. (41). If one of the electrodes is normal, then at $T = 0$, the relative conductance (dI/dV) obtained from Eq. (41) is directly proportional to the relative density of states in the superconducting electrode.

If the density of states in the two metals is constant over the applied voltage range, ohmic behavior is expected [Fig. 16(a)]. If the Fermi level of n (normal) faces the forbidden energy region of s [Fig. 16(b)], a potential across the junction must be raised or lowered by Δ_1, to allow significant tunneling current to flow; i.e., dI/dV will have a maximum at $V = \Delta_1$.

If both metals are superconducting with energy gaps $2\Delta_1$ and $2\Delta_2$ [Fig. 16(c)], the tunnel current can flow to the empty states of the absorber only for $V = (\Delta_1 - \Delta_2)$. With increasing potential, the tunnel current, however, decreases since the density of available states decreases with increasing energy above the gap. As the voltage is further increased to $V = (\Delta_1 + \Delta_2)$ so that the main body of unexcited electrons in one electrode is brought opposite the empty states of the other, the current starts increasing rapidly. If $\Delta_1 = \Delta_2$ [Fig. 16(d)], the initial rise of the current is observed only when $V = 2\Delta_1$.

7.1 Experimental Results

(1) General. It is clear from the preceding discussion that the shape of the tunneling curves provides an accurate, unambiguous, and remarkably easy method to determine the energy gap of a superconductor and the variation in the density of electron states.

Experimentally, a tunnel junction is usually formed by depositing a thin metal film, oxidizing it in air for several minutes, and then superimposing another thin film of a metal in a typical capacitor configuration. Instead of an oxide film, thin insulator films of other materials such as evaporated Al_2O_3, SiO_2, and barium stearate may be used. A typical junction is ~ 1 mm^2 in area, and its resistance may be a few ohms. The dc I-V characteristics are obtained directly on an x-y recorder using a suitable constant-current source. The kinks in the I-V characteristic are easily discernible in the dynamic conductance dI/dV. The ratio $(dI/dV)_{ns}/(dI/dV)_{nn}$ as a function of the voltage measures the relative change in the density of states of the superconductor. The dynamic resistance is measured by superimposing a small alternating current on the dc bias current and then measuring the ac voltage developed across the junction. Circuit details of such a resistance plotter are given by Giaever et al. [178, 179]. Circuits to obtain the second derivative of tunnel currents are given in the literature [180].

Since the exciting discovery of superconductive tunneling, this subject has received extensive experimental and theoretical attention and has been reviewed by Douglass and Falicov [7], and by Schmid [181]. Giaever et al. [177-179] and other workers [182-188] used this technique to measure the energy gap Δ and density of states of various superconducting metals. The values of Δ for some typical materials listed in Table I are derived from such tunneling studies and are in good agreement with those obtained by other techniques and with the predictions of the BCS theory. The temperature dependence of Δ, shown in Fig. 17, is in good accord with Eq. (13) of the BCS theory.

Giaever and Megerle [178] observed that the energy gap of a thin Al film decreased smoothly as the magnetic field applied parallel to the film increased to its critical value H_{cF}. Douglass [182] interpreted this behavior in terms of the G-L theory using Gorkov's result that the order parameter of the G-L theory is proportional to Δ. If the field is present on both sides of the film, this theory predicts a continuous decrease of Δ to zero as $H \to H_{cF}$ (i.e., a second-order phase transition) for films with $t < \sqrt{5}\lambda$. When $t > \sqrt{5}\lambda$, Δ decreases slowly with increasing field and then drops discontinuously to zero (first-order transition) as $H \to H_{cF}$. Douglass calculated the field dependence of Δ numerically. In the limiting case $t \ll \lambda$, it is given by

$$\frac{\Delta(H)}{\Delta(0)} = \left[1 - \left(\frac{H}{H_{cF}}\right)^2\right]^{1/2} \tag{43}$$

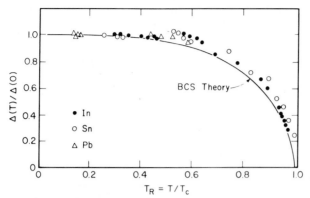

FIG. 17.   Reduced-energy gap of Pb, Sn, and In films as a function of the reduced temperature. Values were determined from tunneling characteristics of sandwich structures. The solid curve is derived from the BCS theory. (*Giaever and Megerle* [178])

If the external field is present on only one side of the film, the critical gap $\Delta(H_c)$, the value of the gap just prior to the abrupt drop as $H \rightarrow H_{cF}$, remains finite at all thicknesses (i.e., the phase transition is always of first order).

Collier and Kamper [188] obtained one-dimensional numerical solutions of the order parameter in the G-L equation for the parallel field. The application of the BCS theory also yields a similar field dependence of Δ. For $T \rightarrow 0$, Mathur et al. [189] obtained a dependence given by

$$\frac{\Delta(H)}{\Delta(0)} = 1 - \left(\frac{H}{H_{cF}}\right)^2 \tag{44}$$

Bardeen [129] predicted that thin films which at higher temperatures display a second-order transition should exhibit a first-order transition at low temperatures. However, no experimental verification of this prediction exists so far.

The temperature, thickness, and magnetic field (parallel) dependence of the energy gap of Al in Al-Al$_2$O$_3$-Pb tunnel junctions was thoroughly investigated by Douglass [182, 183] and Douglass and Meservey [184, 185]. For the same film thickness, the ratio t/λ can be changed by varying the temperature. The field variation of Δ for different values of

t/λ is shown in Fig. 18(a) to illustrate the gradual change in the order of the transition. For $t < \lambda$, the observed values of Δ are too low for all values of the field up to H_{cF}. The critical field is also significantly lower than predicted. The discrepancy may be due to the assumed position-independent energy gap in a thin film. Nevertheless, the t/λ dependence of the critical gap [Fig. 18(b)] provides a striking verification of the predicted behavior for $t > \sqrt{5}\lambda$. The departure of the results from the theoretical curve for high values of t/λ is probably due to the nonlocal effects in Al becoming important in the $t/\lambda > 2.8$ range.

The experimental results of Collier and Kamper [188] on the field variation of the energy gap of clean Sn films are consistent with the G-L theory for the parallel field. The results for the transverse field can be understood in terms of the Abrikosov-Tinkham vortex model.

The changes in the thermal conductivity of superconducting Sn, In, and Pb films upon application of a magnetic field in the plane of the film measured by Morris and Tinkham [190] provide an independent and striking verification of the theory of field dependence of Δ. At $T_R = 0.65$, the results agree with Eq. (43), whereas at $T_R = 0.35$, a more satisfactory fit is obtained with Eq. (44).

The relative conductance $(dI/dV)_{ns}/(dI/dV)_{nn}$ is a measure of the density-of-states function. An example of this function for Hg obtained from $Al\text{-}Al_2O_3\text{-}Hg$ tunnel function is shown in Fig. 19(a). A comparison with the theoretical curve calculated from the BCS density-of-states function shows good qualitative agreement.

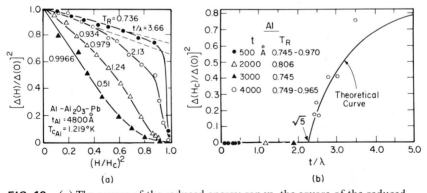

FIG. 18. (a) The square of the reduced-energy gap vs. the square of the reduced magnetic field for Al film data obtained from tunneling characteristics of $Al\text{-}Al_2O_3\text{-}$ Pb junction. (*Meservey and Douglass* [185]). (b) The square of the reduced-energy gap at the critical field as a function of the film thickness normalized by the penetration depth. Theoretical curve is shown for comparison. (*Douglass* [182])

Any structure in the energy spectrum of a superconductor should be reflected in the conductance curve. Bermon and Ginsberg [187] observed a structure in the conductance curve of Fig. 19(*a*) at energies beyond the gap. This structure becomes markedly evident as negative peaks in a d^2I/dV^2 plot as shown in Fig. 19(*b*). These peaks have not been definitely identified as yet but are thought to be due to the structure in the Hg phonon spectrum. Rowell et al. [191] also observed a similar fine structure in Pb and attributed the set of peaks around 4.5 and 8.5 mV in the d^2I/dV^2 plot to the transverse and longitudinal phonons, respectively. Scalapino and Anderson [192] discussed theoretically the possible relationship of the singularities in d^2I/dV^2 plots to singularities* in the phonon spectrum of the superconductor. A correlation between the two is indicated and thus such studies represent a powerful tool for phonon spectroscopy.

Temperature-independent as well as temperature-dependent currents in excess of those expected theoretically have been reported by Taylor and Burstein [193] in Pb-oxide-Pb or Al tunnel junctions for voltages (either polarity) less than the sum of the two energy gaps of the electrodes. The temperature-independent contribution has been ascribed by Schrieffer and Wilkins [194] to the occurrence of double-particle tunneling. The temperature-dependent part may be ascribed to phonon-assisted tunneling processes. More careful experimental results are required to establish these mechanisms.

(2) Tunneling in Superimposed Films. Tunneling studies using a superimposed metal-film couple as one of the electrodes of a tunnel junction have been utilized to verify the existence of the superconductive proximity effects. Such studies should obviously also yield quantitative information on the energy gap induced by the proximity effect and thus give an insight into its origin.

A number of studies on tunnel junctions of the type Al-Al$_2$O$_3$-M-Pb (where M stands for the normal metal under investigation) are reported in the literature [101-104, 107]. The energy gap induced into M = Al, Cd, or Pt is found to be a function of its film thickness. Its value approaches $2\Delta_{Pb}$ for sufficiently thin M films and decreases rapidly with increasing film thickness. The quantitative variation is in agreement with

---

*The density of states (or distribution) functions for actual periodic lattices have singularities known as van Hove singularities. These arise from critical points where the group velocity is zero.

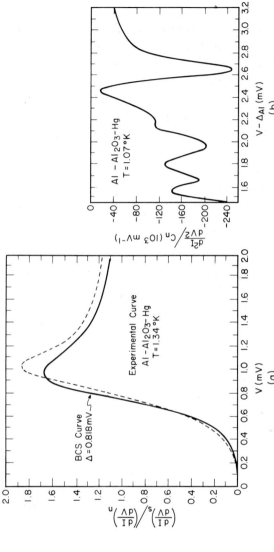

FIG. 19. (a) Comparison of the experimental (dotted) and theoretical (solid) relative differential conductance curves of Al-Al$_2$O$_3$-Hg tunnel junction plotted against bias voltage. The theoretical curve was calculated using the BCS density-of-states function with a value of Δ_{Hg} determined from $I-V$ characteristics. (b) The second derivative of the tunnel current normalized by C_n (C_n = average value of the normal-state conductance) as a function of applied bias voltage minus Δ_{Al}. The energy positions of the extrema in the curve are given by subtracting Δ_{Hg} (1.07°K) = 0.822 mV. (*Bermon and Ginsberg* [187])

581

the theoretical calculations of McMillan [123] for pure films ($t < \xi$) where scattering is assumed to take place at the boundary of the M-Pb couple as a result of the mismatch of the density of states of the two adjoining metals. The experimental results, however, disagree with the predictions of the Saint-James theory [122] according to which a gap can exist only if M possesses an attractive electron-electron interaction. Further, contrary to the observations, this theory asserts that, if M is also a superconductor, the effective value of the gap of the couple will equal that of the low-gap superconductor.

(3) Gapless Superconductivity. When one of the superimposed metals is magnetic, gapless (i.e., no gap between the excitation spectrum of the superconductor) superconductivity and a smeared density of states occur. Gapless superconductivity has also been observed in nonmagnetic couples, such as Pb-Pt. Hauser [106] studied the degree of gapless superconductivity, i.e., the density of states in the Pb gap, in Pb-Ni and Pb-Pt couple tunnel junctions, by measuring the voltage dependence of the dynamic resistance for various thicknesses of the Pb films. He found that the degree of gapless superconductivity is increased by applying a longitudinal magnetic field. His results are similar to those observed in bulk type II superconductors and are in agreement with the theoretical predictions of Maki [120]. These studies also demonstrated that a ferromagnetic Ni film introduces more states into the Pb gap than a nonmagnetic Pt film.

7.2 Supercurrent (Josephson) Tunneling

The tunneling process discussed in the preceding section involves the incoherent transfer of single electrons through the barrier. Josephson [195, 196] predicted that if the tunneling barrier between two superconductors is very thin (~10 Å) and the magnetic field in the barrier is small (a few gauss or less), a flow of "Cooper pairs" of electrons should take place. This supercurrent, or Josephson tunneling of pairs, is characterized by the absence of any potential drop across the tunnel junction for currents less than a certain maximum I_J [Fig. 20(a)].

The barrier in Josephson tunneling behaves as a weak superconducting link between the two superconductors. The phases of the superconducting wave function become locked, and in effect the two superconductors become a single coherent quantum system. An energy ($hI_J/2e$) is associated with the coupling of the pair wave functions. The system is stable against thermal fluctuations if the barrier is very thin

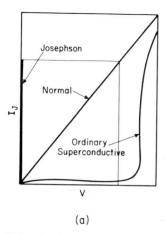
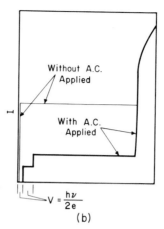

(a)

(b)

FIG. 20. Schematic representation of (*a*) normal, superconductive, and dc Josephson tunneling *I–V* characteristics of a symmetrical tunnel junction; (*b*) *I–V* characteristics of a Josephson junction across which an ac voltage of frequency ν is applied.

(~10 Å). Theoretical discussion of the effect is given by Josephson [196] and Anderson [197].

The predicted supercurrent tunneling was first observed by Anderson and Rowell [198]. The observed temperature dependence agreed with Josephson's prediction. The most striking characteristic of Josephson tunneling is the sensitivity of the supercurrent to small magnetic fields (the earth's field is sufficient to decrease I_J by a factor of 100 in some junctions), which is undoubtedly one of the reasons why the effect remained obscure so long.

The longitudinal magnetic-field dependence of the Josephson current is found [199-202] to be periodic in *H* (Fig. 21). The observed $\sin(\pi H/H_0)/(\pi H/H_0)$ dependence is expected since the field dependence of the critical current is given by the Fraunhofer diffraction-pattern formula for an optical aperture of the same shape as the tunnel barrier. This periodic behavior strongly indicates that the currents observed are indeed tunneling through the barrier and not going through isolated metallic bridges. The minimal current values are obtained whenever the junction contains integral numbers of flux units. The magnetic field interval H_0 between minima is $H_0 = \phi_0/A$, where *A* is the area through which flux threads and ϕ_0 is the fundamental flux quantum [Eq. (15*a*)]. *Thus the flux through the junction increases by one flux quantum between tunneling minima.*

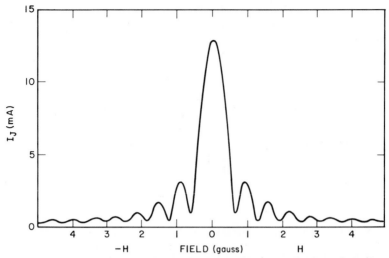

FIG. 21. Recorder tracing of zero-voltage Josephson current through Sn/Sn parallel-strip junction as a function of applied magnetic field in the plane of films parallel to one edge. The magnetic-field interval between minima is the flux quantum per unit area of the junction. (*Fiske and Giaever* [202])

The supercurrent I_J between 1 and 2 superconductor for zero voltage is given by

$$I_J = \frac{\pi}{R_{nn}} \frac{\Delta_1 \Delta_2}{\Delta_1 + \Delta_2} \qquad (45)$$

where R_{nn} is the normal tunneling resistance. Thus, the supercurrent can be ~10 to 100 mA for a typical rectangular junction 0.02 cm on a side. To maintain the flow of Josephson current against thermal and circuit fluctuations, the value of I_J should be as large as possible. This can be done by keeping R_{nn} small and choosing superconductors with large Δ. The junction area cannot be increased indefinitely, however, since the smaller dimension of the junction should remain $\leq 2\lambda_J$, where λ_J is the Josephson penetration distance. This requirement is a consequence [200] of the self-magnetic field generated by the Josephson current which keeps the current flowing to within a distance $2\lambda_J$ from the edges of the junction. The expression for λ_J is

$$\lambda_J = 0.36 (\lambda_L J_J)^{-1/2} \qquad (46)$$

where J_J is the tunneling supercurrent density in A/cm^2 and λ_L is the London penetration depth in Å. For typical values of J_J = 20 A/cm^2, and λ_L = 500 Å, λ_J = 0.004 cm. Thus, it is not practicable to get larger values of J_J by making junctions larger than 0.05 cm on a side.

Josephson also predicted that an alternating supercurrent will appear across the junction whenever a dc potential is developed across it. The ac amplitude may be as large as the maximum dc; its frequency is given by the Josephson relation

$$\nu = \frac{2eV}{h} = 4.84 \times 10^8 \text{ cps}/\mu V \tag{47}$$

where V is the voltage difference across the barrier. Thus for V = 100 μV (which is about $\Delta/10$ for Sn), ν = 48.4 Gc/sec.

The ac Josephson effect involves the transfer of Cooper pairs accompanied by the emission of photons. The direct observation of microwaves emitted from a Josephson tunnel junction was reported by a number of investigators [203, 204]. The spectral-response studies of photon emission suggested the existence of frequency-dependent Josephson current amplitudes which peak in the vicinity of the energy gap. This effect forms the basis of the use [205] of Josephson junctions as sensitive and high-speed emitters (and detectors) of microwaves. Langenberg et al. [206] observed non-Josephson radiations from Sn-oxide-Sn junctions which probably result from the nonlinear mixing of the rf fields driven by the ac Josephson current with the rf fields generated by a suitable parametric mechanism.

The first verification of the existence of an ac Josephson current was actually provided in an indirect experiment due to Shapiro [207], which was verified later by others [208]. In this experiment, a supercurrent was caused to flow by pumping microwave power into the junction, thus effecting stimulated emission of electron pairs from one side of the junction to the other. As the microwave field intensity in the cavity increased from zero, first one, and then additional steps occurred in the dc I-V characteristics so that the current in the Al-Al$_2$O$_3$-Sn junction situated in the cavity flowed at a constant voltage [Fig. 20(b)]. Each step was separated from the next by the voltage $V = h\nu/2e$, as predicted by Josephson. The steps came about from the frequency modulation of the Josephson currents by rf voltages driven by the applied radiation. Whenever one of the modulation sidebands occurs at zero frequency, a current step appears in the I-V characteristic.

The behavior of microwave-induced current steps at constant voltage observed by Shapiro was advantageously exploited by Parker et al. [209] to determine accurately the value of $2e/h$ from various junctions of different superconductors. These authors obtained a value of 483.5912 ± 0.0030 Mc/μV-sec, which differs by 21 ppm from the presently accepted value. This difference is, however, very significant since it removes the present discrepancy between the theoretical and experimental values of the hyperfine splitting in the ground state of atomic hydrogen, one of the major unsolved problems of quantum electrodynamics. This further illustrates the important contribution of a thin film phenomenon to solid-state physics.

Quantum Interference Effects. Since the Josephson effect arises because of the phase coherence between the two superconductors, it is reasonable to expect that a two-junction device, consisting of a pair of Josephson tunnel junctions connected by superconducting links, should exhibit interference behavior directly related to the long-range coherence. It can be shown for the geometry of a double junction that, in addition to the diffraction effects due to the identical single junctions, the maximum current will be modulated by an interference factor with a period corresponding to a change in the flux of ϕ_0 enclosed by the loop. Such interference effects have been observed and studied in detail by Jaklevic et al. [210, 211]. The observation of direct velocity modulation on the phase variation offers itself as a physical probe for the study of current distribution and field penetration into superconducting films.

These interference experiments certainly provide direct and striking observations of the macroscopic quantum behavior of electron pairs in a superconductor, which may be exploited to measure accurately the flux quantum $2e/h$, and the deBroglie wavelength of the electron pair. Further, the two Josephson junctions acting as a quantum interferometer constitute an extremely sensitive magnetometer capable of detecting magnetic-field changes in the nanogauss range.

8. INFRARED TRANSMISSION THROUGH THIN FILMS

At very low frequencies a superconductor behaves as a perfect conductor with zero surface resistance and reflects completely all incident radiation. As the frequency of radiation is increased so that the photon energy $h\nu$ becomes comparable with the energy gap of a superconductor, marked absorption characteristic of a resistive behavior starts abruptly, thereby allowing a direct determination of the energy gap. Since the energy gap $\Delta \sim 10^{-3}$eV, which corresponds to a wavelength of ~0.1 cm, such studies must be carried out in the far infrared.

Tinkham and coworkers [212, 213] studied transmission and absorption through thin superconducting films of Pb, Sn, Tl, In, and Hg using electromagnetic radiation between 0.1- and 6-mm wavelength. The measurements yield the ratio of the power transmitted when the film is superconducting to that when the film is normal. This ratio is then used to determine the real part σ_1 and the imaginary part σ_2 of the conductivity in the frequency-dependent complex conductivity given by the well-known Kramers-Kronig relations. The observed frequency dependence of the ratio σ_1/σ_n (σ_n = normal state conductivity) for a superconducting In film is shown in Fig. 22. The onset of absorption when the radiation energy equals the energy gap provides the most direct experimental method for demonstration, and the determination of the energy gap. This technique has yielded reasonable values of about 3.5 kT_c for the energy gap in films as thin as 100 Å. A precursor absorption peak observed [213] in Pb and Hg films at a frequency somewhat below that of the main absorption edge has now been shown [214] to be an artifact. The observed absorption behavior (Fig. 22) is in good agreement with that expected on the basis of the BCS theory.

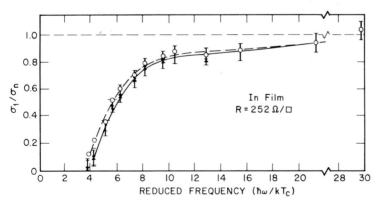

FIG. 22. The observed frequency dependence of the reduced conductivity (real part) of an In film. The solid curve is calculated using the measured film resistance, the broken curve using the calculated value. The outer flags refer to errors in transmission data, the inner to the area removed by gap. (*Ginsberg and Tinkham* [213])

9. SUPERCONDUCTIVE THIN-FILM DEVICES

The unique electromagnetic properties of superconductors offer numerous device applications. This section presents a brief description of *only* those applications which utilize the superconducting behavior of thin films. A detailed analysis of the various bulk as well as thin-film superconductive devices is contained in the reviews by Bremer [10], and by Newhouse [11].

9.1 Cryotrons

The fact that zero electrical resistance of the superconducting state changes reversibly to a finite value on transition to the normal state may be utilized to switch or divert current between different electrical paths. Several arrangements of switching devices, called cryotrons, based on this principle are described below.

(1) Wirewound Cryotron. A magnetically operated current switch is shown in Fig. 23(*a*). The Ta "gate" wire, which is operated slightly below its T_c (4.4°K), acts as the current-diverting element and can be switched to the normal resistive state by the magnetic field generated by

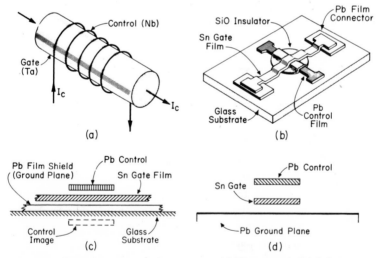

FIG. 23. Several variations of cryotrons: (*a*) Wirewound. (*b*) Schematic of an unshielded crossed-film cryotron (CFC). Tin gate-film cross section: 3 mm × 0.3; Pb control-film cross section: 1 × 15; insulating-film thickness: 0.4. (*Newhouse and Bremer* [217]). (*c*) Cross section of a shielded CFC. (*d*) Cross section of an in-line cryotron.

passing current through the Nb "control" coil. The T_c for Nb being about $9°K$, the control remains superconducting in operation. The first switch of this type was described by Casimir-Jonkers and de Haas [215] using a Pb control wrapped around a gate of Pb-Tl alloy.

Buck [216] demonstrated that the device of Fig. 23a, which he called a "cryotron," could be used as a digital switching element in computer circuits. The term cryotron is now generally used to represent any super-conductive four-terminal device in which the impedance between the two output terminals is controlled magnetically by means of a current passed through the two input terminals.

The cryotron speed, that is, the speed at which the resistance can be inserted into the gate, depends on the superconducting to normal-state transition time and is small enough ($\sim 10^{-10}$ sec) not to be a serious limiting factor. On the other hand, the switching time from one current path to another is determined by the ratio L/R, where L is the inductance of the superconducting loop made up of the current paths, and R the resistance introduced by an opened gate. The wirewound cryotron is severely limited by the fact that this time is no less than 10^{-5} sec. Significant improvement by three to five orders of magnitude in this time can be obtained by using two-dimensional thin-film cryotrons of a suitable geometry. Cryotrons with a time constant of $\sim 2 \times 10^{-9}$ sec have been successfully operated.

(2) Crossed-film Cryotron (CFC). A crossed-film cryotron [217] [Fig. 23(b)] consists of a gate film superconductor such as Sn, crossed by a much narrower control Pb film. The CFC is operated at a temperature just below the T_c of the gate and far below that of the control. The variation of gate resistance with control current at constant gate current exhibits a critical value of the control current above which the gate resistance appears. These characteristics are of course very sensitive to the presence of external magnetic fields.

(3) The Shielded CFC. The operating speed of the CFC circuits can be increased by reducing the inductance of the connecting elements and controls. This can be accomplished by depositing the whole CFC circuit on an insulated superconducting "shield" or "ground plane" [218] [Fig. 23(c)]. The image effect of the shield plane is to make the currents zero everywhere except in the region between the conductor and the shield plane, thereby drastically reducing the self-inductance of all conductors situated close to the shield plane. This geometry increases the gate current in the quenching characteristics because of the cancellation of the normal-field component of the gate current by the shield. The control current at which the resistance appears is approximately the same in the shielded as well as in the unshielded CFC.

The gain of the shielded CFC at a given temperature is defined as $G = I_g/C_c$, where I_g is the critical gate current at zero control current and C_c is the control current sufficient to restore half the maximum gate resistance, at small gate current. Expressing the currents in terms of the fields so produced, we get

$$H_{I_G} = \frac{4\pi I_g}{10W} \quad \text{and} \quad H_{C_c} = \frac{4\pi C_c}{10w} \tag{48}$$

where W and w are the widths of the gate and the control films, respectively. Thus,

$$G = \frac{H_{I_g}}{H_{C_c}} \frac{W}{w} \tag{49}$$

Related expressions for voltage and power gain have been derived by Smallman et al. [219].

For a thick film with $t > \lambda$, $H_{I_g}/H_{C_c} \sim 1$. For thinner films, $H_{I_g}/H_{C_c} < 1$, and the "crossing ratio" W/w has to be made large to achieve a gain larger than unity. Thinner films have a higher gate resistance and, therefore, a higher operating speed. For $t \ll \lambda$, using Eqs. (20), (34), and (35), we can rewrite Eq. (49) as (since $H_I \equiv H_{I_g}$)

$$G \approx \frac{1}{9} \frac{W}{w} \left(\frac{t}{\lambda}\right)^2 \tag{50}$$

Owing to the temperature dependence of λ, G increases rapidly as the temperature decreases below T_c and approaches a constant value at $T/T_c \ll 1$.

The time constant of the shielded CFC can be shown to be

$$\tau = \frac{L}{R} = \frac{8\pi tx}{\rho}\left(\frac{W}{w}\right)^2 \times 10^{-9} \text{ sec} = \frac{8\pi tx}{\rho} G^2 \left(\frac{H_{C_c}}{H_{I_g}}\right)^2 \times 10^{-9} \text{ sec} \tag{51}$$

where x is the effective distance between control and shield. The dependence of τ on t and ρ is complex since these parameters affect H_{I_g}/H_{C_c}. Ittner [220] has shown that τ goes through a minimum of the order of 10^{-8} sec for values of t ranging from 0.3 to 0.5 μ, depending on the film resistivity.

(4) In-line Cryotron. If the control field is parallel, rather than crossed, to that produced by the gate current, the device [Fig. 23(d)] is called an "in-line cryotron" [221]. Here, the control is also influenced by the gate current, and its current polarity relative to the control is thus important. This geometry overcomes the disadvantage of the reduced amplitude of the effective critical current (and hence gain in a CFC) resulting from the use of a bias to increase the operating speed. The quenching characteristics of an in-line cryotron are asymmetrical because of its field distribution. By biasing to the steepest part of the asymmetrical characteristics, it is possible to obtain high-speed operation with a gain above unity. The pulse switching time [221] is limited by the thermal time constant and is $\sim 10^{-8}$ sec.

Logic circuits utilizing cryotron units are analogous to those employing other switching devices, e.g., triodes or transistors. In fact, many circuit techniques that make use of relays and telephone switching and formal network-synthesis techniques can be adapted to cryotron circuits and systems. It should be pointed out that one significant difference between cryotron circuits, as they are presently designed, and those for relays is the necessity for providing an alternative current path when the gate becomes resistive. Thus at least two parallel paths are provided for current, one for each logic situation so that the current flows only in the superconducting gates. This "two-line logic" is the simplest cryotron circuit and forms a basic unit for complex circuitry.

(5) Multicrossover Cryotron (MCC). A high-gain amplifier using cascaded CFCs cannot be operated below 1 kc because of low-frequency noise from refrigerant liquid-helium bubble formation. If the requirement of minimum input inductance is not stringent, a single CFC stage with many crossovers can be used as a single-stage high-gain amplifier. Newhouse et al. [222] used an MCC with In gate $\sim 3,000$ Å thick and control $\sim 7,500$- to 10,000-Å-thick films arranged in an array of 512 crossovers to obtain transimpedances of up to 10 mV/mA with a bandwidth of 1 Mc/sec. A pulsed gate bias allows the amplification of dc signals, and futhermore, helium bubble formation responsible for the low-frequency noise is inhibited. A schematic diagram of an MCC with dc bias and its control characteristic are shown in Fig. 24(a) and (b).

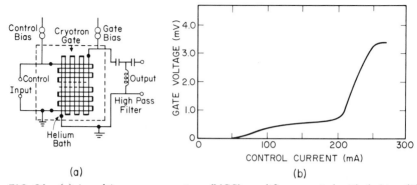

(a)   (b)

FIG. 24. (*a*) A multicrossover cryotron (MCC) amplifier operated with dc bias. (*b*) A typical control characteristic of an MCC consisting of 512, 0.2- by 0.2-mm crossovers with 3,000-Å In gate film carrying 1 mA at $T - T_c = 0.013°$K. (*Newhouse et al.* [222])

(6) Ferromagnetic Cryotron Configuration. The magnetic-field intensity associated with a $180°$ domain wall at a distance of $\sim 10^4$ Å above the wall may be estimated to be ~ 400 Oe. It is therefore reasonable to expect magnetic domain walls in close proximity to the superconducting films to influence the transition from the superconducting to the normal state. This consideration was utilized by Artley et al. [223] in studying a cryotron configuration in which a Permalloy (80 percent Ni, 20 percent Fe) film was deposited between an In gate and a Pb control film. The Permalloy film deposited with an easy direction of magnetization was insulated from the gate and the control by means of an SiO film. A current in the control film producing a field of approximately 1.0 Oe switched the Permalloy film. The resulting field intensity of ~ 400 Oe realized above the domain wall of the ferromagnetic film then switched the superconducting gate film. Thus, a potential gain of ~ 400 is obtained.

Cryotrons are ideal switching devices because they are small, light, low-power ($\sim$ microwatts, three orders of magnitude smaller than that of other conventional high-speed devices), easy to make and interconnect, without input-output interaction, level and polarity-independent, suitable for storage and logic, and have no need for auxiliary components such as resistors, diodes, transformers, or capacitors. Nonvolatile information storage can be performed directly in the device itself (because of its zero resistance) without the need for holding circuitry. The low impedance, low operating temperature, and various fabrication problems are some of the drawbacks. These are, however, by no means serious in view of the rapid advances in technology in this direction.

9.2 Computer Memory Devices

The thin-film cryotron may be used as a storage device in two ways: by causing it to divert current from one superconducting path to another, or by using it to cause persistent currents to flow in a superconducting loop. The switching in these cases is induced by the magnetic field caused by the control current. Storage devices in which switching is caused by an inductively coupled current-induced transition in the film have been studied. These devices may be used for large-scale storage in a superconductive computer or replace magnetic-storage devices in a conventional computer.

The superconducting memory devices make use of the fact that a current induced in a superconducting ring will persist indefinitely. Since the current can circulate either way, one has the possibility of a two-state memory, storing one bit of information with no dissipation of power other than that required to maintain the low temperature. The following variants have been devised.

(1) Crowe Cell. The first storage device of the inductively coupled type was suggested by Crowe [224]. The Crowe cell [Fig. 25(a)] basically consists of a thin superconducting film (say, Pb) with a small hole, a few millimeters in diameter, which has a narrow crossbar running across it. The crossbar is the switched element. Currents are induced in

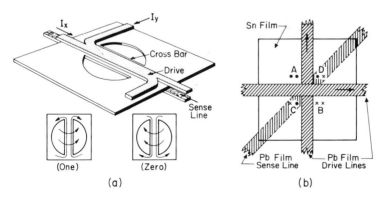

FIG. 25. (a) The "Crowe"-type persistent-current cell and its ONE and ZERO stored modes. (*Crowe* [224]). (b) Continuous-sheet memory cell. Flux penetration may occur at A and B owing to the vectorial sum of drive currents and provides signals at C and D. (*Buchold* [229])

the crossbar by the magnetic coupling from the two electrically insulated drive lines placed directly over and parallel to the crossbar. The crossbar current returns in the edges of the hole, half on one side and half on the other. An insulated sense line is deposited adjacent and parallel to the crossbar, on the opposite side of the drive lines.

The current is stored in the two D-shaped loops formed by each half of the hole and the crossbar. The sum of the currents induced by the two drive lines either adds to or subtracts from the stored current. For the latter case, the crossbar stays superconducting so that the shielded sense lines receive no signal (zero state). In the former case, if the total current exceeds the critical current of the crossbar, the crossbar becomes resistive and allows flux to penetrate through the hole and induce voltage in the sense line (one state).

Studies [224-226] of the Crowe cell show that the maximum operating speed in the adiabatic mode is determined by Joule heating and may be improved by operating closer to the critical temperature. Using a 600 Å by 0.1 mm by 2 mm Pb crossbar deposited on various substrates, Broom and Simpson [225] measured the electrical time constants of their cells to be between 10 and 20 nsec. Rhoderick [226] obtained 3 nsec for the electrical time constant, about 30 nsec for the thermal time constant, and an output voltage of ~50 mV for his cells.

(2) Persistor, Persistatron, and Continuous-film Memory (CFM). The "Persistor" suggested by Crittenden [227] and the "Persistatron" suggested by Buckingham [228] are somewhat similar in operation. These digital storage devices are a persistent current loop consisting of a permanently superconducting inductance connected in parallel with a superconductive film of negligible inductance. A "write" pulse quenches the low-inductance element and thereby diverts a part of the current through the loop inductor, which provides an output signal.

The Persistor, like the Crowe cell, has physical holes in a superconducting plane. The edges of the holes play a significant role in determining the electrical behavior. If large memory arrays are to be made, it may be difficult to obtain devices of matched electrical behavior.

Buchold [229] proposed a continuous-film memory (CFM) to overcome the matching problem. An individual storage cell of a CFM geometry is illustrated in Fig. 25(b). It has no holes, but it consists of drive lines intersecting over a continuous film. The vector sum of the fields produced by drive currents would "punch" holes of normal metal in this continuous film at a point where critical current is obtained. This normal-state transition would yield a signal in the sense line located on the other side of the film. The sense line may be replaced by a single continuous film of silver. The eddy currents in the Ag film produced by the normal hole are large enough to be easily detected. Since film imperfections are expected to trap some of the penetrating flux, this undesirable effect is tolerable only if the film imperfections are homogeneously distributed.

Burns et al. [230] investigated the CFM scheme. The feasibility of a CFM was demonstrated using an In-cell array 1 sq in. in area and conductors of nominal width of 10^{-3} in. It was operated with currents of 30 to 80 mA, depending on the temperature. Pulse rise times were varied between 200 and 20 nsec, giving a line-switching voltage between 0.8 and 8 mV, respectively. According to Burns et al., the heating during a switching cycle is much less here than in other devices and the total "read-write" time is also reduced. Because of its simpler configuration, the CFM is expected to be smaller than other storage cells.

Superconductive memory devices appear to be promising replacements for magnetic-storage devices. In particular, the CFM cell has about the same drive requirements and logic structure and presents about the same output as a magnetic-core memory, yet may be much smaller, faster, and cheaper. A reasonable fabrication yield and acceptable parameter variation from cell to cell are some of the present technical difficulties. This drawback, however, should not discourage further work in view of the rapid advances in thin film technology. The present-day devices require the use of the liquid-helium bath. Rose-Innes [231] has calculated that a Crowe-cell memory of 1 million bits would occupy only about 30 liters and need about 2 liters of liquid helium per hour for refrigeration. This capacity is at present routinely available to many research and development laboratories.

9.3 Superconducting Magnets

A significant application of the superconductivity phenomenon is the development of superconducting magnets capable of providing magnetic fields in excess of 100 kG. Because of the zero ohmic resistance, no steady-state power is required to sustain a magnetic field provided by superconducting coils. The size of a superconducting solenoid is several orders of magnitude smaller than that of a conventional magnet yielding comparable magnetic field. The primary requirement [232] for the superconducting solenoid-winding material is that the combined effect of the field due to the solenoid as a whole and that due to the current be insufficient to render any segment of the solenoid normal. Type II superconducting materials with β-W structures, such as Nb_3Sn ($18°K$), Nb_2Zr ($10°K$), MoN ($12.0°K$), V_3Ga ($16.8°K$), and V_3Si ($17°K$), with high transition temperatures (given in the parentheses), high critical fields, and high current densities in the presence of high fields, are best suited for superconducting-magnet applications. These materials are, however, generally brittle and therefore present severe problems for wire fabrication.

Since the current in a superconductor is carried by a surface layer (in type II materials, it is carried by about 100-Å-thick filaments), the bulk materials could be replaced by thin films. Thin films offer the advantage of the convenience of preparation of single or multicomponent superconductors in single or multiple layers. These films can be deposited onto a flexible plastic ribbon for winding solenoids. Pyrolytically prepared thick Nb_3Sn films [25] are presently being exploited commercially. Magnets wound with such superconducting tapes or ribbon and yielding magnetic fields of up to about 150 kG have been tested. As discussed in Sec. 5, the high current-carrying capability of suitably prepared thin-film hard superconductors is also comparable with that of the corresponding bulk materials (see Fig. 14). In view of the simplicity of thin film techniques for preparing multicomponent materials of controllable structure, further work in search of new superconducting materials with higher current capabilities at still higher fields and hopefully also at higher operating temperatures may prove very rewarding.

9.4 Low-frequency Devices

Thin superconducting films have been used as indicators, amplifiers, oscillators, modulators, and mixers, etc. A cryotron, described already, is essentially an amplifier circuit. Rosenberger [233] described an oscillator based on a thermal-relaxation process. The current in a superconducting Pb film was allowed to exceed the critical value so that the resistive transition occurred. The current was diverted to a shunt copper wire. After cooling, the Pb film became superconducting again and the current diverted back to the film. Oscillations at about 100 kc/sec were observed.

Superconducting cavities and waveguides [234, 235] possess a high Q value because of the vanishingly small resistance. Temporary storage of microwave energy, extremely efficient waveguides, and stable microwave oscillators can be obtained using superconducting cavities or cavities coated with superconducting films. Values of Q as high as 8×10^6 at 10 Gc/sec have been realized [235]. At higher frequencies, when the penetration depth of the superconducting material is comparable with the skin-effect depth, resistive losses occur in the cavity. This takes place at a frequency given by $h\nu \sim \Delta$, which corresponds to a wavelength of about 4 mm for tin.

Woodford and Feucht [236] described an rf mixer operating near 1 Gc/sec. Their device consisted of input and output loops on either side of a 2,000-Å-thick Sn film. The transition in and out of the superconducting state was obtained at a frequency of 2ω by a current at a frequency ω. By mixing the input signal with the 2ω frequency, both the sum and the difference frequencies may be obtained. Using a switching frequency of 900 Mc/sec, the transmission of an rf signal separated by 40 Mc/sec from the switching frequency was achieved. Sherrill and Rose [164] demonstrated the generation of third harmonics at 30 Gc/sec in 100-Å Sn and Pb films exposed to a 10 Gc/sec signal.

The nonlinear switching behavior of a superconductive thin film can also be used for parametric amplification [237-240]. Parametric amplification with a gain of up to 12 dB and side-band oscillations at 7.2 Gc/sec have been observed [239] in a slotted structure of a superconducting Sn film. The ultrathin film, which behaved like a type II superconductor, was placed on one face of a TiO_2 dielectric resonator. Discrete changes in the absorption took place at definite power levels, probably because of the entrance of quantized vortices or flux bundles into the film. The slits cause the induced microwave currents to be spatially inhomogeneous. The mechanism leading to gain and oscillations is believed to be associated with discontinuities in the pump power vs. frequency response of the films and with the slits (geometrical discontinuities) cut into the films.

Parametric amplification was also obtained by Bura [238] by inserting a fin-structured superconducting film radially between the inner and the outer conductor of a coaxial line. Using a 4,000-Å Sn film with radial fins, a gain of 27 dB at 2.3°K and a pump power of 2×10^{-6} W was measured at 2 Gc/sec. Here too, the nonlinear variation of magnetization with the "pump" magnetic field is believed to be responsible for the parametric amplification.

9.5 Bolometers—Radiation Detectors

A radiation detector, whose operation depends on the change of resistance with temperature of a conductor exposed to the heating effect of incident radiations, is called a bolometer. Superconductive bolometers employ a thin film or foil of a superconductor as the detecting element. The film is biased into the intermediate state and held at a temperature corresponding to the steepest part of the resistive transition curve. Any radiation incident on the film which raises the temperature of the film increases the resistance markedly. Because of the narrow superconducting transition, the operating temperature of the superconducting bolometer must be controlled to within 10^{-4}°K or better. The measurement of the resistance of the detector by sensing the penetration of the flux through a thin film rather than by passing a current should help realize the full sensitivity of the device.

Thin-film bolometers have been studied by a number of workers [241-245]. Andrews et al. [243] used Ta and NbN film bolometers. Superconducting bolometers are able to detect a radiation power of $\sim 10^{-12}$ W with a time constant of several seconds. Alpha particles with a calculated energy $\sim 5.2 \times 10^{-13}$ J can be detected [243] with such devices. If used in conjunction with cryotron output amplifiers, even greater detection sensitivities should result in the bolometers. It is estimated that an ideal thermal detector may be able to detect a signal as low as 10^{-16} W with a time constant of about 1 sec.

Superconducting bolometers may find applications as imaging devices, for the detection of microwaves, infrared radiation, and α-particles, etc. Niobium nitride bolometers have been used to detect and demodulate radio waves of frequencies up to 16 Mc/sec.

9.6 Tunnel Devices

In terms of the energy-band picture, a superconductor is represented as an intrinsic semiconductor with a temperature-dependent energy gap equal to 2Δ. The current in tunneling junctions (Sec. 7) results in the injection of electrons and holes into the conduction, and filled bands, respectively, of the superconductor. Burstein et al. [246] suggested by analogy with semiconductors that optical excitation could also inject carriers across the energy gap and thereby make tunnel junctions quantum detectors of microwave and submillimeter-wave radiation. The long wavelength limit of these photodetectors will be determined by the energy gap and is 3.9 mm for Al and 0.46 mm for Pb. These authors analyzed the concept and showed the feasibility of such a device.

The negative-resistance region in the tunnel $I-V$ characteristic may be utilized to operate a bistable element, an oscillator, or an amplifier. Miles et al. [247] used an Al-Al oxide-Pb junction at a temperature of $\sim 1^\circ$K as an amplifier with a power gain of 23 dB at 50 Mc/sec.

A Josephson tunnel junction can serve as a logic element analogous to a cryotron, having the advantages of higher output level and switching at a much smaller field. A typical Sn-oxide-Sn junction carrying a current of 10 mA will be switched by a field of 0.5 G, from a voltage drop of zero to about 1 mV. By contrast, a typical cryotron will require a switching field 10 to 100 times larger, and produce an output voltage around 10 times smaller.

Josephson tunnel junctions made of bulk materials have been used as emitters as well as sensitive and fast detectors of millimeter and submillimeter radiation. It should be possible to realize similar devices with thin-film junctions.

Two Josephson junctions connected in parallel act as an electron wave (or de Broglie) interferometer. Because of the extreme sensitivity of the interference phenomenon to the presence of a magnetic field, the interferometer can be used as a magnetometer in $\sim 10^{-9}$ G region.

10. CONCLUDING REMARKS

Despite the valuable experimental contributions of thin-film specimens to the understanding of the concepts and theories of the superconducting behavior of materials, it is rather surprising to note that workers in this field have paid little attention to the influence of the internal structure of films. Indeed, films have been considered as ideal materials of parallel plane geometry. It is fortunate that such a naive attitude has not been a serious drawback to many experiments. It is, however, abundantly clear now that more sophisticated size-effect studies and other experiments aimed at studying fundamentally important deviations (such as in the tunneling characteristics) from the expected behavior must be performed on structurally well defined films, in particular readily available single-crystal epitaxial films.

The excitingly promising phenomenon of enhancement of superconducting transition temperature and the critical magnetic field in certain fine-grained films is of considerable technical and fundamental interest. None of the theories proposed so far to explain the enhancement of T_c is consistent with the experimental facts. At the same time, experimental facts are too meager to yield a definite relation between enhancement, the internal structure of films, and other possible effects. Careful experiments together with structural characterization of the films are urgently needed.

A search for new superconducting materials with still higher T_c's and H_c's occupies the attention of many workers. The powerful methods of thin film technology may be exploited to prepare easily a variety of normal and abnormal multicomponent alloys and compounds. Some of these materials are impossible to obtain by any other known technique. Some unusual developments in obtaining high T_c thin-film superconducting materials may be expected in the future.

REFERENCES

1. D. Shoenberg, "Superconductivity," 2d ed., Cambridge University Press, New York, 1960.
2. F. London, "Superfluids," vol. 1, 2d ed., Dover Publications, Inc., New York, 1961.
3. E. A. Lynton, "Superconductivity," John Wiley & Sons, Inc., New York, 1962.
4. P. G. de Gennes, "Superconductivity of Metals and Alloys," Benjamin Inc., New York, 1966.
5. J. Bardeen and J. R. Schrieffer, in "Progress in Low Temperature Physics" (C. J. Gorter, ed.), vol. III, p. 170, Interscience Publishers, Inc., New York, 1961.
6. B. T. Matthias, in "Progress in Low Temperature Physics" (C. J. Gorter, ed.), p. 138, Interscience Publishers, Inc., New York, 1957.
7. D. H. Douglass, Jr., and F. M. Falicov, in "Progress in Low Temperature Physics" (C. J. Gorter, ed.), vol. IV, p. 97, Interscience Publishers, Inc., New York, 1964.
8. B. Serin, in "Handbuch der Physik," (S. Flügge, ed.), vol. 15, Springer-Verlag OHG, Berlin, 1956.
9. J. M. Blatt, "Theory of Superconductivity," Academic Press Inc., New York, 1964.
10. J. W. Bremer, "Superconductive Devices," McGraw-Hill Book Company, New York, 1962.
11. V. L. Newhouse, "Applied Superconductivity," John Wiley & Sons, Inc., New York, 1964.
12. H. Mayer, "Physik dünner Schichten," vol. 2, Wissenschaftliche Verlag, Stuttgart, 1955.
13. R. E. Glover III, in "Thin Films," p. 173, American Society for Metals, Metals Park, Ohio, 1964.

14. W. B. Ittner III, in "Physics of Thin Films" (G. Hass, ed.), vol. 1, p. 233, Academic Press Inc., New York, 1963.
15. N. Byers and C. N. Yang, *Phys. Rev. Letters,* **7**:46 (1961).
16. L. Onsager, *Phys. Rev. Letters,* **7**:50 (1961).
17. J. Bardeen, *Phys. Rev. Letters,* **7**:162 (1961).
18. B. S. Deaver, Jr., and W. M. Fairbank, *Phys. Rev. Letters,* **7**:43 (1961).
19. R. Doll and M. Näbauer, *Phys. Rev. Letters,* **7**:51 (1961).
20. For example, R. Frerichs and C. J. Kircher, *J. Appl. Phys.,* **34**:3541 (1963).
21. P. Fowler, *J. Appl. Phys.,* **34**:3538 (1963).
22. C. A. Neugebauer and R. A. Ekvall, *J. Appl. Phys.,* **35**:547 (1964).
23. D. Gerstenberg and P. M. Hall, *J. Electrochem. Soc.,* **111**:936 (1964).
24. C. A. Neugebauer, *J. Appl. Phys.,* **35**:3599 (1964).
25. J. J. Hanak, "Metallurgy of Advanced Electronic Materials," vol. 19, p. 161, AMIE, 1962; also J. J. Hanak et al., in "High Magnetic Fields" (H. Koln et al., eds.), p. 592, John Wiley & Sons, Inc., New York, 1962.
26. H. C. Theuerer and J. J. Hauser, *J. Appl. Phys.,* **35**:554 (1964).
27. J. Edgecumbe, L. G. Rosner, and D. E. Anderson, *J. Appl. Phys.,* **35**:2198 (1964).
28. E. T. S. Appleyard and A. C. B. Lovell, *Proc. Roy. Soc. London,* **A158**:718 (1937).
29. E. T. S. Appleyard et al., *Proc. Roy. Soc. London,* **A172**:530, 540 (1939).
30. A. I. Shalnikov, *Zh. Eksperim. i Teor. Fiz.,* **8**:763 (1938); **9**:255 (1939); **10**:630 (1940).
31. E. Armi, *Phys. Rev.,* **63**:451 (1943).
32. N. V. Zavaritskii, *C. R. Acad. USSR,* **68**:665 (1951).
33. H. E. Vogel, Thesis, University of North Carolina, 1962.
34. A. M. Toxen, *Phys. Rev.,* **123**:442 (1961); **124**:1018 (1961).
35. I. S. Khukhareva, *Soviet Phys. JETP English Transl.,* **16**:828 (1963).
36. M. Strongin, A. Paskin, O. F. Kammerer, and M. Garber, *Phys. Rev. Letters,* **14**:362 (1965).
37. M. Strongin, O. F. Kammerer, and A. Paskin, *Phys. Rev. Letters,* **14**:949 (1965); in "Basic Problems in Thin Film Physics" (R. Niedermayer and H. Mayer, eds.), p. 505, Vandenhoeck and Ruprecht, Göttingen, 1966.
38. T. K. Hunt and J. E. Mercereau, *Phys. Rev. Letters,* **18**:551 (1967).
39. R. H. Blumberg and D. P. Seraphim, *J. Appl. Phys.,* **33**:163 (1962).
40. N. V. Zavaritskii, *Dok. Akad. Nauk SSSR,* **82**:229 (1952).
41. W. Buckel and R. Hilsch, *Z. Physik,* **138**:109 (1954).
42. B. G. Lazarev, A. I. Sudovtsev, and A. P. Smirnov, *Soviet Phys. JETP English Transl.,* **6**:816 (1958).
43. Y. G. Mikhailov, E. I. Nikulin, N. M. Reinov, and A. P. Smirnov, *Zh. Tekhn. Fiz.,* **24**:931 (1959).
44. J. C. Suits, *Phys. Rev.,* **131**:588 (1963).
45. W. Buckel and R. Hilsch, *Z. Physik,* **131**:420 (1952).
46. W. Buckel, *Z. Physik,* **138**:136 (1954).
47. W. Rühl, *Z. Physik,* **138**:121 (1954).
48. H. Richter and S. Steeb, *Naturwiss.,* **45**:461 (1958).
49. S. Fujime, *Japan. J. Appl. Phys.,* **5**:59, 764 (1966).
50. W. L. Bond, A. S. Cooper, K. Andres, G. W. Hull, T. H. Geballe, and B. T. Matthias, *Phys. Rev. Letters,* **15**:260 (1965).

51. O. F. Kammerer and M. Strongin, *Phys. Letters,* **17**:224 (1965); in "Basic Problems in Thin Film Physics" (R. Niedermayer and H. Mayer, eds.), p. 511, Vandenhoeck and Ruprecht, Göttingen, 1966.
52. K. L. Chopra, *Bull. Am. Phys. Soc.,* **II**(12):57 (1967); *Phys. Letters,* **25A**:451 (1967).
53. K. L. Chopra, M. R. Randlett, and R. H. Duff, *Phil. Mag.,* **16**:261 (1967).
54. B. Abeles, R. W. Cohen, and G. W. Cullen, *Phys. Rev. Letters,* **17**:632 (1966).
55. B. Abeles, R. W. Cohen, and W. R. Stowell, *Phys. Rev. Letters,* **18**:902 (1967).
56. R. W. Cohen, B. Abeles, and G. S. Weisbarth, *Phys. Rev. Letters,* **18**:336 (1967).
57. M. H. Cohen and D. H. Douglass, Jr., *Phys. Rev. Letters,* **19**:118 (1967); M. Strongin, O. F. Kammerer, D. H. Douglass, Jr., and M. H. Cohen, *Phys. Rev. Letters,* **19**:121 (1967).
58. W. A. Little, *Phys. Rev.,* **134A**:1416 (1964).
59. R. A. Ferrell, *Phys. Rev. Letters,* **13**:330 (1964).
60. C. J. Thomson and J. M. Blatt, *Phys. Letters,* **5**:6 (1963).
61. R. H. Parmenter, *Phys. Rev.,* **154**:353 (1967).
62. E. A. Shapoval, *Soviet Phys. JETP Letters English Transl.,* **5**:57 (1967).
63. B. A. Tavger and V. Ya Demikhovskii, *Soviet Phys. JETP English Transl.,* **21**:494 (1965).
64. V. Ya Demikhovskii, *Soviet Phys.-Solid State English Transl.,* **7**:2903 (1966).
65. V. Z. Kresin and B. A. Tavger, *Soviet Phys. JETP English Transl.,* **23**:1124 (1966).
66. V. L. Ginzburg, *Phys. Letters,* **13**:101 (1964).
67. D. A. Kirzhnits and E. G. Maksimov, *Soviet Phys. JETP Letters English Transl.,* **2**:274 (1965); V. L. Ginzburg and D. A. Kirzhnits, *Soviet Phys, JETP English Transl.,* **19**:269 (1965).
68. W. Silvert, *Phys. Letters,* **16**:238 (1965); **19**:93 (1965); *Phys. Rev. Letters,* **14**:951 (1965).
69. W. Silvert, *Physics,* **2**:153 (1966).
70. M. L. Cohen, *Phys. Rev.,* **134**:A511 (1964).
71. J. Appel, *Phys. Rev. Letters,* **17**:1045 (1966).
72. V. B. Sandomirskii, *Soviet Phys. JETP Letters English Transl.,* **2**:248 (1965).
73. R. H. Parmenter, *Phys. Rev. Letters,* **7**:274 (1961).
74. J. M. Lock, *Proc. Roy. Soc. London,* **A208**:391 (1951).
75. H. A. Notarys, *Appl. Phys. Letters,* **4**:79 (1964).
76. P. M. Hall, *J. Appl. Phys.,* **36**:2471 (1965).
77. W. B. Ittner III and J. F. Marchand, *Phys. Rev.,* **114**:1268 (1959).
78. R. A. Connel and D. P. Seraphim, *Phys. Rev.,* **116**:606 (1959).
79. D. P. Seraphim, *Solid-State Electron.,* **1**:368 (1960).
80. H. L. Caswell, *J. Appl. Phys.,* **32**:105, 2641 (1961).
81. A summary is given by K. Schwidtal, *Z. Physik,* **158**:563 (1960); **169**:564 (1962).
82. E. Wasserman and G. V. Minningerode, Ref. 73, p. 516.
83. A. A. Abrikosov and L. P. Gorkov, *Soviet Phys. JETP English Transl.,* **12**:1243 (1961).
84. R. E. Glover III and M. D. Sherrill, *Phys. Rev. Letters,* **5**:248 (1960).
85. For example, R. J. Gayley, Jr., E. A. Lynton, and B. Serin, *Phys. Rev.,* **126**:43 (1962) and references cited here.

86. D. M. Ginsberg and J. S. Shier, in "Basic Problems in Thin Film Physics" (R. Niedermayer and H. Mayer, eds.), p. 543, Vandenhoeck and Ruprecht, Göttingen, 1966.
87. W. Rühl, Z. *Physik*, **159**:428 (1960); **176**:409 (1963).
88. W. Rühl, Z. *Physik*, **186**:190 (1965).
89. R. E. Glover III, S. K. Ghosh, and W. E. Daniels, Jr., in "Basic Problems in Thin Film Physics" (R. Niedermayer and H. Mayer, eds.), p. 536, Vandenhoeck and Ruprecht, Göttingen, 1966.
90. A. D. Misener, H. G. Smith, and J. O. Wilhelm, *Trans. Roy. Soc. Can.*, **29**:13 (1935).
91. H. Meissner, *Phys. Rev.*, **117**:672 (1960), and other references cited.
92. R. Hilsch and G. V. Minningerode, in "Basic Problems in Thin Film Physics" (R. Niedermayer and H. Mayer, eds.), p. 521, Vandenhoeck and Ruprecht, Göttingen, 1966.
93. P. Hilsch and R. Hilsch, *Naturwiss.*, **48**:549 (1961).
94. P. Hilsch, Z. *Physik*, **167**:511 (1962).
95. P. Hilsch and R. Hilsch, Z. *Physik*, **180**:10 (1964).
96. P. H. Smith, S. Shapiro, J. L. Miles, and J. Nicol, *Phys. Rev. Letters*, **6**:686 (1961).
97. J. L. Miles and P. H. Smith, *J. Appl. Phys.*, **34**:2109 (1963).
98. H. Meissner, *IBM J. Res. Develop.*, **6**:71 (1962).
99. R. J. Duffy and H. Meissner, *Phys. Rev.*, **147**:248 (1966).
100. F. Steglich, Z. *Physik*, **195**:239 (1966).
101. H. Tsuya, *J. Phys. Soc. Japan*, **21**:1011 (1966).
102. S. M. Marcus, *Phys. Letters*, **20**:467 (1966).
103. T. Claeson and S. Gygax, *Solid State Commun.*, **4**:385 (1966).
104. J. J. Hauser, H. C. Theuerer, and N. R. Werthamer, *Phys. Rev.*, **136A**:637 (1964).
105. J. J. Hauser, H. C. Theuerer, and N. R. Werthamer, *Phys. Rev.*, **142**:118 (1966).
106. J. J. Hauser, *Physics*, **2**:247 (1966).
107. F. Reif and M. A. Woolf, *Rev. Mod. Phys.*, **36**:238 (1964).
108. M. A. Woolf and F. Reif, *Phys. Rev.*, **137**:A557 (1965).
109. J. P. Burger, G. Deutscher, E. Guyon, and A. Martinet, *Phys. Letters*, **17**:180 (1965).
110. J. J. Hauser and H. C. Theuerer, *Phys. Letters*, **14**:270 (1965); **18**:222 (1965).
111. G. Bergmann, Z. *Physik*, **187**:395 (1965).
112. L. N. Cooper, *Phys. Rev. Letters*, **6**:689 (1961).
113. R. H. Parmenter, *Phys. Rev.*, **118**:1173 (1960).
114. P. G. de Gennes and D. Saint-James, *Phys. Letters*, **4**:151 (1963).
115. P. G. de Gennes and E. Guyon, *Phys. Letters*, **3**:168 (1963).
116. P. G. de Gennes, *Rev. Mod. Phys.*, **36**:225 (1964).
117. N. R. Werthamer, *Phys. Rev.*, **132**:2440 (1963).
118. W. Silvert, *Rev. Mod. Phys.*, **36**:251 (1964).
119. K. Maki, *Progr. Theor. Phys.*, **31**:731 (1964).
120. K. Maki, *Physics*, **1**:21 (1964).
121. P. Fulde, *Phys. Rev.*, **137**:A783 (1965); P. Fulde and K. Maki, *Phys. Rev. Letters*, **15**:675 (1965).
122. D. Saint-James, *J. Phys. Radium*, **25**:899 (1964).

123. W. L. McMillan, *Phys. Rev.,* **167**:331 (1968).
124. C. Chiou and E. Klokholm, *Acta Meta.,* **12**:883 (1964).
125. E. Klokholm and C. Chiou, *Acta Met.,* **14**:565 (1966).
126. A. C. Rose-Innes and B. Serin, *Phys. Rev. Letters,* **7**:278 (1961).
127. V. L. Ginzburg and L. D. Landau, *Soviet Phys. JETP English Transl.,* **20**:1064 (1950).
128. L. P. Gorkov, *Soviet Phys. JETP English Transl.,* **10**:593, 998 (1960).
129. J. Bardeen, *Rev. Mod. Phys.,* **34**:667 (1962).
130. A. M. Toxen, *Phys. Rev.,* **127**:382 (1962); *Rev. Mod. Phys.,* **36**:308 (1964).
131. A. M. Toxen and M. J. Burns, *Phys. Rev.,* **130**:1808 (1963).
132. F. Odeh and W. Liniger, *Proc. 9th Intern. Conf. Low Temp. Phys.,* p. 575, Plenum Press, Columbus, Ohio, 1964.
133. A. I. Shal'nikov, *Soviet Phys.JETP English Transl.,* **10**:630 (1940).
134. N. E. Alekseyevskii, *Soviet Phys. JETP English Transl.,* **10**:1392 (1940).
135. N. E. Alekseyevskii, *J. Phys. USSR,* **4**:401 (1941).
136. N. V. Zavaritskii, *Dokl., Akad. Nauk SSSR,* **78**:665 (1951).
137. N. V. Zavaritskii, *Dokl. Akad. Nauk SSSR,* **85**:749 (1952); **86**:501 (1952).
138. D. H. Douglass, Jr., and R. H. Blumberg, *Phys. Rev.,* **127**:2038 (1962).
139. I. S. Khukhareva, *Soviety Phys. JETP English Transl.,* **14**:526 (1962).
140. A. A. Abrikosov, *Dokl. Akad. Nauk SSSR,* **86**:489 (1952); *Soviet Phys. JETP English Transl.,* **5**:1174 (1957).
141. B. B. Goodman, *IBM J. Res. Develop.,* **6**:63 (1962).
142. D. Saint-James and P. G. de Gennes, *Phys. Letters,* **7**:306 (1963).
143. M. Tinkham, *Phys. Rev.,* **129**:2413 (1963); in "Basic Problems in Thin Film Physics" (R. Niedermayer and H. Mayer, eds.), p. 499, Vandenhoeck and Ruprecht, Göttingen, 1966.
144. P. G. de Gennes and M. Tinkham, *Physics,* **1**:107 (1964).
145. J. J. Hauser and H. C. Theuerer, *Phys. Rev.,* **134**:A198 (1964).
146. R. E. Glover III, *Z. Physik,* **176**:455 (1963).
147. G. K. Chang and B. Serin, *Phys. Rev.,* **145**:274 (1966).
148. K. Maki, *Ann. Phys. N.Y.,* **34**:363 (1965).
149. E. H. Rhoderick, *Proc. Roy. Soc. London,* **A267**:231 (1962); R. F. Broom and E. H. Rhoderick, *Proc. Phys. Soc. London,* **79**:586 (1962).
150. W. DeSorbo and V. L. Newhouse, *J. Appl. Phys.,* **33**:1004 (1962).
151. D. Saint-James, *Phys. Letters,* **16**:218 (1965).
152. J. P. Burger, G. Deutscher, E. Guyon, and A. Martinet, *Phys. Rev.,* **137**:A853 (1965).
153. E. H. Rhoderick and E. M. Wilson, *Nature,* **194**:1167 (1962).
154. L. A. Feigin and A. I. Shal'nikov, *Soviety Phys. "Doklady" English Transl.,* **1**:377 (1956).
155. N. I. Ginzburg and A. I. Shal'nikov, *Soviet Phys. JETP English Transl.,* **8**:243 (1962).
156. J. E. Mercereau and T. K. Hunt, *Phys. Rev. Letters,* **8**:243 (1962).
157. J. E. Mercereau and L. T. Crane, *Phys. Rev. Letters,* **9**:381 (1962).
158. R. E. Glover III and H. T. Coffey, *Rev. Mod. Phys.,* **36**:299 (1964).
159. H. H. Edwards and V. L. Newhouse, *J. Appl. Phys.,* **33**:868 (1962).
160. J. W. Bremer and V. L. Newhouse, *Phys. Rev.,* **116**:309 (1959).
161. F. W. Schmidlin, A. J. Learn, E. C. Crittenden, Jr., and L. N. Cooper, *Solid-State Electron.,* **1**:323 (1960).

162. J. I. Gittleman and S. Bozowski, *Phys. Rev.*, **135**:A297 (1964).
163. F. B. Hagedorn, *Phys. Rev. Letters*, **12**:322 (1964).
164. M. D. Sherrill and K. Rose, *Rev. Mod. Phys.*, **36**:312 (1964).
165. J. A. Mydosh and H. Meissner, *Phys. Rev.*, **140**:A1568 (1965).
166. R. B. Flippen, *Phys. Rev.*, **137**:A1822 (1965).
167. W. H. Meiklejohn, *Rev. Mod. Phys.*, **36**:302 (1964).
168. R. Burton, *Cryogenics*, **6**:144 (1966).
169. H. London and G. R. Clark, *Rev. Mod. Phys.*, **36**:320 (1964).
170. A. B. Pippard, *Proc. Roy. Soc. London*, **A203**:210 (1950).
171. T. E. Faber and A. B. Pippard, in "Progress in Temperature Physics" (C. J. Gorter, ed.), vol. 1, p. 159, Interscience Publishers, Inc., 1955.
172. R. F. Broom and E. H. Rhoderick, *Brit. J. Appl. Phys.*, **11**:292 (1960); *Solid-State Electron.*, **1**:314 (1960).
173. J. W. Bremer and V. L. Newhouse, *Phys. Rev. Letters*, **1**:282 (1958).
174. W. H. Cherry and J. I. Gittleman, *ONR Symp. Rept.* ACR-50, p. 75, 1960.
175. I. Giaever, *Phys. Rev. Letters*, **5**:147 (1960).
176. J. Nicol, S. Shapiro, and P. H. Smith, *Phys. Rev. Letters*, **5**:461 (1960).
177. I. Giaever, *Phys. Rev. Letters*, **5**:461 (1960).
178. I. Giaever and K. Megerle, *Phys. Rev.*, **122**:1101 (1961).
179. I. Giaever, H. R. Hart, Jr., and K. Megerle, *Phys. Rev.*, **126**:941 (1962).
180. J. W. T. Dabbs, *Proc. 9th Intern. Conf. Low Temp. Phys.*, p. 428, Plenum Press, Columbus, Ohio, 1964. D. E. Thomas and J. M. Rowell, *Rev. Sci. Instr.*, **36**:1301 (1965).
181. A. Schmid, *Phys. Stat. Solidi.*, **7**:3 (1964).
182. D. H. Douglass, Jr., *IBM J. Res. Develop.*, **6**:44 (1962).
183. D. H. Douglass, Jr., *Phys. Rev. Letters*, **6**:346 (1961); *Phys. Rev. Letters*, **7**:14 (1961).
184. D. H. Douglass, Jr., and R. Meservey, *Phys. Rev.*, **135**:A19 (1964).
185. R. Meservey and D. H. Douglass, Jr., *Phys. Rev.*, **135**:A24 (1964).
186. P. Townsend and J. Sutton, *Phys. Rev.*, **128**:591 (1962).
187. S. Bermon and D. M. Ginsberg, *Phys. Rev.*, **135**:A306 (1964).
188. R. S. Collier and R. A. Kamper, *Phys. Rev.*, **143**:323 (1966).
189. V. S. Mathur, N. Panchapakesan, and R. P. Saxena, *Phys. Rev. Letters*, **9**:374 (1962).
190. D. E. Morris and M. Tinkham, *Phys. Rev.*, **134**:A1154 (1964).
191. J. M. Rowell, P. W. Anderson, and D. E. Thomas, *Phys. Rev. Letters*, **10**:334 (1963).
192. D. J. Scalapino and P. W. Anderson, *Phys. Rev.*, **133**:A921 (1964).
193. B. N. Taylor and E. Burstein, *Phys. Rev. Letters*, **10**:14 (1963).
194. J. R. Schrieffer and J. W. Wilkins, *Phys. Rev. Letters*, **10**:17 (1963).
195. B. D. Josephson, *Phys. Letters*, **1**:251 (1962); *Rev. Mod. Phys.*, **36**:216 (1964).
196. B. D. Josephson, *Advan. Phys.*, **14**:419 (1965).
197. P. W. Anderson, *Rev. Mod. Phys.*, **38**:298 (1966).
198. P. W. Anderson and J. M. Rowell, *Phys. Rev. Letters*, **10**:230 (1963).
199. J. M. Rowell, *Phys. Rev. Letters*, **11**:200 (1963).
200. R. A. Ferrell and R. E. Prange, *Phys. Rev. Letters*, **10**:479 (1963).
201. M. D. Fiske, *Rev. Mod. Phys.*, **36**:221 (1964); D. D. Coon and M. D. Fiske, *Phys. Rev.*, **138**:A744 (1965).
202. M. D. Fiske and I. Giaever, *Proc. IEEE*, **52**:1155 (1964) (review).

203. R. E. Eck, D. J. Scalapino, and B. N. Taylor, *Phys. Rev. Letters,* **13**:15 (1964).
204. I. K. Yanson, V. M. Svistunov, and I. M. Dmitrenko, *Soviet Phys. JETP English Transl.,* **21**:650 (1965).
205. C. C. Grimes, P. L. Richards, and S. Shapiro, *Phys. Rev. Letters,* **17**:431 (1966).
206. D. N. Langenberg, W. H. Parker, and B. N. Taylor, *Phys. Letters,* **22**:259 (1966).
207. S. Shapiro, *Phys. Rev. Letters,* **11**:80 (1963).
208. D. N. Langenberg, W. H. Parker, and B. N. Taylor, *Phys. Rev.,* **150**:186 (1966).
209. W. H. Parker, B. N. Taylor, and D. N. Langenberg, *Phys. Rev. Letters,* **8**:287 (1967).
210. R. C. Jaklevic, J. Lambe, A. H. Silver, and J. E. Mercereau, *Phys. Rev. Letters,* **12**:159, 274 (1964).
211. R. C. Jaklevic, in "Basic Problems in Thin Film Physics" (R. Niedermayer and H. Mayer, eds.), p. 528, Vandenhoeck and Ruprecht, Göttingen, 1966.
212. R. E. Glover III and M. Tinkham, *Phys. Rev.,* **108**:243 (1957).
213. D. M. Ginsberg and M. Tinkham, *Phys. Rev.,* **118**:990 (1960).
214. S. L. Norman and D. H. Douglass, Jr., *Phys. Rev. Letters,* **18**:339 (1967).
215. J. M. Casimir-Jonkers and W. J. de Haas, *Physica,* **2**:935 (1935).
216. D. A. Buck, *Proc. IRE,* **44**:482 (1956).
217. V. L. Newhouse and J. W. Bremer, *J. Appl. Phys.,* **30**:1458 (1959).
218. V. L. Newhouse, J. W. Bremer, and H. H. Edwards, *Proc. IRE,* **48**:1395 (1960).
219. C. R. Smallman, A. E. Slade, and M. L. Cohen, *Proc. IRE,* **48**:1562 (1960).
220. B. W. Ittner II, *Solid-State Electron.,* **1**:239 (1960).
221. A. E. Brenneman, J. J. McNichol, and D. P. Seraphim, *Proc. IEEE,* **51**:1009 (1963).
222. V. L. Newhouse, J. L. Mundy, R. E. Joynson, and W. H. Meiklejohn, *Rev. Sci. Instr.,* **38**:798 (1967).
223. J. L. Artley, G. Buckley, C. A. Willis, Jr., and W. F. Chambers, *Appl. Phys. Letters,* **9**:429 (1966).
224. J. W. Crowe, *IBM J. Res. Develop.,* **1**:295 (1957).
225. R. F. Broom and O. Simpson, *Brit. J. Appl. Phys.,* **11**:78 (1960).
226. E. H. Rhoderick, *Proc. Roy. Soc. London,* **A267**:231 (1962).
227. E. C. Crittenden, Jr., *Proc. 5th Intern. Conf. Low Temp. Phys.,* Madison, 1958, p. 232.
228. M. J. Buckingham, *Proc. 5th Intern. Conf. Low Temp. Phys.,* Madison, 1958, p. 229.
229. T. A. Buchold, *Sci. Am.,* **202**:74 (1960); *Cryogenics,* **1**:203 (1961).
230. L. L. Burns, Jr., G. A. Alphonse, and G. W. Leck, *IRE Trans.,* **EC-10**:438 (1961).
231. A. C. Rose-Innes, *Brit. J. Appl. Phys.,* **10**:452 (1959).
232. See, for example, J. E. Kunzler, *Rev. Mod. Phys.,* **33**:501 (1961).
233. G. B. Rosenberger, *IBM J. Res. Develop.,* **3**:189 (1959).
234. C. J. Grebenkemper and J. P. Hagen, *Phys. Rev.,* **86**:673 (1952).
235. E. Maxwell, in "Advances in Cryogenic Engineering" (K. D. Timmerhaus, ed.), **6**:154 (1961).
236. J. B. Woodford and D. L. Feucht, *Proc. IRE,* **46**:1871 (1958).

237. A. S. Clorfeine, *Appl. Phys. Letters,* **4**:131 (1964); *Proc. IEEE,* **53**:388 (1965).
238. P. Bura, *Appl. Phys. Letters,* **8**:155 (1966).
239. R. V. D'Aiello and S. J. Freedman, *Appl. Phys. Letters,* **9**:323 (1966).
240. H. Zimmer, *Appl. Phys. Letters,* **10**:193 (1967).
241. A. Goetz, *Phys. Rev.,* **55**:1270 (1939).
242. D. H. Andrews, R. M. Milton, and W. DeSorbo, *J. Opt. Soc. Am.,* **36**:518 (1946).
243. D. H. Andrews, R. D. Fowler, and M. C. Williams, *Phys. Rev.,* **76**:154 (1949).
244. J. A. Hulbert and G. O. Jones, *Proc. Phys. Soc. London,* **B68**:801 (1955).
245. D. H. Martin and D. Bloor, *Cryogenics,* **1**:159 (1961).
246. E. Burstein, D. N. Langenberg, and B. N. Taylor, *Phys. Rev. Letters,* **6**:92 (1961).
247. J. L. Miles, P. H. Smith, and W. Schönbein, *Proc. IEEE,* **51**:937 (1963).
248. F. R. Gamble and H. M. McConnel, *Phys. Letters,* **26A**:162 (1968).
249. J. W. Garland, K. H. Bennemann, and F. M. Mueller, *Phys. Rev. Letters,* **21**:1315 (1968); J. M. Dickey and A. Paskin, *Phys. Rev. Letters,* **21**:1441 (1968).

X FERROMAGNETISM IN FILMS*

Mitchell S. Cohen,†

Lincoln Laboratory, Massachusetts Institute of Technology

1. INTRODUCTION

Research on thin ferromagnetic films has proved rewarding from two standpoints: First, the study of this specialized branch of magnetism has deepened knowledge of the fundamental nature of magnetism in general. Second, because ferromagnetic properties are highly structure-sensitive, the ferromagnetic behavior of films can provide insight into the structure and properties of *all* thin films; indeed, many subtle variations in the physical structure of ferromagnetic films, although clearly manifested in ferromagnetic behavior, prove to be very difficult to detect by nonmagnetic means.

Although magnetic films of insulators such as ferrites [1], garnets [1], and Eu chalcogenides [2] have been studied, the overwhelming

*This work was sponsored by the U.S. Air Force.

†This chapter is contributed by Dr. Cohen at the request of the author.

proportion of the research in this field has been carried out on metallic ferromagnetic films. Films of the ferromagnetic metals Ni, Fe, Co, and Gd and their alloys with ferromagnetic and nonferromagnetic metals have been investigated. Films with reproducible ferromagnetic behavior can be made by all the standard deposition techniques (Chap. II): vacuum deposition by evaporation [3] (the most popular technique), sputtering [4], electroplating [5, 6], chemical reduction [7] ("electroless" deposition), and thermal decomposition of metal-organic vapors [8]. Single-crystal and polycrystalline films have been prepared under a wide variety of conditions by these methods, and the resultant magnetic and associated physical properties have been investigated.

The largest portion of the research effort has been devoted to polycrystalline films of nickel-iron alloys, particularly the alloys near the "nonmagnetostrictive" composition 81 percent Ni, 19 percent Fe (sometimes known as "Permalloy"), deposited on flat, smooth substrates. Such films exhibit the following unique magnetic properties: The magnetization $\mathbf{M}$ is confined to the plane of the film by shape anisotropy; a uniaxial magnetic anisotropy can be induced with an easy axis (EA) in a predetermined direction in the film plane [in the minimum-energy state, $\mathbf{M}$ is equivalently directed in *either* sense of this EA (Sec. 3)]; and under proper circumstances $\mathbf{M}$ can reverse its sense by a rotational process in a few nanoseconds under application of magnetic fields of a few oersteds. These films thus have two equivalent stable states between which a fast transition can be made for low driving fields, while large arrays of such films can be manufactured cheaply, with a favorable geometry. The films are therefore attractive for use as memory elements in digital computers; this application has served as the technological impetus behind the intensive research on ferromagnetic films in the past decade (Sec. 12).

A typical film of this variety is vacuum-deposited, at normal incidence of the vapor, on a polished soft-glass substrate heated to perhaps 250°C. An 83 percent Ni, 17 percent Fe melt is evaporated by rf induction at about 10^{-5} Torr* from an alundum crucible, resulting in an 81 percent

---

*Except for certain properties (Secs. 2 and 4.4), the magnetic behavior of ferromagnetic films is not grossly sensitive to the pressure during deposition, so that an oil diffusion-pump system is usually adequate. Of the residual gases in the vacuum chamber, the magnetic parameters are most sensitive to oxygen, which must have a partial pressure greater than about 10^{-5} Torr to have detectable effects [9]. On the other hand, the magnetic parameters are very sensitive to substrate temperature, deposition rate, and composition, as discussed below.

Ni, 19 percent Fe film, perhaps 1,000 Å thick and 1 cm in diameter, having randomly oriented crystallites about 100 Å in diameter. In all deposition techniques, during deposition (and subsequent cooling to room temperature) a dc field is applied in the film plane to define the *EA* direction. The resulting magnetic parameters (see below) are typically $H_K \simeq 3$ Oe, $H_c \simeq 1$ Oe, and $\alpha_{90} \simeq 2°$. With the exception of films with nonplanar magnetization (Sec. 10), the films discussed in this chapter will have roughly the foregoing description unless otherwise specified. This is not a severe restriction, however, because similar properties are exhibited by NiFe films made by other procedures, and by films composed of different ferromagnetic metals.

In the past decade, over 1,500 publications on ferromagnetic films have appeared. In this short review there is space enough only to indicate the major aspects of the field; details of specimen preparation, experimental techniques, and mathematical derivations may be found in the original literature, while the standard modern texts [10, 11, 12] are recommended for discussions of the basic concepts of ferromagnetism. Several reviews of the rapidly advancing field of ferromagnetic films have already been published [3, 13-20]; the present chapter represents an attempt to include the most recent work, particularly the recent micromagnetic ripple theories, which have not been adequately emphasized in the older reviews.

Gaussian cgs units are used throughout this chapter.

2. MAGNETIZATION VS. THICKNESS [15, 21, 22]

The spin on an atom at the surface of a uniformly magnetized ferromagnetic film is less rigidly constrained to the average direction of **M** than the spin on an interior atom simply because of the absence of exchange bonds on the exterior side of the surface atom. Thus, below a certain value of film thickness t, the ratio of surface atoms to interior atoms will be high enough so that a decrease in the saturation magnetization and Curie temperature T_c from bulk values will be detectable. In the limit, in fact, calculations show that a monatomic sheet is not ferromagnetic at all [15]. The problem of the dependence of M and T_c on t (and the dependence of M on temperature T for a given t) has received much experimental and theoretical attention, because it provides a key to the understanding of magnetism as a cooperative phenomenon.

This problem was first attacked by Klein and Smith [23], who made use of the theory of spin waves, i.e., periodic distributions of reversed spins. These authors, along with later theorists, assumed a perfect monocrystalline film, with plane, parallel boundaries. They postulated the same periodic boundary conditions used for the bulk case, even in the direction normal to the film plane. Also, in the summation over all spin waves to obtain M, the term associated with the uniform mode ($k = 0$) was suppressed because it led to an awkward divergence not found in the three-dimensional case. Good agreement was obtained between this theory and early experiments [15, 21], which showed a rapid decrease in M with decreasing t, for t below 150 Å. Later work showed that this agreement was fortuitous, however, since both the theoretical and experimental results were incorrect.

The spin-wave theory was improved by Döring [24], who recognized the unsuitability of periodic boundary conditions for spin waves propagating in the direction normal to the film plane. The correct boundary conditions must reflect the fact that the spins at the interfaces may be either free or subject to a constraint in their motion ("pinned") depending on the magnetic anisotropy at the interface layers. Döring also showed that the divergence originating from the $k = 0$ term disappeared if a small external or anisotropy field was present, as is always true in practice. This theory was extended to cover any arbitrary crystal structure or orientation in the film [25] (still considered as a perfectly flat monocrystal). The improved spin-wave theory predicts that M will not decrease with decreasing t until t is less than ~30 Å, not 150 Å as calculated by Klein and Smith (Fig. 1).

While progress was being made on the spin-wave approach, Valenta [15] applied the Heisenberg molecular-field formulation to thin films. To do this he divided the film into sheets of atoms parallel to the substrate, and considered the magnetic interaction of a given atom in one sheet with nearest neighbor atoms in the other sheets. This theory also indicated little change of M with t except for the thinnest films (Fig. 1). Valenta's theory was extended to include magnetic anisotropy [26] and was used to calculate M as a function of depth within the film [27]; the theory was also generalized to the Heisenberg-trace method [28]. The molecular-field method permits calculations of M at high temperatures (and hence calculations of the Curie temperature) where the spin-wave theories are invalid.

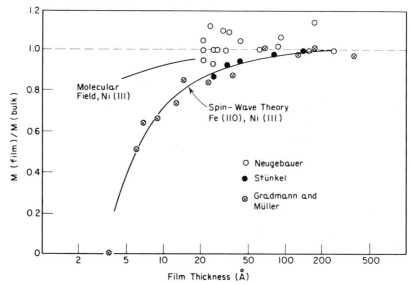

FIG. 1. Dependence of M on film thickness, where M is normalized to the bulk value. Comparison of spin-wave [25] (at 200 Oe) and molecular-field [26] theoretical values with experimental results for polycrystalline Ni (Neugebauer [33]), polycrystalline Fe (Stünkel [35]) and epitaxial 48 percent Ni 52 percent Fe (Gradmann and Müller [43]). (*After Gradmann and Müller* [43])

Most recently the sophisticated Green's-function method has been applied [29, 30]; this approach offers the advantage of being valid over the entire temperature range, but the mathematics involved are not easily tractable. It should also be noted that the development of successful calculations of M in thin films by electron-band theory has begun [31].

Two reasons have been given for the disagreement of Crittenden and Hoffman's early M vs. t data [32] with more recent data: (1) The original films were prepared at 10^{-6} Torr and/or measured (in air) under conditions such that oxidation occurred. (2) Conditions were such that the films were not completely magnetically saturated during measurement. The most reliable recent data comprise those taken from films deposited at pressures $\lesssim 10^{-8}$ Torr and measured directly with high fields in the vacuum system. Thus M vs. t for a given T, and M vs. T for a given t were measured for Ni films with a torque-magnetometer technique [33, 34], and for Fe films with a microbalance [35] technique. In both cases, at room temperature, bulk values of M were found down to $t \simeq 30$ Å (Fig. 1), and in both cases evidence of superparamagnetic behavior was found in the thinnest films. In another experiment on Fe, the films were prepared at $\sim 10^{-8}$ Torr, carefully coated with SiO to protect against oxidation, then measured in air using the Mössbauer effect [36]; bulk values of M were again found down to $t < 30$ Å.

Clear evidence of the importance of oxidation was provided by experiments which gave the $t = 30$ Å limit for Fe films deposited and measured at 10^{-6} Torr, but the $t \sim 150$ Å limit (similar to the original M vs. t data [32]) after admission of air into the vacuum chamber [37]. Similarly, experiments on Ni films showed that oxidation upon admission of air led to an appreciable decrease in magnetic output [34]. On the other hand, following the technique of Crittenden and Hoffman [32], Gondō et al. [38] prepared their Ni films at 10^{-5} Torr and measured them in air after overcoating with a protective SiO layer. In contrast with Crittenden and Hoffman, however, they found the bulk value of M for t down to 16 Å as long as fields high enough to saturate the samples were used; sometimes fields in excess of 3×10^3 Oe were required. They therefore emphasized the requirement of large saturating fields during M measurements rather than good vacuum conditions.

The need for prevention of oxidation and the need for high saturating fields are related, however. Thinner films have a more granular character, with larger spacing between crystallites; in magnetic terms thinner films assume the character of a collection of superparamagnetic particles, which would require a higher saturating field. The granularity of ultrathin Fe films was clearly demonstrated by measurements of electrical resistance as a function of film thickness [35], while the superparamagnetic nature of ultrathin films was shown by observations of magnetic-viscosity effects in thin Ni films (Sec. 7.2). Now besides causing the true magnetic thickness to be less than the net physical thickness, oxidation can promote superparamagnetic behavior because it breaks the intercrystalline exchange bonds. This was shown by the work of Ratajczak [39], who found that Fe films deposited at temperatures above 100°C retained bulk values of M for much smaller values of t than did films deposited at room temperature. Further, he demonstrated, by monitoring the electrical resistivity as a function of time, that this difference in behavior was caused by greater susceptibility to oxidation of the low-temperature (and hence presumably fine-grained) films.

In summary, the best experimental work indicates, in agreement with the best theoretical analyses, that M does not decrease with decreasing t until t is less than 20 or 30 Å. Ultrathin films are granular and superparamagnetic in character, however, and therefore do not fit the flat, planar geometry postulated by the theories. Thus a detailed experimental fit of the theories is not really meaningful for $t \lesssim 50$ Å. For this reason Hellenthal [40, 41] has attempted the calculation of M for a granular film using a sublattice model similar to that of Valenta [15]. This is a formidable theoretical problem, however, especially since the details of the crystallite shapes and the degree of mutual coupling are unknown.

A more promising approach may be that taken by Gradmann [42, 43], who tried to keep his films flat, monocrystalline, and planar in the ultrathin range. He reported that Ni and NiFe films, when grown epitaxially on Cu films which were in turn epitaxially grown on mica, showed no superparamagnetic behavior even for thicknesses as small as 7 Å. Further, reflection electron-diffraction spots were narrow and elongated, which indicates flatness. This conclusion is also supported by the work of Hellenthal [44] who showed that, for the same average thickness, the Curie temperatures of epitaxially grown Ni films were closer to the bulk value than those of polycrystalline films.

Using a pendulum magnetometer, Gradmann was able to measure M as a function of t for films as thin as 5 Å. His results are gratifying, although they may be subject to the objection that even if the films are flat, the theoretical conditions are not met because the magnetic material is bounded not by vacuum, but by Cu. The importance of the change in boundaries is not clear, however.

3. STONER-WOHLFARTH MODEL [3] (A FIRST APPROXIMATION)

If the free energy of a ferromagnetic body depends upon the direction of **M**, the body is said to possess magnetic anisotropy. Directions of **M** associated with energy minima (maxima) are designated as easy (hard) directions. The anisotropy energy of a uniaxial ferromagnetic film, assuming **M** is confined to the film plane (Sec. 4.1), is given by

$$E = K_u \sin^2 \varphi_0 \tag{1}$$

where K_u is the uniaxial anisotropy constant and φ_0 is the angle between **M** and an easy *axis* (*EA*) whose direction within the plane of the film is determined during deposition and remains fixed subsequently. Note that Eq. (1) defines two states of equivalent minimum energy $\varphi_0 = 0$ and π, i.e., either sense of **M** along the *EA* is an equivalent equilibrium state;

states of maximum energy are achieved when **M** is along the hard axis (*HA*), i.e., $\varphi_0 = \pi/2$ or $3\pi/2$. Higher-order anisotropy (biaxial, triaxial, etc.) may be discussed similarly.

The magnetization may be pulled out of its equilibrium position (along an *EA*) by application of a field **H** in the film plane, which exerts a torque $\mathbf{H} \times \mathbf{M}$ on **M**. The field energy is $-\mathbf{H} \cdot \mathbf{M}$, which is a minimum when **M** is pulled parallel to **H**. If **M** in a uniaxially anisotropic film is assumed to stay in the plane of the film and to remain perfectly uniform, the position of **M** can be predicted by the Stoner-Wohlfarth (S-W) model of coherent magnetization reversal [3] for any arbitrary field **H** applied in the plane of the film at an angle β to the *EA*.

Referring the angles to the *EA* (Fig. 2*a*), the total energy is the sum of the anisotropy and field terms:

$$E = K \sin^2 \varphi_0 - HM \cos(\beta - \varphi_0) \tag{2}$$

It is convenient to define the normalized quantities* $\epsilon \equiv E/MH_K$, $h \equiv H/H_K$, $h_{\parallel} \equiv H \cos\beta/H_K$, and $h_{\perp} \equiv H \sin\beta/H_K$. Here $H_K \equiv 2K/M$ is known as the *anisotropy field*, and the subscripts $\parallel$ and $\perp$ refer to the

*Fields normalized to H_K are denoted by lowercase letters, i.e., $h \equiv H/H_K$, $h_c \equiv H_c/H_K$.

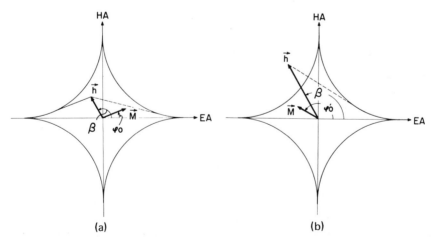

(a)   (b)

FIG. 2. Illustrating the graphical determination of the azimuth of **M** using the Stoner-Wohlfarth astroid. Field **h** applied at an angle β. (*a*) **M** at angle φ_0 before switching. (*b*) **M** at angle φ_0' after switching due to growth in magnitude of *h*.

"longitudinal" and "transverse" field components parallel and perpendicular to the EA, respectively. Then Eq. (2) becomes

$$\epsilon = \frac{1}{2} \sin^2 \varphi_0 - h_{\parallel} \cos \varphi_0 - h_{\perp} \sin \varphi_0 \tag{3}$$

Energy extrema are found by setting $\partial \epsilon / \partial \varphi_0 = 0$, which is equivalent to the statement that the net torque Λ_0 on $\mathbf{M}$ vanishes. Thus

$$\Lambda_0 = \frac{\partial \epsilon}{\partial \varphi_0} = \frac{1}{2} \sin 2\varphi_0 + h_{\parallel} \sin \varphi_0 - h_{\perp} \cos \varphi_0 = 0 \tag{4}$$

Values of φ_0 satisfying Eq. (4) represent stable equilibrium (minimum energy states) only if $\partial^2 \epsilon / \partial \varphi_0^2 > 0$, where, from Eq. (4),

$$\frac{\partial^2 \epsilon}{\partial \varphi_0^2} = \frac{\partial \Lambda_0}{\partial \varphi_0} = \cos 2\varphi_0 + h_{\parallel} \cos \varphi_0 + h_{\perp} \sin \varphi_0 \equiv \Lambda_1 \tag{5}$$

Here $\Lambda_1 = \cos 2\varphi_0 + h \cos(\varphi_0 - \beta)$ is known as the "uniform effective" (or single-domain) field.

From Eq. (4) the quasi-static hysteresis loop, i.e., the integrated flux picked up by a coil whose axis is parallel to an ac driving field, may be calculated for any arbitrary azimuth β of that field. Consider the important special case where $\mathbf{H}$ is parallel to the EA, i.e., $\beta = 0$. Then the solutions of Eq. (4) for $h_{\perp} = 0$ are $h_{\parallel} = -\cos \varphi_0$, and $\varphi_0 = 0, \pi$. From Eq. (5) it is seen that the former and latter solutions are unstable ($\Lambda_1 < 0$) and stable states ($\Lambda_1 > 0$), respectively, and that the transitions between the stable states $\varphi_0 = 0$ and $\varphi_0 = \pi$ occur (at $\Lambda_1 = 0$) at $h = \pm 1$. The resulting hysteresis loop obtained by plotting $M \cos \varphi_0$ vs. h is shown in Fig. 3a; it is rectangular with a coercivity H_c of H_K. Another important case is the hard-axis loop, given by plotting $M \sin \varphi_0$ vs. h. For this case, solutions of Eq. (4) (with $h_{\parallel} = 0$) are $h_{\perp} = \sin \varphi_0$, and $\varphi_0 = \pm \pi/2$; from Eq. (5) stable solutions are given by the former expression for $|h| \leq 1$ and the latter expression for $|h| \geq 1$ (Fig. 3b). Physically, $\mathbf{M}$ rotates with increasing $h_{\perp}$ until at $h_{\perp} = 1$ it is parallel to the HA, where it remains for still greater values of $h_{\perp}$. Hysteresis loops for azimuths other than along the EA and HA may be similarly determined [3].

Experimental hysteresis loops generally follow the theoretical predictions, although there are some important differences in detail [3, 45]. In

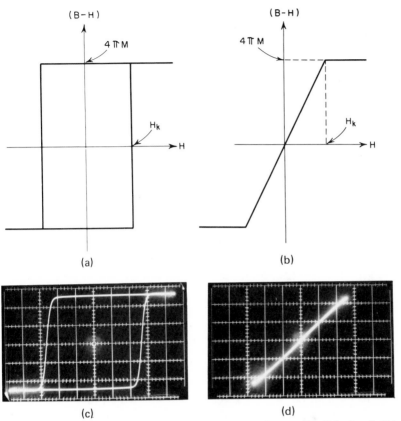

FIG. 3. *M-H* hysteresis loops. (*a*) Calculated *EA* loop. (*b*) Calculated *HA* loop. (*c*) Experimental *EA* loop. (*d*) Experimental *HA* loop. In loops in (*c*) and (*d*), one horizontal division per oersted. ((*c*) *and* (*d*) *courtesy of T. S. Crowther*)

particular, if care is taken to maintain a single-domain state, the *HA* loop (Fig. 3*d*) is similar to the curve of Fig. 3*b*; in fact, since the slope of the *HA* loop is theoretically proportional to the reciprocal of H_K (in non-normalized units) this loop is often used in hysteresigraph determinations of H_K (Sec. 6.3). On the other hand, while the experimental *EA* loop is similar to the curve of Fig. 3*a*, its sides are not infinitely steep and the coercivity is usually much less than H_K (Fig. 3*c*).

Assume now that **M** is pointing to the right along the *EA*. If now an increasing field **h** (Fig. 2*a*) is applied at an arbitrary obtuse angle β, **M** will rotate clockwise. If *h* is less than a certain threshold field h_s, **M** will relax back to its original position upon removal of the field. However, when h exceeds h_s, an irreversible transition process will occur in which **M** rapidly rotates into a position in the second quadrant between **h** and the *EA* (Fig. 2*b*). The film is now said to have been "switched," for upon setting **h** = 0, **M** is now found to point *left* along the *EA*. The switching transition occurs when an energy extremum changes from a minimum to a maximum, or when $\partial \epsilon / \partial \varphi_0 = \partial^2 \epsilon / \partial \varphi_0^2 = 0$ The solution of these conditions is [Eqs. (4) and (5)] $h_{\parallel} = -\cos^3 \varphi_0$, $h_{\perp} = \sin^3 \varphi_0$, or suppressing the signs,

$$h_{\parallel}^{2/3} + h_{\perp}^{2/3} = 1 \tag{6}$$

which represents a mathematical figure known as an astroid (Fig. 2). Equivalently, the uniform effective field Λ_1 vanishes at the astroid [Eq. (5)].

Using the astroid, the orientation of **M** may be found by a simple graphical construction [3] which is helpful in visualizing switching. If a straight line is drawn through the tip of **h** so that it is tangent to the astroid, then **M** is parallel to this line (Fig. 2*a*). It may be noted, however, that when the tip of **h** lies within the astroid it is intersected by tangents to *both* upper branches of the astroid. Which tangent gives the position of **M** depends on the magnetic history; assuming **M** pointed right (left) along the *EA* at **h** = 0, the solid (dotted) tangent would be used. If **h** is increased in magnitude so that its tip touches the astroid, by the definition of the latter a switching threshold is achieved. A further increase in h gives a state where only one tangent representing equilibrium can be drawn (Fig. 2*b*).

To a first approximation, the behavior of uniaxially anisotropic magnetic films is describable on the S-W model. Thus, under certain conditions, experimental hysteresis loops taken at arbitrary orientations β of the applied field **h** are similar to those theoretically predicted, although wide deviations from theoretical behavior are often observed. In a similar fashion, hysteresis loops and critical switching curves have also been calculated for films having *biaxial* (Sec. 4.2) and combinations of biaxial and uniaxial anisotropy [46, 47]. Again, experimental observations often show large deviations from the predicted behavior.

4. MAGNETIC ANISOTROPY

The magnetic properties of a ferromagnetic film are largely deter-mined by the magnitudes of the magnetic anisotropies present in the film. These anisotropies can originate in any of several different mechanisms which depend upon the film structure. Several anisotropy mechanisms can act simultaneously; the resultant magnetic behavior will depend on the energy contribution from each mechanism, and is not predictable simply from a vectorial addition of the anisotropies. Thus, simple trigonometry shows that the net anisotropy resulting from the presence of uniaxial anisotropies from two different mechanisms (e.g., M induced and oblique incidence) with noncollinear EA's leads to another uniaxial anisotropy of a predictable magnitude with an EA in a predictable direction [48, 49].

4.1 Shape Anisotropy [10]

If a ferromagnetic body is uniformly magnetized, magnetic poles will be created at the surfaces. If the body has an ellipsoidal shape, the "demagnetizing" field H_d which is generated by these poles will be antiparallel to **M**. In an ellipsoid of revolution $H_d = -N_1 M$ or $H_d = -N_2 M$, when **M** is parallel or perpendicular to the axis of revolution, respectively. Since the magnetostatic energy is given by $-\frac{1}{2} \int H_d \cdot M \, dV$, where the integration extends over the volume of the ellipsoid, the difference in energy density between the two states is given by $\Delta E = \frac{1}{2}(N_1 - N_2) M^2$.

A thin film may be considered a very flat, oblate ellipsoid of revolution. Then $N_1 \simeq 4\pi$, while $N_2 \simeq 4\pi t/(t + L)$, where L is the film diameter. The demagnetizing field perpendicular to the film plane is then $-N_1 M = -4\pi M$, which is about 10^4 Oe for a NiFe film, and is usually sufficiently large to keep **M** in the film plane. Alternatively, the film may be said to have a shape magnetic anisotropy, of energy density $K_\perp = \Delta E = 2\pi M^2$, with a hard direction normal to the film and an easy plane in the film plane. The demagnetizing field in the plane of the film is then $-N_2 M \simeq 4\pi t M/L$, which is only ~ 0.1 Oe for the film dimensions given in Sec. 1. Measurements of the external demagnetizing fields above a film are in good agreement with predictions of this simple ellipsoidal model [50].

Consideration of the shape anisotropy of a very elongated prolate ellipsoid of revolution will be useful later. Here $N_1 \simeq 0$, while $N_2 \simeq 2\pi$.

In this case the ellipsoid has an easy axis along the long axis of the ellipsoid, and an anisotropy energy density $K = \Delta E = \pi M^2$, which for NiFe is about 2×10^6 ergs/cm^3.

4.2 Magnetocrystalline Anisotropy

The free energy of a ferromagnetic single crystal depends upon the orientation of **M** with respect to the crystallographic axes because of magnetocrystalline anisotropy [10]. Neglecting higher-order terms, for a cubic crystal with direction cosines α_1, α_2, α_3 of **M**,

$$E_{\text{cub}} = K_1 \sum_{i>j} \alpha_i^2 \alpha_j^2 + K_2 \alpha_1^2 \alpha_2^2 \alpha_3^2 \tag{7}$$

For a cubic monocrystalline film with the [100] axis normal to the plane, if the demagnetizing field constains **M** to lie in the film plane, Eq. (7) becomes $E_{\text{cub}} = K_{\text{bi}} \sin 4\theta$; i.e., a biaxial anisotropy is developed.

Single-crystal films of Fe, NiFe, Co, and MnBi have been made by epitaxy [51-53]. The conditions for obtaining good single crystals are the same as those for nonmagnetic films (Chap. IV). In particular, the importance of providing a high density of nucleation sites on the single-crystal substrate, e.g., by pre-exposure to water vapor, has been demonstrated [54-56]. Measurements of the anisotropy constants (usually only K_1) are generally in good agreement with bulk data [57-59], while room-temperature measurements of anisotropy constants of cubic Co and hexagonal Ni phases, both unstable in bulk form, have been made [60, 61]. Departures from bulk results and the presence of uniaxial anisotropy components have been attributed to a strain-magnetostriction mechanism (Sec. 4.3) which has been verified by experiments in which the strain was released by stripping from the substrate [62].

The only true anisotropy of order higher than uniaxial originates in magnetocrystalline anisotropy. It should be pointed out, however, that for small applied fields, the magnetic behavior may be interpreted in terms of pseudo-biaxial, triaxial, or n-axial anisotropy if two or more regions (or films) are coupled magnetostatically or otherwise [46, 63, 64]. However, under the true test condition of large applied fields, only uniaxial anisotropy is admissible, since the addition of uniaxial anisotropies always yields another uniaxial anisotropy (see above).

4.3 Strain-Magnetostriction Anisotropy

When a ferromagnetic crystal is strained, the magnetocrystalline anisotropy is changed. Therefore, in order to minimize the magnetocrystalline anisotropy energy, an unconstrained crystal will exhibit a spontaneous strain which depends on the direction of **M** relative to the crystal axes; the strain measured along the [100] or [111] directions is defined as the "magnetostriction" constant λ_{100} or λ_{111}, respectively, if **M** is parallel to the strain. If a uniform stress S is applied to a sample consisting of randomly oriented crystals, it can be shown [65] that the energy of the resulting strained state is

$$E = \frac{3}{2} \lambda S \sin^2 \varphi_0 \qquad (8)$$

where φ_0 is the angle between **M** and the stress direction, and λ is an "average" magnetostriction $\lambda \equiv \frac{3}{5}\lambda_{111} + \frac{2}{5}\lambda_{100}$. (If there is crystallographic texture in the sample, λ_{100} and λ_{111} will have different coefficients in this equation.) Thus tension (compression) creates an EA (HA) along the strain direction for positive λ.

This mechanism can thus explain the uniaxial magnetic anisotropy which is created in a film subjected to anisotropic strain in the film plane [compare Eqs. (1) and (8)]. It is convenient to define the magnetoelastic coefficient $\eta \equiv dH_K/de$, which gives the change in H_K occurring upon application of a given strain e. It is found that η is a function of composition [66-68] (Fig. 4); in fact η is often used as a convenient index of the composition (Sec. 11.2).

Deposited films are often found in a state of high strain (Sec. 4.7), so that if the net strain is *anisotropic*, the resulting magnetoeleastic anisotropy, when properly combined with the **M**-induced anisotropy (Sec. 4.4), can produce a "skew" of the net EA from the intended direction. Furthermore, *microscopic*, randomly oriented strains can cause local anisotropy fluctuations which contribute to the magnetization ripple (Sec. 5). Since both these effects are usually considered undesirable, films of compositions near the zero magnetostriction composition (ZMC), about 81 percent Ni, 19 percent Fe (Fig. 4), are often made. It should be noted, however, that at the ZMC, λ_{111} and λ_{100} although small, do not separately vanish, but only the average magnetostriction vanishes, so that there can be a *local* magnetoelastic contribution to the ripple at the ZMC.

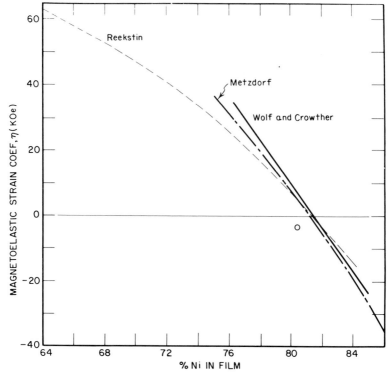

FIG. 4. Magnetoelastic strain coefficient $\eta \equiv dH_K/de$ vs. composition for NiFe films. Metzdorf and Reekstin, vacuum-deposited films; Wolf and Crowther, electroplated films. (*After Wolf and Crowther* [66]; *Metzdorf* [67]; *Reekstin* [68])

It has been recently pointed out [67] that for films of constant composition, variations of η equivalent to a composition shift of several percent can occur, depending on substrate temperature and deposition rate. The origin of this effect probably lies in changes in η with crystalline-defect concentration, as indicated by annealing studies. It has also been noted [68] that a concentration gradient through the film thickness, as often observed in electroplated [69] and some vacuum-deposited [70] films, can cause changes in η.

4.4 *M*-induced Uniaxial Anisotropy [71]

If a saturating magnetic field is applied in the plane of a polycrystalline film during its deposition, a uniaxial anisotropy is induced with the EA parallel to the applied field. This phenomenon, which has been observed in Ni, Fe, Co, Gd, and alloy films, provides the basis for much of the scientific and technological interest in ferromagnetic films. It is therefore ironic that the origin of induced uniaxial anisotropy is still very poorly understood.

This anisotropy should be called magnetization-induced, rather than field-induced anisotropy [72, 73]. Films deposited in the absence of a field still show anisotropy with local easy axes in the directions which **M** happened to take during the deposition; the function of the applied field is simply to ensure that **M** (and hence the resulting EA) is uniformly aligned in a desired direction.

The value of K_u obtained is strongly dependent upon deposition conditions, notably the substrate temperature, deposition rate, and any subsequent annealing treatment. The importance of the inclusion of impurities from the ambient medium is not clear. Changing the residual pressure over several orders of magnitude during vacuum deposition [9, 74] does not strongly affect K_u, but exposure to air after vacuum deposition [75] can affect the magnetic annealing characteristics of K_u. A reported influence on K_u of impurities from the walls of the crucible [76] was not confirmed [77]. In any case, experience has shown that care must be taken to maintain reproducibility of deposition conditions in order to reduce the scatter in experimental values of K_u to reasonable limits.

In Fig. 5, K_u for vacuum-deposited NiFe films is shown as a function of Ni concentration in the film [77-80]. (This concentration is not the same as the melt concentration because of fractionation effects [81].) It should be noted that the true substrate temperature may be very different from that of the substrate support [82]; Wilts [77] is the only author who has taken care to give the true substrate temperature.

An examination of bulk NiFe phenomena should be helpful in the search for the origin of the M-induced anisotropy in films. It is well known that a heat treatment of bulk ferromagnets in a magnetic field (a "magnetic anneal") will sometimes induce magnetic anisotropy [83]. The accepted interpretation is that short-range directional order is induced by atomic rearrangements during a high-temperature anneal; i.e., because of magnetic interactions a preferred orientation of like-atom pairs is developed [84]. This directional order is frozen in when the sample is cooled, thus creating uniaxial anisotropy. In NiFe, the

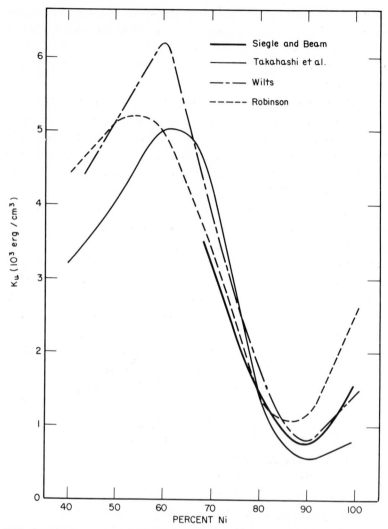

FIG. 5. Uniaxial anisotropy constant K_u vs. NiFe composition. Substrate temperatures: Robinson [78], 240°C; Siegle and Beam [79], 250°C; Takahashi et al. [80], 300°C; Wilts [77], 250°C.

anisotropy due to Ni-Ni (or Fe-Fe) pairs is theoretically given by [83, 84]

$$K_u = \frac{\alpha c^2 (1 - c)^2 [B(T_a)]^2 [B(T_m)]^2}{T_a} \tag{9}$$

where c is the atomic concentration of Ni, T_a is the annealing temperature, T_m is the temperature at which the measurement is taken, $B(T)$ is the Brillouin function, and α is a constant. Only nearest-neighbor interactions were considered in the derivation of Eq. (9), while the composition dependence is based on the assumption of ideal solubility. Experiments on bulk binary alloys are in general agreement with the pair-ordering theory [84].

Even if pair-ordering anisotropy is operative in ferromagnetic films, it is clear from the simple fact that uniaxial anisotropy can be induced in Ni, Fe, and Co films that at least one additional anisotropy mechanism must be involved. For this reason several authors [78] have considered the following strain-magnetostriction mechanism (Sec. 4.3): At a temperature T, a film which is free from its substrate will sustain a temperature-dependent strain $\lambda(T)$, which is the saturation magnetostriction suitably averaged for a polycrystalline film. Assume that, as the film is cooled after deposition, a critical temperature T_0 is reached below which the atoms lose their mobility, so that the film is constrained by the substrate to maintain the strain $\lambda(T_0)$. Thus a stress $S = Y\lambda(T_0)$, where Y is Young's modulus, is exerted by the substrate on the film, and is tensile (compressive) if $\lambda(T_0)$ is positive (negative). At the measuring temperature T_m the saturation magnetostriction is $\lambda(T_m)$; since the film is under the stress S a uniaxial anisotropy is created

$$K_u = \frac{3}{2}\lambda(T_m)S = \frac{3}{2}Y\lambda(T_0)\lambda(T_m) \tag{10}$$

It is difficult to estimate T_0; presumably it is within the range between T_m and the deposition temperature T_d.

West [85] has pointed out that K_u calculated by Eq. (10) is fundamentally incorrect, for rather than the magnetostriction, the magnetoelastic *energy* should be averaged over all the crystallographic directions. Using the correct formulation he obtained values of K_u of 4.2, 1.8, and 42×10^3 ergs/cm^3 for Ni, Fe, and Co, respectively; these are maximum values since it was assumed that $T_0 = T_m =$ room temperature. Experimental results are roughly consistent with these estimates, but it must be emphasized that the strength of the bond between the film and the substrate determines the constraint which the substrate can exert on the film, and hence the effective value of T_0. This bond can change with substrate cleanness, deposition temperature, and background pressure (see Chap. IV), thereby explaining the large scatter in K_u values found for films having high values of λ.

Robinson [78] attempted to fit his K_u data for NiFe (Fig. 5) with a pair-strain model by using *both* Eqs. (9) and (10). The constant $\mathcal{C}$ was fixed by assuming that the strain contribution vanished at the zero magnetostriction composition (81 percent Ni, 19 percent Fe), while T_a and T_0 were set equal to T_d. This procedure gave a fair fit to the data, but the theoretical peak value of K_u was less than the experimental value and occurred at a Ni concentration above the experimental value. However, West [85] showed that his reformulation of Eq. (10) led, for Ni concentrations $<$ 65 percent, to a significant increase in the strain contribution to K_u, while the theoretical peak occurred at a Ni concentration (about 50 percent) near that found experimentally. Siegle and Beam [79] also successfully fitted their K_u vs. composition data (Fig. 5) with Eqs. (9) and (10) both before and after magnetic-annealing treatments changed the values of K_u; the annealing or deposition temperatures were used for T_a and T_0 in these equations.

However, these fits of the pair-strain model to the K_u data are only semi-quantitative because of the doubt about the values of T_0 and T_a, and ignorance of the correct concentration dependence of Eq. (9) and the temperature dependence of the parameters of Eq. (10). Also, the experimental data in Fig. 5 may not be pertinent for a theoretical fit for Ni concentrations less than 40 percent because of a γ-to-α phase change below this Ni concentration; upon proper extrapolation it is found that if the γ phase were retained no peak in the K_u vs. Ni concentration curve would be found at all [86].

Furthermore, determinations of K_u as a function of temperature [87, 88] (*EA* field applied during heating) or as a function of magnetization [89] (by adding Cr) are not in accord with Eq. (9). Thus a better insight into the origins of K_u is needed; this insight is provided by magnetic-annealing studies. Upon application of a *HA* field during the anneal, H_K decreases monotonically; for a sufficiently high annealing temperature H_K can even assume negative values, i.e., an interchange of the *EA* and *HA* occurs. A subsequent *EA* anneal can restore the anisotropy to nearly its original value, and the procedure can be repeated.

Studies of the kinetics of such annealing experiments show that in nonmagnetostrictive 81 percent Ni, 19 percent Fe films (for which the strain-magnetostriction mechanism is nearly inoperative [85]), *several* mechanisms with different activation energies are active [88, 90, 91].

One of these experiments [91] involved deposition at various substrate temperatures in an EA field, annealing in an EA, HA, or rotating field for short times, then quenching to room temperatures; in this way five different anisotropy contributions were found. Still another approach to the study of anisotropy kinetics is provided by studies of the changes of K_u in films subjected to radiation damage by high-energy nuclear particles [92].

However, the experiments giving the best insight into anisotropy sources are those which yielded the full spectrum of annealing activation-energies and decay times. In such experiments [75, 88, 93] the film was subjected to a HA anneal at a temperature T_a shortly after deposition without breaking vacuum, and either continuous or periodic measurements of H_K were made by magnetoresistive techniques. The H_K vs. time t data were then analyzed by assuming that for n annealing processes

$$H_K(t) = \sum_{j=1}^{n} H_{K_j} \left[2 \exp\left(-\frac{t}{\tau_j}\right) - 1 \right] \tag{11}$$

where τ_j is the jth relaxation time with activation energy E_j,

$$\tau_j = \tau_{0j} \exp\left(\frac{E_j}{kT_a}\right) \tag{12}$$

Using Eq. (11), several relaxation times τ_j were obtained for each value of T_a; the experiment was then repeated, after returning the EA to its original position, for different values of T_a. In this way curves of τ_j vs. T_a were obtained which in turn gave values of E_j. Results are shown in Table I; note that the results of the two groups of investigators agree for processes 3 and I.

The activation-energy values in Table I are considerably lower than the Ni-Ni pair-ordering energy of 3 eV found in bulk material [88]. Further, if additional anisotropy sources are sought in the directional order of some of the structural defects (interstitials or vacancies) which are found in high concentration in thin films (Chap. IV), poor agreement with activation energies and relaxation times for either interstitial atomic pairs or vacancy doublets is found [94]. Other processes involving directional order of defects are conceivable; alternatively it is possible that the

Table I.   Anisotropy Processes*

Process	E_j,eV	τ_{0j}, sec	Fraction of total H_K
(Smith et al. [75]):			
3	0.15	2	0.10
1	0.18	7×10^{-2}	0.08
5	0.26	3×10^{-2}	0.05
2	0.34	1.5×10^{-3}	0.07
4	0.48	6×10^{-6}	0.08
(Kneer and Zinn [88]):			
I	0.15	1.5	0.20
II	0.4	3×10^{-2}	0.08
III	0.4	3×10^{-1}	0.16
IV	1.5	2×10^{-8}	0.56

*The process-numbering scheme of Smith et al. is used. Considerable experimental uncertainty is associated with processes 4 and 5. Note agreement between processes 3 and I.

anisotropy originates only in Ni-Ni pairs, but the presence of defects can affect the reorientation of these pairs. Kneer and Zinn [93] favor the latter interpretation* for the predominant anisotropy process (process IV, Table I), after finding that the major activation energy is raised to 2.1 eV following a heat treatment which recrystallized the film. It is hoped that an experimental approach which excites only individual anisotropy processes by a rotating anneal technique [95] will permit more accurate measurements of τ_j and E_j, and hence permit a better identification with atomic processes. Another promising method is that of introducing a specific defect and studying its annealing behavior [96, 97].

Annealing experiments can also be performed on films having nonvanishing magnetostriction, so that the strain-magnetostriction mechanism is operative. The EA in films of NiFe, Ni, or Co,

*The recent annealing experiments of Iwata and Hagedorn (T. Iwata and F. B. Hagedorn, J. Appl. Phys., **40** (1969)) were also explained on this assumption. These authors believe that the annealing out of defects causes behavior of the anisotropy which may be *misinterpreted* [Eqs. (11) and (12)] as changes of relaxation frequency with annealing time.

vacuum-deposited at room temperature, can be reoriented into the annealing field direction by room-temperature anneals of several days duration [74, 96, 98], while films deposited at several hundred degrees Celsius show no room-temperature annealing. This behavior is explainable on the strain-magnetostriction model if it is assumed that the film-substrate bond is weak for a room-temperature deposition so that the constraint of the substrate on the film may be changed by a magnetic anneal. Attempts to measure the activation-energy spectrum in Ni films were not very successful because of wide scatter in experimental results [93], while curves of H_K vs. temperature (measured in an *EA* field) showed small steps which presumably corresponded to sudden strain-relief processes. Kneer and Zinn [93] interpreted these data as showing that the constraint originates not in stress exerted on the film by the substrate [78, 85], but in internal anisotropic microstresses associated with anisotropic atomic-defect structures. This question is still open, however.

4.5 Oblique-incidence Uniaxial Anisotropy

In view of the considerations of Sec. 4.1, it is not surprising that geometric inhomogeneities in a film, i.e., departures from the ideal planar shape, can affect the magnetic anisotropy. Thus unidirectional grinding of the substrate [99] can induce uniaxial shape-anisotropy in the magnetic film deposited on it. Also, small-scale *random* geometric anisotropy scattered throughout the substrate surface causes local randomly oriented magnetic shape anisotropy, which increases the magnetization dispersion (Sec. 5) and hence affects other magnetic properties [100, 101]. For these reasons attempts are usually made to provide a smooth, clean substrate [102] in order to obtain films having uniform *M*-induced anisotropy with low magnetization dispersion.

Shape anisotropy can play a decisive role, however, even if the substrate is perfectly smooth and clean. If a ferromagnetic film is vacuum-deposited with the vapor beam not normal to the substrate, as tacitly assumed above, but at oblique incidence, uniaxial anisotropy will be induced either in the absence of an applied field or in a rotating field [103]. In Fig. 6a, H_K is given as a function of incidence angle θ (measured from the film normal) for 81 percent Ni 19 percent Fe films [104]. Here positive (negative) values of H_K mean the *EA* is perpendicular (parallel) to the plane of incidence. Similar results have been obtained by many workers for NiFe, Fe, Co, and Ni films [105, 106].

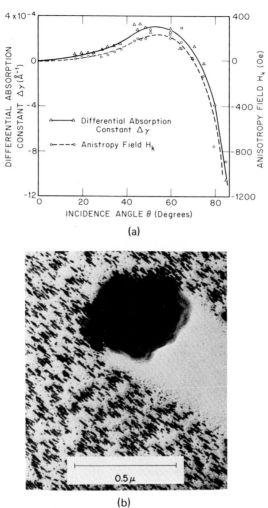

(a)

(b)

FIG. 6. (*a*) Anisotropy field H_K and differential absorption coefficient $\Delta\gamma$ for polarized light vs. incidence angle θ for oblique incidence 81 percent Ni, 19 percent Fe films deposited at 200°C substrate temperature. (*After Cohen* [104]) (*b*) Electron micrograph of an 81 percent Ni, 19 percent Fe film deposited at 85° incidence angle and 200°C substrate temperature. The shadow cast by the large dirt particle reveals the vapor-beam direction. (*After Cohen* [104])

The values of H_K can become very large, ~300 Oe at $\theta \simeq 50°$ for perpendicular anisotropy and ~−1,000 Oe at $\theta \simeq 85°$ for parallel anisotropy. As expected, oblique-incidence effects are not observed in electrodeposited or sputtered films because the incidence directions of the incoming atoms are randomized in these processes (Chap. IV).

Oblique-incidence anisotropy of either sign can be explained on the basis of shape anisotropy. For low values of $\theta (H_K > 0)$, a self-shadowing mechanism is invoked according to which regions behind growing crystallites do not receive incoming vapor because they are in the "shadows" of the crystallites; as the crystallites grow in size they will then tend to agglomerate into parallel crystallite "chains" elongated normal to the plane of incidence and separated from each other by voids.* The shape anisotropy of these chains provides the magnetic anisotropy.

Evidence for this shape-anisotropy interpretation comes not only directly from electron micrographs [103, 108] which show the crystallite chains, but from observations of anisotropic electrical resistance [103] and anisotropic absorption of polarized light [103, 104] (dichroism). The dichroism was investigated by measuring the differential absorption constant $\Delta \gamma \equiv \gamma_\| - \gamma_\perp$, where $\gamma_\|$ and $\gamma_\perp$ are the optical-absorption constants when the plane of polarization is respectively parallel and perpendicular to the axis of greatest absorption. In Fig. 6a, $\Delta \gamma$ is plotted against θ; it is seen that, as anticipated, the optical-anisotropy and magnetic-anisotropy dependences on θ are in agreement. The production of crystallite chain structure by oblique incidence of the vapor is not confined to magnetic materials; its presence in nonmagnetic materials has been detected by electron microscopy [109] and observations of dichroism [103]. Therefore, if NiFe films are deposited at normal incidence [103] or by electrodeposition [110] on underlayers of nonmagnetic metals which were previously deposited at oblique incidence, uniaxial anisotropy with the *EA* normal to the incidence plane is produced; i.e., the geometric anisotropy of the underlayer is "replicated" by the NiFe.

For high values of $\theta (H_K < 0)$ a different geometric anisotropy mechanism is involved; the crystallites tend to elongate in the direction of the incoming vapor beam to form "crystallite columns," as is seen from the micrograph of Fig. 6b. Thus there is a competition between the shape anisotropy of the crystallite columns, contributing to parallel anisotropy, and that of crystallite chains (which can also be seen in Fig. 6b), contributing to perpendicular anisotropy. The columnar growth

---

*An alternative mechanism based on inhibition of atomic mobility by chemisorbed oxygen has recently been offered to explain the anisotropic shape of the nuclei in the *earliest stages* of film growth [107].

overbalances the chains at higher angles. That this geometric interpretation is correct is shown by the agreement of the Δy and the H_K dependences on θ over the entire θ range.

The work just described was carried out at relatively low values of T_d ($\lesssim 250°C$). Other work [108] showed that in the $300°C < T_d < 400°C$ range the magnitude of oblique-incidence anisotropy decreased, presumably because enhanced atomic mobility at higher substrate temperatures allowed filling of the voids. Further, the negative portion of the H_K vs. θ curve (Fig. 6a) shifted toward lower values of θ (at $T_s \simeq 350°C$ negative H_K values could be observed at θ values of 20 to 30°); i.e., at higher values of T_d the chain-column balance shifts in favor of columns [111, 112]. This interpretation was again verified by good correlation with dichroism measurements. Annealing treatments of films having perpendicular oblique-incidence anisotropy also emphasized the delicate nature of the chain-column balance, for it was found that under certain circumstances annealing treatments converted perpendicular into parallel anisotropy, i.e., the annealing suppressed the chains, but not the columns [112, 113].

It should be mentioned that explanations of oblique-incidence anisotropy in terms of fiber axes or strain-magnetostriction models were generally rejected [103, 106] as being inadequate, except in the case of NiFe films with a Ni content >90 percent. These films showed parallel anisotropy under conditions where 81 percent Ni 19 percent Fe films showed perpendicular anisotropy. A strain-magnetostriction mechanism associated with tension of the chains was therefore postulated for these films [103].

More recently, it has been found that if alloy films are produced by simultaneous deposition from two separate sources, one evaporating a ferromagnetic and the other a nonferromagnetic metal (Ag, Al, Cu, Au, Mo, Pb, Sn, or Zn), perpendicular oblique-incidence anisotropy at remarkably low incidence angles is produced [114, 115]. Thus at a (81 percent Ni, 19 percent Fe) vapor angle of only 6°, an addition of 8 percent Cu is sufficient to give a perpendicular anisotropy contribution of about [115] 1/2 Oe. Electron-microscopy, electrical-resistivity, and optical-dichroism measurements did not reveal the crystallite chain structure discussed above, while electron diffraction and studies of M vs. Cu concentration failed to reveal evidence of inclusions of second phases [115]. Therefore, a model was proposed explaining codeposition oblique-incidence anisotropy in terms of a very thin grain-boundary layer which was enriched in the nonmagnetic metal. The thickness of this interfacial layer around a given crystallite would vary with azimuth, depending upon the directions of the incoming vapors, thus generating magnetic anisotropy.

4.6   Unidirectional Anisotropy

It is possible to prepare special films whose lowest-energy magnetic state is achieved when **M** is parallel (but not antiparallel) to a specified easy *direction* (not axis) in the film plane [116]. Such unidirectional anisotropy stands in contrast with uniaxial anisotropy, where neither of the two senses of the *EA* is energetically preferred. Unidirectional anisotropy, which is also found in bulk material, depends on the exchange interaction between a ferromagnet and an antiferromagnet which are in intimate contact; it is therefore sometimes called "exchange anisotropy." The antiferromagnetic component is assumed to have a high uniaxial anisotropy and a Néel temperature lower than the Curie temperature of the ferromagnet. The *EA* direction in the antiferromagnet is then fixed, through ferromagnetic-antiferromagnetic exchange, by the position of **M** in the ferromagnet as the system is cooled through the Néel temperature. The unidirectional anisotropy is thereafter maintained by the exchange interaction; it cannot be changed below the Néel temperature since an external field cannot act directly on the antiferromagnet.

Unidirectionally anisotropic films in the systems Co-CoO, Fe-FeS, Ni-NiO, NiFe-NiFeMn, NiFe-Cr, and NiFe-Cr_2O_3 have been prepared [117-119]. These systems show one or more of the three properties characteristic of ferromagnetic-antiferromagnetic coupling: a unidirectional ($\sin \theta$) component in the torque curve at high fields, a $B-H$ loop displaced from the origin along the H axis, and rotational hysteresis at high fields. A good quantitative theory explaining the measurements does not exist; in particular it is not even understood how these three characteristics can be simultaneously present at the same value of the applied field. Further, some unexplained time-dependent or magnetic-viscosity effects have been noted [119].

4.7   Perpendicular Anisotropy [120]

If **M** is inclined at an angle θ to the film normal, and if anisotropy within the plane is ignored, the resultant energy may be written as

$E_\perp = -K_\perp \sin^2\theta$ where $K_\perp$ is the perpendicular anisotropy constant. This equation implies, for $K_\perp > 0$, that the film normal is a hard axis while the film plane is an easy plane. If the only contribution to $K_\perp$ is given by the shape demagnetizing factor of the film (Sec. 4.1), then $K_\perp = 2\pi M^2$. However, ferromagnetic resonance [121] (Sec. 8.1) and torque-magnetometer measurements [57] have sometimes demonstrated the existence of an additional structural contribution $K_\perp^S$ (which is often negative so as to tend to maintain **M** normal to the film plane); in that case $K_\perp = 2\pi M^2 + K_\perp^S$. [It is useful to define a perpendicular *anisotropy field* $H_K^\perp \equiv 2K^\perp/M = 4\pi M + (H_K^\perp)_S$, where $(H_K^\perp)_S \equiv 2K_\perp^S/M$].

Thin films are often in states of high strain (Chap. V), either because of the difference in thermal-expansion coefficients of the substrate and films (for films deposited at an elevated temperature and subsequently cooled) or because of the "intrinsic" strain in the film caused by defects induced during deposition. As opposed to the strain considered in Sec. 4.3, this strain is considered isotropic within the plane of the film. Strain measurements made on NiFe films demonstrate [122, 123] that the strain, which is tensile at low substrate temperatures, decreases in magnitude as T_d is increased and becomes compressive at sufficiently high values of T_d ($T_d \gtrsim 100°C$). The associated stress S, when coupled to the magnetostriction λ, provides a contribution of $3\lambda S/2$ to $K_\perp^S$ for polycrystalline films (Sec. 4.3). The strain-magnetostriction contribution to $K_\perp^S$ is thus strongly dependent on T_d; it is also dependent on composition because of the strong dependence of λ on composition.

A clear demonstration of the strain-magnetostriction contribution to $K_\perp^S$ is afforded by the observation [124] that the measured value of $K_\perp$ of epitaxially grown Ni films reverted to $2\pi M^2$ upon stripping from the substrate, while the strain, as measured by x-rays, disappeared. Further, the value of $K_\perp^S$ calculated from λ and S was consistent with the measured value of $K_\perp^S$. By measuring $K_\perp, S$, and M, it was also shown [122] that the only contribution to $K_\perp^S$ for polycrystalline Ni films deposited at various substrate temperatures came from strain-magnetostriction.

On the other hand, it is difficult to accept strain-magnetostriction as the sole contributing mechanism in the interpretation of the variation of $K_\perp$ with T_d for films of compositions associated with values of λ considerably *different* from that of Ni. Thus, an appreciable contribution to $K_\perp^S$ is also found [125] for films of 80 percent Ni, 20 percent Fe, which have $\lambda \simeq 0$, while such films can also exhibit "stripe domains" [126] for which an appreciable value of $K_\perp^S$ is required (Sec. 10.1). The additional contribution to $K_\perp^S$ is furnished by the shape anisotropy of crystallite columns (Sec. 4.5), where the columns now grow normal to the substrate under conditions of low T_d and slow evaporation. Estimates [127] of the shape anisotropy of the columns are consistent with measured values of $K_\perp$, while such columns have been observed in Ni and NiFe films by low-angle electron diffraction [128], and by electron microscopy in Al films [129]. It is also found that $K_\perp^S$ increases as the pressure during deposition increases [130]. This effect has been associated with the strain-magnetostriction mechanism, particularly for Ni films [122], but for some compositions there are indications that [125, 131] high pressure may enhance the columnar growth, or possibly generate still another anisotropy mechanism.

The columnar-growth contribution to $K_\perp^S$ is clearly related to oblique-incidence parallel anisotropy (Sec. 4.5). Thus, it has been shown that anisotropy nearly parallel to the direction of the columnar growth is generated in oblique-incidence films [132]; the parallel oblique-incidence anisotropy discussed in Sec. 4.5 is that part of the microscopic shape anisotropy which is effective in the plane of the film.

Another contribution to $K_\perp^S$ can be furnished by magnetocrystalline anistoropy, if the films are either monocrystalline or textured with the proper orientation. Thus [133, 134] stripe domains have been observed in textured Co films and in single-crystal foils of Ni, Fe, and Co.

5. MAGNETIZATION RIPPLE [135]

It is well established that **M** is not uniform in direction throughout a real uniaxially anisotropic film, but in the film plane it suffers small quasi-periodic, local angular deviations (a few degrees) from its average direction. Such deviations, known as "magnetization dispersion" or "magnetization ripple," are important because many of the experimentally observed departures from the S-W theory are associated with them.

The local perturbations in magnetization direction originate in local random magnetic inhomogeneities, usually thought to be associated with the randomly oriented crystallites composing the film. The resulting ripple is known to be primarily longitudinal rather than transverse (Fig. 7); i.e., the "wavefronts" of the ripple are perpendicular rather than parallel to the average **M** direction. Longitudinal ripple is energetically preferred since, although both configurations have the same exchange

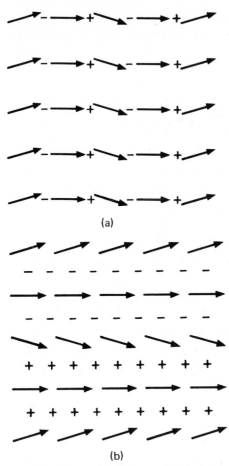

FIG. 7. Schematic diagram of magnetization ripple. Mean magnetization direction is horizontal. (*a*) Longitudinal ripple. (*b*) Transverse ripple. (*After Harte* [155])

and anisotropy energies, transverse ripple has higher values of $\nabla \cdot \mathbf{M}$ and hence higher magnetostatic energy.

Because of the scale of the ripple, only Lorentz microscopy or very high resolution magneto-optic microscopy [136] (Sec. 11) offer direct methods for observation (see Fig. 17, for example). Using the geometric-optics interpretation of Lorentz microscopy, analyses have been made of the predominant ripple wavelengths [137-140]. However, in view of the fact that it is known that only wave and not geometric optics is valid for such analyses (Sec. 11), this work is not totally acceptable. Nevertheless, the rough estimates that the ripple wavelengths and angles are of the order of 1μ and $1°$, respectively, are probably correct [141]. It is hoped that future Lorentz-microscopy investigations utilizing wave optics will provide more information on the ripple spectrum.

A useful *macroscopic* index of the magnetization dispersion in a given film is furnished by the measurement of the splitting angle α_{90} (Sec. 6.2), or by a measurement of the normalized wall coercive force $h_c \equiv H_c/H_K$ which increases monotonically with α_{90} (Sec. 7.2). Low-dispersion films are characterized by $\alpha_{90} \sim 1°$ and $h_c \sim 0.2$, while high-dispersion films can have α_{90} in tens of degrees and $h_c > 1$.

Many attempts have been made to determine whether the random perturbing anisotropies originate in strain-magnetostriction or magneto-crystalline anisotropy.* Thus, several workers have measured α_{90} vs. Ni content in NiFe films [73, 144, 145]. A broad, flat minimum around the composition 81 percent Ni, 19 percent Fe, the composition for zero average magnetostriction, is revealed (Fig. 8a), but this curve does not permit a choice between the two mechanisms because the magnetocrystalline constant K vanishes at 74 percent Ni, 26 percent Fe. Furthermore, strain-magnetostriction could be important even if the *average* magnetostriction vanished, since the *local* magnetostriction and the *local* strain, which may be spatially dependent, are important (Sec. 4.3).

Lorentz-microscopy investigations have given conflicting conclusions regarding the origin of ripple [59, 146]. On the other hand, Faraday-effect observations of *HA* domain splitting [147] (Sec. 6.2) support the strain-magnetostriction model. Perhaps the most incisive study is that of Fujii et al. [148], who illuminated small (a few hundred microns diameter) areas of their films in order to obtain local Faraday-effect measurements of α_{90} (Sec. 11.2). Both long-range (many millimeters) and local (presumably $\sim 1\mu$) contributions to α_{90} were found. The long-range contribution originated in strain-magnetostriction, while the local portion originated in *both* strain-magnetostriction and magneto-crystalline anisotropy. The latter conclusion was based on studies of local α_{90} vs. composition at various values of T_d (Fig. 8b), which showed

*The contribution from the spatial variations in the M-induced anisotropy caused by the random orientations of the crystallites is small [142], while an early report that the film composition is spatially dependent has not been confirmed [143].

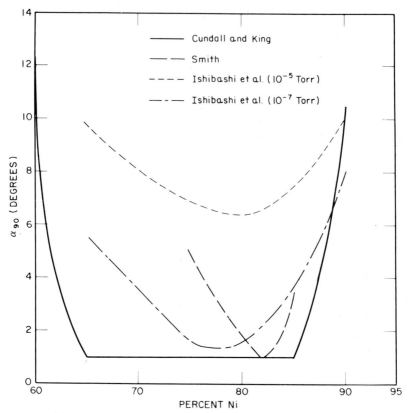

FIG. 8. (*a*) Dispersion angle α_{90} vs. composition of NiFe films.   Substrate temperature 250°C.   (*After Smith* [73]; *Cundall and King* [144]; *Ishibashi et al.* [145])

a minimum at 75 percent Ni, 25 percent Fe (the composition for vanishing magnetocrystalline anisotropy) only when T_d was high enough to minimize local strains; otherwise the minimum was at 81 percent Ni, 19 percent Fe (the composition for vanishing magnetostriction).

5.1 Anisotropy-dispersion Model [3]

The first model proposed to explain ripple was based on a "dispersion of the anisotropy." The film was postulated to consist of a collection of *noninteracting* regions, each region having an *EA* and K_u value which differed slightly from that of its neighbors in direction and magnitude, respectively. Measurements which were sensitive to the presence of ripple (Sec. 6) were interpreted as showing that low-ripple films had statistical angular dispersions "α_{90}" of *EA* directions of a few degrees and

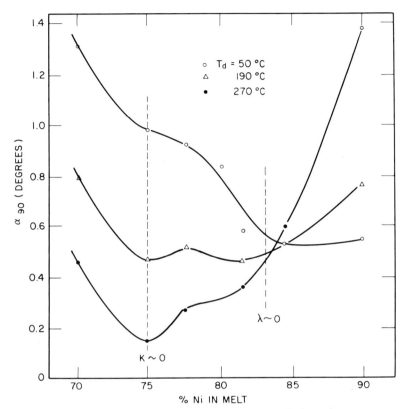

FIG. 8. (*b*) Dispersion angle α_{90} vs. composition of NiFe films for various substrate temperatures. Measurements were made with Faraday effect over 0.1 mm^2 area so as to detect local variations only. (*After Fujii et al.* [148])

magnitude dispersions "Δ_{90}" of K_u of about 10 percent. This model has the advantage of relative computational simplicity since the S-W model applies separately to each region; the astroid associated with each region has an appropriate tilt with respect to the average *EA* direction and represents the H_K value of the region. In case the regions are considered to have higher-order anisotropy than uniaxial, e.g., biaxial (Sec. 4.2), the astroids are simply generalized to more complex critical curves. The anisotropy-dispersion model was not designed primarily to describe ripple itself, e.g., as revealed by Lorentz microscopy, but to aid in discussing certain ripple-dependent experiments (Sec. 6).

5.2 Micromagnetic Model

The anisotropy-dispersion model of ripple does permit explanations of many experimental results, but the model is unsatisfying because ad hoc assumptions are often required upon application to particular problems, and because the model either ignores intercrystallite magnetic interactions or at best handles them in a superficial manner. To meet these objections, the problem has been approached using a "micromagnetic" model [149].

To apply micromagnetics, it is assumed that **M** is independent of the coordinate normal to the film plane, so that the desired distribution of **M** is two-dimensional. Further (in the static case), **M** is everywhere parallel to the film plane because of the shape demagnetizing field. It is also assumed that uniaxial anisotropy, uniform in magnitude and direction, has been induced throughout the film. However, within each randomly oriented crystallite, a local anisotropy is superimposed upon the uniaxial anisotropy. It is this spatially randomly varying local anistoropy which causes the perturbations of the direction of **M** which are known as magnetization ripple (Fig. 7). Now **M** cannot closely follow the anisotropy perturbations because exchange and magnetostatic forces prevent abrupt change in **M**. Thus the magnetization distribution in the film must be determined by consideration of the many components of the local total energy E_{tot}. Using the coordinate system of Fig. 9, these components are

External field energy $\qquad E_H \quad = -HM \cos(\beta - \phi)$ $\qquad\qquad$ (13a)

Uniaxial anisotropy energy $\quad E_{K_u} \quad = K_u \sin^2\phi$ $\qquad\qquad$ (13b)

Exchange energy $\qquad\qquad E_{ex} \quad = A(\nabla\phi)^2$ $\qquad\qquad$ (13c)

Local anisotropy energy $\qquad E_K \quad = Kf(\phi, x, y)$ $\qquad\qquad$ (13d)

Magnetostatic energy $\qquad E_{mag} = -\frac{1}{2}\mathbf{H}_{mag}\cdot\mathbf{M}$ $\qquad\qquad$ (13e)

In these equations, E_H is the energy due to the action of the external field **H** on **M**; E_{K_u} is the uniaxial anisotropy energy; E_{ex} represents the "stiffness" of **M**, i.e., its tendency to avoid sharp gradients; E_K is the local perturbing anisotropy energy which is associated with a "generalized crystallite" of diameter D and is described by the product of the anisotropy constant K and the trigonometric function f; and E_{mag} is the energy originating in the magnetostatic (or stray) field created by internal magnetic poles, which are caused by fluctuations in **M**. Here a

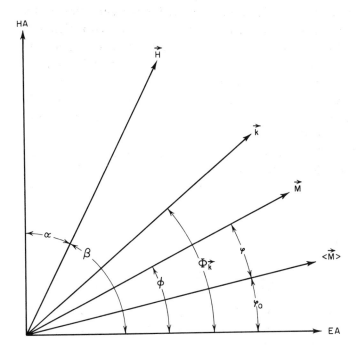

FIG. 9. Coordinate system for micromagnetic theories. In Hoffmann's theory the x axis is along $\mathbf{M}$, and the y axis is perpendicular to $\mathbf{M}$; in Harte's theory, the x and y axes are parallel to the EA and HA, respectively.

generalized crystallite means a region over which the local perturbing anisotropy is constant (either an actual crystallite if K originates in magnetocrystalline anisotropy or perhaps a larger region if K originates in strain-magnetostriction anisotropy). The first two terms are also present in the S-W model (Sec. 3), while the last three terms originate in the magnetization deviations caused by the local anisotropy. It is the E_{mag} term which causes the greatest mathematical difficulties in the micromagnetic approach; indeed, several early theories are unsatisfactory because this term was treated too crudely [137, 150]. However, Hoffman [151-153] and Harte [154, 155] have independently developed micromagnetic theories* which successfully treat E_{mag}.

Hoffmann showed that to find ϕ at a given point in the film, only magnetostatic interactions originating within a small "coupling region"

*It has also been demonstrated that the problem can be attacked with the Green's-function method [156].

surrounding the point were important, so that the magnetic state of the coupling region could be found by micromagnetics as if that region were isolated. This results in an enormous simplification over the micromagnetic problem of the entire film. On the other hand, Harte did consider the entire film by expanding the total torque $\delta E_{tot}/\delta\phi$ in Fourier series and then solving for each Fourier component of the torque. The perturbing anisotropies which cause the ripple were treated by both authors on a statistical basis.

(1) **Hoffmann's Theory.** Hoffmann [151-153] expressed the direction of **M** at any point $\phi(\mathbf{r})$ as the sum of the azimuth φ_0 of the mean magnetization $\langle M \rangle$ (independent of position) and the angular deviation from it $\varphi(\mathbf{r})$:

$$\phi(\mathbf{r}) = \varphi_0 + \varphi(\mathbf{r}) \tag{14}$$

where $\mathbf{r} = \mathbf{i}x + \mathbf{j}y$ is in the film plane and is defined by a coordinate system which moves with $\langle M \rangle$ (Fig. 9). In order to find $\varphi(x, y)$ as a function of position, he then rewrote Eqs. (13) by expansions in terms of $\varphi(x, y)$. Thus the local anisotropy energy is given by the expression

$$E_K = K f_0(\varphi_0, x, y) + K\varphi f_1(\varphi_0, x, y) + \frac{K\varphi^2}{2} f_2(\varphi_0, x, y) + \cdots \tag{15}$$

while the magnetostatic energy is given by

$$E_{mag}(\mathbf{r}) = \frac{1}{2}\mathbf{M} \int \frac{(\mathbf{r}' - \mathbf{r})}{|\mathbf{r}' - \mathbf{r}|^3} \, \text{div}' \, \mathbf{M}(\mathbf{r}') \, dV' \tag{16}$$

where div' implies operation on the primed variables only, and the integration extends over the entire volume of the film. The expansion of $\mathbf{M}(\mathbf{r}')$ about point $\mathbf{r}$ and substitution of the result into Eq. (16) gives E_{mag} in terms of φ and the "magnetostatic" constants C_x and C_y, which depend on the volume V^* surrounding the point $\mathbf{r}$ within which magnetostatic fields are large enough to affect $\mathbf{M}(\mathbf{r})$ appreciably.

In the S-W theory (Sec. 3), equilibrium was specified by the condition that the total torque vanishes, $d\epsilon/d\varphi_0 = 0$; here the total torque includes the "microtorques" from the ripple, and its vanishing is expressed by the vanishing of the variation of the integral of ϵ_{tot}, $\delta \int \epsilon_{tot} \, dV = 0$. (As in Sec.

3, ϵ is the energy normalized to MH_K, $\epsilon \equiv E/MH_K$.) If ϵ_{tot} is expressed in terms of φ as just discussed, an Euler equation results:

$$\left[h \sin(\beta - \varphi_0) - \frac{1}{2} \sin 2\varphi_0 - \frac{Kf_1}{2K_u} \right]$$

$$+ \left[-\varphi\Lambda_1 - \frac{\varphi Kf_2}{2K_u} + \frac{A\partial^2\varphi}{K_u \partial x^2} + \left(\frac{2A + MC_y^2}{2K_u} \right) \frac{\partial^2\varphi}{\partial y^2} \right] \qquad (17)$$

$$+ \left\{ \frac{M^2 C_x \varphi}{2K_u} \left[\varphi \frac{\partial^2\varphi}{\partial x^2} + \left(\frac{\partial\varphi}{\partial x} \right)^2 \right] \right\} = 0$$

Here Λ_1 is defined by Eq. (5). Terms containing $\partial\varphi/\partial y$ and some terms of order φ^2 and higher were suppressed in Eq. (17) because they are negligible for typical films.

If terms only up to first order in φ and its derivatives are retained (the first two brackets), Eq. (17) can be solved for φ in closed form. From the zero-order S-W approximation [Eq. (4)] the first two terms vanish, and upon consideration of its average over the many crystallites in the volume V^*, the term containing f_2 may be neglected in comparison with Λ_1 if Λ_1 is not extremely small (the first-order solution diverges at $\Lambda_1 = 0$, i.e., at the astroid). The solution is expressed in terms of the modified Bessel function $K_0(\zeta)$. Identifying the volume V^* with the volume over which the argument ζ of $K_0(\zeta)$ is less than unity ensures that the approximations made are self-consistent, and allows determination of the constants C_x and C_y and the size of V^*. The result of this linear theory is

$$\varphi(\varphi_0, x, y) = \frac{-K}{4\pi AW} \int_{film} f_1(\varphi_0, \xi, \eta) K_0$$

$$\times \left\{ \left(\frac{K_u \Lambda_1}{A} \right)^{1/2} \left[(x - \xi)^2 + \frac{(y - \eta)^2}{W^2} \right]^{1/2} \right\} d\xi\, d\eta \qquad (18)$$

where

$$W^2 = 8tM^2 (AK_u \Lambda_1)^{-1/2} \qquad (19)$$

Thus if the distribution of local anisotropies is known throughout the film, Eq. (18) will yield φ at any arbitrary point in the film.

Since $K_0(\zeta)$ decreases rapidly as its argument increases, the integral in Eq. (18) may be taken over V^* instead of the entire film. Thus only *within* the coupling region V^* surrounding the point (x, y) will the local random anisotropy cause appreciable contribution to $\varphi(x, y)$. Within V^*, φ may be taken as approximately constant. This coupling region is elliptical in shape with a semiminor axis in the x direction [parallel to $\langle M \rangle$] of length $(A/K_u \Lambda_1)^{1/2}$ determined by exchange forces, and a semimajor axis in the y direction [perpendicular to $\langle M \rangle$] which is W times as long [from Eq. (19), for $\Lambda_1 \sim 1$, $W \sim 30$] and is determined by magnetostatic fields. This interpretation is consistent with Lorentz micrographs of ripple (Fig. 17) which indeed show structure elongated in a direction perpendicular to $\langle M \rangle$. Again in correspondence with ripple, V^* is not anchored in the film; its size and orientation and the resulting $\varphi(x, y)$ depend upon the applied field through Λ_1.

The magnetization dispersion δ is defined quantitatively as the value of φ^2 averaged over the entire film.† For randomly distributed uniaxial anisotropies, it can be shown from Eqs. (18) and (19) that

$$\delta = \frac{0.17S}{t^{1/4} M^{1/2} (AK_u \Lambda_1)^{3/8}} \tag{20}$$

where the structure constant $S \equiv KD\sigma_1/\sqrt{n}$ characterizes those properties of the film which determine the ripple magnitude. Here n is the number of crystallite layers in the film and σ_1 is the standard deviation describing the statistical distribution of the perturbing anisotropies (for randomly distributed uniaxially anisotropic crystallites, $\sigma_1 = 1/\sqrt{2}$).

The linear theory just presented is valid only for unsatisfactorily small values of δ (less than 1 or 2°), i.e., far from the astroid (large Λ_1), for which local stability of the ripple is achieved. To discuss stability for larger δ, an effective field $h_{eff}(\beta)$ is defined in terms of the second variational derivative of the total energy [Eq. (13)] integrated over the film, in analogy with the S-W theory where the uniform effective field $\Lambda_1 \equiv \partial^2 \epsilon / \partial \phi^2$. This field, which is a function of position within the film, is

$$h_{eff}(\beta) = \Lambda_1 + \frac{Kf_2}{2K_u} - \frac{M^2 C_x}{K_u} \left[\bar{\varphi} \frac{\partial^2 \varphi}{\partial x^2} + \frac{1}{2} \left(\frac{\partial \varphi}{\partial x} \right)^2 \right] \tag{21}$$

---

†A *non*-squared average over the film vanishes.

The last two terms represent nonlinear magnetostatic fields $h_1(\beta)$ and $h_2(\beta)$ which are respectively parallel and antiparallel to $\langle M \rangle$. Again in analogy with the S-W theory, where stable or unstable equilibrium is achieved according as Λ_1 is positive or negative [Eq. (5)], it is seen from Eq. (21) that $h_1(\beta)$ acts to stabilize the ripple while $h_2(\beta)$ tends to make it unstable. From Eq. (21) if the condition $h_{\rm eff}(\beta) > 0$ holds in regions where $\varphi = 0$, it will hold a fortiori in regions where $|\varphi| > 0$. Thus, since the second term is usually negligible, the condition for *stability* throughout the film is

$$\Lambda_1 > h_2(\beta) = \frac{M^2 C_x}{2K_u}\left(\frac{\partial \varphi}{\partial x}\right)^2_{\varphi = 0} \tag{22}$$

where C_x is known. By finding $\partial \varphi/\partial x$ from Eq. (18), this condition may be written as $\Lambda_1 > \Lambda_1^b$ where

$$\Lambda_1^b = \frac{1}{AK_u}\left(\frac{2^{1/2}Md^2 K^2 t^{1/2}}{16\pi n}\right)^{4/5} \tag{23}$$

is called the "blocking field."

Since $|h_1(\beta)| \simeq |h_2(\beta)|$, Eq. (22) is *also* the condition for neglect of the higher-order terms in Eq. (17). However, when $\Lambda_1 > \Lambda_1^b$, the linear theory may be used to estimate $h_{\rm eff}(\beta)$ [Eq. (21)], which may then be substituted for Λ_1 in Eq. (20) to give a corrected δ. Further discussion of Λ_1^b is given in Sec. 6.2.

(2) Harte's Theory. Harte's micromagnetic theory [154, 155] is based upon a Fourier-analysis approach which, although further abstracted from physical intuition than Hoffmann's theory, is more satisfactory for treating the magnetostatic energy. Thus Harte's formalism handles nonlinear influences on the ripple more easily than Hoffmann's formalism, but on the other hand the latter approach permits closer study of *local* ripple-instability conditions. For this reason, although Harte's calculations of ripple amplitude are better than those of Hoffmann, Hoffmann has investigated quasi-static ripple phenomena (Sec. 6) which are not readily accessible to Harte's theory.

Harte expanded $\varphi(\mathbf{r})$ [Eq. (14)] in the Fourier series

$$\varphi(\mathbf{r}) = \sum_{\mathbf{k} \neq 0} \varphi_{\mathbf{k}} e^{i\mathbf{k}\cdot\mathbf{r}} \qquad (24)$$

where the coordinate system is now *fixed* in the film plane (Fig. 9). Here $\mathbf{k}$ is a wave vector in the film plane (at an angle $\Phi_{\mathbf{k}}$ to the x axis), defined by $\mathbf{k} = \pi(n_x/L_x, n_y/L_y)$ where n_x, n_y are integers and L_x and L_y are the lateral dimensions of the rectangular film. (Later the limit $L_x, L_y \to \infty$ is taken.) It is desired to find $\varphi_{\mathbf{k}}$ for all $\mathbf{k}$. To do this, the various interactions are expressed in terms of torques which are expanded about φ_0, then Fourier-analyzed, i.e.,

$$T_i(\mathbf{r}) = \frac{\delta E_i}{\delta \phi} = \sum_{\mathbf{k}} (t_{\mathbf{k}})_i e^{i\mathbf{k}\cdot\mathbf{r}} \qquad (25)$$

Here i represents either u [uniform, from addition of Eqs. (13a) and (13b)], ex [exchange, Eq. (13c)], K [local anisotropy, Eq. (13d)], or m [magnetostatic, Eq. (13e)]. Since, in analogy with Eq. (4), the total torque $\sum_i T_i(\mathbf{r})$ vanishes at equilibrium, its Fourier components vanish independently, or

$$(t_{\mathbf{k}})_u + (t_{\mathbf{k}})_{ex} + (t_{\mathbf{k}})_K + (t_{\mathbf{k}})_m = 0 \qquad (26)$$

The torque components $(t_{\mathbf{k}})_K$ are expressed in terms of $p_{\mathbf{k}}$ and $q_{\mathbf{k}}$ which are coefficients in the Fourier expansion of trigonometric functions describing the (random) orientation of the local anisotropies, and hence are random variables. It is possible to express $(t_{\mathbf{k}})_{ex}$ exactly in terms of $\varphi_{\mathbf{k}}$, but the components $(t_{\mathbf{k}})_m$ and $(t_{\mathbf{k}})_u$ are expressed by expansions in terms of the form $\varphi_{\mathbf{k}} \varphi_{\mathbf{k}'} \varphi_{\mathbf{k}''}$ (and higher order). Making these substitutions into Eq. (26) yields

$$p_0 - \left(\mathcal{H}_0\right)_1 - 2 \sum_{\mathbf{k} \neq 0} q_{-\mathbf{k}} \varphi_{\mathbf{k}} + \left(t_0\right)_2 + \left(t_0\right)_3 + \cdots = 0 \qquad (27a)$$

$$p_{\mathbf{k}} - \left(\mathcal{H}_{\mathbf{k}}\right)_1 \varphi_{\mathbf{k}} - 2 \sum_{\mathbf{k} \neq 0} q_{\mathbf{k}-\mathbf{k}'} \varphi_{\mathbf{k}'} + \left(t_{\mathbf{k}}\right)_2 + \left(t_{\mathbf{k}}\right)_3 + \cdots = 0 \qquad (27b)$$

where the $(t_0)_j$ terms are of the jth order in φ_k and originate in uniform and magnetostatic torques and where the "first-order effective field"

$$\left(\mathcal{H}_k\right)_1 \equiv \Lambda_1 + \frac{2Ak^2}{MH_K} + \frac{4\pi M}{H_K}\widetilde{\chi}_k \sin^2(\Phi_k - \varphi_0) \tag{28}$$

while $\left(\mathcal{H}_0\right)_1 = \Lambda_1$. Here the magnetostatic function $\widetilde{\chi}_k$ is a function only of $k \equiv |k|$ (for small k, $\widetilde{\chi}_k \sim tk/2$), while Φ_k is the azimuth of k (Fig. 9). Equation (27a) governs the uniform ($k = 0$) mode, while Eq. (27b) describes nonuniform ($k \neq 0$) modes.

From Eq. (27a), if $\varphi_k = 0$ for all k [$\varphi(r) = 0$], $\Lambda_0 = p_0$. Since averaging over an ensemble of films gives $\langle p_0 \rangle = 0$, in first approximation $\Lambda_0 = 0$ [Eq. (4)]; i.e., the S-W value of φ_0 may be used in Eq. (27b). To the first order in φ_k only the first two terms in Eq. (27b) are utilized, giving

$$\left(\varphi_k\right)_1 = \frac{p_k}{\left(\mathcal{H}_k\right)_1} \tag{29}$$

i.e., the components φ_k are attenuated by the effective field* $\left(\mathcal{H}_k\right)_1$. Thus Eqs. (24) and (29) provide the first-order solution, and are equivalent to Hoffmann's equation (18). Further, in agreement with Hoffmann's theory, magnetostatic fields [$\widetilde{\chi}$ in Eq. (28)] make this attenuation largest in the transverse direction ($\Phi_k - \varphi_0 = \pi/2$), while in the longitudinal direction ($\Phi_k = \varphi_0$), φ_k is attenuated only by exchange. Note that *on* the astroid ($\Lambda_1 = 0$), Eq. (29) diverges for small k [see Eq. (28)] as did Eq. (20).

To obtain a *higher*-order solution, Harte assumed in analogy with Eq. (29) that

$$\varphi_k = \frac{p_k}{\mathcal{H}_k} \tag{30}$$

where the effective field $\mathcal{H}_k \equiv \left(\mathcal{H}_k\right)_1 + V_k$, i.e., it consists of a linear [Eq. (28)] and a nonlinear field, respectively. Since V_k is not a small

---

*Note that this result implies that the ripple state depends on the applied field only through Λ_1. For example, the ripple states for fields along either the *HA* or *EA* are identical as long as the *HA* field exceeds the *EA* field by $2H_K$ [see Eq. (5)].

correction to $\left(\mathcal{H}_{\mathbf{k}}\right)$ (a fact overlooked in early work), to find $\mathcal{H}_{\mathbf{k}}$ is a formidable nonlinear problem. By substituting Eq. (30) into Eq. (27b), manipulating, and taking an ensemble average, Harte derived a difficult integral equation for $V_{\mathbf{k}}$, which involved $\mathcal{H}_{\mathbf{k}}$ but which was dependent only upon the magnitude k. Harte showed that $V_k = V_0$ for $k < k_v$ and $V_k = V_\infty$ for $k > k_v$ was an acceptable approximate solution of this equation, where V_0 and V_∞ are the respective nonlinear effective fields at $k = 0$ and $k = \infty$, and k_v is a critical wave number $\sim 1/t$. After defining the magnetostatic and exchange lengths

$$R_m = \frac{\pi M^2 t}{K_u \mathcal{H}_0} \quad \text{and} \quad R_e \equiv \left(\frac{A}{K_u \mathcal{H}_0}\right)^{1/2} \tag{31}$$

where $\mathcal{H}_0 = \Lambda_1 + V_0$, he achieved a solution for V_0 and V_∞ for certain ranges of R_e and R_m, and hence could calculate $\mathcal{H}_{\mathbf{k}}$ for those ranges; he was thus enabled to calculate the ripple spectrum ($\varphi_{\mathbf{k}}$ vs. $\mathbf{k}$) from Eq. (30). He then found the dispersion δ from Parseval's theorem $\delta^2 = \sum_{\mathbf{k} \neq 0} |\varphi_{\mathbf{k}}|^2$.

It is seen from the first column of Table II, which gives the resulting functional dependence of δ, that δ is governed both by the scale of the inhomogeneities with respect to R_e and R_m, and by the magnitude of KD. Thus for fine inhomogeneities the ripple is attenuated by magnetostatic

Table II. Magnetization Dispersion δ for Thin Films*

Condition	Linear	Nonlinear	Longitudinal coherence length
$t, D \ll R_e$; fine-scale inhomogeneity	$\dfrac{0.17 KD (AK_u \Lambda_1)^{-3/8}}{t^{1/4} M^{1/2}}$ $\xrightarrow{\ 3^0\ }$	$0.29\left(\dfrac{KD}{tM^2}\right)^{2/5}$	R_e
$t, R_e \ll D \ll R_m$; coarse-scale inhomogeneity	$\dfrac{0.41 KD^{1/4} (K_u \Lambda_1)^{-3/4}}{t^{1/4} M^{1/2}}$ $\xrightarrow{\ 6^0\ }$	$0.29\left(\dfrac{KD}{tM^2}\right)^{2/5}$	D

*Magnetization dispersion δ and longitudinal coherence length for various conditions on the scale of inhomogeneity D, exchange length R_e, and magnetostatic length R_m according to Harte's theory. Arrows indicate transition from linear to nonlinear theory as ripple amplitude increases, while the transition δ is given below the arrow. Certain cases, e.g., thick films, have been omitted [155].

and exchange fields, but for coarse inhomogeneities exchange is not important. In examining Table II it must be remembered that R_e, R_m, and Λ_1 are dependent upon the external field; in particular, as the astroid is approached ($\Lambda_1 \rightarrow 0$), δ increases and the nonlinear $(KD/tM^2)^{2/5}$ dependence is valid rather than the linear dependences on the left. For a 500-Å film with $\Lambda_1 = 6$ Oe and $M_0 = 800$ G, the transition δ is $3°$ for fine but $6°$ for coarse inhomogeneities, where $D = 1\mu$ was assumed in the latter case. The linear approximation for fine inhomogeneities in Table II agrees with Hoffmann's equation (20) for $\sigma_1 = 1/\sqrt{2}$.

Variations of **M** in distances smaller than the critical lengths R_e and R_m [Eq. (31)] are suppressed because of the action of exchange and magnetostatic fields, respectively; the effective field $\mathcal{H}_k$ [Eq. (30)], which is expressible in terms of R_e and R_m, suppresses components $\varphi_\mathbf{k}$ associated with large k values in the longitudinal and transverse directions (thus determining *short*-wavelength cutoffs). Then since the autocorrelation function of $\varphi(\mathbf{r})$ is simply the Fourier transform of the ripple spectrum [155], there will be corresponding *maximum* correlation distances, or coherence lengths in these directions (Table II). For the thin film, fine-inhomogeneity, linear case, the longitudinal and transverse coherence lengths are found to be R_e and $(R_e R_m)^{1/2}$, respectively, which are approximately the dimensions of Hoffmann's coupling region V^* (see above). Thus for that case Harte's results agree in all respects with those of Hoffmann.

6. QUASI-STATIC MANIFESTATIONS OF RIPPLE

Magnetization ripple has appreciable influence in many magnetic film phenomena. In the phenomena discussed below, however, its role is decisive. The best theories for these phenomena are based on Hoffmann's micromagnetic theory (Sec. 5.2), but because of the difficulties associated with nonlinear effects in Hoffmann's theory, his predictions must be approached with caution, particularly at field points near the astroid.

6.1 Rotational Hysteresis

The rotational hysteresis W_r is defined as the work done when a field **H** applied in the film plane is rotated $360°$. Thus, since the torque T exerted by **H** upon **M** is given by $HM \sin(\beta - \varphi_0)$,

$$W_r = \int_0^{2\pi} T \, d\varphi_0 = HM \int_0^{2\pi} \sin(\beta - \varphi_0) \, d\varphi_0 \qquad (32)$$

where the angles β and φ_0 are defined in Fig. 9. This work W_r is a consequence of irreversible processes occurring in the film during the rotation of H; W_r is lost to the lattice as heat. Rotational hysteresis is most conveniently measured with a torque mangetometer by integrating the measured torque over a 360° rotation [Eq. (32)].

According to the S-W theory [3], rotational hysteresis occurs only for applied fields in the range $0.5 < h < 1$, since irreversible processes should occur only within that range. Experimental results [157] show, however, that even in films with small ripple, i.e., small values of h_c (e.g., $h_c < 0.2$), nonzero values of W_r appear for values of h less than 0.5 and greater than 1 (Fig. 10a). Furthermore, films with $h_c > 0.5$ exhibit [158] "unidirectional hysteresis," i.e., a hysteresis loss with a rotational period of 2π in fields $h < 0.5$, while films which have even larger magnetization dispersion, i.e., high values of h_c or RIS properties (Sec. 6.4), exhibit [159] a tail of the W_r vs. h curve which extends to very high values (several hundred) of h (Fig. 10b).

The departure from the S-W theory exhibited in Fig. 10a was initially interpreted as an indication of a noncoherent reversal mechanism, while the results for films with higher values of h_c were first explained on the anisotropy-dispersion model (Sec. 5.1). Thus independent, irreversible rotation processes in regions of large randomly oriented anisotropy (assumed to be scattered throughout a matrix of uniaxial anisotropy) were thought to contribute to W_r. The wide range of nonzero W_r would then be explained by the wide range of H_K values of these regions.

A more plausible explanation for W_r involves ripple hysteresis, i.e., irreversible local rearrangements of the ripple in separate regions scattered throughout the film. Indeed, such ripple hysteresis has been directly observed by Lorentz microscopy [160] upon application of a changing field $h < 0.5$ under conditions similar to those employed in obtaining Fig. 10a. Hoffmann [138, 161] has developed a theory of ripple hysteresis based on his micromagnetic theory. He noted that, as φ_0 changes under the action of an external field, $\bar{\epsilon}_K$, the local anisotropy energy averaged over a coupling region, changes because the coupling region changes its size and orientation and hence includes different sets of crystallites. In effect, then, a random perturbation $\bar{\epsilon}_K(\varphi_0)$ is

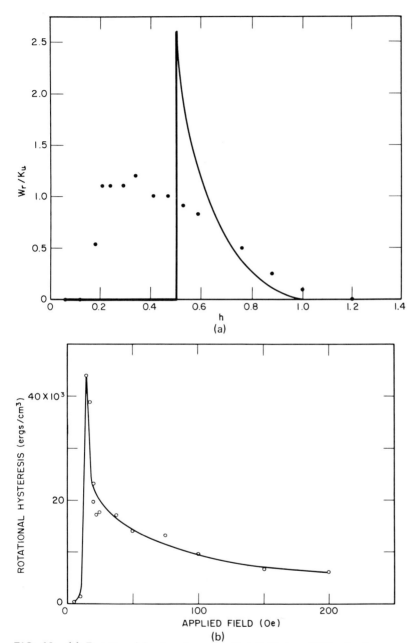

FIG. 10. (*a*) Rotational hysteresis vs. applied field $h = H/H_K$ for low-dispersion film ($h_c = 0.2$). Solid curve calculated from Stoner-Wohlfarth theory. (*After Doyle et al.* [157]). (*b*) Rotational hysteresis vs. applied field for high-dispersion (RIS) film. (*After Cohen* [159])

superimposed on the smooth ϵ_{K_u} vs. φ_0 curve, so that the net curve contains small hills and valleys. When **M** is caught in one of these local energy minima (which occurs preferentially near the *EA* or *HA*), ripple hysteresis results and **M** lags behind its hysteresis-free position. Hoffmann's theory was based only upon his linear-ripple theory, however, and will have to be modified to account for blocking effects near $\Lambda_1 = 0$.

6.2 HA Domain Splitting (HA Fallback)

If a field $h \gg 1$ is applied parallel to the *HA*, the film is saturated in the field direction and the ripple is suppressed. As the value of h is reduced, the ripple grows in magnitude and in average wavelength until at a certain value of h the phenomenon of *HA* domain splitting or "fallback" begins [162]. In this process, as h decreases, in regions where **M** is locally deviated to the "right" ("left") of the *HA*, **M** rotates clockwise (counterclockwise) toward the *EA*. Finally an array of approximately equal-spaced domain walls (Sec. 7) appears parallel to the *EA* (Fig. 11). If the experiment is repeated with **h** applied at an angle α to the *HA* (Fig. 9), a smaller density of "splitting domains" is observed, and domains containing **M** directed along the *EA* sense closest to **h** are wider than oppositely directed domains. Thus the *EA* remanence (at $h = 0$), which is 0 at $\alpha = 0$, increases monotonically with α.

The S-W model predicts only uniform rotations of **M** and thus cannot describe *HA* domain splitting. The first attempt to explain this phenomenon was based on the anisotropy-dispersion model (Sec. 5.1). Since, according to that model, ripple-inducing regions having local easy axes distributed on either side of the average easy axis are equally

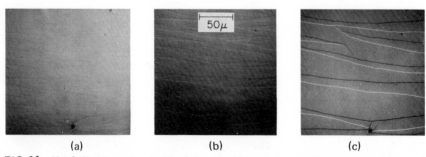

(a)	(b)	(c)

FIG. 11. *HA* fallback as revealed by Lorentz microscopy. Sample was saturated along the *HA* (vertical), after which the applied field was slowly reduced to the following values: (*a*) 3.8 Oe, (*b*) 2.2 Oe, (*c*) zero.

probable, upon *HA* domain splitting an equal number of counterclockwise and clockwise domains are produced, giving a vanishing *EA* remanence. However, when **h** is applied at an angle α to the right of the *HA* only those regions with easy axes which deviate from the mean *EA* direction by an angle greater than α can support local counterclockwise **M** rotation; in all other regions **M** rotates clockwise, thus giving an *EA* remanence to the right. As α increases, the splitting-domain density decreases and the *EA* remanence increases, until, when α is greater than the maximum local *EA* dispersion angle, **M** rotates *uniformly* clockwise toward the *EA*.

Measurements of the remanence as a function of α are useful in characterizing magnetic films [48, 163]; the "domain-splitting angle" α_r, i.e., the field angle α giving r percent remanence, is often quoted; typically α_{90} is a few degrees. However, the newer micromagnetic theory shows that α_{90} is *not*, as often stated in the older literature, a measurement of the angular dispersion of the *EA* direction but is an index of the *magnetization* dispersion in the film. On the other hand, the average ripple angle δ cannot be directly identified with α_{90}; the quantitative connection between these two parameters is complicated.

Hoffmann has developed an explanation of domain splitting based on his micromagnetic theory [138, 164]. He pointed out that as **h** and hence the uniform effective field Λ_1 is reduced from saturation, at first the ripple will reorient "freely" so that (Fig. 12a) δ and the longitudinal ripple period R_e will increase monotonically as predicted by the linear theory [Eq. (20) or Table II]. When the blocking field Λ_1^b is reached [Eq. (23)], however, the ripple becomes unstable in regions where $\varphi = 0$ (Fig. 12b), thus creating negative magnetostatic fields $h_2(\beta)$ for still lower Λ_1 (Fig. 12c). The ripple can then no longer reorient freely and is said to be blocked, although a small additional field can affect the blocked configuration. If Λ_1 is further reduced below Λ_1^b, although the periodicity remains fixed approximately at $R_e\left(\Lambda_1^b\right)$, **M** within each region rotates toward the nearest *EA* direction. The small-angle walls in the regions of instability thus gradually become high-angle walls, and the blocked state is transformed into a "locked" state which is stable in the presence of small additional fields.* In reality, blocking does not occur simultaneously throughout the film, but various values of Λ_1^b should be

---

*The widths of the splitting domains should thus be $\frac{1}{2} R_e\left(\Lambda_1^b\right)$.

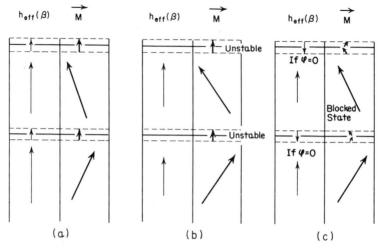

FIG. 12. Schematic diagram showing initiation of the blocking condition, as the uniform effective field Λ_1 is decreased below the blocking field Λ_1^b. Right side shows **M** configuration; left side shows effective field $h_{\mathrm{eff}}(\beta)$. (a) $\Lambda_1 > \Lambda_1^b$, ripple is free. (b) $\Lambda_1 = \Lambda_1^b$, blocking initiates at regions of instability, where $\varphi = 0$. (c) $\Lambda_1 < \Lambda_1^b$, ripple is no longer free, but blocked. (*After Hoffmann* [153])

associated with different regions; further development of the theory will be needed to take this fact into account.

On the basis of this theory, Hoffmann deduced that $\alpha_{90} \propto S^2$. A series of films of various thicknesses, but prepared at the same substrate temperature, followed this relationship closely, while from the slopes of the curves it was possible to deduce that K was too large to be explained by magnetocrystalline anisotropy and must originate in strain-magnetostriction anisotropy. A later study [147] of the frequency distribution of splitting-domain widths (which originates in the distribution of Λ_1^b throughout the film) was also in accord with this conclusion.

The phenomena of ripple blocking and locking can occur for *any* azimuth of the applied field, as long as reverse domains do not sweep out the locked structure; *HA* fallback is simply a special case of ripple blocking and locking. The above considerations should thus be generally valid.

6.3 Initial Susceptibility

The initial susceptibility χ is measured by applying, in the plane of the film, a dc bias field **H** at an angle β to the *EA* (Fig. 9) and a small ac excitation field, while detecting magnetic-flux changes in the ac field direction with a pickup coil. If the ac field is parallel (perpendicular) to H, then $\chi_{l,\beta}$, the longitudinal ($\chi_{t,\beta}$, the transverse) susceptibility, is studied, while either the in-phase component χ' or the $\pi/2$ out-of-phase component χ'' (corresponding to an energy loss) can be measured.

Susceptibility measurements are extremely sensitive to ripple influences. Thus, on the S-W model (Sec. 3) for the condition that M and H are parallel, $\chi'_{l,\beta} = \chi''_{l,\beta} = 0$, so that any measurable susceptibility must originate in ripple. For transverse susceptibility under the same condition, on the S-W model a small excitation field dH perpendicular to H will cause a rotation $d\varphi_0$ of M and a concomitant change in torque Λ_0, which must vanish for equilibrium. Then

$$d\Lambda_0 = MH_K \frac{\partial \Lambda_0}{\partial \varphi_0} \frac{d\varphi_0}{dH} dH + M\,dH = 0$$

or, since the susceptibility $\chi'_{t,\beta} = dM_t/dH = M(d\varphi_0/dH)$,

$$\chi'_{t,\beta} = \frac{M}{H_K(\partial \Lambda_0/\partial \varphi_0)} = \frac{M}{H_K \Lambda_1} = \frac{M}{H_K[h\cos(\beta - \varphi_0) + \cos 2\varphi_0]} \quad (33)$$

where Eq. (5) has been used. Measurements of $\chi'_{t,\beta}$ are most commonly made with the ac field along the *EA* and *HA*, in which case Eq. (33) becomes

$$\chi'_{t,0 \atop t,\pi/2} = \frac{M}{H_K(h \pm 1)} \quad (34)$$

It is seen from Eq. (34) that, in the absence of a dc field, $H_K = M/\chi'_{t,0}$; this equation is the justification for the commonly used hysteresigraph determination of H_K (Sec. 11.2). It is also noted that as $h \to \infty$, $\chi'_{t,\beta} \to 0$, while $\chi'_{t,0}(\chi'_{t,\pi/2})$ becomes infinite at $h = -1$ ($h = +1$). Furthermore, Eq. (34) states that the $\chi'_{t,0}$ and $\chi'_{t,\pi/2}$ vs. h curves are congruent but displaced by 2, in H_k normalized units.

Experimental curves of $\chi'_{t,0}$ and $\chi'_{t,\pi/2}$ are shown in Fig. 13a and b, respectively [165]. At high fields good agreement was found with the S-W theory, although the congruency conditions were not satisfied over the complete range of h because of the introduction of reverse domains at fields $h < -h_c$ in the $\chi'_{t,0}$ measurements. However, the agreement of $\chi'_{t,\pi/2}$ with the S-W theory was poor for $h \sim 1$; e.g., the field h_p for the

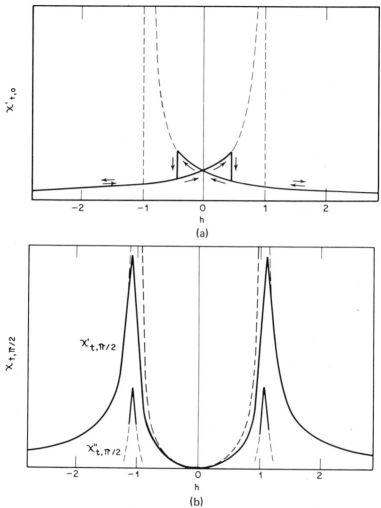

FIG. 13. Initial susceptibility X vs. applied field $h \equiv H/H_K$ for film with $h_c = 0.45$. Dotted line is theoretical curve according to Stoner-Wohlfarth theory. (a) $X'_{t,0}$; dc field along EA. Arrows indicate history of dc field magnitude. (b) $X'_{t,\pi/2}$ and $X'_{t,\pi/2}$; dc field along HA. After saturation, the magnitude of the dc field was slowly decreased. (*After Feldtkeller* [165])

peak value of $X'_{t,\pi/2}$ was not 1, and $X'_{t,\pi/2}$ was of course not infinite. This disagreement with the S-W theory is again caused by magnetization dispersion; as the magnetization dispersion increases, the departure from the S-W theory (e.g., $h_p - 1$) increases [17, 165]. It is also noted from Fig. 13b that nonzero values of $X''_{t,\pi/2}$ are detected. On the S-W theory, such loss should occur [166] only over that miniscule range of h near 1 which permits the net ac and dc field vector to cross the astroid near its tip, and hence correspond to switching of **M**.

These results were again first discussed on the anisotropy-dispersion model, although far more attention was given to [166, 167] $\chi_{t,\beta}''$ than to $\chi_{t,\beta}'$. The anisotropy dispersion was postulated to correspond to a collection of astroids (each associated with a region of deviant anisotropy), randomly distributed in angular orientation about the EA and representing various values of H_K. Then [166] the range of angle about the HA at the dc bias field and the range of the magnitude of h about 1, which lead to appreciable values of $\chi_{t,\beta}''$, were interpreted as the angular and magnitude dispersion of the anisotropy, α_{90} and Δ_{90}, respectively. To explain the failure of an anticipated relationship between α_{90} and Δ_{90}, the random anisotropy regions were later [167] postulated to have not a uniaxial but a complex, uniaxial plus biaxial character. It was claimed that the latter postulate could also explain the occasional observations of $\chi_{t,0}''$, which was alternatively interpreted as being caused by the presence of scattered regions with negative anisotropy [168], i.e., regions with a local EA perpendicular to the average EA.

The micromagnetic theory explains the experimental facts, at least qualitatively, in a much more natural way. A first attempt to construct the theory on the linear micromagnetic [138, 169] model failed, however; comparison of predicted curves of $\chi_{t,\beta}'$ vs. h showed [170] the necessity of considering the nonlinear theory. Hoffmann [171] later pointed out that $\chi_{t,\beta}'$ could be found simply by replacing Λ_1 in Eq. (33) by $h_{\mathrm{eff}}(\beta)$ [Eq. (21)]. However, it is necessary to average over the entire film, while taking account of the fact that certain portions of the film are blocked and other portions are free; this difficult calculation has not yet been performed. In any case it is clear that $\chi_{t,\beta}'$ is reduced by the ripple to a value below the S-W value.*

*Thus, exaggerated values of H_K may result from measurements of $\chi_{t,0}'$ by the common hysteresigraph technique [Eq. (34)], which is a well-known experimental fact [172]. However, since the ripple state depends on the applied field through Λ_1 in the nonlinear (Sec. 5.2) as well as linear theory, the statement that $\chi_{t,0}'$ and $\chi_{t,\pi/2}'$ are congruent but displaced by 2 is true not only on the S-W theory but in the presence of ripple. This leads to a susceptibility technique for measuring H_K, which is insensitive to the presence of ripple [172].

The agreement of the micromagnetic theory with experimental measurements of $\chi'_{l,\beta}$ is more satisfactory than that for measurements of $\chi_{t,\beta}$. Since $\chi'_{l,\beta}$ is proportional to δ^2/Λ_1, then [Eq. (20)] $\chi'_{l,\beta} \propto \Lambda_1^{-7/4}$; this relationship has been experimentally verified [173, 174] and provides strong support for the micromagnetic theory. Measurements of $\chi'_{l,\beta}$ have also been used to deduce δ by integrating $\chi'_{l,\beta}$ as a function of the dc field [175].

Ripple hysteresis [176] is the origin of χ'', as was demonstrated directly by observations of isolated local ripple rearrangements in susceptibility measurements by Lorentz microscopy [160]. However, the treatment of ripple hysteresis offered by Hoffmann (Sec. 6.1) was based on his linear theory and is therefore not yet satisfactory. Nevertheless, the superiority of this explanation to the one based on the anisotropy-dispersion theory has been clearly demonstrated by Harte et al. [177]. These workers measured $\chi''_{t,0}$, $\chi''_{t,\pi/2}$, and $\chi''_{t,\pi/4}$, using dc field configurations so that for a given applied field h, the quantity Λ_1 was the same in all three configurations. Since the ripple state depends only upon Λ_1, the $\chi''_{t,\beta}$ curves should be, and were, congruent for all three configurations. Films with high values of δ were shown to exhibit nonzero values of $\chi''_{t,0}$ and $\chi''_{t,\pi/4}$ only because the accompanying high values of h_c prevented the introduction of reverse domains. These results, and results obtained with similar methods by other authors, are difficult to explain on the anisotropy-dispersion theory.

6.4 Effects of High KD/K_u

As the random perturbing anisotropy KD (or the structure constant S) in a film is increased, the uniaxial character becomes increasingly masked, and the magnetic behavior is increasingly determined by the perturbing anisotropies alone. This fact was directly demonstrated by a Lorentz-microscopy study of the effects of a series of annealing treatments given to a film [178]. It was found that as the annealing treatment proceeded, the crystal size D increased, as did the ripple intensity, h_c, and α_{90}. Splitting domains became more pronounced upon *EA* reversal (*EA* locking), and finally "rotatable initial susceptibility" (RIS) films were produced. These RIS films [not to be confused with the RIS properties of stripe-domain films (Sec. 10.1)] have such high magnetization dispersion that **M** remains in any arbitrary direction in which the film has been previously saturated; the large local anisotropies prevent **M** from relaxing back to the nearest *EA*. The initial susceptibility in the absence of a dc field is then largest in the direction perpendicular to **M**, so that the initial susceptibility can be "rotated" at will [159]. While it is clear that the high dispersion is the determining factor in establishing the RIS property, the exact mechanisms of RIS in ac fields are still under discussion [179].

7. DOMAIN WALLS

7.1 Structure of Walls [180, 181]

Thus far only a state of uniform magnetization (subject to ripple perturbations) has been considered. Under certain conditions, however, films may contain many domains, with the magnetization uniform within each domain but differing in direction from domain to domain (see Fig. 17, for example). The narrow transition region between domains is known as a "domain wall." The domain configuration of a film depends upon both its physical structure and its previous magnetic history because a given configuration is commonly not a state of lowest energy, but a metastable state.

The nucleation and growth of domains, i.e., movement of domain walls, will be discussed in Sec. 7.2. Here the magnetic structure of the wall itself, i.e., the distribution of **M** inside the wall, will be considered.

(1) Bloch Walls. In bulk material the magnetization transition inside a domain wall is accomplished by rotation of **M** about an axis normal to the plane of the wall. In such a wall, known as a Bloch wall, the component of **M** normal to the wall is continuous so that no volume magnetic poles are present [10]. For a large specimen the wall shape will then be determined by the balance between anisotropy energy E_{K_u} [Eq. (13b)] and exchange energy E_{ex} [Eq. (13c)].

Consider such a Bloch wall parallel to the EA in a uniaxially anisotropic film (Fig. 14a). The wall shown separates two oppositely directed domains, so that **M** rotates 180° in the wall. At the intersection of the wall with the surface of the film, magnetic poles will be created so that in addition to E_{K_u} and E_{ex}, magnetostatic energy E_{mag} [Eq. (13e)] must be taken into account in the energy balance. As the film thickness t decreases, E_{mag} will predominate over E_{K_u} and E_{ex}. It is the presence of E_{mag} which makes the calculation of the magnetization distribution within the wall difficult, for in order to find E_{mag} the magnetization distribution must first be known. The problem is thus properly handled by variational methods, but it then becomes tractable only under certain approximations.

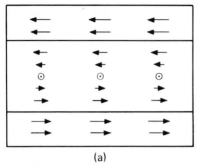
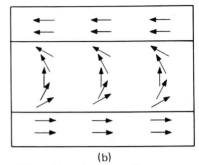

(a)   (b)

FIG. 14. Schematic diagram showing direction of **M** within a domain wall. (*a*) Bloch wall. **M** turns *out* of plane of film. (*b*) Néel wall. **M** turns *in* plane of film. (*After Middelhoek* [180])

Néel [180] approached this problem by assuming the simple distribution

$$\phi = \frac{-\pi}{2}, \quad x \le \frac{-a}{2}$$

$$= \frac{\pi x}{a}, \quad \frac{-a}{2} \le x \le \frac{a}{2} \tag{35}$$

$$= \frac{\pi}{2}, \quad x \ge \frac{a}{2}$$

where x is the coordinate normal to the wall, ϕ is the angle measured in the plane of the wall between **M** and the normal to the film, and a is the wall width (Fig. 14*a*).

The average anisotropy energy density in the wall is then [Eqs. (13*b*) and (35)] $E_{K_u} = K_u/2$, while the exchange energy density [Eqs. (13*c*) and (35)] $E_{ex} = A(\pi/a)^2$. To estimate the magnetostatic energy density the wall is approximated by a cylinder of elliptical cross section, where t is the major axis of the ellipse while a is the minor axis, giving (Sec. 4.1) $E_{mag} = \pi a M^2/(a + t)$. The energy per unit area of the wall is found by multiplying each of the above expressions by a, then adding them:

$$\gamma_B = \frac{a K_u}{2} + \frac{\pi^2 A}{a} + \frac{\pi a^2 M^2}{a + t} \tag{36}$$

It is seen from Eq. (36), as expected on physical grounds, that an increase in a causes an increase in the anisotropy and magnetostatic energies, but a decrease in the exchange energy. The value of a giving the *minimum* energy is found by setting $dy_B/da = 0$, while the value of y_B corresponding to this equilibrium value of a is found by substituting back into Eq. (36). In Fig. 15a, y_B is plotted as a function of t assuming $K_u = 10^3$ ergs/cm^3, $M = 800$ G, and $A = 10^{-6}$ ergs/cm, while in Fig. 15b, a is plotted vs. t. For these typical parameters y_B consists primarily of exchange and magnetostatic energies only. The rapid increase of y_B with decreasing t is thus caused by an increase in the magnetostatic energy.

(2) **Néel Walls.** Néel pointed out that this magnetostatic energy could be decreased by a different magnetization distribution in the wall, in which **M** rotated *in the plane of the film* (Fig. 14b). In the Néel wall there are no poles at the surfaces of the film, but instead volume poles are created within the wall since now the component of **M** normal to the wall is not continuous. The dependence of y_N and a upon t may be calculated in the same manner as in the Bloch-wall case (Fig. 15). Note the rapid increase in a as t decreases because of the decrease in the magnetostatic energy, which predominates in y_N for $t < 400$ Å.

(3) **Crosstie Walls.** As is seen from Fig. 15a, a Néel wall is energetically more favorable than a Bloch wall for values of $t < 350$ Å. It is found experimentally, however, that as the film thickness decreases a Bloch wall does not suddenly change into a Néel wall at a certain value of t. Instead an intermediate structure, known as the crosstie wall, is observed (Fig. 16).

A crosstie wall may be regarded as a Néel wall having alternating intervals of oppositely directed senses of rotation of the magnetization within the wall. This arrangement provides some flux closure in the wall. At the boundaries of the intervals are small ($\sim$100 Å diameter [183]) transition regions, known as "Bloch lines," in which the magnetization is normal to the plane of the film.* (The crossties are pieces of Néel walls associated with alternate Bloch lines.) The calculation of the energy of a crosstie wall is not as straightforward as the Bloch and Néel cases [185]. The estimate [180] given in Fig. 15a shows a Bloch-to-crosstie transition at 850 Å and a crosstie-to-Néel transition at 200 Å; the transition thickness will, of course, depend on the material constants [186].

---

*In thick films, the Bloch-wall analog of a crosstie wall has been found, namely, a Bloch wall having alternating intervals of oppositely directed rotation senses separated by *Néel* lines [184]. This arrangement also serves to decrease the demagnetizing energy by flux closure.

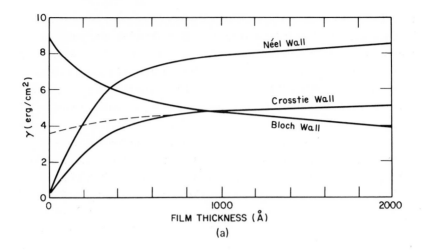

(a)

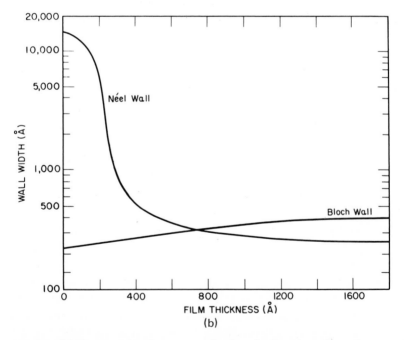

(b)

Fig. 15. (*a*) Theoretical surface energy density of a domain wall as a function of thickness. Assumed $A = 10^{-6}$ ergs/cm, $M = 800$ G, $K_u = 10^3$ ergs/cm^3. The dashed curve shows the estimate of crosstie-wall energy density when the energy of Bloch lines is included. (*After Middelhoek* [180]) (*b*) Theoretical widths of Bloch and Néel walls as a function of film thickness. Assumed $A = 10^{-6}$ ergs/cm, $M = 800$ G, $K_u = 10^3$ ergs/cm^3. (*After Middelhoek* [180])

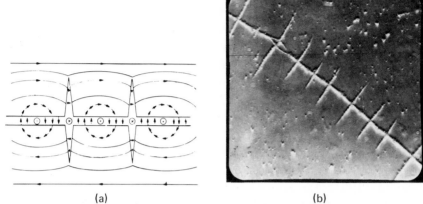

(a)   (b)

FIG. 16. (*a*) Schematic diagram of magnetization distribution around a crosstie wall. Magnetization is normal to the film plane at Bloch-line positions. (*After Huber* et al. [182]) (*b*) Bitter pattern of crosstie wall. (*After Moon* [182])

Walls separating domains whose **M** vectors lie at other than 180° are found in monocrystalline films and can be created in uniaxially anisotropic polycrystalline films by the application of fields in the plane of the film normal to a wall (i.e., a *HA* field). In this latter case, considerations similar to the above show that, for both crosstie and Bloch walls, when the wall angle is decreased below a certain value the wall transforms into a Néel wall.

(4) Other Wall Calculations. Theoretical studies of walls have been made using more sophisticated approaches than the one presented above. These studies were confined, as was the previous discussion, to one-dimensional variations of **M** for computational ease, so that only Bloch or Néel walls were investigated. As indicated above, the primary computational difficulty lies in the calculation of the magnetostatic energy, which leads to a difficult variational problem.

A fruitful approach has involved the use of the Ritz method, where a parametric form of the magnetization distribution is assumed, then the value of the parameter yielding the minimum value of the energy $E_{K_u} + E_{ex} + E_{mag}$ is found. Dietze and Thomas [187] used $\cos \phi = a^2/(a^2 + x^2)$ as the parametric form for both the Bloch and Néel 180° walls, and calculated the surface energy density γ and wall width a as a function of t. Later Feldtkeller [188] found that the increased freedom provided by an expression with two additional parameters predicted lower values of γ than did the single-parameter assumption. It is clear that that parametric form giving the lowest value of γ is the best, but the Ritz method gives no indication of which class of functional forms to choose for computation.

Brown and LaBonte [189] bypassed this difficulty by assuming a general one-dimensional magnetization distribution which was a discrete rather than continuous function of x. The magnetostatic energy was computed explicitly; then a digital computer was used to find the minimum energy state. This procedure showed that at the outer edges of a Bloch wall, M is reverse-rotated to give a component normal to the plane of the film opposite to that obtaining at the wall center. This unexpected configuration serves to lower the magnetostatic energy. Aharoni [190] also avoided the Ritz approach, but by solving the general variational problem after making a tractable approximation for the magnetostatic energy.

All these methods [187-190] for Bloch walls give values of γ_B and a which are in fair agreement and which are not too far from the results deduced on the simple model of Eq. (35). Further, these methods obey the Aharoni necessary self-consistency criterion fairly well [191]. However, the agreement among Néel-wall calculations is not so satisfactory, and the Aharoni criterion is far from satisfied for the extant Néel-wall calculations. Nevertheless these calculations [188-190] all show that a central region in the wall with a high value of $d\phi/dx$ is surrounded on either side by a region of low $d\phi/dx$; Aharoni [190] found that this behavior fitted the form $\sin\phi = \tanh(x/a)$. The ratio of the widths of these regions [192] has been estimated to be about 30, i.e., there can still be appreciable values of ϕ several microns from the center of a Néel wall, while the entire magnetization distribution of a Bloch wall should be confined to a region a few hundred angstroms wide.

It seems probable that significant advances in wall calculations will occur only when at least *two* dimensional variations in the magnetization within the wall (i.e., allowing variation throughout the film thickness) are permitted [193].

(5) Experimental Results. The experimental methods for studying microscopic magnetization distributions (Sec. 11.1) give results in general agreement with the theory discussed above. Using the Bitter technique, the transition from Bloch through crosstie to Néel walls as the film thickness decreases has been confirmed [194], with the transitions at roughly the thicknesses indicated in Fig. 15a. The magnetization distribution in a crosstie wall, first studied by the Bitter technique, has been confirmed by Lorentz microscopy [195]. Furthermore, measurements of wall lengths in special domain configurations (assumed to be in equilibrium) have yielded values of wall energy density γ in fair agreement with theoretical predictions [196]. However, none of the three microscopic techniques has the resolution to permit detailed experimental study of the magnetization distribution within a Bloch or Néel wall. Lorentz microscopy appears to confirm the presence of the long tails of the magnetization distribution in a Néel wall, but previous deductions of the detailed magnetization distribution are questionable, primarily because of the limitations of geometric electron optics in the interpretation of these results (Sec. 11.1).*

In contradiction to the wall theory presented above, the transitions between the various wall types are experimentally found to occur gradually, rather than abruptly [194, 197]. For values of $t > 1,000$ Å, the lowest-energy state will be that of a Bloch wall, with sections of opposite polarity separated by Néel lines. At $t < 900$ Å Néel walls begin to be of energy comparable to Bloch walls, so that various intervals of the wall are made of Néel-wall segments of alternating polarity. As t decreases, the proportion of Néel to Bloch character increases, until around 600 Å the wall has become a crosstie wall. The spacing between Bloch lines decreases as t continues to decrease, until a pure single-polarity (Bloch-line-free) Néel wall is produced for $t < 300$ Å. Existing wall theories do not predict this behavior.

Another limitation of the wall theories discussed above lies in their failure to predict walls having *mixed* Néel and Bloch components [198]. In such walls, **M** rotates across the wall so as to produce magnetostatic fields simultaneously normal to and in the plane of the film. On theoretical grounds such mixed walls occur for wall angles $\theta_c < \theta < 180°$, where θ_c is a critical angle, with the ratio of Bloch to Néel character increasing as θ increases; for $\theta < \theta_c$ pure Néel walls occur, while for $\theta = 180°$ pure Bloch walls can exist. (Mixed 180° walls cannot occur [189, 190].) Magnetization distributions and magnetostatic fields above mixed walls were calculated [199] with a computer technique similar to the one used by Brown and LaBonte [189].

---

*An "inversion" technique which converts an experimental electron intensity distribution into the domain wall profile has recently been demonstrated and justified with wave-optics for high resolution Lorentz microscopy of domain walls [K. J. Harte and M. S. Cohen, *J. Appl. Phys.,* **40** (1969)].

An illustration of the role of Bloch lines is furnished by the study of 360° walls, where the magnetization is parallel on both sides of the wall [200, 201]. In thicker films 360° walls are unstable because moving Bloch lines cause the unwinding of the spins forming such a wall. In thin (< 300 Å) films, however, Bloch lines do not move easily, so that 360° walls may be formed either by application of a high HA field opposed to the magnetization at the center of a 180° wall, or by the trapping by an imperfection of a Bloch line in a moving 180° wall under the application of an EA field to form a "perturbation" 360° wall.

The experimental studies described above have been made on polycrystalline, uniaxially anisotropic films. Monocrystalline, biaxially anisotropic films prepared by epitaxy (Chap. IV) have the same basic domain-wall structure. However, in these films the magnetocrystalline anisotropy causes the frequent occurrence (especially in thinner films) of 90° walls in a checkerboard pattern [202, 203].

7.2 Wall Motion

(1) Domain Nucleation and Growth. If, after saturation, a *slowly increasing* reverse field is applied along the EA of a uniaxially anisotropic film, the magnetic state will change not by the uniform rotational process predicted by the S-W theory, but noncoherently by the nucleation and growth of reverse domains. Two threshold fields are involved: that for nucleation of a reverse domain H_n and that for growth of the nucleated domain (movement of domain walls) H_0. It is found that nucleation preferentially occurs at inhomogeneities like scratches or holes, but especially at the film edges; subsequently the nucleated domains grow in length and width by movement of the domain walls in a series of "Barkhausen" jumps, until the film is saturated in the reverse direction (Fig. 17). The threshold value of the reverse field necessary for the full reversal process is called the "wall coercive force" H_c, although slightly different definitions of H_c are possible within the narrow range of field necessary for complete reversal. Thus, H_c should equal H_n or H_0, whichever is larger. For films about 1 cm in diameter and 1,000 Å thick, usually $H_c = H_0$; the coercive force for wall movement will therefore be called H_c in the following.

Since a magnetic film is not quite a perfect ellipsoid (Sec. 4.1), the demagnetizing field originating from poles at an edge is greater at that edge than in the interior. A lower-energy state than one of uniform magnetization may therefore be achieved if **M** near the edge is parallel to

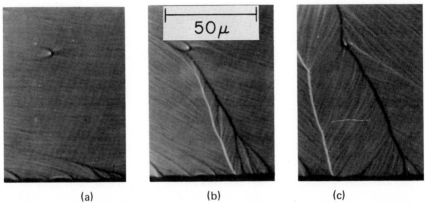

(a)   (b)   (c)

FIG. 17. Growth of reverse domains as revealed by Lorentz microscopy. Magnetization first saturated along *EA* (vertical), then reverse fields applied of (*a*) 0 Oe, (*b*) 1.3 Oe, (*c*) 1.9 Oe.

the edge; the transition region between the interior of the film (where **M** is perpendicular to the edge) and the edge depends on both t and the applied field. As the field **H** (perpendicular to the edge) is decreased in magnitude from a value giving saturation, a $90°$ "curling" of **M** will appear in the transition region [204]; reverse domains can nucleate in the interface between two peripheral regions of opposing curl senses [205] at reverse fields H_n whose magnitude decreases [206] with an increase in t. However, the detailed nucleation process, which must involve magnetization rotation [207], is not well understood theoretically, even for a ripple-free film.

The importance of ripple in determining H_c is clearly demonstrated by the phenomenon of "inversion," i.e., $h_c > 1$, found in films with high magnetization dispersion. The occurrence of such films is unexpected, for their reversal by magnetization rotation would be anticipated at $H = H_K$ before H_c was reached [73]. Observations of inverted films show, however, that before ordinary reverse domains can be introduced, an *EA* locking process occurs [73] which is precisely the same as the *HA* fallback process in a low-dispersion film (Sec. 6.2). Now the sense of **M** in the center of the Néel locking walls is antiparallel to the applied field. In order to create reverse domains, this sense must be reversed; this is accomplished by the nucleation and subsequent separation of two Bloch lines on each wall [162].

Noncoherent reversal is also observed [200, 208, 209] when the reversing field **H** is applied at an appreciable angle β to the *EA*. When β is less than a critical angle β_c, reversal occurs when* $h > h_w$ by the growth of domains nucleated at the film edges, as just described, but if $\beta > \beta_c$ and $h > h_q$ (β), a parallel array of long, slender, inclined reverse domains appears (Fig. 18). If h is increased beyond h_q for moderate angles ($\lesssim 60°$) the reversal is completed at $h \gtrsim h_w$ (β) by edge domains sweeping through, while for high angles ($\gtrsim 60°$) the slender domains grow in width before the edge domains come in. The angular dependence of the threshold fields h_q and h_w may be displayed by plotting these quantities in field space, where they may be compared with the astroid (Fig. 19). It is seen from such plots that for $\beta < \beta_c$, $h_w \cos \beta \sim h_c$, i.e., application of h at an angle β is essentially equivalent to EA reversal, and that as h_c increases β_c decreases and the curves move away from the origin. Since the value of h_c is correlated with the magnetization dispersion, it may be inferred that these phenomena originate in a dispersion-dependent phenomenon.

The character of the high β-reversal mechanism has been controversial. Smith [200] (and Metzdorf [209]) observed *propagation* of the slender inclined domains from the film edge, and called them labyrinth domains because of the labyrinthine path of the magnetization within the domains. Labyrinths were postulated to grow from their tips by sequential switching of adjacent regions by magnetostatic fields from the tips, although because of the ripple the orientation of **M** was so unfavorable in certain regions during labyrinth propagation, that they did not switch. Upon removal of the field the labyrinthine appearance is enhanced by magnetostatic fields created when the magnetization relaxes toward the EA.

On the other hand, Middelhoek [208] postulated a process called "partial rotation," in which the slender domains are created by *simultaneous* reversal without propagation features. In this process the magnetization throughout the film rotates uniformly as h is increased, until at h_q switching occurs in bands which happen to have **M** favorably oriented because of the ripple. The ensuing magnetostatic fields prevent the unswitched bands from also switching as h is further increased.

The controversy between labyrinth and partial-rotation generation was resolved by Cohen [178], who showed that labyrinths occurred in a film

---

*The threshold field for reversal by motion of domains nucleated at the film edges is defined as H_w for a reverse field applied at an angle β to the EA. At $\beta \simeq 0°$, $H_w = H_c$.

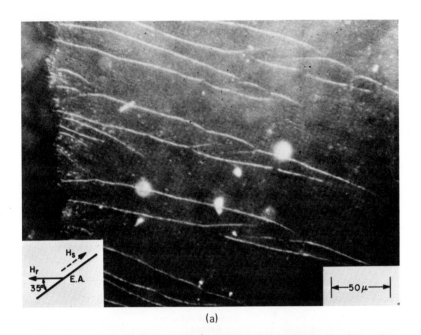

(a)

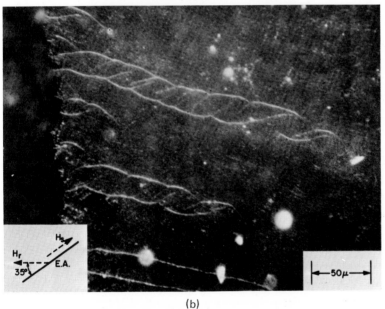

(b)

FIG. 18.   Labyrinths as shown by Bitter patterns. Fields applied in horizontal direction. (*a*) Saturated along *EA*, then reverse field of 1.3 Oe applied at 35° to *EA*. (*b*) Field reduced to zero, showing intensification of inner structure. (*After Smith and Harte* [200])

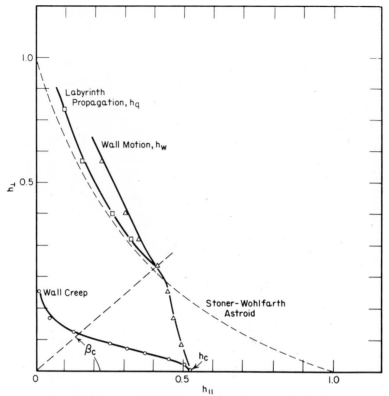

FIG. 19. Critical curves for wall motion, labyrinth propagation, and wall creeping as compared with astroid. (*After Middelhoek and Wild* [231])

with low dispersion, but upon annealing and increasing the dispersion in that film, reversal proceeded by partial rotation. (Middelhoek's films generally had high dispersion values [208].) The aid of the magnetostatic field from a labyrinth tip is apparently not needed if the magnetization dispersion is large enough.

Studies similar to those described above have been carried out for biaxially anisotropic, monocrystalline films prepared by epitaxy. Just as for uniaxially anisotropic films, magnetization reversal does not proceed by simple coherent rotation [210], but the reversal process usually includes reverse-domain nucleation and growth [211, 212]. The considerations involved are similar to those pertinent to uniaxially anisotropic films, but the biaxial (and any admixed uniaxial) anisotropy strongly influences the reverse-domain growth.

(2) Coercive Force [213]. An infinitesimally small field would be sufficient to sweep a previously nucleated reverse domain completely across a perfectly uniform film. The fact that a field equal to or greater than a threshold field H_c must be applied for wall motion may be explained by assuming that inhomogeneities cause the energy of the domain wall to be a random function of position. Then the energy provided by the field H_c is just large enough to overcome the potential hills; i.e., for an infinitesimal displacement dx of the wall parallel to itself $2H_c Mlt\, dx = d(\gamma lt)$, where l is the length of the wall and γ is the wall energy density. Thus

$$H_c = \frac{1}{2Mt}\left[\frac{d(\gamma t)}{dx}\right]_{max} \tag{37}$$

for a wall of constant length.

Now the potential barriers are randomly distributed in position and height [the subscript max in Eq. (37) refers to the steepest barrier]. An applied field may thus be sufficient to sweep a wall through a small region until the wall is stopped by a prohibitively steep barrier; a slight increase in field causes another Barkhausen jump to new barrier positions, etc.* There is thus a small range of fields needed to move walls completely across the film (thereby saturating the film in a sense opposite to its original sense). This fact causes some ambiguity in measurements of H_c, but since the hysteresis loop is nearly rectangular the several possible definitions of H_c yield nearly the same result. Visual measurements of the probability p of occurrence of Barkhausen jumps of a single wall as a function of jump distance x yielded a good fit to the law $p = \exp(-\lambda x)$, where λ is the stopping probability; the deviations from this law found in flux-pickup measurements are caused by the simultaneous measurement of many different walls with that technique [214, 215].

The inhomogeneities which can contribute to H_c [Eq. (37)] include spatial variations in film thickness, magnetic anisotropy, magnetization, exchange constant, and the magnetization ripple. Thus, by assuming that

---

*It should be noted that the diffusion-aftereffect contribution to H_c found by Lambeck (Sec. 7.2) does not fit this conventional interpretation of the Barkhausen effect.

a particular spatial variation of thickness was alone responsible for H_c, Néel deduced, for a Bloch wall, that $H_c \propto t^{-4/3}$. However, although experimental measurements of H_c as a function of t show large scatter and the results of different authors vary greatly, it is nevertheless clear that the $-4/3$ power law is not valid [216, 217]. On the other hand, peaks in the H_c vs. t curves have been associated with transitions between the various wall types [217], which is reasonable because different types of walls have different values of $[d(\gamma t)/dx]_{max}$ [Eq. (37)].

A good quantitative theory which predicts H_c does not yet exist. A fruitful approach [218] is the consideration of a domain wall embedded in the magnetization ripple; the periodically distributed poles thereby produced cause the necessary $d(\gamma t)/dx$ [Eq. (37)]. In this connection it is again noted that h_c increases with an increase in magnetization dispersion, and films prepared under conditions (e.g., high T_d) such as to give high dispersion also have large h_c values [178, 219]. Thus it is well known [100, 220] that substrate inhomogeneities can cause an increase in H_c; a thorough cleaning of the substrate and/or a precoating with SiO will lower H_c. It may also be noted that if defects in the film are large enough, a wall may become caught on them directly [205].

(3) Wall Velocity. When a field **H** in excess of H_c is applied parallel to a 180° domain wall during a Barkhausen jump, the wall will acquire a velocity v, the magnitude of which depends upon $H - H_c$. Kerr-effect measurements show [221], at least for low values of $H - H_c$, that

$$v = m(H - H_c) \tag{38}$$

where m is defined as the wall mobility.

Since the motion of the wall represents a transition from a higher- to a lower-energy state, the speed of the transition, i.e., the mobility m, will be determined by an energy-loss mechanism. One mechanism which has been proposed for transformation of the magnetic energy to heat energy of the lattice is eddy-current generation, which is the predominant mechanism in bulk metals. However, theory [222] shows that in films, m calculated on the basis of this mechanism is approximately proportional to $1/t$, so that below a certain film thickness some kind of intrinsic damping will predominate. This damping mechanism may be the same spin-lattice interaction which operates in coherent switching (Sec. 9); it may be noted that a correlation has been established [223] between the temperature dependence of m and that of λ determined from FMR (Sec. 8.1).

Mobility measurements have been made as a function of film thickness (Fig. 20). The data indicate no eddy-current contribution for $t < 2,000$ Å, while the minimum in the m vs. t curve is associated with the transition from crosstie to Bloch walls. Explanations of the shape of the curve in terms of spin-lattice interactions have been made using Galt's approach [221]. Galt showed that for a shape-invariant, moving, 180° wall

$$ m = \frac{2\gamma}{\alpha} \left(\int_0^\pi \frac{d\phi}{dx} \, d\phi \right)^{-1} \tag{39} $$

where ϕ is the azimuth of $\mathbf{M}$ within the wall at a position x measured normal to the wall, γ is the gyromagnetic ratio, and α is the damping parameter in the Landau-Lifschitz equation (Sec. 8); the value of m thus depends on the details of wall shape. However, because of the scatter in measurement of m, the questionability of Lorentz-microscopy determinations of wall shape (Sec. 7.1), and doubts of the validity of shape

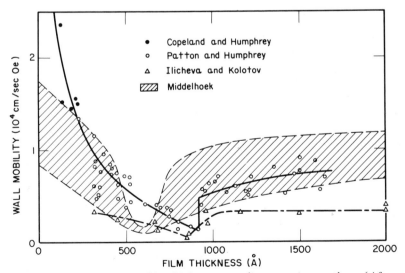

FIG. 20. Wall mobility vs. film thickness according to various authors. (*After Feldtkeller* [213])

invariance of a moving wall, convincing agreement between theory and experiment has not been achieved.

Middelhoek [224] has indicated that much of the scatter in measurements of m as a function of t (Fig. 20) may originate in variations of h_c. He found that m decreased with increasing h_c, which may indicate that ripple plays a role in the damping process. Finally, it should be noted that an unexplained sharp decrease in m for high fields has been observed in thin films.

(4) Magnetic Viscosity. As explained above, the coercive force h_c may be interpreted as the value of the external field which gives a moving domain wall the energy necessary to surmount the potential barriers it encounters. Under certain circumstances, however, part of this energy may come from thermal fluctuations, so that very slow, temperature-dependent wall motion may be observed in the presence of a dc field. Phenomena of this type are known as "magnetic-viscosity effects" [225], and should not be confused with creep (Sec. 7.2) since they occur in the absence of ac fields.

Stacey [226] presented a simple statistical theory to explain these results. The probability per unit time that a segment of a wall will be reversed in a field H with the aid of the thermal activation of a volume V of the film is $p = C \exp[-(E_0 - 2HMV)/kT]$, where E_0 is the barrier height and C is a constant. If all barriers are assumed to be of equal height E_0 and spaced a distance l apart, then the wall velocity may be written as

$$v = pl = Cl \exp\left[\frac{2MV(H - H_v)}{kT}\right] \qquad (40)$$

where $H_v \equiv E_0/(2MV)$ and $H < H_v$. Since H_v is the field necessary for wall motion at $T = 0°K$ (no thermal activation), H_v may be identified with H_c. Further, the fact that V appears in the exponent means that for large V, thermal-activation effects will be unobservable since values of H impractically close to H_v would be required, while for small V wall motion should occur at $H \sim 0$. Thus films having only a small range of particle size will exhibit magnetic-viscosity effects.

Several workers have observed magnetic viscosity in magnetic films [227, 228], but only in selected films because difficulty was encountered in finding films having sufficient uniformity for meaningful measurements (spatial variations of H_v can cause Barkhausen jumps between intervals of slow wall motion). Hellenthal [229] found that magnetic-viscosity effects became more pronounced in Ni films as t decreased ($t < 300$ Å). Such behavior is not surprising since the increasing crystallite separation found in very thin films means increasing magnetic independence of the crystallites, so that the sample may be viewed as a planar aggregate of crystallites with relatively weak mutual magnetostatic interactions, instead of a continuous film. Thermal activation may thus reverse **M** in some crystals at the domain wall. Stacey's simple theory makes no mention of how the volume V in thick, continuous films is determined, i.e., how the exchange and magnetostatic forces are broken in a thermal-activation process. In this connection it is interesting to note that Cohen [115] reported that the addition of Cu to 500-Å-thick NiFe films caused the appearance of magnetic-viscosity phenomena. The Cu was postulated to be preferentially concentrated at the grain boundaries, thus breaking exchange bonds and making the crystallites magnetically more independent.

Lambeck [230] has recently reported a presumably unrelated thermal-activation effect in Fe films. He found that as the temperature was decreased from 20 to $-175°$ C, the probability of Barkhausen jumps diminished, while slow wall motion was more pronounced. By observing that H_c increased with the time a domain wall was allowed to remain in any given position, Lambeck concluded that his observations were interpretable in terms of a diffusion-aftereffect [225]. This phenomenon, never before observed in films, involves thermal activation through atomic-diffusion processes, as opposed to the magnetic-viscosity phenomena discussed above, where the magnetization itself is directly subject to thermal activation.

(5) Creep [213, 231, 232]. As discussed above, a threshold curve for domain-wall motion can be determined for films subjected to dc EA and HA fields. It is found, however, that if a HA ac or pulse field and EA dc field $H < H_c$ are simultaneously applied, a film which is initially domain-wall free will become slowly demagnetized for fields well within the dc threshold curve by a process known as "creep." Creep is of great technological interest, because it means that information stored in a film can be slowly lost under the *repeated* action of pulse fields whose dc values would be too small to reverse the film. Creep is usually characterized by curves associated with the velocity v_c of individual walls, or curves showing the fields necessary to cause a specified amount of demagnetization in the film for a specified number of HA pulses (Fig. 19).

A *changing HA* field is necessary for creep, which does not occur under a dc *HA* and an ac *EA* field; each time the *HA* field pulse is applied, the wall shifts its position by a few microns. These shifts occur during the rising or falling edges of the *HA* field pulses, so that only the *HA* field pulse-height and total number of pulses are important, but not the repetition rate and pulse length. The creep threshold is critically dependent upon the structure of the wall [233]; films thin enough to contain only Néel walls do not exhibit creep under unipolar *HA* pulses, although some creep is seen with bipolar *HA* pulses. It would thus appear that creep is associated with the Barkhausen jumps in the Bloch-line positions caused by the *HA* pulse field $H_\perp$. This correlation has been directly observed by Lorentz microscopy [160].

Telesnin et al. [234] found that plots of ln v_c vs. *H* yielded nearly straight lines, so that the relation

$$v_c = v_0 e^{\alpha(H - H_b)} \tag{41}$$

was followed by their data, where v_0, α, and H_b are constants dependent upon $H_\perp$. Equation (41) is of the form of Eq. (40). This does not imply that magnetic viscosity is important in creep, but that a statistical process is present in both cases; in creep, the activation is provided not thermally, but by the changing *HA* field. When H_b was identified as the threshold field for wall motion for a dc $H_\perp$, Eq. (41) yielded a value of v_0, which upon substitution into Eq. (38) gave an estimate of the *equivalent EA* field H' generated by the changing *HA* field. Telesnin et al. found $H' \sim 10^{-2}$ to 10^{-1} Oe, while Green et al. [235] found $H' = 0.6$ Oe in an experiment which visually matched the effect on a wall of *EA* and *HA* ac fields.

The specific mechanism which generates this equivalent field is not well understood, however. When a *HA* pulse causes a local change in the magnetization distribution of a wall (e.g., a Bloch-line movement) a consequent change in the local potential-energy distribution is also made, so the wall may be enabled to move a short distance before it is again in a local potential-energy minimum. Specific reasons for such "local mobility" transformations may lie in the differences in mass or stiffness between segments of different wall types and/or, for thinner films, **M** distribution differences between Néel segments of opposite polarity. In another interpretation, attention is focused on the moving Bloch or

90° lines (either in crosstie walls or separating Bloch and Néel segments, respectively), the local transient fields from which are postulated to surpass the local values of H_c. Finally, a totally different suggestion, ignoring local differences in wall types, makes local reorientations of the wall, which occur to avoid poles on the walls, responsible for creep. There are fundamental objections to several of these mechanisms, and none of them is on a firm theoretical or experimental footing.

(6) Wall Streaming. If HA, short rise-time (≤ 30 nsec) pulses of amplitude about 0.3 H_K are applied to a demagnetized film (having several walls parallel to the EA), which is thick enough (≥ 800 Å) to support Bloch walls, the walls are observed to move back and forth in a streaming or "worm" motion, even though no EA field is applied [213, 236, 237]. The thinner the film, the shorter the rise time which is required for such wall streaming, but films thin enough for crosstie or Néel walls do not exhibit streaming no matter how short the rise time.

Streaming cannot be explained on the basis of the mechanisms discussed above. Rather, upon subjection to a short rise-time HA pulse, the spins in the wall precess, thus causing wall motion by an "unwinding" process. Segments of the wall of opposite senses will move in opposite directions, so that pulse-induced movement of the Bloch lines separating these segments causes the characteristic oscillating or streaming motion.

8. RESONANCE

A ferromagnet which is subject to both dc and rf fields will exhibit motion of the magnetization; for certain rf frequencies and related dc field values, the system will be in resonance. Experimental determination of the values of those parameters corresponding to resonance yields important ferromagnetic information. Indeed, insights into the basic mechanism of ferromagnetism can thereby be gained. Two types of resonance behavior are observed: ferromagnetic resonance (FMR) in which **M** moves uniformly and coherently throughout the sample, and spin-wave resonance (SWR) in which inhomogeneous magnetization distributions (spin waves) are excited.

8.1 Ferromagnetic Resonance [15, 10, 18]

Consider [10] an atomic system having a total angular momentum $J\hbar$ with an associated magnetic moment μ. Then $\mu = g\mu_B J$, or alternatively,

$\mu = \gamma \hbar \mathbf{J}$, where g is the Landé factor, $\mu_B \equiv e\hbar/(2mc)$ is the Bohr magneton, and $\gamma \equiv g\mu_B/\hbar$ is the gyromagnetic ratio. Speaking in terms of classical physics, if a field $\mathbf{H}$ is applied to the system, a torque $T = \mu \times \mathbf{H}$ will act upon it, or from the previous equation since $T = \mathbf{J}\hbar$, where the dot indicates time differentiation,

$$\dot{\mu} = \gamma \mu \times \mathbf{H} \qquad (42)$$

Equation (42) represents a *precession* of μ about $\mathbf{H}$ with an angular frequency, called the "Larmor frequency," of γH.

By replacing μ by $\mathbf{M}$, Eq. (42) may be used to describe the behavior of the magnetization in a *macroscopic* ferromagnetic body (a film), if $\mathbf{M}$ in the body is perfectly uniform. To incorporate a small amount of damping, which is experimentally found, a phenomenological damping term in either the Landau-Lifshitz or (for small damping) equivalent Gilbert formulation is added to the right side of Eq. (42). Thus

$$\dot{\mathbf{M}} = \gamma \mathbf{M} \times \mathbf{H} - M^{-2}\lambda \mathbf{M} \times (\mathbf{M} \times \mathbf{H}) \quad \text{or} \quad \dot{\mathbf{M}} = \gamma \mathbf{M} \times \mathbf{H} - M^{-1}\alpha \mathbf{M} \times \dot{\mathbf{M}}$$

$$(43)$$

where λ and α are the Landau-Lifshitz and Gilbert damping parameters, respectively. The scalar product of $\mathbf{M}$ with both sides of Eq. (43) vanishes, thus showing that the magnitude $|\mathbf{M}|$ is conserved since $\dot{\mathbf{M}}$ is perpendicular to $\mathbf{M}$; the damping term merely causes $\mathbf{M}$ to spiral in toward $\mathbf{H}$. However, if an rf field at the precession frequency γH is applied perpendicular to $\mathbf{H}$, energy will be taken from this field to maintain the precession against the damping. Such a system is said to be in ferromagnetic resonance.

For practical application of Eq. (43), $\mathbf{H}$ must be treated as an effective field which contains not only the applied field $\mathbf{H}_r$ but demagnetizing fields and anisotropy fields which act upon $\mathbf{M}$. Now $\mathbf{H}_r$ is usually applied normal to the film plane ("perpendicular" FMR) or in the film plane ("parallel" FMR), while the rf field is always perpendicular to $\mathbf{H}_r$ and in the film plane. It is easily shown from Eq. (43) (with the damping term omitted) that for perpendicular FMR, if H_K is the ordinary in-plane uniaxial anisotropy field and $H_K^\perp$ is the perpendicular anisotropy field (Sec. 4.7),

$$\omega_\perp = \gamma \left[\left(H_r - H_K^\perp - H_K \right) \left(H_r - H_K^\perp \right) \right]^{1/2} \qquad (44)$$

For parallel FMR,

$$\omega_\parallel = \gamma \left\{ \left[H_r + H_K^\perp + \frac{H_K}{2}(1 \pm 1) \right] (H_r \pm H_K) \right\}^{1/2} \tag{45}$$

where the $\pm$ signs refer to $\mathbf{H}_r$ perpendicular or parallel to the in-plane EA, respectively. The S-W model gives the position of $\mathbf{M}$ in the film plane for any arbitrary $\mathbf{H}_r$, thus making it possible to find $\omega_\parallel$ for any azimuth of $\mathbf{H}_r$.

Ferromagnetic resonance has been experimentally observed in films in the frequency range from [238, 239] 100 kc/sec up to 70 Gc/sec; most of the work reported has employed microwaves in the centimeter range. For films less than about 5,000 Å thick, the skin depth is larger than the film thickness (for frequencies less than 10 Gc/sec) so that, in contrast with FMR in bulk materials, the rf field is constant throughout the depth of the film and eddy-current damping can be neglected [240]. At the lower frequencies, standard vhf or uhf techniques are used, and the resonance condition is detected either by flux-pickup techniques or by monitoring the Q of a resonant circuit. At the higher frequencies the film is placed in a microwave cavity, and the reflected wave from the cavity is monitored. In all arrangements it is more convenient to fix the frequency and vary the magnitude of the dc field H_r while monitoring the output of the apparatus.

Parallel FMR in the range from about 100 kc/sec to 1 Gc/sec may be used to measure H_K. In that range, H_r and H_K are small enough to be neglected within the first bracket of Eq. (45), so that for the same frequency $\omega_\parallel$, the resonance peaks are separated by [241] $2H_K$. If then $\omega_\parallel^2$ is plotted against H_r for $\mathbf{H}_r$ parallel to both the EA and HA, two straight lines separated at a given $\omega_\parallel^2$ by $2H_K$ result (Fig. 21). From Eq. (45), the line representing the average of these two lines should pass through the origin. Experiments show that instead it intercepts the field axis at a distance H_i to the right of the origin [241, 242], indicating the presence of an unexplained internal field $\mathbf{H}_i$ parallel to $\mathbf{M}$. This internal field, which may attain values of several oersteds, increases monotonically with magnetization dispersion [243] but is nevertheless probably not directly related to the magnetostatic fields predicted by nonlinear ripple theory.*

In the higher-frequency range, large enough values of H_r are required so that H_K can be ignored in Eqs. (44) and (45). If parallel *and*

*See footnote on page 720.

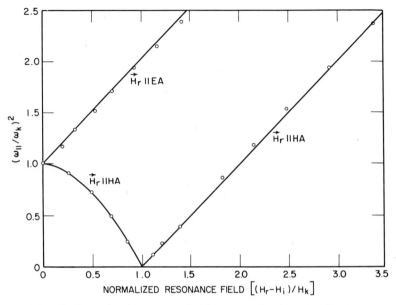

FIG. 21. Normalized resonance field vs. normalized applied field. Here $\omega_K \equiv \gamma(4\pi MH_K)^{1/2}$. Internal field H_i is about 1 Oe. (*After Ngo* [242]

perpendicular FMR experiments are performed, simultaneous solution of Eqs. (44) and (45) will yield γ and hence g; for NiFe, g values of about 2.1 are usually found [15]. In addition, $H_K^{\perp}$ can be determined, thus yielding a measurement of the magnetization if $K_{\perp}^S$ is small enough so that $H_K^{\perp} = 4\pi M$ (Sec. 4.7).

If the damping term in Eq. (43) is included in the considerations leading to the above equations, the width of the resonance peak is calculable. Such a calculation is conveniently carried out by substituting $\mathbf{T} = -\mathbf{r} \times \nabla E$ into Eq. (43), where E is the sum of the anisotropy, magnetostatic, and applied field energies, and $\mathbf{T} = \mathbf{M} \times \mathbf{H}$. For parallel FMR it is found that, if the frequency $\omega_{\parallel}$ is fixed while H is varied,

$$\Delta H_r = \frac{2\lambda\omega_{\parallel}}{\gamma^2 M} = \frac{2\alpha\omega_{\parallel}}{\gamma} \qquad (46)$$

where ΔH_r is the field interval between half-power points. It is found experimentally that ΔH_r can range from a few to several hundred oersteds, depending on the film. However, Eq. (46) is not satisfied experimentally, but rather $\Delta H_r = (\Delta H_r)_0 + 2\alpha\omega_\|/\gamma$; i.e., there is a residual broadening even at zero frequency. This residual broadening is attributable to magnetization dispersion which causes different parts of the film to come into resonance at slightly different fields. The residual broadening has even been used as an index of magnetization dispersion (although identified as "magnitude anisotropy dispersion"); $(\Delta H_r)_0$ increases monotonically with increasing magnetization dispersion of films as measured by susceptibility techniques [243].

Finally, it may be noted that in addition to the forced oscillations excited by FMR, free oscillations of **M** in magnetic films are observable [244, 245]. If a step-function pulse field is applied to the film (with *no* rf field), free oscillations can be observed which decay in a few nanoseconds. The field dependency of the oscillation frequency and the decay constants (about 1 or 2×10^8/sec) are in good agreement with the results obtained from FMR.

8.2 Spin-wave Resonance [246-248]

Spin waves (periodically spaced spin reversals) were first postulated to explain the decrease of magnetization with increasing temperature in bulk material. Their consideration led to the celebrated Bloch $T^{3/2}$ temperature dependence of M, which is valid at low temperatures. Later the reality of spin waves, or "magnons," was dramatically proved by the demonstration of resonance of spin waves in thin films.* The conditions for resonance of the nonuniform magnetization distribution represented by spin waves are similar to those for resonance of the *uniform mode* in FMR, so that the same experimental techniques are used for both FMR and SWR.

If the wavelengths involved are long, a continuum model in which the spins are treated as classical vectors may be used to treat spin-wave dynamics. Exchange forces resist spatial variations in **M(r)** and hence tend to suppress spin waves; exchange is manifested in the equation of

---

*Usually "perpendicular" SWR is studied, i.e., $\mathbf{H}_r$ is normal to the film plane, although it is also possible to excite spin waves in "parallel" SWR, i.e., with $\mathbf{H}_r$ parallel to the plane of the film. However, parallel SWR intensities are usually much weaker than perpendicular SWR intensities; parallel SWR is observed only when film inhomogeneities are such that favorable boundary conditions result.

motion of $M(r)$ [Eq. (43)] by an additional equivalent field of $2A\nabla^2 M(r)/M^2$. Thus Eq. (43) becomes (neglecting damping)

$$\dot{M}(r) = \gamma M(r) \times \left[H_{eff} + \left(\frac{D}{\gamma M}\right) \nabla^2 M(r) \right] \qquad (47)$$

where H_{eff} includes applied, demagnetizing, and anisotropy fields, as before, and where $D \equiv 2\gamma A/M$. Let $M(r) = M + m(r)$, where $m(r)/M \ll 1$. Then, if a saturating field H_r is applied along i_z normal to the plane of the film, a spin wave m_k [here $m(r) = \sum_k m_k$ with a propagation vector k in the i_z direction may be written as

$$m_k = m(i_x + ii_y) \exp[i(\omega t - kz)] \qquad (48)$$

This equation describes a spin wave of frequency ω traveling in the z direction. Upon substitution into Eq. (47), the dispersion relation

$$\omega = \gamma H_{eff} + Dk^2 + \cdots \qquad (49)$$

is found.

If rf energy of frequency ω satisfying Eq. (49) could be coupled to the spin wave, the latter would be maintained in resonance. However, this condition can be realized only if the spin waves are in a coherent state, such as represented by *standing* spin waves. Such standing spin waves are set up by reflections at the film interfaces. If the magnetic anisotropy at the interfaces (coming either from the action of an oxide layer or from Néel surface anisotropy [116]) is strong enough to immobilize or "pin" the spins at the interfaces,* the problem is analogous to that of a vibrating string clamped at both ends. The boundary conditions then demand a standing-wave solution of the form $\sin kz$, where $k = p\pi/t$ and p, the mode number, is an integer (Fig. 22a). Since the rf field is uniform within the film, in order to excite the spin wave, p must be an *odd* integer, for when p is even the instantaneous nonuniform moment vanishes upon integration over t, so that the rf field cannot couple to the magnetization.

As in FMR, it is convenient to keep ω fixed and vary H_r, which results [Eq. (49)] in a linear dependence of the applied field for resonance H_r on p^2. Such an experimental SWR spectrum is shown in Fig. 22b, where it is indeed seen that the even modes are not excited. From the slope of H_r vs. p^2, the exchange constant A can be found if t, γ, and M are

---

*The physical origin of the pinning action remains largely unexplained.

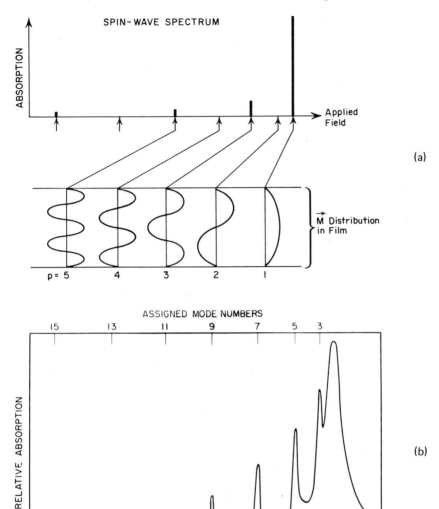

FIG. 22. (*a*) Schematic diagram of theoretical spin-wave spectrum under assumption of perfect pinning in a uniform film. Because of the symmetries of the **M** distribution corresponding to the even-numbered modes, the rf field cannot couple with them, so that only odd-numbered modes are excited. (*After Wolf* [247]) (*b*) Experimental spin-wave spectrum taken at 9.18 Gc/sec from a carefully prepared 4,100 Å thick 85 percent, Ni, 15 percent Fe film. Peaks follow quadratic-dispersion law over the entire range. (*Courtesy of R. Weber*)

known (the latter two may be determined from uniform mode resonance, Sec. 8.1). Determinations of A by SWR constitute a major application of this technique; values of A thus found compare favorably with values measured by other means.

Large deviations from the quadratic p dependence [Eq. (49)] have been observed in many films, particularly at low p values where linear p dependences have sometimes been found. To explain the SWR spectra of such films, an effective variation of M through the thickness of the film, in either a parabolic or other (even linear) fashion, has been postulated instead of pinning at the interfaces as discussed above; alternatively, variation of magnetic anisotropy normal to the film, caused, for example, by variation of strain through the film thickness, can produce the same effect. Such postulates lead to an eigenvalue problem similar to the quantum-mechanical problem of an electron in a potential well. Consideration of such results has led to the viewpoint that a necessary condition for SWR is the presence of inhomogeneity in the film.

In complete contradiction to this viewpoint, other authors [248, 249] have found that a p^2 spectrum dependence could be reproducibly obtained for films made with careful control of substrate temperature, deposition rate, and vacuum, so the films were as *homogeneous* as possible. These authors concluded that pinning originated in thin ($\sim$30 Å) layers, giving an effective magnetic anisotropy at the film interfaces. Inhomogeneous films can also support SWR with anomalous p dependences, e.g., linear, but if the films are too inhomogeneous, e.g., highly magnetostrictive films of Co or Ni, the SWR spectrum may be suppressed.

If in Eq. (49), $H_{eff} = H_r - 4\pi M$, then that dispersion relation becomes $H_r = \omega/\gamma + 4\pi M - D\pi^2 p^2/t^2$. The range of field for SWR is $4\pi M < H_r < 4\pi M + \omega/\gamma$ (for $H < 4\pi M$ the film is no longer saturated). Thus more SWR peaks in the spectrum are available and thinner films can be studied at higher frequencies. This was demonstrated by experiments at 70 Gc/sec; for example, Weber and Tannenwald [239] recorded a spectrum with over 20 peaks for a 7,000-Å film and studied spin wavelengths as short as 225 Å. Upon proper correction of the uniform-mode peak for "exchange" and "electromagnetic" shifts, a very precise ($\sim$1 percent) fit to a quadratic p dependence was observed for the low modes in support of the homogeneous-film viewpoint. Indeed, Weber and Tannenwald's accuracy in this experiment was high enough to permit determination of the coefficient F for an additional Fk^4 term in the dispersion relation [Eq. (49)]; interpretation of this result in terms of the nearest-neighbor Heisenberg model implies that the range of the exchange interaction is substantially greater than merely to the nearest atomic neighbors. Related SWR studies of the temperature dependence of D and F have also been made [250]. Such investigations are of interest to students of the fundamental mechanisms of ferromagnetism.

In magnetostrictive films, interactions between phonons and spin waves (including the uniform mode) have been observed [251]. Ultrasonic phonons can be generated by SWR in such films, or spin waves can be induced by the proper ultrasonic phonons [252].

8.3 Energy-loss Mechanisms

The uniform mode and spin waves are not independent, normal modes, but they interact among themselves and with the lattice. This is demonstrated by the observation of nonvanishing FMR and SWR line widths ΔH_r. The origin of this line broadening [aside from the $(\Delta H_r)_0$ term, Sec. 8.1] is highly controversial, for both bulk and films. Energy-loss mechanisms which have been proposed to explain the damping include (1) eddy-current damping, important only for low-numbered modes [253]; (2) exchange interaction between "magnetic" d and "conducting" s electrons [254]; (3) magnon-magnon scattering at inhomogeneities, i.e., one or more different spin waves are excited at the expense of the original excitation [255]; (4) magnon-phonon scattering, i.e., interaction of spin waves with lattice vibrations through magnetostriction, just discussed in Sec. 8.2.

Various experimental techniques have been developed to study the energy-loss mechanisms, including studies of ΔH_r vs. film thickness [256], ΔH_r vs. the angle of H_r with the film normal [255], high-power FMR effects [257, 258], and parallel pumping [259]. The data do not permit an unambiguous choice of loss mechanism, however.

9. SWITCHING [260]

9.1 Coherent Rotation Theory

Switching is defined as an irreversible rotation of the magnetization which occurs upon the application of sufficiently high fields. The threshold fields necessary for switching were discussed in Sec. 3 on the assumption of coherent rotation according to the S-W model. However, the S-W theory presumes static conditions and thus gives only the equilibrium positions of **M**; the theory must be extended to a dynamic theory [14, 241] in order to study the details of the switching transition.

In investigations of switching, a dc field $H_\perp$ is usually applied along the *HA*, and a step-function pulse field is applied along the *EA*. The coordinate system of Fig. 23 is used. In addition to the energy terms already considered in the S-W theory [Eq. (2)], it is necessary to include a term describing the magnetostatic energy generated by a component of **M** normal to the plane of the film, for it is found that during switching **M** lifts slightly (a few degrees) [261] out of the film plane. The total normalized energy is then

$$\epsilon = \tfrac{1}{2} \sin^2 \varphi_0 - h_\parallel \sin \theta \cos \varphi_0 - h_\perp \sin \theta \sin \varphi_0 + 2\pi M^2 \cos^2 \theta \quad (50)$$

The torque **T** is given by $\mathbf{T} = -\mathbf{r} \times \nabla \epsilon$. After transformation into spherical coordinates and substitution of **T** in this form into Eq. (43), equating the $\mathbf{i}_{\varphi_0}$ and $\mathbf{i}_\theta$ components results in a pair of simultaneous equations for φ_0 and ψ:

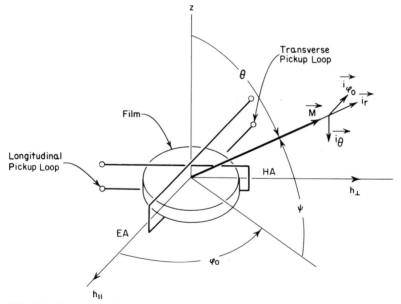

FIG. 23. Schematic diagram of detection arrangement used in switching experiments, along with a coordinate system used in the theoretical analysis. (*After Smith* [14])

$$\dot{\varphi}_0 M \cos\psi = -\gamma \frac{\partial\epsilon}{\partial\psi} - \frac{\lambda}{M \cos\psi} \frac{\partial\epsilon}{\partial\varphi_0}$$

$$\dot{\psi}M = \frac{\gamma}{\cos\psi} \frac{\partial\epsilon}{\partial\varphi_0} - \frac{\lambda}{M} \frac{\partial\epsilon}{\partial\psi} \qquad (51)$$

By differentiating with respect to time and using the assumption, justified a posteriori, that $\psi \ll 1$, these equations may be combined into

$$\ddot{\varphi}_0 + 4\pi\lambda\dot{\varphi}_0 + 4\pi\gamma^2 \frac{\partial\epsilon}{\partial\varphi_0} = 0 \qquad (52)$$

To obtain a solution of Eq. (52), a digital computer must be used. The results show that shorter switching times are achieved for higher values of $h_\|$ and/or $h_\perp$, as expected, and that the solutions can be highly oscillatory, depending critically on the magnitude of λ. As in the case in FMR, the origin of the damping is not well understood; it is clear only that for thin ($t \lesssim 5{,}000$ Å) films the eddy-current contribution should be small [14].

The physical interpretation of the switching process is that the torque exerted on **M** by $h_\|$ initially causes **M** to precess about $h_\|$ until a ψ value of a few degrees is attained, then **M** precesses in the φ_0 direction about the normal demagnetizing field thus created. This process is completed faster for larger $h_\perp$, since the initial torque is then larger.

9.2 Experimental Results

Step-function fields $h_\|$ with rise times less than 0.5 nsec have been generated by switching a charged coaxial cable with a mercury relay, then feeding the resulting current pulse into a strip line surrounding the film. The outputs of the pickup loops surrounding the film (Fig. 23), which are proportional to $\dot{M}_\|$ and $\dot{M}_\perp$ for the "longitudinal" and "transverse" loops, respectively, can be displayed with a response time of less than 0.5 nsec on a sampling oscilloscope [262, 263]. The resulting waveforms (Fig. 24a and b) depend critically upon the magnitude of $h_\perp$ and $h_\|$ and upon the magnetization dispersion.

The only case which is consistent with the coherent-rotation theory is that of low-dispersion films with high values of $h_\perp$ and $h_\|$. Fair agreement between theory [Eq. (52)] and experiment (Fig. 24a) can be obtained [241, 245, 263], but it is usually necessary to assume values of λ several times the resonance or free-oscillation value [14, 245], although it has

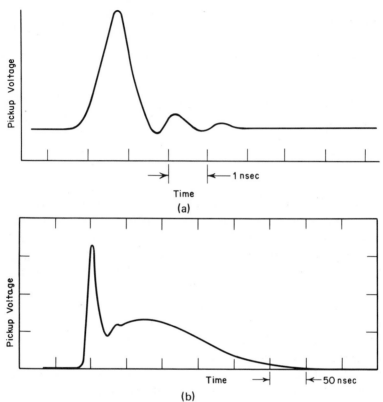

FIG. 24. Voltage waveforms from longitudinal pickup loops. (*a*) $h_\perp = 0.5$, $h_\parallel = 1$. Entire waveform is attributed to coherent rotation. (*After Hoper* [263]) (*b*) Small value of $h_\perp$. Initial spike is attributed to coherent rotation, while remainder of waveform is from wall motion. (*After Dietrich* [263])

been pointed out [263] that this discrepancy may originate in the neglect of the distortion in the sensing system. Smaller fields $h_\perp$ and $h_\parallel$ cause the appearance of a long tail following an initial spike (Fig. 24*b*), which indicates incoherent processes are occurring after an initial coherent process. Waveforms from high-dispersion films do not agree with the coherent-rotation theory no matter what values of $h_\parallel$ and $h_\perp$ are used.

From the longitudinal waveform (Fig. 24*a,b*), a switching time τ can be defined by any of several definitions, e.g., the interval necessary for the logitudinal signal to diminish to 10 percent of its peak value. Then $1/\tau$ may be plotted [264] vs. $h_\parallel$, with $h_\perp$ as a parameter (Fig. 25). If h is

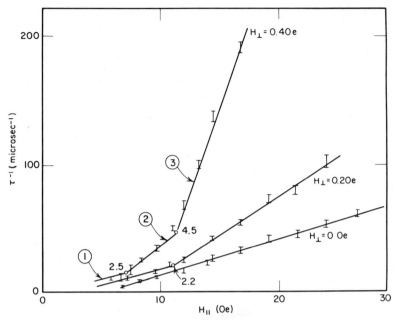

FIG. 25. Reciprocal switching time τ^{-1} vs. $H_{\parallel}$ for various values of $H_{\perp}$. Values of the products $H_{\parallel} \cdot H_{\perp}$ are given at the inflection points. The switching regions 1, 2, and 3 are indicated on the $H_{\perp} = 0.4$ Oe curve. (*After Telesnin et al.* [264])

sufficiently large, these curves are seen to consist of three regions 1, 2, and 3 which have been [265] associated with wall motion, incoherent rotations, and coherent rotation, respectively, separated by inflection points. [Note that a field point considerably outside the astroid is needed for switching in region 3; if $h_{\perp}$ is not large enough (Fig. 25), region 3 or even 2 may not be attained for *any* value* of $h_{\parallel}$.] However, this interpretation of the transitions seen in Fig. 25 may be incorrect, since, by using careful experimental techniques, Stein [262] showed that his data agreed with coherent rotation [Eq. (52)] only for high fields and for $\tau \sim 1$ nsec. In any case, the important question is that of determining the mechanism of breakdown from full coherent rotation to some noncoherent processes.

First, it is clear that dispersion is a very important factor in determining switching behavior. Thus the switching coefficients S_w, i.e.,

---

*Recently, however, Kolotov et al. [266] have reported an unexplained sharp increase in slope of the switching curve when $H_{\parallel}$ exceeds a threshold of 10 to 30 Oe, in the absence of any $H_{\perp}$ field.

the reciprocal of the slope of the switching curve in region 3 (Fig. 25), is directly proportional to the dispersion [267]. Further, it is found [264] that the product B of the field $h_{\parallel}$ at the inflection points and the corresponding field $h_{\perp}$ is a constant for each film, and that B is also directly proportional to the dispersion. Also, measurements show that the fraction of the magnetization which is irreversibly switched does not suddenly jump to unity at the astroid, but it increases gradually with applied fields $h_{\perp}$ and $h_{\parallel}$ and attains unity only beyond the astroid [262, 263]; as the dispersion increases higher fields are needed to attain unity.

Details of the switching process can be learned by the correlation of the integrals of the longitudinal and transverse signals, for since these integrals are proportional to $M_{\parallel}$ and $M_{\perp}$, respectively, $\mathbf{M}(t)$ can thereby be determined. It is illuminating to display [262, 263, 268] the results in a plot of the trajectory of $\mathbf{M}$ vs. time (Fig. 26). The initial angles of $\mathbf{M}$ are indicated for two different values of $h_{\perp}$, while the corresponding trajectories, which result after the application of an $h_{\parallel}$ pulse, are plotted at 0.5-nsec intervals. For coherent rotation the trajectory should be the arc of a circle; this condition is well fulfilled for the initial portions of both the $h_{\perp} = 0.15$ and 0.60 curves. After a certain time interval, however, the angular velocity $\dot{\varphi}_0$ and the value of $|\mathbf{M}|$ decrease rapidly by a factor of 10 to 100 (Fig. 26), indicating the initiation of incoherent behavior. The shape of the $\mathbf{M}$ trajectory depends not only upon $h_{\perp}$ and $h_{\parallel}$,

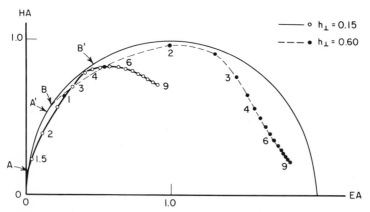

FIG. 26. Vector locus of net magnetization during switching for $h_{\parallel} = 1.92$ and two different values of $h_{\perp}$. Time in nanoseconds after application of $h_{\parallel}$ pulse marked on each locus. The initial directions of $\mathbf{M}$ are indicated at A and B, while the directions corresponding to switching according to the Stoner-Wohlfarth theory are indicated at A' and B', for $h_{\perp} = 0.15$ and $h_{\perp} = 0.60$, respectively. (After Sakurai et al. [268])

but upon the magnetization dispersion, e.g., high-dispersion films can show conservation of $|M|$ even after the initiation of incoherence as revealed by a marked decrease in $\dot{\varphi}_0$.*

Further insight into the switching process is furnished by "interrupted-switching" experiments [262, 263]; i.e., $h_{\parallel}$ is turned off before the switching process is completed. Analysis of the transverse and longitudinal waveforms showed [263] that even in the "coherent" region 3, coherent switching was never completed if $h_{\parallel}$ was interrupted before the full switching time; instabilities occurred, and M collapsed into an incoherent configuration. Presumably the same incoherent processes begin soon after an (uninterrupted) initial coherent rotation for switching at low fields (region 2). In principle, the dynamics of the incoherent processes could be investigated by following the incoherent switching process with Kerr or Lorentz microscopy, but these techniques have not been made sufficiently fast.† Investigators have thus been forced to the somewhat questionable procedure of inferring the magnetization configuration existing *during* switching from Kerr or Bitter observations made just *after* interrupted switching. They conclude that the magnetization breaks up into long, slender domains (similar to labyrinth or partial-rotation configurations), which grow with time during switching.

9.3 Interpretation

In spite of the experimental results given above, the detailed mechanism of incoherent switching is not yet clear. Several proposals have been made, however. Stein [262, 269] suggested that the ripple initially rotates nearly coherently until angles are attained at which the torque from the field can no longer maintain this coherence, and the ripple is then left in a locked state because of its magnetostatic fields.

---

*On the coherent-rotation theory the maximum value $\left(\dot{\varphi}_0\right)_{max}$ will be given by Eq. (52) with the first term omitted; using this relation and the value taken from plots of $\dot{\varphi}_0$ similar to Fig. 26, Sakurai et al. [268] found a value of λ of 1.5×10^8/sec, in good agreement with resonance and free-oscillation results.

†Such measurements have been recently made by Kryder and Humphrey (M. H. Kryder and F. B. Humphrey, *J. Appl. Phys.,* **40** (1969)) with a Q-switched laser. The resulting Kerr photographs show rotational processes during high-speed switching, but reversal by propagation of "diffuse" boundaries at low speeds. In addition, these authors cast further doubt on the interrupted-switching technique by demonstrating large changes in magnetization distribution upon interruption of the switching field.

Further reversal occurs by the motion of high-velocity, broad "dynamic walls." However, the torque may not in fact have the time dependence required by this theory [260], and it is difficult to obtain direct information on the behavior of such dynamic walls.

Harte [270] ignored the later stages of the incoherent state, and used his ripple theory (Sec. 5.2) to discuss the conditions under which incoherence *began*, i.e., the conditions for instability of the coherent state. He noted that if, at any instant during switching, the ripple pattern for some reason suddenly became immobile while the uniform mode continued to rotate, the ripple state would change from longitudinal to transverse. The transverse ripple would then exert on the uniform mode a magnetostatic "reaction torque," which could become large enough to overbalance the torque exerted by the applied field and thus "lock" the rotation. The ripple will be immobilized when it grows large enough during switching so that the associated magnetostatic fields cause violation of the Hoffmann stability criterion [Eq. (22)]. From these considerations Harte predicted the beginning of incoherence; he was also able to derive the relation $h_{\parallel} h_{\perp} = B$ found experimentally by Telesnin et al. [264] (Sec. 9.2), and he derived a magnetization-dispersion dependence of B in agreement with Telesnin's experimental results. However, as noted above, the inflection points in Telesnin's curves may not actually correspond to the initiation of coherent rotation.

This theory assumes that the relaxation time of the ripple is short so that the ripple, if it is not immobilized by becoming unstable, can reach equilibrium with the uniform mode in a time small compared with the switching time. In an early theory Harte [271] assumed the opposite extreme, i.e., a relaxation time far longer than the switching time, so that the reaction torque would be created by the *initial* ripple state. That the assumption of a short relaxation time is better was shown by Hoper [272], who, in studies of magnetization rotations through small angles, found ripple relaxation times of about 1 nsec, which was consistent with values of the free-oscillation damping constant.

Another theory, that of "bisensical rotation," has been offered [273] for reversal with $H_{\perp} = 0$.

10. COMPLEX MAGNETIZATION CONFIGURATIONS

So far, this chapter has been confined to the discussion of flat, planar films in which, because of the large demagnetizing field $4\pi M$, **M** was always constrained to lie in the film plane, except for the small transient tilt experienced during switching (Sec. 9). While the majority of investigations in ferromagnetic films have been concerned with this flat, in-plane magnetization configuration, some studies of variant configurations have been made.

10.1 $\vec{M}$ Normal to Film Plane

(1) **MnBi Films** $(K_\perp < 0)$. Manganese-bismuth crystals are hexagonal and have an easy axis along the c axis with a very high [274] ($\sim 10^7$ ergs/cm^3) associated magnetocrystalline anisotropy. If films of MnBi are prepared with the c axes of the constituent crystallites normal to the film plane, the magnetocrystalline-anisotropy field $\left(H_{K}^{\perp}\right)$ will be high enough ($\sim 3 \times 10^4$ Oe) to overcome the $4\pi M$ demagnetizing field ($\sim 7 \times 10^3$ Oe) so that $K_\perp < 0$ (Sec. 4.7); these MnBi films thus have the unique property that $\mathbf{M}$ in the remanent state is normal to the film plane. Such MnBi films have been successfully made by interdiffusion of Mn and Bi layers successively deposited on glass. However, the Mn-Bi chemical reaction which yields ferromagnetic MnBi is peritectic and hence slow and unreliable; the reaction seems to propagate from nucleation sites of previously reacted magnetic material [275]. Production of single-crystal MnBi films by epitaxy on mica appears to be a more reproducible procedure [53].

While $\mathbf{M}$ is everywhere normal to the MnBi film plane, the sense of $\mathbf{M}$ may vary across the film; i.e., domains of oppositely directed $\mathbf{M}$ may be present. Indeed, a domain configuration consisting of adjacent slab-shaped domains, alternately magnetized normal to the film plane, results in a state of lower total energy because the magnetostatic energy will thereby be reduced. Thus, calculations show that in equilibrium the domain width depends on the film thickness [276]. Although, owing to the presence of local energy barriers, such equilibrium states are not found, it is possible to induce reverse domains by application of fields normal to the film. Such domain structure can be conveniently studied by the Faraday or Kerr [275] effect, thus permitting detailed investigations of domain nucleation and growth [53].

(2) **Stripe Domains** $(K_\perp > 0)$. A structural anisotropy $K_\perp^S$ tending to maintain $\mathbf{M}$ normal to the film plane is again assumed, but now under the condition that $\left(H_{K}^{\perp}\right)_S < 4\pi M$. An equilibrium state of slab-shaped domains alternately magnetized normal to the film plane again obtains for thick films, while below a certain thickness $\mathbf{M}$ will lie in the film plane.

However, under these conditions a transition configuration for intermediate thicknesses has been predicted [277] for a thickness greater than a critical thickness t_c. In this configuration **M** is not alternately magnetized normal to the film plane but is *tilted* out of the film plane in

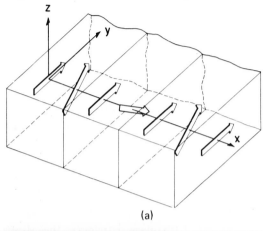

(a)

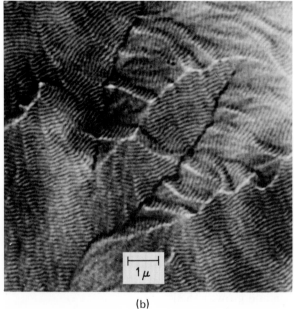

(b)

FIG. 27. (*a*) Schematic diagram of magnetization configuration in a stripe-domain film. Film is in the xy plane, and the mean magnetization is in y direction. (*After Fujiwara et al.* [284]) (*b*) Lorentz micrograph of stripe-domain film. (*Courtesy of S. Chikazumi. See Ref. 281*)

alternate senses in adjacent domains (Fig. 27*a*), with perhaps some curling of **M** at the film interfaces to provide flux closure [278, 279].

This "stripe-domain" configuration has actually been observed in vacuum-deposited films [280, 281] having a large value of $|K_\perp^S|$ (Fig. 27*b*). Because of the tilting of **M**, the hysteresis loop of such films has a characteristic shape, sometimes called a "transcritical" shape [282], indicating low remanence in zero field. (Early workers called films with stripe domains "mottled" films because the stripes in the Bitter patterns, which were not fully resolved, presented a spotted or mottled appearance [283].) It is found that the characteristics of the hysteresis loop and the width of the stripe domains depend on the film thickness, and that, as expected on theoretical grounds, below a critical thickness t_c (usually several thousand angstroms) the stripe domains and transcritical loop vanish, and the film becomes normal with **M** in the film plane. On the other hand, it has been reported that in the thickness range 2 to 10μ, while the transcritical loop is maintained, the magnetization configuration involves domains alternately magnetized *normal* to the film [279, 282], with closure domains at the film surfaces. Films thicker than 10 or 12μ revert to bulk magnetization configurations without an associated transcritical loop.

Stripe-domain films can also display [284] RIS properties as shown by hysteresigraph and torque-magnetometer investigations. The explanation offered for this phenomenon (differing from that of Sec. 6.4) is that while **M** between the stripe walls can rotate under the application of small fields, the walls themselves remain immobile until larger fields are applied; "resetting" the wall direction with a high field thus changes the initial susceptibity characteristics. A relaxation phenomenon was also observed in stripe-domain films; when a small ac field was applied at a large angle to a previously applied high ac field, the hysteresis loop showed hysteresis which increased with time in a time interval which depended on the magnitude of the applied field. It was shown that this effect was caused by a creeplike (Sec. 7.2) growth of regions of stripe domains oriented along the new field direction [285].

Calculations of t_c, stripe-domain width, and the shape of the transcritical loop were made on the basis of the model of Fig. 27*a*, and fair agreement with experiment was obtained. In order to obtain still better agreement with experiment, the model has been generalized [278] to permit more freedom in the angular deviations of **M**. Stripe domains have also been observed in single-crystal platelets [286] and thinned-down foils [134]; for those cases the calculations must take account of the fact that the easy axis may be inclined to the film normal.

It is found that t_c and the domain width depend upon film composition and substrate temperature because of the dependence of $K_\perp^S$ on these quantities. Originally $K_\perp^S$ and hence stripe domains were attributed to a strain-magnetostriction mechanism, i.e., either tension in negative-magnetostriction films or compression in positive-magneto-striction films [287]. It is now realized that other mechanisms can also contribute to $K_\perp^S$ and thus aid in the formation of stripe domains (Sec. 4.7). Thus, the properties of stripe-domain films seem to be strongly affected by the incidence angle during depositon, presumably because of the inclination from the film normal of the out-of-plane *EA* for films made at oblique incidence [132, 288].

10.2 Multilayer Films [289, 290]

(1) Interaction Mechanisms. Many investigators have given consider-able attention to the magnetic interactions between superposed magnetic films, and to the resulting **M** distributions in the films. Such films (with the usual in-plane magnetization configuration) may or may not be separated by nonmagnetic intermediate layers. For a two-film system, the simplest interactions originate in the effect of the demagnetizing field of one film on **M** in the other film and, for contiguous films, the exchange interaction between films. The former interaction tends to make **M** in the two films antiparallel, while the latter tends to keep the magnetization vectors parallel. Assuming the absence of domain walls, calculations of the **M** positions in both films, along with switching thresholds, can be made for these interactions, or indeed for a generalized interaction, in analogy with the S-W calculations for single films (Sec. 3). In these calculations [291-293], the azimuth of **M** is usually allowed to vary as a function of distance in the normal direction (thus requiring consideration of exchange interactions), while t, M, H_K, and the *EA* directions of the films are regarded as known quantities. The values of any or all of these parameters may differ between the films, which can necessitate involved calculations for **M** positions and switching thresholds. When such coupled-film systems have been experimentally investigated, good correlation with the theory has usually been found.

These calculations assume idealized film structures with perfectly planar interfaces. However, a vacuum-deposited sandwich made of two magnetic films separated by a nonmagnetic intermediate film, e.g., SiO, contains small-scale interface roughnesses because of the nonzero crystallite size. Magnetic charges are thus created at both intermediate-layer interfaces. The resulting magnetostatic energy will be minimized when the magnetization vectors in both layers are parallel; this interaction has been demonstrated by Lorentz microscopy [294].

A more subtle coupling mechanism is associated with two superposed vacuum-deposited films separated by a thin nonmagnetic metallic intermediate layer [295, 296]. The magnetic films are usually of different compositions, e.g., NiFe and NiFeCo, so they have markedly different values of H_c. The strength of the interaction can be measured by several means [297]; perhaps the most striking is a shift along the field axis of the *M-H* loop associated with the magnetically softer film. Such measurements demonstrate the presence of an equivalent coupling field of several oersteds which tends to align **M** in both films. This coupling field, which can be ten times the field expected from the above-mentioned surface-roughness model [295], decreases with increasing thickness of the intermediate layer. An early interpretation of this phenomenon involved an indirect exchange mechanism based on spin polarization of the conduction electrons in the intermediate layer [296]. However, later investigations involving heat and oxidation treatments indicated that the effect originated in interdiffusion in Pd intermediate layers [297], while for intermediate layers of Cr, Au, and presumably other metals, the origin lay in the production of tiny bridges of ferromagnetic material penetrating through small pores in the intermediate layers [298].

(2) Domain Wall Interactions. Interesting magnetostatic interactions have been observed between two domain walls which are in the separate magnetic films of a couple having a nonmagnetic intermediate layer [290], e.g., SiO. The nature of these interactions depends upon the thickness of the intermediate layer in relation to the domain-wall width, and the relative coercive forces of the magnetic films. Under favorable circumstances Néel walls having opposing rotational senses can be directly superposed, thus providing good flux closure. The wall energy (Sec. 7.1) is thereby reduced so much, in fact, that it is experimentally found that the Néel-to-crosstie transition occurs for much thicker films than in single-layer films [299]. It is not even necessary to have a full Néel wall in the second film to provide this flux closure, since a Néel wall in one film can induce in the other film a "quasi-Néel" wall (Fig. 28),

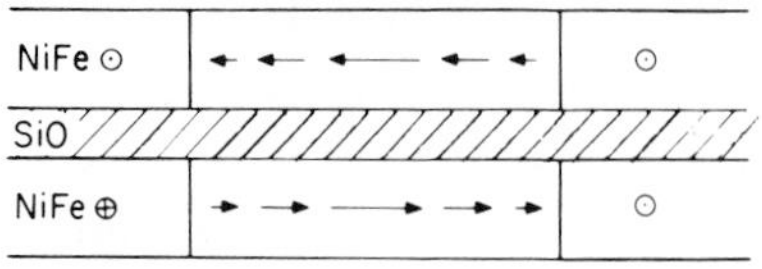

NiFe ⊙	← ← ← ── ← →	○
SiO /////	/////////////	//
NiFe ⊕	→ → ── → → →	○

FIG. 28. Cross section of a double film showing a juxtaposed Néel (bottom layer) and quasi-Néel wall (top layer). (*After Middelhoek* [290])

which provides the same flux closure but is not a real wall since the senses of **M** on both sides of the wall are parallel. Calculations of domain-wall energy, similar to those discussed in Sec. 7.1, show that the Néel quasi-Néel configuration is indeed the lower-energy state for sufficiently thin intermediate layers and magnetic-film thicknesses [290, 300]. The theory also predicts that wall widths in coupled films will be considerably larger than single-film wall widths because of the low magnetostatic energy originating in flux closure for the former case. These predictions are generally confirmed by Lorenz microscopy [294].

Not only the wall behavior but also the domain behavior in such coupled films can be unusual [290]. Domains can begin growing in one film of the couple and then, after encountering a high H_c region, continue their growth in the other film. Such behavior is again permitted by the flux-closure feature of coupled films.

Since walls in coupled films have a lower energy than single-film walls, it is not surprising that H_c is lower for coupled-film walls [298] (Sec. 7.2.2); the value of H_c depends upon the thickness of the magnetic films and the intermediate layer [290]. It might be anticipated [Eq. (39)] that the wall mobility would be higher in coupled than in single films, but published measurements are contradictory on this point [301, 302]. A report that low-field pulse switching is significantly faster for NiFe laminated films has also been challenged [303]. On the other hand, all authors agree that the creep threshold (Sec. 7.2) is raised in coupled films over that for a single film of the same thickness for unipolar, but not for bipolar pulses [297]. The change is caused simply by the retention of the less creep-sensitive Néel wall for larger film thicknesses.

10.3 Cylindrical Films

Films of NiFe (typically about 1μ thick) have been electroplated on metal wires in a manner such that a circumferential *EA* is established. The primary difference between the resulting cylindrical films and flat films is that a closed flux path can be established in the former films in the remanent state. Effects due to torsional strain of the wire [304] and the influence of the substrate surface must be considered [305], but the measured [306] properties of cylindrical films are essentially the same as those of flat films [307]. However, following the behavior of all electroplated films, these properties change slowly ("age") with time, although they can be stabilized by an annealing treatment [308].

11. MEASUREMENT TECHNIQUES [3, 309]

Many techniques have been developed for measurements of properties associated with either the local or the average magnetization distribution in a film. These techniques are based on various interactions with either the inner or external magnetic fields of a film, at frequencies ranging from dc to optical frequencies. The apparatus may employ sensing of either magnetic-field (pickup probes or Bitter technique), mechanical (torque magnetometer), electrical (Hall effect or magnetoresistance), or light (Kerr or Faraday effect) phenomena.

11.1 Microscopic Observations [225]

(1) Bitter Technique. While the external fields just above the surface of a magnetic film can be studied by direct measurements with a miniaturized Hall-effect probe [310], the Bitter technique offers better spatial resolution and more convenience for this purpose. In this technique a colloidal suspension of fine (about 1,000 Å diameter) ferromagnetic particles, e.g., Fe_3O_4, is allowed to flow over the film. A glass cover slip is placed over the colloid, which is then examined by a light microscope, most conveniently in the dark-field mode. The particles will be attracted to regions of high fields which originate in areas of high magnetization divergence, e.g., domain walls, thereby inducing easily visible particle aggregation at those areas (Fig. 16b).

The resolution of this technique is limited by the questions of contrast and particle size, although some improvement in this regard is possible by making electron micrographs of dried colloidal suspensions. The technique does not permit the detection of small divergences of **M**, like ripple, unless the ripple is extraordinarily large as in RIS films, and fast magnetization changes cannot be followed since several seconds are required for the colloid to rearrange. Also, the technique is applicable only within the narrow temperature range within which the colloidal suspension is stable.

(2) Kerr and Faraday Magneto-optical Effects. When plane-polarized light is passed through a magnetic film, the plane of polarization is rotated by an angle (about $1/3°$) which is proportional to the component of **M** parallel to the direction of propagation of the light. Thus, to obtain an appreciable "Faraday" rotation, the light may be passed at normal incidence through a film having **M** normal to the film plane, but the light must pass obliquely through a film having **M** confined to the film plane. If the film is then viewed through an analyzing Nicol prism set for extinction of the light traversing a given domain, light traversing an adjacent domain of opposing magnetization sense will not be extinguished owing to its opposite Faraday rotation. The Faraday (or Kerr) effect thus provides contrast between *domains*, as opposed to the Bitter technique, which distinguishes domain *walls*.

To utilize the Kerr effect, polarized light is *reflected* from the surface of a film, and viewed through an analyzing prism. There are several categories of Kerr effect, depending upon the relative orientations of **M** and the plane of polarization, but only for those in which the electric vector has a component perpendicular to **M** do rotations of the polarization plane occur. Probably the longitudinal meridional configuration is the most commonly used, for which the optimum incidence angle for maximum rotation is about $60°$. It is evident that the Kerr and Faraday effects have the advantages of being able to follow rapid changes in magnetization and of being usable over a wide temperature range. Furthermore, measurements of the major magnetic parameters may be made with this method (Sec. 11.2).

While the Kerr effect is usually used for low-magnification applications, the full resolution of light optics ($\sim 1\mu$) is attainable if a high-numerical-aperture objective lens is used [311]. However, care must be taken to maximize the contrast, which is inherently low. Aside from precautions taken in instrumentation, this goal can be approached by "blooming" the film with a dielectric overlayer which is designed to enhance the Kerr effect by multiple reflections. Optimum thicknesses and indices of refraction of a dielectric superstructure designed to provide the greatest enchancement of the Kerr and Faraday effects have been calculated [312, 313].

(3) Lorentz Electron Microscopy. With the proper instrumental adjustments, magnetic structure can be viewed by transmission electron microscopy. In the *defocused* mode of operation (Fig. 29), the objective lens is focused at an image plane at some distance (typically a few centimeters) from the film. Now, because of the Lorentz force $q\mathbf{v} \times \mathbf{B}$ acting on them, the electrons are deflected on passage through the film, in senses depending on the **M** direction. Then, as seen from Fig. 29*b*, at the image plane an increased electron intensity will result below some walls and a decreased intensity below others, depending on the orientation of the electron beam relative to the senses of **M** in the adjacent domains. The defocused mode of operation thus detects divergences of **M**, e.g., domain walls, in analogy with the Bitter technique. However, if the microscope is operated in the less-popular Foucault mode, for which an aperture is critically positioned in the back focal plane of the objective lens, contrast between different domains is furnished, in analogy with the Kerr effect.

Lorentz microscopy offers the advantage of high resolution, e.g., it can easily detect magnetization ripple, while measurements of the important macroscopic magnetic parameters (Sec. 11.2) can be made in the microscope [160]. Furthermore, the mean direction of **M** at an arbitrary point in a film may be deduced from a Lorentz micrograph by noting that **M** is always perpendicular to the ripple. However, the detailed interpretation of Lorentz micrographs is not straightforward because the geometric-optical theory of contrast formation presented above is only the first approximation to a more complicated wave-optical theory [314]. Since the geometric-optical theory is only marginally valid

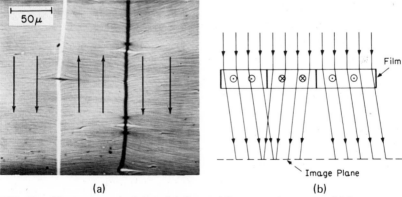

(a)   (b)

FIG. 29. Illustrating principles of defocused Lorentz microscopy. (*a*) Lorentz micrograph, with **M** directions indicated by arrows. (*b*) Side view showing electron trajectories which correspond to the image shown in (*a*). (*After Cohen* [160])

for handling the imaging of domain walls, calculations of domain-wall shapes based on the geometric-optical approach must be treated with caution; a fortiori, magnetization ripple can be quantitatively studied by Lorentz microscopy only if the wave-optical approach is employed.

11.2 Macroscopic Measurements

The macroscopic parameters which characterize the magnetic behavior of a magnetic film are size (thickness and lateral dimensions), composition, magnetization, magnetostriction constant, K_u or H_K, H_c, and α_{90} (in lieu of the ripple spectrum or average magnetization dispersion). The thickness and composition are measured by the usual techniques applicable to all thin films (Chap. III), while specialized methods have been developed to measure the other quantities. Katz et al. [315] have proposed the standardization of some of these methods for measurements required in engineering applications, but many of their proposed methods suffer in accuracy from ripple influences and thus are inappropriate for precise work.

(1) Mechanical Detection (Torque Magnetometer). The most reliable measurements of magnetic anisotropy can be made with a torque magnetometer. The sample is suspended from a fiber in such a way that the plane of the film is horizontal, and a dc field **H** is slowly rotated in the horizontal plane. As **H** rotates, a varying torque **T** is exerted on the fiber, tending to twist it. In modern instruments, any incipient fiber twist is detected automatically and a counter-torque is electronically applied to the fiber so that it maintains its original position; the strength of the electric counter-torque signal, which is proportional to **T**, is then plotted as a function of the azimuth of **H**. The torques involved are very small ($\sim 10^{-2}$ dyne-cm), so that special precautions must be taken to build a sensitive instrument while ensuring that it does not respond to vibrations [316].

Assuming that **H** is large enough to keep the film saturated at equilibrium, the torque exerted by **H** on **M** is always balanced by the torque $T = \partial E_K/\partial\varphi_0$ exerted by the sample on **M**, where E_K is the anisotropy energy and φ_0 is the angle between **M** and the *EA* (Fig. 9). If H is large enough, φ_0 is approximately equal to the field angle β, so that the magnetometer records a signal proportional to $\partial E_K/\partial\beta$ as a function of β. For a uniaxially anisotropic film this signal is thus proportional to

$K_u \sin 2\beta$. Then, after calibration of the instrument, K_u can be determined from the full excursion of the $\sin 2\beta$ trace and a knowledge of the film thickness (since the signal is also proportional to the sample volume). Since the value of H can be made high enough to suppress the ripple, ripple cannot influence the measurement. On the other hand, if measurements are made at values of H for which ripple is present, rotational hysteresis originating in the ripple can be studied (Sec. 6.1).

If the sample is suspended so that the film plane is vertical, application of a small field **H** in the horizontal plane at an angle θ to the film plane induces a torque $MVH \sin \theta$, where V is the film volume, thus permitting the determination [316] of M. If H is now increased so that **M** is pulled out of the film plane, $K_\perp$ (Sec. 4.7) can be measured [57]. Still keeping the same geometry, with proper orientations of applied fields it is possible to measure [316] H_c, or the entire *B-H* loop may be traced out [317].

(2) Flux-pickup Detection. Experimentally, the simplest scheme for detection and interpretation of the macroscopic magnetic state of a film is probably the measurement of the voltage induced in a pickup coil adjacent to the film. Since, by Faraday's law of induction, this voltage is proportional to the rate of *change* of magnetic flux enclosed by the coil, either relative motion between the coil and film must be provided, or the magnetic state of the film must be changed by an applied field. Although the former approach has been used successfully in a "vibrating-sample" magnetometer [318], the latter approach is experimentally easier and more often employed.

Because of its experimental simplicity and its ability to measure most of the pertinent magnetic parameters, the hysteresigraph [3] is the most commonly used instrument for routine examination of magnetic films. In this instrument a pickup coil either surrounds or is adjacent to the film, so that the coil encloses an appreciable fraction of the magnetic flux originating from magnetic poles at the film edges. The magnetization in the film changes with time in response to an ac excitation field which is produced in the plane of the film by a pair of Helmholz coils whose axis is parallel to the axis of the pickup coil. A voltage is then induced in the pickup coil which is proportional to the rate of change of the total flux linking it. This total flux includes a large spurious component due to the mutual inductance of the Helmholz and pickup coils; this component is customarily canceled by the voltage induced in an auxiliary "bucking" coil, which is within the Helmholz coil but far from the film. To obtain a hysteresis loop, the compensated pickup voltage, after integration and amplification, is fed to the vertical plates of an oscilloscope, while a voltage proportional to the excitation field is fed to the horizontal plates. In designing the instrument, care must be taken to balance possible phase shifts between various signals, to use a wideband amplifier in order to reproduce the *M-H* loop with good fidelity, and to use a good pickup-coil design in order to obtain the best signal-to-noise ratio [319].

Since the height of the loop is proportional to both **M** and *t*, upon proper calibration either of these quantities may be measured. By applying the excitation field along the *EA*, H_c may be determined by noting the minimum field for which hysteresis can be detected. If a small excitation field is applied along the *HA* of a film in the single-domain state, H_K may be taken as the field at which the extrapolation of the resulting closed loop cuts the high-drive saturation value of *M* (Fig. 3*d*). However, this technique of measuring H_K is valid only for low-dispersion films, since the procedure is equivalent to determining the reciprocal of the initial susceptibility, which is ripple-dependent (Sec. 6.3).

The *HA* fallback determination of α_{90} (Sec. 6.2) may be made by repositioning the pickup coil so that its axis is perpendicular to the excitation field; the pickup coil now detects the changing flux component normal to the excitation field. A large excitation field (greater than H_K) is then applied along the *HA*. As the sample is rotated to give a gradually increasing angle α between the *HA* and the excitation field, or alternatively as a gradually increasing dc *EA* bias field $\mathbf{H}_{\parallel}$ is applied, a signal is recorded which increases to a saturation value when **M** everywhere in the film rotates in the same sense. The value of α at which *r* percent of the saturation signal is detected is designated as α_r; the results of the bias-field method are defined similarly, but now $\alpha \equiv \sin^{-1}(H_{\parallel}/H_K)$. The results of the two measurements are generally different because of the difference between the ripple states which are involved.

The hysteresigraph has also been used in the determination of the magnetoelastic coupling constant $\eta \equiv dH_K/de$ (Sec. 4.3). The film, on a thin (~0.2 mm) glass substrate, is bent between three or (for better strain uniformity) four [67] knife-edges, while the value of H_K as a function of the strain is monitored by the hysteresigraph technique described above.

With the provision of a dc field perpendicular as well as parallel to the excitation field, and with appropriate amplification of the pickup signal (preferably using a lock-in amplifier), initial susceptibility studies can also be made with the hysteresigraph apparatus (Sec. 6.3).

High-speed switching and FMR experiments may also be said to utilize a flux-pickup technique; the apparatus has been briefly described above. Schemes using flux pickup to study the magnetization behavior of local regions have been suggested, but such schemes have not attained the spatial resolution afforded by the microscopic methods of Sec. 11.1 [320, 321].

(3) Magneto-optic Detection The output of a photomultiplier tube which replaces the ocular of a Kerr or Faraday apparatus (Sec. 11.1) can be made proportional to that component of **M**, averaged over the entire film, which is parallel to the direction of light propagation; then a hysteresigraph based not upon flux pickup but upon magneto-optic detection can be constructed, so that H_K, H_c, and α_{90} can be measured [322]. This technique offers the advantage of insensitivity to electromagnetic pickup from the driving field at the sensing element, and the ability to explore small film areas by using a narrow pencil of light. The technique may also be used instead of flux pickup in sensing fast rotational switching [323].

In addition to the ferromagnetic metals [324], magneto-optic effects in ferromagnetic insulator films, which have large Faraday rotation but small light absorption, have been explored [325].

(4) Magnetoresistive or Hall-effect Detection. If θ is the angle between **M** and the direction of a current I in the plane of the film, then [75, 326] the resistance R is given by* $R/R_0 = 1 + (\Delta R/R_0) \cos^2 \theta$, where $\Delta R/R_0 \sim 10^{-2}$. Thus, if a dc current is applied along the EA while a HA field **H** is applied, the S-W theory predicts a linear dependence of R on H^2. If an ac HA field is used, R vs. H can be conveniently displayed on an oscilloscope, and thus H_K can be determined analogously to the flux-pickup hysteresigraph method. Critical fields for initiation of wall motion can also be determined by magnetoresistive detection.

If electric and magnetic fields are applied in the film plane, a *planar* Hall-effect voltage $V = \frac{1}{2}IR_q \sin 2\theta$ is produced perpendicular to the current, where R_q is a constant characterizing the film. For example, for an 81 percent Ni 19 percent Fe film, 500 Å thick, of length 8 mm and

*Here ΔR is the magnetoresistance constant.

width 1 mm, R_q is found to be about 0.1 Ω when the current is along the length of the strip [326]. This effect may also be used [327] to measure H_K and H_c.

(5) Ripple-independent H_K Determinations. As already discussed, the determination of H_K from the *HA* initial susceptibility suffers from inaccuracy because of ripple influences. Although torque-magnetometer measurements are free from this objection, they are relatively inconvenient. Several methods have therefore been devised to permit a non-torque measurement of H_K which is experimentally easy but yet is free of ripple influences. In principle, these methods can employ either flux-pickup, magneto-optic, magnetoresistive, or even Lorentz-microscopy sensing, but except for method (*b*) they were all originally designed for flux-pickup detection.

(*a*) The Feldtkeller initial-susceptibility method was introduced in Sec. 6.3.

(*b*) If a large ac excitation field is oriented perpendicular to the axis of a pickup coil, no voltage will be induced in the latter when **M**, in a single-domain film, is held parallel to the excitation field. For any orientation of the film, the orientation of **M** giving this null condition will be attained for a unique value of an auxiliary dc field applied perpendicular to the excitation field, and hence perpendicular to **M**; the auxiliary field is thus proportional to a torque exerted on **M** and, when plotted against the azimuth of the film in the range (0, 2π), yields a "pseudo-torque" curve. A second dc field parallel to the excitation field ensures maintenance of the single-domain state; the value of this field must be large enough to saturate the film, but it does not otherwise affect the results. Real and pseudo-torque curves of magnetic films are in good agreement [328].

(*c*) A variation of method (*b*) has been suggested [329] in which the *EA* is placed at 45° to the *EA*; in this case the null condition is satisfied when the value of the auxiliary dc field is $H_K/2$.

(*d*) A saturating dc field is rotated in the plane of the film, and the output of a pickup loop adjacent to the film is Fourier-analyzed in terms of harmonics of the rotation frequency. Upon proper analysis, these harmonics yield not only the value of the uniaxial anisotropy, but the value of higher-order anisotropies and the value of *M* as well [330].

12. APPLICATIONS

The primary application of magnetic films has been in the technology

proposed for use as the coatings on magnetic tape [331], while controlled domain-wall motion in ferromagnetic films has been suggested for logic elements in digital computers [332]. However, most attention has been given to the use of magnetic films as elements in a high-speed digital-computer memory, or store. Magnetic films are attractive for this application because they can be produced with inherent bistable states (uniaxial anisotropy) and can be switched from one stable state to the other in a few nanoseconds by small drive fields. In addition, large high-density arrays of film "bits" can be made in a favorable geometry and at low cost.

12.1 Flat-film Memory

A schematic diagram of four bits in a flat-film memory [333] is shown in Fig. 30a. Such a memory can be constructed by fabricating individual bits by suitable masking techniques during deposition, or by selective etching of a continuous film. Alternatively, the memory plane may be completely covered with ferromagnetic material, while the bits are defined by domain walls surrounding the region of the drive-line crossings. The ferromagnetic film is typically 81 percent Ni, 19 percent Fe, about 1,000 Å thick, and deposited on a smooth, flat substrate (e.g., glass) in a magnetic field so as to give a well-defined EA.

A "one" ("zero") is stored in a given bit when that bit is uniformly magnetized say left (right) along the EA. The memory must be designed to enable interrogation of each bit, so that the stored information may be "read out," and provision must also be made to change the magnetization state at will, or to "write in" information. It is possible to operate the memory in the "coincident-current" mode so that these reading and writing operations are carried out only on each selected bit of interest. However, since such operation puts severe requirements on the tolerances of the film parameters, the "word-organized" mode (Fig. 30a) is more commonly used. For operation in this mode the bits are arranged in a rectangular array, while electrical conductors are placed parallel to the EA ("word" lines) and HA ("digit" and "sense" lines) of each bit. The electrical lines may be mechanically positioned wire or may be fabricated by deposition techniques involving masking or subsequent etching.

To interrogate a given bit, a current pulse is passed through the word line corresponding to the row (word) in which the bit is contained, thus creating a HA (or "transverse") field pulse on all the bits in the word. The magnetization in each bit of the word will then uniformly rotate to-

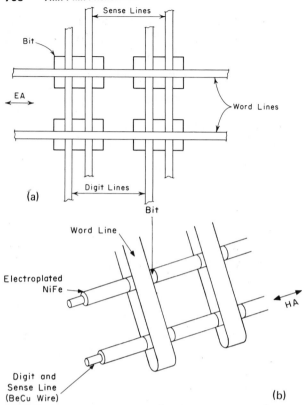

FIG. 30. Schematic diagram of wiring arrangement for a word-organized memory using uniaxially anisotropic ferromagnetic (*a*) flat films, (*b*) electroplated cylindrical films.

ward the *HA*, thereby generating a pulse in its associated sense line of different polarity according as a one or a zero was stored in that bit. Upon termination of the word pulse, the information in the bit will be destroyed by the *HA* domain-splitting process (Sec. 6.2); the type of operation just described is thus known as "destructive" readout (DRO). The information must therefore be rewritten into each bit in the word upon (or before) termination of the interrogating word pulse.

To write information into a given bit, a pulse is again applied to the word line. Before termination of that pulse, another pulse is applied to the digit line (which may be physically the same conductor serving as the sense line) corresponding to the bit of interest, thereby generating an *EA* (or "longitudinal") field in the bit. The magnetization in the bit will then relax back to the *EA* sense corresponding to a one or a zero according to the sense of the digit pulse.

The speed of such a memory is limited not by the switching time of the bits (Sec. 9) but by the transient times of the associated circuitry. These transient times include not only the rise times of the pulses but also the recovery time of the sense amplifier after being subjected to the pulse induced in the sense line by coupling from the digit line. In a large memory, the "cycle time" of a write, read, rewrite operation may thereby be extended to several hundred nanoseconds. [The cycle time may be reduced if more complicated nondestructive readout (NDRO) configurations are used which obviate the need of the rewrite operation [334].]

The detailed design problems are complex. Thus, for example, the sense amplifier must detect the read signal from the interrogated bit through the noise from the capacitive coupling between the word and sense lines (since these lines are orthogonal the inductive coupling between them is low). The requirements on the drive currents are determined by the tolerance on the film parameters; i.e., the digit and word pulses must be large enough to change the state of any desired bit in the array, but not so large that an "unselected" bit is inadvertently reversed. Particular attention must be paid to creep effects (Sec. 7.2), for bits in an unselected word may be inadvertently reversed by creep because of HA pulses from neighboring word lines in the presence of EA digit pulses, which are simultaneously applied to all bits in a given column during writing. In practice, considerations of cycle time, bit density (which influences signal output and creep sensitivity), memory size, and cost (including the necessary circuitry) are interrelated [335]. Memories containing of the order of a million bits with bit diameters and spacings of the order of tenths of a millimeter are currently practical.

Many variations on the configuration of Fig. 30a have been proposed. To reduce the effect of creep, specially shaped [336] or multilayer [337] bits have been suggested, while various proposals [337, 338] for bits having flux closure instead of the open flux structure of Fig. 30a have been made.

12.2 Cylindrical-film Memory [339]

In recent years the cylindrical-film (Sec. 10.3) memory has become increasingly popular. Such a memory typically consists of a parallel array of specially prepared BeCu wires onto which 81 percent Ni, 19 percent Fe has been electroplated, with a set of word lines placed perpendicular to the wires (Fig. 30b). The bits are defined by the intersections of the wires and the word lines, and have circumferential easy axes owing to the passage of current through the wires during the electrodeposition process. The operation of this memory is completely analogous to that of the flat-film memory of Fig. 30*a*.

A major advantage of this type of memory over the flat-film memory is the closed flux path of the bits; for the same bit size the absence of demagnetizing fields permits a greater film thickness (e.g., about 1μ) and hence a larger readout signal. In addition, this geometry provides close coupling between the bits and the sense-digit wire. A disadvantage of the cylindrical-film memory is that fringing fields from the word lines are larger than in the flat-film case, which necessitates bit densities at least an order of magnitude less than those practical with flat films. It is thus difficult to see how this geometry could be used in future ultra-high-density memories.

12.3 Future Developments

A slow, steady improvement in the design of both flat-film and cylindrical-film memories can be anticipated; special schemes for amelioration of the creep, signal-to-noise ratio, bit density, and speed problems will be developed. However, several workers have reached the conclusion that the conventional schemes for reading and writing presented above must be abandoned in order to permit a truly revolutionary improvement in memory size to the gigabit level. For this purpose bit diameters in the micron range have been proposed, with "Curie-point" writing by means of a controlled electron or laser light beam which heats any selected bit [340, 341], and with reading by the Kerr effect (Sec. 11.2).

Acknowledgments

T. S. Crowther, K. J. Harte, H. Hoffmann, D. O. Smith, and R. Weber are thanked for helpful criticisms of the manuscript of this chapter.

REFERENCES

1. G. R. Pulliam, *J. Appl. Phys.*, **38**:1120 (1967).
2. K. Y. Ahn and J. C. Suits, *IEEE Trans. Magnetics*, **MAG-3**:453 (1967).
3. M. Prutton, "Thin Ferromagnetic Films" Butterworth & Co., Ltd., Washington, 1964.
4. W. M. Mayer, *IEEE Trans. Magnetics*, **MAG-2**:166 (1966).

5. I. W. Wolf, *J. Appl. Phys.*, **33**:1152 (1962).
6. R. Girard, *J. Appl. Phys.*, **38**:1423 (1967).
7. L. D. Ranson and V. Zentner, *J. Electrochem. Soc.*, **111**:1423 (1964).
8. T. Matcovich, E. Korostoff, and A. Schmeckenbecher, *J. Appl. Phys.*, **32**:93S (1961).
9. M. C. Paul and M. M. Hanson, *J. Appl. Phys.*, **37**:3743 (1966).
10. E. Kneller "Ferromagnetismus," Springer-Verlag OHG, Berlin, 1962.
11. "Magnetism" (G. T. Rado and H. Suhl, eds.), vols. 1-4, Academic Press Inc., New York, 1963-1966.
12. S. Chikazumi, "Physics of Magnetism," John Wiley & Sons, Inc., New York, 1964.
13. J. B. Goodenough and D. O. Smith, in "Magnetic Properties of Metals and Alloys," p. 112, American Society for Metals, Metals Park, Ohio, 1959.
14. D. O. Smith in "Magnetism" (G. T. Rado and H. Suhl, eds.), vol. 3, p. 465, Academic Press Inc., New York, 1963.
15. W. Andrä, Z. Frait, V. Kamberský, Z. Málek, U. Rösler, W. Schüppel, P. Šuda, L. Valenta, and G. Vogler, *Phys. Stat. Solidi*, **2**:99, 112, 136, 345, 941, 1227, 1241, 1417 (1962); **3**:3 (1963).
16. E. W. Pugh in "Physics of Thin Films" (G. Hass, ed.), vol. 1, p. 277, Academic Press Inc., New York, 1963.
17. E. Feldtkeller, *Z. Angew. Phys.*, **17**:121 (1964).
18. R. F. Soohoo, "Magnetic Thin Films," Harper & Row, Publishers, Incorporated, New York, 1965.
19. R. Jaggi, S. Methfessel, and R. Sommerhalder in Landolt-Börnstein, "Zahlenwerte and Funktionen," vol. 2, part 9, p. I-141, Springer-Verlag OHG, Berlin, 1962.
20. H. Chang and G. C. Feth, *IEEE Trans. Commun. Electron.*, **83**:706 (1964); H. Chang and Y. S. Lin, *IEEE Trans. Magnetics*, **MAG-3**:653 (1967) (a bibliography).
21. R. Abbel in "Basic Problems in Thin Film Physics" (R. Niedermayer and H. Mayer, eds.), p. 375, Vandenhoeck and Ruprecht, Göttingen, 1966.
22. A. Corciovei, *IEEE Trans. Magnetics*, **MAG-4**:6 (1968).
23. M. J. Klein and R. S. Smith, *Phys. Rev.*, **81**:378 (1951).
24. W. Döring, *Z. Naturforsch.*, **16a**:1008, 1146 (1961).
25. R. J. Jelitto, *Z. Naturforsch.*, **19a**: 1567, 1580 (1964).
26. A. Corciovei, *Czech. J. Phys.*, **10**:568, 917 (1960).
27. J. J. Pearson, *Phys. Rev.*, **138**:A213 (1965).
28. A. Corciovei and Ghika, *Czech. J. Phys.*, **12**:278 (1962).
29. A. Corciovei and G. Ciobanu, in "Basic Problems in Thin Film Physics" (R. Niedermayer and H. Mayer, eds.), p. 381, Vandenhoeck and Ruprecht, Göttingen, 1966.
30. W. Brodkorb and W. Haubenreisser, *Phys. Stat. Solidi*, **16**:577 (1966).
31. S. Szczeniowski and L. Wojtczak, *J. Appl. Phys.*, **39**:1377 (1968).
32. E. C. Crittenden, Jr., and R. W. Hoffman, *J. Phys. Radium*, **17**:270 (1956).
33. C. A. Neugebauer, *Phys. Rev.*, **116**:1441 (1959).
34. C. A. Neugebauer, *Z. Angew. Phys.*, **14**:182 (1962).
35. D. Stünkel, *Z. Physik*, **176**: 207 (1963).
36. E. L. Lee, P. E. Bolduc, and C. E. Violet, *Phys. Rev. Letters*, **13**:800 (1964).
37. N. Morita, *J. Phys. Soc. Japan*, **17**:1155 (1962).

38. Y. Gondō, H. Konnō, and Z. Funatogawa, *J. Phys. Soc. Japan,* **16**:2345 (1961).
39. H. Ratajczak, *Acta Phys. Polon.,* **25**:675 (1964).
40. W. Hellenthal, *Z. Angew. Phys.,* **13**:147 (1961).
41. W. Hellenthal, *Z. Physik,* **170**:303 (1962).
42. U. Gradmann, in "Basic Problems in Thin Film Physics" (R. Niedermayer and H. Mayer, eds.), p. 485, Vandenhoeck and Ruprecht, Göttingen, 1966.
43. U. Gradmann and J. Müller, *Phys. Stat. Solidi,* **27**:313 (1968).
44. W. Hellenthal, *Z. Physik,* **194**:323 (1966).
45. F. B. Hagedorn, *J. Appl. Phys.,* **38**:263 (1967).
46. E. J. Torok, H. N. Oredson, and A. L. Olson, *J. Appl. Phys.,* **35**:3469 (1964).
47. I. S. Edelman, *Fiz. Metal. i Metalloved.,* **20**:683 (1965); *Phys. Metals Metallog. USSR English Transl.,* **20**(5):42 (1965).
48. T. S. Crowther, *J. Appl. Phys.,* **34**:580 (1963).
49. R. V. Telesnin, I. M. Saraeva, and A. G. Shishkov, *Izv. Akad. Nauk SSSR, Ser. Fiz.,* **29**:586 (1965); *Bull. Acad. Sci. USSR Phys. Ser. English Transl.,* **29**:590 (1965).
50. R. E. Matick, *J. Appl. Phys.,* **35**:3331 (1964).
51. H. Sato, R. S. Toth, and R. W. Astrue, in "Single Crystal Films" (M. H. Francombe and H. Sato, eds.), p. 395, Pergamon Press, New York, 1964.
52. V. G. Pynko and R. V. Sukhanova, *Izv. Akad. Nauk SSSR, Ser. Fiz.,* **30**:43 (1966); *Bull. Acad. Sci. USSR, Phys. Ser. English Transl.,* **30**:45 (1966).
53. D. Chen, *J. Appl. Phys.,* **37**:1486 (1966); **38**:1309 (1967).
54. J. W. Matthews, *Appl. Phys. Letters,* **7**:255 (1965).
55. S. Shinozaki and H. Sato, *J. Appl. Phys.,* **36**:2320 (1965).
56. L. V. Kirenskii, V. G. Pynko, N. I. Sivkov, G. P. Pynko, R. V. Sukhanova, and M. A. Ovsyannikov, *Phys. Stat. Solidi,* **17**:243 (1966).
57. S. Chikazumi, *J. Appl. Phys.,* **32**:81S (1961).
58. Y. Gondō, S. Usami, K. Itoh, H. Konnō, and Z. Funatogawa, *J. Appl. Phys.,* **34**:1081 (1963).
59. S. Tsukahara, H. Kawakatsu, and T. Nagashima, *Electrochem. Soc. Japan,* **27**:48 (1963).
60. W. D. Doyle, *J. Appl. Phys.,* **35**:929 (1964).
61. J. Goddard and J. G. Wright, *Brit. J. Appl. Phys.,* **16**:1251 (1965).
62. J. F. Freedman, *IBM J. Res. Develop.,* **6**:449 (1962).
63. H. Chang, *J. Appl. Phys.,* **35**:770 (1964).
64. A. Yelon, *J. Appl. Phys.,* **38**:325 (1967).
65. C. Kittel, *Rev. Mod. Phys.,* **21**:541 (1949).
66. I. W. Wolf and T. S. Crowther, *J. Appl. Phys.,* **34**:1205 (1963).
67. W. Metzdorf, *IEEE Trans. Magnetics,* **MAG-2**:575 (1966).
68. J. P. Reekstin, *J. Appl. Phys.,* **38**:1449 (1967).
69. W. D. Doyle, *J. Appl. Phys.,* **38**:1441 (1967).
70. T. C. Penn and F. G. West, *J. Appl. Phys.,* **38**:2060 (1967).
71. J. C. Slonczewski, *IEEE Trans. Magnetics,* **MAG-4**:15 (1968).
72. W. Andrä, Z. Málek, W. Schüppel, and O. Stemme, *J. Appl. Phys.,* **31**:442 (1960).
73. D. O. Smith, *J. Appl. Phys.,* **32**:70S (1961).
74. C. D. Graham, Jr., and J. M. Lommel, *J. Phys. Soc. Japan,* **17**(Suppl. B-I):570 (1962).

75. D. O. Smith, G. P. Weiss, and K. J. Harte, *J. Appl. Phys.*, **37**:1464 (1966).
76. P. Astwood and M. Prutton, *Brit. J. Appl. Phys.*, **14**:48 (1963).
77. C. H. Wilts, in "Basic Problems in Thin Film Physics" (R. Niedermayer and H. Mayer, eds.), p. 422, Vandenhoeck and Ruprecht, Göttingen, 1966.
78. G. Robinson, *J. Phys. Soc. Japan*, **17** (Suppl. B-I):588 (1962).
79. W. T. Siegle and W. R. Beam, *J. Appl. Phys.*, **36**:1721 (1965).
80. M. Takahashi, D. Watanabe, T. Kōno, and S. Ogawa, *J. Phys. Soc. Japan*, **15**:1351 (1960).
81. R. J. Heritage, A. S. Young, and I. B. Bott, *Brit. J. Appl. Phys.*, **14**:439 (1963).
82. B. M. Abakumov, I. G. Orlov, and N. A. Yarkina, *Izv. Akad. Nauk SSSR, Ser. Fiz.*, **31**:383 (1967); *Bull. Acad. Sci. USSR, Phys. Ser. English Transl.*, **31**:374 (1967).
83. C. D. Graham, Jr., in "Magnetic Properties of Metals and Alloys," p. 288, American Society for Metals, Metals Park, Ohio, 1959.
84. J. C. Slonczewski, in "Magnetism" (G. T. Rado and H. Suhl, eds.), vol. 1, p. 205, Academic Press Inc., New York, 1963.
85. F. G. West, *J. Appl. Phys.*, **35**:1827 (1964); **35**:2787 (1964).
86. T. Suzuki and C. H. Wilts, *J. Appl. Phys.*, **38**:1356 (1967).
87. M. Takahashi, *J. Appl. Phys.*, **33**:1101 (1962).
88. G. Kneer and W. Zinn, in "Basic Problems in Thin Film Physics" (R. Niedermayer and H. Mayer, eds.), p. 437, Vandenhoeck and Ruprecht, Göttingen, 1966.
89. M. S. Cohen, *J. Appl. Phys.*, **35**:834 (1964).
90. W. Metzdorf, *Z. Angew. Phys.*, **18**:534 (1965).
91. D. O. Smith and G. P. Weiss, *J. Appl. Phys.*, **36**:962 (1965).
92. C. M. Williams and A. I. Schindler, *J. Appl. Phys.*, **37**:1468 (1966).
93. G. Kneer and W. Zinn, *Phys. Status Solidi*, **17**:323 (1966).
94. A. Seeger, H. Kronmüller, and H. Rieger, *Z. Angew. Phys.*, **18**:377 (1965).
95. K. J. Harte, D. O. Smith, R. M. Anderson, and R. C. Johnston, *J. Appl. Phys.*, **39**:749 (1968).
96. W. Andrä and K. Steenbeck, *Phys. Stat. Solidi*, **17**:191 (1966).
97. F. E. Luborsky, *J. Appl. Phys.*, **38**:1445 (1967); F. E. Luborsky and W. D. Barber, *J. Appl. Phys.*, **39**:746 (1968).
98. Z. Málek, V. Kamberský, and W. Schüppel, *Phys. Stat. Solidi*, **1**:K147 (1961).
99. R. J. Prosen, Y. Gondō, and B. E. Gran, *J. Appl. Phys.*, **35**:826 (1964).
100. H. -E. Wiehl, *Z. Angew. Phys.*, **18**:541 (1965).
101. T. Iwata, R. J. Prosen, and B. E. Gran, *J. Appl. Phys.*, **38**:1364 (1967).
102. B. I. Bertelsen, *J. Appl. Phys.*, **33**:2026 (1962).
103. D. O. Smith, M. S. Cohen, and G. P. Weiss, *J. Appl. Phys.*, **31**:1755 (1960).
104. M. S. Cohen, *J. Appl. Phys.*, **32**:87S (1961).
105. D. E. Speliotis, G. Bate, J. C. Alstad, and J. R. Morrison, *J. Appl. Phys.*, **36**:972 (1965).
106. A. Kamberská and V. Kamberský, *Phys. Stat. Solidi*, **17**:411 (1966).
107. J. G. W. van deWaterbeemd and G. W. van Oosterhout, *Philips Res. Rept.*, **22**:375 (1967).
108. A. J. Hardwick, *J. Appl. Phys.*, **34**:818 (1963).
109. H. König and G. Helwig, *Optik*, **6**:111 (1950).
110. M. Lauriente and J. Gabrowski, *J. Appl. Phys.*, **33**:1109 (1962).

111. W. Metzdorf and H. -E. Wiehl, *Phys. Stat. Solidi*, **17**:285 (1966).
112. T. S. Crowther and M. S. Cohen, *J. Appl. Phys.*, **38**:1352 (1967).
113. P. Weber and W. Ruske, *Phys. Stat. Solidi*, **17**:185 (1966).
114. J. Eckardt, *Z. Angew. Phys.*, **14**:189 (1962).
115. M. S. Cohen, *J. Appl. Phys.*, **38**:860 (1967).
116. I. S. Jacobs and C. P. Bean, in "Magnetism" (G. T. Rado and H. Suhl, eds.), vol. 3, p. 271, Academic Press Inc., New York, 1963.
117. J. H. Greiner, *J. Appl. Phys.*, **37**:1474 (1966).
118. A. A. Glazer, A. P. Potapov, R. I. Tagirov, and Ya. S. Shur, *Phys. Stat. Solidi*, **16**:745 (1966).
119. D. Paccard, C. Schlenker, O. Massenet, R. Montmory, and A. Yelon, *Phys. Stat. Solidi*, **16**:301 (1966).
120. H. Fujiwara and Y. Sugita, *IEEE Trans. Magnetics*, **MAG-4**:22 (1968).
121. M. Kuriyama, H. Yamanouchi, and S. Hosoya, *J. Phys. Soc. Japan*, **16**:701 (1961).
122. E. Klokholm and J. F. Freedman, *J. Appl. Phys.*, **38**:1354 (1967).
123. J. M. Gorres, M. M. Hanson, and D. S. Lo, *J. Appl. Phys.*, **39**:743 (1968).
124. J. F. Freedman, *J. Appl. Phys.*, **33**:1148 (1962).
125. T. Koikeda, S. Fujiwara, and S. Chikazumi, *J. Phys. Soc. Japan*, **21**:1914 (1966).
126. D. S. Lo and M. M. Hanson, *J. Appl. Phys.*, **38**:1342 (1967).
127. H. Fujiwara, *J. Phys. Soc. Japan*, **20**:2092 (1965).
128. R. H. Wade and J. Silcox, *Appl. Phys. Letters*, **8**:7 (1966).
129. C. Kooy and J. M. Nieuwenhuizen, in "Basic Problems in Thin Film Physics" (R. Niedermayer and H. Mayer, eds.), p. 181, Vandenhoeck and Ruprecht, Göttingen, 1966.
130. S. Usami, H. Nagashima, and H. Aoi, *J. Phys. Soc. Japan*, **22**:877 (1967).
131. Y. Sugita, H. Fujiwara, and T. Sato, *Appl. Phys. Letters*, **10**:229 (1967).
132. M. Yamanaka and R. Ueda, *J. Phys. Soc. Japan*, **23**:241 (1967).
133. A. Bourret and D. Dautreppe, *Phys. Stat. Solidi*, **13**:559 (1966).
134. J. P. Jakubovics, *Phil. Mag.*, **14**:881 (1966).
135. H. Hoffmann, *IEEE Trans. Magnetics*, **MAG-4**:32 (1968).
136. H. Boersch and M. Lambeck, *Z. Physik*, **165**:176 (1961).
137. P. Schnupp, *Ann. Physik*, **14**:1 (1964).
138. H. Hoffmann, "Habilitationsschrift," University of Munich, 1965.
139. A. Baltz and W. D. Doyle, *J. Appl. Phys.*, **35**:1814 (1964).
140. H. B. Callen, R. L. Coren, and W. D. Doyle, *J. Appl. Phys.*, **36**:1064 (1965).
141. H. W. Fuller and M. E. Hale, *J. Appl. Phys.*, **31**:238 (1960).
142. M. Roth, *IEEE Trans. Magnetics*, **MAG-2**:563 (1966); *Phys. Stat. Solidi*, **14**:115 (1966).
143. K. Y. Ahn and W. R. Beam, *J. Appl. Phys.*, **37**:240 (1966).
144. J. A. Cundall and A. P. King, *Phys. Stat. Solidi*, **16**:613 (1966).
145. S. Ishibashi, K. Hida, M. Masuda, and S. Uchiyama, *Elec. Eng. Japan*, **86**:23 (1966).
146. E. Fuchs, *Z. Angew. Phys.*, **13**:157 (1961).
147. H. Hoffmann and M. Okon, *Z. Angew. Phys.*, **21**:406 (1966).
148. T. Fujii, S. Uchiyama, E. Yamada, and Y. Sakaki, *Japan. J. Appl. Phys.*, **6**:1 (1967).
149. W. F. Brown, Jr., "Micromagnetics," Interscience Publishers, Inc., New York, 1963.

150. H. Rother, *Z. Physik,* **179**:299 (1964).
151. H. Hoffmann, *J. Appl. Phys.,* **35**:1790 (1964).
152. H. Hoffmann, in "Basic Problems in Thin Film Physics" (R. Niedermayer and H. Mayer, eds.), p. 386, Vandenhoeck and Ruprecht, Göttingen, 1966.
153. H. Hoffmann, *IEEE Trans. Magnetics,* **MAG**-2:566 (1966).
154. K. J. Harte, *Proc. Intern. Conf. Magentism, Nottingham* 1964; *J. Appl. Phys.,* **37**:1295 (1966).
155. K. J. Harte, *J. Appl. Phys.,* **39**:1503 (1968).
156. U. Krey, *Phys. Kondens. Mater.,* **6**:218 (1967).
157. W. D. Doyle, J. E. Rudisill, and S. Shtrikman, *J. Phys. Soc. Japan,* **17** (Suppl B-I):567 (1962).
158. W. D. Doyle, J. E. Rudisill, and S. Shtrikman, *J. Appl. Phys.,* **33**:1162 (1962).
159. M. S. Cohen, *J. Appl. Phys.,* **33**:1968 (1962).
160. M. S. Cohen, *IEEE Trans. Magnetics,* **MAG**-1:156 (1965).
161. H. Hoffmann, *Phys. Stat. Solidi,* **7**:89 (1964).
162. E. Feldtkeller, *Elektron. Rechenanl.,* **3**:167 (1961).
163. J. I. Raffel, T. S. Crowther, A. H. Anderson, and T. O. Herndon, *Proc. IRE,* **49**:155 (1961).
164. H. Hoffmann, *Z. Angew. Phys.,* **18**:499 (1965).
165. E. Feldtkeller, *Z. Physik,* **176**:510 (1963).
166. E. J. Torok, R. A. White, A. J. Hunt, and H. N. Oredson, *J. Appl. Phys.,* **33**:3037 (1962).
167. E. J. Torok, *J. Appl. Phys.,* **36**:952 (1965).
168. D. O. Smith, *Appl. Phys. Letters,* **2**:191 (1963).
169. H. Hoffmann, *Phys. Stat. Solidi,* **7**:383 (1964).
170. K. D. Leaver, M. Prutton, and F. G. West, *Phys. Stat. Solidi,* **15**:267 (1966).
171. H. Hoffmann, *Phys. Stat. Solidi,* **17**:K105 (1966).
172. E. Feldtkeller, *Phys. Letters,* **7**:9 (1963).
173. C. S. Comstock, Jr., A. C. Sharp, R. L. Samuels, and A. V. Pohm, *IEEE Trans. Magnetics,* **MAG**-4:39 (1968).
174. K. D. Leaver, *J. Appl. Phys.,* **39**:1157 (1968).
175. C. S. Comstock, A. C. Sharp, D. J. Frantsvog, and A. V. Pohm, *J. Appl. Phys.,* **38**:1339 (1967).
176. E. Feldtkeller, *Proc. Intern. Conf. Magnetism, Nottingham,* 1964, p. 837.
177. K. J. Harte, M. S. Cohen, G. P. Weiss, and D. O. Smith, *Phys. Stat. Solidi,* **15**:225 (1966).
178. M. S. Cohen, *J. Appl. Phys.,* **34**:1841 (1963).
179. H. Fujiwara, Y. Sugita, N. Saito, and S. Taniguchi, *J. Appl. Phys.,* **39**:736 (1968).
180. S. Middelhoek, in "The Use of Thin Films in Physical Investigations" (J. C. Anderson, ed.), p. 385, Academic Press Inc., New York, 1966; *J. Appl. Phys.,* **34**:1054 (1963).
181. E. Feldtkeller, in "Basic Problems in Thin Film Physics" (R. Niedermayer and H. Mayer, eds.), p. 451, Vandenhoeck and Ruprecht, Göttingen, 1966.
182. E. E. Huber, Jr., D. O. Smith, and J. B. Goodenough, *J. Appl. Phys.,* **29**:294 (1958); R. Moon, *J. Appl. Phys.,* **30**:82S (1959).
183. E. Feldtkeller and H. Thomas, *Phys. Kondens. Mater.,* **4**:8 (1965).
184. S. Shtrikman and D. Treves, *J. Appl. Phys.,* **31**:147S (1960).
185. A. Aharoni, *J. Appl. Phys.,* **37**:4615 (1966).
186. M. Mohiuddin, *Brit. J. Appl. Phys.,* **17**:789 (1966).

187. H. -D. Dietze and H. Thomas, *Z. Physik*, **163**:523 (1961).
188. E. Feldtkeller, *Z. Angew. Phys.*, **15**:106 (1963).
189. W. F. Brown, Jr., and A. E. LaBonte, *J. Appl. Phys.*, **36**:1380 (1965).
190. A. Aharoni, *J. Appl. Phys.*, **37**:3271 (1966).
191. A. Aharoni, *J. Appl. Phys.*, **39**:861 (1968).
192. H. Kronmüller, *Phys. Stat. Solidi*, **11**:K125 (1965).
193. A. Aharoni, *J. Appl. Phys.*, **38**:3196 (1967).
194. S. Methfessel, S. Middelhoeck, and H. Thomas, *IBM J. Res. Develop.*, **4**:96 (1960).
195. E. Fuchs, *Z. Angew. Phys.*, **14**:203 (1962).
196. J. M. Daughton, G. E. Keefe, K. Y. Ahn, and C. -C. Cho, *IBM J. Res. Develop.*, **11**:559 (1967).
197. E. Feldtkeller and E. Fuchs, *Z. Angew. Phys.*, **18**:1 (1964).
198. E. J. Torok, A. L. Olson, and H. N. Oredson, *J. Appl. Phys.*, **36**:1394 (1965).
199. A. L. Olson, H. N. Oredson, E. J. Torok, and R. A. Spurrier, *J. Appl. Phys.*, **38**:1349 (1967).
200. D. O. Smith and K. J. Harte, *J. Appl. Phys.*, **33**:1399 (1962).
201. E. Feldtkeller and W. Liesk, *Z. Angew. Phys.*, **14**:195 (1962).
202. H. Sato, R. W. Astrue, and S. S. Shinozaki, *J. Appl. Phys.*, **35**:822 (1964).
203. F. Birãgnet, J. Devenyi, P. Escudier, R. Montmory, D. Paccard, and A. Yelon, in "Basic Problems in Thin Film Physics" (R. Niedermayer and H. Mayer, eds.), p. 447, Vandenhoeck and Ruprecht, Göttingen, 1966.
204. R. M. Hornreich, *J. Appl. Phys.*, **34**:1071 (1963).
205. R. H. Wade, *Phil. Mag.*, **10**:49 (1964); **12**:437 (1965).
206. Ye. N. Il'yicheva, N. G. Kanavina, and A. G. Shishkov, *Fiz. Metal. i Metalloved.*, **21**:21 (1966); **22**:351 (1966). [English translation in *Phys. Metals Metallog.*, **21**:19 (1966); **22**:30 (1967)].
207. S. Methfessel, S. Middelhoeck, and H. Thomas, *J. Appl. Phys.*, **32**:294S (1961).
208. S. Middelhoek, *IBM J. Res. Develop.*, **6**:394 (1962).
209. W. Metzdorf, *Z. Angew. Phys.*, **14**:412 (1962).
210. A. Yelon, O. Voegeli, and E. W. Pugh, *J. Appl. Phys.*, **36**:101 (1965).
211. L. V. Kirenskii, V. G. Pynko, G. P. Pynko, R. V. Sukhanova, N. I. Sivkov, G. I. Rusov, and I. S. Edelman, *Phys. Stat. Solidi*, **17**:249 (1966).
212. D. S. Lo, *J. Appl. Phys.*, **37**:3246 (1966).
213. E. Feldtkeller, in "Magnetismus," p. 215, Deutscher Verlag für Grundstoff-industrie, Leipzig, 1967.
214. V. F. Ivlev and V. S. Prokopenko, *Izv. Akad. Nauk SSSR, Ser. Fiz.*, **25**:606 (1961); *Bull. Acad. Sci. USSR, Phys. Ser. English Transl.*, **25**:616 (1961).
215. G. M. Rodichev and P. D. Kim, *Izv. Akad. Nauk SSSR, Ser. Fiz.*, **25**:610 (1961); *Bull. Acad. Sci. USSR, Phys. Ser., English Transl.*, **25**:620 (1961).
216. K. H. Behrndt, *J. Appl. Phys.*, **33**:193 (1962).
217. K. Y. Ahn, *J. Appl. Phys.*, **37**:1481 (1966).
218. H. Rother, *Z. Physik*, **168**:283 (1962).
219. R. W. Olmen and S. M. Rubens, *J. Appl. Phys.*, **33**:1107 (1962).
220. A. G. Lesnik, G. I. Levin, and S. N. Kaverina, *Izv. Akad. Nauk SSSR, Ser. Fiz.*, **29**:591 (1965); *Bull. Acad. Sci. USSR, Phys. Ser. English Transl.*, **29**:594 (1965).
221. C. E. Patton and F. B. Humphrey, *J. Appl. Phys.*, **37**:4269 (1966).

222. C. E. Patton, T. C. McGill, and C. H. Wilts, *J. Appl. Phys.*, **37**:3594 (1966); 37:4301 (1966).
223. C. E. Patton and F. B. Humphrey, *J. Appl. Phys.*, **39**:857 (1968).
224. S. Middelhoek, *IBM J. Res. Develop.*, **10**:351 (1966).
225. D. J. Craik and R. S. Tebble, "Ferromagnetism and Ferromagnetic Domains," North Holland Publishing Company, Amsterdam, 1965.
226. F. D. Stacey, *Australian J. Phys.*, **13**:599 (1960).
227. L. F. Bates, D. J. Craik, and S. Rushton, *Proc. Phys. Soc. London*, **80**:768 (1962).
228. D. J. Craik and M. J. Wood, *Phys. Stat. Solidi*, **16**:321 (1966).
229. W. Hellenthal, *Z. Physik*, **184**:39 (1965); **194**:323 (1966).
230. M. Lambeck, *J. Appl. Phys.*, **39**:741 (1968).
231. S. Middelhoek and D. Wild, *IBM J. Res. Develop.*, **11**:93 (1967).
232. W. Kayser, *IEEE Trans. Magnetics*, **MAG-3**:141 (1967).
233. S. Middelhoek, *Phys. Letters*, **13**:14 (1964).
234. R. V. Telesnin, E. N. Ilecheva, N. G. Kanavina, and A. G. Shishkov, *Phys. Stat. Solidi*, 14:363 (1966).
235. A. Green, K. D. Leaver, R. M. Livesay, and M. Prutton, *Proc. Intern. Conf. Magnetism, Nottingham*, 1964, p. 807.
236. T. Kusuda, S. Konishi, and Y. Sakurai, *IEEE Trans. Magnetics*, **MAG-3**:286 (1967).
237. E. Feldtkeller and K. U. Stein, *J. Appl. Phys.*, **38**:4401 (1967).
238. T. E. Hasty and L. J. Boudreaux, *J. Appl. Phys.*, **32**:1807 (1961).
239. R. Weber and P. E. Tannenwald, *Phys. Rev.*, **140**:A498 (1965).
240. K. Goser, *Arch. Elek. Übertragung*, **19**:384 (1965).
241. D. O. Smith, *J. Appl. Phys.*, **29**:264 (1958).
242. D. T. Ngo, *J. Appl. Phys.*, **36**:1125 (1965); 37:453 (1966).
243. R. H. Nelson, *J. Appl. Phys.*, **35**:808 (1964).
244. P. Wolf, *Z. Physik.*, **160**:310 (1960); *J. Appl. Phys.*, **32**: (1961).
245. G. Matsumoto, T. Satoh, and S. Iida, *J. Phys. Soc. Japan*, **21**:231 (1966).
246. C. F. Kooi, P. E. Wigen, M. R. Shanabarger, and J. V. Kerrigan, *J. Appl. Phys.*, **35**:791 (1964).
247. P. Wolf, in "Basic Problems in Thin Film Physics" (R. Neidermayer and H. Mayer, eds.), p. 392, Vandenhoeck and Ruprecht, Göttingen, 1966.
248. R. Webber, *IEEE Trans. Magnetics*, **MAG-4**:28 (1968).
249. M. Nisenoff and R. W. Terhune, *J. Appl. Phys.*, **36**:732 (1965).
250. R. Weber and P. E. Tannenwald, *J. Phys. Chem. Solids*, **24**:1357 (1963).
251. R. Weber, *Phys. Rev.*, **169**:451 (1968).
252. M. H. Seavey, Jr., and W. J. Kearns, *J. Appl. Phys.*, **36**:1205 (1965).
253. P. Pincus, *Phys. Rev.*, **118**:658 (1960).
254. R. F. Soohoo, *Proc. Intern. Conf. Magnetism, Nottingham*, 1964, p. 852.
255. P. E. Wigen, *Phys. Rev.*, **133**:A1557 (1964).
256. C. E. Patton, C. H. Wilts, and F. B. Humphrey, *J. Appl. Phys.*, **38**:1385 (1967).
257. J. B. Comly, T. Penney, and R. V. Jones, *J. Appl. Phys.*, **34**:1145 (1963).
258. A. -J. Berteaud and H. Pascard, *J. Appl. Phys.*, **36**:970 (1965).
259. J. B. Comly, Thesis, Harvard University, 1965.
260. F. B. Hagedorn, *IEEE Trans. Magnetics*, **MAG-4**:41 (1968).
261. E. Tatsumoto and M. Nomura, *Japan. J. Appl. Phys.*, **5**:1119 (1966).

262. K. -U. Stein, *Z. Angew. Phys.,* **18**:528 (1965); **20**:36 (1965).
263. J. H. Hoper, *IEEE Trans. Magnetics,* **MAG**-3:166 (1967); W. Dietrich, *IBM J. Res. Develop.,* **6**:368 (1962).
264. R. V. Telesnin, E. N. Ilicheva, O. S. Kolotov, T. N. Nikitina, and V. A. Pogozhev, *Phys. Stat. Solidi,* **14**:371 (1966).
265. F. B. Humphrey, *J. Appl. Phys.,* **29**:284 (1958).
266. O. S. Kolotov, T. N. Letova, V. A. Pogozhev, and R. V. Telesnin, *Phys. Stat. Solidi,* **16**:K121 (1966).
267. R. V. Telesnin and T. N. Nikitina, *Fiz. Tverd. Tela,* **6**:1234 (1964); *Soviet Phys.-Solid State English Transl.* **6**:957 (1964).
268. Y. Sakurai, T. Kusuda, S. Konishi, and S. Sugatani, *IEEE Trans. Magnetics,* **MAG**-2:570 (1966).
269. K. -U. Stein, *Z. Angew. Phys.,* **20**:323 (1966).
270. K. J. Harte, *J. Appl. Phys.,* **38**:1341 (1967).
271. K. J. Harte, *Proc. Intern. Conf. Magnetism, Nottingham* 1964, p. 843; *J. Appl. Phys.,* **36**:960 (1965).
272. J. H. Hoper, *J. Appl. Phys.,* **39**:1159 (1968).
273. O. A. Vinogradov, *Phys. Stat. Solidi,* **15**:377 (1966).
274. C. Guillaud, *J. Phys. Radium,* **12**:492 (1951).
275. L. Mayer, *J. Appl. Phys.,* **31**:346 (1960).
276. Z. Málek and V. Kamberský, *Czech. J. Phys.,* **8**:416 (1958).
277. J. Kaczer, M. Zelený, and P. Šuda, *Czech. J. Phys.,* **B13**:579 (1963).
278. Y. Murayama, *J. Phys. Soc. Japan,* **21**:2253 (1966).
279. G. S. Krinchik, A. N. Verkhozin, and S. A. Gushina, *Fiz. Tverd. Tela,* **9**:2314 (1967).
280. N. Saito, H. Fujiwara, and Y. Sugita, *J. Phys. Soc. Japan,* **19**:1116 (1964).
281. T. Koikeda, K. Suzuki, and S. Chikazumi, *Appl. Phys. Letters,* **4**:160 (1964).
282. L. S. Palatnik, L. I. Lukashenko, and A. G. Ravlik, *Fiz. Tverd. Tela,* **7**:2829 (1965); *Soviet Phys. Solid-State English Transl.,* **7**:2285 (1966).
283. E. E. Huber, Jr., and D. O. Smith, *J. Appl. Phys.,* **30**:267S (1959).
284. H. Fujiwara, Y. Sugita, and N. Saito, *Appl. Phys. Letters,* **4**:199 (1964).
285. Y. Sugita and H. Fujiwara, *J. Phys. Soc. Japan,* **20**:98 (1965).
286. R. W. DeBlois, *J. Appl. Phys.,* **36**:1647 (1965).
287. R. J. Spain, *Appl. Phys. Letters,* **6**:8 (1965).
288. A. Baltz, *J. Appl. Phys.,* **37**:1485 (1966).
289. E. Feldtkeller, *J. Appl. Phys.,* **39**:1181 (1968).
290. S. Middelhoek, *J. Appl. Phys.,* **37**:1276 (1966).
291. E. Goto, N. Hayashi, T. Miyashita, and K. Nakagawa, *J. Appl. Phys.,* **36**:2951 (1965).
292. E. Fulcomer, H. Chang, and L. A. Finzi, *J. Appl. Phys.,* **37**:4451 (1966).
293. W. Andrä, *IEEE Trans. Magnetics,* **MAG**-2:560 (1966).
294. E. Feldtkeller, E. Fuchs, and W. Liesk, *Z. Angew. Phys.,* **18**:370 (1965).
295. J. C. Bruyère, O. Massenet, R. Montmory, and L. Néel, *IEEE Trans. Magnetics,* **MAG**-1:10 (1965).
296. J. C. Bruyère, G. Clerc, O. Massenet, R. Montmory, L. Néel, D. Paccard, and A. Yelon, *J. Appl. Phys.,* **36**:944 (1965).
297. J. C. Bruyère, G. Clerc, O. Massenet, D. Paccard, R. Montmory, L. Néel, J. Valin, and A. Yelon, *IEEE Trans. Magnetics,* **MAG**-1:174 (1965).

298. O. Massenet, F. Birägnet, H. Juretschke, R. Montmory, and A. Yelon, *IEEE Trans. Magnetics,* **MAG-2**:553 (1966).
299. F. J. Friedlaender and L. F. Silva, *J. Appl. Phys.,* **36**:946 (1965).
300. J. C. Slonczewski, *IBM J. Res. Develop.,* **10**:377 (1966).
301. C. E. Patton and F. B. Humphrey, *J. Appl. Phys.,* **37**:1270 (1966).
302. S. Middelhoek and D. Wild, *Nature,* **211**:1169 (1966).
303. K. U. Stein, *IEEE Trans. Magnetics,* **MAG-2**:781 (1966); F. B. Humphrey and R. Hasegawa, *IEEE Trans Magnetics,* **MAG-2**:781 (1966).
304. H. J. Kuno, *Proc. IEEE,* **53**:1754 (1965).
305. W. D. Doyle, *Phys. Stat. Solidi,* **17**:K67 (1966); *IEEE Trans. Magnetics,* **MAG-2**:206 (1966).
306. H. S. Belson, *IEEE Trans. Commun. Electron.,* **83**:317 (1964).
307. D. B. Dove, T. R. Long, and J. E. Schwenker, *IEEE Trans. Magnetics,* **MAG-1**:180 (1965).
308. J. T. Chang, U. F. Gianola, and M. W. Sagal, *J. Appl. Phys.,* **35**:830 (1964).
309. F. B. Humphrey, *J. Appl. Phys.,* **38**:1520 (1967).
310. B. Kostyshyn, J. E. Brophy, I. Oi, and D. D. Roshon, Jr., *J. Appl. Phys.,* **31**:772 (1960).
311. J. Kranz and A. Hubert, *Z. Angew. Phys.,* **15**:220 (1963).
312. D. O. Smith, *Opt. Acta,* **12**:13, 193 (1965); **13**:121 (1966).
313. R. P. Hunt, *J. Appl. Phys.,* **38**:1652 (1967).
314. D. Wohlleben, *Phys. Letters,* **22**:564 (1966).
315. H. W. Katz, A. H. Anderson, W. Kayser, U. F. Gianola, H. P. Louis, J. H. Glaser, C. D. Olson, J. Hart, and I. W. Wolf, *IEEE Trans. Magnetics,* **MAG-1**:218 (1965).
316. F. B. Humphrey and A. R. Johnston, *Rev. Sci. Instr.,* **34**:348 (1963).
317. E. J. Torok, D. C. Agouridis, A. L. Olson, and H. N. Oredson, *Rev. Sci. Instr.,* **35**:1039 (1964).
318. P. J. Flanders and W. D. Doyle, *Rev. Sci. Instr.,* **33**:691 (1962).
319. H. J. Oguey, *Rev. Sci. Instr.,* **31**:701 (1960).
320. C. J. Bader and D. M. Ellis, *Rev. Sci. Instr.,* **33**:1429 (1962); **34**:1188 (1963).
321. R. F. Soohoo, *J. Appl. Phys.,* **33**:1276 (1962).
322. W. R. Beam and K. Y. Ahn, *J. Appl. Phys.,* **34**:1561 (1963).
323. D. A. Thompson and H. Chang, *Phys. Stat. Solidi,* **17**:83 (1966).
324. D. O. Smith, in "Optical and Electro-optical Information Processing" (J. T. Tippett, D. A. Berkowitz, L. C. Clapp, C. J. Koester, and A. Vanderburgh, Jr., eds.), p. 523, The M.I.T. Press, Cambridge, Mass., 1965.
325. S. Methfessel, M. J. Freiser, G. P. Pettit, and J. C. Suits, *J. Appl. Phys.,* **38**:1500 (1967).
326. G. Kneer, *IEEE Trans. Magnetics,* **MAG-2**:747 (1966).
327. Wu ding Ke, *Izv. Akad. Nauk SSSR, Ser. Fiz.,* **29**:576, 580 (1965); *Bull. Acad. Sci. USSR, Phys. Ser. English Transl.,* **29**:581, 585 (1965).
328. W. R. Beam and W. T. Siegle, *Rev. Sci. Instr.,* **36**:641 (1965).
329. V. V. Kobelev, *Fiz. Metal. i Metalloved.,* **13**(3):467 (1962). [English translation in *Phys. Metals Metallog.,* **13**(3):146 (1962).]
330. F. B. Hagedorn, *Rev. Sci. Instr.,* **38**:591 (1967).
331. D. E. Speliotis, *J. Appl. Phys.,* **38**:1207 (1967).
332. R. J. Spain, *IEEE Trans. Magnetics,* **MAG-2**:347 (1966).

333. J. I. Raffel, T. S. Crowther, A. H. Anderson, and T. O. Herndon, *Proc. IEEE*, **49**:155 (1961).
334. A. V. Pohm, T. A. Smay, and W. M. Mayer, *IEEE Trans. Magnetics*, **MAG-3**:481 (1967).
335. J. I. Raffel, *J. Appl. Phys.*, **35**:48 (1964).
336. P. D. Barker and E. J. Torok, *J. Appl. Phys.*, **37**:1363 (1966).
337. K. U. Stein and E. Feldtkeller, *IEEE Trans. Magnetics*, **MAG-2**:184 (1966).
338. J. E. Eide, *J. Appl. Phys.*, **37**:1365 (1966).
339. I. Danylchuk, A. J. Perneski, and M. W. Sagal, 1964 Proc. Intermag. Conf., New York, p. 5-4-1.
340. H. J. Kump and J. T. Chang, *IBM J. Res. Develop.*, **10**:255 (1966).
341. D. O. Smith, *IEEE Trans. Magnetics*, **MAG-3**:433 (1967).

Footnote for page 679:

*Recent work (J. H. Hoper, to be published) shows that the internal field has been previously measured incorrectly both in magnitude and in *sign*; in past experiments the low-frequency resonance criterion was erroneously assumed to be maximum signal rather than *vanishing phase shift* between the rf excitation and the induced pickup voltage. Upon application of the latter criterion, H_i was indeed found to be interpretable as a magnetostatic field which originates in ripple and is antiparallel to **M** (Sec. 5.2). Furthermore, below a certain cutoff frequency it is not possible to observe true resonance; the resonance "peaks" previously reported (see below) were in reality peaks in the initial susceptibility (Sec. 6.3).

XI OPTICAL PROPERTIES OF THIN FILMS

1. INTRODUCTION

The phenomenal growth of thin film research and development owes much to the stimulus provided by the early utilitarian interest in the application of optical films in mirrors and interferometers. The considerable theoretical and experimental investigations on the optical behavior of thin films deal primarily with optical reflection, transmission, and absorption properties, and their relation to the optical constants of films. As a result of these studies, complex multilayer optical-device systems with remarkable reflection, antireflection, interference, and polarization properties have emerged for both laboratory and industrial applications. Moreover, the reflection, transmission, and interferometric properties of thin films have made it possible to determine the optical constants conveniently. The absorption studies have led to a variety of interesting thin film optical phenomena which have thrown considerable light on the electronic structure of solids.

This chapter reviews the optical behavior of metal and nonmetal films, optical-absorption phenomena in thin films and their relation to the electronic structure of solids, and the major applications of optical coatings. The emphasis is on the basic understanding and the significance of the subjects covered rather than their detailed analysis. A good deal of the information in this chapter has been obtained from the various review articles and treatises on the optical properties of thin films [1-10] and their applications [1, 3, 6, 11-14].

2. THIN FILM OPTICS

This section outlines the basic relations for the reflection, transmission, and absorption of light by single and multilayer films. Thin film optics may be developed more or less completely from several independent approaches. The classical approach of Airy [15] of summing multiple reflections is lacking as a fundamental theory but is still very useful in approximate methods of calculation. The most rigorous treatment is derived from the electromagnetic theory where the problem of the determination of the reflectance, transmittance, and absorptance of a single film or a multilayer film reduces basically to the solution of the boundary-value problem, namely, the determination of the steady-state amplitudes of the electric- and magnetic-field vectors at the interfaces arising from an incident light wave of certain assumed characteristics. This problem is treated extensively in standard textbooks [16].

For our present purpose, it suffices to quote the results for the amplitude and state of polarization of a light beam reflected and transmitted by a thin film in terms of the Fresnel coefficients of reflection and transmission at the interfaces between the media and as a function of the optical constants, the film thickness, and the angle of incidence of the light beam. It must be pointed out that these results are based on the assumptions that a thin film is a homogeneous, isotropic, and plane-parallel layer of thickness t, which is comparable with the wavelength of light λ. The film is characterized by a refractive index n and, if absorption occurs, by $(n - ik)$, where k is the extinction coefficient. Note that when $t \sim \lambda$, the path difference introduced between the multiply-reflected and -transmitted beams is small compared with the coherence length of any reasonably monochromatic source. Further, the total lateral displacement of the beam is small compared with its breadth. Thus, we may regard the multiply-reflected and -transmitted beams as combining coherently. Moreover, the lateral dimensions perpendicular to the film thickness are taken as infinite (that is, large compared with λ) so that the multiply-reflected amplitudes may be summed to infinity without significant error. This approximation obviously breaks down in systems in which total reflection occurs at opposite faces of the film, as in the frustrated total-reflection filter. The same is true if light amplification takes place in the medium.

2.1   Reflection and Transmission at an Interface

The reflection and transmission of electromagnetic radiation at a plane interface between two isotropic media of uniform refractive indices n_0 and n_1 are dealt with in standard textbooks on optics. If a beam of light is incident at an angle ϕ_1 at the n_1/n_0 interface separating the media of refractive indices n_1 and n_0, and is refracted at an angle ϕ_0 in the second medium, then the ratios reflected/incident (r_1) and transmitted/incident $(\hat{t}_1)$ amplitudes are given by the following relations:

$$r_{1p} = \frac{n_0 \cos\phi_1 - n_1 \cos\phi_0}{n_0 \cos\phi_1 + n_1 \cos\phi_0} \tag{1}$$

$$r_{1s} = \frac{n_1 \cos\phi_1 - n_0 \cos\phi_0}{n_1 \cos\phi_1 + n_0 \cos\phi_0} \tag{2}$$

$$\hat{t}_{1p} = \frac{2n_1 \cos\phi_1}{n_0 \cos\phi_1 + n_1 \cos\phi_0} \tag{3}$$

$$\hat{t}_{1s} = \frac{2n_1 \cos\phi_1}{n_1 \cos\phi_1 + n_0 \cos\phi_0} \tag{4}$$

These ratios are called the Fresnel reflection and transmission (amplitude) coefficients for plane-polarized components in which the electric vector lies parallel (p) and perpendicular (s) to the plane of incidence. Equations (1) and (2) can be rewritten by using Snell's law $n_0 \sin\phi_0 = n_1 \sin\phi_1$. For normal incidence, these equations are simplified since the distinction between planes of polarization vanishes.

The reflectivity (or reflectance) R of an interface is defined as the ratio of the reflected energy to the incident energy. Since the reflected and incident beams are in the same medium,

$$R_p = R_s = r_{1p}^2 = r_{1s}^2 \tag{5}$$

Similarly, the transmissivity (or transmittance) T may be defined as the ratio of transmitted to the incident energy. But here the two beams are in different media so that

$$T_p = \frac{n_0}{n_1} \hat{t}_{1p}^2 \quad \text{and} \quad T_s = \frac{n_0}{n_1} \hat{t}_{1s}^2 \tag{6}$$

For normal incidence at n_1/n_0 interface,

$$R_p = R_s = \left(\frac{n_1 - n_0}{n_1 + n_0}\right)^2 \quad \text{and} \quad T_p = T_s = \frac{4 n_0 n_1}{(n_1 + n_0)^2} \tag{7}$$

If the substrate is an absorbing medium, the refractive index n_0 should be replaced by $n = n_0 - i k_0$ so that

$$R_p = R_s = \frac{(n_1 - n_0)^2 + k_0^2}{(n_1 + n_0)^2 + k_0^2} \tag{8}$$

2.2   Reflection and Transmission by a Single Film

For a parallel-sided, isotropic film of refractive index n_1 between media of indices n_0 and n_2 (Fig. 1), the amplitude reflectance and transmittance are readily obtained by summing multiply-reflected beams. The result is

$$R = \frac{r_2 + r_1 \exp(-2i\delta_1)}{1 + r_2 r_1 \exp(-2i\delta_1)} \tag{9}$$

and

$$T = \frac{\hat{t}_1 \hat{t}_2 \exp(-i\delta_1)}{1 + r_2 r_1 \exp(-2i\delta_1)} \tag{10}$$

where $r_1, r_2, \hat{t}_1,$ and $\hat{t}_2$ are the Fresnel coefficients at the n_0/n_1 and n_1/n_2 interfaces (see Fig. 1), $\delta_1 = (2\pi/\lambda) n_1 t \cos \phi_1$ is the phase thickness of the

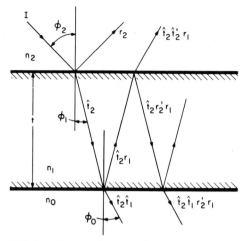

FIG. 1. The ray diagram of a light beam incident at an angle ϕ_2 on a plane, parallel-sided film of thickness t and refractive index n_1 supported on a substrate of index n_0. The index of the first medium is n_2. The r's and $\hat{t}$'s are the Fresnel coefficients of reflection and transmission, respectively.

film, and λ is the wavelength in vacuo. Equations (9) and (10) are valid for either the p or s directions of polarization provided the correct values of r and $\hat{t}$ are used consistently.

The reflectivity and transmissivity are given by

$$R = \frac{r_1^2 + r_2^2 + 2r_1 r_2 \cos 2\delta_1}{1 + r_1^2 r_2^2 + 2r_1 r_2 \cos 2\delta_1} \tag{11}$$

$$T = \frac{n_0}{n_2} \frac{\hat{t}_1^2 \hat{t}_2^2}{1 + 2r_1 r_2 \cos 2\delta + r_1^2 r_2^2} \tag{12}$$

These general expressions can be written in terms of the refractive indices but are too long to write out in full. A tractable form obtained only for the special case of normal incidence and transparent media (real n) is

$$R = \frac{\left(n_0^2 + n_1^2\right)\left(n_1^2 + n_2^2\right) - 4n_0 n_1^2 n_2 + \left(n_0^2 - n_1^2\right)\left(n_1^2 - n_2^2\right)\cos 2\delta_1}{\left(n_0^2 + n_1^2\right)\left(n_1^2 + n_2^2\right) + 4n_0 n_1^2 n_2 + \left(n_0^2 - n_1^2\right)\left(n_1^2 - n_2^2\right)\cos 2\delta_1} \tag{13a}$$

$$T = \frac{8n_0 n_1^2 n_2}{\left(n_0^2 + n_1^2\right)\left(n_1^2 + n_2^2\right) + 4n_0 n_1^2 n_2 + \left(n_0^2 - n_1^2\right)\left(n_1^2 - n_2^2\right)\cos 2\delta_1} \quad (13b)$$

The normal reflectance exhibits oscillatory variation as illustrated in Fig. 2 by the plot of R vs. the film thickness normalized by the wavelength of the light. The reflectance at the positions of the maxima and minima are as follows:

For $n_0 \lessgtr n_1 \lessgtr n_2$,

$$R_{max} = \left(\frac{n_2 - n_0}{n_2 + n_0}\right)^2 \qquad \text{at } \lambda_{max} = \frac{2n_1 t}{m} \qquad (14a)$$

and

$$R_{min} = \left(\frac{n_1^2 - n_0 n_2}{n_1^2 + n_0 n_2}\right)^2 \qquad \text{at } \lambda_{min} = \frac{4n_1 t}{2m + 1} \qquad (14b)$$

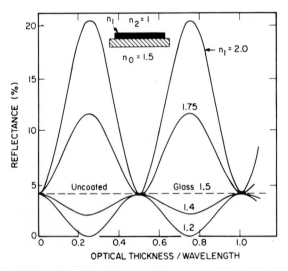

FIG. 2. Theoretical variation of reflectance R (at air side) with normalized thickness for films of various refractive indices on a glass substrate of index 1.5. (*Heavens* [1])

For $n_0 \lessgtr n_1 \gtrless n_2$,

$$R_{max} = \left(\frac{n_1^2 - n_0 n_2}{n_1^2 + n_0 n_2}\right)^2 \quad \text{at } \lambda_{max} = \frac{4 n_1 l}{2m + 1} \tag{15a}$$

and

$$R_{min} = \left(\frac{n_2 - n_0}{n_2 + n_0}\right)^2 \quad \text{at } \lambda_{min} = \frac{2 n_1 l}{m} \tag{15b}$$

where m is an integer. It is clear from these expressions that, by a suitable choice of the film index, we can either enhance (reflection coating) or diminish (antireflection coating) the reflectance of a given substrate (see Sec. 5). Further, the oscillatory behavior of R may be exploited as a spectrophotometric method for monitoring the thickness of single or multilayer films during deposition, as well as for determining the optical constants of thin films (see Sec. 3.1.4).

Equations (11) and (12) for R and T are quite general and are valid for any type of the film and substrate media. The algebraic expressions are, however, cumbersome for different combinations of absorbing (complex n) and nonabsorbing (real n) media and are given by Heavens [1]. A point of interest is that the p and s components of a reflected beam, in general, have different amplitudes for nonnormal incidence. For reflection from a transparent material, the p component goes through a minimum of amplitude and a phase change of π at an angle of incidence greater than ϕ_p. This angle is called the "polarizing angle" because an unpolarized light will be reflected at this angle to become plane-polarized. At this angle ϕ_p, we have

$$n_0 = n_1 \tan\phi_p \quad \text{(Brewster law)} \tag{16}$$

For reflection from an absorbing surface, both the amplitudes and the phases of the s and p components are different; thus, the resultant reflected light is elliptically polarized. The ratio of the amplitudes of reflectance [Eq. (9)] of the p and s component can be written in a generalized form due to Drude [17] as

$$\frac{R_p}{R_s} = e^{i\Delta} \tan\psi \tag{17}$$

where Δ is the relative phase difference and Ψ is the azimuth, the ratio of the reflection amplitudes of the p and s components. Since Δ and Ψ are related to the optical constants, a determination of the latter is, in principle, possible from the measurements of Δ and Ψ. This method of measurement forms the basis of the polarization spectrometry or ellipsometry—a name first given to it by Rothen [18]. Ellipsometry has assumed an important role in the study of surfaces only in the last decade. A good account of the subject is given in the proceedings of a 1963 symposium [19] on ellipsometry.

The explicit relationship between Δ and Ψ and the optical constants n_0, k_0 of a substrate and n_1 of the medium are given by

$$n_0^2 - k_0^2 = n_1^2 \sin^2 \phi_1 \left[1 + \frac{\tan^2 \phi_1 (\cos^2 2\psi - \sin^2 2\psi \sin^2 \Delta)}{(1 + \sin 2\psi \cos \Delta)^2} \right] \qquad (18)$$

$$n_0 k_0 = \frac{n_1^2 \sin^2 \phi_1 \tan^2 \phi_1 \sin 2\psi \cos 2\psi \sin \Delta}{(1 + \sin 2\psi \cos \Delta)^2} \qquad (19)$$

2.3 Anisotropic and Inhomogeneous Films

The conditions of a homogeneous and isotropic medium assumed in the preceding discussion are met when the optical constants are independent of the position and direction within a film. Multicomponent films are expected to show anisotropic and inhomogeneous behavior. The theoretical treatment of such films is quite complex, and generally no single method exists for directly evaluating their overall behavior in all cases. Several authors [20] have attempted to tackle this problem. The results are preliminary in nature. The interested reader should refer to the original papers.

2.4 Multilayer Films

A multilayer thin film is a finite combination of single layers having different optical constants and film thicknesses. Because a wide variety of applications of thin film optics are based on a multilayer thin film, the calculations of reflection, transmission, and absorption properties of a multilayer as a function of wavelength of light for any arbitrary angle of incidence is an important problem. The treatment of such a problem is obviously tedious, and various mathematical formulations have been devised for computer computations. This subject has been reviewed in detail by Berning [11]. In view of the complexity of the problem of a multilayer film and the necessity for performing trial-and-error calculations in the design of a multilayer device, the use of a modern electronic digital computer is indispensable. Although very little use has so far been made of an analog computer, its continuous-data-display capability is expected to increase its popularity.

In addition to the several analytical schemes of computation, various graphical methods exist, some of which have considerable value either as aids to hand computing or in offering a degree of visualization useful in design problems. These methods, which include vector diagram, Smith or admittance chart, Kard's calculator, and Male's nomogram, have been reviewed by Berning [11]. This review should be consulted for references to the original works.

2.5 Optical Absorption

Radiation absorption occurs in the majority of media. The intensity of radiation is generally attenuated in an exponential form of the type e^{-Kt}, where K is the absorption coefficient and is related to the imaginary part k of the refractive index by $K = 4\pi k/\lambda$. Absorption is a phenomenon of fundamental interest because of its relation to the dynamics of the electrons and ions of the medium under the influence of electromagnetic radiation. An absorbing medium is characterized by a complex dielectric constant $\epsilon = \epsilon_1 - i\epsilon_2 = 1 + 4\pi\alpha$, and a complex refractive index $n - ik$. Here, α is the sum of the polarizabilities resulting from free carriers, bound carriers, interband transitions, etc. It follows from the Maxwell equations that $\quad (n - ik)^2 = \epsilon_1 - i\epsilon_2 \quad$ (for the free-electron case)

so that

$$\epsilon_1 = n^2 - k^2 \quad \text{and} \quad \epsilon_2 = 2nk = \frac{4\pi\sigma}{\omega} \tag{20}$$

where ω is the angular frequency of the incident light and σ is the static (dc) electrical conductivity.

If absorption takes place primarily because of conduction or free electrons, as in the case of metals, then the Drude (also called Drude-Lorentz) theory [21] yields the following relations for the dielectric constant in terms of the optical constants:

$$\epsilon_1 = n^2 - k^2 = 1 - \frac{\omega_p^2}{\omega^2 + \omega_0^2} \tag{21}$$

and

$$\epsilon_2 = 2nk = \frac{\omega_p^2 \omega_0}{\omega\left(\omega^2 + \omega_0^2\right)} \tag{22}$$

where

$$\omega_p = \left(\frac{4\pi n_e e^2}{m^*}\right)^{1/2} \quad \text{(in free space)} \tag{23}$$

and

$$\omega_0 \equiv \frac{1}{\tau_0} = \frac{n_e e^2}{m^* \sigma} \tag{24}$$

Here ω_p is the plasma frequency of collective oscillations of the free electrons (see Sec. 4.3 of this chapter), ω_0 is the Lorentz-Sommerfeld frequency of a free-electron gas (metal) of conductivity σ, and m^* is the effective mass and n_e the density of electrons (the subscript e is used in order to distinguish n_e from the refractive-index symbol in this chapter).

At long wavelengths in the far infrared, $\epsilon_1 \to 0$ and $\epsilon_2 \to \infty$, and thus Eq. (20) yields

$$n = k = \sqrt{\frac{\epsilon_2}{2}} = \sqrt{\frac{2\pi\sigma}{\omega}} \tag{25}$$

This relation is expected to be valid in the region of high damping or absorption so that $\omega < \omega_0$ $(\omega\tau_0 < 1)$ is satisfied. The corresponding wavelength for typical metals is $\sim$10 to 20 μ. In this spectral range, the reflectance from a metal surface in air [Eq. (8)] reduces to

$$R \approx 1 - \left(\frac{2\omega}{\pi\sigma}\right)^{1/2} \tag{26}$$

which is the well-known Hagen-Rubens relation.

If absorption is due to bound electrons with which a natural frequency of vibration of ω_n may be associated, then one can show that

$$\epsilon_1 = 1 + \omega_p^2 \frac{\omega_n^2 - \omega^2}{\left(\omega_n^2 - \omega^2\right)^2 + \omega_0^2 \omega^2} \tag{27a}$$

and

$$\epsilon_2 = \omega_p^2 \frac{\omega_0 \omega}{\left(\omega_n^2 - \omega^2\right)^2 + \omega_0^2 \omega^2} \tag{27b}$$

It is clear that a maximum in ϵ_2 and hence in the absorption occurs when $\omega = \omega_n$. This absorption band occurs in the ultraviolet for most materials. If interband transitions take place, the above relations will be modified considerably.

The preceding discussion suggests that studies of the optical constants in the spectral regions where strong absorption takes place would yield fundamental information on the density and the effective optical mass of the carriers.

If the free electron gas is weakly damped so that $\omega \tau_0 \gg 1$ is satisfied, then $\epsilon_2 \to 0$ and the dielectric constant is obtained from Eq. (21) as

$$\epsilon_1 = n^2 = 1 - \frac{\omega_p^2}{\omega^2} \tag{28a}$$

The physical significance of this relation and the value of ω_p is given in Sec. 4.3 of this chapter. If the conduction electrons in a system give rise to a dielectric constant ϵ_∞ at optical frequencies (sufficiently large compared to the ionic resonance), then the static dielectric constant is given by

$$\epsilon = n^2 = \epsilon_\infty \left(1 - \frac{\omega_p^2}{\omega^2}\right) \tag{28b}$$

$$= \epsilon_\infty - \frac{4\pi n_e e^2}{m^* \omega^2} \quad \left(\text{since } \omega_p^2 = \frac{4\pi n_e e^2}{m^* \epsilon_\infty}\right)$$

Thus, the intercept and the slope of the straight line plot of n^2 versus λ^2 in the region $\omega > \omega_p$ should yield values of ϵ_∞ and n_e/m^*, respectively.

We should point out that when the optical-penetration depth of the incident light is comparable with the mean free path of the conduction electron, the field in which electrons move between collisions is no longer constant. Consequently, a correction to Drude's theory, called an anomalous skin-effect correction, is necessary (Sec. 3.4).

3. OPTICAL CONSTANTS OF THIN FILMS

The optical behavior of a material is generally utilized to determine its optical constants, n, k. Films are ideal specimens for reflectance, transmittance, and interferometric types of measurement. A review of these methods of measurement is given by Heavens [1, 3]. Several of these methods actually determine the effective optical thickness from which the metrical thickness of the film can be derived, provided the optical constants are known, or vice versa. These methods are discussed elsewhere (Chap. III, Sec. 1.5).

Some of the commonly employed techniques to determine the optical constants and the results obtained therefrom will now be described.

3.1 Experimental Methods

(1) Reflection Methods. The optical constants of an isotropic medium can, in principle, be derived from a pair of reflectance measurements using linearly polarized radiation at nonnormal incidence. It is required to make two measurements which may be selected in a number of ways. Humphreys-Owen [22] described the various measurement possibilities and gave a thorough discussion of the relative merits and sensitivities of each method.

One class of methods is comprised of two reflectance measurements— of either p, s, or both components, at one angle of incidence; or one reflectance measurement at each of two angles of incidence. The reflectance measurement of unpolarized light at two angles of incidence (Simon's method [23]) is not too sensitive but is useful at wavelengths at which polarization is difficult, as, for example, in the ultraviolet region. Avery's method [24] employs the ratio R_p/R_s at two angles of incidence and has the obvious advantage that only the ratio and not the absolute value of reflectance is required. At an angle of incidence $45°$,

Abelès has shown that $R_s = R_p^2$ is generally valid for unpolarized light, provided the film is homogeneous. This condition has been exploited by Robin [25] to determine n and k of evaporated films of Cr, Au, and Pb from simple measurements of R_s at 0 and 45° in the ultraviolet region.

The second class of measurements involves determination of the Brewster angle ϕ_B, the angle of incidence for which R_p is a minimum, and a reflectance measurement at this angle. A special case arises for a transparent medium ($k = 0$) when ϕ_B is equal to the polarizing angle at which $R_p = 0$ and the reflectance amplitude changes phase by 180°. One obtains the Brewster law $\tan \phi_B = n_1/n_2$ [Eq. (16)], which forms the basis of the Abelès method [26] of determining n_1.

Humphreys-Owen concluded that all reflectance methods show an apparent decrease in sensitivity even at very low k. Although no one method is overwhelmingly more suitable than the others, the methods involving measurements of ϕ_B and R_s or R_p/R_s seem best for low k. Accuracies of a few percent for n and k are obtained in most of these methods.

Normal Incidence.   The reflectance of an absorbing surface of index (n,k) in air for normal incidence is given by Eq. (8) as

$$R = \frac{(n - 1)^2 + k^2}{(n + 1)^2 + k^2}$$

If $k^2 \ll (n - 1)^2$, n can be determined from a measurement of R. If this condition does not hold, a second measurement at another angle of incidence would be required to determine n and k. Robinson [27] deduced n and k in vacuum ultraviolet by measuring normal-incidence reflectance together with a determination of the phase change on reflectance from the reflectance data as a function of wavelength.

Abelès' Method [26].   We have already mentioned this method. For a film of index n in air, $\tan \phi_B = n$ determines the index. This result is independent of the film thickness and the optical constants of the substrate. For measurements on a transparent substrate, an accuracy to ± 0.002 is attainable in films of suitable thickness, provided their index lies within ± 0.3 of that of the substrate. For films of optical thickness equal to an even multiple of $\lambda/4$, the reflectance curve is nearly the same as that for the bare substrate. This condition is clearly satisfied for only one angle of incidence. Since monochromatic light is used, no errors from dispersion effects arise.

The experimental method involves photometric measurements to compare the light intensities reflected from the bare substrate surface and the film-coated surface. A mechanized photometric arrangement [28] may be used. Phase-sensitive detection gives a higher sensitivity; precision of ± 0.01 is estimated for films of index about 2. This method has been applied [28] to oxide films and is quite useful in investigating the homogeneity and anisotropy of the films by comparing the reflectance vs. angle-of-incidence curve in the neighborhood of ϕ_B with the measured values. In the presence of slight absorption, the reflectance of the filmed surface will be below that for a transparent film, giving rise to a shift in the apparent Brewster angle and thus an error in the value of the refractive index.

(2) Reflectance and Transmittance Methods. Separate determinations of n and k can be made by measuring reflectance and transmittance of the same specimen. The transmission data allow an accurate determination of k since the absorption coefficient $K = 4\pi k/\lambda$. This method, used by Schulz and Tangherlini [29] for metal films, requires that the thickness of the film must be such that the effects of multiple reflections are suppressed. For metal films, a lower limit of ~300 to 400 Å is imposed on the film thickness. Under this condition the transmittance of a film of index $n_1 - ik_1$ and thickness t is given by [4]

$$T = \frac{16n_0 \left(n_1^2 + k_1^2\right)^2 \exp\left(-4\pi k_1\, t/\lambda\right)}{\left[(n_1 + 1)^2 + k_1^2\right]\left[(n_0 + n_1)^2 + k_1^2\right]} \qquad (n_1 > n_0) \qquad (29)$$

where the index of the substrate is n_0 and the ambient is assumed to be air. According to this relation, the intercept as well as the slope of the log T vs. t straight-line plot would give k_1.

Neglecting interference and multiple reflections, T and R are related by $T = (1 - R) \exp\left(-4\pi k_1\, t/\lambda\right)$. If reflection at the film-substrate interface $(n_0 < n_1, k_0 = 0)$ is also considered, then $T = (1 - R)^2 \exp\left(-4\pi k_1\, t/\lambda\right)$. This relation together with Eq. (8) offers the most convenient method for determining n_1 and k_1 from R and T data on the same film.

The absorption coefficient may be obtained by using a double-beam spectrophotometer with films of two different thicknesses, one in the sample and the other in the reference beam. This method automatically cancels out the surface-reflection effects. The absorption is then due only to the difference in the film thicknesses.

Murmann's Method [30]. In this method the reflectance at each side of the film is measured. From the value of the film thickness and that of the wavelength to be used, the values of the reflectances R on the air side and R' on the substrate side are calculated for a range of n_1 and k_1 values, and plots of R and R' vs. n_1 for different k_1 values are obtained. Plotting against n_1, the values of k_1 which give the values of R and R' obtained with the film in question then yield curves which give two possible values for n_1, k_1. The correct one may be distinguished from the set.

Malé's Method [31]. If, in addition to the reflectances on both sides of the film, the transmittance is also measured, then n_1, k_1, and t can be determined by laborious computations as given by Malé.

Wolter [32] *and David* [33] deduced approximate relations for R, R', and T in terms of the optical constants for films whose thickness is small when compared with the wavelength. These relations may be used to determine optical constants of ultrathin films.

(3) Interferometric Methods. Fochs [34] used the Michelson interferometer to determine film thickness and refractive index. The minimum thickness required is of the order of one to two wavelengths. By covering part of the Michelson mirror with the film and part with the film and a reflecting razor blade, three types of fringes are obtained: (*a*) normal fringes from the uncovered area, (*b*) the fringe pattern affected by the film thickness, and (*c*) the channeled spectrum arising from multiple reflections in the film. By displacing the interference mirror so that the (*a*) and (*b*) fringes are complementary, the film thickness is determined and then n is deduced from the optical-thickness value.

Variable-angle Monochromatic-fringe Observation (VAMFO). Pliskin and Conrad [35] have described a variable-angle monochromatic-fringe observation (VAMFO) arrangement to measure the thickness and the refractive index of transparent films deposited on reflective substrates. This interferometric technique, shown schematically in Fig. 3, uses a rotating stage, attached to an xy stage, for the sample so that the reflected light from both sides of the sample is observed by a microscope at a magnification of about 10. If Δm is the number of fringes observed between the angles of refraction θ_2 and θ_1, the film thickness is given by

$$t = \frac{\Delta m \lambda}{2n(\cos \theta_2 - \cos \theta_1)} \tag{30}$$

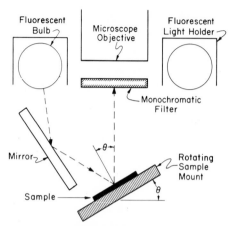

FIG. 3. Schematic of the essential parts of a variable-angle monochromatic fringe observation (VAMFO) arrangement. The rotating stage is attached to an xy stage so that the reference point under examination (generally at $10 \times$) can be maintained in the center of the field of view. (*Pliskin and Conrad* [35])

If t is known, n can be calculated, and vice versa. This simple and convenient technique is best suited for films thicker than 1 μ, although measurements on SiO films as thin as 900 Å have been made. An accuracy of 0.2 percent in the determination of n for thick films is claimed.

(4) Spectrophotometric Methods. *Jacquinot's Method* [36]. When white light is reflected from a thick film ($t > \lambda/4$), the reflectance shows maxima and minima at various wavelengths in the spectrum [Eqs. (14) and (15)]. The positions of such turning points, which are determined by the effective optical thickness of the film relative to the wavelength of light, allow the determination of the optical constants of the film (or conversely, the thickness of the film if the index is known). If dispersion takes place in the film, the turning values no longer occur exactly at half- and quarter-wave optical thicknesses, so that dispersion correction must be taken into account. In practice, the dispersionless equations applied at two or more maxima produce a first approximation to the values of n and the dispersion $dn_1/d(1/\lambda)$. These values are then used for further refinements by successive approximations.

It is difficult to determine precisely the wavelengths corresponding to the extrema in the spectrophotometric method. Giacomo and Jacquinot [37] have used a vibrating-mirror technique to measure the rate of change of reflectance, rather than the reflectance itself, thereby increasing the sensitivity of detecting the extrema. The ac signal further allows automatic electronic measurements with a high rejection ratio for the background. For further discussion of the experimental details of this method see Chap. III, Sec. 1.5. The method is clearly applicable to films of thickness greater than $\lambda/4$. For thinner films, Abelès [38] suggested that two reflection measurements at normal incidence at two different wavelengths may be used to compute the index as well as the thickness of the film.

Absorbing Films. For absorbing films on a transparent substrate, the general expressions [Eqs. (11) and (12)] for R and T at normal incidence may be used to calculate the indices. Hadley [39] greatly simplified this process by preparing a comprehensive set of curves by calculating R and T as a function of t/λ for a range of values of n, producing one set of curves for each of a range of values of k and a given substrate index. One such set of curves for $k = 0$ is shown in Fig. 4. By

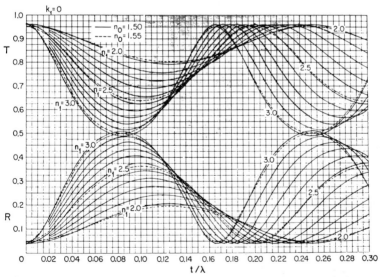

FIG. 4. One of Hadley's curves of normal incidence R and T vs. t/λ (t is film thickness, λ is wavelength) for the film refractive index $n_1 = 3.0, 2.5$, and 2.0, and the extinction coefficient $k_1 = 0$, on transparent substrates of indices 1.50 and 1.55. (*Courtesy of Hadley* [39])

measuring R and T, one can thus read off the values of n for each value of k at one wavelength. The k value at which the two n values are the same determines the true n and k. This can be determined graphically from the intersection of the plots of n's against k. An optimum accuracy is obtained if the thickness is an odd multiple of $\lambda/4$.

(5) Critcial-angle Method. The phenomenon of total reflection, which occurs at an angle arcsin (n_0/n_1), where n_1 is the medium of incidence and $n_1 > n_0$, has been commonly used to determine n_0 of nonabsorbing media. If absorption takes place in either medium, the transition from zero to total reflectance is gradual and no well-defined angle exists. The critical-angle method has been applied by Hunter [40] to measure the optical constants of Al, Mg, and In films. The total-reflection method is applicable in ultraviolet because $n < 1$ and k is sufficiently small. The latter condition also results in interference fringes at angles of incidence below the critical angle. The value of n may be determined from the position of the interference maxima and minima.

(6) Polarimetric Methods (Ellipsometry). A plane-polarized light reflected from an absorbing substrate at nonnormal incidence assumes elliptical polarization (Sec. 2.2). The ellipticity (the ratio of minor to major axis) of the reflected beam is determined by the relative phase difference Δ and the azimuth ψ. The optical constants of the reflecting surface can be determined from the values of Δ and ψ [Eqs. (18) and (19)].

The experimental arrangement utilized to measure Δ and ψ is called an "ellipsometer." A schematic arrangement of a typical ellipsometer is shown in Fig. 5. Analysis of the ellipticity of the reflected beam is carried out by any of the standard methods. A commonly used arrangement [41] employs a plane-polarized incident light with its plane of polarization at 45° to the plane of incidence. The reflected light passes through a compensator and an analyzer which are rotated to give the smallest intensity of the reflected light. The vibration direction of the compensator determines the angle χ between the great semiaxis of the ellipse and the plane of incidence. The difference between the position of the compensator and the analyzer gives the ellipticity angle $\bar{\gamma}$. The values of ψ and Δ can be determined from the relations

$$\tan 2\chi = \tan 2\psi \cos \Delta \tag{31}$$

and

$$\sin 2\bar{\gamma} = \sin 2\psi \sin \Delta \tag{32}$$

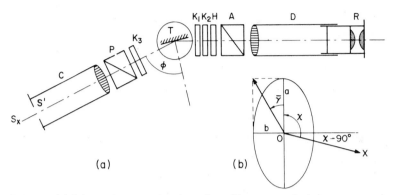

FIG. 5. (*a*) Schematic representation of an ellipsometer (polarization spectrometer): S_x = source of monochromatic light, S' = slit; C = collimator; P = polarizer; T = rotating table; K_1, K_2, K_3 = compensators; H = Nakumara's double plate; A = analyzer; D = telescope with eyepiece R for measurements in a dark field—one may use an eyepiece for half-shade measurements. (*Vašiček* [41] (*b*) The $\bar{y}$ and x coordinates of the sllipse.

One can also set the analyzer at 45° and rotate the polarizer and the compensator (on the analyzer side) until the detector gives the minimum of the reflected light. The angle of incidence for this case is called the principal angle of incidence $\bar{\phi}_1$, and the angle between the restored plane of polarization and the plane of incidence is the principal azimuth $\bar{\psi}$. Under this condition, the major and minor axes of the polarization ellipse of the reflected light lie in and perpendicular to the plane of incidence. Equations (18) and (19) then simplify to

$$n_0^2 - k_0^2 = n_1^2 \sin^2 \bar{\phi}_1 \left(1 + \tan^2 \bar{\phi}_1 \cos 4\bar{\psi}\right) \tag{33}$$

$$n_0 k_0 = n_1^2 \sin^2 \bar{\phi}_1 \tan^2 \bar{\phi}_1 \sin 2\bar{\psi} \cos 2\bar{\psi} \tag{34}$$

The general Drude equations [Eqs. (18) and (19)] as such do not allow direct calculations of the refractive indices and thickness from the values of Δ and ψ. Vašiček [41] developed a solution for a transparent film on glass and obtained a table of values of ψ and Δ for films of thickness up to 1 to 1.5μ and for refractive indices of 1.2 to 2.75 for a substrate of index 1.5163 and for λ = 5,890 Å. Computations of the optical constants are conveniently handled by an electronic computer. Archer [42] computed and prepared a chart of Δ and ψ values of a range of thicknesses and indices of films on a silicon substrate. Thus, a given point in the (Δ, ψ) chart uniquely defines a film thickness and index.

McCrackin and Colson [43] have also described computational methods for both dielectric and absorbing films.

The refractive index of a transparent substrate may be obtained easily by determining the angle of incidence ϕ_p at which the reflection of unpolarized light produces plane-polarized light [ϕ_p = polarizing angle— see Eq. (16)]. In this case, $n_0 = n_1 \tan \phi_p$, where n_1 is the index of the medium.

The phase changes in the reflected as well as transmitted beam have been utilized by Schopper [44] for determining the optical constants of Sb_2S_3 films. Försterling [45] discussed the ellipticity of light reflected at each side of the film and transmitted by the film and used the results as a basis for determining the optical constants of evaporated-metal films.

While the general Drude equation is cumbersome for calculations, a tractable form results for ultrathin surface films. Drude's solution for surface films much thinner than the wavelengths of light was completed by Tronstad and Feachem [46]. For both absorbing and nonabsorbing surface films with $t < \lambda$, the changes in Δ_0 and ψ_0 of a substrate due to the presence of a film characterized by Δ and ψ are given by

$$\Delta - \Delta_0 \approx \frac{-Ct}{\lambda} \tag{35}$$

$$2\psi - 2\psi_0 \approx \frac{C't}{\lambda} \tag{36}$$

where C and C' are constants determined by various optical parameters. The linear change of Δ and ψ with film thickness is generally valid to thicknesses below 100 Å. Extension of the Drude equations (35) and (36) to include second-order terms in the expansion of exp $2i\delta$ has been made by Hauschild [47]. Further extensions to thicker films have been analyzed by Leberknight and Lustman [48] and Winterbottom [49].

Some useful applications of ellipsometry are: (1) In addition to the optical constants of dielectric films, the technique can be used to determine the optical constants of metals in the infrared [50]. (2) The strong influence of a surface layer on the state of polarization of light reflected from an absorbing surface makes ellipsometry a very sensitive technique. Vašíček has claimed that it is possible to detect ~1-Å-thick surface films and a change in the refractive index of ~0.002. Fane et al. [235] used an ellipsometer to detect oil- and mercury-contamination films of less than 5 Å average thickness. (3) An ellipsometer may be mechanized [51] to monitor continuously surface phenomena such as oxidation and corrosion [52, 53], epitaxially induced strains in oxide films, and electrochemical surface reactions [54].

(7) Summary. A comparative summary of the commonly employed methods is given in Table I. We should mention that there also exist various specialized methods to measure the optical constants. Among these, a method [55, 56] useful in the ultraviolet region based on the measurement of the energy loss of energetic electrons through thin films deserves mention. The optical constants are derived from the complex dielectric constant.

3.2 Results on Optical Constants

(1) General Remarks. The optical properties, like other properties of thin films, show profound sensitivity on the film microstructure. Since the various deposition parameters affect the microstructure, we expect, and indeed observe, strong dependence of the optical properties of ultrathin films on the deposition conditions. The results are further complicated by aging effects such as may arise because of recrystallization, surface and volume oxidation, etc., of a film. Consequently, violent disagreements between the results of different authors employing different—generally poorly or not at all specified—deposition parameters have been commonly reported. As a result of these uncertainties, the possible size-dependent optical phenomena in ultrathin films remain obscure.

With increasing film thickness, the effect of ·the initial granular structure on the optical properties is decreased but not eliminated completely. Therefore, thickness dependence is still observed, although the general behavior of the optical parameters follows that of the bulk, at least qualitatively. A reproducible and nearly bulk behavior is expected in carefully prepared epitaxial films of optical materials. This optimism is well supported by the results on epitaxial films of CdS [57], and PbS, PbSe, and PbTe [58].

The optical constants of thin films of most of the useful optical materials have been measured and reported in the literature. The primary aim of these studies has been to obtain data for optical-coating applications *without* regard to the influence of the film microstructure. The effects of deposition rate, substrate temperature, ambient conditions, etc., on the optical properties have been described in various

Table I. Summary of the Commonly Used Methods for the Measurement of Optical Constants of Transparent (T) and Absorbing (A) Films

Method	Quantities measured	Film	Substrate	Measurement gives	Remarks
Abelès	Brewster angle	T	T, A	n	Simple
Schulz	Reflectance and transmittance	A	T	n,k	Convenient
Malé	Reflectance and transmittance	T,A	T	n,k,t	Simultaneous n,k,t
Hadley	Reflectance and transmittance at several wavelengths	T,A	T	n,k	Convenient, charts available
Spectrophotometric	Wavelength of interference maxima and minima	T,A	T,A	n,t	Simple, dispersion problems
VAMFO	Fringe displacement for two angles of incidence	T	T	n,t	Simple
Ellipsometer	Δ, ψ	T,A	T,A	n,k,t	Valuable for ultrathin films (<50 Å). Indirect, tables and charts required

reviews [3, 13, 14]. The single important conclusion one may draw from these reviews is that the published optical data should be interpreted with prudence. Some general results of basic and applied interest are described in the following sections.

(2) Metal Films. The optical constants of thin films (<100 Å) show a marked thickness dependence. Nearly dielectric behavior is exhibited by ultrathin discontinuous metal films. A change to metallic behavior takes place when the film becomes physically continuous and electrically conducting with increasing thickness. The value of n is initially much higher than that of the bulk material and decreases smoothly to approach a steady value for thick films. The extinction coefficient k starts from a low value and rises, either monotonically or through a maximum, to a limiting value which may be close to the bulk value. This type of variation of n and k is illustrated in Fig. 6 for Ag films. Also

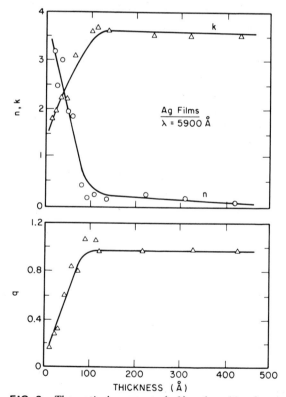

FIG. 6. The optical constants (n,k) and packing fraction (q) of thin silver films at a wavelength of 5,900 Å. (*Ishiguro and Kuwabara* [70])

displayed in this figure is a typical thickness dependence of the packing q, defined as the fractional volume occupied by the metal in a discontinuous film. The correlation between the two curves is discussed later.

The wavelength dependence of the optical constants is complicated and depends markedly on the deposition parameters. It is generally agreed that the dispersion curve for thick continuous films is qualitatively similar to that of the bulk material. This is illustrated in Fig. 7 for a 211-Å-thick Au film compared with the bulk material.

The observed optical constants and the calculated reflectance of thick films of Au, Ag, Cu, and Al for several wavelengths are listed in Table II. Representative works on the optical constants of continuous metal films may be found in the following references: Au [29, 59-69], Ag [24, 29, 63, 70-75], Cu [24, 29, 59, 61, 68], Al [29, 65, 75-79], Pd [80], Pb [81, 82], Fe, Ni, and Co [83], Bi [84], Sn [69, 82, 85, 86], In [87], Ta, W, and Zr [87, 88], Mg, Be, Ca, Ba, Sr, La, Mn, and Ce [79], and Na, K, Cs, and Rb [89].

Joos and Klopfer [68] studied the dependence of the optical constants on temperature down to $20°$ K for Au, Ag, and Cu films. A small decrease in absorption was observed at lower temperatues. The frequency of the absorption edge of Ag increased by about 10^{11}/sec per degree decrease in temperature. This magnitude of the temperature

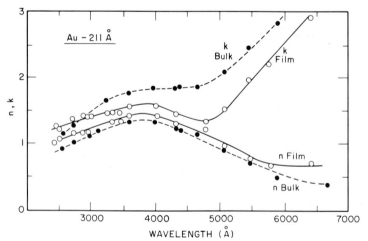

FIG. 7. The optical constants of a 211-Å-thick gold film and bulk gold as functions of the wavelength. (*Philip* [63])

Table II. Optical Constants of Thick Annealed Films of Ag, Au, Cu, and Al (Schulz [29]) The films were annealed until, by the $R_s^2 = R_p$ criterion, they are homogeneous. The infrared data are taken from (a) Beattie [75], (b) Padalka and Shklyarevskii [60], (c) Shklyarevskii and Padalka [61]

Wavelength, μ	Silver n	Silver k	Gold n	Gold k	Copper n	Copper k	Aluminum n	Aluminum k
0.40	0.075	1.93	1.45	1.88	0.85	2.20	0.40	3.92
0.45	0.055	2.42	1.40	1.84	0.87	2.42	0.49	4.32
0.50	0.050	2.87	0.84	2.37	0.88	2.42	0.62	4.80
0.55	0.055	3.32	0.34	2.97	0.72	3.07	0.76	5.32
0.60	0.060	3.75	0.23	3.50	0.17	3.65	0.97	6.00
0.65	0.070	4.20	0.19	3.97	0.13	4.17	1.24	6.60
0.70	0.075	4.62	0.17	4.42	0.12	4.62	1.55	7.00
0.75	0.080	5.05	0.16	4.84	0.12	5.07	1.80	7.12
0.80	0.090	5.45	0.16	5.30	0.12	5.47	1.99	7.05
0.85	0.100	5.85	0.17	5.72	0.12	5.86	2.08	7.15
0.90	0.105	6.22	0.18	6.10	0.13	6.22	1.96	7.70
0.95	0.110	6.56	0.19		0.13		1.75	8.50
2.0	0.48(a)	14.4	0.54(b)	11.2			2.30(a)	16.5
3.0					1.22(c)	17.1		
4.0	1.89	28.7	1.49	22.2			5.97	30.3
6.0	4.15	42.6	3.01	33.0			11.0	42.4
7.0					5.25	40.7		
8.0	7.14	56.1	5.05	43.5			17.0	55.0
10.0	10.69	69.0	7.41	53.4			25.4	67.3

effect can be explained by the lattice-expansion coefficient and the temperature dependence of the Fermi distribution function.

(3) Abnormal Absorption Phenomenon. The structure-sensitive optical absorption of films is considerably enhanced if the films are deposited at an oblique angle of incidence. Further, the absorption depends on the direction of polarization relative to the direction of incident vapor. These effects have been studied for Au and Al films by various authors [90] and may be ascribed to the anisotropic, columnar growth of films (see Chap. IV, Sec. 2.3 for further discussion).

A typical absorption curve for bulk metals consists of absorption peaks corresponding to the interband electronic transitions, superimposed on the absorption by the conduction electrons. The absorption by conduction electrons for thin discontinuous and poorly conducting metal films ($t \ll \lambda$) must be small. However, a marked abnormal absorption maximum appears at a position which varies with the thickness of the film (more precisely with the crystallite size). This abnormal optical-absorption phenomenon has been the subject of numerous experimental [65, 72-74, 91-97] and theoretical investigations [95, 98-100]. The results have been reviewed by Rouard [8, 96, 97].

The abnormal absorption phenomenon has been studied [65, 72-74, 91, 97] in Au and Ag films. The absorption of a film in air is given by the product nkt, since [5]

$$ nkt \sim \frac{\lambda}{4\pi} n_0 \frac{1 - R - T}{T} \tag{37} $$

where n_0 is the refractive index of the substrate, and R and T are the reflectance and transmission of the film at normal incidence. Absorption curves (given by the spectral dependence of nkt) for Ag films of different thicknesses are shown in Fig. 8. The absorption maxima shift toward longer wavelengths as the film thickness increases. A similar trend is observed when the film is exposed to air, indicating the sensitivity of this absorption phenomenon to structural changes. Note that the absorption peak at $\sim$2,600 Å for all films is due to the transition between the electron energy bands of Ag.

Studies [92] on films of transition elements (Fe, Co, Ni) also yield an abnormal absorption region when absorption curves are normalized with respect to the bulk value. The curves indicate the three wavelength regions of distinct behavior. At long wavelengths the absorption is below

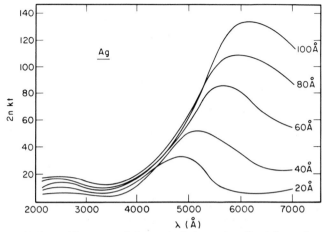

FIG. 8. Variation of the optical absorption ($2\,nkt$) as a function of wavelength for silver films of different thicknesses. The crystallite size is unknown, but it is expected to increase with the increasing thickness. (*Payan and Rasigni* [91])

the bulk value and is followed by a region of bulklike behavior. At still shorter wavelengths, the absorption shows a maximum which in Co films takes place at 2,600 Å. Since the position of the maximum is found to be independent of the film thickness, the original explanation in terms of crystallite size effect may not be correct.

Abnormal absorption has also been studied [93, 94] in other granular films such as In and Ga with results similar to those of Ag films.

(4) Maxwell-Garnett Theory of Abnormal Absorption. An explanation of the abnormal optical properties of ultrathin films was sought by Maxwell-Garnett [98] on the basis of structural discontinuities in films. He assumed that the film consists of a large number of spherical metal particles of diameter small compared with the wavelength of the light used, and also the film dimensions, and further, these particles, embedded in a dielectric medium, have bulklike values of the optical constants. The electric field E of the electromagnetic wave will polarize the spherical crystallites, thus producing an extra field $4\pi P/3$. Here, P is the electric moment per unit volume of the medium and is proportional to N or $1/q$, where N is the number of crystallites per unit volume and is thus inversely proportional to the packing factor q. From the well-known Clausius-Mosotti relation [21] we obtain (since $\epsilon = n^2$)

$$\frac{\epsilon - 1}{\epsilon + 2} = \frac{n^2 - 1}{n^2 + 2} = \frac{4}{3}\pi P = AN = \frac{B}{q} \tag{38}$$

where A and B are constants and n is the refractive index of the spheres.

We can rewrite Eq. (38) in terms of an effective index n^* of the film as

$$q\frac{n^2 - 1}{n^2 + 2} = \frac{n^{*2} - 1}{n^{*2} + 2} = \text{constant} \tag{39}$$

Thus, the effective index of the film can be calculated in terms of n and q.

Malé [99] made use of the Maxwell-Garnett theory [Eq. (39)] and calculated n and k as functions of q. In general, q increases with the film thickness. As q tends to zero, n tends to unity and k to zero, i.e., the optical constants of air. If $k(\lambda)$ is greater than $n(\lambda)$, both the curves of n and k as functions of q show a maximum, which occurs at lower q (smaller thickness) for n than for k. No maximum is observed in these curves if $k(\lambda)$ is less than $n(\lambda)$ for the bulk metal. The occurrence of a maximum is physically related to the fact that a system of metal particles acting as a damped oscillator resonates at a particular frequency. The theoretical conclusions obtained by Malé are in qualitative agreement with the experimental results for Au, Ag, and Pd films.

To improve the agreement of the theoretical and experimental results, David [100] modified the Maxwell-Garnett theory to assume two-dimensional, rather than three-dimensional, crystallites of the form of ellipsoids of revolution. By choosing a suitable distribution function for the ratio of the axes of the ellipsoids, values of q can be derived to fit the experimental data. Schopper [10, 44] used a Gaussian distribution to explain his results. Extensive use of David's generalized theory has been made by several investigators to obtain excellent agreement with the experimental results (see Ref. 8) on films of noble metals, Ga, and In.

It is unfortunate that in most cases where abnormal absorption has been studied and theoretically accounted for, little information on the shape and size of the crystallites is available. The crystallite shape and size in a film depend much on the substrate and deposit materials, the deposition parameters, and the film thickness (Chap. IV, Sec. 3). The initial growth stage of a film consists of cap-shaped three-dimensional nuclei. The coalescence of these nuclei with increasing film thickness at relatively low substrate temperatures produces wormlike microstructures. Doremus [95] made a careful study of the structure as well as optical absorption of Au films in the island stage. Electron micrographs of the spherical islands were used to determine q directly. Assuming the Maxwell-Garnett thoery, Doremus made use of the free-electron-theory expressions for the coupled dielectric constant of a metal film and showed that, with decreasing size of the spherical particles, the position of the absorption maximum is not affected whereas the width of the maximum increases. If electrical conductivity is limited by the size of the particle, the dielectric constant should exhibit size effect and hence result in widening of the absorption maximum. This conclusion agreed with the observations for particles of diameter less than 90Å. The position of the maximum was found to be in excellent agreement with that calculated from the Maxwell-Garnett theory for spherical particles. The results on broadening are necessarily qualitative, however, since a spectrum of particle sizes is generally present in the films.

(5) **Dielectric and Semiconducting Films.** The importance of dielectric and semiconducting films in optical applications has brought about voluminous literature in this field. The optical constants and the transmission range of films of several materials of interest are listed in Table III. Let us emphasize that the parameters given depend strongly on the deposition conditions, nonstoichiometry, inhomogeneity, and anisotropy of the deposited films. Now we shall examine some results on commonly used materials.

Table III. Refractive Indices of Some Transparent Materials in Thin Film Form
In most cases, the films are formed by thermal evaporation. The values of the indices vary with the deposition conditions and should therefore be taken as only representative. Values are given for one wavelength only—the center of the visible spectrum (v) in most cases. (Taken mostly from Heavens' compilation [3])

Substance	n	λ, μ	Range of transparency, μ
Al_2O_3	1.60	0.55	8
	1.55	2.0	
AlF_3	1.38–1.39	0.546	
CaF_2	1.23–1.26	0.546	$<0.15-15$
CeF_3	1.65	0.55	
	1.59	2.0	
CdS	2.5–2.7	0.5	$0.52-> 14$
CeO_2	2.35	v	
CsBr	1.8	v	0.23–40
CsI	2.0	uv	0.25
GaAs	3.6	2.0	

Table III. Refractive Indices of Some Transparent Materials in Thin Film Form (*Continued*)

In most cases, the films are formed by thermal evaporation. The values of the indices vary with the deposition conditions and should therefore be taken as only representative. Values are given for one wavelength only—the center of the visible spectrum (v) in most cases. (Taken mostly from Heavens' compilation [3])

Substance	n	λ, μ	Range of transparency, μ
Ge	4.4	2.0	1.7–20
InAs	3.4	3.8	> 3.8
InP	3.7	2.0	
InSb	4.0	7.3	> 7.3
Irtran 2	2.2	10	
KBr	1.56	0.589	0.60–30
LiF	1.36–1.37	0.546	< 0.11–10
	1.39–1.40	0.546	
MgF_2	1.35	2.0	0.23–10
Na_3AlF_6 (cryolite)	1.35–1.39	0.57	< 0.20–10
NaCl	1.54	0.589	< 0.20–> 15
PbS	4.0	3.0	> 3
PbSe	4.52	5.1	
PbTe	5.5	3.8	
RbI	2.0	uv	0.25
Si	3.3	2.0	> 1.0
SiO	1.55–2.0	0.55	
	1.5–1.8	2.0	0.50–8
SiO_2	1.44	0.55	
	1.46	2.0	0.2–5
SnO	1.96	v	> 10
Ta_2O_5 (anodic)	2.2	0.590	
Te	4.9	6.0	3.4–20
$ThOF_2$	1.36	v	
	1.5	0.632	
TiO_2 (anodic)	2.4	0.546	
TiO_2 (thermal)	2.73	0.546	
WO_3	1.8	v	> 10
ZnS	2.35	0.55	
	2.2	2.0	0.4–14
ZnSe	2.6	0.63	
ZrO_2	2.1	v	
	2.0	2.0	

The high value of n and robust mechanical properties of CeO_2 [101] films make them very suitable for high-reflecting multilayers. Among other widely used dielectric materials for optical devices are cryolite (Na_3AlF_6) [102], MgF_2 [102-108], LiF [109], and CaF_2 and NaF

[106], which have been the subject of several studies. The fine-grained MgF_2 films deposited on glass exhibit [108] negligible absorption in the visible. The absorption rises steeply in the ultraviolet below 2,000 Å and shows a marked dependence on the film thickness and deposition parameters.

The dispersion of SiO [110] films (Fig. 9) is sensitive to oxidation since higher oxides (SiO_2 [111, 112]) have different absorption behavior. Evaporated SiO films exhibit a wide range of refractive indices depending on the deposition conditions. Indices from 1.55 to 2.00 may be obtained in the 5,000- to 7,000-Å wavelength range. The mechanical properties of SiO, as well as of other dielectric materials such as MgF_2, are satisfactory in thin films, but stress-relief cracks are commonly observed in thicker films. Exposure to water vapors changes the tensional stress to compressional stress, thus generating interesting stress-relief patterns (see Chap. V, Sec. 3.5).

Studies of optical constants of thermally evaporated TiO_2 [113, 114] films, anodized Al_2O_3 [115-117] and Ta_2O_5 [28, 118, 119] films, and reactively sputtered Fe_2O_3, InO, and CdO films [120] have also been reported. These films are generally amorphous. The measured dispersion of TiO_2 follows that of the bulk. The observed dispersion for evaporated Al_2O_3 films and anodic tantalum oxide [118] films is shown in Fig. 10(a) and (b), respectively. The index values for Al_2O_3 depend markedly

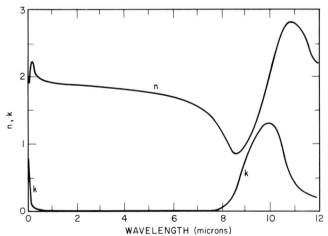

FIG. 9. The dispersion of the refractive index and extinction coefficient of vacuum-evaporated SiO films. (*Hass and Salzburg* [110])

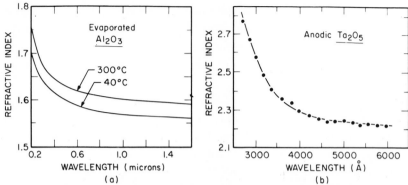

FIG. 10. The dispersion of the refractive indices of (*a*) vacuum-evaporated films of Al_2O_3 deposited at 40 and 300°C (*Cox et al.* [117]); (*b*) anodic films of tantalum oxide. (*Young* [118]).

on the substrate temperature during deposition. At $\lambda = 0.5\ \mu$, an average value of $n = 1.6$ is well below that of the bulk value (1.76) for Al_2O_3. A gradient of the index throughout the film thickness is shown by the anodic Ta_2O_5 films.

Zinc sulfide is a popular material for high-index films in multilayer technology and has therefore been studied extensively [121-125]. The wavelength dependence of n and k is shown in Fig. 11. The values obtained for films are slightly below (~0.1) those for the bulk crystal; large differences, however, occur below $\lambda = 4,000$ Å. The enhanced absorption observed in the neighborhood of $\lambda = 3,300$ Å may be due to an exciton band. The ZnS films used in these studies were deposited on glass substrates at room temperature and are expected to have a fine-grained structure.

Gonella [124] studied absorption in thin, crystalline, β-ZnS films. Abnormal absorption attributed to the resonance of discrete particles in ultrathin metal films described already was also observed in these films in the neighborhood of $\lambda = 3,000$ Å. This absorption effect for films of comparable thickness but of different crystallite sizes is shown in Fig. 12. The absorption peak shifts toward longer wavelengths with increasing film thickness; the width of the maxima increases with increasing crystallite sizes. The latter behavior is quite the opposite to that observed by Doremus [95] on Au films.

Optical films with refractive indices varying over an extended range are highly desirable for certain applications. This result may be achieved by preparing films of mixtures of different substances. Encouraging

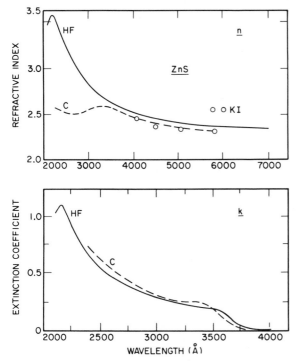

FIG. 11.   The dispersion of the refractive index and extinction coefficient of thick ZnS films. *HF*, Hall and Ferguson [107]; *C*, Coogan [122]; and *KI*, Kuwabara and Ishiguro [125].

results have been obtained with evaporated films of dielectric mixtures such as CeO_2 + SiO [126], CeO_2 + CeF_3 [127], and CeF_3 + ZnS [128].

The optical properties of semiconducting films are of considerable interest for optical devices. Studies of films of Si [87], Ge [129], CdS [107, 130], InSb [131], InP [132], GaAs [133], and CdTe [133] have been reported in the literature. Generally, these (and other) studies have been carried out on polycrystalline films of unknown microstructure. Zemel et al. [58] carried out extensive measurements of the spectral and temperature dependence of the optical constants of epitaxial films of PbS, PbSe, and PbTe. Their results are shown in Fig. 13 to emphasize two points. First, the values of the optical constants obtained on carefully prepared epitaxial films are in good agreement with the bulk values. Secondly, significant fundamental information may be derived from the conveniently available data using the readily prepared clean,

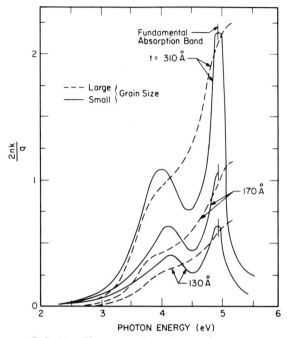

FIG. 12. The energy dependence of the absorption parameter $2nk/q$ (q is the packing fraction) for evaporated films of β-ZnS of different thicknesses with small (<100 Å), and medium (100 to 200 Å) size grains. The abnormal absorption region is pronounced for smaller grain size and shifts toward smaller energies with increasing film thickness. The fundamental absorption band occurs at smaller energies for smaller-grain-size films. (*Gonella* [124]).

smooth, single-crystal films. The data in Fig. 13 show a peak in the index (of each material) which is associated with the rapid change in the absorption coefficient in the fundamental absorption edge. As the temperature decreases, this peak shifts to lower energies. This shift corresponds to the temperature dependence of the energy gap. The position of the peaks yields a direct estimate of the energy gap. By extrapolating the linear plot of n^2 vs. λ^2, the high-frequency dielectric constant is obtained as a function of temperature.

The use of suitable deposition techniques makes it possible to obtain a material in thin film form with different microstructures and, in some cases, with different crystal structures (Chap. IV, Sec. 4.3). Any changes in the electronic structure of the material would be reflected in its optical behavior. Such studies have been performed on the amorphous

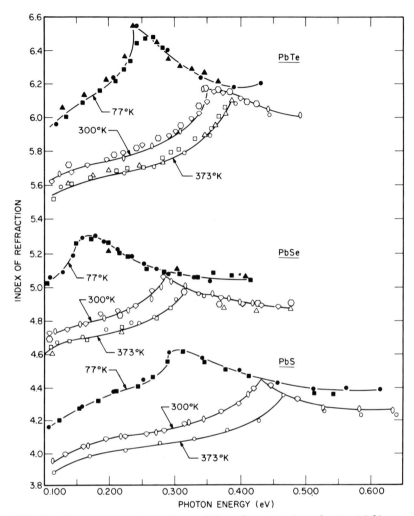

FIG. 13. Photon-energy dependence of the refractive index of epitaxial films of PbS, PbSe, and PbTe at 77, 300, and 373°K. (*Zemel et al.* [58])

and crystalline structures of Ge [134], Se [135], and GeTe [136]. When deposited at or near room temperature, the films of these materials are amorphous. Either deposition at elevated temperatures, or subsequent annealing of the amorphous film yields a crystalline structure. Tauc et al. [134] studied infrared absorption in amorphous and crystalline (annealed) Ge films. Their results for 3,100-Å-thick films are shown in Fig. 14. The absorption spectra for the crystalline film are sharper but are very similar to those of the amorphous film. These spectra correlate

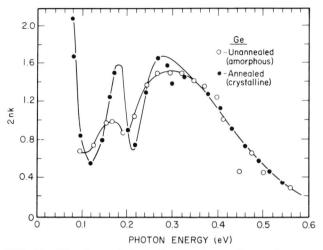

FIG. 14. The observed absorption spectrum of amorphous (unannealed) and crystalline (annealed Ge films). The solid curves are calculated assuming the film spectra to be similar to bulk Ge and arising from direct transitions of holes between three branches of the valence band of bulk Ge. (*Tauc et al.* [134])

with those of bulk *p*-type Ge, except that the peaks and the absorption edge in films are shifted by ~0.1 eV toward smaller photon energies. This correlation is further shown by the excellent agreement of the experimental data with the theoretical curves, which were calculated by assuming that the film spectrum is similar to that of bulk Ge and arises because of direct transitions of holes between the three branches of the valence band. One may thus conclude that the electronic structure of the valence band is similar in both the amorphous and crystalline structures of Ge and that a momentum vector sufficiently well defined to keep the three branches of the valence band is apparently retained in the amorphous state of Ge (the conclusion is valid *only* near *k* = 0).

The optical absorption edge has been found [136, 236] to be nearly the same for both crystalline and amorphous GeTe films. According to the band model of crystalline GeTe (degenerate semiconductor) proposed by Tsu et al. [237], the observed optical gap (~0.8 eV) is the sum of the direct energy gap (~0.1-0.2 eV), the Fermi level (~0.4 eV) in the valence band, and the energy in the conduction band corresponding to the *k*-vector at the Fermi level. In view of the markedly different electrical behavior of the crystalline and amorphous forms of GeTe (see Chap. VII, Sec. 3.2), the optical and electrical data can be reconciled within the same band structure only if we assume that the electrical conductivity is dominated by the trapping effects of the localized levels (due to fluctuation potential) above the Fermi surface.

3.3 Size Effects in Optical Properties

The optical properties of a thin film generally differ from those of the bulk. The differences are generally attributed to the microstructure of the films. One may ask a natural question as to whether the optical constants are the same for a film as for the bulk material. Let us discuss this question separately for nonabsorbing (nonmetallic) and absorbing (metallic) films.

(1) Semiconductor and Dielectric Films. In an infinite crystal, the electron energy is a multivalued continuous function of the quasi-momentum. In a thin film specimen, the quasi-momentum assumes discrete values along the thickness dimension. The energy spectrum represents a system of discrete levels with separation between them given by the uncertainty principle as

$$\Delta E \approx \frac{\hbar^2}{2m^* \, t^2} \quad \text{[Eq. (13), Chap. VII]}$$

where t is film thickness and m^* is the effective mass of the carriers. The effect of the discrete energy levels on the forbidden gap of a semiconductor has been considered by Sandomirskii [137] and others. According to Sandomirskii, all levels of the energy spectrum of a semiconductor will be shifted by ΔE in a thin film specimen, provided smearing of energy levels by temperature and diffuse scattering of the carriers at the film surfaces are not significant. This shift will increase the band gap and thus affect the optical behavior of semiconducting films. A significant effect should be observable in films of semimetals with overlapping conduction and valence bands since the overlap may be diminished or completely eliminated, as in the case of thin Bi [138] films (Chap. VII, Sec. 3.2).

The effect of the reduction of the overlap between the conduction and valence bands has been observed in the electrical conductivity of thin Bi, Sb, and As films (Chap. VII, Sec. 3.2). A thickness-dependent change in the optical band gap has been observed by Stasenko [238] in evaporated CdS films and has been attributed to the aforementioned effect. Clearly, the effect is very small and very precise measurements of the absorption edge are required to verify the quantum size relation given above.

Soonpaa [139] studied optical transmission of single-crystal films of $Bi_8Te_7S_5$ obtained by cleaving of bulk crystals. This semiconducting material cleaves at a multiple of quintuple layers, each corresponding to a film thickness of 9.8173 Å. Soonpaa found no shift in the energy gap of thin films as compared with that of the bulk material. The absorption coefficient, however, showed an abrupt *decrease* as the film thickness increased. This has been explained on the basis that, because of the limited number of momentum vectors available along the thickness dimension of the film, momentum-conserved direct optical transitions are only possible in ultrathin films in contrast to the direct optical transitions which take place in the bulk. Thus, the absorption is reduced in thinner films.

Another type of size effect may arise in ultrathin films owing to the influence of the substrate. When two different media are brought together, electrochemical equilibrium between the phases in contact must take place by the transfer of charges across the interfacial boundary. The interface, therefore, is no longer a discontinuous boundary, which is commonly assumed in calculating the optical (or other physical properties) behavior of surface films. Two chemically distinct phases form an electric double layer, which is sharp and localized at the interface for low-dielectric-constant media but is diffuse for high-dielectric media. Since a diffuse layer would influence the electrical and optical properties of the interface, ultrathin dielectric films deposited on metallic substrates may exhibit abnormal values of optical constants. This explanation was proposed by Plumb [140] to account for his observations of the abnormal values of optical constants of ultrathin films. To demonstrate the validity of his argument, he studied optical absorption in barium stearate layers deposited by the Blodgett-Langmuire technique onto Au surfaces. The absorption index, measured by ellipsometry, decreased with increasing film thickness and followed an approximately inverse exponential law. This observation was shown to be consistent with the assumed existence of a diffuse double layer. Further clear-cut experiments are needed to support this explanation and may be provided by a study of the spectral dependence of the absorption index.

(2) Metal Films. The situation in metal films is different. The optical properties of a metal depend ultimately on the behavior of conduction electrons in the metal in response to the incident electromagnetic radiation. When the specimen thickness is comparable with the mean free path (mfp) of the conduction electrons, the scattering of electrons at the surface reduces the conductivity. This size effect depends on the ratio of the specimen thickness to the mfp of electrons, and the surface scattering coefficient p (Chap. VI, Sec. 3.1). A size effect in the electrical conductivity would produce a size effect in the optical constants of a metal film, particularly in the infrared region where optical absorption is dominated by the conduction electrons and the indices are proportional to $\sqrt{\sigma}$ [Eq. (25)].

If the penetration depth of the incident radiation of frequency ω is much larger than the film thickness, and the surface scattering of conduction electrons is diffuse, the electrical conductivity, and hence also optical reflectivity, have been shown by Flechsig [141] and Kao [142] to exhibit oscillatory behavior. The maxima of these oscillations will occur when $(\omega/\omega_0)\gamma = 1, 7, 13, 19, \ldots$, where ω_0 is the Lorentz-Sommerfeld frequency [Eq. (24)], and $\gamma = t/l$ is the ratio of thickness to the mfp. These oscillations amount to about several percent deviations from the classical Drude theory for bulk material. This predicted oscillatory behavior should be observable in films of semimetals and degenerate semiconductors.

When the incident radiation is absorbed in a distance much smaller than the mfp of the conduction electrons, a correction, called the "anomalous skin-effect correction," needs to be applied to the simple Drude theory. This correction arises because now the conduction electrons will travel a considerable distance free from the applied electromagnetic field. The regions of the spectrum most affected are the near infrared and the visible, where absorption due to conduction electrons dominates. The anomalous skin correction has been theoretically treated by Reuter and Sondheimer [143], Dingle [144], and Holstein [145]. The correction obviously depends on the nature of surface scattering of conduction electrons. If the surface scattering is diffuse, the correction may be significant. For specular scattering, the correction is, however, negligible. Dingle has derived explicit formulas for the optical properties of metals in the near infrared, visible, and ultraviolet, for both specular and diffuse surface scattering of electrons. Theoretical reflectance curves for Ag calculated from the simple and the corrected Drude theory for $p = 0$ (diffuse scattering) and $p = 1$ (specular scattering) are shown in Fig. 15. It is clear that, in this frequency range, the differences between the different curves are very small. Therefore, extremely careful and accurate (better than a fraction of a percent) measurements of reflectance are needed to verify the validity of the Drude theory and its size-effect modifications.

Investigations on Au films of thickness 100 Å or more have been reported by Weiss [59], Hodgson [69], Schulz [146], and Bennett et al. [147]. The results of Schulz and Hodgson are in reasonable agreement with the anomalous skin effect corrected Drude theory for $p = 0$. On the other hand, very precise measurements of Bennett et al. on opaque, polycrystalline Au and Ag films deposited on supersmooth quartz agree with the simple Drude theory and also the corrected theory for $p = 1$. The data for Ag films are shown in Fig. 15 and are compared with the various calculated curves using the measured film conductivity. It is clear that the data above about 8μ show a good agreement with the corrected Drude theory for $p = 1$. Above about 15μ, better agreement is obtained with the simple Drude theory. The same holds true for Au films above 3μ. This observation is important since it is not only a verification of the simple Drude theory but also means that the optical constants in the far infrared may be obtained by direct calculations from this theory. The decline of the reflectance below $\sim 5\mu$ may be explained as due to the frequency dependence of the relaxation time τ in this wavelength region.

Bennett et al. also observed a rapid decrease of the reflectance of Ag films when the roughness of the films exceeded about 45 Å rms. Roughness was produced by depositing films onto the substrate coated

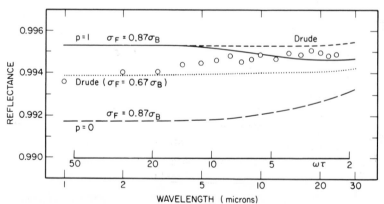

FIG. 15. Calculated and measured infrared reflectance of evaporated Ag films. The experimental points are averaged for several films. The solid $(p = 1)$ and the long dashed $(p = 0)$ curves are calculated from the corrected Drude theory for a conductivity of 87 percent of bulk value. The short dashed and the dotted curves are calculated from the simple Drude theory for given values of the film conductivity. The parameters used are $\sigma_B = 4.85 \times 10^{17}$/sec, $\tau = 1/\omega_0 = 2.99 \times 10^{-14}$ sec and $n_e = 1.09$ electrons/ atom. (*Bennett et al.* [147])

with discontinuous CaF_2 films. This result was interpreted in terms of a change from $p = 1$ to $p = 0$ scattering behavior of the films. This point is further discussed in Sec. 3.4 of Chap. VI.

4. THIN FILM ABSORPTION AND PHOTOEMISSION PHENOMENA

4.1 Infrared Absorption

(1) Optical Modes. An ionic crystal subjected to electromagnetic radiation exhibits two optical-absorption modes due to the dielectric-dispersion relation. Between the frequencies ω_T of the transverse optic mode (TO) and ω_L of the longitudinal optic mode (LO), a forbidden region of high reflectivity, called the "restrahlen" or "residual-ray" region, exists. The frequencies* ω_L and ω_T are determined by the dynamics of the crystal lattice and are related to each other by the Lyddane-Sachs-Teller relation [148]

$$\omega_L = \omega_T \frac{\epsilon_0}{\epsilon_\infty} \tag{40}$$

where ϵ_0 and ϵ_∞ are the static (low-frequency) and the optical dielectric constants, respectively.

Because of the strong interaction of transverse electromagnetic waves with the long-wavelength TO mode, resonant absorption is produced at ω_T which is easily detected by reflection or transmission studies. Such studies have been carried out on thin films of cubic ionic crystals of SiC [149], PbSe [150], and CdS [151]. The observed values are in good agreement with the data from other sources.

The longitudinal mode in a finite crystal may be excited only as a result of its strong coupling with the transverse mode. Berreman [152], however, showed that the longitudinal mode can be excited in a

---

*Note that the forbidden region is not related to the periodicity of the lattice. In this region of frequencies, the dielectric constant is negative, so that electromagnetic waves cannot propagate in an infinite crystal. The upper bound of the forbidden range is denoted by ω_L; so $\omega_L > \omega_T$ and $\epsilon_0 > \epsilon_\infty$. It is interesting to observe that if $\omega_T \rightarrow 0$, $\epsilon_0 \rightarrow \infty$, that is, a ferroelectric behavior is expected. Physically, this implies that the effective restoring force caused by the polarization of the lattice on passage of a transverse optical phonon is zero and thus the lattice is unstable.

thin-film geometry by a nonnormal incidence of p-polarized radiation. Berreman utilized this fact to measure ω_L of thin LiF films by transmission and reflection spectra. The reflection spectra of a 3,250-Å-thick LiF film evaporated onto a collodion film for s- and p-polarized radiation incident at an angle between 26 and 34° is shown in Fig. 16. The known absorption band at 14.97μ due to atmospheric CO_2 allows a very precise measurement of ω_L which occurs at 14.95μ. The absorption at 32.6μ is due to the transverse mode. The position of ω_L depends on deposition conditions; the value of ω_L shifts toward longer wavelengths for films deposited on cold substrates.

The data on epitaxially grown PbS and PbSe [150] films have been used to calculate ϵ_0 and ϵ_∞. The values for PbS at room temperature are ϵ_∞ = 174.4 and ϵ_0 = 17.3; for PbSe, ϵ_∞ = 200. The high-static dielectric constant in lead salts is also expected from the mobility data.

(2) Surface Modes. In thin films, surface-absorption effects become important. The occurrence of a surface or a boundary means that a certain number of atomic forces are set to zero, which results in lowering of the stiffness of the lattice and, therefore, a lowering of some or all of

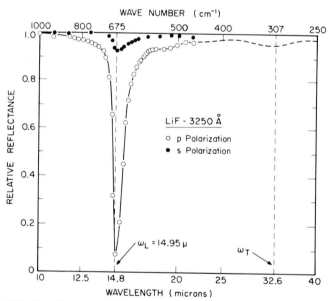

FIG. 16. Observed reflectance of s- and p-polarized radiations by silvered glass slides with LiF films (3,250 Å thick) deposited on them at 265°C, relative to that by silvered glass alone. Radiation was incident from 26 to 34°. (*Berreman* [152])

the normal-mode frequencies. A shift of some of the normal-mode frequencies out of the allowed frequency range and into the forbidden frequency gap may also occur. The modes corresponding to these frequencies may be called "localized" surface modes. Their amplitude is expected to be maximum at the surface, decreasing exponentially with increasing distance from the surface.

Surface vibrational waves were discussed theoretically first by Rayleigh [153]. More recently Gazis et al. [154] and Wallis [155, 156] investigated the surface modes of a diatomic lattice. The surface modes of vibration of a diatomic lattice were shown to exist with frequencies in the forbidden gap between the optical and acoustical branches. Because of the surface modes, a net electric-dipole moment would exist at the surface, so that absorption in the infrared for frequencies lying below the transverse optical mode is expected.

Because of very high absorption of the transverse optical mode compared with the surface mode, the detection of surface modes may be possible only in a thin film deposited on a metallic mirror, so that multiple reflections enhance their absorption effect. There is at present no direct verification of the existence of surface modes.

If a foreign atom is adsorbed on the surface, localized surface modes should occur provided the mass of the adsorbed atom is small enough or the bonding of the surface is sufficiently strong. Theoretical examination of surface impurities in a two-dimensional lattice by Kaplan [157], and Hori and Asahi [158] established that surface-localized modes may arise in the forbidden frequency gap on either side of the allowed normal optical modes. The observed [159] infrared absorption of the Pt surface covered with H_2 may be due to the surface modes of the adsorbed H_2.

4.2 Magneto-optical Absorption

In the presence of a large external magnetic field, the continuum of states in the valence and conduction band of a semiconductor are quantized into a set of semidiscrete Landau levels which interact with the incident radiation to produce characteristic absorption spectra. Measurements of interband magneto-optical absorption give information on the band gap, degeneracy of the bands, the reduced effective mass, and g factors for the individual bands. This information forms the basis of a powerful technique for determining the band structure.

In order to perform these experiments in regions of high interband optical absorption, a few microns thick specimens are required. High mobilities are also required in order to satisfy the condition for cyclotron orbits, e.g., $\omega_i \tau_i > 1$, where ω_i is the cyclotron frequency and τ_i is the relaxation time for the ith band. Both these conditions are conveniently obtained in epitaxially grown films. Palik et al. [160] studied the magneto-optical spectrum in PbS films of 2 to 4μ thickness, grown epitaxially by depositing on rock salt at temperatures of $\sim$100 to 300°C. These data were used by Dimmock and Wright [161] to calculate the band structure of PbS. A detailed discussion of the interpretation of the interband magneto-optical absorption data in terms of the band parameters is given by Mitchell [162].

Magneto-optical effects are small in all but ferromagnetic materials. The ellipticity introduced in the transmitted (and the reflected) beam when a plane-polarized light is incident on a magnetized ferromagnetic film is known as the Faraday effect—generally referred to as a rotation of the plane of polarization. This effect provides a powerful tool for studying the micromagnetic behavior of ferromagnetic films (Chap. X, Sec. 11.2).

4.3 Plasma-resonance Absorption

The free electrons in a metal are capable of collective oscillations [163] (called plasma oscillations) owing to their coulomb interaction with the ions of the lattice. The plasma frequency ω_p of a free electron gas is given by Eq. (23)

$$\omega_p = \left(\frac{4\pi n_e e^2}{m^*} \right)^{1/2} \qquad (\epsilon_\infty = 1 \text{ for free space})$$

where n_e is the density of electrons. This frequency lies in the ultraviolet for metals (also for Ge, if plasma oscillations occur in the high carrier density valence band), and in the infrared for typical semiconductors. Note that the above relation for ω_p is valid for small wave vectors ($\lambda \to \infty$). The wavelength dependence is given by

$$\omega \approx \omega_p \left(1 + \frac{3}{10} \frac{v_F^2}{\lambda^2 \omega_p^2} \right) \qquad (v_F = \text{Fermi velocity})$$

Plasma oscillation may be viewed as a kind of longitudinal optical phonon in which the electron gas plays the part of the negative ions. Since gas has no shear elastic modulus, $\omega_T = 0$ and $\omega_L = \omega_p$. If plasma oscillations are confined to the surface subject to the boundary condition that the tangential component of electric field be continuous at the boundary, then one can show that the plasma frequency is reduced by a factor $1/\sqrt{2}$.

If plasma oscillations are weakly damped so that $\omega\tau_0 \gg 1$, then from Eqs. (21) and (22) for the dielectric constant of a metal, we note that ϵ_1 is positive only for $\omega > \omega_p$. Thus, the electromagnetic waves can propagate in such a medium only for $\omega > \omega_p$, while below this cut-off frequency an exponential decay occurs and the waves are totally reflected. This criterion makes it convenient to measure ω_p experimentally and thereby determine (n_e/m^*). When free electron absorption is dominant and $\omega\tau_0 \gg 1$, Eqs. (21) and (22) reduce to

$$\epsilon_1 = n^2 = 1 - \frac{\omega_p^2}{\omega^2} \qquad \epsilon_2 = 0 \qquad\qquad (41)$$

An excellent general review of this subject as applied to thin films has been given by Steinmann [164]. The results obtained with the alkali-metal films are in agreement with the theoretical value of ω_p. The free-electron theory fails for most other cases. For example, the observed ω_p for Ag is 3.75 eV, rather than 9 eV as calculated from Eq. (23), and is the result of a transition of $4d$ electrons into the conduction band.

Plasma oscillations can be excited externally only by high-energy charged particles or photons since the energy of plasmons is higher than the Fermi energy. The plasma oscillations are damped because of various interactions. Ferrell [165] showed that in a thin film the plasma oscillations are attended by the emission of electromagnetic radiation. This plasma radiation is similar to the Ginzburg-Franck [166] prediction of the so-called "transition radiation" emitted during the transfer of a charged particle from one dielectric medium to another. Plasma-resonance emission in thin films has been the subject of much theoretical [167] and experimental [168-173] investigations. In general, excellent agreement has been obtained between theory and experimental results on films of Ag, Al, and K.

On the basis of Ferrell's physical picture of plasma-radiation emission, an inverse process is also expected. That is, plasma oscillations should occur when electromagnetic radiation of the correct frequency and correct state of polarization (*p* state) is incident at a nonnormal angle on a thin metal film. The occurrence of plasma oscillations is attended by an absorption peak, the position of which determines ω_p. Plasma-resonance absorption has been treated theoretically by several workers [174-176]. The effect has been observed in thin Ag [177, 178] films at $\hbar\omega_p$ = 3.8 eV, in K [179] films at about 3.9 eV (λ = 3,200 Å), and in Al [180] films at 14.8 eV. The resonance absorption of *p*-polarized radiation in K films of various thicknesses observed by Deichsel [179] is shown in Fig. 17. The angle of incidence of radiation is 55°. As expected, the *s* polarization does not exhibit the effect. As the film thickness increases, absorption processes other than those due to plasma resonance dominate so that the transmission peak becomes more and more diffuse.

The plasma-resonance phenomenon in thin metal films is a simple but very powerful tool for a precise determination of the plasma frequency

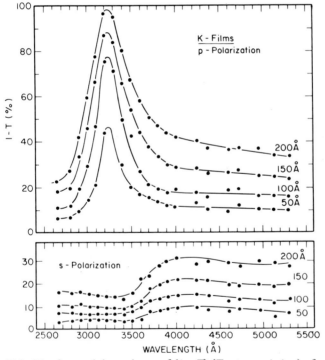

FIG. 17. Spectral dependence of $(1 - T)$ (T = transmission) of *p*- and *s*-polarized radiation in thin potassium films of different thicknesses. The radiation is incident at an angle of 55°. (*Deichsel* [179])

and the dielectric constant in the vicinity of the plasma frequency. The width of the transmission dip and its magnitude, respectively, determine the slope of the real part of the dielectric constant and the value of the imaginary part.

4.4 Ultraviolet Absorption

Absorption bands due to bound electrons in most materials occur in the ultraviolet. The absorption spectrum in the ultraviolet may be conveniently studied in specimens in thin film form. The data can be used to extract information on the electronic structure of the energy bands. Several investigators [181-185] have measured the absorption spectra of various materials. The positions of the absorption edges agree well with those obtained from the known higher-energy x-ray transitions.

Owing to the occurrence of plasma oscillations of free electrons in a metal, a critical wavelength λ_p exists which corresponds to the plasma frequency ω_p [Eq. (23)], so that radiations of wavelengths greater than λ_p are completely reflected while the shorter wavelengths penetrate into the medium (i.e., the medium becomes transparent). This critical wavelength (and hence ω_p) can be determined easily by reflectance measurements in the ultraviolet. Experimental results for Al and Ge films available in literature are presented in Fig. 18. Also shown in the figure are the theoretical curves calculated for several metals on the basis of the Drude model. The assumed values of τ and λ_p are given in the figure caption.

4.5 Photoemission from Metal Films

The thickness and spectral dependence of photoemission from thin metal films are expected to depend on the escape depth of photoelectrons and the mechanisms by which they lose energy during escape. Although the earliest studies in this connection date back as far as 1903, reliable measurements on K, Cs, and Na films have been obtained only recently [186-188]. The results show that photoelectrons are generated in the bulk of the film and their number reaches a constant value at a critical film thickness termed the "escape depth." The escape depth is found to decrease with increasing energy of the photoelectrons. The decrease observed in K and Cs films is very marked and has been interpreted as being caused by the onset of energy loss by emission of plasmons. For Na [188] films, the observed decrease is not as rapid and is ascribed to the fact that the plasma threshold for Na is beyond the

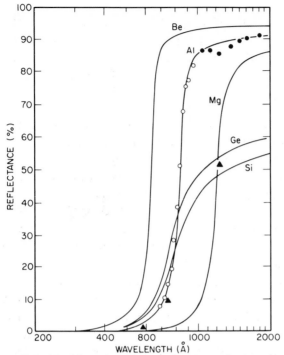

FIG. 18. Normal incidence reflectance of various
metals in the extreme ultraviolet. Data points are for
Al (solid and open circles) and Ge (solid triangles)
films. The solid curves are calculated using a Drude-
type model with the following parameters:

	Be	Al	Mg	Ge	Si
λ_p, Å	639	837	1198	743	762
τ, 10^{-15} sec	1.3	1.1	1.1	0.16	0.14

(*Data compiled by Hunter* [181]) Note that the
calculations for Ge and Si refer to their respective
valence bands.

frequency range used in the experiment. The energy dependence of the
escape depth is qualitatively similar to that predicted theoretically [189,
190] for the energy dependence of the mfp for electron-electron
collisions (see Chap. VI, Sec. 5).

Photoresponse studies of thin film metal-barrier-collector structures
provide valuable information on both the mfp of hot electrons in the
metal electrode (Chap. VI, Sec. 5) and the barrier potentials (Chap. VIII,
Sec. 4.3) over which the hot electrons are transported to the collector
electrode. By plotting a square root of the photoresponse as a function

of the photon energy (Fowler plot), a straight line is obtained whose intercept yields the barrier potential. The film-thickness dependence of the photoresponse yields the mfp of hot electrons.

Photoemission measurements by Garron [191] on thin films of Al, Ag, Au, and In showed a pronounced thickness-dependent maximum which was ascribed to a thickness-dependent minimum of the work function of films. Extensive measurements of the work function of annealed and unannealed evaporated films of Ti, Cr, Mn, Fe, Co, Ni, Cu, Zn, Ge, Pd, Ag, Sn, W, Pt, Pb, and Bi were made by Suhrmann and Wedler [192]. The films deposited at 77 or 90°K were subsequently annealed to 273°K. The work function was measured from the spectral response of photoemission. Generally speaking, the work function increased on annealing while that of the unannealed specimens was lower by as much as 0.72 eV for Cu films. If structural defects vary with film thickness, an apparent thickness dependence of the work function may be observed [193, 194].

Photoelectron energy-distribution and quantum-yield measurements provide considerable information on the density of states and electron scattering within the metal for electron energies not generally accessible. This powerful technique has been exploited by several workers, and the results are reviewed by Spicer [195]. For example, Wooten et al. [196] studied photoemission from about 2,000-Å-thick Al films. The energy distribution of photoelectrons from Al was found to be a near replica of the theoretical density of occupied states. By fitting data to the Monte Carlo calculations [189] of the energy distribution and quantum yield, a value for the mean free path for electron-electron scattering of about 1,000 Å for electrons of energy 6 eV above the Fermi energy was obtained.

5. MULTILAYER OPTICAL SYSTEMS

Partly transparent, highly reflecting metal films have been used in optics for a considerable time. The use of dielectric films has led to striking developments in optical systems for high transmission (antire-flection), and high reflection over a wide or a narrow band of well-defined radiation wavelengths. In this section we shall survey briefly the various optical applications of single and multilayer films, empha-sizing the physical basis rather than the technical details involved in these applications. The techniques for deposition and monitoring of multilayer films are described in Chaps. II and III, respectively.

5.1 Antireflection Coatings

When light is incident on a surface separating media of differing refractive indices, part of the light is reflected. The reflection coefficient R given by Eq. (13a) is, for example, 4 percent for glass of index 1.5 and 36.9 percent for Ge of index 4.1 in the infrared. Most of the light reflected at the surface is lost to the transmitted beam. With an increasing number of surfaces and media of different indices, as may be required in an optical system, the transmittance loss increases considerably. The loss becomes intolerable for elements of high value of the index. In addition to this loss in the useful intensity of radiation, the multiple reflections between the surfaces of various transmitting optical elements cause unwanted light to fall on the image plane, thereby reducing the contrast and definition of the image. The reduction of the reflectance, and enhancement of the transmittance may be obtained by the use of suitable surface coatings called "antireflection coatings." This antireflection process is also termed "blooming" and is now a commonplace industrial process that is customarily applied to many optical components.

The reduction of reflectance of polished crown-glass plates with aging was noted as early as 1887 by Rayleigh [197] and was attributed to the presence of a surface layer. Bauer [198] was the first to explain the reduction due to an interference phenomenon. Extensive experimental and theoretical work on single or multilayer reflection-reducing coatings has been published in literature. The subject has been thoroughly reviewed by Heavens [1, 3] and by Cox and Hass [13]. A brief survey of the film materials for such optical coatings is given by Hass and Ritter [199]. Computational methods for multilayer optics are discussed by Berning and Berning [200], and Miyake [201].

(1) Inhomogeneous Films. The reflection coefficient R at normal incidence at a surface of index n in air is given by Eq. (7) as $[(n - 1)/(n + 1)]^2$. If it were possible to coat the surface with a film whose index varied continuously from n to unity over a distance not small compared with the wavelength of light used, then zero R would be obtained at all wavelengths. This ideal medium cannot be realized in practice since no transparent solid materials with sufficiently low indices are known. A low-index layer on glass can be produced by removing heavier elements (e.g., Pb, Ba) in dense glass by using ion-beam sputtering. In the case of materials of high index, even a small matching by step-graded index coating gives considerable improvement in the transmittance.

The production and properties of inhomogeneous thin films produced by codeposition of materials of different indices have been examined by several workers [202-204]. Jacobsson [203] has shown that for a given index profile, the optical properties may be calculated by numerical integration of the wave equation.

(2) Homogeneous Single Films. For a nonabsorbing thin film of index n_1 deposited on a substrate of index n_0 and exposed to a medium of index n_2, the normal incidence reflectance is given by Eq. (13a). If the film thickness is one-quarter wavelength, or odd multiples thereof (that is, $n_1 t = \lambda/4,\ 3\lambda/4,\ \cdots$, and therefore, $\delta = \pi/2,\ 3\pi/2,\ \cdots$), then Eq. (13a) yields

$$R_{\lambda/4} = \left(\frac{n_1^2 - n_0 n_2}{n_1^2 + n_0 n_2} \right)^2 \tag{42}$$

A universal curve for R vs. $n_1/\sqrt{n_0 n_2}$ based on Eq. (42) is shown in Fig. 19. Thus $R = 0$ if

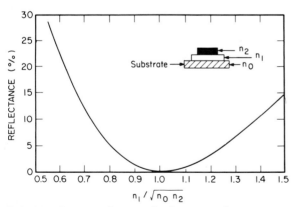

FIG. 19. A universal curve of reflectance of a transparent $\lambda/4$ film of index n_1 as a function of $n_1/\sqrt{n_2 n_0}$ (n_2 and n_0 are indices of the medium and the substrate, respectively).

$$n_1^2 = n_0 n_2 \tag{43}$$

If Eq. (43) is not satisfied, then Eq. (42) will indicate either a minimum or a maximum, depending upon whether $n_1 < n_2$ or $n_1 > n_2$, respectively. The position of the minimum is determined by the optical thickness $n_1 t$. Therefore, by changing film thickness, we can position the reflectance minimum at any wavelength. If the ideal condition [Eq. (43)] is not satisfied, the value of R increases rapidly for deviations from this condition.

Another interesting property that follows from Eq. (43) is that there are two values of n_1 for any given n_0 and n_2 which give the same R. The two values coincide at $n_1 = \sqrt{n_0 n_2}$ and yield $R = 0$. The two n_1's satisfy the relation $n_1' n_1'' = n_0 n_2$. The spectral reflectances for these two values of n_1 are identical, as pointed out by Osterberg and Smith [205]. Thus, in practice, we may choose a film of higher or lower index to obtain the same R as long as the $n_1' n_1'' = n_0 n_2$ condition is satisfied.

It should be mentioned that for a film which is an even multiple of quarter-wavelength thick, the reflectance is the same as that of the uncoated substrate. If $n_1 < n_0$, a $\lambda/2$ film gives a maximum in reflectance. This means that such an antireflection coating will have a spectral reflectance which never exceeds the reflectance of the uncoated substrate. On the other hand, for $n_1 > n_0$, the minimum in R will be that of the uncoated substrate, and the maximum is given by Eq. (13a).

To satisfy Eq. (43), for crown glass ($n_0 = 1.51$) in air, the film index would have to be 1.23. A durable material with such a low index is not available. The best compromises are MgF$_2$ ($n_1 = 1.38$), and cryolite ($n_1 = 1.34$) $\lambda/4$ films which yield reflectance of 1.33 and 0.75 percent, respectively, compared with 4.13 percent for the reflectance of an uncoated glass surface. The computed spectral reflectance of a single $\lambda/4$ layer ($n_1 = 2.0$) on a Ge substrate ($n_0 = 4.0$) for various angles of incidence is shown in Fig. 20. The minimum placed at 2μ is matched at the given angle. The reflectance, taken as the average of s and p components for nonnormal incidence, rises steeply around the minimum. The rise is much steeper for an antireflected, high-index material than for a low-index material.

(3) **Multilayer Films.** The effectiveness of a single-layer antireflection coating is limited by the available materials. Moreover, zero reflectance can be obtained at only one wavelength. These difficulties can be largely overcome by using coatings with two or more layers. As

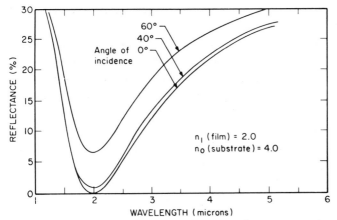

FIG. 20. Calculated reflectance as a function of wavelength of a single-layer coating ($n_1 = 2.0$) on Ge ($n_0 = 4.0$) for 0, 40, and 60° angle of incidence (matched at the given angle of incidence). (*Cox and Hass* [13])

the number of layers increases, the mathematical analysis becomes enormously complicated. Cox and Hass [13] have given a comprehensive discussion of two- and three-layer systems. They computed the reflectance of these systems with an IBM 650 or RCA 301 digital computer using programs developed by Berning and Berning [200]. Some of the essential results are illustrated in this section.

If n_0, n_1, n_2, n_3, are the indices of the substrate, first film, second film, and so on, then for two layers of equal optical thickness, so that $n_1 t_1 = n_2 t_2 = \lambda/4, \; {}^3/_4 \lambda, \; \cdots$, the reflectance is given by

$$R_{\lambda/4} = \left(\frac{n_2^2 n_0 - n_1^2 n_3}{n_2^2 n_0 + n_1^2 n_3} \right)^2 \tag{44}$$

A zero reflectance is obtained if

$$n_0 n_2^2 = n_1^2 n_3 \tag{45}$$

Equation (45) gives a universal curve identical to that of Fig. 19 if $n_1/\sqrt{n_0 n_2}$ is replaced by $n_2 \sqrt{n_0}/n_1 \sqrt{n_3}$.

The condition of Eq. (44) gives more flexibility in obtaining single- or double-zero reflectance with durable film materials on glass. The calculated reflectance of quarter-quarter coatings on Ge, satisfying the ideal conditions of single zero, double zero, and a compromise condition, are shown in Fig. 21.

A $\lambda/4$ layer of SiO ($n_1 = 1.7$) next to glass followed by a $\lambda/4$ layer of MgF_2 yields nearly zero R at $\lambda = 1\mu$ and is thus useful for near-infrared viewing devices. The superiority of this double layer over the single layer is limited to a narrow spectral range since R rises more rapidly with λ in the former case.

It should be pointed out that zero reflectance can also be obtained by double-layer combinations consisting of high-index films thinner than $\lambda/4$ and low-index films thicker than $\lambda/4$. These non-quarter-wave coatings show only a slightly steeper spectral reflectance than the quarter-wave coatings. Such coatings of ($MgF_2 + CeO_2$) and ($MgF_2 + SiO$) on glass have been prepared to yield zero reflectance at 1μ.

A broader region of low reflectance in a double layer may be obtained [206, 207] by using a coating consisting of a $\lambda/2$ high-index (n_1) film *next* to the glass (n_0) and a $\lambda/4$ low-index (n_2) film as the *outer* layer. Zero reflectance for this system is obtained if either

$$n_2^2 = n_3 n_0 \qquad (46a)$$

or

$$n_1^3 - \frac{1}{2} \frac{n_1 n_0}{n_3 n_2} \left(n_3^2 + n_2^2 \right) \left(n_2 + n_1 \right) + n_2 n_0^2 = 0 \qquad (46b)$$

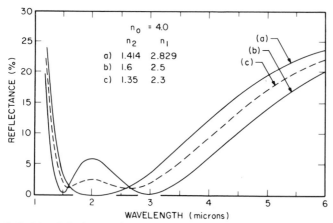

FIG. 21. Calculated reflectance as a function of wavelength for step-down double-layer $\lambda/4 - \lambda/4$ coatings on Ge ($n_0 = 4.0$). The coatings are (*a*) single zero, $n_2 = 1.414$, $n_1 = 2.829$; (*b*) double zero, $n_2 = 1.6$, $n_1 = 2.5$; and (*c*) intermediate or compromise case, $n_2 = 1.35$, $n_1 = 2.3$. (*Cox and Hass* [13])

The first solution gives less effective coatings than a single layer. The second case yields a maximum in reflectance at $\lambda_0 = 2n_1 t$, but the maximum is independent of n_1. Two zeros in reflectance are obtained, one on each side of λ. A quarter-half coating of $(MgF_2 + SiO)$ on glass yields reflectance minima of 0.5 percent at about 1.5 and 2.3μ. This coating has a lower reflectance than a single MgF_2 layer over a wavelength region 1.2 to 2.6μ.

A lower and extended spectral region with lower and more uniform reflectance can be produced by using triple-layer coatings [208-209]. If each layer is $\lambda/4$ (or odd multiples thereof), three zeros in reflectance are obtained when the following condition is satisfied:

$$n_3 n_1 = n_2^2 = n_4 n_0 \qquad (47a)$$

If $n_4 = 1$, this condition is satisfied only by infrared transparent materials. We can rewrite Eq. (47a) in a more general form

$$n_1 n_3 = n_2 \sqrt{n_0 n_4} \qquad (47b)$$

which is a less restrictive condition and yields only one zero at the wavelength where each layer is one-quarter wavelength thick.

If the three-layer arrangement is a substrate-quarter-half-quarter (or three-quarter) system, then only a single zero in reflectance is obtained for

$$n_3^2 n_0 = n_1^2 n_4 \qquad (47c)$$

Figure 22 gives a comparison of the spectral reflectance of three types of triple-layer coatings on glass: (a) quarter-quarter-quarter satisfying Eq. (47b), (b) quarter-half-quarter [Eq. (47c)], and (c) quarter-half-three quarters [Eq. (47c)]. The curve (b) has a broader low-reflectance region, and thus quarter-half-quarter construction is the best choice. Such coatings with $(MgF_2 + ZrO_2 + CeF_3)$, $(MgF_2 + CeO_2 + CeF_3)$, and $(MgF_2 + SiO + CeF_3)$ on glass have been prepared by thermal evaporation. The experimental results for a $(MgF_2 + CeO_2 + CeF_3)$ coating on glass are also shown by the dotted curve (d) in the same figure. A reasonable qualitative agreement with the theory is thus obtained.

As the number of layers increases, the mathematical analysis becomes more complicated. An equivalent concept proposed by Epstein [210] is

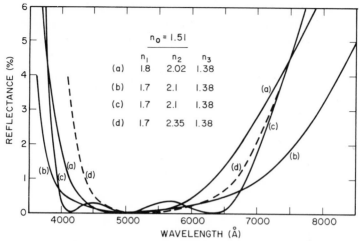

FIG. 22. Calculated reflectance as a function of wavelength of three types of triple-layer coatings (indices given inside the figure) on glass ($n_0 = 1.51$): (a) quarter-quarter-quarter; (b) quarter-half-quarter; (c) quarter-half-three quarters. The observed reflectance of a quarter-half quarter coating consisting of MgF_2 ($n_3 = 1.38$) + CeO_2 ($n_2 = 2.35$) + CeF_3 ($n_1 = 1.7$) is shown by the dotted curve (d). (*Cox and Hass* [13])

valuable for designing symmetrical multilayers. According to this concept, a symmetrical combination of thin films is "equivalent" at any given wavelength to a single film characterized by an equivalent index and thickness.

(4) Infrared Antireflection Coatings. The high refractive indices of semiconducting materials used in the infrared cause large reflection losses. Consequently antireflection coatings are essential in obtaining high transmittance and proper imaging with optical elements in the infrared. Single-layer evaporated $\lambda/4$ SiO coatings have been used for Si, Ge [211], and InAs [212] for wavelengths up to about 8μ, beyond which SiO becomes strongly absorbing. In the 8- to 15-μ region, ZnS films are most suitable [211]. Antireflection coatings for InSb consisting of $PbCl_2$ or As_2S_3 have been used [213]. Antireflection coatings of MgF_2, SiO_2, Al_2O_3, and didymium fluoride are suitable for infrared film material, Eastman Kodak Irtran-2.

Double-layer (quarter-quarter) coatings of (MgF_2+CeO_2) [213] and (MgF_2+Ge) [214] on Si, of (SiO+ZnS), and (MgF_2+Ge) on Ge, and of (MgF_2+SiO) [207] on Irtran-2 have also been employed for antireflection. The transmittance of the uncoated substrates is ~50 percent or less; that of the coated substrates is increased to above 90 percent in most cases. A triple quarter-layer [13] coating of (Sb$_2$S$_3$+SnO+ThF) on Ge increases the transmittance to above 90 percent over the 1.3- to 3.3-μ wavelength range.

5.2 Reflection Coatings

As seen in Sec. 2.2, the surface reflectance may be considerably enhanced by the deposition of a $\lambda/4$ optical thickness film having an index greater than that of the substrate (Fig. 2). Under this condition, the beams reflected from the air/film and film/substrate interfaces are in phase. The reflectance in this case is given by Eq. (13a). The higher the index of the film, the greater is its reflectance.

(1) Metal Mirrors. Highly reflecting surfaces are commonly made of metals; Ag, Al, and Rh evaporated-film mirrors have been used. The reflectance of a good evaporated-film mirror is always higher than that of a polished surface of the same material. Because of the extensive use of mirror coatings, the reflectance data [14] over a wide spectral range for several coatings is of interest and is listed in Table IV.

Table IV. Percent Reflectance at Normal Incidence of Freshly Evaporated Mirror Coatings of Aluminum, Silver, Gold, Copper, Rhodium, and Platinum from the Ultraviolet to the Infrared (after Drummeter and Hass [14])

λ, μ	Al	Ag	Au	Cu	Rh	Pt
0.220	91.5	28.0	27.5	40.4	57.8	40.5
0.240	91.9	29.5	31.6	39.0	63.2	46.9
0.260	92.2	29.2	35.6	35.5	67.7	51.5
0.280	92.3	25.2	37.8	33.0	70.7	54.9
0.300	92.3	17.6	37.7	33.6	73.4	57.6
0.315	92.4	5.5	37.3	35.5	75.0	59.4
0.320	92.4	8.9	37.1	36.3	75.5	60.0
0.340	92.5	72.9	36.1	38.5	76.9	62.0
0.360	92.5	88.2	36.3	41.5	78.0	63.4
0.380	92.5	92.8	37.8	44.5	78.1	64.9
0.400	92.4	94.8	38.7	47.5	77.4	66.3
0.450	92.2	96.6	38.7	55.2	76.0	69.1
0.500	91.8	97.7	47.7	60.0	76.6	71.4
0.550	91.5	97.9	81.7	66.9	78.2	73.4
0.600	91.1	98.1	91.9	93.3	79.7	75.2
0.650	90.3	98.3	95.5	96.6	81.1	76.4
0.700	89.9	98.5	97.0	97.5	82.0	77.2
0.750	88.0	98.6	97.4	97.9	82.6	77.9

Table IV. Percent Reflectance at Normal Incidence of Freshly Evaporated Mirror Coatings of Aluminum, Silver, Gold, Copper, Rhodium, and Platinum from the Ultraviolet to the Infrared (after Drummeter and Hass [14]) (*Continued*)

λ, μ	Al	Ag	Au	Cu	Rh	Pt
0.800	86.3	98.6	97.7	98.1	83.1	78.5
0.850	86.0	98.7	97.8	98.3	83.4	79.5
0.900	89.0	98.7	98.0	98.4	83.6	80.5
0.950	91.8	98.8	98.1	98.4	83.9	80.6
1.0	94.0	98.9	98.2	98.5	84.2	80.7
1.5	96.8	98.9	98.2	98.5	87.7	81.8
2.0	97.2	98.9	98.3	98.6	91.4	81.8
3.0	98.0	98.9	98.4	98.6	95.0	90.6
4.0	98.2	98.9	98.5	98.7	95.8	93.7
5.0	98.4	98.9	98.6	98.7	96.4	94.9
6.0	98.5	98.9	98.6	98.7	96.8	95.6
7.0	98.6	98.9	98.7	98.7	97.0	95.9
8.0	98.7	98.9	98.8	98.8	97.2	96.0
9.0	98.7	98.9	98.9	98.8	97.4	96.1
10.0	98.7	98.9	98.9	98.9	97.6	96.2
15.0	98.9	99.0	99.0	99.0	98.1	96.7
20.0	99.0	99.2	99.0	99.2	98.3	97.0
30.0	99.2		99.0			
40.0			99.1			

Mirror coatings may be protected and their reflectance enhanced over a part of the spectrum by the use of suitable thicknesses of transparent dielectric films. A significant increase in reflectance in the visible spectrum is obtained [215] for Al mirrors coated with $\lambda/4$ films of Al_2O_3, $Al_2O_3+TiO_2$, or MgF_2+CeO_2. Silicon monoxide films form the most suitable protective coating for Al mirrors. If SiO films are partially oxidized by evaporation in poor vacuum, the protected Al mirrors show high absorption in the ultraviolet. Several hours of exposure of these mirrors to ultraviolet radiation from a quartz mercury arc eliminates completely the undesired ultraviolet absorptance of partially oxidized SiO but has very little effect on the reflectance of Al mirrors coated with stoichiometric SiO. Ultraviolet irradiation appears to change the structure of SiO molecules, and also increase the oxygen content in the films [216].

(2) **All-dielectric System.** When metal mirrors are used for partial reflection and transmission, a considerable loss of light by absorption

takes place. Highly reflecting systems with very small absorption are useful in various optical devices. Such systems are made possible by the use of dielectric coatings. Considerable improvement in the calculated reflectance of glass coated with $\lambda/4$ layers of various dielectric materials is shown by the data in Table V.

Since the high reflectance of a $\lambda/4$ layer is due to the phase agreement of the beams reflected at the two surfaces, the same relation can be maintained in a stack of alternately high (H) and low (L) index $\lambda/4$ films. With such a stack of odd number of dielectric films, reflectances very close to 100 percent and with negligible light loss can be expected. The usual arrangement has high-index $\lambda/4$ layers next to the substrate in odd-number positions and is denoted as $HLHLHLHL$....Epstein [210] and Turner [217] analyzed this system. The calculated maximum reflectance at $\lambda/4$ for z number of layers in air is given by

$$R = \left(\frac{n_H^{z+1} - n_L^{z-1}n_0}{n_H^{z+1} - n_L^{z-1}n_0}\right)^2 \qquad (48)$$

where n_H, n_L, and n_0 are the high, low, and substrate indices, respectively. A comparison of Eq. (48) with Eq. (15a) shows that the k layers are equivalent to one layer of an effectively high index $= \sqrt{n_H^{z+1}/n_L^{z-1}}$.

The calculated reflectance of different layers of various HL stacks is given in Table VI. It is clear that, even for a fairly small number of HL pairs, the high reflectance approaches closely to that of an infinite stack. As the number of HL pairs increases, the transmittance decreases and the bandwidth increases.

Table V. Calculated Normal-incidence Reflectance of a Surface of Index 1.50 When Covered by Various $\lambda/4$ Films (Single) (from Heavens [1])

Film material	Refractive index	Wavelength, μ	Reflectance, %
ZnS	2.30	0.546	31
TiO$_2$	2.6	0.546	40
Sb$_2$S$_3$	2.7	1	43
Ge	4.0	2	69
Te	5.0	4	79
Uncoated	1.50	0.546	4

Table VI.   Calculated Normal-incidence Reflectance at Maximum of Quarter-wave Stacks (from Heavens [1])

System	No. of layers	CeO_2 + cryolite, $\lambda = 0.55\,\mu$	Sb_2S_3 + CaF_2, $\lambda = 1\mu$	Ge + cryolite, $\lambda = 2\mu$	ZnS + cryolite, $\lambda = 0.589\mu$
DH	1	0.35	0.447	0.69	
DHLH	3	0.73	0.835	0.96	0.695
DHLHLH	5	0.91	0.900	0.99	0.891
DHLH	7	0.97	0.977		0.964
DHLH	9		0.995		0.988
DHLH	11		0.999		0.988
		$n_D = 1.52$	$n_D = 1.45$	$n_D = 1.45$	$n_D = 1.50$

D = dielectric substrate; $H = \lambda/4$ high-index layer; $L = \lambda/4$ low-index layer.

Ring and Wilcock [218] obtained a 98.3 percent reflectance and 0.5 percent absorption for a seven-layer ZnS+cryolite stack at λ = 4,300 Å. The absorption loss is probably by scattering rather than by true absorption. A reflectance close to 100 percent may be obtained [219] by using 25 or more layers. A transmittance of 0.05 percent and a reflectance of up to 99.8 percent was obtained by Behrndt and Doughty [220] with 21 layers of ZnS+$ThOF_2$ (Fig. 23). They obtained a still better performance by using ZnSe+$ThOF_2$ layers.

By using suitable materials (see Table III), similar $\lambda/4$ *HL* stacks can be used in the ultraviolet and infrared. For instance, $PbCl_2$+MgF_2 [221] yields 94 percent reflectance down to λ = 3,000 Å; CsI with either cryolite or MgF_2 [222] has been used in the 250- to 280-mμ region with 2.5 to 4 percent absorption for 11 and 13 layers. Germanium and cryolite [223] are satisfactory beyond 1.7μ; Te+NaCl has been used by Greenler [224] to obtain 80 percent reflectance with negligible absorption in the neighborhood of 12 to 15μ.

The *HL* $\lambda/4$ stack yields maximum reflectance at one particular wavelength and its side bands. The bandwidth over which high reflectance is obtained may be broadened by the use of staggered layer thicknesses so that several consecutive maxima in the spectral reflectance are observed. Baumeister and Stone [225] discussed the design of a staggered-thickness 15 layers of ZnS and cryolite to cover the entire visible region.

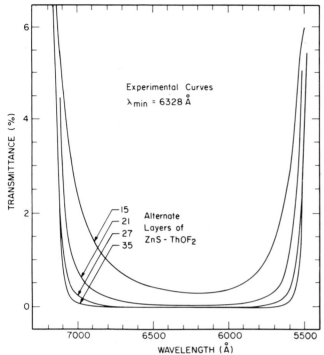

FIG. 23. Experimental curves of transmittance vs. wavelength for varying numbers of alternate $\lambda/4$-*HLHL* layers of ZnS (n = 2.32) and ThOF$_2$ (n = 1.5). The minimum transmission occurs at 6,328 Å. (*Behrndt and Doughty* [220])

The high reflectance by all-dielectric reflection systems is of interest in interferometry, laser technology, optical-delay line, storage, etc. A common application is in the fabrication of a cold mirror for a cinema projector to avoid overheating of the film. A semiconducting substrate (Ge, for example) which is transparent to infrared when used as a substrate for a $\lambda/4$ *HL* stack reflects in the visible spectrum. This construction forms a cold mirror since it reflects the nonheating visible light and transmits the "heating" infrared. An average transmittance of about 80 percent is obtained with a Ge mirror in the region 0.8 to 2.7μ.

5.3 Interference Filters

(1) Reflection Filters. The cold mirror mentioned above is an example of a selective-reflection filter. Selective reflection over a limited spectral region may be obtained by depositing interference combinations of dielectric and metal films on an opaque mirror, or a stack of all-dielectric $\lambda/4$ *HL* layers. If semiconducting films, which are transparent to infrared, are used, the mirror combination has negligible reflection in the visible and high reflection in the infrared. Such a system, called a "dark mirror," is of much interest in applications where it is necessary to obtain selective energy absorption for solar-powered thermal systems in the visible without increasing the operating temperature considerably. Hass et al. [226] used an Al-Ge-SiO coating to obtain 2 percent reflectance in the visible and 90 percent at $\lambda = 1.2\mu$ and above. At temperatures above 200°C, the dark mirror loses its dark appearance because of the diffusion of Al into the Ge film. A 10- to 20-Å-thick film of SiO placed between Al and Ge layers prevents this diffusion up to temperatures of more than 400°C without seriously affecting the reflection characteristics. The equivalent optical thickness of the Al-SiO combination is half wavelength.

Another dark-mirror coating of Al-SiO-Cr-SiO used by Hass et al. [226] (Fig. 24) has a normal solar absorptance of about 80 percent, a thermal emittance of about 5 percent, and reflection of over 90 percent at 5μ. The maximum temperature to which this coating can be heated before being destroyed because of crystallization is ~450°C. Coatings of the type Pt-SiO$_2$-Pt-SiO$_2$ can be used at temperatures as high as 600°C. Reasonable vacuum stability has been obtained for Mo-Al$_2$O$_3$-Mo dark mirrors at 530°C.

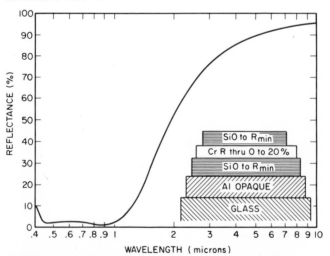

FIG. 24. The construction and the reflectance curve of an Al-SiO-Cr-SiO dark-mirror coating. Layer deposition controlled at $\lambda = 5,500$ Å. (*Drummeter and Hass* [14])

The $\lambda/4$ *HL* all-dielectric stack discussed in the preceding section has a broad central spectral-reflectance maximum with sharper secondary maxima. The bandwidth of the central maximum decreases with increasing number of units in the stack. This essentially acts as a reflection interference filter. The central-reflectance peak can also be sharpened by using a combination of several filters of different orders. If three filters with first-order peaks at wavelengths $\lambda_0, \lambda_0/2$, and $\lambda_0/3$ were combined to form a system, very narrow reflectance peaks at $\lambda_0, \lambda_0/2$, and $\lambda_0/3$ with a reflectance band comparable in width with one of the maxima of the highest-order filter result. The side maxima are suppressed by the lower-order filters. Billings [227] used a three-filter combination using CaF_2 and Ag_2S coatings. The multiple $\lambda/4$ *HL*-dielectric stack filter enables higher peak reflectances to be obtained than with the dielectric-plus-metal system.

(2) Transmission Interference Filters. The antireflection coating (already described) is a sort of broadband-transmission interference filter. If we arrange two selectively reflecting stacks such that their central-reflectance maxima are suitably displaced, a filter with a high transmission (>95 percent) and fairly narrow bandwidth may be produced. These stacks may be metal-dielectric interference-filter type or all-dielectric $\lambda/4$ *HL* type. Dufour [228] used a five-layer system of the type $Ag-MgF_2-Ag-MgF_2-Ag$. The all-dielectric stacks improve the performance with negligible absorption losses. For stacks of four layers each, a filter with a steep-sided band 250 Å wide, centered at $\lambda = 560$ mμ and with a peak transmission of 97 percent may be obtained. On either side of the band the transmission remains below about 4 percent in the $\lambda = 470$ to 680 mμ range.

The Fabry-Perot interference filter consists of two, partly reflecting, Ag films separated by a dielectric spacer layer. Transmission maxima occur for $\lambda_0 = 2nt$, where the product nt is the effective optical thickness of the spacer. The maximum transmittance of the filter is limited by the absorption of silver layers. The absorption effects may be reduced by combining the silver layers with suitable multilayer dielectric stacks. The high reflecting $\lambda/4$ *HL* all-dielectric stacks could be used to improve transmittance and bandwidth considerably.

The spectral dependence of the transmittance of a 15-layer filter of the type (*HL HL HL H 2L HL HL HL H*) observed by Polster [229] is shown in Fig. 25 for $H = \lambda_0/4$ ZnS, $L = \lambda_0/4$ cryolite, and $\lambda_0 = 550$ mμ. A peak transmission of 80 percent, and a pass-band half-width of 65 Å as

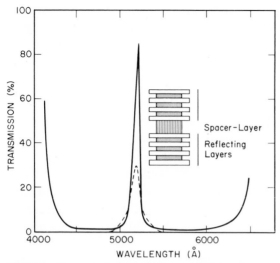

FIG. 25. Transmission of a 15-layer all-dielectric Fabry-Perot filter with a first-order pass band. Broken curve is for a second-order filter of comparable half-width but with silver reflecting layers. (*Polster* [229])

compared with the theoretical value of 83 Å was obtained. Ring and Wilcock [218] also studied a similar filter at $\lambda = 460$ mμ and obtained a half-width of 22 Å with a peak transmission of 70 percent. The 30 percent light loss is apparently due to scattering. As the number of layers is increased, the half-width of the filter should decrease. For the filter at $\lambda_0 = 550$ mμ, the calculations give a half-width of 0.9 Å for 31 layers. Whether such small half-widths can be achieved in practice depends entirely on our deposition technology and knowledge of scattering at the interfaces between layers. To obtain sufficient transmission in a filter of 20 Å half-width centered at $\lambda_0 = 500$ mμ, the thickness of a homogeneous film must be known to an accuracy better than 0.4 percent, a difficult though not impossible experimental challenge.

A very narrow pass band can be obtained by using systems of the type *HLLHLHLLH* (due to Smith [230]), and *HHLHHLHH* (due to Turner [231] and Smith [230]. The first is called the double half-wave system and the second the treble half-wave, denoted by Turner as *WADI* (wideband all-dielectric interference filter). The double half-wave system shows a fairly steep rise in transmittance at twice the pass-band wavelength and continues thereafter at long wavelengths.

(3) Frustrated Total-reflection Filter. A high reflectance with no absorption and a minimum of scattering may be obtained by using total reflection from a surface covered by a low-index film called the "frustrating layer." Turner [232] devised such a system with a spacer layer sandwiched between two surfaces coated with frustrating layers. The transmission of this filter is adjusted by controlling the thickness of the frustrating layer. A frustrating layer of MgF_2 is suitable on a $60°$ prism of dense flint glass ($n = 1.72$), the critical angle for the combination being about $53°$.

Because of a phase change on reflection at the spacer-frustrated layer interface depending on the state of polarization of the incident light, this filter exhibits two transmission bands when used with unpolarized light. Billings [233] used a birefringent spacer layer obtained by evaporation of some organic materials such as uric acid to bring the two transmission bands together. Birefringent frustrating layers could also be used for the same purpose if suitable materials were available.

5.4 Absorptance and Thermal Emittance of Coatings

The fact that the reflectance (and hence absorptance) characteristic of a surface can be drastically modified by suitable coatings has important technical applications in the field of space technology. The absorptance ($A = 1-R$) and thermal emittance e (the ratio of the power emitted by a surface area to the power emitted by a blackbody with the same physical conditions) of the surface of an object in space ultimately determine the equilibrium temperature by balancing the power absorbed from sunlight (or other sources) to the power radiated from the surface. The higher the ratio A/e, the higher the surface temperature will be. In other applications such as a solar cell, an efficient absorption of the useful spectral range is required at a low operating temperature. That is, the solar-cell surface should have high specific spectral absorptance and high thermal emittance. If, on the other hand, solar energy is first to be transformed into heat, then a high solar absorptance and a low thermal emittance are required for the development of high equilibrium temperature.

It is clear from our discussion in the preceding sections that, with the help of surface coatings, desired absorptance characteristics can be obtained. Some of the examples are the cold mirror and the dark mirror. Because of strong absorption bands of SiO_2 in the infrared, the thermal emittance of $Al+SiO_2$ mirrors can be adjusted to any value between 2 percent and more than 70 percent by controlling the thickness of the SiO_2 film. Similarly, Al_2O_3 coatings on Al mirrors may be used to increase thermal emittance without changing the solar absorptance. The experimental and theoretical analysis of solar absorptance and thermal emittance is thoroughly reviewed by Drummeter and Hass [14].

5.5   Thin-film Polarizers

The high degree of polarization produced on reflection at the Brewster angle has been exploited to obtain polarizers. Efficient, large-aperture polarizers may be obtained in a simple fashion by the use of a few thin-film elements. Thin-film elements of suitable effective optical thickness also allow the possibility of suppression of one of the polarized components while enhancing the reflectance of the other component. This principle of a polarizer is of particular interest for ultraviolet and infrared regions and awaits further explotiation.

Abelès [103] studied the polarizing properties of TiO_2 films deposited on glass. For a glass plate with a TiO_2 film on each face, a 0.92 polarization value was obtained. A polarization of 0.997 was obtained for two such plates placed parallel to one another. The superiority of this polarizer over the conventional pile of plates is clear from the fact that 12 such plates would be necessary to obtain a polarization of 0.96.

Lostis [234] examined the use of transparent films deposited on a totally reflecting prism face for the purpose of quarter- and half-wave plates. A single reflection at the hypotenuse of a $45°$ prism ($n = 1.52$) coated with a $\lambda/3$ layer of CeO_2 ($n = 2.1$) produces a quarter-wave plate. A film of $0.92\ \lambda$ thickness of TiO_2 ($n = 2.6$) produces a half-wave plate. The phase variation over the visible spectrum is $\pm 5°$ in the latter case.

The foregoing description suggests that multilayer films deposited on a prism face may be used to produce highly polarized reflecting or transmitting beams. This arrangment, called a "polarizing beam splitter," does not suffer from the absorption losses that accompany a metal-film partial reflector. Banning [113] used two $60°$ prisms, with ZnS-cryolite-ZnS coatings, cemented together to obtain an average polarization of about 0.98 for white light incident over a range of $\pm 5°$ from the normal.

6.   CONCLUDING REMARKS

Thin films are ideally suited for optical reflection and absorption studies. A number of examples have been presented in this chapter to show the usefulness of thin films in studying a variety of optical phenomena and in deriving valuable information on the electronic structure of solids. Most of these studies have not considered the microstructure of a film, which has considerable influence on the results obtained. It is hoped that future investigations will be carried out on films of well-defined structures, and possibly on single-crystal epitaxial films.

By codepositing a number of substances, a large variety of homogeneous and inhomogeneous thin-film optical materials may be produced which are hardly possible to obtain in bulk form. These normal and abnormal materials await extensive exploitation both for optical coatings and for fundamental investigations of the electronic structure of solids.

REFERENCES

1. O. S. Heavens, "Optical Properties of Thin Solid Films," Dover Publications, Inc., New York, 1965.
2. A. Vašíček, "Optics of Thin Films," North Holland Publishing Company, Amsterdam, 1960.
3. O. S. Heavens, *Rept. Progr. Phys.,* 23:1 (1960) (review).
4. O. S. Heavens, in "Physics of Thin Films" (G. Hass and R. E. Thun, eds.), vol. 2, p. 193, Academic Press Inc., New York, 1964.
5. H. Wolter, "Optik dünner Schichten" (in German). in "Handbuch der Physik" (S. Flügge, ed.), vol. 24, p. 461, Springer-Verlag OHG, Berlin, 1956.
6. H. Mayer, "Physik dünner Schichten," vol. 1, Wissenschaftliche Verlag, Stuttgart, 1950.
7. G. V. Rozenberg, "Optics of Thin Film" (in Russian), Moscow, 1958.
8. P. Rouard and P. Bousquet, in "Progress in Optics" (E. Wolf, ed.), vol. 4, North Holland Publishing Company, Amsterdam, 1965.
9. P. Rouard, Propriétés Optiques (and Applications) des Lames Minces Solides, *Mem. Sci. Phys. Paris,* 55:1952.
10. H. Schopper, *Fortschr. Physik,* 2:275 (1954).
11. P. H. Berning, in "Physics of Thin Films" (G. Hass, ed.), p. 69, Academic Press Inc., New York, 1963.
12. R. P. Madden, in "Physics of Thin Films" (G. Hass, ed.), p. 123, Academic Press Inc., New York, 1963.
13. J. T. Cox and G. Hass, in "Physics of Thin Films" (G. Hass and R. E. Thun, eds.), p. 239, Academic Press Inc., New York, 1964.
14. L. F. Drummeter, Jr., and G. Hass, in "Physics of Thin Films" (G. Hass and R. E. Thun, eds.), p. 305, Academic Press Inc., New York, 1964.
15. G. B. Airy, *Trans. Cambridge Phil. Soc.,* 4:409 (1832).
16. For example, J. A. Stratton, "Electromagnetic Theory," McGraw-Hill Book Company, New York, 1941.
17. P. Drude, *Ann. Phys. Leipzig,* 34:489 (1888).

18. A. Rothen, *Rev. Sci. Instr.*, **16**:26 (1945).
19. "Ellipsometry in the Measurement of Surfaces and Thin Films," 1963 Symposium Proceedings (E. Passaglia, R. R. Stromberg, and J. Kruger, eds.), National Bureau of Standards, Washington, D.C., 1964.
20. See Ref. 11; also, F. Abelès, Ref. 19, p. 41.
21. For example, see M. Garbuny, "Optical Physics," Academic Press Inc., New York, 1965.
22. S. P. F. Humphreys-Owen, *Proc. Phys. Soc. London*, **77**:949 (1961).
23. I. Simon, *J. Opt. Soc. Am.*, **41**:336 (1951).
24. D. G. Avery, *Proc. Phys. Soc. London*, **B65**:425 (1952).
25. S. Robin, *J. Phys. Radium*, **13**:493 (1952); *Compt. Rend.*, **236**:674 (1953).
26. F. Abelès, *J. Phys. Radium*, **19**:327 (1958).
27. T. S. Robinson, *Proc. Phys. Soc. London*, **B65**:910 (1952).
28. O. S. Heavens and J. C. Kelly, *Proc. Phys. Soc. London*, **72**:906 (1958); Opt. *Acta*, **6**:339 (1959).
29. L. G. Schulz and F. R. Tangherlini, *J. Opt. Soc. Am.*, **44**:362 (1954); L. G. Shulz, *J. Opt. Soc. Am.*, **44**:357 (1954).
30. H. Murmann, *Z. Physik,* **80**:161, 1933; **101**:643 (1936).
31. D. Malé, *J. Phys. Radium*, **11**:332 (1950).
32. H. Wolter, *Z. Physik,* **105**:269 (1937).
33. E. David, *Z. Physik,* **106**:606 (1937).
34. P. D. Fochs, *J. Opt. Soc. Am.*, **40**:623 (1950).
35. W. A. Pliskin and E. E. Conrad, *IBM J. Res. Develop.*, **8**:43 (1964).
36. P. Jacquinot, *Rev. Opt.*, **21**:15 (1942).
37. P. Giacomo and P. Jacquinot, *J. Phys. Radium*, **13**:59A (1952).
38. F. Abelès, *Ann. Phys. Paris*, **5**:750 (1950).
39. L. N. Hadley, cited by Heavens, Ref. 4.
40. W. R. Hunter, *J. Opt. Soc. Am.*, **54**:15 (1964).
41. A. Vašíček, *J. Opt. Soc. Am.*, **37**:145 (1947); **37**:979 (1947).
42. R. J. Archer, *J. Opt. Soc. Am.*, **52**:970 (1962).
43. R. L. McCrackin and J. P. Colson, Ref. 19, p. 61.
44. H. Schopper, *Z. Physik,* **131**:215 (1952).
45. K. Försterling, *Ann. Physik,* **30**:745 (1937).
46. L. Tronstad and C. G. P. Feachem, *Proc. Roy. Soc. London*, **145**:115 (1934).
47. H. Hauschild, *Ann. Physik,* **63**:816 (1920).
48. C. E. Leberknight and B. Lustman, *J. Opt. Soc. Am.*, **29**:59 (1939).
49. A. B. Winterbottom, *Trans. Faraday Soc.*, **42**:487 (1946).
50. A. B. Winterbottom, Ref. 19, p. 97.
51. S. Roberts, Ref. 19, p. 119.
52. M. A. Barrett, Ref. 19, p. 213.
53. J. V. Cathcart and G. F. Petersen, Ref. 19, p. 201.
54. A. K. N. Reddy and J. O'M. Bockris, Ref. 19, p. 229.
55. H. Mendlowitz, *Proc. Phys. Soc. London*, **75**:664 (1960); *J. Opt. Soc. Am.*, **50**:739 (1960).
56. R. E. Lavilla and H. Mendlowitz, *J. Phys. Radium*, **25**:114 (1964).
57. I. H. Khan and K. L. Chopra, Electrochemical Society Meeting, Washington, D. C. (1964).
58. J. N. Zemel, J. D. Jensen, and R. B. Schoolar, *Phys. Rev.*, **140**:A330 (1965).
59. K. Weiss, *Z. Naturforsch.*, **3a**:143 (1948).

60. V. G. Padalka and I. N. Shklyarevskii, *Opt. i Spektroskopiya,* **11**:527 (1961).
61. I. N. Shklyarevskii and V. G. Padalka, *Opt. i Spektroskopiya,* **6**:78 (1959).
62. R. Philip and J. Trompette, *Compt. Rend.,* **241**:627 (1955).
63. R. Philip, *Compt. Rend.,* **247**:1104, 2322 (1958); **248**:3418 (1959); **249**:1343 (1959).
64. R. Philip, *J. Phys. Radium,* **24**:120 (1963).
65. J. Richard, *Compt. Rend.,* **255**:662, 670 (1962).
66. S. Robin-Kandaré, Thesis, Paris, 1959.
67. J. E. Davey and T. Pankey, *J. Appl. Phys.,* **36**:2571 (1965).
68. G. Joos and A. Klopfer, *Z. Physik,* **138**:251 (1954).
69. J. N. Hodgson, *Proc. Phys. Soc. London,* **B68**:593 (1955).
70. K. Ishiguro and G. Kuwabara, *J. Phys. Soc. Japan,* **6**:71 (1951).
71. J. R. Beattie, *Physica,* **23**:898 (1957).
72. G. Rasigni and P. Rouard, *J. Phys. Radium,* **23**:211 (1962).
73. G. Rasigni, *Rev. Opt.,* **41**:373, 566, 625 (1962).
74. G. Rasigni et al., *Compt. Rend.,* **254**:2325 (1962).
75. J. R. Beattie, *Phil. Mag.,* **46**:22 (1955).
76. R. Gerharz, *Z. Angew. Phys.,* **9**:95 (1958).
77. G. Hass, W. R. Hunter, and R. Tousey, *J. Opt. Soc. Am.,* **46**:1009 (1956); **47**:1070 (1957).
78. G. Hass and J. E. Waylonis, *J. Opt. Soc. Am.,* **51**:719 (1961).
79. H. M. O'Bryan, *J. Opt. Soc. Am.,* **26**:122 (1936).
80. D. Malé and J. Trompette, *J. Phys. Radium,* **18**:128 (1957).
81. J. Trompette, *Compt. Rend.,* **248**:207 (1959).
82. G. P. Motulevich and A. A. Shubin, *Opt. i Spektroskopiya,* **2**:633 (1957).
83. F. Bueche, *J. Opt. Soc. Am.,* **38**:806 (1948).
84. R. Burtin, *Compt. Rend.,* **250**:1998 (1960).
85. R. Burtin and F. Dorel, *J. Phys. Radium,* **22**:33S (1961).
86. P. L. Clegg, *Proc. Phys. Soc. London,* **B64**:774 (1952).
87. D. Fabre and J. Romand, *Compt. Rend.,* **242**:893 (1956).
88. C. Mouttet and P. Gravier, *Rev. Opt.,* **43**:245 (1964).
89. H. E. Ives and H. B. Briggs, *J. Opt. Soc. Am.,* **26**:122 (1936); **27**:181, 395 (1937).
90. L. Reimer, *Optik,* **14**:83 (1957); H. Koenig and G. Helwig, *Optik,* **6**:111 (1950); L. Holland, *J. Opt. Soc. Am.,* **43**:376 (1953).
91. R. Payan and G. Rasigni, *J. Phys. Radium,* **25**:92 (1964).
92. M. Belzons, *J. Phys.,* **26**:259 (1965).
93. C. Wesolowska, *Acta Phys. Polon.,* **25**:323 (1964).
94. A. J. Ejczis and G. P. Skorniakow, *Opt. i. Spektroskopiya,* **16**:85, 159 (1964).
95. R. H. Doremus, *J. Appl. Phys.,* **37**:2775 (1966).
96. P. Rouard, *Appl. Opt.,* **4**:947 (1965).
97. P. Rouard, in "Basic Problems in Thin Film Physics" (R. Niedermayer and H. Mayer, eds.), p. 263, Vandenhoeck and Ruprecht, Göttingen, 1966.
98. J. C. Maxwell-Garnett, *Phil. Trans. Roy. Soc. London,* **A203**:385 (1904).
99. D. Malé, *Compt. Rend.,* **235**:1630 (1952); *Ann. Phys. Paris,* **9**:10 (1954).
100. E. David, *Z. Physik,* **114**:389 (1939).
101. G. Hass, J. B. Ramsey, and R. Thun, *J. Opt. Soc. Am.,* **48**:324 (1958).
102. P. Bousquet, *Opt. Acta,* **3**:153 (1956); *Ann. Phys. Paris,* **2**:5 (1957).
103. F. Abelès, *J. Phys. Radium,* **11**:310, 403 (1950).

104. L. G. Schulz and E. J. Scheibner, *J. Opt. Soc. Am.*, **40**:761 (1950).
105. O. S. Heavens and S. D. Smith, *J. Opt. Soc. Am.*, **47**:469 (1957).
106. D. Fabre, J. Romand, and B. Vodar, *J. Phys. Radium*, **25**:55 (1964).
107. J. F. Hall and W. F. C. Ferguson, *J. Opt. Soc. Am.*, **45**:74, 714 (1955); J. F. Hall, *J. Opt. Soc. Am.*, **46**:1013 (1956).
108. N. Morita, *J. Phys. Soc. Japan*, **11**:975 (1956); **12**:1142 (1957).
109. L. G. Schulz, *J. Chem. Phys.*, **17**:1153 (1949).
110. G. Hass and C. D. Salzburg, *J. Opt. Soc. Am.*, **44**:181 (1954).
111. W. A. Pliskin and R. P. Esch, *J. Appl. Phys.*, **36**:2011 (1965).
112. A. Revesz and K. H. Zaininger, *J. Phys. Radium*, **25**:66 (1964).
113. M. Banning, *J. Opt. Soc. Am.*, **37**:688, 792 (1947).
114. G. Hass, *Vacuum*, **2**:331 (1952).
115. W. Weisskirchner, *Z. Naturforsch.*, **6a**:509 (1951).
116. L. Harris, *J. Opt. Soc. Am.*, **45**:27 (1955).
117. J. T. Cox, G. Hass, and J. B. Ramsey, *J. Phys. Radium*, **25**:250 (1964).
118. L. Young, *Proc. Roy. Soc. London*, **A244**:41 (1958).
119. T. Thach Lan, F. Naudin, and P. Probbe-Bourget, *J. Phys. Radium*, **25**:11 (1964).
120. L. Holland and G. Siddall, *Vacuum*, **3**:375 (1953).
121. K. V. Shalimova, *Dokl. Akad. Nauk SSSR*, **80**:587 (1951).
122. C. K. Coogan, *Proc. Phys. Soc. London*, **B70**:845 (1957).
123. L. Huldt and T. Staflin, *Opt. Acta*, **6**:27 (1959).
124. J. Gonella, in "Basic Problems in Thin Film Physics" (R. Niedermayer and H. Mayer, eds.), p. 280, Vandenhoeck and Ruprecht, Göttingen, 1966.
125. G. Kuwabara and K. Ishiguro, *J. Phys. Soc. Japan*, **7**:72 (1952).
126. H. Vonarburg, *Optik*, **20**:43 (1963).
127. S. Fujiwara, *J. Opt. Soc. Am.*, **53**:1317 (1963).
128. S. Fujiwara, *J. Opt. Soc. Am.*, **53**:880 (1963).
129. F. Lukes, *Czech. J. Phys.*, **10**:59 (1960).
130. S. Martinuzzi, M. Perrot, and J. Fourny, *J. Phys. Radium*, **25**:203 (1964).
131. E. Hahn and L. Sagi, *J. Phys. Radium*, **25**:14 (1964).
132. R. P. Howson, *J. Phys. Radium*, **25**:212 (1964).
133. H. Merdy, S. Robin-Kandare, and J. Robin, *Compt. Rend.*, **257**:1526 (1963); *J. Phys. Radium*, **25**:223 (1964).
134. J. Tauc, R. Grigorovici, and A. Vancu, *Phys. Stat. Solidi*, **15**:627 (1966).
135. F. R. Kessler and E. Sutter, *Z. Physik*, **173**:54 (1963).
136. K. L. Chopra and S. K. Bahl, Abstracts 13th Natl. Vacuum Symp., 1967, p. 33.
137. V. B. Sandomirskii, *Sov. Phys. JETP English Transl.*, **10**:1630 (1963); B. A. Tavger and V. Ya. Demikhovskii, *Soviet Phys.-Solid State English Transl.*, **5**:469 (1963).
138. V. N. Lutskii, *Soviet Phys. JETP Letters English Transl.*, **2**:245 (1965).
139. H. H. Soonpaa, in "Basic Problems in Thin Film Physics" (R. Niedermayer and H. Mayer, eds.), p. 289, Vandenhoeck and Ruprecht, Göttingen, 1966.
140. R. C. Plumb, *J. Phys. Radium*, **25**:69 (1964).
141. W. Flechsig, *Z. Physik*, **176**:380 (1963); **162**:570 (1961); **170**:176 (1962).
142. Yi-Han Kao, *Phys. Letters*, **18**:16 (1965).
143. G. E. H. Reuter and E. H. Sondheimer, *Proc. Roy. Soc. London*, **A195**:336 (1948).
144. R. B. Dingle, *Physica*, **19**:311, 348, 729 (1953).
145. T. Holstein, *Phys. Rev.*, **88**:1427 (1952).
146. L. G. Schulz, *J. Opt. Soc. Am.*, **44**:540 (1950).

147. H. E. Bennett, J. M. Bennett, E. J. Ashley, and R. J. Motyka, *Phys. Rev.,* **165:** 755 (1968); H. E. Bennett and J. M. Bennett, in "Optical Properties and Electronic Structure of Metals and Alloys" (F. Abelès, ed.), p. 175, North Holland Publishing Company, Amsterdam, 1966.
148. R. Lyddane, R. Sachs, and E. Teller, *Phys. Rev.,* **59:**673 (1941).
149. W. G. Spitzer, D. A. Kleinman, and C. J. Frosch, *Phys. Rev.,* **113:**133 (1959).
150. J. N. Zemel, *Proc. 7th Intern. Conf. Semicond. Phys.,* 1964, p. 1061.
151. M. Balkansky, in "The Use of Thin Films in Physical Investigations" (J. C. Anderson, ed.), p. 347, Academic Press Inc., New York, 1966.
152. D. W. Berreman, *Phys. Rev.,* **130:**2193 (1963).
153. Lord Rayleigh, *Proc. London Math. Soc.,* **17:**4 (1887).
154. D. C. Gazis, R. Herman, and R. F. Wallis, *Phys. Rev.,* **119:**533 (1960).
155. R. F. Wallis, *Phys. Rev.,* **116:**302 (1959).
156. R. F. Wallis, *Surface Sci.,* **2:**146 (1964).
157. H. Kaplan, *Phys. Rev.,* **125:**1271 (1962).
158. J. Hori and T. Asahi, *Progr. Theoret. Phys. Kyoto,* **31:**49 (1964).
159. W. A. Pliskin and R. P. Eischens, *Z. Phys. Chem.,* **24:**11 (1960).
160. E. D. Palik, D. L. Mitchell and J. N. Zemel, *Phys. Rev.,* **135:**A763 (1964).
161. J. O. Dimmock and G. B. Wright, *Phys. Rev.,* **135:**A821 (1964).
162. D. L. Mitchell, in "The Use of Thin Films Physical Investigations" (J. C. Anderson, ed.), p. 363, Academic Press Inc., New York, 1966.
163. D. Pines, "Elementary Excitation in Solids," Benjamin Inc., New York, 1963.
164. W. Steinmann, in "Basic Problems in Thin Film Physics" (R. Niedermayer and H. Mayer, eds.), p. 363, Vandenhoeck and Ruprecht, Göttingen, 1966.
165. R. A. Ferrell, *Phys. Rev.,* **111:**1214 (1958).
166. V. L. Ginzburg and I. M. Franck, *Zh. Eksperim. i Teor. Fiz.,* **16:**15 (1946).
167. E. A. Stern, *Phys. Rev. Letters,* **8:**7 (1962).
168. H. Boersch, C. Radeloff, and G. Sauerbrey, *Z. Physik,* **165:**464 (1961).
169. P. von Blanckenhagen, H. Boersch, D. Fritsche, H. G. Seifert, and G. Sauerbrey, *Phys. Letters,* **11:**296 (1964).
170. E. T. Arakawa, R. J. Herickhoff, and R. D. Birkhoff, *Phys. Rev. Letters,* **12:**319 (1964).
171. R. J. Herickhoff, E. T. Arakawa, and R. D. Birkhoff, *Phys. Rev.,* **137:**A1433 (1965).
172. J. Brambridge and H. Raether, *Z. Physik,* **199:**118 (1967).
173. J. Bösenberg and H. Raether, *Phys. Rev. Letters,* **18:**397 (1967).
174. R. A. Ferrell and E. A. Stern, *Am. J. Phys.,* **30:**810 (1962).
175. N. Matsudaira, *J. Phys. Soc. Japan,* **18:**380 (1963).
176. M. Hattori, K. Yamada, and H. Suzuki, *J. Phys. Soc. Japan,* **18:**203 (1963).
177. S. Yamaguchi, *J. Phys. Soc. Japan,* **18:**266 (1963).
178. A. J. McAlister and E. A. Stern, *Phys. Rev.,* **132:**1599 (1963).
179. H. Deichsel, *Z. Physik,* **174:**136 (1963).
180. A. Ejiri and T. Sasaki, *J. Phys. Soc. Japan,* **20:**876 (1965).
181. W. R. Hunter, *J. Phys. Radium,* **25:**154 (1964).
182. W. C. Walker, O. P. Rustgi, and G. L. Weissler, *J. Opt. Soc. Am.,* **49:**471 (1959).
183. O. P. Rustgi, *J. Opt. Soc. Am.,* **55:**630 (1965).
184. W. R. Hunter, D. W. Angle, and R. Tousey, *Appl. Opt.,* **4:**891 (1965).
185. W. R. Hunter, in "Optical Properties and Electronic Structure of Metals and Alloys" (F. Abelès, ed.), p. 136, North Holland Publishing Company, Amsterdam, 1966.

186. H. Thomas, *Z. Physik,* **147**:395 (1957).
187. H. Mayer and W. Wiegrefe, in *Symp. Elec. Magnetic Properties Thin Metallic Layers, Louvain,* 1961, p. 9.
188. F. J. Piepenbring, Ref. 185, p. 325.
189. R. N. Stuart, F. Wooten, and W. E. Spicer, *Phys. Rev.,* **135**:A495 (1964).
190. J. J. Quinn, *Phys. Rev.,* **126**:1453 (1962).
191. R. Garron, *Compt. Rend.,* **254**:4278 (1962); **255**:1107 (1962); *Ann. Phys. Paris, 10,* **(13)**:(1965); Ref. 97, p. 332.
192. R. Suhrmann and G. Wedler, *Z. Angew. Phys.,* **2**:70 (1962).
193. S. M. Bryla and C. Feldman, *J. Appl. Phys.,* **33**:774 (1962).
194. E. N. Clarke and H. E. Farnsworth, *Phys. Rev.,* **85**:484 (1962).
195. W. E. Spicer, in "Optical Properties and Electronic Structure of Metals and Alloys" (F. Abelès, ed.), p. 296, North Holland Publishing Company, Amsterdam, 1966.
196. F. Wooten, T. Huen, and R. N. Stuart, as in Ref. 195, p. 333.
197. Lord Rayleigh, *Proc. Roy. Soc. London,* **A41**:275 (1887).
198. G. Bauer, *Ann. Physik,* **19**:434 (1934).
199. G. Hass and E. Ritter, *J. Vacuum Sci. Tech.,* **4**:71 (1967).
200. J. A. Berning and P. H. Berning, *J. Opt. Soc. Am.,* **50**:813 (1960).
201. K. P. Miyake, *J. Phys. Radium,* **25**:255 (1964).
202. G. Tricoles, *J. Phys. Radium,* **25**:262 (1964).
203. R. Jacobsson, *J. Phys. Radium,* **25**:46 (1964).
204. R. Jacobsson and M. O. Martensson, *Appl. Opt.,* **5**:29 (1966).
205. H. Osterberg and L. W. Smith, *J. Opt. Soc. Am.,* **50**:494 (1960).
206. P. Leurgans, *J. Opt. Soc. Am.,* **39**:639 (1949).
207. J. T. Cox, G. Hass, and R. F. Rowntree, *Vacuum,* **4**:445 (1954).
208. L. B. Lockhart and P. King, *J. Opt. Soc. Am.,* **37**:689 (1947).
209. J. T. Cox, G. Hass, and A. Thelen, *J. Opt. Soc. Am.,* **52**:965 (1962).
210. L. I. Epstein, *J. Opt. Soc. Am.,* **42**:806 (1952); **45**:360 (1955).
211. J. T. Cox and G. Hass, *J. Opt. Soc. Am.,* **48**:677 (1958).
212. J. T. Cox, G. Hass, and G. F. Jacobus, *J. Opt. Soc. Am.,* **51**:714 (1961).
213. S. D. Smith and T. S. Moss, *J. Sci. Instr.,* **35**:105 (1958).
214. J. T. Cox, *J. Opt. Soc. Am.,* **51**:1406 (1961).
215. G. Hass, *J. Opt. Soc. Am.,* **45**:945 (1955).
216. A. P. Bradford and G. Hass, *J. Opt. Soc. Am.,* **53**:1096 (1963).
217. A. F. Turner, *Bausch and Lomb Tech. Rept. (Multilayer Films),* No. 5, 1951.
218. J. Ring and W. L. Wilcock, *Nature,* **171**:648 (1953).
219. M. Laikin, *Appl. Opt.,* **4**:1032 (1965).
220. K. H. Behrndt and D. W. Doughty, *J. Vacuum Sci. Tech.,* **3**:264 (1966).
221. S. Penselin and A. Steudel, *Z. Physik,* **142**:21 (1955).
222. A. Steudel and S. Stotz, *Z. Physik,* **151**:233 (1958).
223. O. S. Heavens, J. Ring, and S. D. Smith, *Spectrochim. Acta,* **10**:179 (1957).
224. R. G. Greenler, *J. Opt. Soc. Am.,* **45**:788 (1955); **47**:130 (1955).
225. P. W. Baumeister and J. M. Stone, *J. Opt. Soc. Am.,* **46**:228 (1956).
226. G. Hass, H. H. Schroeder, and A. F. Turner, *J. Opt. Soc. Am.,* **46**:31 (1956).
227. B. H. Billings, *J. Phys. Radium,* **11**:407 (1950).
228. C. Dufour, *Vide,* **16-17**:480 (1948).
229. H. D. Polster, *J. Opt. Soc. Am.,* **39**:1038 (1949).
230. S. D. Smith, *J. Opt. Soc. Am.,* **48**:43 (1958).

231. A. F. Turner et al., Infrared Transmission Filters, *Bausch and Lomb Tech. Rept.,* Nos. 1-6, 1953.
232. A. F. Turner, *J. Phys. Radium,* **11**:444 (1950).
233. B. H. Billings, *J. Opt. Soc. Am.,* **40**:471 (1950).
234. P. Lostis, *Compt. Rend.,* **240**:2130 (1955); *Rev. Opt.,* **38**:1 (1959).
235. R. W. Fane, W. E. J. Neal, and R. M. Rollanson, *Appl. Phys. Letters,* **12:265** (1968).
236. S. K. Bahl and K. L. Chopra, *J. Appl Phys.* (1969).
237. R. Tsu, W. E. Howard, and L. Esaki, *Phys. Rev.,* **172**:779 (1968).
238. A. G. Stasenko, *Soviet Phys. Solid State* (English translation), **10**:186 (1968).

AUTHOR INDEX

Reference numbers are in italic type followed by page numbers (in roman) on which the references appear in the text. A colon separates reference numbers and page numbers, and a semicolon indicates the separation of chapters. The names of the authors are given in the References compiled at the end of each chapter. Each chapter has an independent list of references. Only some author names are mentioned in the text.

Mikhailov, Y. G., *43*:539, 601
Mikkola, D. E., *179*:121, 135
Miksik, M. G., *117*:459, 463
Milazzo, G., *131*:43, 79
Miles, J. L., *117*:42, 79; *96, 97*:550, 603, *96*:
   551, *247*:599, 607
Milgram, A. A., *17*:141, 254
Miller, A., *139*:44, 80
Miller, D. P., *304*:249, 263
Miller, K. J., *162*:49, 80
Miller, N. C., *191*:388, 390, 431
Miller, R. E., *122*:42, 79
Mills, J. C., *193*:125, 136
Milton, R. M., *242*:598, 607
Minamiya, A., *149*:47, 80; *285*:232, 236, 262
Minden, H. T., *65*:453, 462
Minn, S. S., *8*:330, 335, 337, 342, 426
Minnaja, N., *8*:468, 523
Minningerode, G. V., *82*:548, 602, *92*:550,
   630
Misener, A. D., *90*:550, 603
Miservey, R., *184, 185*:577, 578, 579, 605
Mitchell, D. F., *64*:152, 256
Mitchell, D. L., *37*:271, 274, 322, 324; *160,
   162*:764, 791
Mitchell, J. M., *116*:41, 79
Mitchell, J. W., *155*:188, 259, *291*:194, 239,
   246, 263
Mitra, S. S., *70*:98, 132
Miyake, K. P., *201*:770, 792
Miyashita, T., *291*:696, 718
Moak, C. D., *101*:40, 79
Moazed, K. L., *82*:154, 256, *334*:154, 155,
   157, 201, 264
Moest, R. R., *170*:50, 81
Mohiuddin, M., *186*:661, 715
Moll, A., *15*:470, 523
Moll, J. L., *195*:63, 82; *237*:411, 413, 414,
   415, 432, *238*:411, 412, 413, 432
Mönch, W., *208, 209*:198, 260; *148*:382, 430
Montgomery, M. D., *16*:441, 445, 448, 461
Montmory, R., *119*:633, 714, *203*:666, 716,
   295, 296:697, 718, *297*:697, 698,
   718, *298*:697, 698, 719
Moodie, A. F., *193*:125, 136
Moon, R., *182*:663, 715
Moore, A. J. W., *30*:147, 255
Moore, C. P., *304*:249, 263
Moore, W. J., *43*:23, 77
Morita, N., *37*:613, 711; *108*:750, 751, 790
Morris, C. E., *92, 93*:492, 493, 494, 495, 497,
   525, *122*:503, 504, 526
Morris, D. E., *190*:579, 605
Morris, E., *111*:375, 429

Morris, W. C., *151*:47, 80
Morrison, J. R., *105*:629, 713
Morrison, R. D., *267*:423, 433
Morritz, F. L., *139*:44, 80
Moser, J., *1*:328, 425
Moses, S., *130*:314, 326
Moss, M., *193*:197, 260
Moss, T. S., *60*:452, 462, *75*:454, 462; *213*:
   776, 792
Mostovetch, N., *18*:335, 426
Mott, N. F., *16*:335, 426, *161*:385, 388, 430,
   251:369, 414, 417, 432; *58*:449,
   462; *6*:468, 523
Motulevich, G. P., *82*:744, 789
Motyka, R. J., *109*:371, 375, 428; *147*:760,
   791
Moubis, J. H. A., *86*:37, 78
Moulton, C., *128*:43, 79
Mountuala, A., *268*:423, 433; *33*:445, 446,
   461
Mouttet, C., *88*:744, 789
Mueller, F. M., *249*:544, 607
Mullen, D. C., *44*:477, 524
Mullendore, A. W., *27*:18, 19, 76
Müller, E. K., *21*:18, 76; *326*:235, 264
Müller, E. W., *140, 141*:114, 134, *148*:115,
   134; *79*:485, 525
Müller, J., *43*:612, 614, 712
Mundy, J. L., *222*:591, 592, 606
Murayama, Y., *278*:695, 718
Murbach, H. P., *3*:267, 271, 275, 276, 283,
   288, 323
Murmann, H., *30*:735, 788
Murphy, C. S., *108*:41, 79
Murphy, E. L., *78*:485, 525
Murray, L. A., *96*:102, 133
Musatov, A. L., *172*:497, 518, 527
Mydosh, J. A., *165*:568, 570, 571, 605

Näbauer, M., *19*:535, 537, 601
Naber, C. T., *200*:64, 82; *56*:448, 462
Nagakura, S., *25*:18, 76
Nagashima, H., *130*:635, 714
Nagashima, T., *59*:620, 637, 712
Nagasima, K., *51*:275, 279, 324
Nahrwold, R., *1*:11, 76
Naik, Y. G., *81*:293, 325; *166*:387, 430
Nakagawa, K., *291*:696, 718
Nakai, J., *101*:455, 463
Nakaya, S., *51*:97, 131
Naudin, F., *119*:751, 790
Navinsek, B., *130*:379, 429
Neal, W. E. J., *235*:740, 793

SUBJECT INDEX

Abelés' method of measuring optical
constants, 733
Abnormal absorption of discontinuous films,
746
Maxwell-Garnett theory, 747
Absorptance, definition of, 785
Absorption of gases in films, 109, 378
effect on electrical conductivity, 378, 379
Absorption coefficient (*see* Optical constants)
Adatom, mobility of, 140
Adhesion of films, 313
measurement techniques of, 314
removal of film, 314
scratch test, 315
x-ray topography, 314
origin of: interface charge effects, 321,
322
due to van der Waals forces, 321
results of, 316
sputtered versus evaporated Au films,
319, 320
scratch-test critical load, table of, 318
Adion, 44
Adsorption:
and desorption processes, 138
on films: effect on condensation
process, 138, 151
effect on electrical conductivity, 379
study of, surface area, 186
Agglomeration, 165
Alloy superlattices, 223
Alloying of superimposed films (*see* Diffusion)
Amorphous films:
annealing of, 197
optical absorption of, 756
structure of, 198
superconductivity of, 539
Amorphous structures in films:
amorphous-crystalline transformation,
metastable structures, Ostwald's
rule, 199

Amorphous structures in films *(Cont'd)*
amorphous versus disordered structures,
195, 196
binary alloy films: electrical behavior of,
385
table of, 202-207
codeposit-quenched, 197
deposition-parameter controlled, 206-212
impurity stabilized, 196
optical absorption of amorphous Ge,
GeTe, 756
pure elements, 196
table of materials, 200, 201
vapor-quenched, 196
(*See also* Structure and microstructure of
films)
Angle of incidence effects (*see* Obliquely
deposited films)
Anisotropy field:
definition of, parallel and perpendicular
components, 615
determination of, 706
Annealing behavior of films:
changes in electrical conductivity, metal
films, 381-387
magnetic, 623
structural changes, 185
Annealing temperature, 383
Anodic oxidation (*see* Electrodeposition)
Anomalous skin effect, skin depth, and skin
resistivity, 405-408
Antireflection coatings:
infrared, 776
inhomogeneous, 770
multilayer, multizero reflectance
condition, 772-776
single, 771
Area, surface (*see* Surface area of films)
Astroid, 615
Atomic mobility (*see* Mobility, atomic)
Auger-electron spectroscopy, 114